BIOLOGY OF ANIMAL BEHAVIOR

Common cranes *(Grus grus)* congregated during migration at Lake Hornborga in southern Sweden. Some of the birds are displaying the courtship dance.
Photograph courtesy Sture Karlsson, Mariestad, Sweden.

BIOLOGY OF ANIMAL BEHAVIOR

JAMES W. GRIER

Professor of Zoology,
North Dakota State University,
Fargo, North Dakota

With a contribution by William W. Beatty,
North Dakota State University

Original artwork by Barbara Bradley,
Tucson, Arizona

With 866 illustrations

Times Mirror/Mosby
College Publishing

St. Louis
Toronto
Santa Clara

1984

To my own mate and closest kin

Editor   Diane Bowen
Assistant editor   Susan Dust Schapper
Manuscript editor   Connie Povilat
Design   Jim Buddenbaum
Production   Kathleen L. Teal, Judy England, Jeanne Bush

Copyright © 1984 by Times Mirror/Mosby College Publishing

A division of The C.V. Mosby Company
11830 Westline Industrial Drive
St. Louis, Missouri 63146

Printed in the United States of America

Library of Congress Cataloging in Publication Data

Grier, James W.
   Biology of animal behavior.

   Bibliography: p.
   Includes index.
   1. Animal behavior. I. Title.
QL751.G836 1984   591.51   83-5419
ISBN 0-8016-1971-8

TS/VH/VH 9 8 7 6 5 4 3 2 1   01/D/068

PREFACE

Animal behavior as a subject of study covers a large, diverse, and poorly defined array of topics within the fields of biology and psychology. Different authors and teachers vary greatly in the material they include and the manner in which they present it. Because the topics I wanted to include in an introductory course on animal behavior were so scattered through the literature, I found myself assigning six textbooks, many readings, and several additional references to the class. Students were buying only one of the six and sharing with other students. This illustrated some of the advantages and disadvantages of sharing certain resources, but it created many problems, and I was also getting weary of carrying around so many books and papers.

I was wrestling with this situation when a sales representative from Times Mirror/Mosby College Publishing walked into my office. The outcome of his visit was analogous to what used to happen when I was a kid and several of us and our fathers went fishing: Dad always said that anyone who complained about the cooking could fix the next meal. Thus I wrote this book for my own teaching purposes, but I hope that others can use it as well.

Intended Audience

Biology of Animal Behavior was written for an introductory course in behavior or for persons wishing to learn about the subject on their own. This book should serve to present the subject both for those who want an introduction only and for those who may wish to delve into the topic deeper, take more advanced courses, or simply be able to read the original or more technical literature in animal behavior. I borrowed the working assumptions of Dawkins (1976) "that the (reader) has no special knowledge, but I have not assumed that he is stupid." I also assumed that one should not be afraid to tackle topics such as inclusive fitness and optimality theory. Rather, these concepts should be made understandable although they cannot be treated at an advanced level in this book. Aside from having little or no previous exposure to animal behavior and the subject's contemporary topics, however, the reader should have some background in general biology—ideally at least a year of introductory biology in college or a good high school course.

Because different people (including those graduating from the same educational institution) generally have remarkably different backgrounds and understandings of biology, I have attempted to identify and include critical connecting links of basic information that are commonly assumed to be understood but almost as commonly are not. In addition, the reader should have access to a good general biology or zoology text, such as one of those listed at the end of Chapter 4, for occasional reference.

Approach and Organization

My basic goal was to integrate (1) the structure and function of behavior and (2) the topics of ethology, comparative psychology, and neurobiology. A major guideline in my writing was to consider how well ideas and conclusions fit the real, rather than the academic, world of animal life before such material went into the book.

Biology of Animal Behavior is organized in four parts: (1) an introduction including history, methods, genetics, and evolution, (2) behavior as observed output, including ecological and further evolutionary aspects, (3) the internal neural-endocrine aspects, and (4) additional peripheral topics. I arrived at this structure by using reviewer input and personal trial and error, by presenting this material in different ways to introductory classes, and by observing how others taught it. This book was tested, in manuscript form, in two introductory courses, and it has worked for me.

The structure, in essence, first establishes basic, necessary principles; then the book presents observed forms of behavior, the main topic of interest. Next, internal mechanisms are discussed. This provides a thread of continuity or something of a plot. By the time students reach the neural topics in Part 3, they generally are curious and ready for them.

Many persons prefer to cover the neural and learning topics first and thus get the internal mechanisms in place from the start. Instructors who want to present the material in this sequence can cover Part 3 before Part 2. This book is modular and includes numerous cross-references to material elsewhere in the book.

To describe further what this book is, it may be easier to say what it is not. First, it is not an encyclopedia of animal behavior, which would require a large book or many volumes and would still cover only a small part of what is known and published about behavior. Instead this book is intended to give a broad perspective of the subject through a sampling of details from the available mountain of literature.

Second, to my own frustration, it is not continuously changing with the subject. New books and reviews of topics covered in this book appeared constantly as I was writing about them. The subject of behavior is moving and changing fast. Trying to write about it reminded me of my attempts to catch armadillos that were running toward their holes: I could run fast enough to keep up with them, but as soon as I would stop to reach down and grab one, it would get ahead and away from me. So it was with this book. I had to quit long enough to write something down. The material in this book is as up to date as I could get it at the time it was written, but the subject will continue to run forward and will have to be tracked with future editions.

Third, the book is not about people—very much. Humans *are* animals and share many of the same mechanisms and behavioral characteristics of other animals. Furthermore, some of the best information on behavior, particularly that concerning the central nervous system, is about human behavior: this is included where appropriate. But this book is oriented toward behavior in general in all animals.

Finally, this book is obviously not a living animal and, as such, is a poor substitute for one. To appreciate and understand animals properly, one must spend a lot of time with the real thing. Good films on animal behavior are the next best route. My only caution on films is that they show behavior as interpreted by the filmmaker and commentator. Many have an overly anthropomorphic view. I hope all readers will have access to both good films and live animals.

Distinctive Features

At the end of each chapter, the reader will find a summary encapsulating the major points and principles and a list of recommended additional readings that will guide him or her to resources for further investigation of selected topics. The Appendices contain useful supplementary information, including suggestions for observing live animals. The Instructor's Manual contains test questions, further suggestions for field trips and in-class demonstrations, and a list of recommended films.

Reviewers

I wish to extend my profound thanks to the following reviewers. Their helpful criticism was invaluable in improving and bringing this book to completion, and if I have not always followed their advice, I hope they will forgive me.

Russell D. Fernald   *University of Oregon*
Stephen J. Gaioni   *Washington University*
H. Carl Gerhardt   *University of Missouri at Columbia*
Thomas C. Grubb, Jr.   *Ohio State University*
Herbert E. Hays   *Shippensberg State College*
Stephen H. Jenkins   *University of Nevada at Reno*
John A. King   *Michigan State University*
Randall Lockwood   *State University of New York at Stony Brook*

Valuable input and criticism were also provided by many students on whom earlier drafts of the book were tried and by my wife, who served as first critic and editor of rough drafts.

Other Acknowledgments

Much, if not most, of the credit for the content of this book must go to many persons who contributed greatly to my understanding of animals. The following were particularly significant and influential teachers and advisors:

V. Sponseler, A. Potter, B. Eyestone, C. McCollum, R. Goss, R. Winier, V. Dowell, R. Tepaske, M. Grant, J.T. Emlen, W. Burns, J.F. Crow, J. Nees, M. Konishi, T.J. Cade, S. Emlen, A. vanTienhoven, H. Ambrose, and T. Eisner. Several senior colleagues from whom I have learned much include W. Dilger, D.

Peakall, J. Wiens, H.C. Mueller, J.J. Hickey, F. and F. Hamerstrom, and J. Hailman. I particularly thank Fran Hamerstrom, who constantly reprimanded me as a student to write in "plain English." Several influential peers who shared mutual interests in animal behavior include several personal friends in Ontario, Iowa, Wisconsin, Minnesota, and New York, as well as a large number of fellow undergraduate and graduate students at the University of Northern Iowa, University of Wisconsin at Madison, and Cornell, my own present and past graduate and undergraduate students, many falconer friends and eagle/raptor enthusiasts, several faculty members of the Zoology and Psychology Departments at North Dakota State University and University of North Dakota, and other biologists in the state and nearby, particularly at the Northern Prairie Wildlife Research Center, North Dakota Game and Fish Department, Ontario Ministry of Natural Resources, Canadian Wildlife Service, and U.S. Fish and Wildlife Service.

Additional sources of my understanding have been many observers, thinkers, and writers with whom I have not had direct contact but who helped synthesize the subject for everyone. Whether this amounts to a form of altruism on the part of those persons or just successful replication and recombination of some "memes" (see Dawkins 1976), I do not know; without the aid of many review articles and texts, an overall review of the subject would be nearly impossible. I find that on any given topic, for example, sociobiology or the brain, one or a few well-written items stand out from among the many available. I thank the authors of these materials and have tried to acknowledge them as recommended further readings.

In addition to W.W. Beatty, who wrote Chapter 17, I would like to especially thank J.D. Brammer, J.F. Cassel, J.W. Gerst, C.R. Gustavson, and W. Maki at North Dakota State University for discussions and clarifications concerning various topics in this book.

I thank Susan Abrams and Diane Bowen, Editors, Susan Dust Schapper, Assistant Editor, Connie Povilat, Manuscript Editor, Peter Wold, who initially contacted me, and others at Times Mirror/Mosby College Publishing for their constant encouragement and support during the preparation of this book. Barbara A. Bradley was responsible for most of the line drawings. Additional illustrations were done by Karen Schuler. Dean Grier did much of the typing and entering of revisions into the word processing system and helped greatly with numerous other tasks. Additional typing and greatly appreciated help were provided by Karlene and Joyce Grier, Joyce Mortensen, Sheila Kath, Charlotte Meester, and George Allen. Previously unpublished photographs were graciously donated by Tom Brakefield, Ed Bry, and Gerald Holt. Thomas P. Freeman, Kathy L.H. Iverson, and Gary Fulton of the North Dakota State University Electron Microscope Laboratory prepared several EM photographs for the book. Several other persons, indicated in figure credits, also contributed photos. Sture Karlsson kindly donated the cover photograph. I thank Cindy Lowes, Florissant, Missouri, for doing most of the time-consuming, complex, and often frustrating task of obtaining permissions and originals of previously published materials.

I was permitted a part of a year's leave of absence from North Dakota State

University to work on this book, for which I thank the administration and the people of North Dakota and northwestern Minnesota.

Last, but certainly not least, those in my immediate family who have both tolerated and shared in my interests with animals have been of utmost importance. I particularly wish to thank my parents, Mr. and Mrs. P.H. Grier, my wife, Joyce, my parents-in-law, Mr. and Mrs. E. Petersen, and my children, Karlene and Dean. They have had to put up with such things as opposums in clothes baskets, birds flying around the house, rattlesnakes in empty wash buckets, leaking aquaria full of invertebrates, numerous unmentionable items in the refrigerator and on the kitchen table, and my frequently being lost in thought when they wished to chat about other things.

James W. Grier

CONTENTS

PART ONE

INTRODUCTION TO BEHAVIOR AND ITS CAUSES

1 Behavior and biology, 3

2 History: ethology, comparative psychology, and neurobiology, 25

3 Observation and measurement of behavior, 53

4 Genetics and ecology of behavior, 87

5 Evolution of behavior: consequences of genetic and ecological variability, 109

PART TWO

BEHAVIOR AS OBSERVED OUTPUT: ETHOLOGICAL AND ECOLOGICAL ASPECTS

6 Basic and maintenance behavior of individuals, 147

7 Finding a place to live: habitat selection, homing, and migration, 187

8 Introduction to sociobiology: the phenomenon of grouping and who's who, 225

9 Communication: modes, mechanisms, and ecological considerations, 261

10 Behavior and reproduction: courtship, parental care, and other considerations, 303

11 Interspecific behaviors: symbioses and predation, 341

12 Play, 385

PART THREE

INTERNAL CONTROL OF BEHAVIOR: PHYSIOLOGICAL AND PSYCHOLOGICAL ASPECTS

13 Nervous systems, neurons, and muscles, 403

14 Different sensory worlds, 443

15 Pathways through the brain, 473

16 Hormones, neurochemicals, and behavior, 507

17 Neural mechanisms and feeding: a case in point, 527
   William W. Beatty

18 Ontogeny of behavior, 553

19 Learning and memory, 579

PART FOUR

ADDITIONAL TOPICS

20   Abnormal behavior, 621

21   Applied behavior, 637

Appendix A   Suggestions for observing live animals, 657

B   Gene frequencies in populations, 663

C   Nervous system response times, 667

Literature cited, 670

DETAILED CONTENTS

PART ONE

INTRODUCTION TO BEHAVIOR AND ITS CAUSES

1   Behavior and biology, 3

Bat and moths: a behavioral case in point, 4
    Echolocation and hunting behavior in bats, 7
        Properties of sound, 8
        Use of echolocation by bats, 10
        Little brown bats and noctuid moths, 13
    Noctuid moth hearing and behavior, 15
So what? 20
Summary, 23

2   History: ethology, comparative psychology, and neurobiology, 25

Early beginnings, 29
    Medical anatomy and physiology, 30
    The evolutionary foundation, 30
    The development of different disciplines, 31
Ethology, 33
    Early ethological concepts and terms, 33
    Problems with the concepts, 39
Comparative psychology, 43
    Cognitive versus reinforcement theories, 45
    Behaviorism, 46
Neurobiology, 48
Since the sixties, 50
Summary, 50

3   Observation and measurement of behavior, 53

Behavior and scientific methods, 54
Becoming familiar with the species, 55
Observing without interfering, 56
Locating and identifying individual animals, 62
Ingenuity and technology, 65
Describing, recording, and cataloging behavior, 68
    The units of behavior, 68
    Sampling behavior, 73
    Data recording, 76
Interpreting and presenting behavioral data, 77
    Correlations and the "trained cricket," 81
    Complex analyses and computers, 83
Suggestions for learning the methods of behavioral study, 84
Summary, 85

4   Genetics and ecology of behavior, 87

An overview of genetics, 87
    DNA, RNA, and protein, 87
    Interactions with the environment, 90
    Reproduction, 92

Mendelian genetics, 93
Single genes and behavior, 94
Physical versus statistical (inferential) bases of heredity, 95
    Quantitative genetics, heritability, and behavior, 96
    Additional evidence for genetic control of behavior, 100
Concluding comments on the genetics of behavior, 102
Ecology and behavior, 103
Summary, 107

5 **Evolution of behavior: consequences of genetic and ecological
   variability,** 109

Methods of studying evolution, 109
The process of evolution: changes in the gene pool, 111
    The classic Darwinian view, 111
    Shifts in gene frequencies, 111
    Origins of new species and characteristics, 114
    "Selfish" genes and the level at which selection operates, 117
        Group selection and behavior, 117
        Selfish genes and behavior, 119
        Shifting balance theory, 121
Evolution versus other considerations, 122
    Special creation—creationism, 123
    Extraterrestrial origin of species, 124
    Lamarckianism, 124
    Neutralism, 124
    Recent evolutionary considerations, 125
The role of future events in behavior, 125
More on survival and reproduction: a recap, 127
The evolution of behavior, 128
    An example: the evolution of crane vocalizations, 131
    Evolutionary characteristics of behavior, 140
Summary, 143

PART TWO

BEHAVIOR AS OBSERVED OUTPUT:
ETHOLOGICAL AND ECOLOGICAL ASPECTS

6 **Basic and maintenance behavior of individuals,** 147

Activity and inactivity: rhythms and sleep, 148
    Rhythms, 150
    Sleep, 155
        Sleep: need or compulsion? 158
Locomotion and intention movements, 162
Maintenance and related behavior, 164
Feeding and optimal foraging, 172
    Description and examples of optimal foraging, 174
        Optimal patch choice and method, 175
        Optimal food type, 177
        The role of learning in optimal foraging, 179
        Optimal foraging and search image, 179
        Optimal behavior in general, 180
        Problems with optimality theory, 180
    Caching behavior, 181
Shelter seeking and construction, 181
Summary, 184

7 **Finding a place to live: habitat selection, homing, and migration,** 187

The monarch butterfly: a case in point, 187
Simple forms, 190
Habitat use and selection, 191
  Home range and familiar area, 199
Complex systems of orientation and navigation, 200
  Ultimate factors in migration, 200
  Proximate factors in migration and homing, 201
  A brief phylogenetic survey of migration, orientation, and navigation, 202
    Aquatic plankton, 202
    Molluscs and crustaceans, 202
    Arachnids, 202
    Insects, 203
    Fish, 206
    Amphibians, 209
    Reptiles, 209
    Birds, 210
    Mammals, 220
Summary, 222

8 **Introduction to sociobiology: the phenomenon of grouping and who's who,** 225

Elements of social behavior and who's who, 226
Causes of grouping, 231
Maintenance of social structure and spacing, 237
Kin selection and the evolution of social behavior, 243
Evolutionarily stable strategy, 247
The sociobiological controversy, 252
Summary, 258

9 **Communication: modes, mechanisms, and ecological considerations,** 261

What is communication? 261
Methods of studying communication, 268
Modes and mechanisms of communication, 269
  Signal characteristics, 269
  Comparison of major communication modes and channels, 271
    Chemical communication, 271
    Auditory communication, 274
    Tactile communication, 275
    Visual communication, 276
    Electrical communication, 276
  Communication by multiple channels and antithesis, 276
Evolution of communication, 280
  Specific advantages and functions, 280
  Evolutionary origins of displays, 282
  Phylogenetic comparisons of communication, 282
  Advanced forms of communication, 287
    Cephalopods, 287
    Honeybees, 288
    Birds, 293
Tapping the lines, 294
A major controversy: who has language and who does not? 294
Summary, 300

10   Behavior and reproduction: courtship, parental care, and other considerations, 303

The key club: who is allowed to mate and why, 306
   Sequences of behavior and basic biological considerations, 306
   Environmental considerations and timing, 310
Mating patterns in reproduction, 313
   Differential investment by the sexes, 316
   Spatial distribution of resources, 322
   The evolution of mating patterns, 324
   Communal (arena or lek) displays, 326
   Nonmodal forms of reproductive behavior, 328
Phylogenetic comparisons of reproductive behavior, 328
Parental care, 330
   Parent-offspring conflicts, 335
Other cooperation in reproduction and care of young, 337
Summary, 339

11   Interspecific behaviors: symbioses and predation, 341

Aggregations/chance encounters, 343
Commensalism, 344
Mutualism, 346
Parasitism, 348
   Tapping the communication lines, 353
Predator and antipredator behavior, 357
   Evolution of predatory behavior, 359
   Proximate factors of predation, 361
   Problems of finding, recognizing, and selecting prey, 365
      Which prey species are taken? 366
      Prey switching, 366
      Search image and associated notions, 367
      Prey selection, 370
   Hunting and capture methods, 371
   Handling and killing tactics, 374
   Antipredator defense tactics, 375
   Long-term predator-prey relationships, 383
Summary, 383

12   Play, 385

Descriptions of play, 386
   Examples of play behavior, 386
   General attributes of play, 390
Why do animals play? 394
   Surplus energy theory, 394
   Pleasure theory, 394
   Arousal/stimulation theory, 395
   Practice theory, 395
   Exercise theory, 396
Play in a social context, 396
Summary, 399

PART THREE

**INTERNAL CONTROL OF BEHAVIOR:
PHYSIOLOGICAL AND PSYCHOLOGICAL ASPECTS**

13   Nervous systems, neurons, and muscles, 403

Nervous systems, 403
Neurons, 412
   The structures of neurons, 413
   Neurophysiological methods, 416
      Visual techniques, 416
      Recording of bioelectric events, 416
      Stimulation, ablation, and lesioning, 417
      Experimental manipulation and analysis, 418
   The operations of neurons, 418
      Discrimination by neurons, 422
      Transmission of decisions, 422
      Maintenance of neurons: recycling and recharging, 427
      Nonneuronal components of the nervous system, 428
Examples: moth ears and cockroach kickers, 428
   Moth ears, 428
   Cockroach cerci, 432
Muscular output, 436
Summary, 439

14   Different sensory worlds, 443

Methods of investigating senses, 445
Sensory attributes, 446
   Sensitivity, 447
   Spatial or temporal pattern discrimination, 448
   Ability to localize the source, 448
Brief survey of the senses, 449
   Mechanoreception, 449
   Inertial senses, 450
   Vibrational senses, 451
   Vision, 454
      Color vision, 455
      Night vision, 458
      Polarized light detection, 460
      Acuity and form vision, 460
   Infrared and heat detection, 461
   Electric senses, 462
   Magnetic senses, 462
   X-ray and radio wave sensing, 463
   Chemical senses, 463
Emitted-energy senses, 466
Combinations of sensory input, 468
Mechanisms of transduction, 469
Summary, 470

15   Pathways through the brain, 473

Vertebrate neural pathways with emphasis on humans, 475
   Reflexes, 475
   Mapping the brain, 481
   Abstraction and pattern recognition: vision as a case in point, 484

Motor centers in the brain, 491
Language pathways, 493
Right versus left sides of the cerebrum: brain lateralization, 497
Contributing components and other pathways, 500
Tracing the actual circuits: pathways through invertebrate brains, 502
Summary, 503

16 **Hormones, neurochemicals, and behavior,** 507

Insect behavior and hormones, 509
Vertebrate behavior and hormones, 509
Pathways and mechanisms of hormone action, 517
Pleasure, pain, and exogenous mood-altering chemicals, 521
Summary, 524

17 **Neural mechanisms and feeding: a case in point,** 527
William W. Beatty

Is feeding regulated internally? 527
Control of feeding in an insect: the blowfly, 532
    Feeding behavior and patterns, 532
    Control of feeding, 533
        Initiation of feeding, 533
        Cessation of feeding and satiety, 535
Control of feeding in a mammal: the rat, 540
    Feeding behavior and patterns, 540
    Peripheral controls, 541
    The hypothalamus and central mechanisms, 543
    Recent interpretations, 549
Summary, 550

18 **Ontogeny of behavior,** 553

Embryological and neonatal roots of behavior, 554
Maturation and behavioral change, 560
Effects of deprivation and general effects of early experience on the development of behavior, 565
    Environmental enrichment, 568
    Development and the acquisition of specific changes in behavior, 569
        Critical or sensitive periods, 569
        Imprinting, 570
        Development of bird vocalizations, 572
Summary, 576

19 **Learning and memory,** 579

Learning categories and terminology, 579
    Habituation, 580
    Sensitization, 580
    Classical conditioning, 580
    Operant conditioning, 581
    Taste aversion, 583
    Latent or exploratory learning, 584
    Place or spatial learning, 585
    Cultural or observational learning, 585
    Imprinting, 587

Insight learning, 587
Learning-set learning, 587
Overlaps and combinations of types of learning, 588
Phylogenetic survey of learning, 588
Protozoans, 589
Porifera, 589
Coelenterates, 589
Platyhelminthes, 591
Molluscs, 595
Annelids, 596
Arthropods, 597
Echinoderms, 601
Vertebrates, 601
Mechanisms of learning and memory, 603
Habituation and sensitization in *Aplysia*, 603
Memory and the elusive engram, 605
Learning processes versus learning phenomena: a synthesis, 609
Summary, 614

PART FOUR

ADDITIONAL TOPICS

20   Abnormal behavior, 621

Examples of abnormal behavior in animals, 622
Nondomestic animals under natural conditions, 622
Animals in zoos, 623
Abnormal escape or hiding reactions, 623
Refusal to eat, 624
Stereotyped motor reactions, 624
Displacement and redirected behavior, 625
Self-mutilation, 625
Aggressiveness, 625
Miscellaneous abnormal behavior, 626
Captive animals used in research, 628
Domestic livestock, 629
Domestic pets, 633
Causes of abnormal behavior: a review, 634
Preventing, correcting, and curing abnormal behavior, 634
Summary, 635

21   Applied behavior, 637

Management of animals in captivity, 638
General considerations, 638
Legal ramifications and requirements, 638
Housing facilities, 638
Specialized handling and moving techniques and equipment, 638
Nutrition and feeding, 638
Behavioral welfare ("psychological concerns"), 639
Prevention and correction of disease and injury, 639
Special problems, 639
References and specialists, 639
Zoos, 640
Livestock and research animals, 643

Pets, 644
Veterinary management of animals, 645
Management of nondomestic animals, 646
Managing pests and nuisance animals, 646
Game, nongame, and endangered species management, 648
Game and nongame management, 648
Endangered species management, 651
Applied animal behavior in research and photography, 653
Training animals, 654
Summary, 655

Appendix A   Suggestions for observing live animals, 657

B   Gene frequencies in populations, 663

C   Nervous system response times, 667

Literature cited, 670

PART ONE

INTRODUCTION
TO BEHAVIOR
AND ITS CAUSES

CHAPTER 1

Bat searching for prey using echolocation.

BEHAVIOR AND BIOLOGY

A botany student who was enrolled in an ornithology course once remarked that it was her least favorite class. The birds kept moving; she could not get close enough to them or keep them in sight long enough; and she was terribly frustrated. Several zoology students, on the other hand, have complained that botany is dull and boring because the plants do not "do anything." These statements certainly do not express the sentiments of all botanists and zoologists. But they do indicate the central difference between plants and animals, namely, that most animals can actively move, whereas most plants cannot. The movements of animals, what they do, how they do it, and why are all part of animal behavior.

Following Kandel (1976), who based his definition on the tradition of Skinner (1938) and Hebb (1958), behavior is defined as *all observable muscular and secretory responses to changes in an animal's internal or external environment*. This definition is quite broad. It includes simple muscular contractions and secretions as well as higher-order categories such as courtship and communication. What can be seen or otherwise detected about animals, however, must be viewed as only surface phenomena. Thus the *study* of behavior may take one far below the surface to things that cannot be sensed directly. One may consider and make inferences about relatively inaccessible matters such as molecular and evolutionary processes and, perhaps, even mental or cognitive states.

Movement requires mechanisms such as cilia, flagella, pseudopods, or muscles. In many forms of animals movement is aided by hard parts, skeletons, and other structures located either inside or outside the body. Nearly the entire remaining anatomy and physiology of an animal supports or is otherwise affected by the animal's ability to move.

Of particular interest are the systems that control and coordinate movement and other responses such as secretions: the nervous and endocrine systems. Animals do not just move and behave completely at random or by some mysterious invisible force. All one has to do is cut some nerves, remove certain endocrine glands, or damage various parts of the brain or spinal cord to observe striking changes in behavior, including immobility or paralysis. The internal control mechanisms operate behind the scenes and are unknown or taken for granted by most persons, but the underlying machinery is even more fascinating than would be some mysterious invisible force.

Nervous system tissues and organs are soft, delicate, and often extremely complex. Hence neural structures have presented biologists with some of the greatest challenges in science. The precise manners in which information travels through the structural connections and pathways of nervous and endocrine systems also have presented many challenges and have generated much interest. Many behavior patterns, including some that are very complex, appear to be mostly fixed internally. They often are said to be innate or instinctive. Other patterns may be more open to modification, or "learned," as a result of events in the animal's life. A large portion of the study of behavior has been devoted to what goes on inside the animal.

There is more to understanding behavior than just considering what is inside the animal. The various internal mechanisms and accompanying behavior patterns or abilities to learn have evolved in response to various ecological factors and forces. The coordinated movements and actions of animals have become diverse, complex, and sophisticated over the course of many millions of years. Much of the subject has centered on attempts to decipher these ecological and evolutionary forces.

Animal behavior probably has been of interest to humans for as long as they have been curious. But only in the last two centuries has there really been much progress in the understanding of behavior. This progress resulted largely from the application of standard scientific methods, including carefully designed experiments, to behavior on virtually all levels, from molecular to ecological. In addition to the advances in each particular area, the relationships between these areas have become clearer. Formerly distinct topics such as neurophysiology and the ecological aspects of behavior have begun to merge into a more unified subject. Animal behavior as a discipline is maturing and taking on an academic status along with the other relatively recent developments of molecular biology and ecology. Furthermore, to the interest and excitement of some persons and the fears of others, scientists keep noticing that humans are animals, and the findings of animal biologists have begun spilling more into studies of human behavior. Psychologists have been at the forefront of behavioral biology, and some sociologists also have been entering the arena.

When faced with a subject that pervades so many aspects of animal life and modern biology, a textbook writer feels like a mosquito at a nudist camp; one does not know where to start. This book begins arbitrarily with a microcosm of the subject: the interaction between bats and moths. The example has become something of a classic in behavior. The ecological and evolutionary aspects are fairly straightforward, and some of the anatomical and nervous system considerations are as simple and well studied, yet interesting, as one is likely to find.

BAT AND MOTHS: A BEHAVIORAL CASE IN POINT

The animals involved in this frequent life-and-death drama are the little brown bat (*Myotis lucifugus*) and moths of the family Noctuidae. The setting might be an abandoned field near a New England town on a warm summer evening. At first glance the behavior patterns of the bats and moths are not particularly remarkable. They seem to be merely flying about. Bats may be seen briefly overhead as dark forms silhouetted against the dark-blue dusky sky. The moths may be noticed flitting against a window screen, attracted to the lights inside. But on closer inspection the details and mechanics of the behavior patterns are truly impressive.

It is hard to put oneself in these animals' positions, but try to imagine doing just that. To illustrate the differences between human lives and the lives of bats

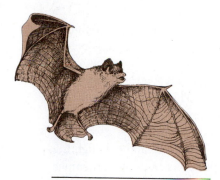

Figure 1-1 Little brown bat.

and moths and to start breaking the habit of looking at the world only from the human viewpoint, stop reading and take a few minutes to think about the following questions. What did you do in the past hour? Where were you, and how did you spend the time last night from 6:00 PM to 6:00 AM? *Why* did you do all of those things? What did you eat for your most recent meal, and how did you *get* that food? Where did the food come from?

Now consider the lives of the bat and moth. These animals are largely inactive during the day and active at night. The little brown bat (Figure 1-1) is a small mammal of the order Chiroptera (hand-wing). Its body is covered with soft fur. The bat's wings are largely composed of long, slender finger bones and naked skin. These bats sleep during the day by holding on with their back legs and hanging headdown in the dark crevices of large trees, caves, mines, and buildings. They are highly social when not hunting and roost in groups, depending in part on their age, sex, and reproductive condition. They hibernate during the winter in more protected sites, hibernacula, which may be several kilometers from their roosts at other times of the year.

Moths normally are overlooked or considered to be nuisances by most persons. The noctuid group includes such notorious pests as armyworms *(Cirphis unipuncta)*, cabbage looper *(Trichoplusia ni)*, corn earworm *(Heliothis armigera)*, cutworms (several genera), and underwings *(Catocala* sp.). These moths are large, soft-bodied insects with a wingspan of approximately 2 to 12 cm, depending on the species. Noctuids (Figure 1-2) are beautiful and intriguing. If picked up, they do not bite, but a dusty material, the fine scales that cover the wings in insects of the order Lepidoptera (scale-winged), may rub off on one's hands.

Noctuid moths go through the typical complete metamorphosis of egg-larva-pupa-adult as do other lepidopterans. During the day they hide inconspicuously in crevices or vegetation. The social life of various noctuids ranges from solitary, except for mating, to highly gregarious.

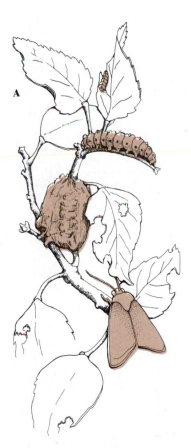

A

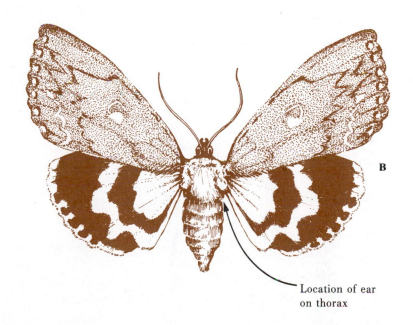

B

Location of ear
on thorax

Figure 1-2 A, Noctuid moth, including egg, larval, pupal, and adult forms. B, Arrow indicates position of ear on the posterior side of the thorax. (See also Figures 1-14, 1-15, 13-16, and 13-19.)

The summer New England evening environment may contain an overwhelming variety of objects, substances, and forms of energy, many of which are associated with humans. There are trees, shrubs, grasses, telephone poles, wires, houses, people, cats, dogs, night-flying birds, mice, moths, a great variety and large number of other insects, bats, vehicles, and much more. The air and environment contain innumerable odors, sounds, light, and other forms of energy. Consider just the sounds and mechanical vibrations: a slight breeze and many associated vibrations such as from rustling leaves, sounds of dogs barking and cats fighting in the distance, children crying, people talking, nighthawks buzzing and snipes winnowing overhead, high-pitched chirps of bats flying about, sounds of televisions and stereos coming from houses, roaring motorcycles, a police siren, various sounds of machines, vehicles and squeaking hinges, buzzing of mosquitoes and droning of large flying beetles, chirping and whirring calls of many species of insects, a few twitters from perched birds, and a buzzing transformer on a power pole. Energy in other forms includes light from the rising moon, starlight, many radio and television signals from near and distant broadcasting stations, microwaves, cosmic and x-rays plus many radio waves and other forms of radiation from space, the earth's magnetic field, streetlights, lights from houses, headlights from vehicles, light from fireflies, children with flashlights, and much more.

Out of this bewildering hodgepodge of substances and energies, a *few* are of extreme significance to the lives of bats and moths. These few may mean the difference between life and death. The remainder are irrelevant. The animals' sensory and nervous systems must, somehow, sort through all of it, tune into the significant items, omit the rest, and then initiate the appropriate behavior. The means by which the animals do this may not be conscious. The biological machinery that does the job has resulted from and been refined by millions of years of natural selection. The items associated with humans have not been around for long, but there have always been many objects and various vibrations, favorable, unfavorable, and neutral, in the environment.

The natural selective pressures resulting from the interactions between bats and moths are easy to understand. The bats and moths that continue living are

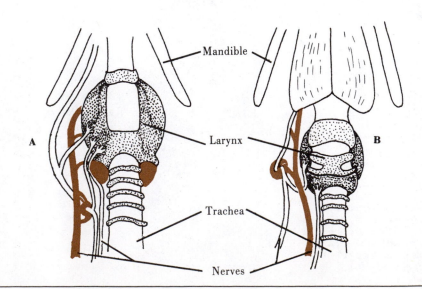

Figure 1-3 Larynxes of bat (*Nyctalus* sp.) (A) and rat (*Rattus* sp.) (B) illustrating relative sizes and development. The animals and their jaws are roughly similar in size, but the bat larynx is more muscular and much larger, extending about three fourths the distance between the ends of the mandible compared to about one third for the rat. Modified from Yalden, D.W., and P.A. Morris. 1975. The lives of bats. Quadrangle/New York Times Book Co., New York.

those which, among other things, eat but are not eaten themselves. Although bats are occasionally killed and eaten by other animals such as hawks, it is their eating that is of interest at the moment. Bats that are able to detect (in this noisy environment) and then capture and eat insects survive and produce new bats in greater numbers than those which do not detect, capture, and eat insects. The only moths, on the other hand, that survive and reproduce are those which either do not encounter bats (among other things) or those which can detect and then avoid the danger. They have to identify bats against the background of all the other potential input. If one is normally unaware of the moths, the ignorance is mutual on the part of the moth; they also usually are tuned to other things.

Both the bats and moths are detecting and identifying each other on the basis of airborne mechanical vibrations or "sounds." Sound waves are common in the environment, are sensed mechanically, and have been used extensively by animals. But the biological machinery, coming from two, long, separated lines of evolution, is vastly different in the two cases.

Echolocation and Hunting Behavior in Bats

Bats use a variety of sounds in many different ways (Griffin 1958, Simmons et al. 1979). Some squeaks and chitters, for example, are uttered when the bats are disturbed. These sounds are often within the range of human hearing. But the sounds of interest at present are those which the bats use for detecting objects, such as obstacles or prey. These sounds are *ultrasonic* or above the range of normal human hearing. All of the calls are produced in the larynx, as in other mammals, but the bat's larynx is more highly specialized (Figure 1-3).

The role of sound and hearing in the orientation of bats was implicated first by Lazzaro Spallanzani in Italy in 1793. He discovered that owls could not fly normally in complete darkness or with their eyes covered but that bats could. Furthermore, bats that he blinded, released, and then recovered later at their roost in the bell tower of the cathedral at Pavia had filled their stomachs with as much insect food as others that he had not blinded. But Spallanzani (in conjunction with a Swiss surgeon, Jurine) found that tight plugging of the ears caused the bats to fly about at random and collide with obstacles. Being ingenious and thorough and wishing to eliminate the possibility that the bats were disoriented by irritation of the ears, Spallanzani had some miniature brass tubes fitted to the bats' ears (the ear canal was less than 1 mm in diameter). Then he plugged and unplugged the tubes. Bats with open tubes could orient almost normally, but with closed tubes they could not.

No one knew of the ultrasonic calls of bats at that time, and Spallanzani was rejected, ridiculed, and nearly forgotten. The more respected Cuvier offered an alternate, armchair opinion that bats avoided obstacles by sense of touch (in spite of the fact that Spallanzani had already tested for that by covering bats' wings with varnish or flour paste and they still flew acceptably). Cuvier's opinion was accepted for 120 years until a Cambridge physiologist, H. Hartridge, contemplated a bat that strayed into his room one night. Hartridge knew of sonar, which had been developed for detecting submarines in World War I, and hypothesized that bats were using ultrasonic sounds.

The matter was not settled until 1938 when a curious undergraduate student at Harvard, D. Griffin, checked some bats with an ultrasonic recorder. The equip-

Figure 1-4 **Variety among different kinds of bats that rely on vision or echolocation to varying degrees. Illustrated are two species from the suborder Microchiroptera: little brown bat** *(Myotis lucifugus)* **(A) and horseshoe bat** *(Rhinolophus ferrumequinum)* **(B), which rely predominantly on echolocation during flight. Two species are from the suborder Megachiroptera: rousette bat** *(Rousettus* **sp.) (C) and the fruit bat** *(Notopteris macdonaldi)* **(D), which rely more on vision. Note differences in the relative sizes of the eyes and ears. Worldwide there are around 800 species of bats in 17 families belonging to these two suborders.**

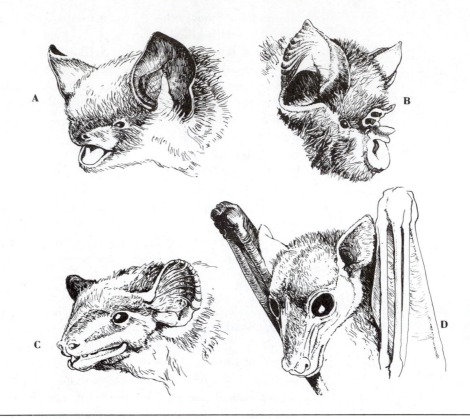

ment had been developed by a Harvard physicist, G.W. Pierce, for studying the high-pitched sounds of insects. Coincidentally and independently at about the same time, a Dutch zoologist with very good hearing, S. Dijkgraaf, listened to bat calls with his own hearing. He concluded that faint sounds were being used by the bats for echolocation.

More recent investigations have shown that different bats vary in their use of sound for echolocation or whether they even echolocate at all. Some of the larger old-world tropical bats of the suborder Megachiroptera, for example, have large eyes and relatively small ears and are thought to rely almost exclusively on vision while flying. Other species are intermediate (Figure 1-4).

Properties of Sound

Before this topic is discussed further, the physics of sound must be considered briefly. Sound waves are vibrations, the to-and-fro movements or displacements of molecules in a given medium such as air, water, or other substances. The displacements of particular molecules affect neighboring molecules, and the movement is passed on; that is, it is propagated and travels. When sound waves encounter an object, depending on its size and characteristics, the waves may be reflected (echoed) or set up vibrations in the object.

A sound wave can be described by several characteristics (Figure 1-5). The **amplitude** or loudness is the maximum intensity of the wave or, in terms of the density of air molecules, the difference in pressure between the maximum com-

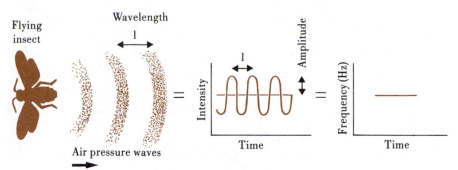

Figure 1-5 Physical characteristics of sound waves. Sound pressure waves can be graphed by intensity or frequency. The relationships between speed *(v)*, frequency *(f)*, and wavelength *(l)* are indicated by a mathematical formula and graphically.

Velocity = v = 344 m/second in air at 20° C
Frequency = f, in cycles/second or Hz; 1 kHz = 1000 Hz
Wavelength = 1, in length/cycle
v = fl, f = v/l
Human hearing: 40 to 20,000 Hz(0.04 to 20 kHz)

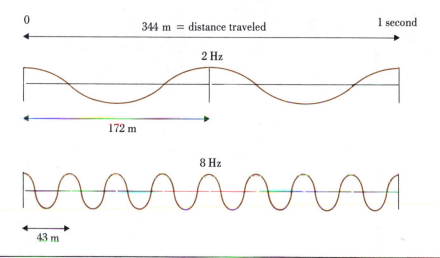

pression and the average density. The amplitude of sound waves is very slight relative to the earth's atmosphere. The amplitude of a loud human shout, for example, is only about a hundred thousandth to a ten thousandth of one atmosphere in pressure. A barely audible whisper would be less than a billionth of an atmosphere.

One *cycle* of sound is the time from the peak of intensity of one wave to the peak of the next. Sound waves travel in air (20° C) at a speed of about 344 m per second (1130 feet/second or roughly 1000 feet/second). In 1 msec (1/1000 second) the sound moves 34.4 cm (about 1 foot), a convenient figure to remember. The length of a cycle can be expressed in terms of either the distance traveled (its *wavelength*) or the time it lasts (how many cycles fit into a second), that is, its *frequency* or pitch. If each cycle lasts 1 msec, then 1000 cycles would occur in 1 second; the frequency would be 1000 cycles per second (hertz, Hz) or 1 kHz. The wavelength would be 34.4 cm. The frequency of middle C in music is 523.25 Hz. Sounds in which all cycles are the same length are pure tones (Figure 1-6), but most sounds encountered in nature are a complex mixture of frequencies.

These principles are important for understanding the acoustical properties of bat calls and subsequent echoes. Objects only reflect wavelengths that are rough-

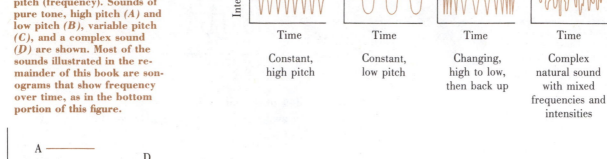

Figure 1-6 Relationship between sound intensity and pitch (frequency). Sounds of pure tone, high pitch *(A)* and low pitch *(B)*, variable pitch *(C)*, and a complex sound *(D)* are shown. Most of the sounds illustrated in the remainder of this book are sonograms that show frequency over time, as in the bottom portion of this figure.

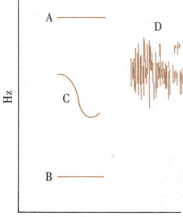

ly the same size as the object or smaller (Figure 1-7). The larger wavelengths break up, *scatter,* and are not reflected intact. This can be easily demonstrated by riding in an automobile at a constant speed with the windows open along a quiet street. Differences in sound result from echoes returning from the complex mixture of noises produced by the vehicle. One usually can hear differences even between areas of roadway with curbs versus areas without curbs, such as at driveways and intersections. The echoes from tree trunks and posts alongside the road sound higher pitched than echoes from the sides of larger objects such as parked cars, trucks, and buildings. This is because the larger-wavelength lower-pitched sounds do not reflect off the smaller objects and, hence, are lost.

One also may notice that the volume or intensity of echoes can change. This results from *interference,* in which the amplitudes of the reflected waves add to (*constructive* interference) or subtract from (*destructive* interference) those of more recently emitted waves. The interference between echoes and newly emitted sounds can be a problem for bats. The problem is avoided by producing calls in short bursts, rather than continuously, so that the calls and echoes are separated.

Another property of sound that may be of use to some (but not all) species of bats is the *Doppler effect.* If the object that is producing (or reflecting) the sound is moving, the start of the sound will be at a different point in space than the finish of the sound. The wavelength of a sound that is moving away is stretched out essentially, which lowers the frequency, and an approaching sound is compressed in length; thus it is raised in frequency. The same effect occurs if the other or receiving end is moving. The effect is familiar to most people as the change in pitch that occurs with an approaching and passing vehicular noise such as a train whistle. The Doppler effect permits the relative motion of two objects to be assessed if the change in frequency can be measured.

Use of Echolocation by Bats

Predominantly insectivorous bats use ultrasonic echolocation for locating prey and obstacles. The brief, loud, high-pitched calls are given in rapid succession, and the frequency of each call is varied or modulated. Each call starts at a high pitch, then rapidly drops (Figure 1-8). If the call is slowed down to bring it within the range of human hearing, it sounds like a "chirp" or "wheeough." Because of the change in pitch, the call is said to be *frequency modulated (FM),* and the bats are called chirping or FM bats. Echoes from FM calls may contain a fair amount

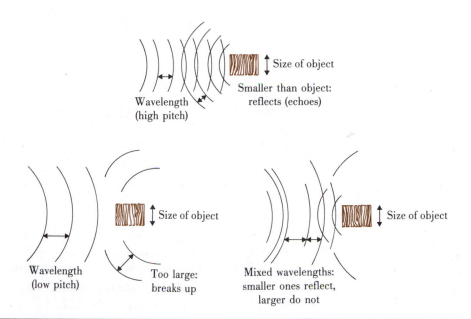

Figure 1-7 Characteristics of echoes. Objects reflect sounds only of wavelengths smaller (higher frequency) than the reflecting dimension of the object. Larger-wavelength sounds break up and scatter: therefore they do not cause an echo. From variable-pitch, complex sounds, only those vibrations of smaller wavelength echo. Thus from a variable sound source such as a bat call, smaller objects such as small insects would reflect higher-pitched echoes than larger ones.

of information, such as the size of objects. Furthermore, freshly reflected echoes retain their chirp quality and can be identified against a noisy background of reverberations from old echoes, other bats, and sounds from other sources.

Still other species of bats, also mostly insectivorous, incorporate calls of a more *constant frequency (CF)*. Echoes from CF calls do not permit recognition of the size of an object, but they permit movement relative to the bat to be detected via Doppler shifts. To use the Doppler effect, the bats do not measure a change in the echo but rather adjust the frequency of their call to achieve an echo of constant pitch (Simmons et al. 1979). The extent that the call has to be adjusted permits the bats to measure the magnitude of the Doppler shift.

Bats using CF calls also may use, to a lesser or greater extent, some FM calls. Some species are relatively fixed in the pattern of CF and FM calls that they emit under different conditions. Other species appear more flexible and vary the patterns of calling in different situations.

Figure 1-8 Types of bat echolocation calls. FM calls (*left panel*) start at a high pitch (about 80 kHz here) and drop rapidly to a lower frequency (40 kHz). CF calls (*right panel*) remain at a constant frequency (around 83 kHz) but may contain an FM element at the end. The limit of human hearing is just below 20 kHz. Time spans are approximately 0.05 second in the left panel and 0.13 second in the right panel.

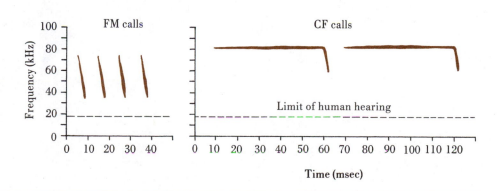

A bat's sensory system operates to accentuate echoes, whereas the human system supresses echoes. The echo suppression of the human system can be demonstrated by tape-recording a series of clicks or short bursts of sounds (such as sounds made by firing a cap pistol) in a room, preferably a large room or hallway with hard walls. When the sound is originally made and when the tape is played in the proper direction, the echoes are perceived only with difficulty or not at all. But if the tape is played backward, the echoes, which now precede the sound, are plainly audible.

The bat's hearing, on the other hand, is suppressed for the original emission of the call. The intensity at the ears is partly reduced by the structure of the mouth and nose, which direct the call forward. Then the hearing is further, actively suppressed by the muscles in the middle ear, which are attached to the ear bones. During the brief emission of the bat's call the muscles clamp down and dampen the transmission of the sound to the inner ear. Then the muscles rapidly relax and permit full amplification of the echo before contracting again to suppress the next emission. Further suppression and amplification of calls and echoes, respectively, may occur by processing in the brain.

Thus many species of bats use a variety of information in returning echoes to sense the environment while flying. Echoes return earlier and louder from close objects than from distant ones. At any one time the incoming sound contains a mixture of echoes from close and distant objects, including perhaps echoes from earlier calls. Bats have an incredible ability to discriminate very small distances on the order of 0.1 mm (Simmons 1979).

The information in echoes is not received continuously by bats because it must be interrupted for sending new calls. The bat's view of the world might be somewhat analogous to the effect produced if one were to walk around with eyes closed and repeatedly open them for a sequence of many brief glimpses of the surroundings. However, in the bat's perception, that is, in the processing of the input by the brain, the flickers may be fused for continuous interpretation, just as one is unaware of the flicker in a projected movie film or the normal blinking of the eyes.

Figure 1-9 External ear of little brown bat. The tragus is believed to help focus sound in the vertical dimension.

Tragus

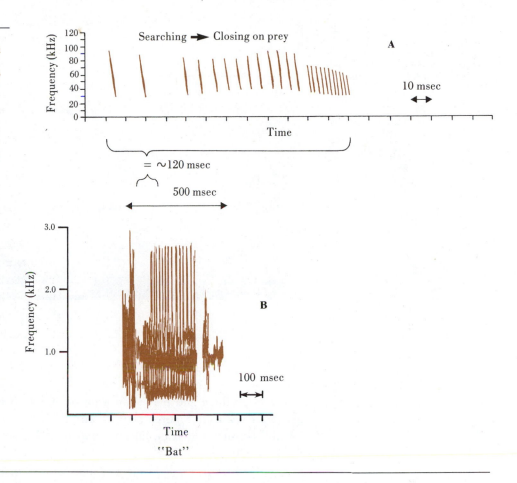

Figure 1-10 Calls of a hunting little brown bat closing on prey (A) contrasted with human speech (B). The bat calls occupy approximately 0.120 second (for 24 separate vocalizations) and span a frequency range of over 60 kHz. By comparison, human vocalization requires nearly 0.50 second per vocalization and spans a frequency range of only 2 to 3 kHz.

Little Brown Bats and Noctuid Moths

The external ears of the little brown bat discussed in the previous bat and moth example are illustrated in Figure 1-9. The ears are cone-shaped funnels that serve to focus the sound, thus accentuating the sound intensity from a given direction and reducing the intensity from other directions. A part of the ear, the tragus, is believed to aid the process of focusing sounds in the vertical dimension.

Little brown bats are FM bats. Figure 1-10, *A*, is a graph, or sonogram, of the calls of a little brown bat. When cruising, the bats emit calls that start at about 90 kHz and drop to 45 kHz and last about 1 or 2 msec. The calls are repeated at rates of 10 to 20 per second. Then when potential prey or small obstacles are encountered, the rate of calling may rise to as many as 250 calls per second, and each call lasts less than a millisecond. Bat detectors are available or can be made that render the calls audible to a human as clicks in an earphone or speaker. If one eavesdrops on bats with such a device, their cruising calls sound like the "putting" of a slow gasoline engine, and the final approach to prey turns into a rising tempo of whining rapid clicks.

To appreciate better the numbers in the preceding paragraph, consider the vocal performances of a human. At best a human can make only about 10 distinct sounds per second. An average short, one-syllable human word takes about ½

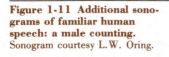

Figure 1-11 Additional sonograms of familiar human speech: a male counting. Sonogram courtesy L.W. Oring.

1 second

second to produce and may span a range of 2 to 3 kHz. Sonograms of human speech are shown in Figures 1-10, *B*, and 1-11. Compared with the needle points of sound produced by bats, human vocalizations are truly low, slow, and slurred (Figure 1-10).

Bats use these sounds to identify and capture insects. Some insects may be detected by the sounds produced by the insect itself. The fact that bats take dead insects (Griffin et al. 1965) and make characteristic call patterns on the final approach indicates quite clearly, however, that bats use their emitted calls as the primary detector. Little brown bats weighing 7 g have been shown to routinely catch 1 g of insects per hour of hunting. A smaller related species of bat studied in the laboratory caught at least 175 mosquitoes in 15 minutes, or one every 6 seconds (Griffin 1959). The smallest prey recorded for a bat was a small gnat weighing 0.0002 g, which was found still in the mouth of a bat that was killed while it was hunting. Noctuid moths are easily detected by bats, and many are captured and eaten.

The "appearances" of a flying moth's echoes were investigated by Roeder (1963), who simulated a bat's calling and receiving system (Figure 1-12). The setup permitted him to photograph simultaneously the actual appearances of the moth with the companion echoes as they were displayed on an oscilloscope screen (Figure 1-13). The results, briefly, were that the echo was strongest when the moth was sideways to the bat and its wing surface was broadside to the sound. It was weakest when the moth was facing toward or away from the bat or when the moth's wings were down. Even when the wings were up and appeared similar visually, there were many variations in the echoes (Figure 1-13). At best the moth would present a flickering sound return to the bat.

The perception of sound in the bat goes beyond the highly specialized ear. The vibrations from the air, having been transformed into nervous system impulses,

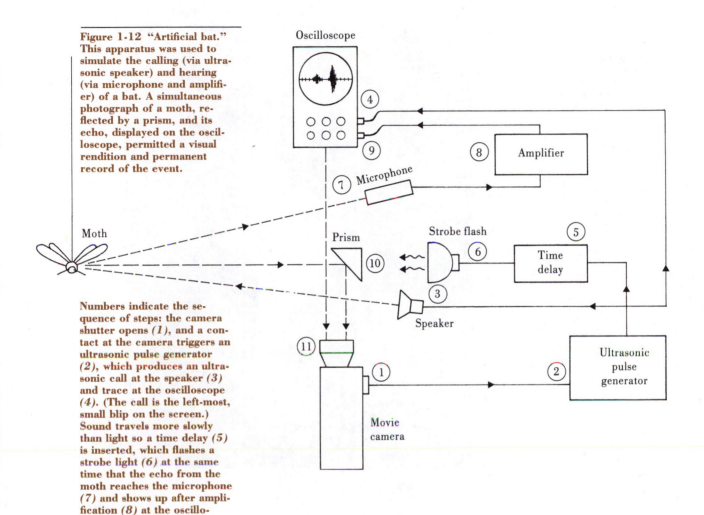

**Figure 1-12 "Artificial bat."
This apparatus was used to
simulate the calling (via ultrasonic speaker) and hearing
(via microphone and amplifier) of a bat. A simultaneous
photograph of a moth, re-
flected by a prism, and its
echo, displayed on the oscilloscope, permitted a visual
rendition and permanent
record of the event.**

**Numbers indicate the se-
quence of steps: the camera
shutter opens (1), and a contact at the camera triggers an
ultrasonic pulse generator
(2), which produces an ultrasonic call at the speaker (3)
and trace at the oscilloscope
(4). (The call is the left-most,
small blip on the screen.)
Sound travels more slowly
than light so a time delay (5)
is inserted, which flashes a
strobe light (6) at the same
time that the echo from the
moth reaches the microphone
(7) and shows up after amplification (8) at the oscilloscope (9) (right-most blip on
the screen). The visual appearance of the moth and image on the oscilloscope reach
the prism (10) and camera
(11) at the same time. Photographs resulting from this setup are shown in Figure 1-13.
Modified from Roeder, K.D.
1967. Nerve cells and insect
behavior. Harvard University
Press, Cambridge, Mass.**

are then transmitted via the auditory nerve to the bat's brain where they are
further sorted, processed, and identified. After the detected object is identified as
edible, the bat moves toward it and attempts to keep it in range and capture it.
But in addition to detecting and tracking the moth, the bat also needs information
on its position in space, its speed and direction, air movement, and many other
facts normally needed by pilots. Furthermore, all the detection, decisions, and
movements have to be *fast*, often in small fractions of a second. If the whole
process took as long as it does to write or read about it, the bat would not only miss
its meal but also would stall and crash into the ground.

**Noctuid Moth Hearing
and Behavior**

Compared with bats, moths have much simpler and different equipment for hearing. The moth's hearing basically is just a simple detector for the bats' calls,
something of an original "fuzz buster" or radar detector. The moth's ears are
located on the thorax (Figure 1-14). Each ear consists simply of a tympanic
membrane connected to only two sensory cells (Figure 1-15). The moth does not
have several thousand sensory cells or an elaborate funnel-shaped ear, and it
does not, as far as is known, use its eyes to detect bats. Yet moths not only detect
bats; as described later, they can even determine the general location and dis-

A   B   C

Figure 1-13 Echo and visual appearances of moths with wings in different positions. Photographs obtained with the apparatus are shown in Figure 1-12. In A the wings are in different positions and reflect quite different echoes. Note in B, however, the variety among echoes even when the wing positions appear similar visually. In C echoes from the rear view of a departing moth provide a minimum of broadside reflecting surface, hence, a more constant and smaller echo, which would be more difficult for a bat to detect. Reprinted by permission of the publishers from *Nerve Cells and Insect Behavior*, by Kenneth D. Roeder, Cambridge, Mass.: Harvard University Press, Copyright © 1963 by the President and Fellows of Harvard College.

Figure 1-14 Noctuid moth showing ear on thorax. Photo by James W. Grier.

To brain

Figure 1-15 Noctuid moth ear on right side of thorax as viewed from above. Modified from Roeder, K.D. 1967. Nerve cells and insect behavior. Harvard University Press, Cambridge, Mass.

tance of the bats and take appropriate action. Furthermore, they do it very rapidly, in fractions of a second and generally in much less time than humans can respond to a stimulus.

Crashing is not as serious for moths as it is for bats because the moth is much lighter and possesses much less momentum. But slow or unlucky moths are eaten. Photographs of the tracks of a moth that won and another that lost in their encounters with bats are shown in Figure 1-16.

Depending on the time available, that is, the distance of the approaching bat, responding moths have been observed to behave in one of two ways. If the bat is far enough away and the moth detects it but the bat may not have detected the moth yet, the moth makes a directed turn away from its present path. This frequently keeps it out of range of detection by the bat. If, however, the bat is closer, within the range where it may have detected the moth's echo and be closing in, then the moth displays a random, wildly gyrating, unpredictable dive. These

Figure 1-16 Flight paths of moths and bats recorded with camera shutter left open and live moths tossed into the air above an illuminated field. A, Moth *(bright wavy track)* **and bat** *(smooth lower track)* **both enter from left. The moth dives, and the paths approach each other, but then the moth turns sharply upward and escapes while the bat leaves picture to right. B, Bat enters from right side of photograph; moth enters in center from below in a looping flight. The paths intersect, indicating that the bat captured the moth.** Reprinted by permission of the publishers from *Nerve Cells and Insect Behavior,* by Kenneth D. Roeder, Cambridge, Mass.: Harvard University Press, Copyright © 1963 by the President and Fellows of Harvard College.

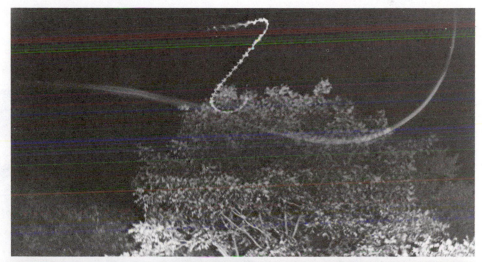

A

B

Figure 1-17 Method for observing moth responses to simulated bat calls. Calls are presented by an ultrasonic loudspeaker mounted on a mast. The area and any moths that enter the scene are lighted and photographed from the side. The camera shutter is left open to record moth flight paths as continuous tracks of light. Results of moth responses are shown in Figure 1-18.

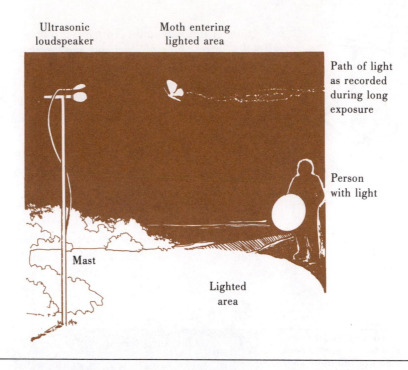

Ultrasonic
loudspeaker

Moth entering
lighted area

Path of light
as recorded
during long
exposure

Person
with light

Mast

Lighted
area

Figure 1-18 Flight responses of moths toward artificially presented, simulated bat calls. Arrows indicate position of moth when calls were initiated. Other objects and tracks in the pictures are from other insects at the edge of the field. A, Power dive. B, Passive dive interrupted by brief period of wing flapping. C, Looping dive. D, Directed upward movement away from sound source. E and F, Directed horizontal moves away from sound. Reprinted by permission of the publishers from *Nerve Cells and Insect Behavior*, by Kenneth D. Roeder, Cambridge, Mass.: Harvard University Press, Copyright © 1963 by the President and Fellows of Harvard College.

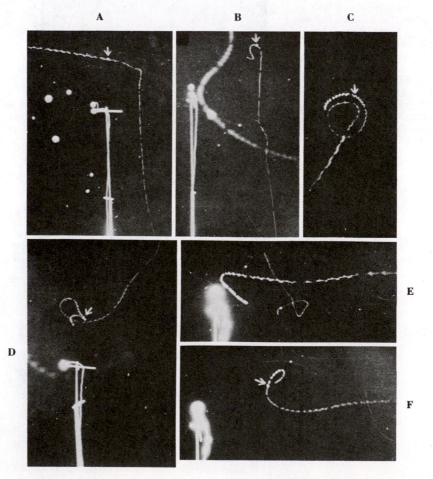

A   B   C

D

E

F

Figure 1-19 Technique used to record directed flight behavior of moths in presence of artificial bat calls in laboratory. The ultrasonic loudspeaker (1) directs calls at one side of the moth, which is mounted by the thorax to a phonograph pickup (2). A thermistor-anemometer (3) detects temperature changes created by differences in airflow as the moth attempts to turn while flying. Changes from the thermistors are balanced by a bridge circuit (4) and recorded as upward or downward deflections, depending on direction of turn, on the oscilloscope (5). Thoracic vibrations and loudspeaker pulses also are recorded on the oscilloscope as indicated. Modified from Roeder, K.D. 1967. Nerve cells and insect behavior. Harvard University Press, Cambridge, Mass.

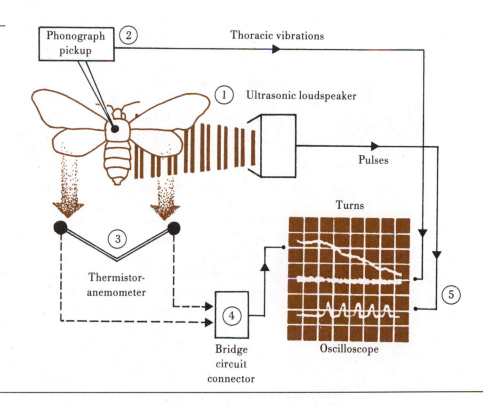

rapidly changing moves are difficult for the bat to track and intercept, and there is a reduced chance that the moth will be caught.

The erratic behavior of noctuid moths in response to high-pitched sounds can be observed fairly easily, such as by jingling keys or making other high-pitched noises near moths at a streetlight. With patience and if there are moths and bats around, the behavior and interactions can be observed at night with the aid of low illumination, such as from a 100 W light bulb, or near yard lights, if the hand is held up to block the light of the lamp from one's eyes. Observations on the erratic behavior as well as information on moth ears were published as early as 1919. With later knowledge of bat sounds, correlations were made between moth behavior and the hunting of bats (see Roeder 1967).

The directed turns and directional escapes, however, are more difficult to observe and were not known or even suspected before the 1960s. The possibility of such behavior was raised when neurophysiological studies (Chapter 13) showed that moths might be able to determine the *direction* of a sound source. Directional, rather than erratic, behavior was reasoned to be more advantageous if a bat were still far enough away.

Roeder (1962) confirmed the presence of two basic evasive behavior patterns in moths with a controlled source of artificial bat calls. An ultrasonic transmitter capable of producing a train of 70 kHz, 5 msec pulses at a rate of 30 per second was placed on a mast at the edge of a field. Intensity of the sound was adjusted to simulate bats at different distances. A floodlight and camera were used to observe and record the behavior of passing moths (Figure 1-17). The camera shutter was opened and kept open to photograph, as a continuous line, the path of a moth

Figure 1-20 Silhouette of Arctiid moth showing position of ear (*oval* to right) and timbal, a click-producing organ (*striated object* to left of ear).

entering the field. After a sufficient length of flight had been recorded, the transmitter was turned on. Many moths behaved as predicted and displayed one of the two behavior patterns, depending on the intensity of the ultrasonic sounds. Other moths did not react at all. After recording the behavior, Roeder attempted to capture the moth with a net to identify its family and species. Figure 1-18 shows a few of the photographs obtained during the course of over 1000 observations.

The tendency of noctuid moths to turn in an oriented manner away from an ultrasonic sound of low intensity was further confirmed in the laboratory by using flying moths fixed to an electronic device that measured their attempts to turn. The moth was attached with a drop of wax on the surface of the thorax to an insect pin in a manner that did not interfere with the flapping of the wings. The pin was inserted into a phonograph pickup for amplification and recording. A differential, or directional, anemometer (a device for measuring wind speed) was placed behind the moth's wings to measure the direction of the wake produced by the movement of the wings. The moths in this setup showed clear directional responses to low-intensity ultrasonic sounds at one side or the other (Figure 1-19). If an ear was deafened, the moths always turned toward the "silent" side.

Some moths from another family (Arctiidae) display yet a different response to the calls of bats. These moths answer back to ultrasonic sounds with a series of rapid high-pitched clicks from a timbal organ on their thoraxes (Figure 1-20). The moths of this group taste bad, and bats were observed to refuse capturing and eating them. The clicking behavior appears to warn bats of this distastefulness.

It may seem reasonable, if not almost obvious at this point, that moth hearing evolved at least partly under the selective pressure imposed by bats. But, until tested, this remains only a hypothesis. Roeder and Treat (1961) attempted to test the hypothesis by observing 402 encounters between moths and bats. Each encounter was scored for whether of not the moth took evasive action and whether or not it survived. For every 100 moths showing evasive behavior that survived, only 60 not showing evasive behavior survived. This demonstrates a clear selective advantage for the evasive behavior and gives some support to the hypothesis. (It also raises the question of how moths with nonevasive behavior continue to exist!)

SO WHAT?

This bat-moth story focuses on predator and antipredator behavior, sound and hearing, and two species of animals. There are, however, many other kinds of behavior and other species of animals, and hearing plays only a small role in the broad, general consideration of behavior. If the bat-moth story were to be an example of something, rather than just a story of its own, what has it illustrated about animal behavior? The important points and questions raised by this example follow:

1. There are events surrounding an animal, such as the sudden encounter between a bat and moth, that may require prompt behavioral responses if the animal is to survive.

The predator-prey interactions considered here comprise only a fraction of the changing events that enter into every animal's life. Some changes occur more slowly, such as seasonal weather changes over a period of weeks, but they occur well within the lifetime of an individual animal and may be considered as relatively immediate, short-term factors. The immediate conditions that cause a particular behavioral response are called **proximate factors.**

2. Behavioral responses to proximate conditions may vary among different animals in similar situations. Some moths, for example, showed evasive behavior, and some did not; arctiids responded differently from noctuids. Different behavior may be shown by the same animal in different situations. Moths responded differently, for example, to close versus distant bats. A given individual may possess the capacity for many different behavior patterns, and different animals may behave quite differently even under similar circumstances.

3. An animal's behavior depends to a large part on the biological structures that the animal possesses. The structures may be quite different in different organisms (e.g., the ears of bats and the ears of moths), but all are subject to the same principles such as the physics of sound waves. The structures may be relatively simple or extremely elaborate and sophisticated. The structures do not tell the whole story, however. How do the senses enter the information into the nervous system? How is it processed, and how are the appropriate movements brought about? Does experience alter subsequent behavior?

4. Each animal has its own sensory and perceived world. This concept often is associated with the term *Umwelt* (German for own or self world) introduced by Jakob von Uexküll in a classic book, *Umwelt und Innenwelt der Tiere* (1909). (Von Uexküll's original use of the term involved more than just sensory and perceptual aspects of an animal's relationship with its environment. For a translation of Von Uexküll's work, see Schiller, 1957. The matter also is discusssed briefly in Chapter 14. The common, contemporary view of *Umwelt* as an animal's perceptual world, however, is adequate for present purposes.) Humans usually take their own perceptual world for granted and assume that what they see is all there is and the only way it is. But it is a serious mistake to think that all other animals sense and perceive the world in the same way as humans do.

5. Further contemplation of the observations suggest that there is more to behavior than just the immediate, proximate conditions. A moth, for example, has to respond appropriately on its first encounter with a bat. It may not get a second chance, and it may not have time to invent a solution to the problem on the spot. Furthermore, although there is some variability in responses, many different moths may give similar responses in similar situations; such consistency would not be likely if each moth were acting completely independently. If an organism possesses particular biological structures and behavioral responses, what is their source?

Behavioral responses are **adaptive**; that is, they are correlated with a function. The moth's responses (such as erratic versus directed flight) were appropriate to the particular situations (bats near or far). Thus it is reasonable to consider the long-term factors, events that occurred during the ancestry of the bats and moths, which led to their present-day adaptations. These long-term evolutionary causes of behavior are called **ultimate factors,** as opposed to the closer, more immediate proximate factors. Ultimate factors are essentially synonymous with selective pressures, but the terminology is less tautological (explaining itself) when discussed in connection with the process of natural selection.

A *function* is what something, such as a leg, wing, or particular behavior, *does*. The current biological interpretation is that functions are the *results* of ultimate factors interacting with biological systems over long periods. A loose and less neutral term for

Figure 1-21 Relationship between ultimate and proximate factors in the life of an individual animal.

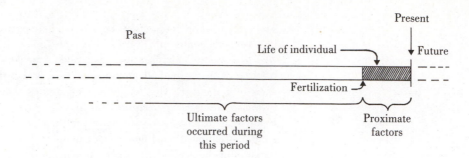

function is *purpose*. Use of the word "purpose" should be avoided, however, because it also may imply a larger design than many biologists are willing to accept or, in other contexts, may imply conscious purpose on the part of the animal. The presence of consciousness in species other than humans thus far has resisted testing and demonstration, although it has generated much debate.

Ultimate factors impose their effects through natural selection operating on the differences among different animals in their inherited abilities to survive and reproduce. Ultimate factors can be distinguished from proximate factors essentially by whether they occur *before* or *during* the life of an individual animal (Figure 1-21). Note, incidentally, that the anticipation of *future* events also could play a potential role in behavior; this topic is discussed in Chapter 5.

What are the ultimate factors in the lives of bats and moths, and how do they operate to produce the observed results? Also, if ears in moths are so valuable, why do not all species of moths have them? Just as one can break the question, "Why does a moth flee from a bat?" into proximate answers ("because the bat calls are sensed, processed in the nervous system, and cause a sequence of contractions in flight muscles that lead to directed flight") and ultimate answers ("because the ancestors of present-day moths did it and, as a result, had higher survival and reproduction"), the distinction between proximate and ultimate factors can be applied to virtually all behavior. Thus one could ask, for example, "Why does a male dog chase after a female in heat?" or, for that matter, "Why do humans engage in sex?" The answers to the latter questions are not just "to produce offspring."

The concept of ultimate causation is critical to an understanding of behavior but, at the same time, often seems circuitous and confusing to many people. For these reasons the general subject of the evolution of behavior will be addressed early in this book. The distinction between ultimate and proximate factors in behavior is important not only for conceptual reasons. It forms a basis for dividing the entire subject of behavior into two major subdivisions. The historical development of interest in ultimate versus proximate causation is discussed in Chapter 2.

6. Even "simple," seemingly uninteresting behavior as bats and moths flying around at night can prove fascinating once one understands what really is taking place. Achieving that understanding may require dedicated effort and special techniques employed by many persons over long periods. One must consider not only the findings and current interpretations of science but also the research history and methodological substrate on which these findings rest. In this example the use of ultrasonic, electronic sound–producing and detection equipment as well as bat-simulating devices and differential anemometers were encountered. Thus understanding of behavior relies heavily on a historical background, various techniques often involving specialized equipment, curiosity, ingenuity, and persistence, along with other very real human emotions and interactions including cooperation or jealousy. Recall, for example,

Spallanzani-Jurine-Cuvier, Griffin, and Roeder. The course of behavioral research is not always smooth; interpretations change; and the end is usually open. The discovery that bats use ultrasonic calls and echolocation took over 120 years, including some confusion and controversy, after Spallanzani first implicated the use of hearing by bats for orientation. Armchair thinking is important, if not necessary, but it sometimes leads one astray, as with Cuvier's explanation of bat orientation. Previous history and scientific methods are extremely important in behavioral biology. They are so important that the next two chapters will be devoted to these topics, and they will be noted throughout the rest of book.

SUMMARY

Even behavior that appears to be quite simple and uninteresting at first glance may be found to be rather complex and extremely fascinating on closer inspection, as illustrated by the behavioral interactions between bats and moths. Yet behavior is not so impossibly complicated that one cannot understand it with patience, persistence, and proper methods. Many of the techniques simply extend one's senses and slow down or stop the action so one has more time to comprehend it or "translate" what is going on into a visual form for easier understanding by humans, who percieve the world primarily through sight. Ultrasonic bat calls, for example, had to be slowed down to bring them into the range of human hearing or illustrated via sonograms; echoes were made visible with the aid of an oscilloscope; and lights and photographs were used to record behavior that occurred in the dark.

This introduction focuses on the structure and function of behavior and the fact that there are both proximate and ultimate factors that contribute to the final outcome. These points are sometimes referred to, respectively, as the "how's" and "why's" of behavior. Which parts or systems within an organism are involved and how? The proximate factors in behavior are mediated primarily (although not exclusively) through the neural and endocrine systems. These systems are considered in Part Three. Ultimate factors interact with living organisms through genetic and evolutionary processes, which are reviewed in Chapters 4 and 5. Ecological factors are discussed along with descriptions of behavior in Part Two. First, in Chapters 2 and 3, respectively, the general history and methods of behavior will be considered.

Recommended Additional Reading

Griffin, D.R. 1958. Listening in the dark. Yale University Press, New Haven, Conn. (Reprinted 1974. Dover Publications, Inc., New York.)

Griffin, D.R. 1959. Echoes of bats and men. Doubleday & Co., Inc., New York.

Roeder, K.D. 1967. Nerve cells and insect behavior. Harvard University Press, Cambridge, Mass.

Simmons, J.A., M.B. Fenton, and M.J. O'Farrell. 1979. Echolocation and pursuit of prey by bats. Science 203:16-21.

Wimsatt, W.A. 1970, 1977. Biology of bats. 3 vols. Academic Press, Inc., New York.

Yalden, D.W., and P.A. Morris. 1975. The lives of bats. Quadrangle, New York.

CHAPTER 2

Charles Darwin.

HISTORY
ETHOLOGY,
COMPARATIVE
PSYCHOLOGY, AND
NEUROBIOLOGY

The historical past is important for two major reasons. First, present understanding rests on a broad foundation of previous ideas, personalities, and techniques. One cannot truly understand today without also understanding the past. Second, although some past ideas are clearly wrong, if not even amusing in some cases, they may nonetheless contain a few elements of truth and remain at least heuristically useful; that is, by knowing what is wrong with the old ideas and why they no longer are accepted, one greatly broadens his or her understanding of what remains today. Otherwise there is little or no context against which to judge current ideas. It is useful, if not sometimes actually humbling, to trace back the roots of "modern" ideas and discover that they often are not so modern after all. Frequently the idea, or an important kernel of it, was present some time ago, but with insufficient data to confirm its validity. Much of what is possessed today is not so much new ideas as the ability, based on improved information, to differentiate between past valid ideas and those which were less correct or wrong.

History, although important, still must be kept in balance. Unless one personally knows the people involved or really appreciates the magnitude of their contributions (which takes more perspective than an introductory course), history is like someone else's home movies or a glass of warm milk; it readily can put one to sleep. Accordingly, this review will cover only some of the highlights, essentially in summary fashion.

Because science is done by scientists, I will take the liberty to include a few names. Persons who are included represent only a small, incomplete sample of some of the familiar names from particular fields as a starting point for discussion. Each person is an individual and often does not fit easily with or between other individuals. Some persons contributed over a long period of time, and often their views changed over the span. Some persons contributed through one or a few important publications, whereas others contributed through a continuing series of many publications and via other (private and public) interactions with their colleagues, students, and adversaries.

Any attempt to view the history of behavioral study should note the exponential increase, if not explosion, in the interest and number of persons involved. An estimate of the number of psychologists in the United States, for example, shows 228 in 1910, slightly over 1000 in 1930, and well over 45,000 in the late 1970s (Baron et al. 1978)! Comparable growth also occurred in other disciplines con-

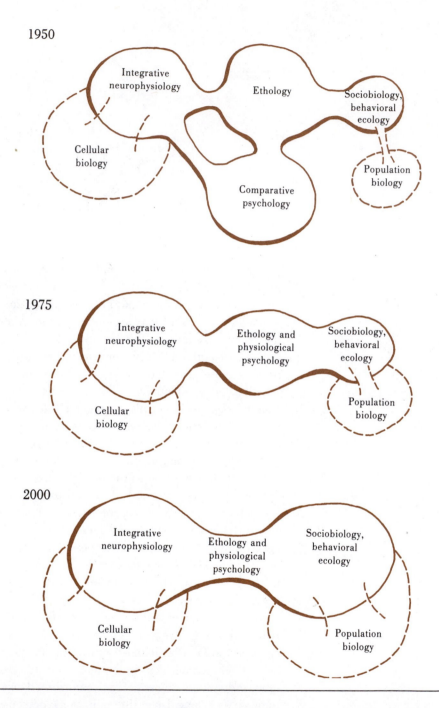

cerned with behavior. Professional meetings, which at one time were attended by a handful of people, suddenly began counting attendance in the hundreds and thousands. A similar explosion has occurred in the literature. In terms of the contributions of individual persons, although one can point to a few important forefathers, it is no longer possible to give equal credit where credit is due. Today there are *many* highly qualified individuals who are making significant contributions on a broad front. Thus, although I will take the liberty to include a few names from the 1970s, it is not meant to give them special recognition; they are, again, just a very small sample.

Given these qualifications, what kind of historical overview can one provide for an understanding of behavior? Present-day understanding of behavior is based on a complex, intertwined splitting and joining of disciplines. The subject has reached into and drawn knowledge from virtually all aspects of animal biology as well as psychology. The historical developments have not been all dull and boring. The subject has had its share of controversy and lively debate with some of the academic scuffling in the bushes even attracting the attention of passersby.

Wilson (1975) provided a subjective impression of changes in behavioral biology from 1950 to 1975 and projected where it might lead by the year 2000 (Figure 2-1). He pictured the subject something as an amoeba drawing its sustenance from cellular biology on one hand and population biology on the other. Behavior at the time (and yet today) commonly is considered to be unified by the central fields of ethology and comparative psychology. Wilson (1975) predicted, however, that these two fields will be "cannibalized by neurophysiology and sensory physiology from one end and sociobiology and behavioral ecology from the other," leading to something of a dumbbell-shaped configuration.

The basic division of academic interest in behavior follows the distinction between ultimate and proximate factors. Bullock (1981) described the difference, somewhat poetically, as "ecology over the eons" versus "physiology of the neurons." Such a division has been characterized also for Russian ethology, although some Russians dispute the characterization (Kovach 1973). Although the two basic branches are quite distinct, a complete understanding of behavior requires that they remain tied together, and one must deal with both, although perhaps not simultaneously.

The picture can now be expanded by tracing and adding the historical descent of knowledge in behavior before 1950 (Figure 2-2). This view is subjective and arbitrary, just as was Wilson's of 1975, and many of the details certainly could be debated. Figure 2-2 is only a rough, two-dimensional abstraction. The dimensions are time (on an exponential scale to represent increased growth and interest) and, more subjectively, fields of interest. An improved diagram would involve three or more dimensions. Neurobiology, for example, is as close to ethology as it is to comparative psychology, and the three topics are more like the legs of a three-legged stool. One person who looked at an early version of the chart offered to program it in several dimensions into a computer so it could be rotated graphically and better represent both the subject matter and the workers. A different early version of the diagram was constructed as a tree with numerous, intertwining branches, but it became too complex to be meaningful. This chart is meant merely to provide an overview and help organize a mass of historical bits and pieces.

The persons listed are a few of those whose names are generally familiar to behavioral biologists. The list does not include well-known behaviorists whose works are oriented primarily toward particular taxonomic groups, such as Schaller with gorillas and large cats, Geist with ungulates, Mech with wolves, several persons with particular groups of birds, and many people with primate studies. Taxonomy would add another dimension to this picture.

The positions of many of the names on the chart are misleading, resulting from the two-dimensional view and other problems. The three-way division breaks down after the 1950s, and the horizontal position is only partially meaningful.

Figure 2-2 Descent of behavioral biology—a subjective impression of the history and relationships of the major disciplines contributing to the subject. This figure is an abstract, two-dimensional shadow of a historical structure that is better represented by three or more dimensions; see text for qualifications.

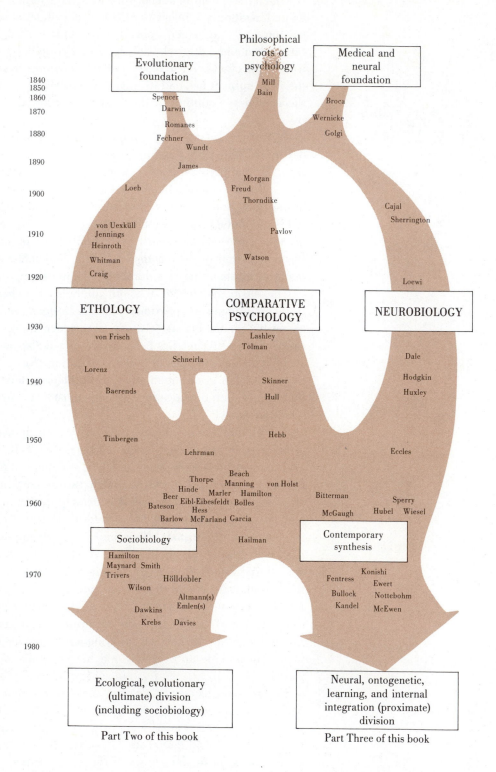

Bitterman, for example, remains quite clearly in the traditional psychology viewpoint. Von Holst, on the other hand, appears to be under psychology in this chart, but his position results, instead, from being between ethology and neurobiology. Many of the persons have or have had multiple interests or otherwise are difficult to pin to a single position.

This chart (Figure 2-2), although it has inadequacies, can be used as the basis for several exercises to help gain a better understanding of historical aspects. For suggestions, one can investigate the contributions of persons listed but not discussed, try to construct improved models of the figure, represent persons by areas rather than by single points, add additional names to the chart, fit one's own interests and background to the figure, and try to find areas that might be improved. This will demolish the figure, but a lot of history will be learned in the process.

In broad outline, then, modern understanding of behavioral biology started from two branches. One major conceptual line is identified with Darwin and evolutionary thinking. The other involves knowledge about neural anatomy and early attempts to understand the human brain. The two fields of evolution versus neurophysiology were somewhat separate from the start, but the initial division might better be described as biology of the natural living world as opposed to human medical interests. Only recently has it emerged as the distinction between ultimate and proximate causation.

EARLY BEGINNINGS

History is a slippery topic because human thought and understanding form a continuity with few clear beginnings and endings. Furthermore, there is much that is arbitrary. Different persons perceive and describe the same events in quite different ways. These problems led Flugel (1933) to begin a book on the history of psychology with the disclaimer that "such a book as this is almost inevitably bad."

Thus it is difficult if not impossible to decide where understanding of behavior began. One can go back to or before the ancient Greek anatomists and philosophers, including Alcmaeon (500 BC) and Aristotle (340 BC). An Egyptian document (around 1700 BC; cited in Bullock et al. 1977) included descriptions of head injuries associated with brain functions. But most of the early ideas and observations were either forgotten or solidified into authoritarian viewpoints not subject to revision and correction during the 400 to 1200 years of the Dark Ages.

The long period of ignorance continued with only a few isolated and mostly unimportant discoveries related to behavior. One of the earliest recorded references to behavior, just recently brought to attention (Kevan 1976), concerns the phenomenon of imprinting. In Thomas More's *Utopia* (1518; translated from Latin in 1551 by R. Robynson) there is the following account:

They brynge vp a great multytude of pulleyne, and that by a meruelous policie.
For the hennes doo not syt vpon the egges: but by kepynge them in a certayne
equall heate, they brynge lyfe into them, and hatche them. The chyckens, assone
as they be come owte of the shell, followe men and women in steade of the
hennes.

This account, however, can be viewed mostly as incidental; it did not, in itself, play a significant role in the history of behavioral science.

Medical Anatomy and Physiology

Perhaps the first small but truly significant beginning for a scientific view of behavior can be identified with A. Vesalius in 1543 with publication of the anatomical work *De Humani Corporis Fabrica*. This work included nothing on behavior per se, but it included detailed figures of the human brain and rekindled interest in original observation as opposed to citing ancient authorities. This was followed by scattered work on the brain and nerves, including investigations by Galvani in 1791 and Helmholtz in 1850 on conduction of nerve impulses in frogs. This path in neural anatomy and physiology formed the beginnings for one of the two foundations of modern understanding of behavior.

The Evolutionary Foundation

The second foundation for modern understanding of behavior arose with Darwin and Wallace's theory of evolution. As with neural anatomy and physiology, psychology and interest in human behavior can be traced back to the ancient civilizations and can be found in works from the sixteenth century onward. But most of the early psychology was more philosophically oriented and not on the same track as modern psychology. Contemporary scientific psychology generally is considered to have started during the nineteenth century and shares its basic beginnings with biological interests in behavior following Darwin. Older, philosophical views of psychology were joined to physiological and other biological aspects during the latter half of the nineteenth century by such workers as J.S. Mill, A. Bain, and W. Wundt. Mill, incidentally, was a brilliant thinker who was well versed in many fields, only one of which was psychology.

Before Darwin, interest in animal behavior, from any standpoint, was largely incidental to other interests such as animal husbandry or training animals for the circus. In 1855 just before the publication in 1859 of Darwin's *The Origin of Species*, Herbert Spencer published the book *Principles of Psychology*. He anticipated some of Darwin's points of evolution but disagreed or was wrong on several others; he believed, for example, in the inheritance of acquired characteristics. One of Spencer's important points, however, was his belief in the continuity of mental states; that is, he thought that there was some kind of continuity in the psychology of lower animals to that of higher animals. Later (1896) he expanded his viewpoint to rank psychic states from simple reflex behavior to volition (free will or choice) and suggested that they graded from one to the other and that the distinction among different animals was quantitative and not qualitative.

Perhaps the most significant starting point in understanding behavior came with Darwin. In *The Origin of Species* he established the general principles of evolution but did not focus on behavior as such. In later books, particularly *Expression of the Emotions in Man and Animals* (1873) and *The Descent of Man* (1871), Darwin did concentrate more on behavior. He applied evolutionary arguments to behavior as well as to other biological traits and made many inferences concerning the internal mechanisms of behavior. The latter included his "principle of serviceable associated habits," which closely resembled later notions of associative learning, his "principle of antithesis," in which animals outwardly express their internal emotions (with opposite signals for opposite emotions, see p. 279), and his "principle of the direct action of the excited nervous system of the body, independently of the will and in part of habit." Darwin provided numerous examples, mostly anecdotal, and proposed several ideas that, after translation to modern language, still are accepted. He also included a few ideas that since have been refuted. In a real sense Darwin started the moves toward progress in understanding behavior by providing the conceptual framework on which further development could take place.

The next person in line was Darwin's student, George John Romanes. Romanes published a comparative analysis of mental function and evolution, *Mental Evolution in Animals* (1884). He later extended similar arguments to humans in *Mental Evolution in Man* (1889). He included unpublished material from Darwin, given to him by Darwin, in both books. One of the theses of these books was that behavior and psychological traits could be studied among different animals and, from that, one could infer phylogenetic relationships just as for morphological traits, a genuinely comparative viewpoint.

Romanes also discussed extensively conscious states and what he called "injective knowledge," that is, inferring what is going on inside some other person (or animal) by observing how he, she, or it reacts to particular circumstances and by knowing how one feels in the same situation. For example, if one experiences fear in a particular dangerous situation and one acts a certain way and if another animal is seen acting similarly in the same situation, then one may infer that it experiences fear also. This view is considered **anthropomorphic** or "of human form." (Anthropomorphism is discussed further in Chapter 5.)

The Development of Different Disciplines

The period immediately around and after the time of Romanes was exceptionally stimulating, and from around 1890 to 1910 many people became involved in the subject. Numerous books and articles were published, and differences of opinion began to develop. These differences of opinion and interest led to significant splits and the development of three major disciplines that largely went their own ways. All were concerned one way or another with the internal control of behavior, but they developed mostly independently of each other and also had other interests beyond the internal control of behavior. The three major branches were **ethology, comparative psychology,** and **neurobiology.** Each of these will be described separately.

The divisions that developed among the three branches were often a result of ignorance or lack of interest in what the other branches were doing and, in some cases, sharp differences of opinion and outright hostility toward one or both of the other branches. Beer (1975) described the division by comparing ethology to a "broad river, fed by numerous tributaries, and braided in its lower reaches through the division and shifting of its channels in the loose gravel that is its bed" but which is "a single fluvial system that is separated from others flowing in its vicinity and through similar terrain."

The word *ethology* (Greek *ēthos*, habit) was in use at least by the seventeenth century, 200 years before Darwin, but it was used in the context of human stage actors or, later, relative to human ethics. It was not until the middle of the nineteenth century that the term was used relative to living animals (as opposed to anatomical specimens), and even then it was used more in an ecological rather than in a behavioral context. It was not until about the 1940s that the term ethology came to be widely used in relation to the study of animal behavior in the animal's natural surroundings.

Ethology developed largely (but not entirely) in Europe, and comparative psychology was concentrated in the United States. Ethology was devoted to the study of animal behavior under natural conditions and focused on consistent species-specific patterns of behavior with relatively little emphasis on learning or human behavior. Ethologists were primarily biologists-zoologists and maintained a fairly broad biological view of behavior, with particular attention to evolution and phylogenetic relationships. Much of their work was anecdotal, and only some of the early ethologists used good experimental design and statistics in their work.

Comparative psychology, on the other hand, minimized, then eventually lost, for the most part, the evolutionary perspective. Comparative psychologists focused more narrowly on principles of associative learning, using laboratory and experimental studies with attention to statistics. Their basic underlying interest was the internal control (via learning) in humans. It was thought to resemble, if not to be identical with (except for capacity), learning in other vertebrates. The work of the comparative psychologists thus concentrated on a few species, particularly white rats, pigeons, dogs, and occasionally other species of rodents and primates.

Neurobiologists came from a basic biological background but were interested in the immediate, proximate mechanisms of the nervous system rather than whole, functional units of behavior or ultimate, evolutionary considerations.

One subdiscipline that was more or less intermediate between psychology and biology, because its members were generally well versed in both areas, was physiological psychology. The physiological psychologists were concerned largely with the proximate causes of behavior and tended to lack or not be interested in evolutionary aspects or adaptations of different animals living in different ecological settings. The underlying focus of the physiological psychologists, as with other types of psychologists, was human behavior. Other species were of interest mainly to the extent that they helped explain human behavior and not because they were of any interest in themselves.

From among these three main branches and all of the individuals involved, one person and his students and associates stood out as an important bridge between the psychological and ethological schools of thought. The person was T.C. Schneirla. In the midst of the period of division between the naturalist ethologists and the psychologists, he, his co-workers, and several productive students used information from both areas and presented a remarkably modern view of learning and the internal integration of behavior. A major publication along this line was the book *Principles of Animal Psychology* (Maier and Schneirla 1935).

Then and during the next two decades Schneirla maintained a distinctly important comparative view by taking the best from both biology and psychology. He considered the internal mechanisms, such as the differences between the nervous systems of different taxonomic groups, and looked broadly at both invertebrates and vertebrates. He related differences among different organisms to their evolutionary backgrounds and to the different demands of surrounding, natural environmental factors. Particularly important for learning theory (discussed further in

Chapter 19) was Schneirla's finding that maze-learning behavior in ants showed major differences from maze learning by rats. Today more details can be added, but his basic viewpoints remain valid and unchanged.

As more people became involved in and aware of what the other disciplines were doing, ethology, comparative psychology, and neurobiology gradually began to reunite. There have been several heated exchanges. Ethologists accused the psychologists of being narrow minded and ignorant of evolutionary perspective, while psychologists often accused ethologists of being too sloppy and anecdotal (Lehrman 1953). Some persons believed the old fields were dying or dead (Lockard 1971). The division between the two camps of ethology and comparative psychology was particularly deep, and much of it exists today, as evidenced by splits that still exist between respective departments on many university campuses. But generally there has been much merging of thought and information, and the two groups are at least speaking more to each other and, in many cases, have genuinely joined, for example, as single departments at several universities.

Beer (1975) described the changed attitudes and atmosphere in a memorial to one of the major opponents, D.S. Lehrman. According to Beer, "The second half of the 1950's and early 1960's was a period of questioning and dismantling as far as ethological instinct theory was concerned." From the mid-1950s on there were several conferences that provided some of the most important interchanges of ideas. At one point Lorenz (p. 35) even helped hold a blackboard for Lehrman, an incident that drew loud applause from the audience. Beer remarked that the "conference was notable . . . for its spirit of harmony, enthusiasm, and optimism."

A modern synthesis seems to be gradually emerging, and the many remaining problems concerning internal control of behavior are being pursued jointly by persons with backgrounds in the three major branches. A brief discussion of the history of each of the branches follows.

ETHOLOGY

Continuing in what was already something of a tradition from earlier workers, two naturalists, Charles Whitman of the University of Chicago and Oskar Heinroth of the Berlin Aquarium, are sometimes considered to be the founders of the modern discipline of ethology (Kandel 1976, Lorenz 1981). Whitman studied the reproductive behavior of closely related species of pigeons, and Heinroth worked with duck behavior. Both came to essentially the same conclusions: the displays of different species are remarkably constant and can be used as taxonomic characters.

Early Ethological Concepts and Terms

One of Whitman's students, Wallace Craig, studied complex behaviors further and noted that they often consist of two components: (1) a steering or *appetitive* component and (2) a *consummatory* part. During the appetitive part the animal actively searches and orients its movements to external stimuli. The appetitive part is variable, whereas the consummatory part is more fixed and stereotyped.

Figure 2-3 Egg retrieval by grey lag goose. The goose uses its beak and neck movements to guide an egg back into the nest after it has accidentally rolled or has been experimentally placed out of the nest. From Lorenz, K., and N. Tinbergen. 1938. Z. Tierpsychol. 2:1-29.

Figure 2-4 Konrad Lorenz. Thomas McAvoy, Life Magazine ©1955 Time, Inc.

Morris (1957) likened the difference to that between an automobile accelerator (more pressure leads to faster acceleration) and the ringing of a telephone (always rings at the same rate no matter how urgent the call).

The classic example in animal behavior involves egg retrieval by birds such as geese (Figure 2-3). If a goose egg rolls out of the nest, the goose reaches out with its beak and brings it back. As the egg wobbles, the goose must correct for the displacement by moving its beak and steering the egg back. Once back at the nest, the egg is tucked under the bird's body in a very stereotyped fashion that seems insensitive to external stimulation. In fact, if the egg is manually removed by an observer before it is all the way under the goose, the goose continues to perform the full sequence.

Further work with different species, particularly in birds, fishes, and insects, was conducted by many naturalists, the most prominent being Konrad Lorenz (Figure 2-4), Karl von Frisch (Figure 2-5), and Niko Tinbergen (Figure 2-6). Their methods are important in three aspects: (1) they observed animals in their

Figure 2-5 Karl von Frisch. From Hickman, C.P., Jr., L.S. Roberts, and F.M. Hickman. 1982. Biology of animals, 3rd ed. The C.V. Mosby Co., St. Louis. Photo by W.S. Hoar.

Figure 2-6 Niko Tinbergen.
From Hickman, C.P., Jr., L.S.
Roberts, and F.M. Hickman.
1982. Biology of animals, 3rd
ed. The C.V. Mosby Co., St.
Louis. Photo by W.S. Hoar.

natural habitat under conditions in which the behavior had evolved; (2) they
studied both proximate and ultimate levels of causation; and (3) they worked with
many different species of both invertebrates and vertebrates (Klopfer 1974). Tinbergen published a concise book, *The Study of Instinct*, in 1951. It summarized
and synthesized the findings to that point and has since become a classic. In 1973
Lorenz, von Frisch, and Tinbergen jointly received the Nobel prize for their
work.

Several important generalizations resulted from the work of these and several
other naturalists. First, the expression of particular behaviors in many species is
remarkably stereotyped or relatively constant among all individuals of the same
age and sex under similar circumstances. This led Lorenz to formulate the notion
of the ***fixed action pattern (FAP).*** More careful quantitative recent studies have
shown that behavior is less constant than Lorenz's FAP suggested. (An early
psychologist, William James, had been closer to the truth in 1890; see p. 43.)
Accordingly, Barlow (1968, 1977) relaxed the requirement of stereotypy somewhat and referred to ***modal action patterns (MAPs).*** Schleidt (1974) provides a
further, detailed discussion of the concept of FAP.

A second important finding from the ethologists was that the stimulus required
to trigger a response could be very simple. This principle was observed among a
wide variety of species; classic examples are fighting responses of sticklebacks
(Figure 2-7), territorial threat responses in European robins (Figure 2-8), begging
responses in newly hatched gulls (Figure 2-9), and the escape responses of birds
triggered by silhouettes of hawks and other birds (Figure 2-10). (More will be said

Figure 2-7 Courtship and fighting in three-spined sticklebacks. A, Two unrestrained males fighting. The fish on the left has assumed the threat posture, which elicits an attack from the fish on the right. **B,** Experimental manipulations of posture by restraining males in glass tubes. The vertical, "threatening" posture elicits an attack, whereas the horizontal, normal one does not. **C,** Experimental manipulations of stimuli (shape, color, realistic overall appearance) via models. All models with red bellies, regardless of overall shape and realistic appearance, elicited attack from male sticklebacks, whereas models without the red, including the most realistic fish, did not. The results shown in B and C suggest that a combination of two stimuli, posture and red belly color, are important in stimulating male fighting behavior. Male courting of female sticklebacks also is stimulated by limited features of the female. **D,** A female in normal posture and shape does not elicit vigorous courting, whereas a crude model with swollen belly or a fish of another species but in the proper posture (E) does. A, B, and **E** from Pelkwijk, J.J. Ter, and N. Tinbergen. 1937. Z. Tierpsychol. 1:193-204. **C** from Tinbergen, N. 1948. Wilson Bull. 60:6-51. **D** from Tinbergen, N. 1942. Biblioth. Biother. 1:39-98.

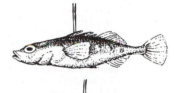

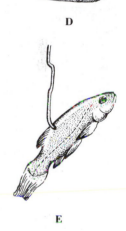

D

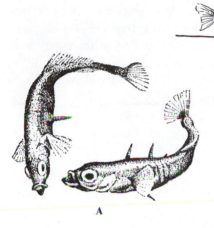

A

B

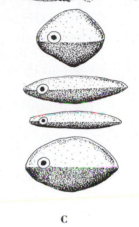

C

E

Figure 2-8 Models used to test the effective stimulus (a red breast) for territorial threat displays in male European robins. The mounted real robin without a red breast elicited much less of a response than a tuft of red feathers. Modified from Lack, D. 1943. The life of the robin. H.F. & G. Witherby Co., Ltd., London.

Figure 2-9 Cardboard models of herring gull heads used to measure the food-begging, pecking response of chicks. Magnitude of the response is shown by the length of the bar by each model. A, Colored spots on mandibles of normal color. B, Spots of varying shade from white to black against a gray mandible. C, Uniformly colored mandibles of different colors. Red elicited the greatest response, and the presence and contrast of a spot on the beak appeared to be important. (Note that the unnatural solid red beak elicited the greatest response. One might ponder why herring gulls do not have solid red mandibles in nature.) Redrawn from Tinbergen, N. 1949. Bijdragen tot de Dierkunde 28:453-465.

about the hawk-response later.) A live bird that lacks the appropriate signal might not trigger a response, whereas a properly colored bundle of feathers on a piece of wire will. Similarly, pieces of cardboard and other simple models shaped and painted properly can stimulate whole patterns of complex behavior. The critical stimulus properties are called *sign stimuli.* Those which are used as genuine signals in communication by a species are referred to as *releasers.* The term releaser sometimes is used synonymously with sign stimulus, but technically releasers are special, intraspecific cases of sign stimuli.

Occasionally responses (FAPs) occur in the absence of the appropriate stimulus. These are called *vacuum behaviors.* Insectivorous birds deprived of flying insects, for example, may fly out and go through all the motions of catching, killing, and eating an imaginary insect. Occasionally vacuum behaviors are performed with inappropriate or minimal stimuli. Rats, for example, may carry and use their own tails in an attempt to build a nest in the absence of suitable materials (Eibl-Eibesfeldt 1958, cited in Lorenz 1981). One of the most remarkable cases of vacuum behavior that I have observed involved our dog burying a bone on the linoleum in a corner of the kitchen. It first dug an imaginary hole, then dropped the bone into it. Next the dog carefully shoveled imaginary dirt up over the bone with its nose and tamped it in place. Then it walked away with the bone lying in full view on top of the linoleum.

Some artificial stimuli were found to elicit responses more effectively than the natural, normal stimuli. A larger than normal egg, for example, may be incubated by some birds in preference over their own eggs (Figure 2-11). Such stimuli are called *supernormal* releasers.

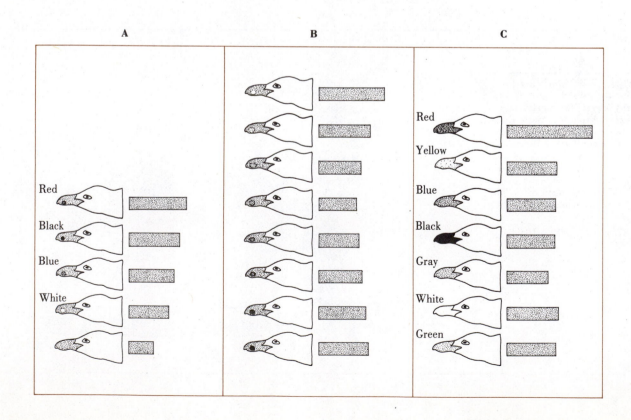

Figure 2-10 Models used to test the escape responses of young gallinaceous birds and waterfowl to flying birds of prey. Models that elicited responses are indicated with a plus sign. A general model that was flown over subjects in one direction resembled a hawk and elicited a response, whereas when it was flown in the opposite direction, it resembled a goose and did not receive an escape response. The important stimulus characteristic that was inferred from these results was "short neck." From Tinbergen, N. 1948. Wilson Bull. 60:6-51.

Putting the sign stimulus together with the FAP led Lorenz to the concept of the ***innate releasing mechanism (IRM).*** The sign stimulus supposedly released the FAP via the IRM (Figure 2-12). The true neurological nature of the IRM was not known, but Lorenz in his famous hydraulic (or "flush toilet") model compared the process to the filling of a reservoir. The water could be released by a valve or might spill over the top. The hydraulic model was not meant to represent exactly what was going on inside the nervous system but only to serve as an analogy. Lorenz (1981) recently resurrected and slightly modified his hydraulic analogy (Figure 2-13). In terms of the nervous system, Lorenz postulated an ***action-specific energy*** to describe a buildup of motivation for particular behavior patterns.

Problems with the Concepts

The notions associated with action-specific energy, IRMs, and similar concepts such as drive have been substantially rejected as too vague and not based on neurological foundations (Lehrman 1953, Hinde 1956, 1970, McFarland and Sibly 1972, Andrew 1974). In discussing drives, such as "thirst," Hinde (1970) pointed out several problems that overlap to some extent:

1. Drives often are not "defined independently of the variations in behavior they are supposed to explain." There may be much circularity, for example, in defining thirst as the tendency to drink.
2. How well drives correlate with behavior may depend on *which aspects of the behav-*

Figure 2-11 Supernormal stimulus. The oyster catcher is attempting to incubate a large artificial egg chosen over a normal egg *(foreground)* or a herring gull egg *(left)*. From Tinbergen, N. 1951. The study of instinct. Oxford University Press, Inc., New York.

Figure 2-12 Relationship of the innate releasing mechanism *(IRM)* **to the sign stimulus and fixed action pattern** *(FAP)* **in the classic ethological viewpoint.**

Sign stimulus   IRM   FAP

ior are measured and *how they are measured*. Different measures of the same aspect of behavior often produce different results.

3. Drives have been defined in different ways by different authors; therefore the terminology is ambiguous.

4. It is not clear which level of behavior drive should be applied to. In nest-building behavior, for example, is there a reproductive drive, a nest-building drive, a stick-carrying drive, or all of these?

5. The use of drive can be an oversimplification and can obscure what really is occurring. One may think one has an answer when really there is only a new name for one's ignorance.

6. Drive can be misleading. One may look for or provide detailed explanations of things that do not even exist.

Hinde concludes "not that drive concepts are always useful or always useless, but that they have a limited range of usefulness and that they can be misleading and dangerous if misused." He notes further, "This, of course, is true of all explanatory concepts."

Figure 2-13 Lorenz's old (A) and new (B) "psycho-hydraulic" models of motivation. The level of water represents the level of "action-specific potential," which may be released by sign stimulus, or the weight at the end of the pulley (in A) or by other input (in B). Modified from Lorenz, K.Z. 1981. The foundations of ethology. Springer-Verlag New York, Inc., New York.

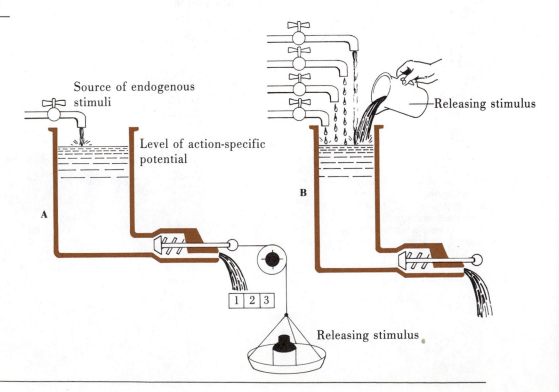

Source of endogenous stimuli

Level of action-specific potential

Releasing stimulus

A

B

Releasing stimulus

Table 2-1 Defining (D) and secondary (S) attributes or characteristics of fixed action patterns as interpreted from 10 sources

Criterion	Lorenz (1932)	Thorpe (1951)	Verplanck (1957)	Hess (1962)	Moltz (1965)	Barlow (1968)	Hinde (1970)	Eibl-Eibesfeldt (1972)	Alcock (1975)	Heymer (1977)
1. Triggered (form independent of environment)	D	D			D	D	D	S	D	
2. Rigid sequence				D		D		S		
3. Unlearned	D		D	D	D			S		D
4. Inherited	D	D		S	S			D		D
5. Species-characteristic	D	D	D					S	S	D
6. Stereotyped			D		D			S	S	D
7. Complex		D		S		S				
8. Unintelligent	D			S				S		
9. Contains imperfections	D									
10. Spontaneous					D					
11. Common causal factors						D				

From Dewsbury, D.A. 1978. Anim. Behav. 26:310-311.

Dewsbury (1978) attempted to clarify the ambiguous definition of FAP. He compared 10 sources on 11 attributes of the concept. His results are shown in Table 2-1. He concluded, "It is apparent that there is little unanimity of opinion. The term 'FAP' cannot presently be used in effective scientific communication."

The matter of innateness or instinct also has been subjected to much scrutiny. Beach (1955) reviewed the concept of instinct in depth. Also, development of food-begging, pecking responses of gull chicks, for example, has been shown by Hailman (1967) to involve learning. Schleidt (1961) demonstrated that turkey chick escape reactions in response to hawk silhouettes involves learning and habituation; they respond to any novel object passing overhead until becoming habituated to familiar forms. In nature hawks fly over rarely enough that the chicks do not habituate to them, but they do to forms of other birds that fly over more commonly. The hawk response story, incidentally, is not finished. Mueller and Parker (1980) showed that naive ducklings *do* have some innate responses, such as more variable heart rates, to hawk versus nonhawk silhouettes. Some of the differences in responses may depend on the species studied. Cassidy (1979) provides a general review of the concept of innateness. Regardless of cases involving learning, many stimulus-response relationships have been shown to involve little learning. Even learned responses often may be triggered by simple cues.

Further observations of the ethologists revealed that during times when two or more stimuli occur simultaneously, the effect of one usually predominates. Occasionally, however, two stimuli are approximately equal, and this produces **conflict behavior.** An example is a male fighting another male at a territorial border where the impulse to flee may be balanced by the impulse to stay and fight. Two major categories of conflict behavior have been identified: **redirected behavior,** in which the proper response is directed at a different object (e.g., pecking the

Figure 2-14 Hierarchical organization of instinct as envisioned by Tinbergen. A, Basic model of organization as shown for reproductive behavior in the male three-spined stickleback. B, Hypothesized application of the innate releasing mechanism (IRM) concept to two levels of an instinct. C, Conjunction of A and B to form a hierarchical system of levels of instincts controlled by levels of IRMs. From Tinbergen, N. 1951. The study of instinct. Oxford University Press, Inc., New York.

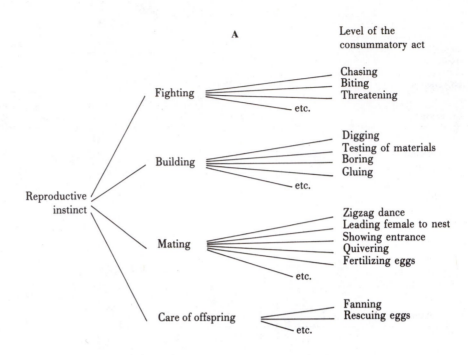

ground instead of the opponent), and ***displacement activity,*** in which an irrelevant response is suddenly given (e.g., preening or eating instead of either fleeing or fighting).

Several other behavioral responses also have been observed in conflict situations. These include inhibition of all but one response, intention movements (Chapter 6), alternation between behaviors, ambivalent behavior, immobility, compromise behavior, autonomic responses (e.g., defecation), and others (Hinde 1970).

In a tentative attempt to interpret some of these observations and somewhat more in line with known neurological mechanisms than Lorenz's IRMs, Tinbergen (1951) proposed hierarchical systems of centers underlying complex behavior. This also would account for higher and lower levels of behavioral organization, for example, reproduction consisting of lower levels such as nest building and mating, which, in turn, consist of lower levels (zigzag dances, etc.) (Figure 2-14). Factors such as hormones were envisioned as stimulating certain centers, and the IRM in this scheme was viewed as a system of inhibitions that, when disinhibited, released lower centers. This view is closer to our modern understanding of internal neural mechanisms (Part Three), but it concentrates too heavily on inhibitions that must be released (as opposed to centers that may be stimulated), and it conceives of an abstract organization, or hierarchy, that may have little real foundation in the nervous system. Furthermore, much of the stimulus filtering takes place peripherally at the sense organs rather than in the central nervous system.

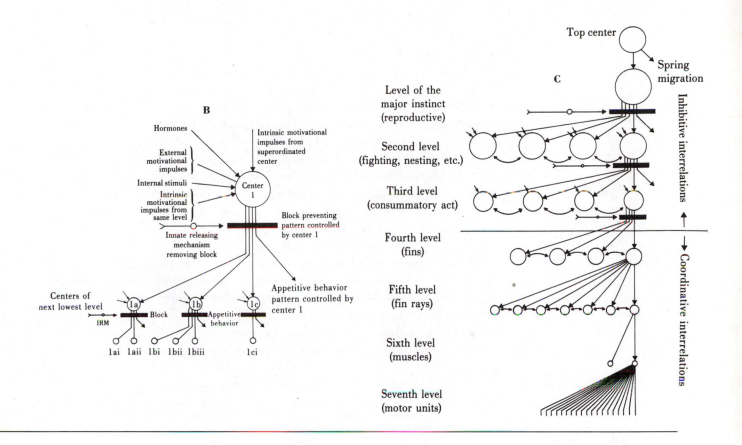

All of the early concepts and attempts to understand what goes on inside the animal, although now largely rejected or replaced by more recent concepts, nonetheless helped organize observations and led to progress and revisions of theory when different sets of information did not match as predicted. The main contributions of the ethologists, aside from describing some of the general characteristics of input-output, have been to increase the catalog of behavior as observed in a wide variety of animals under a wide variety of natural conditions and to emphasize the roles of ecology and evolutionary forces in causing those behaviors.

COMPARATIVE PSYCHOLOGY

Comparative psychology developed initially in close association with the branch that eventually became ethology. One of the early workers who influenced both branches was William James. In 1890 he published the book *Principles of Psychology*, which dealt with instincts or bundles of reflexes. He noted that instincts were not invariable: "What is called an instinct is usually only a tendency to act in a way of which the *average* is pretty constant, but which need not be mathematically 'true'" (1890:391). In addition, he proposed some general principles such as "inhibition by habit," by which the expression of particular behaviors is narrowed down, for example, to a particular mate or particular place to live.

Another principle is the "law of transitoriness," which was an early recognition of what is now called critical periods (Chapters 18 and 19).

The 30-year period following James' publication was active, and *many* people were involved, including Sigmund Freud. There was a profusion of interest in psychology, and numerous works were published at the time. During this period there was much controversy over consciousness and other internal states. Several schools of thought developed, including structuralism (concerned with the "structure" of mental states) and introspectionism (which dealt with looking inwardly toward one's awareness to understand what is going on inside). Efforts to deal with internal mental events were rejected gradually, and a shift was made toward experimental approaches and observing outward expressions of behavior, the foundations of subsequent comparative psychology.

One important person in this movement was C. Lloyd Morgan, who published *Introduction to Comparative Psychology* in 1894. Among his other contributions, he rejected anecdotalism and undisciplined anthropomorphism in the interpretation of behavior in other animals. He called for a principle of theoretical parsimony, which became known as **Morgan's canon:** "In no case may we interpret an action as the outcome of the exercise of a higher psychical faculty if it can be interpreted as the outcome of the exercise of one which stands lower in the psychological scale." A behavior should not be interpreted as resulting from thought, for example, if it can be ascribed to a conditioned reflex instead.

Figure 2-15 One of Thorndike's "problem boxes" (1911) from which animals had to learn to escape. Below the box are plotted the performance curves of a cat learning to escape from two different boxes. From Bitterman, M.E. 1979. Historical introduction. In Bitterman, M.E., et al. Animal learning: survey and analysis. Plenum Publishing Corp., New York.

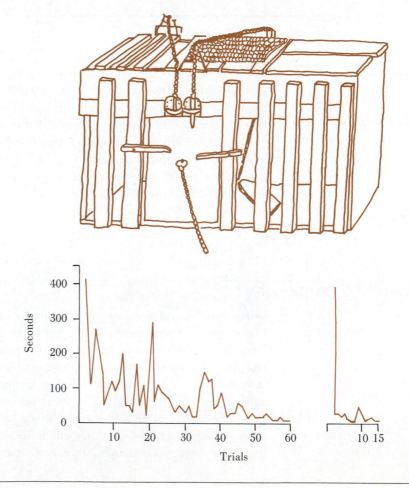

Cognitive versus Reinforcement Theories

Morgan's work did not exclude internal constructs; in fact, his work led to an important class of theories known as ***cognitive theories,*** which were developed further by others such as Tolman (1932) and his students. *Cognition* is difficult to define satisfactorily, but it refers generally to internal mental processes that may exist apart from immediate external stimuli. In latent learning (discussed more fully in Chapter 19), for example, an animal might be turned loose in a maze to roam at will. A few days later, if food is provided in the maze, the animal can quickly return to that place for subsequent food. Learning in this case is said to depend on a ***cognitive map*** of the maze inside the animal's head. The map does not initially involve reinforcement with food rewards. The *anatomical* structure of the map or other cognitive structures may not be known, but, based on the observed behavior, the *organizational* structure or characteristics involving such things as time relationships can be described.

A second branch of theories, known as ***reinforcement theories,*** also developed from Morgan's work. One of the initial architects of this line of thinking was Thorndike, who published a series of papers beginning in 1898. These papers later were collected into a single volume, *Animal Intelligence: Experimental Studies.* One of Thorndike's chief techniques was to put an animal into what he called a *problem box* (Figure 2-15). He plotted performance over subsequent trials. From this work Thorndike proposed his laws of *Exercise* and *Effect.* According to the law of Exercise, performance of a behavior improves with practice. The law of Effect states that the strength of a stimulus to evoke a response increases with pleasant consequences and decreases with unpleasant consequences.

Comparisons of the performance of rats, sparrows, and monkeys in similar sorts of mazes suggested that learning was similar in all species (Figure 2-16). Information from elsewhere, such as from Pavlov's ***classical conditioning*** studies of dog salivation in Russia (1906, translated into English in 1927), seemed to support these conclusions. (Pavlov received a Nobel prize in 1906, but it was for his work on digestive physiology rather than for his work on behavior.) In classical conditioning a normal response, such as salivation, becomes associated with a new stimulus, such as a ringing bell, rather than just the sight of food. The conditioning depends on pairing the unconditioned stimulus (US), in this case sight of food, with the conditioned stimulus (CS), ringing of bell. As a result of the general consensus that learning is similar in all organisms, comparative psychology moved from comparing behavior in a variety of animals and concentrated on comparing the properties of learning under a variety of circumstances.

Figure 2-16 The Hampton Court maze (developed by Small 1901) used in tests of learning ability. Beside the maze are comparative learning curves for three different animals: rat (Small 1901), sparrow (Porter 1904), and monkey (Kinnaman 1902).

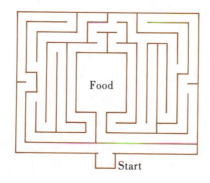

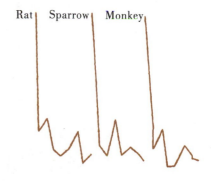

Behaviorism

Thorndike was followed by John B. Watson, who established the school of thought known as **behaviorism** in a series of publications from 1913 to 1930. The last book in the series is entitled *Behaviorism*. Watson's view is that one can only infer the nature of the animal mind from the outward observable behavior of the animal. The outward behavior has sometimes been referred to as "public" events and the internal mental processes as the "private" events beyond reach in an animal's life. Watson attempted to explain human behavior simply as a collection of stimulus-response reactions. Although Watson apparently did not spend much time in the laboratory or academic setting and was considered a bit eccentric, some of his students, such as Lashley, carried on with Watson's ideas.

Perhaps the next most prominent person in this line of thinking was B.F. Skinner (Figure 2-17). In 1938 he published the book *The Behavior of Organ-*

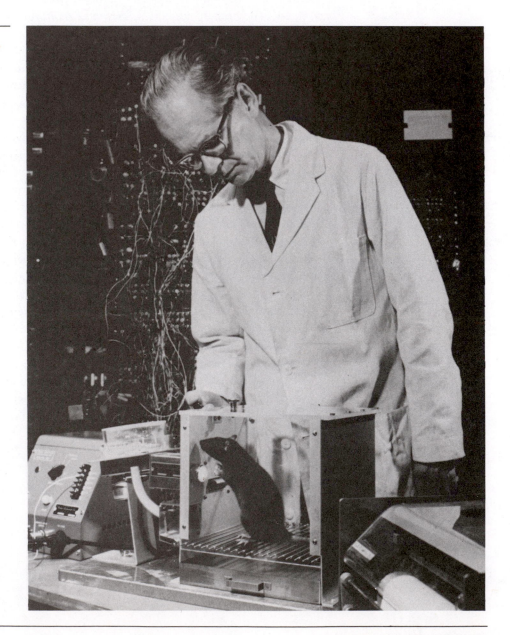

Figure 2-17 B.F. Skinner. A modern automated Skinner box also is shown. When the rat pushes a lever, a food pellet or other form of reinforcement is delivered to the animal. Nina Leen, Life Magazine ©1964 Time, Inc.

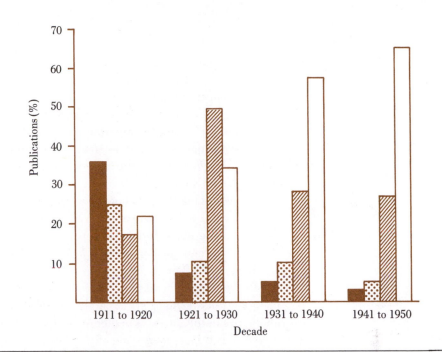

Figure 2-18 Numbers of published learning studies using laboratory rats compared with the use of other types of animals as subjects through 1948. Modified from Beach, F.A. 1950. Am. Psychol. 5:115-124.

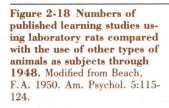

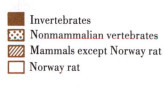

Invertebrates
Nonmammalian vertebrates
Mammals except Norway rat
Norway rat

isms, which is of major importance to psychology and studies of learning. Skinner pursued a class of conditioning that involved reinforcements. (Others had referred to these as *rewards* and *punishments,* but Skinner preferred to avoid implying internal effects.) Responses of animals that may be reinforced are called **operants,** and this type of learning became known as **operant conditioning.** This category also has been called **instrumental learning,** a term that Skinner and his colleagues disliked. Operants include responses such as pressing bars in boxes, and a standard method of studying such learning involved the use of Skinner boxes (Figure 2-17). Operant conditioning differs from classical conditioning in that it modifies the patterns of the response rather than focuses on different stimuli. Conditioned responses are shaped through sequences of reward and punishment.

Skinner took an extreme operational view of learning. Reinforcement was viewed simply as that which alters the probability of behavior, and all that is needed is to discover how one variable (stimulus) affects another (response). No theories are needed, and one need not worry about what is going on inside the animal, including neural or hormonal mechanisms, according to this viewpoint. Much of the concern in comparative psychology with the properties of classical and operant conditioning continued up through the 1950s and is still present. The variations and refinements of the themes have been incredibly diverse. There has been much controversy as to whether the two categories of conditioning are really two or are just one. The increased focus on learning studies using just a few species of subjects, particularly the white laboratory Norway rat, was plotted graphically by Beach in 1950 (Figure 2-18). Nearly 20 years later Hodos and Campbell (1969) observed that the situation had not changed much.

There have been a few notable exceptions to the mainstream focus on conditioning in comparative psychology. Schneirla, as discussed earlier, was one. In addition, several other categories of learning, such as habituation and imprinting (Chapter 19), had to be taken into account and somehow fitted into or alongside conditioning paradigms. Cognitive viewpoints demanded more attention. Garcia and colleagues in the 1960s and 1970s (e.g., Garcia et al. 1974) began quantifying taste aversion forms of learning and found that the results differed significantly from traditional forms of negative conditioning. It seemed that learning theory was developing cracks faster than they could be patched up.

Because of the major importance of learning and the contribution of comparative psychology to an understanding of the internal integration of behavior and because this goes beyond the historical aspects, learning will be discussed further in Chapters 18 and 19.

Before leaving the psychologists, however, and to provide additional contrast with the ethologists, one can note a potential difference in the types of behavior that the two groups study. Some persons (e.g., Denenberg 1972a) have distinguished between *species-specific behaviors,* the stereotyped specific patterns of behavior that often are unique to each species, and *psychological behaviors,* which are more general characteristics such as learning abilities, emotions, and memory traits. Ethologists generally have been interested in the former and psychologists in the latter. Whether or not the distinction is valid, however, remains to be seen. If it is, then some of the differences between ethological and psychological viewpoints may have resulted from the two groups working with different categories of behavior. If the distinction is not valid, however, both it and many of the historical differences between ethology and psychology simply may be artifacts of different conceptual schemes. Whether the differences are in the subject or merely in the processes of study has not been settled.

NEUROBIOLOGY

Neurobiology was divided from the start into two major subdivisions: one concerned mostly with anatomy and the other with physiology. Although the categories are not exclusive and one cannot generalize completely, the anatomists tended to come from the ranks of the human medical sciences, and the physiologists were generally more traditional, academic biologists or zoologists.

The human medical sciences have been heavily and historically involved with the nervous system, particularly the eyes, ears, brain, spinal cord, and neuromuscular interactions. Veterinary sciences also have played some role, but not to the extent that human medical sciences have. Horses and animals with broken legs, to say nothing of nervous system problems, normally are just disposed of. With humans, however, usually almost every effort had been made to keep persons alive and, if possible, repair the problem. Although many nervous system disorders are congenital or disease related, the bulk of human problems probably have come from accidents, such as with vehicles, from falls, and from diving into underwater objects, from strokes, and from injuries encountered in war, domestic quarrels, and other conflicts between people. The changes in behavior resulting from these externally caused problems, the findings associated with subsequent operations, and information from the extent and location of damage documented at autopsy have led to a gradually accumulating body of knowledge about the human nervous system.

One of the most striking medical history cases involving an accident is the story of Phineas Gage. Gage was the unfortunate victim of a blasting accident in 1884. He was foreman of a crew of railroad construction workers. While Gage was tamping the blasting powder with an iron rod into a hole drilled in a rock, a spark ignited the powder. The explosion blew the rod out of the hole and through Gage's head. The rod, over 1 m long and 3 cm in diameter and weighing about 5 kg, landed 50 m away. Amazingly, Gage lived, but his behavior and personality were altered radically as a result of the brain damage. His skull and the iron rod currently are on display in the Harvard Medical School museum.

Important neuroanatomists listed on the chart (Figure 2-2) include two very early ones, Broca and Wernicke, who studied speech problems associated with brain dysfunction, and a recent one, R. Sperry, who is best known for his work with split-brain subjects. The contributions of all three are discussed more fully in Chapter 15.

In addition to collecting information from accidental and similar human cases, medical researchers have deliberately experimented with the nervous systems of nonhuman subjects in an attempt to obtain more insight into what is happening in human systems. Animals used in the latter work primarily have been cats, dogs, and a number of primates, particularly rhesus monkeys. D. Hubel and T. Wiesel, who shared a Nobel prize with R. Sperry, studied the development of visual pathways in cats and primates (Chapters 15 and 18).

The neurophysiological biologists (and psychologists) have approached neural integration primarily from the cellular level. Although thoroughly "modern" and at the forefront of contemporary biology, neurophysiology grew largely from the ranks of traditional, classical physiology. The emphasis was mostly on the components of the nervous system and physiological, molecular considerations, with less attention paid to the whole system or behavior per se. Invertebrates, particularly molluscs (such as squid and a marine snail *Aplysia*) and arthropods (such as crayfish and insects), have been the main subjects of this work. Invertebrates provide systems that are much simpler and easier to reach than vertebrates. Also, for the most part, where larger parts of the system have been investigated, only simple movements and behaviors have been considered. A number of outstanding early investigations, however, were directed at more complex behavior. These included work on moth hearing by Roeder and colleagues (Chapters 1 and 13).

Much of the problem of working with the actual nervous system has been the small size of neurons and their fragility and lack of pigmentation. The first significant progress in looking at nervous tissue had to wait until the initial development of the microscope. But even under a good microscope neurons all blend together in an amorphous mass. With most stains they turn into a stained amorphous mess. In 1875 Golgi discovered a stain that somehow (still not understood) stains only a small fraction of the neurons present. This opened the door to studies of neuroanatomy. A contemporary of Golgi, Ramón y Cajal, used Golgi's new stain and devoted the rest of his life to studying virtually all of the nervous system of humans and other vertebrates. In 1904 he published a monumental book, *Histologie du Système Nerveux de l'Homme et des Vertébrés*. It is still considered the most important single work in neurobiology.

From the 1930s until the present there developed new techniques and several investigations of the operations of individual neurons. Hodgkin and Huxley were prominent pioneers in these developments. With recent developments and the ability to record from several neurons simultaneously, there have been many advances in tracing neural pathways in simple animals and for simple behaviors in more advanced organisms. Some of these findings are described in Chapter 15.

SINCE THE SIXTIES

The classical history of behavioral science probably can be viewed as closing somewhere around the 1960s. The biological study of behavior since the 1960s has been developing along the two major lines indicated at the start of this chapter: ecology and sociobiology on one hand and neurophysiology (including continued attention to learning and memory) on the other hand. Most of the remainder of this book is organized along those lines: Part Two on ecological and evolutionary aspects and Part Three on the internal neurophysiological aspects. The recent history of these topics, such as in behavioral ecology and sociobiology, will be introduced later where appropriate.

SUMMARY

Attention and interest in behavior traditionally followed three main divisions: ethology, comparative psychology, and neurobiology. All three branches had a more or less common origin with Darwin and the foundation provided by evolutionary theory. The three branches then mostly parted company and developed independently of each other. Neurobiology was somewhat separate even before Darwin.

Ethology has been concerned primarily with natural, often stereotyped behavior as shown by a wide variety of animals living under natural or seminatural conditions. A number of concepts developed from these observations; they include sign stimuli, releaser (a type of sign stimulus), fixed (or modal) action patterns, innate releasing mechanisms, vacuum behavior, conflict behavior (redirected or displacement), and the hierarchical organization of behavior. Ethologists retained a strong background in evolution and general biological principles, maintaining interests in the causation of whole units of behavior at both proximate and ultimate levels.

Comparative psychology began by comparing the behavior of different types of organisms but, from initial suggestions that learning was similar among different species, essentially abandoned the comparative approach and concentrated on the properties of associative learning in a few species, primarily to gain insight into human learning. In addition to focusing on learning and a human orientation, psychology has concentrated on proximate factors with little attention, interest, or knowledge of ultimate, evolutionary considerations. Psychologists developed a strong experimental tradition from the start and used careful experimental design and statistics more than did the early ethologists.

A notable exception to the separate paths of psychologists and ethologists was Schneirla, who essentially stood in both camps and who, as early as the 1930s, had a remarkably broad view of animal behavior and its likely internal integration. Schneirla viewed learning as consisting of specific forms in different animals rather than an increase in a general learning ability; he also believed that learning must be viewed from a phylogenetic and environmental standpoint.

The neurobiologists were of two basic types: those working at the anatomical and human clinical level, dealing mainly with neural dysfunction as a result of injury, and those who concentrated on the operations of neurons. The neurobiologists tended to focus on components of behavior and not on functional whole units. They, like the psychologists, were concerned mainly with immediate, proximate factors.

Since the 1950s and 1960s there has been significant merging and rejoining among the three former divisions. Ethologists have become much more conscious of experimentation and statistics, relying less on anecdotal information, and psychologists have begun paying more attention to evolutionary considerations and the differences among animals. Both groups have turned to the neurobiologists for information on and insight into internal mechanisms, and the neurobiologists have become more interested in complex behavior and learning. Neurobiologists have worked out many of the components and now are attempting to put them together into whole functional units.

Contemporary studies of behavior seem to be divided largely between the neurobiological, proximate aspects and the evolutionary, ecological, ultimate aspects. For a good understanding of behavior, one needs to consider both major aspects.

Recommended Additional Reading

Baron, R.A., D. Byrne, and B. Kantowitz. 1978. Psychology: understanding behavior. W.B. Saunders Co., Philadelphia.

Beach, F.A. 1955. The descent of instinct. Psychol. Rev. 62:401-410.

Beer, C.G. 1975. Was Professor Lehrman an ethologist? Anim. Behav. 23:957-964.

Bitterman, M.E. 1975. The comparative analysis of learning. Science 188:699-709.

Bitterman, M.E. 1979. Historical introduction. In Bitterman, M.E., et al., editors. Animal learning, survey and analysis. Plenum Press and NATO, New York.

Boring, E.G. 1957. A history of experimental psychology, 2nd ed. Appleton-Century-Crofts, New York.

Bullock, T.H., R. Orkand, and A. Grinnel. 1977. Introduction to nervous systems. W.H. Freeman & Co., Publishers, San Francisco.

Diamond, S., editor. 1974. The roots of psychology. Basic Books, Inc., Publishers, New York.

Evans, R.L. 1976. The making of psychology. Alfred A. Knopf, Inc., New York.

Flugel, J.C. 1933. A hundred years of psychology. Duckworth & Co., Ltd., London.

Hinde, R.A. 1970. Animal behaviour, 2nd ed. McGraw-Hill, Inc., New York.

Hubel, D.H. 1979. The brain. Sci. Am. 241(3):45-53.

Kandel, E.R. 1976. Cellular basis of behavior. W.H. Freeman & Co., Publishers, San Francisco.

Klopfer, P.H. 1974. An introduction to animal behavior: ethology's first century. Prentice-Hall, Inc., Englewood Cliffs, N.J.

Lorenz, K.Z. 1981. The foundations of ethology. Springer-Verlag New York, Inc., New York.

Manning, A. 1967. An introduction to animal behavior. Addison-Wesley Publishing Co., Inc., Reading, Mass.

Skinner, B.F. 1976. Particulars of my life. Alfred A. Knopf, Inc., New York.

Skinner, B.F. 1979. The shaping of a behaviorist: part two of an autobiography. Alfred A. Knopf, Inc., New York.

Thorpe, W.H. 1979. The origins and rise of ethology. Heinemann Educational Books, Ltd., London.

Tinbergen, N. 1951. The study of instinct. Oxford University Press, Inc., New York.

Watson, R.I. 1971. The great psychologists, 3rd ed. J.B. Lippincott Co., Philadelphia.

CHAPTER 3

C.B. Rippley

A naturalist photographing turtle behavior.

OBSERVATION
AND MEASUREMENT
OF BEHAVIOR

The answer is maybe and that's final.

Anonymous comment on science

Behavior, in brief, has been defined as the observable actions of animals. What one observes, however, and how one does it are intimately tangled up with what one already thinks, hypothesizes, and knows. The role of an underlying theoretical framework or lack thereof, for example, has been implicated in the difficulties of understanding play behavior (Fagen 1981). Advances in theory pertaining to social behavior rapidly pushed forward the observation and understanding of social actions, leading to a new subdiscipline called sociobiology within the past two decades.

But theory without facts can only go so far, and facts require methods and techniques. The advancement of understanding is much like a person trying to move a large heavy object by himself. The facts are at one end and the theory at the other; progress is made by picking up one end, moving it forward and setting it down, then going back to the other end and walking it forward. Behavior is a stream of effector output. How that stream is broken into segments and how those segments are separated and related depend on many things.

The main purpose of this chapter is to permit a deeper appreciation of both the origins and limitations of information in subsequent chapters. Second, this chapter is provided as a practical guide for persons wanting or needing to do more than just casually observe or read about behavior, such as students with class assignments or others intending to become further involved with behavior research. This chapter is not intended to be a cookbook on how to study behavior or a list of projects in need of doing. (Suggestions for exercises to give experience in observing live animals are provided in Appendix A.) Rather, the discussion that follows is for insight into the general problems and solutions that are unique to studying behavior.

For most of human history animals were viewed simply as useful allies or objects of food, danger, competition, or entertainment. Animal behavior has been viewed from a biological perspective, for the most part, only within the past two centuries (Chapter 2). Progress resulted from the application of general biological and scientific methods, systematic observation and recording, critical but open attitudes, formulation of testable hypothesis, and experimentation. In addition, there have been advances in technology and the accumulation of specific knowledge about behavior. The study of behavior shares with other sciences a number of attributes, including general scientific methods and attitudes, much routine work of data gathering, and the need for quiet, uninterrupted contemplation.

BEHAVIOR AND SCIENTIFIC METHODS

Science in general involves a complex body of techniques, approaches, and human activities. It includes not only observation but also various patterns of logic, assumptions, hunches and insights, discrimination and decision making about "facts" and ideas, experiments in some cases, plus a fair amount of luck (frequently known in the trade as serendipity). Although many textbooks make science appear clear cut and purely objective, it is very much an art. What is "beautiful" in art is often said to be "elegant" in science, whether it is an outcome, method, hypothesis, or a whole piece of research.

Much, if not most, science works indirectly by making *inferences* about the subject; these are sophisticated or educated guesses about how or why something is as it seems. There are numerous familiar examples outside behavior. In astronomy, for example, the planet Pluto was inferred to exist through its effects on the planet Uranus before it was seen and known to actually exist. Atoms, the subject of chemistry, had never been seen until recently with powerful electron microscopes. Gregor Mendel never saw the units of inheritance (now called genes or cistrons); he inferred their existence from the way his garden peas grew and reproduced. Examples from behavior are found throughout much of this book.

What most people think of as the scientific method probably is better called the experimental method or hypotheticodeductive method, only one of several methods used in science. Other important methods in science include pure description, as in much of anatomy and embryology, comparative methods (Chapter 5), historical inquiry, and modeling. (For an example of modeling behavior, see Ludlow 1976.) Experiments are based on previous observations (which may or may not have come from earlier experiments) and thinking about what they might mean. The possible meaning is called the **hypothesis.** If the hypothesis is correct, it should permit testable predictions.

To test the predictions, two or more sets of items are arranged to include the possible effect, perhaps at different levels of effect, and one set—the control group—is left without the effect. One could hypothesize, for example, that the hormone testosterone affects fighting behavior in male house cats. Several cats could be castrated (to eliminate the complication of their own internally secreted testosterone) and then be divided into two groups: one group to receive hypodermic injections of a solution containing testosterone and another group to serve as controls. The handling and injection processes themselves may cause effects, perhaps making the cats more likely to fight, so the control cats are handled and injected similarly to the others except that they only receive saline solution with no testosterone, known as a placebo. Each cat in turn, after receiving its injection, can be paired with another male cat and observed for the length and intensity of fighting, if any.

The hypothesis (that testosterone affects fighting behavior) is formally phrased into a **null hypothesis** for testing purposes. In this case the null hypothesis would be that there is no difference between the lengths of times that the cats from the two groups fight. Because of natural variation and perhaps effects of other factors not being tested (such as whether or not the cats came from human homes or off the street), one would not expect all the cats to fight the same length of time even if they received the same treatment. Statistical techniques are employed, and the results are compared against what would be expected by chance.

Because of variation and chance, conducting experiments is much like playing a game of cards. If a deck of cards is shuffled and one is dealt four cards, the chance that one will get four aces is very slim. But it still is possible. Just because a person plays cards day after day all his life and never draws four aces in a row does not mean that it cannot happen. The experimental method is similar. One uses probability theory and an understanding of the "deck" to design the experiment and calculate the odds. Then if the observed outcome turns out to be unlikely, say less than 5% by chance alone, the results are said to be significant, and the null hypothesis is rejected; that is, it is inferred that the effect (of testosterone in this example) is present. The formal, customarily reported expression of this finding is "$p < .05$." If $p > .05$, there is better than a 5% probability of getting these results by chance alone rather than from testosterone. In this case the results would not show a significant difference among groups, and one could not reject the null hypothesis.

In either case one might be wrong. An unlikely outcome might have happened by chance, like a four-ace hand of cards, rather than from testosterone, and the null hypothesis would be rejected mistakenly. This is known formally as a type I error. Or perhaps testosterone really does produce an effect, but this outcome resembled something that could have occurred easily by chance. The null hypothesis would fail to be rejected, and a type II error would be committed.

Although one cannot be certain regardless of outcome, the importance of the technique is that it provides a standardized form of guessing that can lead to at least reasonable inferences. This subject will be returned to in the section about data analysis.

These are the important points for the moment: (1) the experimental method provides an important tool for analyzing and understanding behavior; (2) it is only one of several tools; and (3) it requires understanding of probability and the deck of cards (in this case, cats and hormones) before it can be used properly.

Skills in using and understanding experiments, as well as other scientific methods, also depend on other things such as attitudes. Awareness that one might be making errors in inferences is one of these attitudes. Another attitude is to consider that important other factors, such as the source of the cats, might have been overlooked. Skepticism of results, whether one's own or those of someone else, is a golden quality in science. Skepticism not only helps detect errors, but it also leads to new questions and new hypotheses. Basic scientific attitudes are acquired from the continued study of different sciences, including behavioral research, from the study of history and philosophy of science, and by associating with practicing scientists.

For more detailed information and insights into the application of scientific methods applied to animal behavior, see Altmann (1974), Hazlett (1977), and Lehner (1979). In addition, one should read the original behavioral literature in scientific journals, giving attention to the methods sections.

BECOMING FAMILIAR WITH THE SPECIES

Perhaps the most important, often overlooked, basic consideration for studying animal behavior is how familiar a person is with the overall behavior and biology of the species being studied. Regardless of which aspect of methodology one might consider, the degree of familiarity with the species is critical. In fact, if there is a central dictum for the study of behavior, it is to "know thy animal." A formal list of a species' behavioral repertoire is known commonly as an *ethogram* (discussed later).

Whether or not an ethogram has been compiled, general subjective understanding and familiarity with the species are required. Familiarity is gained by reading about the species and closely related species and by personal observation. Because of variability, the number of behaviors, interactions, and relationships involved and because many behaviors are shown only rarely, many hours of observation generally are required. Full-scale behavioral studies consume much more time than most persons realize. Wilson (1975) cites several examples including a study by Dane et al. of courtship displays in goldeneye ducks that used 22,000 feet of film alone, a study of Serengeti lions by Schaller that involved 2900 hours of observation, and a study of one troop of olive baboons by Ransom with 2555 hours of observation. Lindauer (1952), during one of several studies of bees, logged a total of 176 hours 45 minutes watching a single worker!

Hence, whether involved in the study of behavior oneself or considering another's work, one must ask about the underlying familiarity with the species. The true "experts" generally have spent a considerable portion of their lives observing and working with particular animals.

OBSERVING WITHOUT INTERFERING

Observational or experimental artifacts and unintentional interference with the outcome plague most sciences. These problems are especially acute and pernicious in the study of behavior. The problem is that most animals, via their various senses and nervous systems, may be as aware of an observer as the observer is of them. Most vertebrates and many invertebrates are easily frightened, threatened, distracted or otherwise interrupted from their normal activities. They generally respond to the presence of a human or a change in surroundings by attempting to

Figure 3-1 Cats rubbing a vertical sensor rod in response to the presence of a human observer. During the 1940s similar behavior shown in a puzzle box was interpreted as evidence for learned responses. From Moore, B.R., and S. Stuttard. 1979. Science 205:1032. Copyright 1979 by American Association for the Advancement of Science. Photo courtesy Bruce R. Moore.

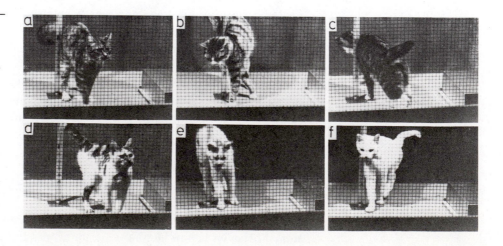

escape or hide, becoming immobile, otherwise acting abnormally, or at least diverting a proportion of their attention to the intrusion.

Moore and Stuttard (1979) uncovered an example of observational interference called "tripping over the cat," which previously had been interpreted, published, and widely cited as a different behavior. Many types of cats, including domestic cats, have a species-typical "greeting" reaction called flank rubbing, head rubbing, or *kopfchengeben*. The response is shown toward conspecifics and humans, which may be treated as conspecifics. The behavior involves arching the back, raising the tail, and brushing past or rubbing the head on the subject being greeted (Figure 3-1). If the subject cannot be reached easily, the cat will redirect the behavior toward a nearby object.

In 1946 Guthrie and Horton published a book titled *Cats in a Puzzle Box*, describing a box in which they supposedly conducted learning experiments. However, the box contained a vertical rod as a "neutral response sensor," which served as an ideal but unrecognized (by Guthrie and Horton) target for redirected rubbing. The observers sat in full view of the cats, which triggered the rubbing. Moore and Stuttard repeated the work but controlled for the presence or absence of the observer and showed that the cat behavior resulted from observer presence rather than the supposed operant conditioning (Figure 3-2).

Two general techniques have been used to solve or reduce the problem of observer interference:

1. Hide or otherwise avoid detection by the subjects
2. Accustom the subject to the presence of the observer, equipment, or holding facilities

Both approaches may be required for animals in captivity.

Hiding can be accomplished in many different ways, such as by using blinds and barriers or by being removed considerable distances. Viewing can be aided with slits or other small openings, one-way glass or plastic, devices such as

Figure 3-2 Results of a replication study of cats in puzzle boxes, but the presence of an observer visible to the cat was controlled. Cats contacted the vertical rod often in the presence of the observer but only rarely in the absence. The few responses in the absence of an observer occurred during pawing or tail chasing rather than from flank rubbing. Results are shown for one of several test subjects. From Moore, B.R., and S. Stuttard. 1979. Science 205:1032. Copyright 1979 by American Association for the Advancement of Science.

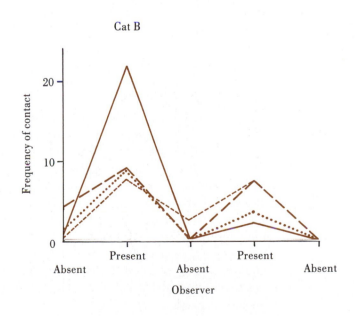

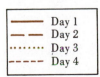

Figure 3-3 Examples of designs for blinds or hides for the observation of animals. From Pettingill, O.S., Jr. 1970. Ornithology in laboratory and field, 4th ed. Burgess Publishing Co., Minneapolis, Minn.

closed-circuit television, or, if at a distance, binoculars or telescopes. For sound there are many physical or electronic recording and "bugging" devices.

Blinds sometimes are quite elaborate, or they may be extremely simple and inexpensive. They may be built of local materials—reeds, branches, grasses, rocks, dirt, or snow—or of items as readily accessible as cardboard boxes, scrap construction materials, or blankets. Old large-appliance boxes and shipping crates work well for temporary blinds. For more permanent blinds there are a variety of custom-built structures, scaffolds, or tentlike devices that can be built or purchased (Figure 3-3). One commonly can use buildings or other structures that are already present. Figure 3-4 was obtained from a house window. Figure 3-5 was taken from a tree blind built from a single bedspread formed into a tentlike structure 10 m from the eagles' nest (Figure 3-6). Vehicles of a type with which the animals are familiar provide a common and mobile form of blind; the use of landrovers for observations in Africa has become almost legendary.

The primary requirement for blinds is simply that they be functional, including attention to the comfort of the observer. Position is important in many respects, including lack of disturbance to animals, good viewing for the observer, access for entering and leaving, and such things as angles and lighting for observation or photography. A good blind also is constructed with provisions for protection of equipment from wind and rain. Blinds generally need to be constructed and in position well in advance of when they are needed to permit time for the animals to become accustomed to them.

Figure 3-4 Wild white-tailed deer fighting. Photo was taken from window of a house used as a blind and from which the deer were not disturbed. Photo by James W. Grier.

Figure 3-5 Photograph of adult bald eagle on nest feeding nestling. Photo was taken from a tree blind located approximately 10 m from nest. Photo by James W. Grier.

Figure 3-6 Diagram of tree blind used to photograph eagles and constructed of readily available materials (bedspread and scrap lumber). A permanent tube made from a painted empty can for housing camera lenses projects from the blind. Note the single row of large nails in the tree used for climbing to the blind. The photograph in Figure 3-5 was obtained from such a blind.

Other means of direct observation of animals without their awareness of being watched include a variety of photographic and electronic video cameras, which are frequently camouflaged or otherwise hidden. Motion picture cameras can be housed in weatherproof boxes and operated at time-lapse speeds for extended observation during the absence of the researcher.

Working openly and often right in the midst of the animals is possible with many species that either ignore humans in the first place (an uncommon situation) or become accustomed to the presence of the observer. This type of observation has received much publicity, for example, Jane van Lawick-Goodall's work (1968) with chimpanzees. Becoming accepted by the subjects to the point where the observer is ignored and the animals go about their routine activities generally requires much time, patience, and expertise.

Signs that an animal is distracted by an observer often are subtle and may depend on the species. Common indications include an animal being unusually still, even stiff, lack of normal response to surrounding natural stimuli, gazing directly toward the observer, stereotyped "camouflage" postures such as the stiff upright posture shown by owls (Figure 3-7), and intention movements (Chapter 6) or actual flight from the area. How disturbed an animal becomes often depends on

**Figure 3-7 Screech owl in
normal undisturbed posture
(A) and in stiff-upright
"freeze" posture (B), indicating that it is aware of the
presence of an observer.**
Photos by James W. Grier.

distance from the observer. Hence, one often can detect disturbance by first observing from a long distance, then gradually moving closer until the animal shows signs of distraction.

The use of captive animals, whether in zoological parks, specialized research holdings, or private facilities, for behavioral observation has become routine and has led to much of the knowledge about animal behavior. In addition to the usual problems of observational interference, however, the conditions of captivity itself may create additional undesirable artifacts. (Depending on the nature of the observations, simplified environments and controls may be intentional, such as for the sake of controlled experiments.)

It is unlikely that few, if any, species other than humans have abstract concepts of freedom that pertain to captivity. Almost all animals, however, require some means of reasonable exercise, minimum housing and health considerations, some environmental diversity, and an absence of fear- and stress-inducing factors. Some species show poorly understood quirks in their captive psychology. House cats, for example, act restless, nervous, and upset in a variety of ways in some caged settings. With a change in the orientation of the cage so that only the view is different or if given certain simple furniture, such as a cardboard box to sit in or a brown paper bag to crawl into, the same cats will become quite contented and seem to accept their captivity much more readily.

Highly social species require either others of their own species or surrogate interactions with humans, which may require many hours of a researcher's or an assistant's time. Various species of mammals and birds, with both highly advanced nervous systems and social behavior, seem to require the most attention. In highly unnatural surroundings, such as may be imposed in captivity, animal behavior can be affected adversely. The two most frequent symptoms are the extremes of (1) excessive activity or (2) inactivity and a very simplified repertoire of behavior.

The holding and care of animals in captivity generally are controlled by various laws and regulations, which are designed primarily to prevent abuses of the animals and ensure proper cleanliness, health care, feeding, and humane treatment. In addition, some professional groups have formulated policies and codes for the housing, care and humane consideration of animals (e.g., Dawkins 1977, Animal Behavior Society Animal Care Committee 1981; also see Chapters 20, 21, and Appendix A).

LOCATING AND IDENTIFYING INDIVIDUAL ANIMALS

The individual identity of animals is perhaps more important to the study of behavior generally than in any other biological discipline because of variability in behavior, the importance of recording data for the correct animal, and the need to know which animals are interacting with which other animals. It is easy to become confused over who is who, particularly in free-ranging situations where different individuals resemble each other.

Techniques to initially locate, identify, and then follow or relocate individual animals seem to be limited mostly by the imagination. The methods that have been used are numerous and often ingenious. Again, much usually depends on the investigator's personal experience and familiarity with the species and its natural life history. Species that are rare or otherwise difficult to find can often be found readily and routinely in their natural habitat by someone who knows how and where to look. But because of the differences between species and the length

Figure 3-8 Individual markings in the ears and horns of dikdiks. From Hendrichs, H. 1975. Z. Tierpsychol. 38:55-69.

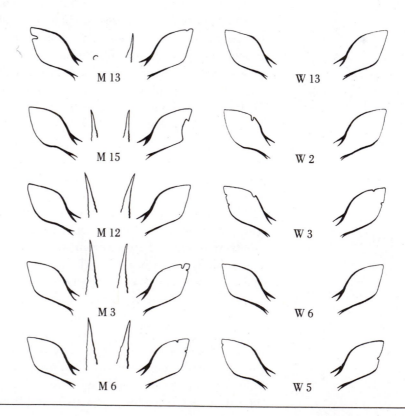

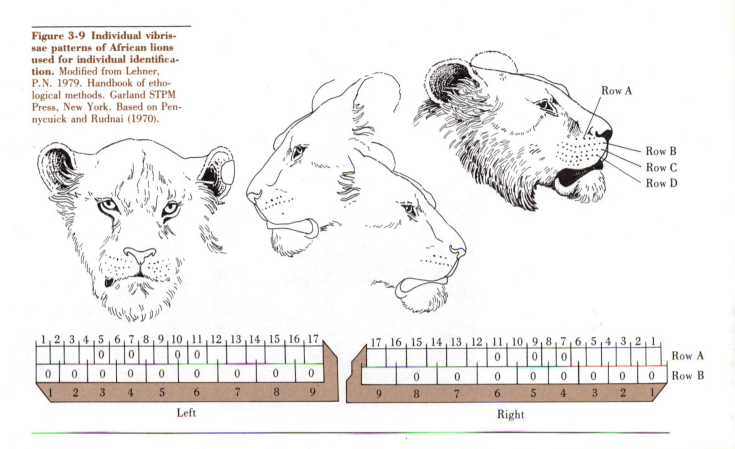

Figure 3-9 Individual vibrissae patterns of African lions used for individual identification. Modified from Lehner, P.N. 1979. Handbook of ethological methods. Garland STPM Press, New York. Based on Pennycuick and Rudnai (1970).

of time required for a person to gain experience, few people ever become intimately acquainted with more than a few species or groups of animals.

Individual animals are recognized by two primary means: their own unique individual characteristics and the use of artificial, applied marks or markers. Examples are numerous in either case; only a few will be given here. Natural individual differences in appearance have been used, for example, in porpoises *(Tursiops truncatus)*, dikdiks (Figure 3-8), and several primates. Individuals may be identified even through differences as subtle as vibrissae patterns in lions (Figure 3-9). Applied markers have included bands, tags, paint, physical alterations—such as toe, tail, ear, scale, shell, hair, and feather clipping—and radio telemetry. Coding has been numerical, via color combinations, or through other, basically digital, forms. Examples of artificial marking are shown in Figure 3-10.

Radio telemetry deserves special mention. Modern electronic technology has provided access to methods for gathering information never before possible. Through radio telemetry (Figure 3-11) one can relocate animals where they otherwise might be impossible to find, and one can follow subjects over long distances and periods of time, often at a considerable distance. In addition to simple location information, sophisticated telemetry instrumentation can provide other data, such as on types and amounts of activity, heart rate, respiration and oxygen consumption, body temperature, and other behavioral and physiological characteristics. Overall, radio telemetry has proved quite useful in the study of animal behavior. Perhaps its biggest drawback is that it generally also is labor intensive;

Figure 3-10 Examples of applying artificial markings to animals for purposes of individual identification. A, Clipping the edges of feathers on the wing of a martial eagle. B, Tagging the wing of a monarch butterfly with a numbered label. A courtesy John Snelling. **B** courtesy Fred A. Urquhart.

A   B

Figure 3-11 Transmitter weighing 3.0 g on the back of a Swainson's thrush (Cath-arus ustulatus). From Demong, N.J., and S.T. Emlen. 1978. Bird Banding 49:342-359. Photo by Stephen T. Emlen.

it requires much time and effort, not to mention cost, to equip and follow the animals.

When animals with large ranges, such as birds, are marked, there is potential for overlap between different researchers and subsequent confusion and conflict. Bird banding on different continents has been largely standardized and coordinated through government agencies. Attempts to coordinate and control auxiliary markers, such as color markers on birds, have been partially successful and depend on the cooperation of the researchers involved. There is only so much information that can be coded reliably and read from applied markers. Researchers need to agree as to who should be coding for what and where in those instances where the animals may travel across the boundaries of research projects. Nonetheless, applied individual markers provide the only means to get some types of information and remain a basic tool in much behavioral research.

It is important to note, however, that not everyone can or should capture and mark animals wherever and whenever they wish. Expertise may be required in the capture, handling, and marking; the animal's health, welfare, and behavior may be involved; some species populations may be threatened or endangered and unable to tolerate additional stresses; and in some areas, such as public parks, the marking of animals may destroy their aesthetic appeal to other persons. Accordingly and depending on the species and situation, a person interested in capturing, marking, and observing animals may have to gain experience first by working with other persons, use discretion, and usually obtain federal, state, and local permits, including permission of local landowners.

INGENUITY AND TECHNOLOGY

Science in general is a creative activity and uses ingenuity and innovations in technology. Both have played important roles in the study of behavior. Techniques to locate and identify individual animals have already been mentioned. In addition, other advances pertaining to behavior involve the extension of one's senses and improved methods for recording data. The bat and moth interactions in Chapter 1 provide an introduction to these principles.

Observation and alertness to detail and different events remain the fundamental method in the study of behavior. But in addition to the unaided use of eyes and ears, there are a great many devices that increase the sensitivity of one's seeing and hearing, shift sights and sounds into the range where one can detect them, render other sensory modes into visual or auditory format, and record sensory information for later or repeated analysis. Seeing is aided by binoculars and various telescopes as well as by a number of night-viewing devices and closed-circuit television. For hearing there are microphones or hydrophones (for underwater), amplifiers, quality headphones and speakers, and intercoms.

Events that occur too rapidly can be stopped by high-speed photographic techniques or by recording them at normal speed on film or magnetic tape and then replaying them at slower speeds. At the other extreme, with events that occur very slowly, the recordings are often speeded up, as in time-lapse photography or the speeded playback of whale songs. Special camera lenses, filters, and film or electronic photosensors allow one to obtain still or motion pictures or televised pictures in the infrared and ultraviolet ranges.

One of the most basic techniques for dealing with nonvisual items is to render them in visual format. Sounds can be converted to sonograms and laid out in a two-dimensional format for the analysis of patterns, as was done in Chapter 1 for bat and human vocalizations. Electrical and magnetic fields can be plotted two or three dimensionally. Chemicals are analyzed with various sorts of chromatographs. Sensory data of many kinds are depicted in line, bar, and other types of graphs.

Not all persons have access to, funding for, or desire to use advanced technology. But nearly everyone has or can develop his or her imagination. In some instances the ingenuity in methods may be as much or more interesting than the actual results of the behavioral research. The ingenuity shown in behavioral research, with either simple or advanced equipment, has been remarkable in many cases.

S. Emlen, for example, devised several creative techniques to study bird migration. In one where flying birds were to be followed at various altitudes, Demong and Emlen (1978) lifted birds in boxes suspended from balloons. The floors of the boxes were released by fuses cut to different lengths, depending on the time the balloon needed to rise to a particular height (Figure 3-12). After the birds were released at that height, they were tracked by specialized radar at the

Figure 3-12 Diagram of box suspended from a balloon used to release experimental birds on migration under controlled conditions. From Demong, N.J., and S.T. Emlen. 1978. Bird Banding 49:342-359.

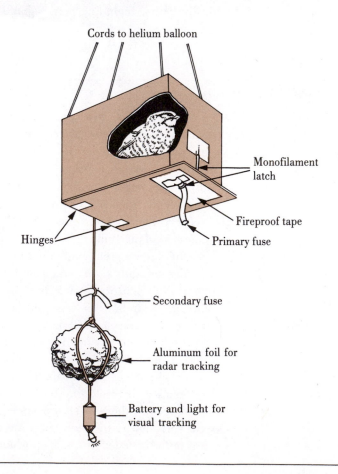

Cords to helium balloon

Monofilament latch

Fireproof tape

Hinges

Primary fuse

Secondary fuse

Aluminum foil for radar tracking

Battery and light for visual tracking

Figure 3-13 Narrow-beam tracking radar at NASA facility, Wallops Island, Virginia. This radar was used to track individual birds released from balloons out to a distance of approximately 16 km. Another radar dish at Wallops Island was used to track individual birds to a distance of 35 km. Recording of data and three-dimensional plotting of results were aided by computer. From Demong, N.J., and S.T. Emlen. 1978. Bird Banding 49:342-359. Photo by Stephen T. Emlen.

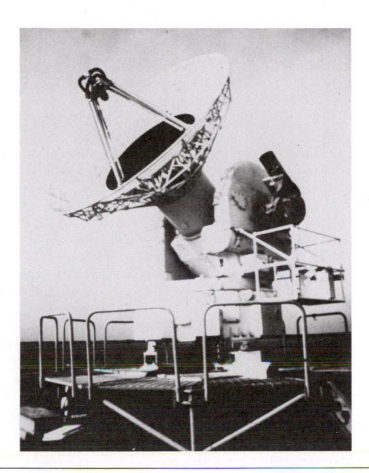

Wallops Island NASA facility in Virginia (Figure 3-13). The arrangement included a small battery and light for visual tracking and a ball of aluminum foil for radar tracking until the bottom fell out of the box at the desired height. Then the radar picked up the bird and followed it. The radar was connected to computer facilities and permitted the flying birds to be tracked precisely in time and three-dimensional space. The radar equipment was available at times when it was not being used in NASA projects. Another classic example of ingenuity in technique involved J.L. Gould's (1975) altering the senses of bees on the presence of the sun to test their use of dance language (Chapter 9).

Wilcox (1979) used several ingenious techniques to study communication in water striders *(Gerris remigis)*. He used masks to block the striders' vision. Masks were created by painting black liquid-silicone rubber on the heads of dead striders, peeling off the cured masks, then using them on live individuals. Wilcox altered the frequency of water surface wave signals produced by the striders through the use of small magnets affixed to the striders' legs. By surrounding the setup with a wire coil, he could induce currents and force the striders to produce wave signals as desired for the experiments. From this, incidentally, he discovered that females and males produce different vibrations, and the male striders can use them to determine the sex of another individual, even when vision is blocked.

DESCRIBING, RECORDING, AND CATALOGING BEHAVIOR

The Units of Behavior

Before one can talk about or analyze parts of behavior, one must be able to describe them and give them names or some other form of identity. An animal's total behavior can be broken into a number of smaller components, much like dissecting the physical body into systems and parts. The process is largely arbitrary, however, much like the naming of species in taxonomy. Dissection and naming of behavior are for man's convenience, not the animal's, and there is not always agreement among behaviorists how to best divide a stream of behavior into parts.

Behavior involves various body parts such as limbs and individual muscles and neurons. From the most reductionist viewpoint, behavior is nothing more than sequences of movements in these individual parts. It is possible, however, to recognize and classify relatively unique patterns of movement. The minimum identifiable units of behavior do not have a widely accepted name. The classical ethologists called them action patterns (Chapter 2). Ellis (1979) named the smallest units *ethons*, a name that has not yet gained widespread use. For simplicity and generality (including learned patterns) and to follow Lehner (1979), these units will be referred to as **behavioral acts.** Sequences of acts usually are called **behavior patterns** or simply **behaviors.** Because of variability of behavior and a degree of arbitrariness in identifying specific behaviors, all persons do not necessarily recognize the same acts in the same animals. Nonetheless, consistency in recognizing units of behavior is generally quite high, particularly among persons

Table 3-1 Interobserver reliability in the measurement of rhesus monkey behavior*

Category	r	Category	r
Infant-mother Behavior		**Approach-withdrawal Play**	
Nonnutritive nipple contacts	.99	Mutual	.99
Ventral contact	.99	Unilateral	.93
Lateral contact	.99	Mixed play	.90
Nonventral contact	.99	Nonspecific contact	.99
Nonspecific contact	.83	Visual exploration	.96
Visual exploration	.64	Oral exploration	.97
Manual exploration	.94	Approach	.97
Oral exploration	.97	Withdrawal	.94
Approach	.90		
		Mother-infant Behavior	
Individual Behavior		Grooming	.99
Self-mouthing	.99	Approach	.95
Cooing	.99	Threatening	.82
Visual exploration	.83	Cradling	> .85
Manual exploration	.85	Punishing	> .85
Oral exploration	.98	Restrain-retrieve	> .85
		Signals to return	> .85
Infant-infant Behavior			
Threatening	.90	**Disturbance Behavior**	
Thrusting	.98	Convulsive jerking	> .85
Rough-and-tumble play		Vocalizations	> .85
Mutual	.99		
Unilateral	.97		

From Fox, M.W., editor. 1968. Abnormal behavior in animals. W.B. Saunders Co., Philadelphia. Based on data from Hansen, E.W. 1966. Behaviour 27:107-149.

*Values of r = interobserver reliability coefficients for different observers watching the same events.

familiar with the behavior. Hansen (1966, cited in Sackett 1968), for example, demonstrated high interobserver reliability coefficients (similar to correlation coefficients) in a study of rhesus monkey behavior (Table 3-1). Inexperienced observers required as little as 7 to 10 days of experience before they could reliably identify rhesus behavioral acts and properly record them.

Hinde (1970) lists two basic methods of describing or naming individual acts: by *physical description* (such as low-pitched whistle, knee-jerk, wing-snap, head-up, tip-up) or by *function or consequence* (such as alarm call, sleep posture, food peck, aggressive jab). Naming by consequence may be more descriptive than its physical expression and may distinguish between similar motor patterns (such as food peck and aggressive jab), but the name includes an element of interpretation. That may inadvertently affect subsequent hypotheses and interpretations, blinding the observer to other alternatives, and it is susceptible to overinterpretation. A "learned peck response," for example, may or may not involve learning.

Once behavioral acts have been named, they can be classified, that is, grouped together with other behaviors. Hinde (1970) lists three ways of classifying: (1) *immediate causation* (e.g., agonistic fighting), (2) *function* (e.g., territorial threats), and (3) *origin ("historical classification")* (e.g., those which originate from similar groups of muscles or from similar evolutionary origins). The first two categories may be quite similar when the cause and function are closely related.

Acts may be grouped into ever-larger collections. Several distinct acts may be involved in sexual behavior. Both sexual and parental categories can be grouped under the heading of reproductive behavior.

The *list* of an animal's entire behavioral repertoire, or at least a major segment of it, is an *ethogram*. The number of acts that one observes in an animal's repertoire generally depends on three major factors: (1) the number the animal actually possesses, (2) how rare particular acts are, and (3) how much time one spends observing. The longer one watches, the more likely rare acts will be seen for the first time. Common, previously recorded acts also will be seen over and over

Figure 3-14 Relationship between behavior observed and time spent observing. The number of different or newly observed behaviors diminishes as time or the total number of observations increases because many acts are repeated and will have been seen previously. Also, different acts vary in their frequency of occurrence; some are rare and some quite common. The most frequently expressed behaviors are those most likely to be recorded at first and during short spans of observation. Observation of less frequent behaviors generally requires much longer periods of surveillance.

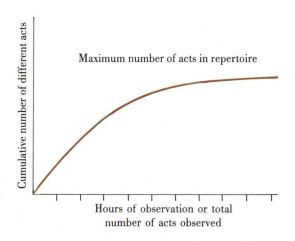

Figure 3-15 A few acts in the behavioral repertoire of the mallard duck. Each act depicted here has been given a specific name. Acts are associated with bathing (A), shaking (B), and nesting (C). A sequence of acts for oiling behavior is shown in D. A, C, and **D** from McKinney, F. 1965. Behaviour 25:120-220. Drawings by Peter Scott. **B** from McKinney, F. 1975. In Hafez, E.S.E., editor. The behaviour of domestic animals, 3rd ed. The Williams & Wilkins Co., Baltimore.

again. The net result of this is seen in Figure 3-14. Fagen and Goldman (1977) provide a discussion and practical techniques for determining the amount of observation required for different species under different conditions.

Behaviors can be grouped, or the ethogram divided, into natural, logical categories based on function. A list of these general categories includes *maintenance*, *feeding*, *orientation* and *navigation*, a number of *inter*specific behaviors (*symbioses*, *predator-prey*, etc.), and several categories of *intra*specific social behaviors that recently have come under the banner of *sociobiology*. The chapters in Part Two of this book are organized on the basis of these major categories. Figure 3-15 illustrates several behavioral acts involved in maintenance behavior of ducks. The entire duck repertoire is too large to illustrate here. A list indicating the size of the ethogram for another group of animals, rodents, is presented on pp. 72 and 73.

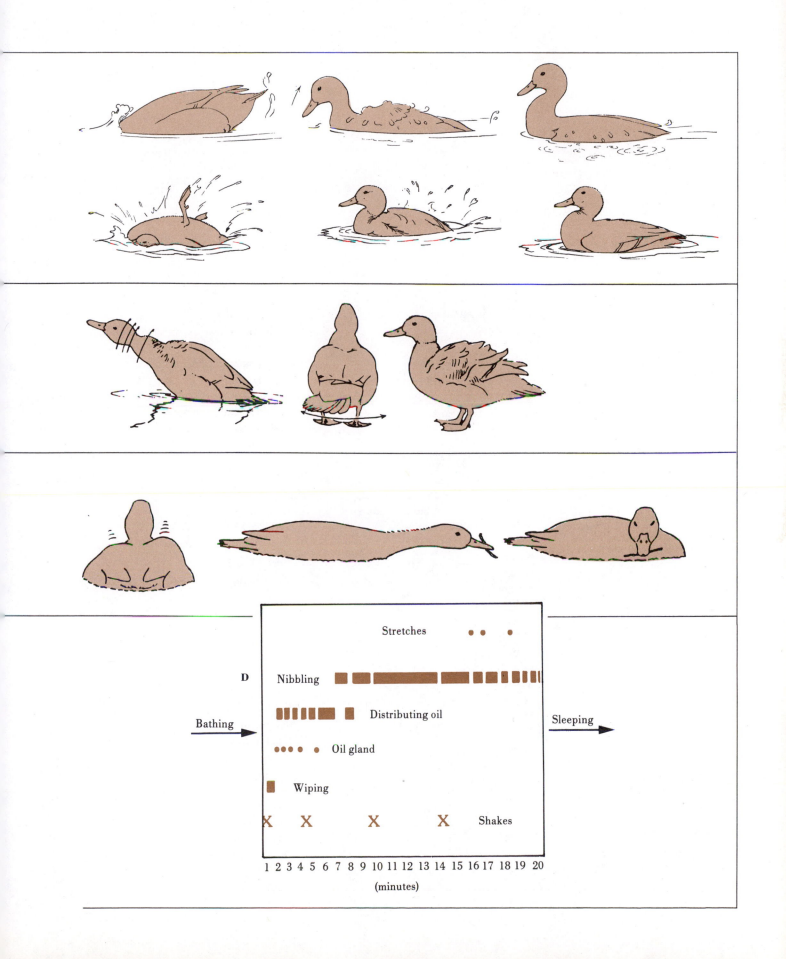

Rodent Ethogram

General Maintenance Behavior

Sleeping and resting
  Curled
  Stretched
    On ventrum
    On back
  Sitting
Locomotion
  On plane surface
    Diagonal
    Quadrupedal saltation
    Bipedal walk
    Bipedal saltation
    Jumping
  Climbing
    Diagonal coordination
    Fore and hind limb alteration
  Swimming
Care of the body surface and comfort
  movements
  Washing
    Mouthing the fur
      Licking
      Nibble
    Wiping with the forepaws
    Nibbling the toenails
  Scratching
  Sneezing
  Cough
  Sandbathing
    Ventrum rub
    Side rub
    Rolling over the back
    Writhing
  Stretch
  Yawn
  Shake
  Defecation
  Urination
  Marking
    Perineal drag
    Ventral rub
Side rub

Ingestion
  Manipulation with forepaws
  Drinking (lapping)
  Gnawing (with incisors)
  Chewing (with molars)
  Swallowing
  Holding with the forepaws
Gathering foodstuffs and caching
  Sifting
  Dragging, carrying
  Picking up
    Forepaws
    Mouth
  Hauling in
  Chopping with incisors
  Digging
  Placing
  Pushing with forepaws
  Pushing with nose
  Covering
    Push
    Pat
Digging
  Forepaw movements
  Kick back
  Turn and push (forepaws and breast)
  Turn and push (nose)
Nest Building
  Gathering
  Stripping
    Biting
    Jerking
    Holding
  Pushing and patting
  Combing
  Molding
  Depositing
Isolated animal exploring
  Elongate, investigatory
  Upright
  Testing the air
  Rigid upright
  Freeze (on all fours)
  Escape leap
  Sniffing the substrate
  Whiskering

By permission of the Smithsonian Institution Press from *Proceedings of the U.S. National Museum*, Vol. 22, No. 3597, "A Comparative Study in Rodent Ethology with Emphasis on Evolution of Social Behavior, I" by J.F. Eisenberg, Smithsonian Institution, Washington, D.C., 1967.

Social Behavior

Initial contact and contact promoting
    Naso-nasal
    Naso-anal
    Grooming
    Head over-head under
    Crawling under and over
    Circling (mutual naso-anal)

Sexual
    Follow and driving
    Male patterns
        Mount
        Gripping with forelimbs
        Attempted mount
    Copulation
        Thrust
        Intromission
        Ejaculate
    Female patterns
        Raising tail
        Lordosis
    Neck grip
    Postcopulatory wash

Approach
    Slow approach
    Turn toward
    Elongate

Agonistic
    Threat (proper) (remains on all four legs)
    Rush
    Flight
    Chase
    Turn away
    Move away
    Bite

Agonistic (continued)
    Locked fighting (mutual)
    Fight (single)
    Defense (on back)
    Side display
        Shouldering
        Sidling
        Rumping
    Uprights
        Class 8 (upright threat)
        Class II
        Locked upright
    Striking, warding
    Sparring
    Tail flagging
    Kicking
    Attack leap
    Escape leap
    Submission posture
    Defeat posture
    Tooth chatter
    Drumming
    Pattering (with forepaws)
    Tail rattle

Miscellaneous patterns seen in a social context
    Sandbathing
    Digging and kick back
    Marking
        Ventral rub
        Side rub
        Perineal drag
    Pilo-erection
    Trembling

Behavior may be described not only via discrete acts but also by measuring continuous aspects of an animal's appearance, such as its posture, angles of one body part to another, positions of various limbs, or gradual changes in color (Figure 3-16).

Sampling Behavior

How does one know how, when, where, and what to watch for in behavior? These depend in part on the underlying research questions being asked. In addition, the how, where, and what depend importantly on one's familiarity with the species, as

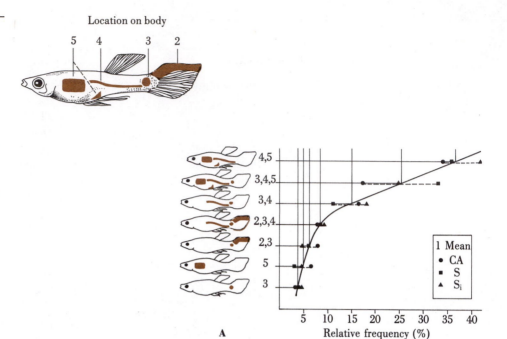

Figure 3-16 Examples of descriptions and measurements of changes in animal behavior along a continuum, as opposed to discrete changes. A, Darkening (from background color to black) in various regions of body of male guppy during courtship. *CA,* Copulatory attempts; *S,* sigmoid postures; S_i, sigmoid intention movements; *1,* overall body darkening. Measurement of body posture changes in drinking ducks **(B)** and hawks **(C). A** from Baerends, G.P., et al. 1955. Behaviour 8:275. **B** from Clayton, D. 1976. Anim. Behav. 24:127. **C** from Grier, J.W. 1968. M.S. Thesis. University of Wisconsin, Madison.

has already been mentioned. Familiarity comes from many hours of casual observation and various forms of more formal observation and by learning what is known of the species from other sources such as publications, meetings, and correspondence or other contact with other persons working with the species. *When,* however, presents a somewhat separate problem. Even the most dedicated observer cannot spend 24 hours a day, 365 days a year watching all subjects.

Until recently the timing of behavioral observation could be described best either as "as much and often as possible" or as "hit-and-miss," depending on such things as convenience and the amount of time an observer has available for observing. The name for such observation is *ad libitum* (Latin, at pleasure). This type of observation, however, presents a number of problems. Animals may display particular behaviors at regular intervals, which may coincide with observation periods in different ways or not at all.

Depending on when observations are conducted, particular behaviors may be underrepresented or overrepresented, studies may not be repeatable, and it may be difficult or impossible to compare the results of one study with those of another. Most persons are familiar, for example, with the general springtime singing of songbirds. The birds are most active, vocal, and visible during the early morning hours around sunrise and again at dusk. An ornithologist who works only 9 AM to 5 PM will miss much of the birds' behavior. A study by someone who works for 2 hours at dawn and another 2 hours at dusk cannot be compared easily with a study by someone else who observes the species at midday, even if the total amount of time (say 4 hours per day) is the same.

Because of the problems and lack of standardization with ad libitum sampling, improved statistical sampling techniques have been developed. They are described and compared by Altmann (1974), perhaps the most important recent methodological publication in the study of behavior.

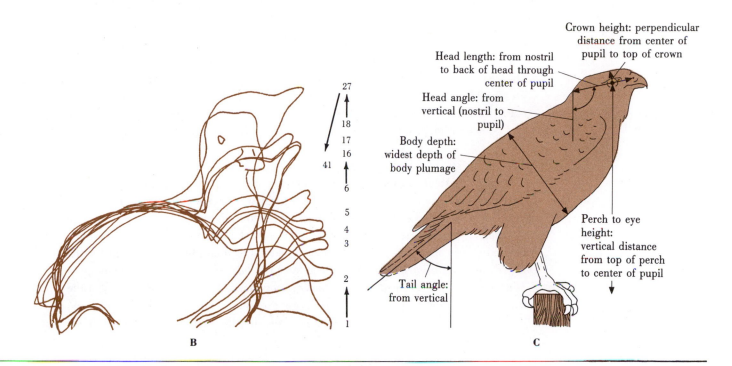

The gist of statistical sampling of behavior is to use random sampling rather than subjective preferences to decide when and which subjects to observe; that is, within the overall time frame (e.g., 1 month or 3 years), one allots the amount of time available for observing into segments that are randomly dispersed over the total time the animals are to be observed. And if there is more than one animal (or family, group, etc.) and all cannot be observed simultaneously, the animal or group to be observed at any one time is chosen at random. This procedure is called *focal animal sampling.*

The random allotments are chosen before observing begins; allotments may involve equal numbers in different cases to provide a balanced design; the initial design should include contingency plans in case problems develop and observations cannot proceed as desired. (Problems seem to be very common when observing behavior! Murphy's law applied to living organisms is that animals always will do as they please.)

With random sampling, as with ad libitum observation, one does not obtain a complete record of every move that all animals make. However, random sampling should give a more *representative* quantitative picture than the ad libitum method. Furthermore, random sampling permits the application of standard statistical procedures. The time units are treated as sampling or experimental units, and one can tally frequencies of particular behavioral occurrences or measure proportions or lengths of times during the sampled time units for statistical purposes.

There are, of course, many ways to divide time—both the total time frame of the study and the time to be spent observing. This has led to several general categories, as reviewed by Altmann (1974). There are some guiding principles, and some techniques offer certain advantages over others. Much of how one proceeds with specific details depends on what is being measured behaviorally, goals of the study, and numerous logistical considerations. Persons wanting further information should refer to Altmann (1974) and recent references (e.g., Dunbar 1976, Altmann and Altmann 1977, Simpson and Simpson 1977, and Tyler 1979). The important point is to remove the subjective preferences of the observer and the possible bias that unconscious and ad libitum choices may create. The goal is to be as objective as possible while at the same time keeping statistics in perspective; statistics and sampling are simply tools and means to an end, not an end in themselves.

Data Recording

The next methodological concern, whether one is making initial descriptive observations or seeking formal experimental answers to specific questions, is how to record the data. Good data recording can be done in many ways, depending on the particulars of the situation and the personal preferences of the observer. Contemporary methods are almost too numerous to mention. But, except for the initial observational stages, there are at least two *bad* methods: relying on memory and using general descriptive notebooks or logbooks. Memory is necessary at the time of an event and for temporary storage of new and unusual observations, but for recording routine data, the weakest ink is better than the strongest memory.

Logbooks also are useful for noting one's thoughts, occasional interpretations, and extraordinary events. But general descriptive notebooks are too cumbersome and inefficient for the storage of routine data, in spite of their widespread use for such work. Trying to dig out the facts for later analysis can be a time-consuming, frustrating, and defeating job. If one is to record data manually, well-planned data forms or tables are far superior. Their construction in the first place focuses one's attention on which data are to be obtained and which may be unnecessary; data are not as likely to be overlooked during the observing process; and specific items are much easier to find later because they have specific locations. Furthermore, it frequently is possible to include numbered boxes or fields on the form for later computer key entry, perhaps by someone such as a computer center keypuncher who is not familiar with the project. Good data forms should be provided with a comments or miscellaneous section for noting the kinds of items for which most persons use general notebooks.

Many, if not most, behavioral events happen too quickly or too many occur simultaneously for manual writing of data. At the other extreme, events may be so infrequent or occur so slowly that it would be a waste of time to sit by and keep track manually. In the first case, if the problem is not too serious, it may be possible to use a tape recorder and orally describe what is occurring, without having to take one's eyes off the subject. Because of the importance of time to behavior, tape recordings, particularly if the stop, pause, fast-forward, or rewind is used, generally need to incorporate time information. A mechanical or spoken time signal can be added to the background or on a separate channel (Wiens et al. 1970). Alternately, the recording can be played back in full sequence at the same speed it was recorded (usually a time-consuming process). Tape recorders, as with all mechanical and electronic devices, are subject to failure and various technical problems with subsequent loss of data. Manual recording can be expedited through the use of symbols or various types of shorthand invented for the occasion. It is important to provide a key somewhere with or near the data in case

Figure 3-17 Automated recording of pigeon pecking responses. In a study in which pigeons were to choose a small "o" amid an array of "x's" displayed on a television screen, responses were detected by a photocell attached to the birds's beak and sent directly to a computer. From Blough, D.S. 1977. Science 196:1013. Copyright 1977 by the American Association for the Advancement of Science. Photo courtesy Donald S. Blough.

someone else has to decipher it or if one may not need or want to go over it until a later time. Things that seem as though they will never be forgotten have an uncanny way of disappearing when they are actually needed; cleverly coded sets of data can turn into a useless mess.

The next levels of sophistication in data recording involve multiple-key paper-chart recorders or direct computer entry. Persons working with the complex multiple interactions among groups of primates have developed this technology to a high level. For examples of sophisticated, computer-compatible recording techniques, see Fernald and Heinecke (1974), Butler and Rowe (1976), and Stephenson and Roberts (1977). The use of photographic, sound, and video tape for storage, replay, and general recording of data was described earlier.

For many events, particularly the very rapid, rare, slow, or routine spread over a long period, visual observation and associated recording may be unnecessary or inefficient. One may be able to turn the whole job over to automatic systems with the aid of mechanical or electronic counters or other digital and analog devices that measure and keep track of the events. Blough (1977), for example, devised an ingenious way to have pigeons record their own behavior directly into the computer (Figure 3-17).

As in sampling, the methodology is only a means to an end, not an end in itself. The most important aspect of data recording is to decide which data are needed *before* they are gathered; then one merely records in a manner that will make the data easiest to extract and analyze. Because of the financial and time costs involved, typical unexpected problems with gadgets and gimmicks, and the constant risk of technical breakdowns, the simpler one can keep the apparatus for a particular job, the better. Again, ingenuity can be a researcher's biggest asset. Emlen and Emlen (1966), for example, recorded the direction and amount of activity by indigo buntings by simply placing stamp pads in their cages and surrounding them with paper to catch the foot prints as they jumped about (Figure 3-18).

INTERPRETING AND PRESENTING BEHAVIORAL DATA

Once the behavioral observations have been made and recorded, what does one do with them? Ideally there was prior planning and understanding of why and how the data were gathered so that analyzing them will be almost mechanical and simply amount to plugging the facts into their proper places. In many cases, perhaps all too frequently, observers proceed without design or planning and end up with answers for which there were no questions, that is, "theory-free" data; such information often presents an analytical quagmire that many statisticians refuse to touch.

Whether one has well-designed experimental results or just some kind of data hodgepodge, however, the methodology of data analysis and most of the statistical treatment are not unique to behavioral research. The techniques are more or less common to other branches of biology and science in general; details are beyond the scope of this book, and only a few general comments will be made.

Figure 3-18 Automated recording of bird activity using a stamp pad to create foot tracks on blotter paper. A, Diagram of technique. B, Photograph of cages with birds (lower left in photo) inside plantetarium. C, Examples of results obtained under experimental conditions. A and C from Emlen, S.T. and J.T. Emlen. 1966. Auk 83:361-367. B courtesy Stephen T. Emlen.

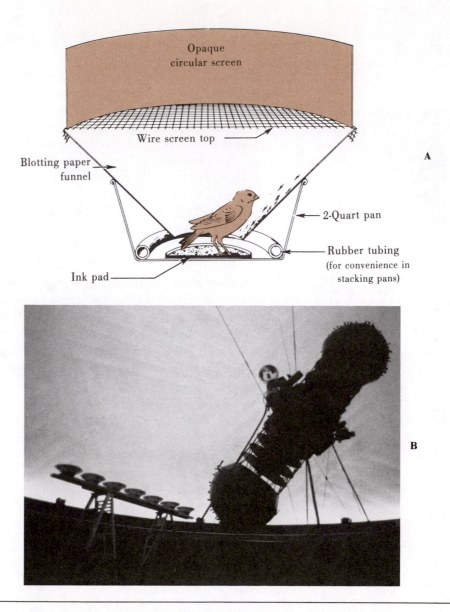

Because biological variability is a fact of life, an understanding of statistics is integral to a deeper understanding and pursuit of biology, including behavior. Most persons reading this introductory text probably will not yet have much, if any, exposure to formal statistics. Thus, in addition to keeping these comments general, this section will be used simply to encourage all readers to learn more about statistics.

There are two general categories of statistics: descriptive and inferential. **Descriptive statistics,** such as the mean and variance, help describe the characteristics of a set of numbers. With **inferential statistics,** however, one attempts to *infer,* with the aid of various assumptions and particular tests (such as *t* tests, analyses of variance, chi square, and many others), how one group relates to another, whether there is any underlying natural structure or order, or perhaps whether the observed results can be accounted for by the inherent variability. If the latter is the case, then similar results might have been obtained by chance

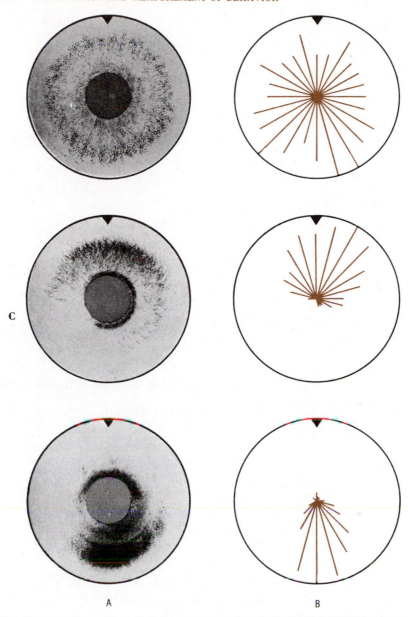

Table 3-2 Aggressiveness in male house cats following testosterone treatment*

Testosterone Treated	Control (Placebo)
158	87
47	22
122	36
89	0
76	14
160	28
143	106
93	0
68	56
114	34

*Hypothetical results of paired fighting tests are shown to illustrate the experimental method. Numbers indicate length of time (in seconds) that a male fought with an opponent under controlled test conditions. Tests were conducted humanely with the aid of specially designed rubber caps placed over claws and teeth to prevent injury. Each cat was tested once.

Table 3-3 Multiple-behavior analysis—simple hypothetical examples representing frequencies of behavior under common situations of interest*

Situation 1: Variations of Behavior Among Different Individuals

Behavior

		A	B	C	D	E	F
	1	12	2	46	0	87	17
	2	18	6	15	2	73	46
Individual	3	22	14	23	0	84	13
	4	16	8	32	0	66	12
	5	9	5	38	1	91	22
	6	11	4	4	0	42	54

Situation 2: Sociometric Analysis of Possible Communication Behavior

Receiver's Behavior

		A	B	C	D	E	F
	A	11	8	6	8	0	13
Sender's	B	9	6	84	10	12	9
Behavior	C	13	0	7	13	11	16
	D	76	12	0	15	12	82
	E	12	96	9	8	0	11
	F	7	5	11	4	6	8

Situation 3: Markov Chains—Sequences of Behavioral Acts

Following Behavior

		A	B	C	D	E	F
	A	3	5	2	47	4	4
Preceding	B	27	7	6	7	2	0
Behavior	C	3	20	4	0	5	57
	D	1	6	38	1	0	30
	E	6	0	5	8	3	5
	F	3	44	2	7	6	4

*For actual examples, see Tables 9-2, 9-3, and 11-2.

alone rather than because of some underlying effect of interest, as discussed earlier. To continue with the cat testosterone example, see the results of the hypothetical experiment in Table 3-2.

A person rarely is interested in only one behavioral act at a time. Usually several must be considered. Three common categories of multiple-behavior analysis and data presentation include (1) comparisons of frequencies of different behavioral acts among different individuals, (2) analysis of behavioral interactions between two (or more) individuals (sociometric matrices), and (3) sequences of behavior over time (Markov chains). Hypothetical sets of data in these categories are shown in Table 3-3. They normally are analyzed with standard chi-square procedures. If sufficiently versed in statistics, one may wish to use these examples as an exercise to see which are significant, consider assumptions, and try to interpret what they mean. For further introductory discussion and additional references, see Lehner (1979). For an expanded exercise, the reader can search for examples of these and related kinds of behavior analyses in the contemporary, original scientific literature.

Complex processes in biology, particularly in behavior, physiology, and ecology, usually include a host of interacting factors; that is, a particular result, such as how an animal responds to a predator or chooses where it will nest, may be an effect of several factors simultaneously and not just one or a few simple ones. Two basic statistical approaches for dealing with these processes are (1) experimental techniques whereby one or a few factors are deliberately varied in a known fashion while all others (it is hoped) are held constant and (2) a whole category of techniques known as multivariate statistics. This raises a problem that will be referred to as the "trained-cricket" problem.

Correlations and the "Trained Cricket"

Many things in nature occur together and are said to be correlated. They may occur together always, nearly always, or just a percentage of the time. When one occurs, the other occurs, and their possible single effects cannot be separated clearly. In an obvious case there is a story about a person who studied crickets and noticed that they could be trained to jump every time he presented a certain sound. Thus he removed all of the legs from a cricket, made the sound, and the cricket no longer jumped. He reasoned that it no longer jumped because it did not hear the noise; therefore the crickets' ears are located on their legs. (He was, in fact, correct but not because of his logic.)

As an illustration of more subtle but just as significant problems, imagine a pollster seeking to understand economic conditions in a community. He has a set of questions about such things as income and purchases. If he goes to the door of a house and asks the husband, he will obtain a certain amount of information. But what happens if he asks the wife the same questions? The answers may not be exactly the same, but it is reasonable to expect that at least some of the information will be the same and that little new has been learned. One would expect the husband's and wife's answers to be highly correlated. If children in the family are asked the same questions, there probably would be less correlation because they know less about the family finances, but there might still be some correlation.

The problem of correlation is pernicious in science in general and behavioral studies in particular when different variables are measured, as in asking different members of a family, or when the wrong variable, such as the sense of hearing, is interpreted as responsible for an effect.

As an example in behavior, consider a study that compares bird migration with measured weather variables. If only one variable, for example, wind direction, is considered, it may appear that migration relates to (depends on or itself correlates with) wind direction. But it could depend on a fall in air temperature instead, and a particular wind direction simply correlates with the decrease in air temperature. If several variables are measured, they all may simply yield similar correlated information. After the passage of a cold front in the northern hemisphere, for example, several things normally occur together: the wind changes to the northwest; temperature drops; cloud cover decreases; humidity drops; and barometric

Figure 3-19 Alternate interpretations of a set of data by two observers. A, Quadratic polynomial performed with aid of a computer. B, Alternate in jest by an observer who believed the quadratic polynomial to be inappropriate. A from Roubik, D.W. 1978. Science 201:1031. **B** from Hazen, R.M. 1978. Science 202:823. Copyright 1978 by the American Association for the Advancement of Science.

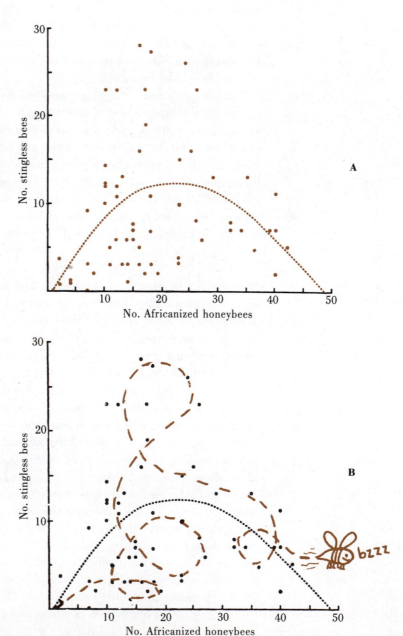

pressure increases. One can have difficulty, depending on the amount of correlation and whether or not appropriate statistical techniques are used, determining which, if any, single factor or set of factors is truly responsible for the bird migration.

The implications are that one always must be on guard for the correlation or trained-cricket problem. When possible it should be avoided in designing research. If it cannot be avoided, however, one must apply proper multivariate statistical techniques and maintain a cautious attitude toward the results.

Complex Analyses and Computers

With increased interest and the aid of high-speed, modern computers, advances have been rapid in both experimental and multivariate analyses, but particularly in multivariates. The new techniques, however, generally are much more complex than those with which people formerly have been familiar, and they require much time and effort to properly learn and understand.

The speed with which the new techniques have proliferated and swept through biology and the ease with which data can be plugged into canned computer programs have led to two major problems. First, the complexity may become bewildering. Many persons have given up before trying to tackle such statistical situations. Second, it may become more difficult to interpret outcomes than to plug data into the computer. Manuals for using complex programs often include warnings such as "A user who specifies this procedure for the analysis of unbalanced data assumes sole responsibility for the validity of the output." Different researchers viewing outcomes of complex analyses may disagree sharply over the validity or interpretation of such results. In one case published in *Science* (Roubik 1978a,b, Hazen 1978), for example, Hazen took issue with Roubik's analysis of competitive interactions between two types of bees: "The rather fanciful curve-fitting of Roubik has prompted me to propose an alternative interpretation of his data." Roubik responded, "I applaud Hazen's skepticism about the validity of the fitted curve. . . [but] judging from the alternative interpretation of my data given by Hazen, it is often most reasonably performed with the aid of a computer." The two different interpretations are shown in Figure 3-19.

Figure 3-20 Graphic presentation of results from a multivariate (principal component) analysis of woodpecker nesting habitat. Eight variables in nesting habitat were measured for each of five different species of woodpeckers. Through analysis of correlations among variables, new abstract variables (the "principal components") were derived. The positions of the five species relative to the first three components are shown in A and to the first two in B. The two-dimensional plot (B) is as if one were looking down from above on the three-dimensional version but illustrates areas of overlap rather than simple points for species. Note that the red-headed woodpecker *(R)* stands apart as unique and the most restricted in its use of habitat. The other species are downy *(D)*, hairy *(H)*, flicker *(F)*, and pileated *(P)*. From Conner, R.N., and C.S. Adkisson. 1977. Wilson Bull. 89:122-129.

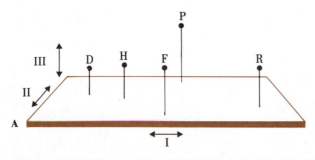

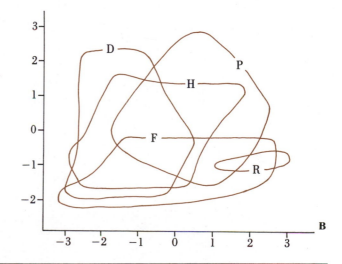

Multivariate statistics are like other statistics only more so; one simply must take time out to sit down, study, and learn them. Anyone contemplating actually *using* multivariate statistics is advised to first take one or more advanced college courses on the subject and consult at least one professional statistician who has experience with multivariate and biological applications. In conjunction with this, one should be prepared to spend much time, effort, and concentration mastering the assumptions and finer details and understanding the computer programs.

After data have been analyzed, the results normally are presented in manners common to many areas of science. These embody a variety of graphs and plots, including three-dimensional forms. Some results of a multivariate study of woodpecker nesting habitat, for example, are shown in Figure 3-20.

SUGGESTIONS FOR LEARNING THE METHODS OF BEHAVIORAL STUDY

This section is not intended for persons wishing merely to read about behavioral studies by others but for those who might be interested or required to do some behavioral observations themselves but do not know where to start. A person who is doing or may be doing research in behavior eventually will be faced with the problem of acquiring new techniques. New methods are learned, as has been stressed earlier, through time and effort. In addition, there are some hints.

1. Be willing to try something new. Accept the uncomfortable feeling of being ignorant and a neophyte in an unfamiliar area; do not let pride get in the way; and simply watch a colleague or teacher and ask for help.
2. Learn to accept, if not relish, problems and chaos rather than avoid or fear them. Without problems there generally is little progress.
3. Be alert to methodology. One should constantly ask oneself how someone else accomplished whatever he or she did. When reading the literature, do not just look at and accept the results and conclusions sections—read the methods sections both for specifics and for valuable ideas and insights.
4. Use imagination. The importance of ingenuity has been stressed repeatedly throughout this chapter. The problem with most people is not that they do not have an imagination but rather that they have not practiced using it.

SUMMARY

Whether one is actively engaged in behavioral research, is a potential researcher, or is merely interested and reading about behavior, the contemporary student of behavior needs an awareness of methodology. This includes knowledge of and, ideally, acquaintance with optics, photography, sound and video equipment, various recording devices, computers, data analysis, and statistical techniques. The depth of one's understanding of behavior will depend in part on how familiar one is with the methods used to derive that knowledge. Practicing researchers not only need to be versed in but versatile with these methods.

Methodology in behavior, in addition to the general scientific methods used in all branches of science, is particularly important in three major categories: (1) techniques and the use of equipment to extend one's senses, (2) finding means of locating and identifying individual animals, and (3) finding ways to observe and, in some cases, hold animals without interfering with the behavior one wishes to study. Familiarity with the species, spending much time in observation, and ingenuity are essential to good behavioral research.

Recommended Additional Reading

Altmann, J. 1974. Observational study of behavior: sampling methods. Behavior 49(3,4):227-265.

Hazlett, B.A., editor. 1977. Quantitative methods in the study of animal behavior. Academic Press, Inc., New York.

Lehner, P.N. 1979. Handbook of ethological methods. Garland Publishing, Inc., New York.

See also Appendix A.

CHAPTER 4

Fruit fly (*Drosophila melanogaster*) laying an egg.

GENETICS AND ECOLOGY OF BEHAVIOR

Why do animals behave as they do? In one way or another this question has motivated much of human interest in behavior. Of particular concern and intrigue have been the detailed inner mechanisms. Contemporary understanding of organisms at the cellular and molecular levels in general, although still far from complete, now provide much insight into the workings of animal behavior and renders superfluous many earlier ideas, such as "drive" and "action-specific energy," as discussed in Chapter 2. When an understanding of organisms at the cellular and molecular levels is combined with an understanding of how organisms interact with their surrounding environments, one is better able to comprehend the various ultimate and proximate causes of behavior.

Therefore the first job is to establish contact with the current state of knowledge at the cellular and molecular levels. The surrounding environment also must be considered. Basic biology and its connection with behavior are considered briefly in this chapter and in Chapter 5. Persons who already possess a good understanding of biology may wish to treat this basic material as a quick review or skip over it. Others should regard it as an introduction. Anyone wanting a further review or introduction may refer to a general introductory text on biology (Recommended Additional Reading). All readers, however, need at least an elementary understanding of this material; therefore it is included in this book. This material provides the basic context for the remainder of the book.

AN OVERVIEW OF GENETICS

Individual animals from the same species and even from the same parents (except for identical twins, etc.) are different from each other. Variation is a rule and a fact of life in biology. If it were not, biology might be much simpler; for one thing, biologists would not have to learn statistics. Biology without variation also would be much less interesting and, perhaps, might not even exist.

DNA, RNA, and Protein

All animals are composed of an immense number of biochemicals. Of these, different proteins are most immediately responsible for the observable differences among individuals, whether in shape, color, or any other trait. Various proteins form the basic structure of organelles, cells, tissues, and organs, including neural and sensory organs. Proteins, in one way or another, also are responsible for the chemical activity within the nervous system. Enzymes, which catalyze and thus guide all of the body's metabolism, are largely protein.

Proteins, or the polypeptide components of proteins, are assembled by the joining of various amino acids into specific sequences. The unique character of any particular protein depends on the specific arrangement of the amino acids.

Figure 4-1 Schematic diagram of the steps of protein synthesis. Study the parts of this figure, labeled to be self-explanatory, in conjunction with the text and perhaps an auxiliary general biology textbook. A, The illustration of transcription is simplified as in prokaryotic organisms to aid understanding. The more complex situation in eukaryotes, which includes all animals, is shown in B. In eukaryotes extra chunks of RNA, the introns, are edited out in the process of constructing the final, functional mRNA.

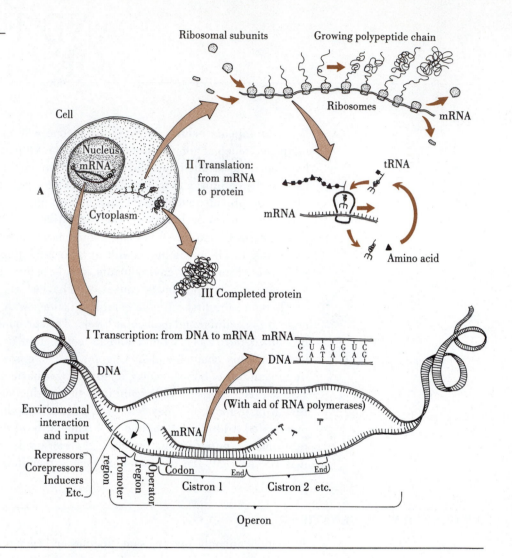

The synthesis of protein requires the interplay of two things: a set of "instructions" and an intracellular environment. The intracellular environment, in turn, is affected by the external environment. Both environments are important for providing raw materials and for determining the rates of metabolism for any given protein.

The basic sets of instructions for protein (and hence the cell's structure and metabolism) reside in the nucleus of each cell in a complex chemical, deoxyribonucleic acid or *DNA*. These instructions are basically analogous to the read-only memory of a computer: they may be copied for use, but they are not modified (except in abnormal situations such as by the ribonucleic acid virus that causes Rous sarcoma in chickens). The original set of instructions stays protected on the shelf, as it were, and is used only for making copies. The copies, molecules of messenger ribonucleic acid *(mRNA)*, leave the nucleus and go out into the cytoplasm of the cell where they direct the assembly of protein. The information travels one way from DNA to protein. It changes hands twice: once in being

Within nucleus of higher (eukaryotic) organisms

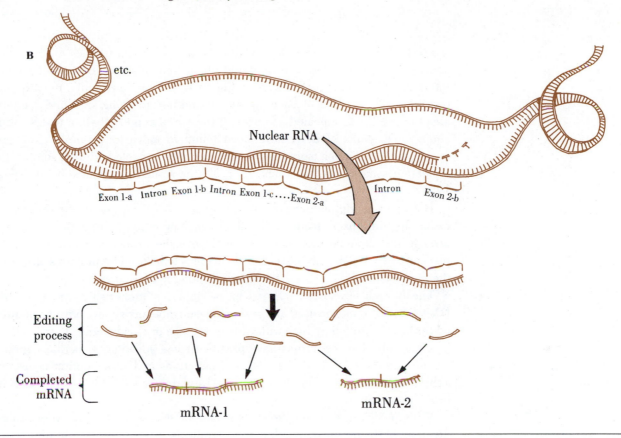

copied *(transcribed)* from DNA to mRNA and once in being *translated* (with the help of another molecule, transfer or *tRNA*) into the final polypeptide chain (Figure 4-1).

Each cell contains the full set of instructions, similar to a complete library, for the entire organism. But any one cell at any one time uses only a small fraction of the available information in the process of manufacturing its protein; that is, different cells are doing different jobs at different times, and apparently they use only the relatively small pieces of information necessary for each job. Thus most of the DNA in an organism is being used for the synthesis of various proteins. But equally important for considering evolution, *there are duplicate, unused sets of instructions that are set aside in the reproductive tissue* (germinal epithelium, ovaries, or testes). Keep this point in mind for later reference.

In the process of transcription in *prokaryotic* organisms (those without nuclei such as bacteria), the making of mRNA copies from the DNA template does not begin at just any site on the DNA. Rather it starts at a point of the DNA called a promoter. The promoter is followed by a segment called the operator, which is then followed by one or more cistrons (segments of DNA responsible for particular polypeptide sequences).

In more complex *(eukaryotic)* organisms that have cell nuclei, which includes all animals, the organization of DNA and processes that lead to mRNA are much more complex than in bacteria. The rapid revolution in the understanding of genetic expression in eukaryotes started around 1977 with the advent of techniques to artificially cut, splice, copy, and read DNA (recombinant DNA techniques). Parts of the DNA that eventually get expressed (the *exons*, from "*ex-pressed*") (Gilbert 1978) are scattered about and interspersed between other, intermediate segments called *introns* (Figure 4-1). Functional units, roughly comparable to the cistrons of bacteria are, in addition to being split and scattered about in the DNA, repeated and present in several copies. If all of that were not enough, segments of DNA have been found to move (or "jump") about and exchange places much more readily than had been expected. During some of this relocation some genes may even get "lost" (the so-called orphons) (Childs et al. 1981).

During transcription in eukaryotes, a nuclear RNA copy of the entire length of exons and introns is made. It is then somehow edited within the nucleus to arrange the exons in proper order and remove the introns to produce a usable piece of mRNA. This mRNA leaves the nucleus to be used for protein synthesis in the cytoplasm of the cell via the familiar route of ribosomes.

The introns have been referred to as "junk-," "gibberish-," or "nonsense-DNA." These nonexpressed and repeated segments may simply have accumulated through evolutionary time with little advantage or disadvantage to the cell. Or it is possible that the introns have provided some advantages, perhaps such as cushioning the DNA from the effects of mutations and various other occasional abberrations that occur in chromosomes. Research and thinking in this area are progressing rapidly.

With an increase in complexity, eukaryotic DNA is likely to show even more diversity in the means by which protein synthesis is regulated than in prokaryotic DNA. Depending on the organism and enzymes involved, there are numerous ways of regulating protein synthesis even in prokaryotes. Some of the prokaryotic systems will be described for the sake of simplicity. A few eukaryotic systems are known or are expected to be similar. Some systems for protein regulation are different and are found only in either prokaryotes or eukaryotes.

Interactions with the Environment

Some mRNA apparently is produced at constant rates, yielding enzymes that are produced at a constant rate and called *constitutive* enzymes. But in other situations the transcription can be modified and either induced or repressed, depending on the situation, by whole compounds, parts of compounds (metabolites), or other compounds, which may be extracellular or intracellular hormones.

These hormones are polypeptides themselves. They require *other* cistrons for their manufacture, and because they have the ability to modify transcription rates, the cistrons that code for them are often referred to as *regulator genes*. Active repressors, which attach to the operator segment and block mRNA synthesis, may be rendered inactive by inducers (Figure 4-2), or there may be different positive control methods. The details get rather complex.

The important point is that this is one of several general ways by which the environment can influence the development and behavior of individual animals. It can lead to some variation of animals even with the same genotype. It also means that genes and environments work *together* to produce animals (and, eventually, their behavior). The roles of genes and the environment cannot be completely divorced. (Another major form of environmental interaction, discussed in

Figure 4-2 Environmental influences in the transcription of DNA and, hence, genetic expression. A, Close-up of the transcription process making mRNA along an operon. The heavy solid arrow represents RNA polymerase. B, A repressed operon in which transcription is blocked but which can be induced by a molecule that inactivates the repressor. There are numerous variations of this system including, for example, complexes of corepressors. C, In this example transcription cannot begin until the promoter region of the operon is covered by a special protein—cyclic AMP complex. These examples merely represent a sample of several known ways in which genetic transcription is influenced by environmental effects.

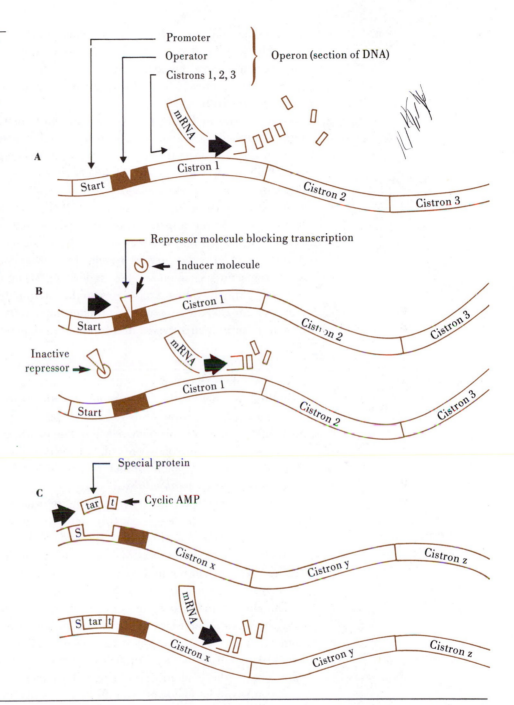

Chapter 5, is in affecting an animal's longevity and ability to reproduce. A third major category of environmental interaction is at the sensory level through the neural and endocrine systems, the subjects of Part Three.)

The living cells of animals may be exposed directly to the external environment in the smaller and simpler organisms, exposed at a few places in the outer layer of larger and more complex animals, or shielded by other layers of cells and buried deep inside the animal. But even the deepest, most protected cells of large complex animals are surrounded by a multitude of physical factors and chemicals. Many of the chemicals are very simple raw materials or by-products. Many, however, are large, complex "information"-containing molecules, such as hormones or pheromones, from other parts of the same organism or from other organisms, respectively.

The nervous and endocrine systems are particularly important sources for the larger "message" molecules. These various substances, by their absence or presence in varying concentrations and often through complex intermediate pathways, can affect the protein synthesis of the nonconstitutive polypeptides in many ways, depending on the particular case.

What does all of this have to do with behavior? Behavior depends on structures, not only the appendages and muscles, but also on an incredibly large number of structures, connections, and chemicals including transmitters and hormones, used in the nervous system. Even when a particular behavior involves a large learning component, the underlying structure and chemicals must be present. Thus the interactions between the environment and the transfer of information from DNA to protein are intimate, numerous, and complex. In addition, how well an animal survives and reproduces depends on how it is built.

Reproduction

DNA, therefore, has two separate roles in an organism: (1) a functional role of providing information for use in protein synthesis for building and operating the animal and (2) a repository role in which "spare" copies of the information are set aside and from which they may be passed on to offspring. In both cases the DNA, as well as the chromosomes on which it resides, has to be copied at some stage. With the first case, where copies are made for the use of each new cell, the process is referred to as *mitosis*.

In *asexual reproduction*, which is found in protozoans and some of the simpler invertebrates, the process is roughly similar to other forms of mitosis whereby a new individual develops from cells that bud off to eventually separate from the old individual. The process is sometimes referred to as *cloning*. The two roles of "using" versus "storing and passing on" information are thus not as distinct in these simpler animals as in other species.

The storing and passing of genetic information via *sexual reproduction*, however, involves a different copying and dividing process called *meiosis*. The doubled DNA is divided not once but twice. The resulting "daughter" cells (gametes)—usually either eggs or sperm—contain only one chromosome from each pair. This is the haploid (from "single") number, or N (e.g., 23 in humans).

The subsequent fusion of two different gametes, in the process of fertilization, restores the paired (2N) number of chromosomes. The resulting cell, a zygote, then divides mitotically. With subsequent divisions, differentiation, growth, and development (often through intermediate or larval stages), a new individual is produced. Some of the newly formed (2N) set of DNA is set aside in the reproductive tissue, and the rest is put to work in its role of assembling polypeptides; hence the cycle begins again.

There may be alterations in base pairs, known as *mutations*, in the original DNA or mistakes in copying, all of which add to variation. The reshuffling of genes through sexual reproduction and perhaps the "jumping" genes and different clustering of genes in eukaryotic organisms can lead to much variation. In addition, the environment can introduce further variation.

Mendelian Genetics

One now can move from molecular considerations to what is easier to see: the end products of metabolism and protein synthesis. The outward appearances and characteristics of an organism are referred to collectively as its *phenotype* (from the Greek, "to show"). The phenotypes were all that our scientific predecessors could observe. From these outward appearances they attempted to *infer* what was going on inside the organism. By the time Gregor Mendel, an Austrian monk, began working with garden peas in the 1860s, Darwin and Wallace had proposed (in 1858) that many traits were inherited. In addition, information on hybridization was already available.

Mendel's work led to two important conclusions. The first, the **law of segregation,** is that inherent factors are **particulate** or discrete. Furthermore, they occur in pairs. Some are **dominant** (which Mendel symbolized with capital letters), and some are **recessive** (symbolized with lowercase letters). Dominant and recessive refer to the expression of the factors. If only one of a pair is dominant, it rather than the other is expressed. For recessive factors to be expressed, both members of a pair must be the same recessive factor. The discrete "factors" that Mendel identified were later named **genes.** Alternative forms of a gene (such as tall versus short or wrinkled versus smooth) are called **alleles.** The terms gene and allele are more or less synonymous.

The second major conclusion of Mendel's work, the **law of independence,** concerns the inheritance of *different pairs of factors*. Although *each* trait follows

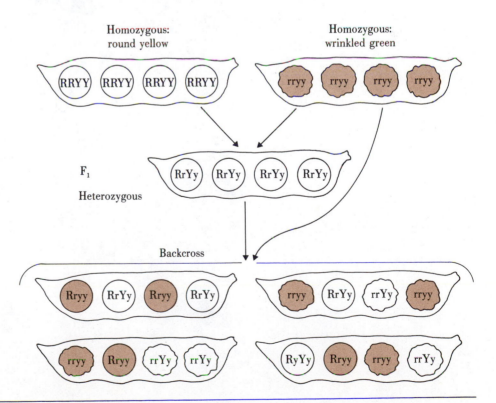

Figure 4-3 Single genes and peas. Results of the classic pea-breeding experiments by Mendel. In the backcross between plants of the F₁ generation and parent type, approximately equal number (55:51:49:52) of peas with the different combinations of shape and color were obtained.

the same pattern of segregation, the different traits (at least those which Mendel chose for study) segregate *independently of each other*. Whether peas are green or yellow, for example, does not depend on whether they also are wrinkled or smooth. All combinations are possible (Figure 4-3).

SINGLE GENES AND BEHAVIOR

Probably the best known and most convincing example of single-gene behavior patterns involves honeybee nest cleaning. A bacterial disease, American foulbrood *(Bacillus larvae)*, kills honeybee larvae. If the bees remove dead larvae

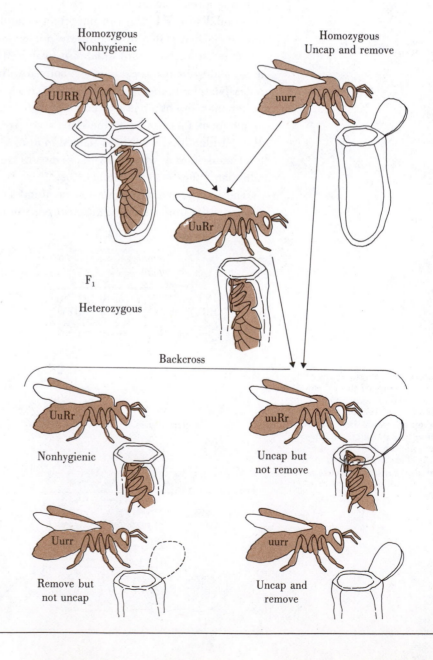

Figure 4-4 Single genes and bees. Breeding was accomplished using drones from known lines, artificial insemination of queens, and experimental inoculation of broods with American foulbrood disease. The ratio of different combinations of uncapping and removal behavior observed in 29 hives resulting from backcrosses was 6:9:6:8. Based on information in Rothenbuhler, N. 1964. Am. Zool. 4:111-113.

from the brood comb, further contamination and disease in the colony may be reduced.

The hygienic behavior of worker bees that show the trait has two components: first the caps are removed from brood cells that contain dead larvae; then the larvae are removed. Rather remarkably, each of these components was found to be associated with an independently segregating recessive allele (Rothenbuhler 1964). The alleles were named *u* for uncapping behavior and *r* for removal behavior. Both are recessive. Only workers that are homozygous recessive for both traits *(uu rr)* show the complete behavior. Workers that are homozygous recessive only for *u (uu Rr* or *uu RR)* uncap but do not remove dead larvae. Bees that are homozygous recessive for *r (rr)* but not for *u (Uu rr* or *UU rr)* do not uncap dead larval cells, but they will remove the larvae if another individual, such as another worker bee or Rothenbuhler himself, first uncaps them. Workers that are heterozygous (e.g., *Uu Rr)* or homozygous for the dominant allele do not perform the cleaning behavior (Figure 4-4). The behavior obviously is adaptive, and it clearly is affected by single alleles as classically defined and measured.

There are several known single-gene morphological traits, such as albinism in mice, that have behavioral correlations. White male mice, for example, were shown to mate much more successfully than black agouti males in one study of competitive mating (Levine 1958). The difference may have resulted from increased aggressiveness by the white males. Thiessen et al. (1970) provide a large number of other behavioral correlations with single-gene effects in mice. Similar, although not single-gene, cases involving behavioral differences among animals with genetically different morphologies include differences in aggressiveness by white-throated sparrows (Ficken et al. 1978).

PHYSICAL VERSUS STATISTICAL (INFERENTIAL) BASES OF HEREDITY

Research since the time of Mendel has shown that not all inheritance is as clear cut as it first appeared. Many traits do not show single-factor segregation, and different traits do not always segregate independently of each other. Many of the complications of segregation during meiosis result from linkage of some genetic material on the same chromosomes and a variety of aberrations, such as crossover, deletions, inversions, translocations, duplications, and breaks, that occur during some stage of copying and division. Furthermore, traits actually result from interactions among numerous genes (i.e., they are polygenically determined), and many, if not most, genes apparently have numerous (pleiotropic) effects. Traits that appear to result from single genes probably are each the products of a large number of cistrons working together but which differ in only one of the cistrons. It is somewhat analogous to an assembly line at a factory where the presence of a single worker who is incompetent or changes what is being done may change the outcome of the final product.

Statistical maps of chromosomes, using rates of recombination to determine which genes are linked and how closely, have turned out to be remarkably similar to physical maps of the same chromosomes based on physical appearances of chromosomes in fruit flies *(Drosophila* sp.). The maps differ mainly in the distances separating the inferred genes (Figure 4-5). These two approaches, molec-

Figure 4-5 Comparison of physical and statistical maps of the inferred locations of genes on their chromosomes. Physical maps are constructed by carefully examining the detailed appearance of chromosomes and comparing differences among individuals that lack or possess particular traits. Statistical maps are constructed by carefully analyzing ratios of combinations of characters from very large numbers of breedings. The statistical inferences are based on the assumption that the closer the genes for two traits are on the same chromosome, the more often the traits will occur together, or, conversely, the less often they will occur separately because of chromosomal crossover and other aberrations. Connecting lines between the two maps indicate the relative positions of genes affecting 11 different traits.

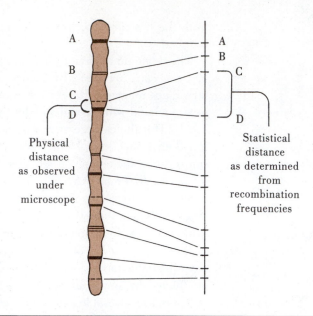

Physical distance as observed under microscope

Statistical distance as determined from recombination frequencies

ular and statistical, have almost met, and the "gene" can almost be equated with "cistron"—but not quite.

The purpose of this discussion has been to show that there is a tenable connection between the statistics and the actual physical substrate of heredity. One can use the term gene with some confidence that we are dealing with the real world but in a somewhat abstract and limited way.

Quantitative Genetics, Heritability, and Behavior

Most behavior patterns that clearly show some aspect of inheritance do not show simple segregation and independence. Many of the examples involve crosses between subspecies or even different species, which would not be expected to show simple outcomes. The animals are too unrelated and the differences in traits involve too many genes to show simple, single-gene segregation. Some examples are shown in Figures 4-6 to 4-8. Descriptions are provided in the figure legends. Hybrids show intermediate or confused forms of behavior in as diverse examples as cricket calls, dove calls, and lovebird nesting behavior.

Genes clearly are involved in behavior. But it is equally clear that this is not the whole story. Genes do not function in isolation from the environment, as was discussed under molecular considerations. Many of the complex interactions between the environment and protein synthesis are mediated via the endocrine and nervous systems. These processes are important in the development (ontogeny) and maintenance of all an individual's traits. Even a leg, for example, may not show gross changes after it has grown and developed, but continued metabolism is necessary for maintenance of a leg.

Behavior is removed a step further from the structures. Behavior involves neural and endocrine input and processing far more than do morphological traits, and it has the potential for much more lability and environmental effect. Some behavior patterns, such as moving to a particular location to obtain food, may routinely require modification and updating throughout the animal's life. Trying

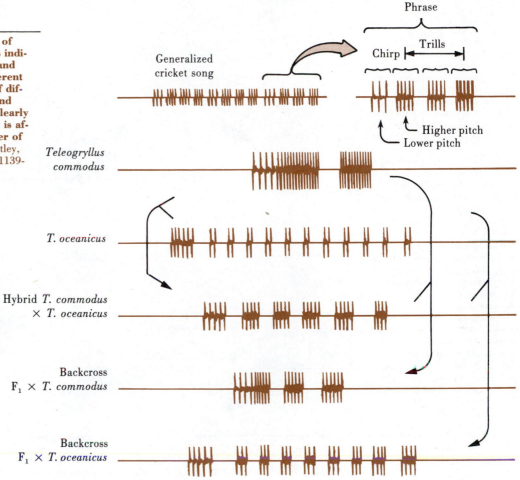

Figure 4-6 Inheritance of cricket song patterns as indicated by hybridization and backcrosses of two different species. A continuum of differences can be seen, and control of the pattern clearly is innate but apparently is affected by a large number of genes. Modified from Bentley, D.R. 1971. Science 174:1139-1141.

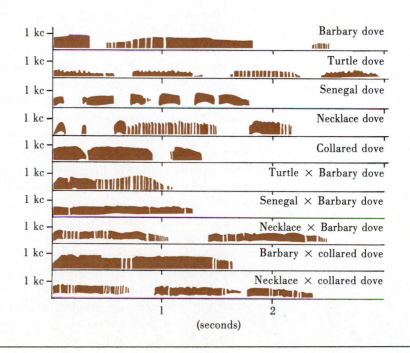

Figure 4-7 Inheritance of cooing song patterns in hybrids between combinations of six different species of doves. The picture is similar to that seen in crickets (Figure 4-6), but the calls are more complex, and the hybrids are less clearly intermediate. Reprinted by permission from *Nature*, Vol. 202, pp. 366-368, Copyright © 1964 Macmillan Journals Limited.

Figure 4-8 Confused carrying behavior in hybrid lovebirds (*Agapornis* sp.). A, Peach-faced lovebirds *(A. roseicollis)* tear up strips of nesting material, reach back and tuck it under rump feathers, then carry it to the nest. B, Fischer's lovebird *(A. fischeri)*, however, inherits a pattern of carrying behavior in which the nesting material is carried directly in the beak. C, Hybrids of these two species have difficulty carrying nesting materials. Initially the hybrids act completely confused and are unable to carry nesting material either way, although they try both. They eventually begin carrying material in the beak, but the seemingly simple behavior takes up to 3 years to perfect. Attempts to tuck material under the rump feathers diminish, but traces of the behavior and incomplete movements persist indefinitely. Figures based on information from Dilger, W.C. 1962. Sci. Am. 206(1):88-98.

to understand and untangle the multitude of interactions and feedbacks, whether at the structural or the behavioral level, has led to a great many biological misunderstandings and controversies, often under the banner of "nature versus nurture" or "instinct" (innate) versus "learning."

The problem arises when one attempts to view the genetic as opposed to the environmental contributions to an animal's structure of behavior as being separate or additive. That is as absurd as saying that a cake is 60% ingredients and 40% baking. The cake is a combination of both, and it is impossible to talk of only a 60% cake or a 40% cake. But there must be some way of separating the contribution of the two factors. Common sense and perhaps experience tell one, for example, that it makes a big difference in a cake whether the sugar is included or left out of the ingredients. Also, the temperature of the oven and the length of baking time obviously are very important. The solution is not to consider individuals (whether cakes or animals) but rather to inspect the *variability among individuals*. If the correct measure of variability is used, it may be possible to separate the internal from the environmental contributions to this variability among individuals. The correct measure of variability is the **variance** of some appropriately measured aspect of the subject of interest.

Variance is extremely useful statistically because it *can* be broken into additive components. *Variance*, briefly, is defined as the sum of the squared differences between each of the measurements and the sample mean, divided by the number in the sample, as in the following formula:

$$S^2 = \frac{\Sigma(x - \bar{x})^2}{N}$$

Variance is roughly like the "average variation." For mathematical reasons, the average variation is not quite correct, and the true average variation is not a useful statistic. It is easier to understand, but it is not useful because it cannot be broken into additive components.

Because variance is additive, the total phenotypic variance (V_t or V_p) in a trait for a given population can be broken into the components of environmental variance (V_e) and genotypic variance (V_g). These are determined for each component respectively from measurements of traits for animals under a variety of environmental and genetic conditions. The conditions can be varied experimentally or allowed to vary naturally. When the variation is not controlled by the investigator, complex multivariate statistical techniques may be required to deal with the data. Interpretation can be difficult.

For a simple hypothetical example that can be understood even without performing the calculations, that is, just by looking at the numbers, see Table 4-1. The behavior, as measured by length of time that male cats fight, can be compared for different genetic (breed) and environmental (rearing) conditions. In this example it is easy to see that both breed and rearing conditions make a difference. Puddy cats fought more than Docile cats, and Macho cats fought even more. In all cases the feral cats fought more than those raised in homes by children. Breed appears to impose greater differences than rearing conditions. As an exercise for persons who have the statistical background, calculations can be carried out on this set of data.

With further information, breeding experiments, and statistical analyses, one can subdivide the genetic variance into a number of components and arrive at a statistic known as heritability (in the "narrow sense"), h^2. It is a measure of genetic variability available for selection and ranges from 0 to 1. Persons interested in details should refer to Ehrman and Parsons (1976).

An example of an actual behavioral trait for which the variance has been broken into environmental and genetic components is mating speed in male fruit flies (Ehrman and Parsons 1976). Remember that differences among members of

Table 4-1 Aggressiveness in males of three hypothetical breeds of house cats*	Breed A (Docile)	Breed B (Puddy)	Breed C (Macho)
Males hand-raised by children in homes	10	30	200
	0	16	250
	5	55	98
	20	72	174
	16	10	210
	3	27	218
Feral (wild) males obtained from city alleys	14	84	280
	8	92	158
	11	0	324
	36	63	253
	32	58	207
	24	94	246

*Hypothetical results of paired fighting tests are shown to illustrate variance. Testing conditions were as in Table 3-3 except that no other treatment was involved. Normal, unneutered cats were used; six different cats in each category were tested once each.

a population are being discussed. Furthermore, the measures are limited to the specific population *and environmental conditions*. If the conditions change, so may the values of the measurements and estimates.

It often is possible to control or infer the variability among genotypes from knowledge about the relationships of the individuals. The animals may have identical genotypes (zero genotypic variance) if they all are derived from the same initial zygote, as in identical twins or via cloning. Or a degree of homozygosity approaching 100% can be obtained from inbreeding. The amount of inbreeding is called the *inbreeding coefficient (F)*.

Once the genetic and environmental variances have been controlled or measured, the genetic variance can be dissected further with appropriate statistical techniques (Ehrman and Parsons 1976). Thus the roles of genes in their final expression must be *inferred* (i.e., determined indirectly) through statistical techniques of experimental or multivariate analysis of populations (not individuals).

Additional Evidence for Genetic Control of Behavior

J.L. Gould (1974) made the statement that there is now "so *much* evidence for [the genetic] control [of behavior] that it is perilous to define precisely what is meant by the phrase." All connections still cannot be traced from the specific DNA sequences to the observed, specific muscular contractions. Also, one rarely knows which portion of the normal variability is contributed by inherited factors. But there is enough circumstantial evidence and enough parts of some connections have been traced to leave little doubt that genetic controls exist for many specific behavior patterns.

In addition to lines of evidence such as mendelian crosses and interspecific hybridization, there now are many other convincing lines. One, which is almost classical, involves the presence of behavioral mutations that can be passed directly to descendants. Breeding the animals, however, often is accomplished with difficulty because many mutations are lethal, near lethal, or otherwise handicapping.

Kung et al. (1975) have isolated over 300 behavioral mutations in the genus *Paramecium*. The paramecia studies were facilitated by the phenomenon of *autogamy* found in these animals. In autogamy the single-celled animal's chromosomes undergo meiotic replication and division followed by haploid mitosis. Subsequently there is self-fertilization, which produces completely homozygous individuals. Some of the classes of mutants include "paranoic," where paramecia move backward for long periods, "ts-paranoic," which are normal when grown at 23° C but paranoic at 35° C (again illustrating the interplay between environment and genetics in determining behavior), "pawns," which cannot swim backward at all, "spinners," which spontaneously spin upon hitting obstacles, "staccatos," which dash back and forth about 5 body lengths when stimulated, and "sluggish," which move slowly or not at all.

Among fruit flies *(Drosophila melanogaster)*, as reviewed by Benzer (1973) with additional examples collected by Gould (1974), there are mutants that get stuck in copulation, others that go to the other extreme ("coitus interruptus"), some that do not climb, others that do not fly (although they appear to have normal wings and muscles), some that have seizures and faint when mechanically jolted, others that emerge from their pupal cases at abnormal times, some that respond abnormally to light and odors, and many more.

As early as 1965 Sidman et al. had cataloged 250 neurological-behavioral mutations in laboratory mice. The list included "weavers" and "waltzers," "tumblers," "shakers," and "quakers." Behavioral mutations that can be passed on to offspring also are numerous in many other species, particularly in domestic animals.

Closely associated with the mutant line of evidence are behavioral traits that are artificially selected over several generations. The case of a one-locus difference of locomotor activity in mice (Abeelen 1979) was obtained through selective breeding. Selection for altered mating behavior has produced numerous examples (reviewed in Spieth 1974 and others). Eoff (1977) was able to artificially select for behavior in two species of *Drosophila* that reduced the sexual isolation of the species. Likewise, numerous selected unique behavioral traits have been obtained in pigeons (e.g., tumblers and pouters) (Nicolai 1976) and in many breeds of dogs, cats, and livestock.

Kovach (1980, Kovach and Wilson 1981) in a series of carefully controlled studies have shown that color preferences in coturnix quail chicks *(Coturnix coturnix japonica)* can be modified by artificial selection. They produced lines of birds that preferred blue or red (Figure 4-9). From their data over several generations and by comparing variances, they estimated that four to eight genes were involved. Their studies furthermore indicated that the effect was not from changes in color perception at the peripheral level but involved some other aspect controlling behavior, apparently in the central nervous system.

Another line of support for the genetic control of specific behavior includes cases where the neural regions controlling the behavior have been isolated and, in some cases, artificially modified. In fruit flies this has been accomplished with the aid of genetic mosaics and gynandromorphs—individuals in which different parts of the body develop from different genotypes, even of different sex type (Hotta and Benzer 1972)—and through direct neuromuscular studies (Levine and

Figure 4-9 Artificial genetic selection of color preference in coturnix quail. Birds were selected over 14 generations for approach choices toward blue or red stimuli. Birds were tested by scoring color choice in 14 trials. Horizontal lines indicate mean number of choices, and vertical bars indicate standard deviations. From Kovach, J.K. 1980. Science 207:549-551. Copyright 1980 by the American Association for the Advancement of Science.

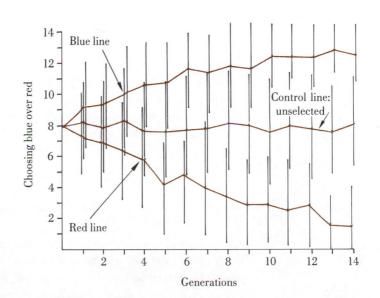

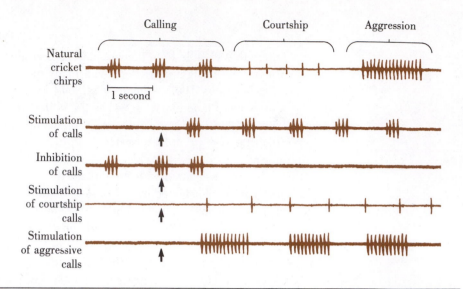

Figure 4-10 Stimulation and inhibition of natural cricket calls by artificial stimulation of different regions of the brain. Specific regions of the cricket brain control the pattern of chirping. Artificial stimulation of these regions triggers or inhibits the full, appropriate pattern as indicated. Arrows indicate start of brain stimulation. Modified from Bentley, D.R., and R.R. Hoy. 1970. Science 170:1409-1411.

Wyman 1973). Direct neuromuscular studies in crickets have shown that neural centers controlling song are located in the thoracic ganglia (Bentley 1971) and inhibited by others in the head (Bentley and Hoy 1970). The researchers were even able to artificially modify the songs by altering the firing rates of the command interneurons (Figure 4-10). Further work of Bentley, Hoy, and others, as reviewed by Bentley and Hoy (1972), suggested the control of these systems to be polygenic, that is, involving several genes, and that some genes regulating some song characteristics were located on the X chromosome. Rossler (1978) was able to change feeding behavior to that of a donor species by transplanting the appropriate parts of the brain in developing amphibian embryos.

Scheller et al. (1982) used recent recombinatorial ("gene-splicing") techniques to identify specific genes and demonstrate that a family of genes control egg-laying behavior in the marine snail *Aplysia*. These effects are mediated through different combinations of peptide hormones, which are controlled by the different genes. Hormones are discussed further in Chapter 16.

Perhaps one of the least reliable, although commonly or casually accepted, lines of evidence for inherited behavioral traits involves the simple constancy or lack of variation often seen in a particular behavior. Hailman (1967) and others have warned of the dangers of concluding that *consistent behavior* in a species indicates a large genetic component. Such constancy may also indicate consistent environmental, developmental, and learning factors. As an obvious illustration, consider the behavior of automobile drivers who all (or almost all) consistently raise their foot and step on the brake on seeing a red stoplight. It would be absurd to conclude that such species-specific and constant braking behavior were completely inherited. The same fallacy applies to many of the consistent behavior patterns seen in other species as well.

CONCLUDING COMMENTS ON THE GENETICS OF BEHAVIOR

All behavior involves both inherited and environmental contributions. The questions of interest are to what *extent* does each of these contribute to variability (mathematically assessed via variance) in specific behavior patterns observed among different, individual animals. Measuring and interpreting the variance are

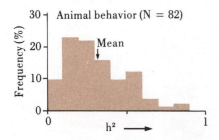

Figure 4-11 Frequency distribution of heritability values measured for 82 instances of animal behavior. Measurements ranged over a variety of taxonomic groups and types of behavior, including reproductive, social, individual behaviors, orientation, learning capacity, and emotionality. Two distinctly separate peaks do not exist, suggesting that all behavior cannot be classified simply as innate or noninnate. Rather, behaviors span a wide range. A few can be viewed as largely innate, a few as largely noninnate, and most are somewhere between. Modified from Jacobs, J. 1981. Z. Tierpsychol. 55:1-18.

extremely difficult because of numerous, complex, and intimate interactions between environmental factors and genes (which themselves are proving to be much more complex than imagined). Also, there are numerous statistical assumptions and requirements that must be taken into account. Yet in many cases it is clear that genetic control is important even in the final, phenotypic patterns of behavior.

In spite of the historical nature-nurture dichotomy, Jacobs (1981) showed with the best variance data available that the proportions of variance accounted for by heritability are mostly *unimodal* and not bimodal (Figure 4-11); that is, there is not one peak for "innate" behavior patterns and another for "learned" patterns, as one would expect if most behavior were one or the other. This suggests, as has so often been argued, that the long-standing dichotomy is not very useful. One should, instead, be concerned with the *degree of heritability* or the *degree of innateness*. But even with that there may be problems, including much random variation ("spontaneous variability" as Jacobs calls it), that is, background "noise" in behavioral variability. Actual heritability may be low. Jacobs suggests that rather than being preoccupied with innate-acquired considerations, one should be looking at other relevant factors such as overall genetic diversity, environmental heterogeneity, and variable selection pressure (Jacobs 1981).

This leads to the subject of environmental variability. The subjects of behavioral genetics and ecological diversity will then be merged in the next chapter in a discussion of the evolution of behavior. The role of inherited, specific patterns of behavior (as opposed to the general underlying, structural substrate) will recur in Chapters 18 and 19 in the discussion of learning.

ECOLOGY AND BEHAVIOR

Variability among individual animals is only one major piece in the puzzle of animal evolution. Another major piece concerns the variability of the external environment.

Animals do not live in a vacuum. Many people are familiar with animals that are relatively isolated in captivity such as pets, livestock, or animals in zoos. Under conditions in which most species are found, however, animals exist in a very complex set of surroundings. The physical and biological conditions vary greatly from place to place and, at any one place, from moment to moment. Some environmental conditions are favorable, such as the presence of food, mates, places to live, suitable temperatures, and supply of moisture. Other conditions may be unfavorable, such as the presence of predators, competitors, and toxic substances. Many conditions may be essentially neutral. Thus the *natural* environment of most species consists of a complex mixture of favorable, hostile, and neutral factors.

The *spatial* distribution of these various conditions often varies in a *patchy* manner; that is, patches of suitable conditions may be interspersed with areas that are unsuitable. The most familiar spatial differences are associated with latitude and altitude (Figure 4-12) and major geographical and climatic landforms (e.g., deserts and tropical rain forests). But even at the local level there may be much spatial heterogeneity (Figure 4-13), often referred to as environmental patchiness.

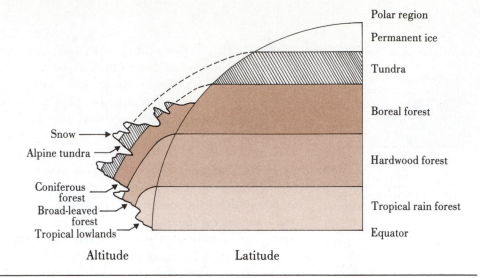

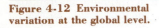

Figure 4-12 Environmental variation at the global level.

Likewise in *temporal* variation, conditions can be favorable or unfavorable. The familiar time changes include night-day and the various seasons. Many changes, such as a moving food supply, involve both time and space. The physical distance or length of time over which something changes can be large or small, occurring over millimeters and seconds or over many kilometers and years. The variations can be continuous or discrete, with sudden changes or sharp boundaries. Also, they can be fixed or regular (hence predictable), as in photoperiod, or very irregular (and unpredictable) as with weather conditions. Irregular conditions tend to vary within certain limits, but those limits may be very broad and rough. Thus the earth's environment is very heterogeneous and extremely dynamic. Some conditions, such as the proportion of oxygen and carbon dioxide in the air and the conditions in deep caves, may remain fairly uniform.

Those animals which survive in the midst of all this variablility and uncertainty are those with suitable structures and behavior patterns. Behavior can be viewed in a sense simply as putting certain structures into action. Particular components or characteristics of the animal are called *adaptations*. Structures (and behavior patterns) are said to be adaptive if they permit an individual to avoid or tolerate unfavorable conditions or if they permit the animal to obtain favorable or necessary surroundings and resources.

There are two broad classes of resources which all animals need and which they must obtain from the environment: energy and materials. These frequently come packaged together as food. Energy is needed because life depends on a high degree of chemical organization and reorganization. The second law of thermodynamics states that, left to themselves, systems will move toward an increase in the amount of disorder, or entropy; that is, it costs energy to organize and keep organized. All living organisms must pay to live by continuously using energy. Plants capture the energy from sunlight and produce their own food internally; hence they are called *autotrophs*. All but a few protozoan animals, however, are *heterotrophs*, which must obtain their food from other organisms.

Shortages of energy and material frequently occur in nature; the problem is by no means unique to humans. Furthermore, it appears almost inevitable, if not one of the laws of nature, that someone else also will attempt to take the resource if it

Figure 4-13 Environmental variation at the local level. A view of habitat with patches of woods, field, ponds and surrounding marsh, and distant hill shown realistically (A) and diagrammatically (B). Neighboring patches even as close as pictured here may be favorable or hostile to various animals, depending on the species and which patch is involved.

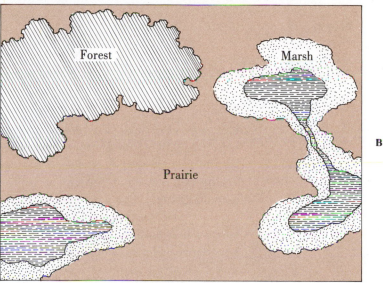

is both limited and concentrated. Different types of competitive interactions, predation, parasitism, various forms of piracy, and theft are common problems in animal life. These problems pose a constant threat to nearly all animals. Various protective measures may reduce or minimize the problem, but it is almost never eliminated. The necessity of protecting one's resources, including the very protein that one is composed of, is an ecological fact of life, and it has many implications for both evolution and behavior.

Because of the diversity and threats in the environment, the diversity seen in animal structures is not surprising. Some structures are better adapted under some circumstances, and other structures are better under others. For any situation there may be several, alternate ways to accomplish the job.

Some animals can tolerate wider variability in living conditions and food than can other animals. An animal's total adaptation is something of an accumulation of compromises. Under two different conditions that an individual may have to

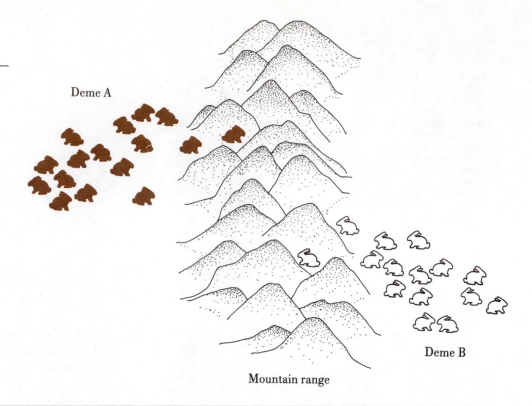

Figure 4-14 Two distinct demes of rabbits, represented here as brown and white, belonging to the same species and same general population. The two groups are separated by an intervening range of unsuitable habitat. Although the rabbits are capable of crossing to the other side and some mixing could occur, it has not happened as shown here, and the two demes are completely separate. In other cases, however, it is easy to imagine mixing to variable degrees. At the extreme, if rabbits crossed the hills and mixed freely and completely, the situation would result in one deme rather than two.

Deme A

Deme B

Mountain range

face, a given structure can be advantageous in one and a handicap in the other. A heavy insulation of feathers, for example, is useful at low temperatures but may lead to heat prostration or even death under hotter conditions. The capture and killing of a moving object may be useful if it is potential food but not if it is a potential mate.

An animal's environment includes other organisms. All organisms of the same species living within a prescribed place and time represent a *population*. The boundaries of place and time for any particular population are defined by the investigator; all populations represent abstract groupings.

The smallest recognized grouping of a population is a *deme:* animals that are reasonably capable of interbreeding with each other because of proximity and other factors that promote access. Individuals from the opposite sides of a species' range could interbreed (by definition of a species) *if* they were to come into contact. But because they are separated widely, they are not likely to come into contact and thus do not belong to the same deme (Figure 4-14). Much abstraction exists even at the level of the deme.

Demes vary in geographical extent. Some demes have clearer boundaries than others. Within a deme most individuals actually mate with only one or a few other individuals. Hence it is the *potential* ability to interbreed that earmarks the definition. All of the students in a class, for example, could potentially breed with all other students of the other sex, but it is not likely that they will actually do so.

Population size at any particular time depends on the size of the population at a previous time plus reproduction and immigration minus death and emigration that occurred in the time period. Very few animals die in the wild from physiological "old age." Those which commonly do belong to species in which reproduction occurs only once (*semelparous* animals, or the so-called big-bang reproducers) and then the adults die soon after. Examples include salmon, octopuses, and the whole order of insects Ephemeroptera (part of the name means "short-lived"). Even in semelparous species, however, most individuals die long before reaching the adult stage.

In addition, there are many ways to die. Death itself may be certain, but the time, place, and manner of death certainly are not. These facts of death—that it usually occurs before the body's clock runs out and that there are many different ways to die—have enormous implications for the long-term or ultimate biology of living organisms.

SUMMARY

Animals are highly variable, with much of the variability inherited. The hereditary mechanism was not known until recently. It can now be stated, in a simplified version, that individual animals carry a unique and elaborate library of information in DNA molecules. The complete library exists in the animal in essentially two sets. One set (the somatic DNA) is *used* by the individual to direct protein synthesis and, consequently, to determine how the animal lives and behaves. The second set (the germinal-line DNA), in a sense, is *set aside and saved*. Depending on how well the working set works in the environmental context in which it exists, that is, to the extent that the individual survives and reproduces, the second spare set of DNA information is passed on to the next generation. In the process the instructions are recombined with similar but not identical instructions from the other parent to form a new complete library. This library is duplicated into two full sets, and the cycle starts again. This simplified version provides a useful starting point. The real picture is more complex and requires further understanding of population genetics.

At least some behavior has been shown to follow the same rules of classic and statistical inheritance as morphological traits. Examples were given in this chapter.

The earth's environment also is highly complex, dynamic, and variable, both temporally and spatially. The heterogeneity is both widespread and local, with unsuitable habitat often neighboring on suitable habitat. The environmental variability imposes differential survival and reproduction on the populations.

Recommended Additional Reading

Crow, J.F. 1976. Genetics notes, 7th ed. Burgess Publishing Co., Minneapolis.

Ehrman, L., and P.A. Parsons. 1976. The genetics of behavior. Sinauer Associates, Inc., Sunderland, Mass.

Hickman, C.P., Jr., L.S. Roberts, and F.M. Hickman. 1984. Integrated principles of zoology, 7th ed. The C.V. Mosby Co., St. Louis.

Keeton, W.T. 1972. Biological science. W.W. Norton & Co., Inc., New York.

Krebs, C.J. 1978. Ecology. Harper & Row, Publishers, Inc., New York.

Srb, A.M., R.D. Owen, and R.S. Edgar. 1965. General genetics. W.H. Freeman & Co., San Francisco.

Watson, J. 1965. The molecular biology of the gene. W.A. Benjamin, Inc., New York.

Wilson, E.O., T. Eisner, W.R. Briggs, R.E. Dickerson, R.L. Metzenberg, R.D. O'Brien, M. Susman, and W.E. Boggs. 1978. Life on Earth. Sinauer Associates, Inc., Sunderland, Mass.

CHAPTER 5

G. Archibald, International Crane Foundation

A pair of Manchurian cranes (*Grus japonensis*) displaying their courtship dance in a marsh.

EVOLUTION OF BEHAVIOR CONSEQUENCES OF GENETIC AND ECOLOGICAL VARIABILITY

The relationship of ultimate, proximate, and future events in the life and behavior of an individual is shown in Figure 1-21. The ultimate causes of behavior come about through a somewhat complicated biochemical system that codes and uses information in conjunction with differential survival and reproduction. The memory of the past, in an indirect manner, is locked into DNA. It is indirect because it is not the events that are remembered but rather the successful solutions for dealing with those events. In the previous chapter DNA was discussed. Now it will be shown how it fits into the broader evolutionary and behavioral picture.

Fortunately the explanation for behavior appears to be basically the same as for the evolution of bones, body shapes, and other biological characteristics. *The way to understand the evolution of behavior is to understand biological evolution in general.* If one understands the basic principles, then the evolution of behavior becomes (almost) self-evident. For this reason one must look closely at the general process of evolution.

The view that behavior can be shaped and changed by evolution in a manner similar to morphological and physiological traits can be traced back to Darwin and other early workers. Elaboration of the principles as applied to behavior received much impetus through the work of Konrad Lorenz, Karl von Frisch, and Niko Tinbergen, for which they received the Nobel prize in 1973.

METHODS OF STUDYING EVOLUTION

The primary method used to make historical inferences about evolution, known as *uniformitarianism,* is based on the belief that physical and biological processes operate in a uniform manner throughout time (Simpson 1970). The underlying principles are that the present is the key to the past, that the laws of the universe have not changed, and that the dimensions of space and time known today have existed for at least longer than life has existed on Earth.

Historical inference in biology, in addition to following uniformitarianism as the guideline, also employs *documentation* (such as fossils and their positions in various geological strata) and *comparative* methods. A fossil or other natural document is described, analyzed, and placed in its historical context. The comparative method describes relationships among existing organisms that can then be added to the historical inferences. Finally, all this information can be added to the insights derived from studies of processes, such as artificial selection, to infer the mechanisms and forces that led to present organisms.

Quantitative predictions and tests, such as the numbers of moths with hearing versus those without that survived bat attacks, may supplement or lend support to the inferences. But as in all of science, including the use of the hypotheticodeductive method, no inferences, hypotheses, or predictions are "proved" for certain. They are only "tested"; that is, one can reject null hypotheses or fail to reject them. Certain outcomes increase confidence in the validity of the inference, and other outcomes lead to a reevaluation of the picture.

The comparative method is based on the premise that closely related organisms will share more traits than will less related organisms. The more traits that are considered, the more complete will be the picture. Chance and convergent evolution, in which certain environmental factors lead to similar characteristics in relatively unrelated organisms, often cause confusing similarities and may lead to improper inferences. Interpreted carefully, however, convergent evolution can provide useful insight into adaptations. The external morphology of dolphins, for example, strongly resembles that of fishes, but they are not closely related; they both merely evolved under the influence of aquatic environments. Aside from the outward shape and appearance, dolphins have hair and mammary glands, bear live young from placentae, have advanced brains typical of mammals, and possess a mammalian skeleton in nearly all aspects, including the limbs, vertebrae, ribs, and skull.

Two terms distinguish between converged and ancestral characteristics. **Analogous** structures have the same function, whereas **homologous** structures have the same ancestral origins. The wings of birds and wings of insects are analogous but not homologous. The wings of birds and arms of humans are homologous but not analogous. The wings of birds and wings of bats are both analogous and homologous.

The environmental conditions that result in convergence in behavior may be more complex, involving interactions with other animals as well as the more inanimate aspects of the environment, and different conditions can lead to specific behaviors that are very similar in expression. Polygyny in mating (Chapter 10), for example, may result from one set of factors in some species and another set in other species.

Animal behavior, being a step removed from the morphological structures, is not easily fossilized. The inferences are more difficult and invoke less confidence. Nonetheless, uniformitarianism may be applied even to behavior. The role of head ornaments of dinosaurs, for example, has been interpreted from the behavior of animals such as deer and certain beetles that have head ornaments today (e.g., Molnar 1977). Likewise, the behaviors that produced certain nest structures, patterns of footprints in the mud, and teethmarks on the bones of other animals in the fossil record are inferred from the behaviors that produce similar patterns and tracks today.

Although behavior may leave few fossil records, it is frequently more complex and richer in detail than most morphological structures (except at the molecular level). Species that appear almost identical morphologically may behave very differently. This often permits the comparative approach to be quite productive, and these methods have been of utmost importance to understanding the evolution of behavior.

THE PROCESS OF EVOLUTION: CHANGES IN THE GENE POOL

**The Classic
Darwinian View**

The assumption that events and processes in the past could be inferred from what is seen today led Darwin and Wallace to the theory of evolution. The main tenets of the Darwin-Wallace theory of evolution, now considered to be the "classic viewpoint," are:

1. *Organisms produce more offspring than survive.* This leads to what Darwin considered a "struggle for existence." But the phrase is somewhat misleading because it may imply that there is an actual physical struggle, which does not always occur, or that the problems are always related to overpopulation and competition, which also do not always occur.
2. *There is variation among individuals, some of which is inherited.*
3. *Some of the survival and subsequent reproduction of organisms may depend on inherited characteristics.* The consequence is **natural selection:** only the fittest survive and reproduce; that is, nature "selects" which ones will die or live and reproduce. (The phrase "survival of the fittest," incidentally, originated a few years before Darwin-Wallace. It was used by the British philosopher Herbert Spencer.)
4. *The occurrence of natural selection over long periods of time* leads to gradual changes in the species and, eventually, the origin *of new species.*

**Shifts in Gene
Frequencies**

Research and observation in biology since publication of Darwin's *Origin of Species by Means of Natural Selection* in 1859 have remained solidly consistent with and supportive of the theory. With further detail from genetics, it is possible to state more explicitly what occurs in evolution. Part of the explanation depends on the relative frequencies of different genes, or alleles, among all of the animals in the population. The **Hardy-Weinberg law** states that gene frequencies in a population remain constant regardless of changes in population size as long as there are no chance fluctuations (known as *genetic drift*) or outside influences. For a further explanation of this principle, see Appendix B.

Under natural conditions there may be genetic drift, possibly other factors such as developmental changes (discussed later), and "outside influences," that is, factors external to the workings of the animals' cells and bodies. The external environmental influences result in **differential survival and reproduction.** Recall that individual animals are not the same in their ability to survive and reproduce. Death comes at different times and in different ways for different individuals. Mortality and reproduction that vary strictly as a matter of chance either result in genetic drift or cancel each other out; hence, they are not of concern. We now are interested only in the extent to which variations in mortality and reproduction result from inherited genetic traits.

A simple example of 32 individuals and 2 alleles, A and a, in equal proportions in the population will be used. The equal proportions of frequencies of

Figure 5-1 Changes of gene frequencies in a hypothetical population as a result of selection. Two alleles *(A* and *a)* initially exist in equal proportions *(p* and *q),* but the ratios change rapidly under the effects of extreme selective pressures as indicated. Also see text.

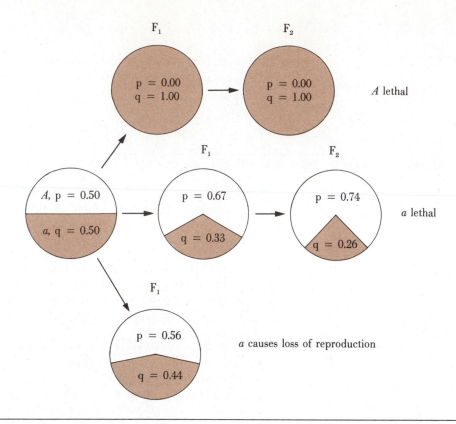

alleles are expressed as p (the proportion of A) = 0.50 and q (the proportion of a) = 0.50 (Figure 5-1). (For expanded treatment, see Appendix B.) Consider the extreme case where A is a dominant lethal gene. Imagine a hypothetical case where the allele would cause an animal not to eat; this will be called the non-feeding gene. All of the AA and Aa individuals die from starvation soon after birth, only aa individuals remain, and the frequencies of A and a become 0.00 and 1.00, respectively (Figure 5-1). If the lethal gene is recessive, aa's die, leaving 8 AA's and 16 Aa's: p = (8 + 8 + 16)/48 = 0.67 and q = 0.33. The 12 pairs will produce (ignoring chance) 144 offspring. Depending on the pairing (exact proportions are no longer possible in this reduced population), the genotypes are either 63 AA, 66 Aa, and 15 aa or 66 AA, 60 Aa, and 18 aa. (For an exercise check the pairing and arithmetic on your own. Use the figure in Appendix B [p. 665] or a copy of it for your work.) The gene frequencies after the 15 or 18 aa young die will be either p = 0.74 or 0.76 and q = 0.26 or 0.24. Either way the average genotype is changing rapidly and will continue to do so with a leveling off as the chance of heterozygotes finding each other by chance, mating, and producing homozygotes becomes reduced.

Maladaptive recessives are nearly impossible to eliminate completely unless there are few of them and they happen to be lost by chance. Selection against an allele, as in the earlier examples, will rapidly reduce the frequency of it or eliminate it. If the selection against a gene is not so harsh, the rate of reduction in gene frequency may be much lower. One could easily modify this example so that only half the aa's rather than all of them are lost. Occasionally a new gene is advantageous and modifies a trait or introduces a new trait so the individual possessing it has a higher chance of survival than other individuals with the previous form of the trait. This has the opposite effect of selection against genes and increases the proportion of new alleles in the population. The changes in gene frequencies depend on the relative advantages and disadvantages of the two (or more) genes in comparison with each other. (The figure on p. 665 or a copy of it may be used to impose different rates of selection and calculate the effects on the hypothetical population.)

Sometimes heterozygotes have an advantage over homozygotes. A classic case involves sickle cell anemia in humans. The condition is caused by a single recessive gene—a single cistron that results in substituting a single amino acid in the approximately 600 that make up the hemoglobin molecule. The result is defective red blood cells that tend to be eliminated by the body; this causes anemia. Persons without the allele do not suffer from this anemia; heterozygotes show only slight effects; and homozygotes with the allele have serious anemia and usually die in childhood. The allele remains common in parts of Africa, however, with as high as 40% of the members of some tribes possessing it, generally as heterozygotes. The reason for the high frequency is that heterozygotes are more resistant to a virulent form of malaria caused by the protozoan *Plasmodium falciparum*. The heterozygotes have an advantage over both homozygotes: the homozygotes with the sickle cell allele die of anemia and the homozygotes without it die of malaria. In the United States, however, where the malarial protozoan is absent, only about 9% of blacks, descended from Africans, possess the allele.

The sickle cell example is interesting but unusual; such striking examples of heterozygote superiority are relatively rare in nature. Heterozygote superiority per se is not especially important for this discussion. But it is a striking example of the *role of the environment* in gene frequencies; this point is of utmost importance. Gene frequencies depend not only on the relative advantages and disadvantages among different alleles but also on the particulars of the surrounding environment.

Survival is only part of the picture. The other part involves reproduction, the bridge between generations. Returning to the simple population, consider that survival is not affected by the genes A and a, but that reproduction is; for example, the allele a causes a loss of reproductive behavior so that a pair with one or both aa individuals produces only half as many young as the other pairs. From the 16 pairs there now are 48 AA, 72 Aa, and 30 aa; hence $p = 0.56$ and $q = 0.44$, and the frequencies have shifted from the initial Hardy-Weinberg equilibrium of $p = 0.50$ and $q = 0.50$ (Figure 5-1).

Thus differential survival and reproduction may change gene frequencies. If the alleles are not lethal and the individuals possessing them remain in the population to be observed and measured, then changed gene frequencies produce changes in the average phenotype for the species.

Because of extensive exchange and recombination of genes within a population, the phenotypic characteristics of a population tend to change throughout the entire population if they change at all. The phenotypes and their substrate genotypes vary among individuals so the previous statement may seem confusing. However, a particular individual is only a small part of most populations. In a population of 10,000 animals, for example, one individual would constitute only a hundredth of 1% of the population. Also, the individual genotypes are usually

Figure 5-2 Random dispersal of genes within a population and differences of average appearance depending on gene frequency. Individual rabbits in this example carry the alleles for either brown or white color, but each rabbit is only a small portion of the total population and, hence, contributes only a small portion toward the average species' characteristic color. Try to view the overall lightness or darkness of these different groups of rabbits by holding the figure at a distance or by viewing it out of focus. Whether rabbits are brown or white was determined at random in approximate proportions of 25% (A), 50% (B), and 75% (C) brown individuals, respectively.

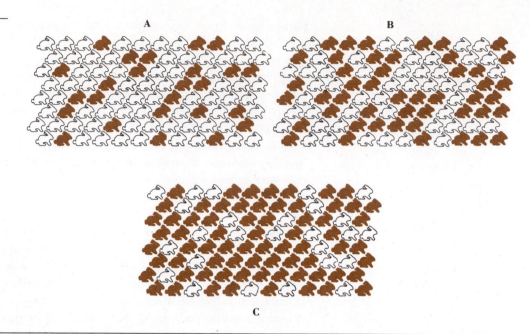

more or less randomly dispersed throughout the population (deme). It is much like a print made up of white and black dots or components interspersed together (Figure 5-2). If the frequency of white items is increased, the overall appearance becomes lighter: one is looking at the total picture and not at the individual dots composing it. On a species level, which may contain several demes with low gene flow between demes, the "grain" of the picture can change, and one could get patches of gene frequencies. In the print analogy they would be patches of white, black, and gray. This may permit such things as *clines* (i.e., gradual differences over a large geographical range) to develop. But within a deme the mobility of individuals and their ability to mate with others (hence, recombine genes) lead to a diffusion and spread of alleles throughout the entire deme.

Origins of New Species and Characteristics

New species are believed to originate in one of two ways: (1) *via gradual accumulation of changes over time* and (2) through *geographical isolation*. The spreading of alleles through a deme, as discussed earlier, is thought to keep the average gene frequency spread more or less uniformly throughout the population. It is somewhat like the level of water being the same among connected vessels. In like manner changing the gene frequencies in some areas but not others is thought to require a barrier to mixing, breeding, and consequent gene flow through the population. Time is one barrier that prevents breeding; that is, animals obviously cannot breed with other animals long since dead. Thus enough changes may accumulate over time in a population so that it becomes different enough to qualify as a new species to the taxonomist (Figure 5-3, *A*). The changes represent a gradual continuum, however, that creates difficulty sometimes in knowing where one species stops and the next starts, such as a change from white through gray to black.

The second common method of speciation in animals results in branching, splitting, and "radiation" of species; that is, more than one species can develop at the same time (Figure 5-3, *B*). This requires a barrier to gene mixing. The daughter population(s) must be cut off and separated from the parent population. This usually is thought to involve geographical isolation. It may happen when a part of a population gets carried across a normal barrier, such as by riding a natural raft across the water to a new location. Or the isolation may occur when environmental changes interrupt a population and separate it into two or more new populations. A large lake full of fish could partially dry up, for example, leaving two smaller lakes, each with its own population of fish. The climate of a large land mass may change and cause intermediate portions of the range to become inhospitable. Other major geological events such as volcanic action, erosion, and the gradual movement of land masses away from each other also are important isolating mechanisms.

Under the new conditions the daughter populations gradually can change over time. Changes are particularly likely if new environments are involved. If the isolation does not last long enough for many genetic changes to accumulate and the populations reunite, the individuals of the two populations will interbreed and gene mixing will resume. If the populations are isolated long enough, morpho-

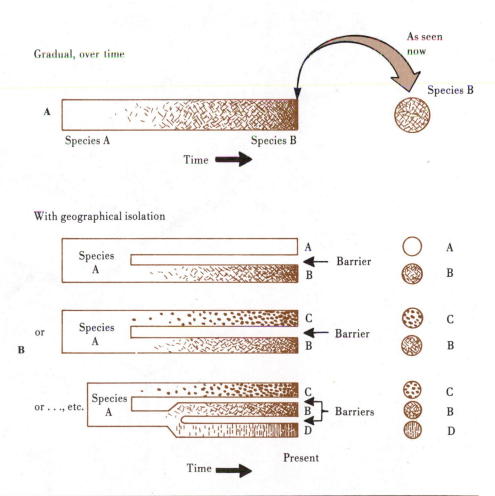

Figure 5-3 Evolutionary origins of new species. Diagram of the commonly accepted basic models. A, The average characteristics of a species change over time until animals from different time periods are sufficiently different to be considered as different species. B, Barriers may prevent mixing of genes so that, in conjunction with genetic drift and selective forces in different locations, different species develop during the same period of time. The latter situation may lead to a diversity of radiations in which the original parent species may or may not continue to exist. At any one point in time, such as the present, it will appear that distinct species exist and their past relationships may not be obvious.

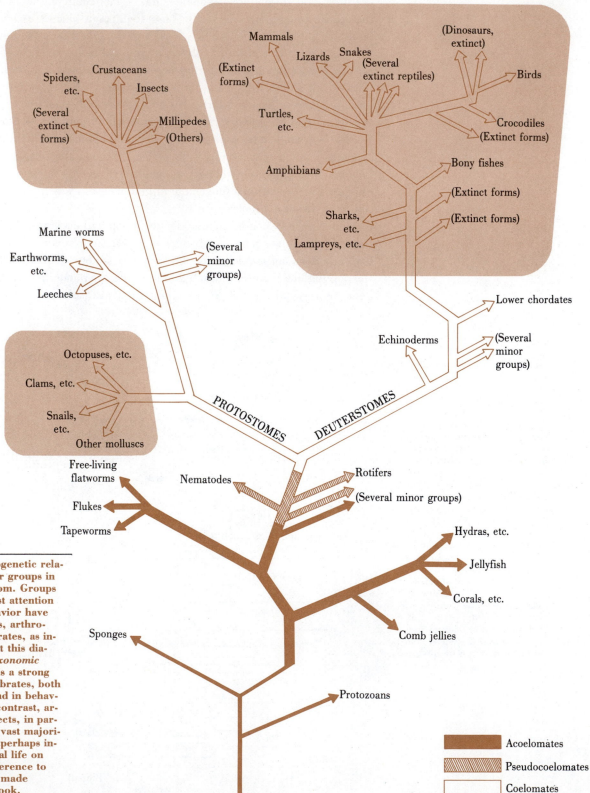

Figure 5-4 Phylogenetic relationships of major groups in the animal kingdom. Groups receiving the most attention in studies of behavior have been the molluscs, arthropods, and vertebrates, as indicated. Note that this diagram is *not to taxonomic scale*, and there is a strong bias toward vertebrates, both in the diagram and in behavioral studies. In contrast, arthropods and insects, in particular, form the vast majority of species and perhaps individuals of animal life on earth. Future reference to this chart will be made throughout the book.

logical, physiological, or behavioral barriers to interbreeding and gene mixing develop. The populations then represent clearly separate species, even if they should come back together in the same location.

The range of possibilities for geographical isolation represents a continuum similar to the time continuum. Although the species concept is related to the real world via the capacity of the animals to interbreed, it also is abstract and somewhat arbitrary. Interesting things may happen when fledgling or closely related sibling species in the taxonomic gray zone come into contact. Sterile or otherwise defective hybrids may be formed. Hybrid vigor or heterosis may occur because of heterozygosity covering maladaptive recessive alleles or from other factors causing heterozygote superiority. Some crosses involve several effects; the mule, for example, is strong and robust but sterile. In some cases the genes of one species essentially overshadow those of the other. When coyotes breed with red wolves, *introgression* occurs; the few differences that marked the two species are largely absorbed into the larger population gene pool, and the "species" in the minority (in this case, the red wolf) is lost.

Another less common source of new species involves sudden, strange genetic events such as polyploidy in which chromosomes may not double or sort properly in meiosis and the new species is essentially a 3N, 4N, etc., of a previous species. Such events are most easily propagated with asexual reproduction and are seen more frequently in plants than in animals.

The properties of variable environments and variable animals, via the reproductive and evolutionary processes outlined earlier, lead to a natural expansion in the diversity of organisms. It is a process that branches through time and space with some of the branches diverging, some branches running parallel, and occasionally some branches growing back, or converging, toward each other.

The basic relationships, as currently understood, for the animal kingdom are pictured in Figure 5-4. The classification of the major dinosaur-bird branch currently is in some dispute (Thomas and Olson 1980). The basic phylogeny and relationships are not in dispute as much as whether dinosaurs should be grouped with reptiles, grouped with birds, or split into a separate new taxonomic group. Dinosaurs, however, are quite different from both birds (by virtue of lacking feathers, having more primitive skulls and tails, and probably having different thermoregulatory physiologies) and other reptiles (by virtue of bipedal ancestry and body size in the majority of species). Hence, I agree with Dodson (1974) that dinosaurs should be placed in a separate group, as indicated in Figure 5-4.

Remember Figure 5-4 for reference throughout this book. The groups of major interest from a behavioral standpoint are the molluscs, arthropods, and vertebrates. Figure 5-4 emphasizes these groups, although not to numerical scale, and illustrates their ancestral relationships.

"Selfish" Genes and the Level at Which Selection Operates

Group Selection and Behavior

There has been academic argument over the point at which natural selection operates. The issues have been discussed extensively in behavioral contexts so they will be considered here in some detail. Darwin believed that selection operates on the individual, but he had no knowledge of genetics. Also, sometimes individual animals do things, such as exhibit "altruistic" behavior, that would seem to be disadvantageous to the individual and thus contradict the concept of individual selection. Worker bees, for example, sacrifice their own reproduction and occasionally even their lives and help another bee, the queen, reproduce instead. Some theorists have considered that selection may operate on whole groups of individuals (e.g., Wright 1955, Dunbar 1960, Wynne-Edwards 1962, Lewontin 1965) or among close relatives within groups ("kin selection," to be discussed later and in detail in Chapter 8).

Wynne-Edwards proposed that somehow entire groups of animals come to possess certain behavioral traits that other entire groups of the same species do not have. He interpreted many behaviors, such as territorial behavior, to be what he called *epideictic* phenomena, "displays, or special occasions, which allow all the individuals taking part to sense or become conditioned to population increase" (Wynne-Edwards 1965); that is, animals could essentially census and then regulate their own populations to avoid overutilization of resources. In particular, according to Wynne-Edwards, animals could adjust their reproduction to the level that resources would permit. Groups that possessed such ability would survive, and those without the traits would perish. This is similar to what Darwin proposed for individuals, but it operates on groups instead. Furthermore, Wynne-Edwards used this explanation for the evolution of virtually all social behavior.

These ideas were hotly disputed among biologists. Wiens (1966), for example, described the difference between Wynne-Edwards' and others' ideas of "group selection" and argued that behaviors which Wynne-Edwards called group selected could be interpreted better via individual selection. (For explanations of territorial behavior, for example, see Chapter 8.) Lack (1966), in disputing Wynne-Edwards, suggested that reproductive characteristics, such as clutch size in birds, may be evolved to long-term environmental averages. But from a proximate viewpoint, according to Lack, birds produce more young than survive long enough to reproduce themselves. Mortality (primarily during the winter) is the mechanism that trims population excesses rather than prevention of excesses through controlled reproduction.

The outcome of the ruckus over Wynne-Edwards' ideas appeared to be twofold: (1) it drew attention to the evolution of social behavior, and (2) it helped focus attention on the *gene* as the basic point at which selection operates, although selection normally affects genes through the individuals that carry them. The theoretical aspects of kin selection and the role of genes as the points of selection were being developed at the same time as or slightly before the Wynne-Edwards controversy. Chief architects of this theory were W.D. Hamilton (1964), J. Maynard Smith (1964), R.L. Trivers (1971), and others. This other important theory on the evolution of social behavior was well under way and probably would have flourished without Wynne-Edwards. But the Wynne-Edwards controversy helped attract attention to the subject.

Wynne-Edwards, incidentally, recently changed from his earlier position in light of the various discussions, theory, and findings. Wynne-Edwards (1977) stated, "The general consensus of theoretical biologists at present is that credible models cannot be devised, by which the slow march of group selection could overtake the much faster spread of selfish genes that bring gains in individual fitness. I therefore accept their opinion." Further details on social aspects are presented in Chapter 8.

Selfish Genes and Behavior

Perhaps the most lucid summary of the recent body of theory concerning genes and kin selection is a popular and very readable book, *The Selfish Gene*, by R. Dawkins (1976). The focus is on the gene but in a context of organisms and environments.

In his popularized account Dawkins describes individual animals as "survival machines." How well a given gene does in the gene pool depends on the phenotypes that result from it. Individual animals, the carriers of genes, are short-lived, but the genes (or rather copies of them) have the potential of being almost immortal and existing for millions of years—if they do a good job of building phenotypes. To paraphrase an old saying, individuals are just DNA's way of making more DNA.

Genes do not operate singly but in concert with other genes. Although the focus is on the individual gene, its survival depends in part on its combination with other genes. Dawkins (1976) describes a clear analogy between genes and oarsmen in a college boat race, but almost any other team sport can be substituted in the analogy. The individual team members are analogous to the genes; the whole team is the gene pool; mutation occasionally provides new potential players; and the combination of members playing during any given time would be analogous to an individual organism. The members are tried in different combinations, and there is some turnover, but usually it is found that some combinations of members tend to win more often than others. Sexual reproduction is the coach, who sends in different team members. Some members are good and remain on the team for a long time, although they occasionally are in a losing race or game. Some members, however, are so bad, regardless of who else they race or play with, that they are dropped from the team. The particular conditions of the home pond or court, audience, and such things as wind and rain that may occur during the game are environmental factors that influence the outcomes of games. Chance also may be an important element.

From this line of reasoning arises the concept of "gene selfishness" and gene fitness. The genes operate as if in a selfish manner, that is, as if for their own promotion and increase. The selfishness is not purposeful, and it obviously is not conscious on the part of the gene; rather it is a *consequence* of the process. Dawkins (1976) goes on to give other excellent analogies to illustrate how genes may "control" phenotypes and behavior in an indirect manner.

Fitness has been viewed in terms of **individual fitness,** which is defined or measured by the reproductive success of an individual during its lifetime, or in terms of the fitness of particular genes. Because a particular gene may be carried by many individuals, gene fitness is called **inclusive fitness.** This important and major concept appears to provide the best explanation for the problem of altruistic behavior mentioned earlier. Inclusive fitness deserves further discussion, but because it is tied so intimately with social behavior, the discussion is postponed to Chapter 8.

It must be emphasized that the fitness of genes is relative to the environment and the other genes present in the total combination. A set of genes for "sharp meat-cutting teeth" is advantageous to a carnivore but might be disadvantageous if it showed up in a cow. Genes out of context can be as worthless as oarsmen on a basketball team. But in a suitable combination, certain genes are better than

Figure 5-5 Simplified differences between the "group selection" theories associated with Wynne-Edwards and Wright. A, As originally proposed (but recently abandoned) by Wynne-Edwards, groups of animals arise that vary in their ability to survive as entire groups. Some groups continue to exist, and others die. One of several problems with the concept concerned how such groups would arise in the first place. B, Wright's shifting balance theory proposes that characteristics become established in small groups in which genetic drift is more likely; then animals disperse into other demes. This permits the average characteristic (such as color) to shift and become established on a larger population and species scale.

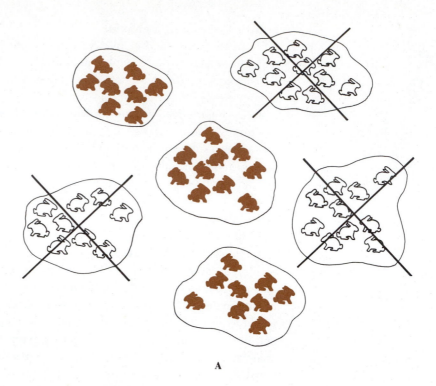

A

others. Their fitness (value as an athlete in the analogy) depends on how many copies eventually result as a consequence of the survival and reproduction of the individual phenotypes they produce.

Dawkins (1978) generalized and carried the concept of inclusive fitness to its logical conclusion—to the elimination even of individual fitness. He generalized the terminology to "replicator selection." He used "replicator" rather than "gene" for two reasons: (1) to encompass *any* (including DNA) entity in the universe that may be copied and undergo evolution by selection and (2) because it is not yet clear that the cistron (gene) is the best or only unit of DNA to view as undergoing replication and selection.

The value of the selfish gene concept to many (but not all) biologists is its generality; it basically covers all cases without being excessively tautological, that is, explaining so much that it explains nothing or is used to explain itself. The principle is the same whether the specific outcome of the gene's action is to cause an animal to capture prey better, escape from being prey, evade harsh environmental conditions such as cold, cause a mate to copulate or permit copulation, or cause a parent to feed young. These are all just different sports in a sense; they are similar in that, by the same analogy, they are team sports. The successful genes are those which, in conjunction with fellow team members, build a better "survival machine," regardless of which particular adaptation one is discussing.

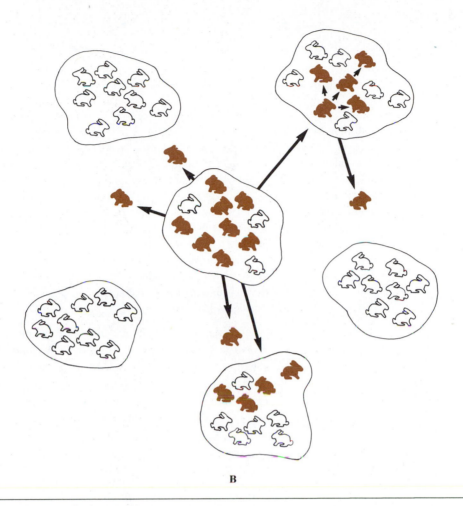

B

Much of an animal's environment consists of other animals possessing their selfish genes. Some of these genes may be shared. The problem of competing selfish genes within groups leads to an important recent topic, the concept of the *evolutionarily stable strategy (ESS)*. Because this subject is largely concerned with interactions among animals, it is discussed further in Chapter 8.

Shifting Balance Theory

The role of groups of animals to selection is acknowledged in the selfish gene viewpoint only in a somewhat indirect manner through the shared possession of certain alleles. Additionally, not all biologists agree with the focus of selection on the level of the gene. Mayr (1963) and Ford (1975) maintained that the individual is the unit of selection (which Dawkins rebutted, 1978). Perhaps at a more serious, deeper, and heavily statistical level is Wright's focus of evolution at the level of local populations (demes) or his "shifting balance theory." This important body of theory and supporting evidence are contained in four massive volumes (Wright 1968-78). A useful review is provided by Wade, who summarizes:

Wright's conviction [is] that "the existence of complex patterns of factor interactions must be taken as a major premise in any serious discussion of population genetics and evolution" (volume 1, p. 105). Evolution by natural selection occurs as a result of the interaction of phenotype and environment. In

Wright's view there are many different kinds of natural selection depending upon the levels of biological organization and the degree of variation existing at any level. The levels of organization discussed by Wright range from single genes and individuals to families and entire demes. He considers the most important evolutionary force to be selection among more or less randomly differentiated local populations. This interdeme selection is brought about by the differential dispersion of those local populations in which random genetic drift and directional mass selection have established a "system of interacting genes of superior fitness." This is Wright's shifting balance theory of evolution, and in volume four he abundantly documents the fact that the genetic differentiation of local populations, necessary to his theory, is widespread in nature. Although many evolutionary biologists and geneticists at the present time accept Wright's major premise regarding the complex relationship of gene and character, they disagree with his views on the levels of selection. . . . I am certain that as the recent trend of evolutionary argument from metaphor and plausibility is reversed [Wright's work] will become a major reference.*

Because Wright's shifting balance theory may not be clear from a verbal description and because it may seem quite similar to Wynne-Edwards' earlier proposal of group selection, which clearly has been rejected as discussed earlier, Wright's view and the difference between the two are illustrated in Figure 5-5.

Much of the whole topic of the point at which selection acts remains in an abstract and academic realm. The main contenders in the controversies may actually agree with each other more than they think or say. For the purposes of this book and in attempt to step aside from the current academic controversies, *selection* is defined as *any process (except random drift) that changes the Hardy-Weinberg equilibrium and alters gene frequencies.* Various alleles in a population are shared more widely than by immediate or near relatives. Thus the shifts of gene frequencies in whole populations, as discussed by Wright, should be kept in mind.

EVOLUTION VERSUS OTHER CONSIDERATIONS

The current biological view of evolution, as discussed earlier, is that the phenotypes of organisms may gradually change, (i.e., evolve) from one generation to the next over long periods, usually thousands or millions of years. The rates of evolution depend partly on heritability, partly on environmental pressures, and partly on other things such as chance. Changes probably occur more slowly in the real world than under artificial selection. Nature works with time. Phenotypic changes are the result of genotypic and subsequent changes in protein synthesis. Except for the earliest organic molecules, all life comes from previous life. Present-day species of animals did not always exist (although many have been around for hundreds of millions of years). Various species arose by changing from previous species.

*From Wade, M.J. 1980. Science 207:173-174. Copyright 1980 by the American Association for the Advancement of Science.

Most biologists consider the existence of gradual change to be *fact* (e.g., Wilson et al. 1978). The *evolutionary theory*, which attempts to explain how these changes occurred, is one of the strongest theories in all of science. But not all persons have understood or agreed with the facts and theory of evolution, and even within the evolutionary framework there are other considerations that should be mentioned.

Special Creation—Creationism

The theory of special creation states that all species were created uniquely with their full sets of characteristics during a short period of time. The theory, however, simply has no supporting evidence in the geological and fossil record. The bones and traces of earlier forms of life exist, can be seen and handled in various museums and laboratories, and must be reckoned with. It seems very unlikely that, as was once proposed, God put those fossils there just to confuse us! The fossil evidence points quite clearly to gradual changes over very long periods of time.

Special creation (or, more recently, creationism) has been closely allied with Judeo-Christian religion, and the biological view of evolution often is considered atheistic. Neither of these views is completely tenable. I would prefer to drop the topic at that, but in my experiences with many students and even some colleagues, the topic refuses to be ignored. An understanding of evolution is critical for an understanding of behavior. So I will briefly address the issue further, then allow each reader to draw his or her own conclusion. The subject of creationism, incidentally, has been receiving much attention. The history and current status of the topic are reviewed by Numbers (1982).

Special creation is a recent historical (probably within the last 1500 years) biblical interpretation. Persons alive when the creation account originally was written and up until the last 1500 years or so apparently were not concerned with scientific accuracy in the same sense that persons are today (Fretheim 1969). A critical reader can see that special creation is only *one interpretation* of the Bible. This interpretation made sense to persons who were not aware of the vast expanses of geological time. The idea that species suddenly appeared from nowhere prevailed both in and out of a religious context. Away from a religious context, for example, one could find a similar notion in the form of *spontaneous generation*. This idea apparently arose from the familiar experience of seeing such things as mice appear in grain sheds and maggots in meat.

In the first chapter of Genesis (any translation, including King James) the Bible states, "Let the earth bring forth . . ." (1:11), "Let the waters bring forth . . ." (1:20), or "[organisms] yielding fruit each according to its kind" (1:11). The "days" in Genesis were not likely 24-hour days. The initial origin of the universe ("Big Bang"), according to modern theory (Gamow 1952), took less than 24 hours and even less than 1 second. Subsequent events, including the development of galaxies, stars, and systems such as our solar system, have taken much longer. Also, the sequence of appearance of different species, such as flowering plants before aquatic creatures, is out of order. But beyond the timing and sequence, the Bible seems to be as compatible with modern evolutionary theory as it is with special creation theory. "Each according to its kind" can be interpreted easily as referring to *reproduction* rather than the original appearance of the species. The story of Adam and Eve can be interpreted as literary prose that describes the relationships among humans and God that apply to all humans rather than as a scientific account of the first man and woman.

One may view the process of evolution as the *means* by which God created organisms. Anyone who does not believe in God is free to believe that it all just happened. The point here is that there are *many* interpretations of the Bible and how it relates to science. Further discussions from diverse viewpoints are provided by Fretheim (1969), Sagan (1979), Fischer (1981), and many others. Old arguments that attempt to draw sides on the issue seem a waste of time whether one is agnostic, theistic, or a devout atheist.

Extraterrestrial Origin of Species

Another theory, recently pushed by Daniken (1977), proposes an extraterrestrial origin of species, particularly humans. This at best only moves the problem to a new location, and there is virtually no biological evidence for it. The morphological similarities of humans (i.e., homologies) with other organisms argue overwhelmingly that human origins, although *Homo* is only about 3 million years old, have been tied to Earth for at least 400 million years. Furthermore, all known animals are adapted to the conditions on this planet and not interplanetary space. If life here came from elsewhere, it most likely would have been of the most primitive form. The arguments and evidence for life starting and evolving on Earth are more plausible than those which state that life came from elsewhere. Even the most plausible support for extraterrestrial contacts has been adequately disputed (Sagan 1979).

Lamarckianism

Ideas commonly associated with Jean Baptiste de Lamarck are that species evolved over many generations from the gradual accumulation of *acquired characteristics* and that there is a natural tendency to evolve from simple to more complex organisms, reaching the peak in humans. These theories have few current followers and are, for the most part, of little but historical interest (although see Lewin 1981a).

Neutralism

Unlike the three previous schools of thought, the neutralist view of evolution merits further consideration in biology. Many, if not all, mutations in the DNA nucleotide bases occur at random. However, random changes in gene frequency (genetic drift) also may play an important role in speciation. There may be more characteristics that are accounted for by this mechanism than formerly was thought by biologists. If changes occur without the pressure of natural selection, without either advantage or disadvantage, they are said to be "neutral" changes. Because of the random nature of neutral changes, the view also is considered occasionally as the "random walk" theory of evolution.

Just how important neutral changes are in the overall process of evolution and just how they mesh with natural selection still remain to be seen. The neutral theory of evolution is a subject of continuing discussion and investigation, and it is meshed in complex statistics and mathematics. For recent references see Kimura (1979) and Jukes (1980). In the opinion of most biologists, random walk is insufficient to account for the *entire* evolutionary process (Felsenstein 1980). There simply is too much evidence for the effects of selection.

Recent Evolutionary Considerations

In addition to all of the preceding considerations, there have been some new worms crawling out of the evolutionary can. In particular, the part of the picture that describes evolution as a smooth, continual accumulation of many genetic mutations has come under fire from several quarters. Also, the findings from recombinant DNA research have raised numerous questions about the role(s) of genes in evolution.

Several persons have raised the possibility that there are important constraints and implications on evolution from *embryonic development*. The bithorax genetic complex of the fruit fly (a genetic complex that has been studied intensively for many years by E. Lewis, see Marx 1981), for example, suggests some evolutionary advances may be developmentally repressed ancestral conditions and may occur in part from shuffling and jumping genes. At a 1981 conference in West Berlin (reported by Lewin 1981b) two major points were stressed: (1) developmental processes limit the influence of natural selection; some changes are developmentally possible, and some are not; and (2) modification of genetic *regulation* that affects early or various developmental stages may result in greater evolutionary differences than simple changes in a large number of other genes. A few slight alterations early in the dynamic process of a developing embryo are known to account for some major morphological differences between certain species of salamanders and may be responsible, for example, for many of the differences between chimpanzees and humans.

Another recent point of contention among evolutionists is concerned with whether evolutionary changes are steady and gradual (i.e., **gradualism**) or occur in spurts *(**punctuated equilibria**)* with possible implications for the classic darwinian theory of evolution. Many (not all) sequences of species show long periods without noticeable change in the fossil record that are interspersed with other periods of (apparently) rapid change. This led S.J. Gould to state (1980a:120), "If Mayr's characterization of the synthetic theory is accurate, then that theory, as a general proposition, is effectively dead, despite its persistence as textbook orthodoxy." This is an unsettling statement. Although there is nothing wrong with being unsettled, Stebbins and Ayala (1981) dismissed it by inspecting the logical framework and concluded that "the extent to which macroevolution is gradual or punctuational remains to be ascertained. Macroevolutionary processes are underlain by microevolutionary phenomena and are compatible with the synthetic theory of evolution. . . . [The] principles are compatible with both gradualism and punctualism."

THE ROLE OF FUTURE EVENTS IN BEHAVIOR

There have been two generally opposed viewpoints, one anthropomorphic (Greek *anthrōpomorphos*, of human form) and the other more mechanistic, on why animals behave in certain ways. At one extreme, the generally popular view is that animals are aware of what they do and the consequences; hence, they behave purposefully. A turkey hen, for example, might be interpreted as taking care of her chicks "for the purpose (as understood by her) of raising them," "to give them love and care," "for the survival of the species," or for whatever purpose one may wish to insert in the statement. This view is known as **anthropomorphism.**

Anthropomorphism is based on the assumption that because much of human behavior is purposeful, based on reason, and often directed by the anticipation of future events, then it must be the same in other species. This assumption usually is accepted without much thought or criticism. The resulting anthropomorphic view has been passed down as folklore among the general public, and it is fostered by much of the cartoon and film entertainment industry. In its fullest expression, anthropomorphism views other animals, although they may not be very smart or able to talk, as little people in fur, feathers, or chitinous exoskeletons.

But some careful observers noticed that not all behavior seems so reasonable or purposeful. The absence of purpose is seen most strikingly when something goes wrong with otherwise adaptive behaviors. Turkey hens normally care for their chicks, apparently as if they understood why they are doing it. If the hen is deafened, however, she will ignore or kill the chicks. Or if a speaker is placed on a kitten so that chick calls can be played back as if they were coming from the kitten, then the turkey hen will attempt to mother the kitten (Schleidt et al. 1960). Thus the behavior might be viewed simply as a more or less mechanical response to the chicks' vocalizations. Even among humans one often can see behavior that occurs without deliberate understanding or in spite of understanding and attempts to act otherwise. Examples include problems many people have with compulsive behavior (e.g., alcoholism, overeating, overwork, and a variety of sexual behaviors).

These observations led to the opposite viewpoint and a rejection of anthropomorphism. In this alternate viewpoint, behavior is seen as being completely mechanical. Insects could be thought of, for example, simply as ganglia on legs. Most, if not all, animals could be considered as being no more aware of why, and perhaps even what, they do than are washing machines or automobiles.

There has been some problem of where to fit humans into this picture. Most people who rejected anthropomorphism in other animals were nonetheless conscious of their own self-awareness. They considered that only the other animals were little living machines. Some persons (e.g., Skinner 1974), however, proposed that even humans, in spite of our capacity for learning and self-consciousness, do not really understand why we behave as we do either. Hence, humans are not much different from the other animals, and people should be viewed simply as animals that are more advanced in respect to some aspects of the nervous system and behavior.

Bringing some moderation to these two extremes, Griffin (1976a,b, 1977)— the same Griffin who worked with echolocation in bats—courageously questioned the view that nonhuman species lack self-awareness and purposeful behavior. He defined "awareness" on the basis of mental images of events that may be remote or near in time and space, a definition that he later (1981) described as compatible with accepted dictionary and philosophical usages of the term. His argument, in brief, is that if biologists are to take an evolutionary view of the structure of brains, nervous systems, and behavior, then all aspects, including awareness and purposefulness, should have some phylogenetic history. At least traces of these characteristics should be found in other species. If one is willing to use other animals such as white rats and pigeons (forms that are not direct ancestors of humans) as models of human learning, then why not also use them as models, at least to a degree, of awareness and purposefulness? In what he called "a possible window on the minds of animals," Griffin proposed that humans could establish dialogues with other species through their species-specific channels of communication (Chapter 9).

Beyond the proposal, however, the argument has not made much progress. There is no universally accepted method of testing animal awareness, and Griffin, as might be anticipated, was criticized from several angles. Persons wishing to pursue further the debate on this and the closely related issue of consciousness should refer to Corben et al. (1974), Mason (1976), Hailman (1978a), Premack and Woodruff (1978), Griffin (1981), and references contained therein. Davidson and Davidson (1980) provide a review of current research and thought on the problem of consciousness. But, as Robinson (1981) commented, even Davidson and Davidson's "major strength is its willingness to acknowledge that psychology's abiding problems did not disappear when they were ignored."

Much of the discussion on these topics bogs down in academic argument and the prior positions of the observers. Different persons frequently draw opposite conclusions from the same or similar results. Gallup (1970), for example, concluded that the ability of chimpanzees to use mirrors to locate objects on their bodies indicated "self-awareness." Epstein et al. (1981) taught pigeons to do the same thing (Figure 9-24) and proposed a "nonmentalistic" explanation, which they then applied also to chimpanzees and humans. Epstein et al. discounted the self-awareness interpretation. Self-awareness, consciousness, and the mind, whatever and wherever they are, remain elusive at best.

Much of the semantic quagmire results from the presence of several different facets: the question of purposefulness of a higher order in the universe, such as from a creator, the question of conscious purposefulness in individual animals, and a common shorthand manner of biological speech using "to" and "for" to describe adaptations and goal-directed behavior. Biologists frequently make statements such as "wings are *for* flying," for example, without necessarily referring to purpose or meaning that wings evolved in anticipation of flying. This shorthand manner of speaking may have caused more problems that it has been worth. One way of describing the *apparent* purposefulness in biological adaptation is through the label **teleonomy**, as opposed to the more religious or mystical *teleology*. The term teleonomy was designated by Pittendrigh (1958, cited in Lorenz 1981).

Regardless of whether or how much an individual animal understands and guides its behavior by the anticipation of *future* events, such as the future welfare of its offspring, it appears that much of why animals behave in specific ways has resulted from evolutionary processes, that is, *past* events. Even the ability or capacity to think about the future depends on the past.

MORE ON SURVIVAL AND REPRODUCTION: A RECAP

For some unknown reason, perhaps as a result of the history of misinterpretation or sloppy biological expression, many persons have difficulty interpreting the proper roles of survival and reproduction. The problem has been particularly pernicious concerning behavior.

Figure 5-6 Conservative nature, or lack of evolutionary change, of vertebrate head scratching. Many birds retain the trait of scratching the head by reaching over the wing ("front leg") as shown by present quadrapeds and presumably as inherited from quadrapedal ancestors. Modified from Lorenz, K.Z. 1958. Sci. Am. 199(6):67-68.

Except for the possibility of some purposeful behavior in humans and occasionally other species, as discussed earlier and as stressed by Fisher (1958) and Hamilton (1963), the reasons most animals behave as they do are *not* "for the preservation of the species," "to survive," or "to reproduce." All of these statements project into the future. Rather, the ultimate reasons lie in the *past:* they are because the animal's *ancestors did* survive and *did* reproduce (Figure 1-21). They possess*ed* suitable genotypes; that is, they *were* adapted to the particular ecological conditions under which they existed. Because those animals survived and reproduced, the characteristics were passed on to their descendants—the animals that are observed today.

Be wary of any causative statement that contains the words "to" or "for." Ultimate factors should not be viewed as projecting into the future. They may lead to or result in improved survival and reproduction in the future (i.e., they are said to have a "function"), but that is a *consequence and not a reason or cause.* Skinner (1981) extended this principle ("selection by consequence") to account for all learned behavior and the development of cultures. Most biologists are not prepared yet to go that far, but most might find Skinner's discussion worth considering and thought-provoking.

Learning and culture aside, both ultimate and proximate factors are generally considered to be environmental. Thus *proximate factors* are those aspects of the external (ecological and social) and internal (physiological and neural) environments that trigger a behavior in the life of a given individual. *Ultimate factors* are those environmental factors which *led* to differential survival and reproduction among the ancestors. These points are very important to a proper interpretation of behavior. If they are not clear yet, one should go back and study this chapter and perhaps a general biology or zoology text again until they are understood.

As a concluding perspective, it still appears that Darwin was not wrong, although there were a lot of details (such as in genetics, molecular biology, and statistics) that he obviously lacked. The modern evolutionary synthesis (e.g., Mayr and Provine 1980) has filled in many of the details. Still much detail is lacking; a more complete and fascinating evolutionary picture continues to develop. Yet the basics, beginning with Darwin and Wallace's main tenets, seem to have survived thus far. These basics will be considered and applied to behavior.

THE EVOLUTION OF BEHAVIOR

On the premise that behavior has a genetic foundation, then various behaviors should be subject to the forces of evolution just as are morphological traits. If heritability is low, selection may be relatively ineffective and require a very long time. But if there is a genetic component at all, then long-term genetic changes should be able to accumulate and cause divergence just as they do for the more familiar morphological and physiological traits.

The comparative method is the most common approach to the evolutionary study of behavior in a particular taxonomic group of animals. The general steps are:

1. Qualitatively describe the behavioral repertoire (i.e., construct an ethogram) (Chapter 3).

2. Compare different species.
3. Formulate an evolutionary, phylogenetic hypothesis for the behavior.
4. Quantitatively test and refine the hypothesis.
5. Evaluate the outcome against other evidence.

The comparative study of behavior differs from other aspects of biology in that most sequences of behavior are shown only infrequently and they may be of very short duration. One can always see a leg or wing, for example, but a particular courtship behavior or act of predation may be seen only rarely. Also, the behavior may appear only under natural or undisturbed conditions. Thus the comparative study of behavior requires *much* time, often years of observation.

Detailed comparative behavioral studies have been conducted on a large number of taxonomic groups. A familiar simple example of homologous behavior involves head scratching in mammals and many birds (Figure 5-6). Quadrapedal animals generally scratch their heads by propping themselves on the front two and one of the back legs and using the other back leg for scratching. In its movement the back leg comes up and over the front leg on that side. Birds also use their back legs for scratching (the forelegs having become wings). Although it would seem "logical" to simply bring the leg forward to scratch the head, only a few birds do it in this way. Instead, most species of birds have retained the primitive quadrapedal pattern and scratch by dropping the wing and bringing the leg, somewhat awkwardly, up and over the top.

Other homologous behavior patterns include many display postures in gulls and courtship actions in waterfowl (Figure 5-7). Different species may use similar

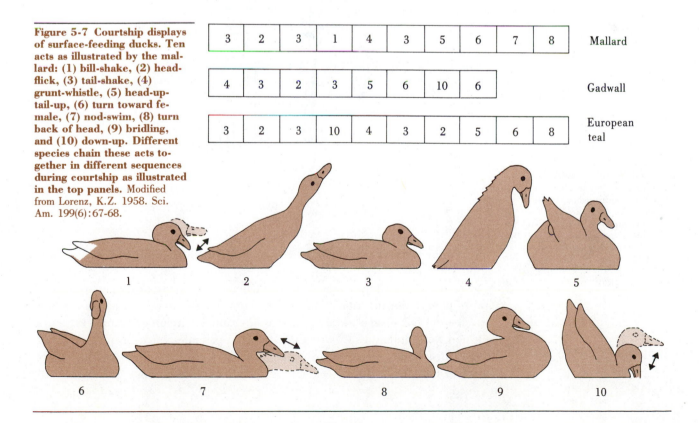

Figure 5-7 Courtship displays of surface-feeding ducks. Ten acts as illustrated by the mallard: (1) bill-shake, (2) head-flick, (3) tail-shake, (4) grunt-whistle, (5) head-up-tail-up, (6) turn toward female, (7) nod-swim, (8) turn back of head, (9) bridling, and (10) down-up. Different species chain these acts together in different sequences during courtship as illustrated in the top panels. Modified from Lorenz, K.Z. 1958. Sci. Am. 199(6):67-68.

| 3 | 2 | 3 | 1 | 4 | 3 | 5 | 6 | 7 | 8 | Mallard |

| 4 | 3 | 2 | 3 | 5 | 6 | 10 | 6 | Gadwall |

| 3 | 2 | 3 | 10 | 4 | 3 | 2 | 5 | 6 | 8 | European teal |

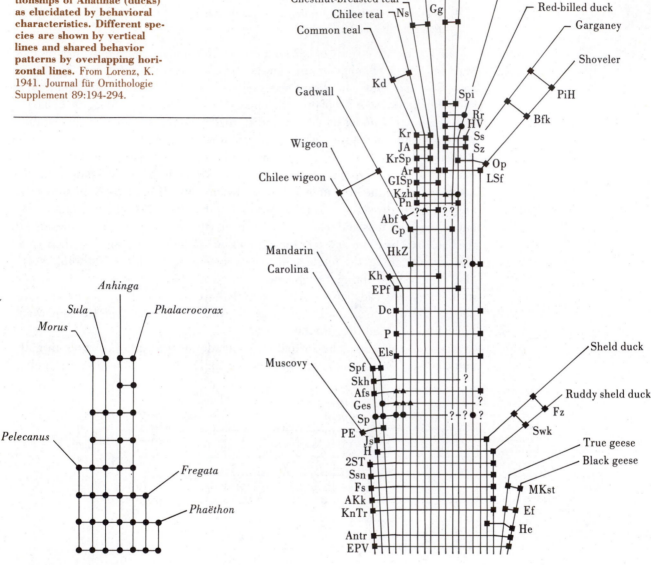

Figure 5-8 Taxonomic relationships of Anatinae (ducks) as elucidated by behavioral characteristics. Different species are shown by vertical lines and shared behavior patterns by overlapping horizontal lines. From Lorenz, K. 1941. Journal für Ornithologie Supplement 89:194-294.

Figure 5-9 Taxonomic relationships of Pelecaniformes (pelicans, anhingas, etc.) as revealed by comparative behavioral analysis. Lines as in previous diagram but with shared characteristics represented by dots. Modified from Tets, G.F. van. 1965. Ornith, Monogr. 2:1-88.

acts, but the sequences are different as in different series of numbers in combination padlocks.

Some studies of large numbers of behavior patterns in particular taxonomic groups permit the determination of the likely ancestral relationships through the apparent homologies. Two of the classic examples include waterfowl (Figure 5-8) and Pelecaniformes (Figure 5-9). Similar examples have been worked out by now for a large number of taxonomic groups. To further illustrate how this is accomplished, results of a comparative study of vocalizations in cranes will be used.

Classification of Cranes*

Family Gruidae

Subfamily	**Balearicinae**	
Genus	*Balearica*	Crowned cranes
Species	1. *pavonina*	West crowned
	B.p. pavonina	Nigerian crowned
	B.p. ceciliae	Sudan crowned
	2. *regulorum*	East crowned
	B.r. regulorum	Southern crowned
	B.r. gibberceps	Kenyan crowned
Subfamily	**Gruinae**	
Genus	*Bugeranus*	
Species	1. *carunculatus*	Wattled
Genus	*Anthropoides*	
Species	1. *virgo*	Demoiselle
	2. *paradisea*	Stanley
Genus	*Grus*	
Species	1. *canadensis*	Sandhill
	G.c. canadensis	Lesser sandhill
	G.c. rowani	Canadian sandhill
	G.c. tabida	Greater sandhill
	G.c. pratensis	Florida sandhill
	G.c. pulla	Mississippi sandhill
	G.c. nesiotes	Cuban sandhill
	2. *antigone*	Sarus
	G.a. antigone	Indian sarus
	G.a. sharpii	Eastern sarus
	3. *rubicunda*	Brolga
	4. *vipio*	White-naped
	5. *leucogeranus*	Siberian
	6. *monacha*	Hooded
	7. *grus*	Common
	G.g. grus	European
	G.g. lilfordi	Lilford
	8. *nigricollis*	Black-necked
	9. *americana*	Whooping
	10. *japonensis*	Manchurian

*In subsequent tables and figures scientific names are abbreviated to the first letters of the genus and species.

An Example: the Evolution of Crane Vocalizations

The family of cranes (Gruidae) consists of 15 living species found in various parts of the world (see boxed material above and Figures 5-10 and 5-11). They are large, long-legged, spectacular birds with populations that vary from common to extremely endangered. Of the two species found in North America, the sandhill crane is now relatively common, whereas the whooping crane is well-known for its endangered status.

Figure 5-10 Different species of cranes found in the world. Although all species are shown together in this illustration, they are widely distributed in nature, and only a few species at most can be found in the same location. Modified from Singer, A. 1978. Audubon 80(2):17-24.

Cranes are monogamous, and nesting females normally lay two eggs. The birds display a well-known "dance" during courtship. The male and female share incubation duties and then give considerable parental care to the chicks. The two chicks, if both hatch and survive, show a period of serious aggression toward each other and are kept separate by one going with each parent until the aggressive period passes and the family rejoins. The family remains together most of the time, at least through the first migration. For further details on the natural history of the various species, see Walkinshaw (1973).

Cranes are very vocal and have a repertoire of around 10 distinct calls, including contact, stress, food-begging, and several others. The crane chick initially has a high-pitched voice, which changes to a lower pitch around age 9 to 11 months. The calls of interest for present purposes are the *unison call* and the closely related *guard calls*. The guard call is a single, short vocalization (approximately 1 second) given in threat or fear contexts when flight is not yet the course of action. It is independent of other vocalizations that may precede or follow it. Sonograms of the guard calls for the 13 species are shown in Figure 5-12.

The unison call is a series of calls lasting several seconds to over 1 minute and given in duet by both members of a pair of cranes. The unison call generally occurs in a courtship or sexual context. It may be accompanied by a variety of movements, particularly of the head, neck, and wings, commonly known as the "dance." Unison calls often incorporate guard calls as components in various manners, depending on the species. Figure 5-13 presents a sonogram of the unison call of the white-naped crane.

The unison calls and associated behaviors of 13 of the species were studied intensively by George Archibald (1976a,b, Wood 1979). This section is derived from Archibald's work (1976a).

For a behavior to be useful in elucidating the evolutionary relationships of a

Figure 5-11 Distribution of crane species throughout the world. Names are located on the continent where the birds nest.

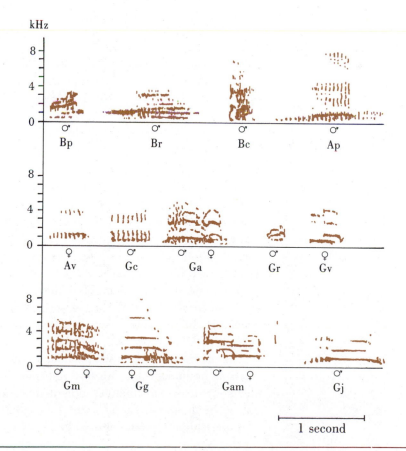

Figure 5-12 Sonograms of the guard calls for 13 species of cranes. For abbreviations used see boxed material on p. 131. Modified from Archibald, G.W. 1976. Ph.D. Thesis. Cornell University, Ithaca, N.Y.

Figure 5-13 Example of the unison call and associated dance movements for one species of crane, the white-naped crane *(Grus vipio)*. The call lasts for 9 seconds, as shown in the 3-second panels. The sequence of calling by both male and female is indicated beneath the sonograms along with diagrams of the associated postures. Modified from Archibald, G.W. 1976. Ph.D. Thesis. Cornell University, Ithaca, N.Y.

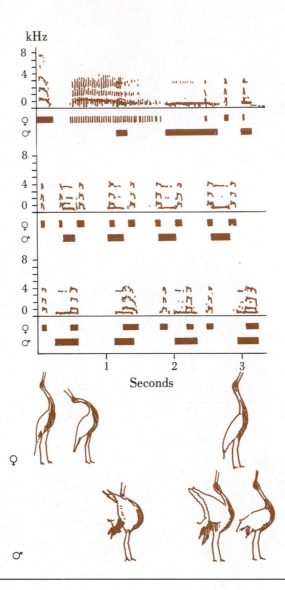

group, it should meet three criteria: (1) be homologous among the members of the group, (2) be largely innate, with little variation resulting from experience, and (3) not be used as a reproductive isolating mechanism (which might create potentially confusing and excessive divergence of the display) (Marler 1957). Behavior, incidentally, plays a frequent and very important role in reproductive isolation, as will be discussed in Chapter 10.

The crane unison calls meet all these criteria. The unison calls are sexual pair displays in all species, and the sonograms show varying degrees of structural homology in the calls: thus they may be reasonably considered as homologous.

Chicks that are incubated, hatched, and reared under unnatural conditions, such as by humans or with foster species, and isolated from their own species develop unison calls that are indistinguishable from those of chicks hatched and raised by their own parents. Thus the calling appears to be innate.

Concerning reproductive isolation, various subspecies of sandhill cranes

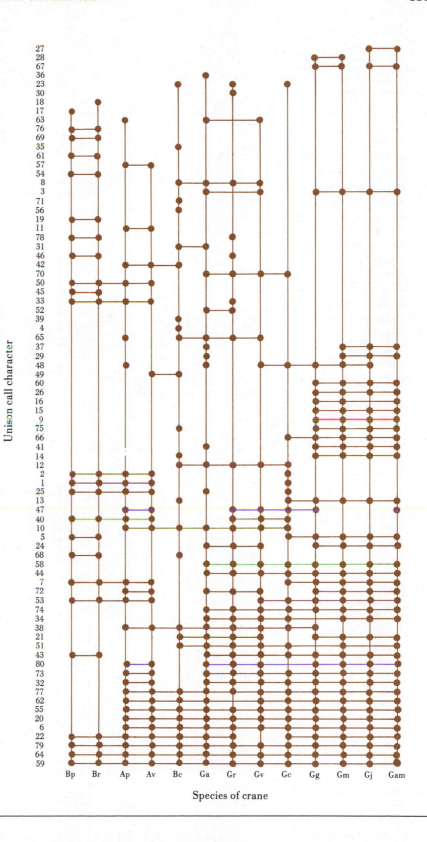

Figure 5-14 A chart of crane unison call dance characters arranged to reveal shared characteristics among 13 of the species of cranes in a fashion similar to Figures 5-8 and 5-9.

winter together and have indistinguishable unison calls but do not breed together; hooded and common cranes have very similar unison calls and winter together in some areas and breed sympatrically in Siberia but have not been observed to hybridize. Eastern sarus cranes, which recently entered Australia, have very

Table 5-1 Characteristics of the unison call dance for 15 species of cranes*

	Bp	Br	Bc	Av	Ap	Gc	Ga	Gr	Gv	Gm	Gg	Gam	Gj	Gl	Gn
1. UC a series of GCs	X	X		X	X	X									
2. GCs often precede UC	X	X		X	X	X									
3. GCs rarely precede Uc							X	X	X	X	X	X	X		X
4. Gcs never precede UC			X											X	
5. Unique introductory calls absent	X	X				X				X	X	X	X		X
6. Unique introductory calls present			X	X	X	X	X	X	X	X	X	X	X	X	X
7. Either sex begins UC	X	X		X	X	X				X	X	X	X	X	X
8. Only ♀ begins UC			X				X	X	X						
9. Usually ♀ begins UC										X	X	X	X		X
10. Both broken and unbroken calls			X	X	X	X	X	X	X						
11. Broken to unbroken to broken to unbroken sequence				X	X										
12. Unbroken to broken to broken sequence			X				X	X	X						
13. Only unbroken calls			X			X				X	X	X	X	X	X
14. Glissando to regular calls			X							X	X	X	X	X	X
15. Several short calls to regular calls										X	X	X	X		X
16. Glissando and several short calls to regular calls										X	X	X	X		X
Regular Elements of the UC															
17. Guard calling mostly	X														
18. Boom calling mostly		X													
19. Boom	X	X													
20. Shrill calls			X	X	X	X	X	X	X	X	X	X	X	X	X
21. Calls generally uniform throughout			X				X	X	X	X	X	X	X	X	X
22. Two types of calls	X	X	X	X	X	X	X	X		X	X	X	X		X
23. Last call low, long, broken			X			X			X						
24. First few regular calls shorter							X	X	X	X	X	X	X		X
25. Guard calls and regular calls	X	X		X	X	X	X								
26. Monosyllabic and disyllabic calls										X	X	X	X		X
27. Usually monosyllabic												X	X		
28. Usually disyllabic										X	X				X
29. Glissandi present							X			X		X	X	X	X
30. Shrill call modified by gular sac								X							
31. Regular calls longer than regular ♂ calls	X	X	X				X								
32. Regular ♂ calls longer than regular ♀ calls				X	X	X	X	X	X	X	X	X	X		X
33. One ♂ call per ♀ call	X	X		X	X		X							X	
34. Two ♀ calls per ♂ call						X	X	X	X	X	X	X	X		X
35. More than two ♀ calls per ♂ call			X												
36. More than one ♂ call per ♀ call							X								
37. ♀ calls between 0.30 and 0.52 second							X			X		X	X	X	
38. ♀ calls less than 0.22 second			X	X	X	X	X	X	X			X			X
39. ♂ calls 0.11 second			X												
40. ♂ calls 0.21 to 0.30 second	X	X		X	X	X		X	X						
41. ♂ calls 0.38 to 0.65 second							X			X	X	X	X	X	X

Data for the first 13 species are from Archibald, G.W. 1976. The unison call of cranes as a useful taxonomic tool. Ph.D. Thesis. Cornell University, Ithaca, N.Y. Details for the last 2 species were obtained later and provided by Archibald.

*X means the character is present in either both male or female. For full species names, see p. 131.

Table 5-1 Characteristics of the unison call dance for 15 species of cranes—cont'd

	Species														
	Bp	Br	Bc	Av	Ap	Gc	Ga	Gr	Gv	Gm	Gg	Gam	Gj	Gl	Gn
42. UC determinate in length			X	X	X										
43. UC indeterminate	X	X				X	X	X	X	X	X	X	X	X	X
44. UC usually between 4 and 40 seconds						X	X	X	X	X	X	X	X	X	X
45. UC can last more than 1 minute	X	X												X	
46. Basal frequency below 0.70 kHz	X	X						X							
47. Basal frequency about 0.70 kHz to 0.90 kHz				X	X	X		X	X		X	X			X
48. Basal frequency about 0.90 kHz to 1.2 kHz						X	X	X	X	X			X		
49. Basal frequency about 1.4 kHz				X	X									X	
50. Only basal frequency with significant amounts of sound energy	X	X		X	X										
51. Much sound energy in several harmonics above the basal frequency			X			X	X	X	X	X	X	X	X	X	X
52. More than 80 dB at 10 feet							X	X							
53. Between 75 and 80 dB at 10 feet	X	X	X	X	X	X		X	X	X	X	X	X	X	X
54. Asynchronous UC	X	X													
55. Synchronous UC			X	X	X	X	X	X	X	X	X	X	X	X	X
56. After introduction, no apparent synchrony				X											
57. Depending on which sex starts UC the other synchronizes with it				X	X									X	
58. After the introduction, ♀ synchronizes her calls with ♂						X	X	X	X	X	X	X	X		X

Visual, Dance Characters Associated with UC

	Bp	Br	Bc	Av	Ap	Gc	Ga	Gr	Gv	Gm	Gg	Gam	Gj	Gl	Gn
59. Stand stationary	X	X	X	X	X	X	X	X	X	X	X	X	X	X	X
60. Usually walk										X	X	X	X	X	X
61. May or may not stand near each other	X	X												X	
62. Usually stand near each other			X	X	X	X	X	X	X	X	X	X	X		X
63. Sexes sometimes touch while UC					X		X	X	X						
64. Wings folded throughout	X	X	X	X	X	X	X	X	X	X	X	X	X	X	X
65. Always raise elbows			X		X		X	X	X						
66. Raise elbows with increased threat						X				X	X	X	X	X	X
67. Lower primaries with increased threat										X	X	X	X	X	X
68. Head lowered to shoulder level at start	X	X	X											X	
69. Head lowered to shoulder level throughout	X	X													
70. Wings pumped						X	X	X	X						
71. Upward neck thrust			X												
72. Neck back beyond vertical at start				X	X			X	X	X	X	X	X		X
73. Neck back beyond vertical throughout				X	X	X		X	X	X	X	X	X		X
74. Neck vertical throughout				X		X	X	X	X	X	X	X	X		X
75. Neck forward beyond vertical throughout			X							X	X	X	X		X
76. Head movement from side to side	X	X													
77. Head held in vertical plane throughout			X	X	X	X	X	X	X	X	X	X	X		X
78. Gular sac inflated	X	X						X							
79. Sexual context	X	X	X	X	X	X	X	X	X	X	X	X	X	X	X
80. Threat context				X	X	X	X	X	X	X	X	X	X	X	X

different unison calls from the native brolga cranes and have been observed to hybridize with the brolgas. There appears to be little reproductive isolation associated with the unison calls. They function, instead, as a sexual display between members of a pair in all species and as a territorial threat display in some.

Archibald broke the unison calls and associated movements into 80 identifiable characters. The list is presented for the 15 species in Table 5-1. The total number of characters in a given species' unison calls ranges from 25 to 40.

The proper analysis and construction of phylogenetic dendrograms ("trees") are best performed with the aid of computers. For techniques and examples see Sneath and Sokal (1973) and Schnell (1970a,b). But for the sake of illustration, an alternate, manual technique will be employed that was introduced by Lorenz (1941) for waterfowl (Figure 5-8). This method renders a visual pattern of all the species simultaneously; it is accomplished by inspecting a graph of species versus behavioral characters and rearranging the characters and species in different sequences until the maximum number of contiguous shared relationships is found. The job can be facilitated by starting with the characters that are found in all or most of the species, then working up to those which are least common. If not done with a computer, one also needs patience, scissors, and a good supply of paper and tape. An outcome of one of these juggling efforts is shown in Figure 5-14.

Figure 5-15 Alternate graphic views of the numbers of shared characteristics in unison call dances among different species of cranes. A, Numbers of shared characteristics are represented by length and thickness of connecting lines. B, Sequentially larger groupings of shared characteristics.

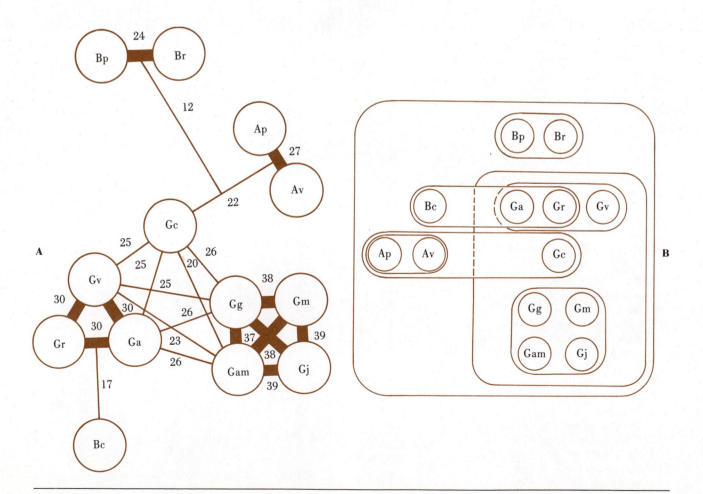

Table 5-2 Number of unison call dance characteristics shared among different cranes*

	Bp	Br	Bc	Av	Ap	Gc	Ga	Gr	Gv	Gm	Gg	Gam	Gj
Bp	25												
Br	24	25											
Bc	6	6	29										
Av	12	12	13	29									
Ap	12	12	13	27	30								
Gc	12	12	15	22	22	34							
Ga	7	7	17	17	19	25	37						
Gr	9	7	17	18	19	25	30	37					
Gv	6	6	15	18	20	25	30	30	32				
Gm	8	8	14	17	17	20	27	22	23	40			
Gg	8	8	15	18	18	26	26	24	25	38	40		
Gam	8	8	14	18	17	20	26	23	23	38	37	40	
Gj	8	8	14	17	17	20	27	22	23	39	37	39	40

From Archibald, G.W. 1976. The unison call of cranes as a useful taxonomic tool. Ph.D. Thesis. Cornell University, Ithaca, N.Y.

*Total number of characters in the dance repertoire of each species is shown in the central diagonal. For species identification of symbols, see p. 131.

Figure 5-16 Inferred family tree of crane species based on shared characteristics of the unison call dance.

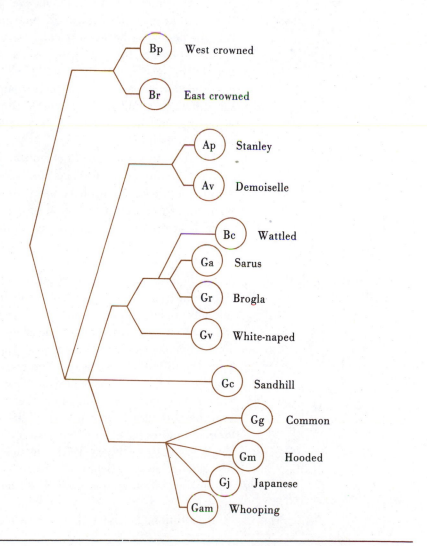

Both Table 5-1 and Figure 5-14 present only two-dimensional views of the relationships. To view them from another dimension, similar to looking at a tree from above, one can construct diagrams based on the number of shared characters (Table 5-2) as in Figure 5-11. The species are connected by lines (Figure 5-15, *A*) that depict by length and thickness the largest numbers of shared characters. Alternately, in Figure 5-15, *B*, one can enclose the species in figures that depict sequentially larger groupings of shared characteristics. Assuming that most closely related species share the largest number of characters and the most distantly related share the fewest, one can use this information to infer the probable phylogenetic tree as shown in Figure 5-16.

The use of the unison calls to infer phylogenetic relationships corroborated the picture obtained from the fossil record, geographical distribution, physiological cold adaptedness, external morphology, and tracheal anatomy (Archibald 1976a). For some of the comparisons it provided better resolution, comparable to that obtained from a detailed morphological analysis (Wood 1979).

The reasons for the specific unison calls and dance behaviors are understood only partially at present (Archibald 1976a). Some of the differences probably are just accumulated genetic drift that developed after the species separated. Some of the differences are fairly clear adaptations to different ecological conditions. The two species with the loudest calls, sarus and brogla, are found in wide open spaces where the birds are widely separated and loud calls more likely would be heard. The species with the quietest call, the wattled crane, is found under closer conditions where the birds are more grouped together. But the reasons for most of the differences and why the most recently evolved species, hooded, common, whooping, and Manchurian cranes, should be so specialized require further study and consideration.

The remaining two species of cranes, the Siberian and black-necked, were not included in the initial study because specimens could not be obtained for observation. But these species have been observed since that time, and the results are shown in Table 5-1. The task of fitting these two species into the analysis performed for the other 13 will be left to the reader as an exercise.

Evolutionary Characteristics of Behavior

When behavior has been described and compared among related species, one finds nearly all of the evolutionary characteristics associated with morphology and physiology. *Adaptive radiation*, in which closely related species living under different environmental conditions show a variety of different traits, can be seen in the behavior of Darwin's finches. The appropriate feeding behaviors accompany the different beak shapes and body sizes. Courtship behavior shows radiation frequently; the cranes provide a good example. *Convergence* can be seen in such behaviors as parental care, other social behaviors of many types, learning, and the gliding behavior of "flying" lizards, frogs, and squirrels. *Parallel evolution* and *clines* have been demonstrated in a number of behavioral traits such as displays of jays (Brown 1963), feeding behavior of garter snakes (Burghardt 1970a), and nesting behavior of some mice (King et al. 1964). Examples of these various evolutionary traits will be encountered throughout the remainder of this book.

In addition, behavior shows some new evolutionary twists such as *ritualiza-*

tion. A particular behavior may evolve away from an original function to become ritualized in a new context. Other, nonbehavioral traits can change functions, but most do not acquire anything similar to a ritualized function unless associated with behavior. Feathers of the wings of birds, for example, gradually may have changed from insulating to insect-catching to flying functions (Ostrom 1979), but all of these are explicit, direct functions, unlike the substituted, more or less indirect function that occurs with ritualization.

Once the principle of ritualization was recognized, it became easier to understand several confusing behaviors. For example, male empid flies of one species give empty silk balloons to the females during courtship. A comparative study (Kessel 1955) revealed that males of some species give nothing and occasionally get eaten themselves. In other species the males give prey wrapped in silk. In yet others the males present the females with prey only. If P = prey and W = wrapping, the reconstructed hypothetical sequence of evolution of the behavior is

$$P-W- \text{ to } P+W- \text{ to } P+W+ \text{ to } P-W+$$

That is, the behavior has become ritualized and greatly changed from its original form. Figure 5-17 arranges the different species by their behaviors in this order. The original function (distraction of the female with edible prey during the time the male is mating) has been replaced by a behavior that distracts without providing food.

There are several sources for the ritualization and evolution of behavior. Many

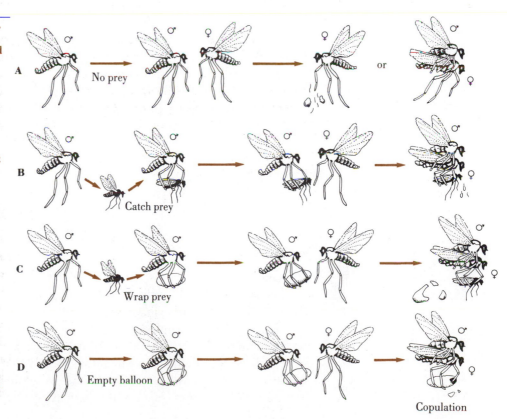

Figure 5-17 Some of the major categories of empid fly courtship behavior, presented in order of inferred evolutionary development. These flies are predatory, normally eating other insects, but they also may kill and eat conspecifics. A, Males of some species go directly to the female and may either be eaten by the female or be successful at copulation. B, Males in these groups of species carry prey when courting and present it to the female. The male copulates while the female eats the item. C, Males capture prey and wrap it in silk balloons before presenting it to the female. The male copulates while the female unwraps and eats the prey. D, Males of at least one species *(Hilara sartor)* do not capture prey first but simply present the female with an empty balloon. Other intermediate stages in the evolution of this behavior also have been identified. Based on information in Kessel, E.L. 1955. Syst. Zool. 4:97-104.

A   No prey

B   Catch prey

C   Wrap prey

D   Empty balloon   Copulation of the sources involve maintenance or other routine behavior (Chapter 6). Some of the commonly recognized sources include:

1. *Thermoregulation.* This includes hair and feather raising and the distribution of blood (as in blushing).
2. *Intention movements for locomotion.*
3. *Protective movements.* Some facial movements of dogs, cats, and primates and head turning in gulls are thought to be examples.
4. *Conflict behavior.* Examples of conflict behaviors as sources for the evolution of new behaviors include displacement or redirected body care (e.g., preening) and eating movements.

Other evolutionary characteristics of behavior include *exaggeration*, for example, of coloration, structures, and movement; *transfer of a display beyond the animal*, for example, bower birds using objects in the environment as part of the display; and *developmental acceleration* of behavior for display purposes, that is, behaviors appearing in some species in younger animals that have other functions when the animal is older.

Although most behavior might be subject to evolution, some categories of behavior seem to be more evolutionarily labile; that is, they show greater change and diversity among species than other types of behavior. Why? Some of the proposed reasons include: (1) many behaviors, such as communication and courtship, may have a species-specific or species-isolating function; (2) they may permit more efficient use of resources; (3) competition may lead to increased specialization; and (4) intense interactions between eaters and eaten (including plants) may lead to highly specialized avoidance or food-obtaining behavior.

It also may be worth considering why some other categories of behavior are more stable and do *not* change. (1) Perhaps the variation has not occurred. (Imagine an empid fly not having a prey item when it went to court.) In other words, evolution simply has not yet had the chance to start. (2) There may be no preadapted behavior (a reason closely related to the first). Preadapted (or "exapted," Gould and Vrba 1982) refers to traits existing first in a different, even nonfunctional, context before they serve the particular role or selective advantage being discussed. Nonpredatory flies, for example, normally do not carry prey. The change needs a foundation or starting point. (3) There may be no advantage to changing. Some changes may not offer a selective advantage; they might even be disadvantageous. In the case of the empid flies, there was a clear selective advantage to males giving females something to eat other than themselves.

•   •   •

In conclusion, the most important point of this chapter is the view that behavior is as subject to the forces of evolution as morphological and physiological characteristics. This view greatly advanced the collective scientific knowledge of behavior. The evolutionary viewpoint also can help significantly to organize one's understanding of the subject. Or, stated in negative terms, one's understanding of behavior is seriously impaired without an evolutionary perspective.

In the broadest sense, the mechanics of evolution via the biochemical processes of DNA-protein synthesis (for the making of individuals) and passing of

DNA in reproduction (by those individuals which survive and reproduce the best) permit one to understand how the long-term ultimate factors affect living organisms. This helps explain why animals behave as they do as a result of *past*, rather than in anticipation of future, events.

In the remainder of this book, the validity of behavioral evolution will be assumed as a given, underlying premise. This is not to say that all behavior is genetically predetermined or fixed. Modification and changes of behavior during the life of an individual will be discussed in Chapters 18 and 19.

SUMMARY

Inherited variability may interact with ecological factors to cause differences in survival and reproduction. The ultimate causes of behavior lie in the past: the ancestors of living animals *did* survive and *did* reproduce. Living animals are believed to behave as they do, for the most part, because they inherited those traits from their ancestors. Previous animals that did not possess those traits either did not survive and reproduce or else they became the ancestors of *different* animals with different characteristics today. Except for some behavior in humans and perhaps in a few other species, animals do *not* behave as they do *to* survive or *to* reproduce. Subsequent or future survival and reproduction are largely consequences of, not causes or reasons for, behavior. This last statement seems difficult for most persons to grasp for some unknown reason, perhaps because they have been misled by previous, subtly different interpretations. If the points in this paragraph are confusing, stop and think and perhaps go back and study this and the previous chapter some more. The genetic basis permits the process of evolution to occur in behavior as in morphology and physiology. The comparative study of behavior has revealed other evolutionary characteristics as well, such as ritualization.

Recommended Additional Reading

Aronson, L.R., E. Toback, D.S. Lehrman, and J.S. Rosenblatt, editors. 1970. Development and evolution of behavior. W.H. Freeman & Co., Publishers, San Francisco.

Crow, J.F. 1976. Genetics notes, 7th ed. Burgess Publishing Co., Minneapolis.

Dawkins, R. 1976. The selfish gene. Oxford University Press, Inc., New York.

Dobzhansky, T. 1951. Genetics and the origin of species. Columbia University Press, New York.

Falconer, D.S. 1960. Introduction to quantitative genetics. Oliver & Boyd, Edinburgh.

Griffin, D.R. 1981. The question of animal awareness (revised and enlarged edition). Rockefeller University Press, New York.

Hecht, M.K., and W.C. Steere. 1970. Essays in evolution and genetics. Appleton-Century-Crofts, New York.

Hickman, C.P., Jr., L.S. Roberts, and F.M. Hickman. 1984. Integrated principles of zoology, 7th ed. The C.V. Mosby Co., St. Louis.

Keeton, W.T. 1972. Biological science. W.W. Norton & Co., Inc., New York.

Mayr, E. 1963. Animal species and evolution. Belknap Press, Cambridge, Mass.

Wilson, E.O., T. Eisner, W.R. Briggs, R.E. Dickerson, R.L. Metzenberg, R.D. O'Brien, M. Susman, and W.E. Boggs. 1978. Life on Earth. Sinauer Associates, Inc., Sunderland, Mass.

PART TWO

BEHAVIOR AS OBSERVED OUTPUT

ETHOLOGICAL AND ECOLOGICAL ASPECTS

CHAPTER 6

James W. Grier

Female mallard *(Anas platyrhynchos)* preening its feathers while standing at the edge of a lake.

BASIC AND MAINTENANCE BEHAVIOR OF INDIVIDUALS

Basic behavior of animals is that which permits individuals to meet the minimum requirements of living. This category of behavior includes those behavior patterns which afford the animal protection from environmental adversities and which keep the body operational. More specifically, there are feeding and drinking movements, basic locomotion, shelter seeking or construction, a host of related maintenance behaviors such as various types of grooming and toilet behavior, and miscellaneous simple body movements. The importance of these behaviors to the animal should be obvious.

An awareness of this somewhat mundane category of animal behavior is important also to anyone interested in behavior for four major reasons. First, basic behavior and inactivity occupy much of an animal's time and behavioral repertoire. Such behavior will, as a result, be that which is observed most commonly.

Second, these common, relatively simple behavior patterns often serve as good subjects for understanding the internal machinery and control of behavior. To understand the role of sensory feedback, for example, one can remove one leg from an insect and then analyze subsequent locomotion changes in the remaining five legs (Hughes 1957).

Third, these behavior patterns in their original function generally are "conservative" and least likely to be modified by either evolution (Chapter 5) or learning (Chapter 19): hence they may provide insight into both of these phenomena. The relatively fixed nature of these behaviors can be seen in a variety of contexts. Classic examples of the conservative nature of maintenance behaviors involve head scratching (Chapter 5) and drinking in birds.

Finally, these behaviors serve as important sources for the evolution of other behaviors and displays. Although the basic behaviors, in their normal context, show little change, they may become incorporated into other behavior patterns, which then show much change and variability.

There are numerous examples of basic behaviors becoming incorporated into more advanced behaviors. The speculum-flashing displays of many ducks (Figure 6-1) seem quite clearly to have been derived from preening movements. Another interesting and possibly derived display in waterfowl, and perhaps in other birds, is tail wagging. Hailman and Dzelzkalns (1974) suggested that tail wagging in waterfowl may be a form of punctuation in their communication. The subjects of animal displays and communication, which so importantly incorporate and elaborate some of the basic movements, are discussed in Chapter 9 and subsequent chapters. The remainder of this chapter is devoted largely to descriptions of basic behavior with comments on evolutionary aspects where appropriate.

Much routine individual behavior, in spite of the importance to the animal and relevance to studies of internal integration or evolution, is not very interesting to most people. Therefore much of this chapter consists of brief surveys that are deliberately superficial and short. Before movement of any kind is discussed, perhaps we need to consider times when there are few if any movements and how these states of inactivity relate to the times when the animal is moving.

Figure 6-1 Speculum-flashing or preen-behind-wing display in the mallard duck. The speculum is the brightly colored patch with contrasting borders on the secondary feathers of the wing. During the display males orient laterally toward the female, raise the secondaries on that side in a preening-type motion, and run the bill behind them, occasionally making a rattling noise. The display clearly seems to have been derived from basic preening movements. From McKinney, F. 1975. The behavior of ducks. In Hafez, E.S.E., editor. The behavior of domestic animals, 3rd ed. Baillière Tindall & Castle, London. Photo courtesy F. McKinney.

ACTIVITY AND INACTIVITY: RHYTHMS AND SLEEP

The amount and timing of activity shown by different species and individuals are highly variable. Anyone who spends time watching animals soon discovers that most animals spend much time doing little or nothing, or they engage in seemingly insignificant activities. Even during the busy reproductive period animals may be inactive for long periods. Over the span of an entire year or lifetime, inactivity is much more common than most persons realize (Table 6-1 and Figures 6-2 to 6-5). Even those species traditionally considered to be most active, such as shrews and bees, spend much time doing little or nothing.

Sometimes activity appears to vary almost randomly and continuously from very inactive to very active. Many birds of prey and other carnivores, when they are not involved otherwise in reproduction, migration, or escaping from danger, show fairly discrete but irregular periods of activity alternating with inactivity.

Table 6-1 Time budget of free-ranging shorthorn cattle in Australia during 24-hour periods throughout the year*

Period	Grazing	Walking	Standing/ Ruminating	Standing/ Resting	Lying/ Ruminating	Lying/ Resting	Drinking
Day	46	9	8	12	15	11	—
Night	38	6	7	9	22	15	—
24-hour period	42	8	8	10	19	13	0.2

Modified from Low, W.A., et al. 1981. Appl. Anim. Ethology 7:27-38.
*Values are percentages based on minutes per day, night, or 24-hour period.

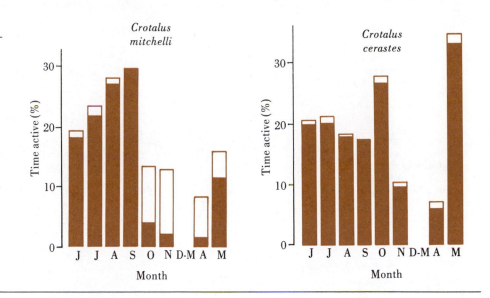

Figure 6-2 Rattlesnake activity patterns throughout the year. During the months of January through March the snakes showed no activity. Activity was measured with the aid of ingested radio transmitters and grids of antennas. Percentages of time active were calculated from numbers of hours active during 2-week sampling periods each month. Night is represented by dark parts of bars, day by open parts. Note that snakes were active less than 40% of the time even during the most active times of the year. Redrawn from Moore, R.G. 1978. Copeia 3:439-442.

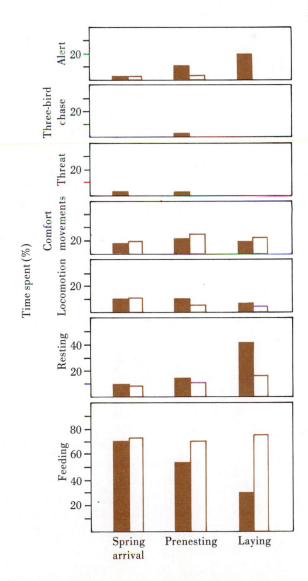

Figure 6-3 Time budgets for gadwall ducks during the breeding season. During this active time of the year most of the time is spent feeding, resting, and with basic locomotion and comfort movements. Males are represented by dark bars, females by open bars. From Dwyer, T.J. 1975. Wilson Bull. 87:335-343.

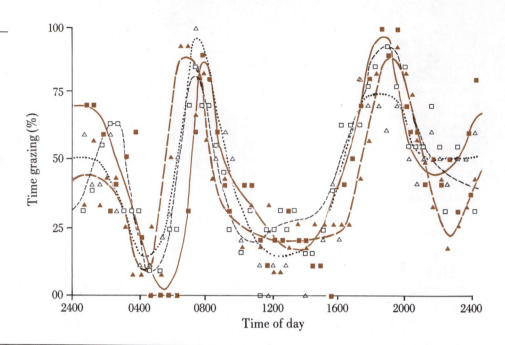

Figure 6-4 Grazing patterns of free-ranging cattle in central Australia. The percentage of time cows spent grazing during 30-minute intervals are remarkably consistent regardless of season. Two sharp peaks occur during morning and late afternoon; there is a moderate amount during the night, with a minimum just before sunrise and a lull during midday. From Low, W.A., et al. 1981. Applied Anim. Ethology 7:11-26.

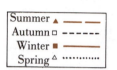

During active periods birds of prey, for example, appear restless and may fly or frequently change perches. Active periods occur whether the birds are satiated or not and whether they are in the wild or in captivity (Grier 1971). Just as noticeable are periods of inactivity. When inactive, birds of prey will sit for long periods in characteristic relaxed postures on one foot, feathers slightly fluffed out, and with an almost hypnotized appearance, although they are still quite alert and move their heads frequently. Eagles, for example, may perch quietly if not disturbed for several hours at a time. But the activity varies. Hawks of the genus *Accipiter* are much more active on the average than those of the genus *Buteo*, and for any one species there is much individual variation. Mueller and Berger (1973) showed that times of activity during the day for such things as migrating flight also may vary among the different genera of hawks and falcons (Figure 6-6).

Rhythms

The alternating periods of animal activity and inactivity often occur in more or less regular patterns or **rhythms.** The most commonly recognized rhythms are those of about a day, that is, **circadian** (*circa*, about and *dies*, day), those of about a year, **circannual,** and those associated with the daily tides and monthly movement of the moon (e.g., "menstrual").

That the rhythms are only approximate is well expressed in the "about" and "circa." They are kept regular, or **entrained,** by environmental cues called **Zeitgebers** (German, timegiver), such as the regular rising and setting of the sun. If the normal *Zeitgeber* is blocked from the animal (such as in a laboratory), it may readily switch to a substitute (such as the sound of a janitor coming in to clean the room next door everyday), and this has created some problems for research on the subject. Not only may the *Zeitgeber* be switched, it can be used to reset the internal rhythms, a characteristic used to advantage in clock-shifting (discussed further in the next chapter).

Figure 6-5 Time budgets for free-ranging female domestic cats. These results were obtained from focal sampling techniques (p. 75) with five subjects and 4692 records (391 hours of observation time). A and B, Different activities. C, Times during which cats were lost to view. D, Average time when cats were active, including times when lost to view *(dashed line).* Modified from Panaman, R. 1981. Z. Tierpsychol. 56:59-73.

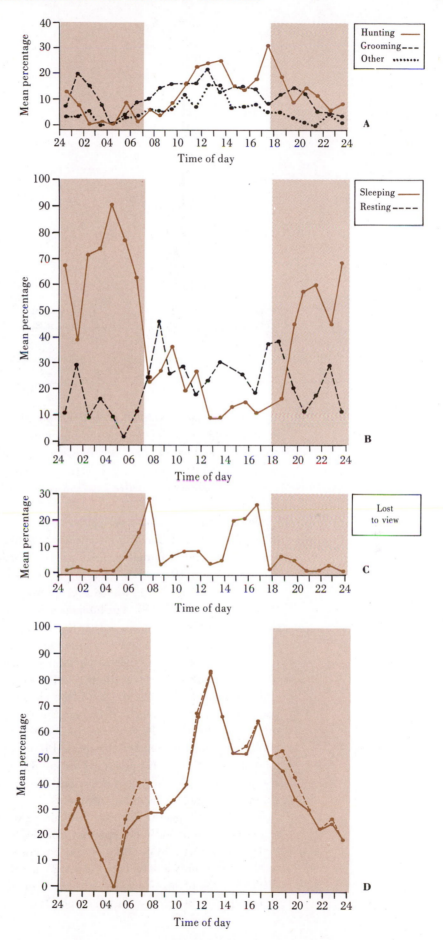

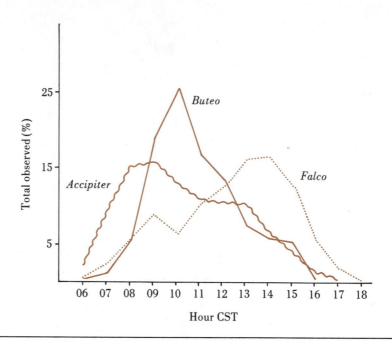

Figure 6-6 Differences in migration activity times among different genera of birds of prey passing along Lake Michigan at Cedar Grove, Wis. Data are based on 7906 birds observed from 1958 to 1961. Redrawn from Mueller, H.C. and D.D. Berger. 1973. Auk 90:591-596. Reprinted by permission of the American Ornithologists' Union.

If the animal is isolated completely so all of the *Zeitgebers* are blocked, the approximate nature of the rhythm becomes very apparent: the activity or other output cycles begin to drift and become out of phase with other individuals that remain in contact with their *Zeitgebers*. Examples of these drifting periods of activity are shown for both circadian (Figure 6-7) and circannual (Figure 6-8) rhythms.

The length of the free-running period (τ) is highly variable. It may range from 22 to 28 hours, depending on species, environmental conditions (particularly temperature and illumination level), and conditions under which an animal was raised (Aschoff 1979). Few clear correlations have been shown between τ and such things as nocturnal versus diurnal species, body size, or taxonomic relationships (Aschoff 1979). For humans τ is approximately 25 hours. The human period of 25 hours explains why it is easier to go to bed and get up later rather than earlier, rotate work to later shifts rather than to earlier shifts, and travel east to west rather than vice versa. Czeisler et al. (1982) applied this principle to improving work conditions under rotating work-shift schedules.

Rhythmic phenomena are so widespread, if not essentially universal, among both plants and animals that two significant ramifications exist. The first is a practical aspect: rhythmicity and periodic differences must be recognized and taken into account or at least suspected in virtually all physiological and behavioral studies. This is discussed in Chapter 3 in reference to sampling techniques. Harcourt (1978) studied rhythms in gorilla behavior and demonstrated that effects extend even into social relationships.

The second aspect is more theoretical: the widespread presence of rhythmicity suggests that it actually is important, that is, highly adaptive. Is it really important, or is it just a by-product or consequence of other processes? Perhaps it is neither and is only a figment of the imagination. Cole (1957), who has questioned the existence of many cycles in nature, demonstrated a diurnal cycle in the metabolic rate of unicorns based on purely random numbers. Most biologists who

Figure 6-7 Circadian rhythms of animal activity during normal light-dark (*LD*) cycles versus periods of continuous light (*LL*) or continuous dark (*DD*). Activity is indicated by dark lines. Under constant conditions the free-running period (τ) is only approximately 24 hours. If less than 24 hours, activity commences earlier and earlier each day, with patterns shifting to the left on the graph. If longer than 24 hours, patterns shift to the right. A, Locomotor activity of two house sparrows. B, Activity of a pig-tailed macaque *(Macaca nemestrina).* C, Awake activity (not constantly active) of a human living alone in a cave with a watch. At first the subject tried to maintain a constant schedule; then he gave up and slept as he felt inclined, after the indicated point. *12,* Noon; *24* midnight. A From Eskin, A. 1971. In Menaker, M., editor. Biochronometry. National Academy of Science, Washington, D.C. B from Aschoff, J. 1979. Z. Tierpsychol. 49:225-249. C modified from Halberg, F. 1973. In Mills, J.N., editor. Biological aspects of circadian rhythms. Plenum Press, New York. Based on data in Mills, J.N. 1964. J. Physiol. 174:217-231.

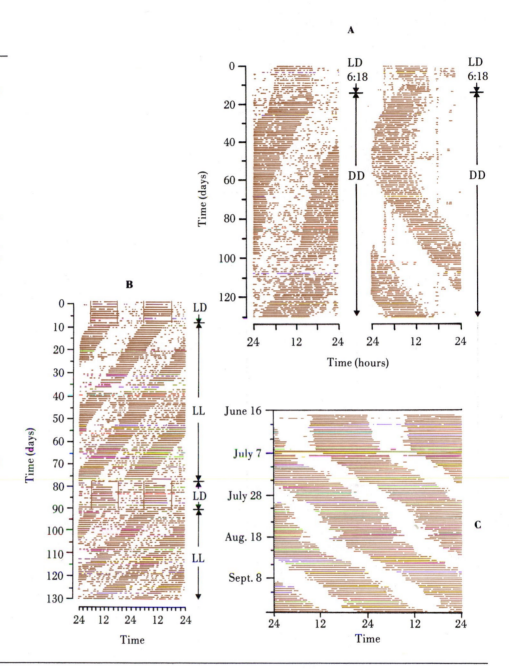

have considered circadian and circannual cycles in the physiology and behavior of animals, however, have little doubt that they exist. If one accepts that they do exist, are they merely consequences of other processes or is timing per se advantageous?

One can argue from a logical basis that rhythm and timing are advantageous. Temporal variation in the suitability of the environment is described in Chapter 4. Some of this variability is highly periodic and regular, such as with effects from the daily rotation and yearly orbit of the earth's movements. The optimal time

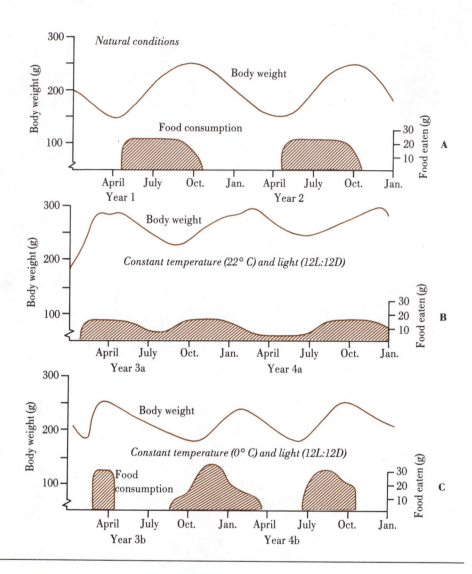

Figure 6-8 A, Natural circannual rhythms of ground squirrels and for those held under constant environmental conditions of 12 hours of light each day and constant temperatures of either 22° C (B) or 0° C (C). Animals continued to show hibernation cycles in spite of constant conditions. Food was constantly available. Squirrels continued to eat, more during the higher temperature, and became occasionally active even while hibernating. Cycles of food consumption and weight change were quite marked. Modified from Pengelley, E.T., and S.J. Asmundon. 1971. Sci. Am. 224(4):72-79.

(optimality is discussed further elsewhere in this chapter) to seek food is when it is most available and not at other times. The safest time to be active is when predators are not active and vice versa. Any biological mechanism that best synchronizes the organism with these and other environmental variables should offer a large selective advantage. The reasonableness of the explanation does not prove it, but it certainly seems difficult to refute.

Whether specific timing mechanisms are consequences or directly advantageous, there is rapidly accumulating evidence for the location and nature of such mechanisms. The diversity of mechanisms suggests strongly that they are analogous rather than homologous. The hormonal changes associated with circannual and seasonal rhythms probably are the best understood and will be encountered again in Chapter 16. Circadian pacemakers involve a variety and number of endocrine and cellular or molecular mechanisms. They appear to be located in such places as at the base of the eye in some molluscs (Block and Wallace 1982), the optic lobes in the protocerebrum of the brain of cockroaches (Page 1982), a part of the vertebrate brain called the suprachiasmatic nucleus in the hypothalamus (Rusak and Groos 1982), and perhaps the pineal gland—long a favorite target of searches for vertebrate biological clocks (Underwood 1977).

The location of the cockroach circadian pacemakers was demonstrated convincingly and remarkably by first rearing cockroaches under different schedules of light and dark to achieve different free-running activity rhythms (22.7- and 24.2-hour periods). Then the optic lobes of different cockroach brains were transplanted to other individuals. After recovery the recipients assumed the free-running rhythm previously shown by the donor (Page 1982). Similar results have been shown in house sparrows by transplanting pineal glands (Zimmerman and Menaker 1979).

How are free-running pacemaker rhythms coupled with external *Zeitgebers?* The exact mechanisms are not yet understood. At least for rhythms involving light, however, the chemical serotonin, which is affected by light, has been implicated in both vertebrates and invertebrates. In the marine snail *Aplysia*, for example, treatment by serotonin, a chemical found naturally in the eye, shifted the phase of the circadian output of optic nerves (Corrent et al. 1978). In birds the pineal gland inside the skull is light sensitive (Deguchi 1981), whereas in mammals light sensitivity for the serotonin pathway uses the main light sensor, the retina of the eye (Moore 1978). For further details on the mechanisms of these biological clocks and how they are thought to be coupled with the *Zeitgebers*, see Moore-Ede et al. (1982) and Takahashi and Zatz (1982).

Sleep

Sleep is the most extreme state of inactivity. It is associated with immobility and change of alertness thresholds (Figure 6-9), sometimes taxon-specific postures, secluded or hidden and protected locations, and characteristic physiological and neural states.

Sleep has been studied and described most extensively in mammals, particularly in humans. There are several categories and stages of sleep, with some having more than one name. The basic mammalian pattern of sleep during a given time period consists of a series of sleep cycles with subsequent cycles becoming shallower and shallower (Figure 6-10, *A*). A typical cycle involves going from the active, awake state through deeper stages classified as stages I to IV, then returning to stage I. Each of the stages of sleep is associated with relatively distinct patterns of brain waves (Figure 6-10, *B*) and different thresholds of arousal. It is more difficult, for example, to awaken from stage IV than from stage I. Following the return to stage I, there is a period associated with *rapid eye movement*, occasional other muscle twitches, and otherwise general muscular immobility. The period of rapid eye movement commonly is called **REM sleep.** This period of sleep also goes by several other names in the literature including *active sleep (AS), paradoxical sleep, rhombencephalic sleep*, and *fast* or *dreaming sleep*. The remaining time during the cycle, before the REM sleep, is called **non-REM (NREM)** or **quiet sleep (QS).** The period of REM sleep normally marks the end of a sleep cycle.

A sleeping mammal generally is immobile, aside from the REM eye movements and muscle twitches, during both REM and most of the NREM sleep. Between REM and NREM periods the individual may change posture, that is, turn over and otherwise move about while remaining asleep. These periods of postural body movements show up strikingly on time-lapse photographs.

Figure 6-9 Sleeping gray seal *(Halichoerus grypus)* while floating in the open sea. A, The seal is awake. B, In quiet sleep. C to F, In rapid eye movement sleep.

From Ridgway, S.D., and P.L. Romano. 1975. Cover photo of Science, Feb. 14, 1975. Copyright 1975 by the American Association for the Advancement of Science.

Figure 6-10 Generalized mammalian (including human) sleep patterns (A) and associated brain waves (B). A modified from Hobson, J.A., et al. 1978. Science 201:1251-1253. **B** modified from Dement, W., and N. Kleitman. 1957. Electroencephalogr. Clin. Neurophysiol. 9:673.

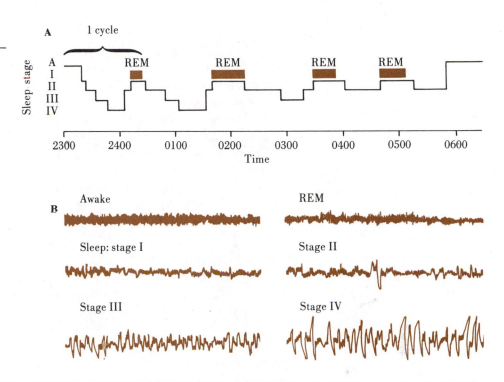

At least in humans, who are able to talk about their experiences, REM sleep is the time of dreaming. For most persons the total amount of dreaming each night is about 90 to 120 minutes (about the duration of a feature-length movie).

Dreams vary among individuals and between the sexes and apparently depend on much of what goes on in an individual's life. Researchers who have systematically awakened subjects during dreams (Cartwright, cited in Kiester 1980) find a more or less standard format to the sequence of a night's dreams. The general sequence involves first dreaming about concerns from the previous day to generalizing about similar previous experiences, to a final imaginary and often tense and vivid conclusion. All components might be quite fanciful. If anything is remembered for long after waking, it usually is only the final scene, as if one saw only the end of a confusing movie.

In the classic Freudian view (reviewed in McCarley and Hobson 1977), dreams were considered to function as censors that disguised repressed wishes to prevent them from waking the sleeping individual. This view has been labeled the ***guardian-censorship*** or ***wish fulfillment—disguise theory*** and, thus, ascribes a psychological purpose to dreaming. This led to much interest in dream analysis, trying to determine which wishes were being repressed, and attempts to use the results in psychotherapy.

Alternate recent interpretations, however, have focused more on the physiological aspects of dreaming. Hobson and McCarley (1977) proposed an ***activation-synthesis theory*** for dreaming based on neurophysiological evidence. According to this theory, parts of the brain periodically and endogenously generate impulses during sleep that activate the forebrain. The forebrain then attempts to make sense of, or synthesize, the hodgepodge of impulses, perhaps using past memories from the real world in an attempt to match the new input. As stated by Hobson and McCarley (1977), "The forebrain may be making the best of a bad job in producing even partially coherent dream imagery from the relatively noisy signals sent up to it from the brainstem."

Thus it may be that dreams are mostly literal representations rather than mystical, symbolic, or repressed thoughts or meanings as was formerly thought. A dream of a banana, for example, is just that and not a dream about a penis according to this view. Hobson and McCarley later commented (in Kiester 1980) that dreaming is essentially just a *consequence* of neurological processes in which "dreams may be the signals made by the system as it steps through a built-in test pattern—a kind of brain tune-up crucial to prepare the organism for behavioral competence."

Sleep: Need or Compulsion?

Whether or not there is a need for dreams or whatever dreaming represents or, more generally, for REM sleep in mammals is not clear. The function, if any, of sleep in general has eluded biologists. Only mammals and birds, with some indications in reptiles, have been shown to possess the brain and muscular characteristics of REM sleep, suggesting that REM sleep is a relatively recent evolutionary phenomenon. However, representatives of virtually all taxonomic groups show periodic immobility and several other characteristics (Table 6-2) of what might be called sleep (Meddis 1975).

Some persons (e.g., Allison and Van Twyver 1970) have tried to correlate sleep habits with general life history patterns, such as predators versus prey species. But more recent information suggests the picture is more complex (Table 6-3). Note that cats and mice sleep similar amounts of time; small animals with high metabolic rates (bats and shrews) are at opposite extremes; large and small animals (elephants and shrews) are different but more similar than bats and shrews; and related cetaceans (Dall's porpoise and bottle-nosed dolphins) are different in their sleeping habits.

Hypotheses on the function of sleep fall into three general categories: no function, recuperation or restoration, and immobility. The first category suggests that there is no function and that the environmental rhythm (of night and day) is so

Table 6-2 Some characteristics of sleep and their phylogenetic distribution

Characteristic	Primates	Other Mammals	Mono-tremes	Birds	Reptiles	Amphibia	Fish	Molluscs	Insects
Prolonged period of inactivity	+	+	+	+	+	+	+	+	+
Circadian organization	+	+	+	+	+	+	+	+	+
Reduced alertness	+	+	+	+	+	+	+	+	+
Specific sleep sites/postures	+	+	+	+	+	+	+	+	+
High-voltage slow brain waves	+	+	+	+	−	−	−	−	−
REM sleep	+	+	−	+	−?	−	−	−	−

From Meddis, R. 1975. Anim. Behav. 23:676-691.

Table 6-3 Total sleep time per 24 hours for various mammals for which figures are available

Hours	Mammal
20	Two-toed sloth
19	Armadillo, opossum, bat
18	
17	
16	Lemur, tree shrew (*Tupaia*)
15	
14	Hamster, squirrel, mountain beaver
13	Rat, cat, mouse, pig, phalanger
12	Chinchilla, spiny anteater [echidna]
11	Jaguar
10	Hedgehog, chimpanzee, rabbit, mole rat
9	
8	Human, mole
7	Guinea pig, cow
6	Tapir, sheep
5	Okapi, horse, bottle-nosed dolphin, pilot whale
4	Giraffe, elephant
3	
2	
1	
0	Dall's porpoise, shrew

From Meddis, R. 1975. Anim. Behav. 23:676-691.

regular and complete that over millions of years nervous systems have somehow locked into the periodicity and now show a proximate but no ultimate need for sleep. Note that the no-function hypothesis does not deny that sleep or cycles exist, as was discussed briefly earlier for circadian rhythms; the hypothesis simply suggests that sleep has no adaptive function.

Recuperation hypotheses propose that sleep may be necessary to somehow "recharge" or reorganize thought or other cerebral processes. Persons engaged in mental activities and a lot of daily decision making seem to require more sleep than those engaged primarily in physical activities. Sleep may permit the more efficient and complete repair of other bodily systems, such as the resting of muscles. It is not clear in this case, however, why an animal should become *so* deeply and physiologically immobilized.

Immobility hypotheses propose that sleep evolved to keep animals quiet and immobile when their full sensory abilities cannot operate (as with vision at night) and when they may be subject to predation or other dangers such as falling (Freemon 1972). It has even been suggested (Freemon 1972) that different parts of the brain take turns recharging, while other parts remain on guard for sudden emergencies requiring arousal and that this leads to the two observed forms of sleep. Sagan (1977) hypothesized that the sleep condition in modern mammals goes back to the period when large and dangerous reptiles roamed the earth along with smaller primitive mammals.

In attempting to distinguish between these categories of hypotheses, however, one runs into the trained-cricket problem (p. 81): there may be unavoidable correlations, or the data can otherwise be interpreted to support any of the alternatives. Among animals that sleep, including humans, there seems to be a proximate need for sleep. Humans and other animals deprived of sleep, for example, seem to become increasingly unable to function normally until sleep is restored. This debilitation and drowsiness under sleep deprivation have been used as one of the main assumptions demonstrating a recuperation function; that is, without repair and recuperation the system breaks down. But it has never been very clear just what is breaking down.

The other two theories state that sleep simply has become fixed into the behavioral repertoire of some species, complete with "appetitive" and "consummatory" components, and the apparent debilitation is nothing more than an increased attempt of the motivation system to find expression. The no-function hypothesis would say it just happens. Immobilization proponents suggest that the selective advantage is to enforce periods of inactivity to prevent wasting energy or avoid predators rather than for any physiological benefit.

There are supportive data and arguments on all sides. Shapiro et al. (1981), for example, show significant correlations between amount of sleep and exercise (Figure 6-11). Meddis (1975), on the other hand, states:

Recuperation theories would seem to imply that a healthy existence without sleep is impossible, at the very least among mammals. . . . [Those mammals] whose sleep requirements are negligible must therefore remain the Achilles' heel of recuperation theories. By contrast, the immobilization hypothesis not only tolerates the possibility but predicts that such conditions should occur among species who have little spare time or who have alternative procedures for obtaining the advantages normally resulting from the sleep instinct.

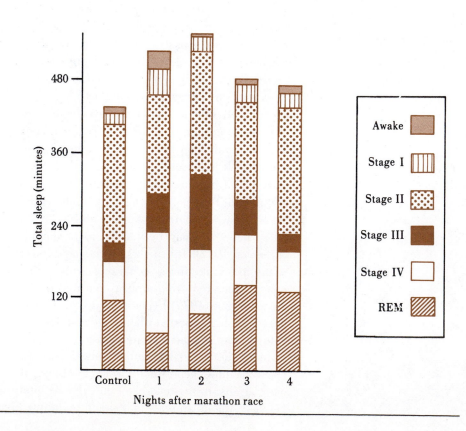

Figure 6-11 Amounts of sleep following a 92 km marathon race. Averages are shown for six subjects. Primary differences following the heavy exercise were increases in sleep of stages 3 and 4 and a reduction of REM sleep. From Shapiro, C.M., et al. 1981. Science 214:1253-1254. Copyright 1981 by the American Association for the Advancement of Science. Figure courtesy Colin M. Shapiro.

Total sleep (minutes)

Nights after marathon race

Figure 6-12 Species-specific differences in activity patterns of sympatric desert rattlesnakes. The two species are active at different times of the day or night during most of the year but merge and both become nocturnal during the hot summer months. Open bars represent *Crotalus mitchelli*, **solid bars** *C. cerastes*. **Arrows indicate sunrise and sunset.** Redrawn from Moore, R.G. 1978. Copeia 3:439-442.

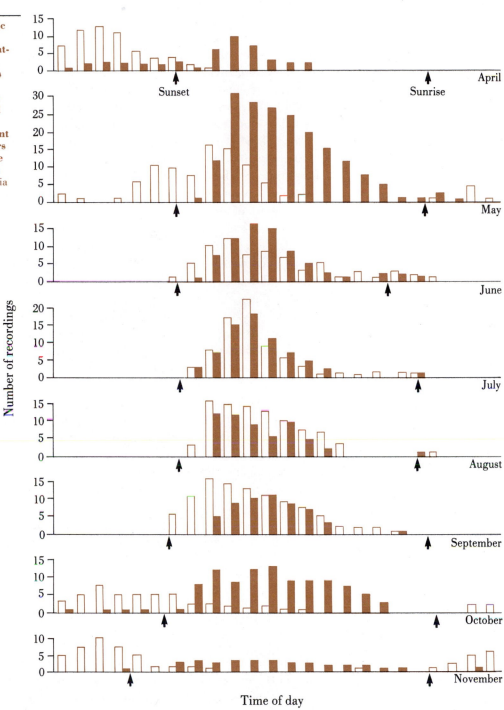

Meddis then notes several species (Table 6-3) and individuals (including bona fide cases involving active healthy humans) that show little or no sleep.

The immobilization hypothesis permits much diversity among sleep patterns depending on different organisms' natural histories. According to Meddis (1975), for example, the cat has few predators and "can sleep long and securely," whereas the mouse has numerous predators but obtains security in a protected nest. In the case of small-bodied, high-metabolism species, "The shrew tackles its energy problem with almost incessant food gathering. The bat adopts the alternative of almost total inactivity and spends a great deal of time asleep." Large herbivores, such as elephants, do not spend much time sleeping because they constantly must be eating. In short, the immobilization hypothesis states that immobility occurs if needed or can be afforded but not if not needed and cannot be afforded.

None of the sleep hypotheses can be eliminated completely, and they are not mutually exclusive. The immobility hypothesis, for example, cannot deny that there still might be some recuperative selective advantages. Different advantages or even no advantages might occur in different situations among different species. The function of sleep remains an intriguing unknown.

About the only safe generalizations about sleep that one can draw are: (1) most animals show periods of inactivity, some of which may be fairly regular, and the animal may become less alert and therefore "sleeps"; and (2) there are considerable differences in activity and sleep patterns among different species and, within a species, even during different seasons. A striking example of the variability is shown by two sympatric, closely related species of desert rattlesnakes (Figure 6-12). The sidewinder *(Crotalus cerastes)* is consistently nocturnal in its activity patterns, whereas the speckled rattlesnake *(C. mitchelli)* is similar to the sidewinder during the summer months but shifts to diurnal patterns the remainder of the year.

Returning to the activities of animals, one may classify them on a functional basis (Chapter 3). A functional category of movements may be represented by simple actions in one species under one set of circumstances or be fairly complex in another case. Movements for thermoregulation, for example, may involve as little as raising or lowering the hair or feathers. In other instances thermoregulation may involve changing the orientation of the body or moving considerable distances to obtain a different microhabitat, such as into or out of sunlight, to warm rocks, or underground. Because of such variability, basic behaviors sometimes defy easy classification; any attempt to define categories will be arbitrary. The sections that follow simply provide a basis for discussion.

LOCOMOTION AND INTENTION MOVEMENTS

Animals move about in many different ways, and these travels result in many different functions. Some functions are readily apparent, such as movements toward food and shelter or away from danger, whereas other activity appears to be largely "exploratory." The apparent advantage of exploration is that it may expose an animal to new opportunities and increase the probability of favorable encounters with food, mates, or other situations. Whether for a specific function or of a more general, exploratory nature, however, locomotion among different animals shows much diversity.

Patterns of locomotion within a given species may be fairly consistent and recognizable. Many types of birds, for example, fly in recognizable ways. Some species fly in a straight line with a steady, specific wing-beat frequency. Many woodpeckers (Piciformes) fly in an undulating pattern, flapping and rising, then not flapping and falling. Hawks of the genus *Accipiter* typically fly by interspersing somewhat regular sequences of flapping and gliding. Different species of birds, in spite of considerable structural similarity, show many differences in terrestial gaits, as reviewed by Clark (1975). Some walk on alternate legs, some hop on both legs, and some alternate between stepping and hopping. In many species there are ontogenetic (developmental) changes in gait patterns. Although most species show stereotyped patterns, including differences in various age, sex, feeding, and social contexts, individuals can compensate and change gait patterns under conditions such as injury. Similar generalizations can be drawn about the gaits of mammals, swimming of fishes, flight and walking or hopping of insects, and locomotion movements of other animals. A person familiar with differences of locomotion patterns often can identify a species at a great distance by a moving silhouette or when other details of an animal's appearance cannot be seen.

A simple list of different types of basic locomotion includes:

1. Movement by pseudopods, cilia, or flagella in protozoans
2. Pumping actions as in coelenterates and scallops
3. Paddling, flipper, and rowing motions for swimming in many forms of invertebrates and vertebrates
4. Other jet-propulsive techniques as in cephalopods
5. Tunneling, burrowing, digging, peristalsis-like motions and other underground movements
6. Gliding, friction, and extension-type movements as in flatworms, gastropods, "inchworms," and snakes
7. Walking
8. Jumping
9. Running, trotting, and cantoring
10. Climbing and related movements in trees and on other vertical surfaces
11. Gliding and various kinds of flight by different invertebrates and vertebrates

The cost of locomotion has received considerable attention (reviewed by Denny 1980). As shown in Figure 6-13, it requires more energy, for a given weight and assuming the appropriate structure is present, to run than to fly and to fly than to

Figure 6-13 Costs of locomotion. From Denny, M. 1980. Science 208:1288-1290. Copyright 1980 by the American Association for the Advancement of Science.

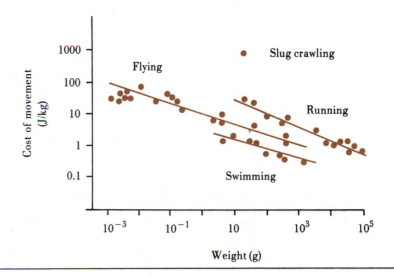

Intention Movements and Other Effector Output as Sources for Visual Displays, as Reported in the Literature

Intention Movements of Skeletal Action Patterns	Autonomic Responses

Maintenance and basic activities
  Locomotion
  Rolling and wallowing
  Scratching
  Bill wiping
  Preening and grooming
  Orientation of sense organs
	Pilomotor actions of fur and feathers
	  Sleeking
	  Fluffing and ruffling

Foraging and feeding	Respiratory responses Yawning

Reproductive behavior
  Mounting and copulation
  Penile erection and lordosis
  Nest building
  Parental care
	Vasoresponses
	  Flushing
	  Blanching

Agonistic behavior
  Fighting and combat
  Fleeing, flight, and protective responses
	Ocular responses
	  Pupillary actions

Antipredator behavior	

From Hailman, J.P. 1977. Optical signals. Indiana University Press, Bloomington.

swim. The costliest form of locomotion, interestingly, appears to be crawling on a mucous track by molluscs because of the production of mucus that holds the mollusc to the substrate (Denny 1980).

For some movements, particularly jumping and flying, there are preparatory, orienting movements. They usually involve a preliminary posture and a readiness to move, with limbs in the proper position for springing or taking off. They are referred to as *intention movements.* In birds, for example, the bird stands or perches on both legs, leans forward, and slightly lowers the body; it may also stretch the neck forward, extend the wings slightly, and direct its gaze in the direction toward which it is ready to move. Intention movements are significant not only as preliminary actions for completed movements but because they frequently are seen by themselves; that is, the subsequent movement is abandoned and not carried out. They also have played an important role in the evolution of many communication displays (Chapter 9). Hailman (1977) includes intention movements from general locomotion along with those from several other categories of movement, plus antonomic responses as major sources for the evolution of visual displays (see box above).

MAINTENANCE AND RELATED BEHAVIOR

Many relatively simple behaviors are associated with respiration and general body care. Maintenance behaviors include elimination, stretching and yawning, grooming, scratching, washing and bathing, drying, thermoregulation, and molting or ecdysis. Most species, including humans, show very characteristic, stereotyped patterns of movement in these basic behaviors.

Respiratory movements include (1) gill movements among osteichthyes (bony fishes), (2) continuous swimming or pharyngeal pumping movements in chondrichthyes (cartilaginous fishes), (3) gill waving among animals with external gills, (4) pumping of the oral and pharyngeal regions in amphibians with positive pressure breathing, in which air is pumped into the lungs by force from the mouth region, and (5) bellowslike sucking movements of the chest or abdomen by amniotic vertebrates and various invertebrates that possess negative pressure (sucking) breathing. Some aquatic species with rudimentary lungs gulp air during periods of low oxygen concentration in the water. Other aquatic but air-breathing organisms, including both invertebrates and vertebrates, direct much of their movement toward the surface to breathe periodically. Some invertebrates and humans have acquired various means for carrying or storing air outside the body for use underwater. Other respiratory or respiratory-related movements include coughing, hiccuping, and sneezing.

Elimination, defecation and urination (micturition), involve characteristic postures (Figure 6-14) that also may differ between the sexes or even with social status. Male and female dogs, for example, urinate quite differently; the males cock one leg and urinate to the side, whereas females usually squat. Sprague and Anisko (1973) described in detail several characteristic elimination behavior patterns in the beagle dog. Cats (Felidae) are well known for their toilet behavior. They dig in a soft substrate, eliminate, then cover the waste by pawing and pulling material over it. Birds usually defecate by leaning forward, raising the tail, and spreading the feathers in the anal area. Some birds eject the defecated material great distances, whereas other species merely drop it. Young birds in the nest may back to the edge to defecate at or over the edge; they may defecate with droppings in a membrane, the package of which is picked up and removed by the parent; or they may simply defecate on the nest material, which, depending on the species and individual, may be picked up with the nest material and removed (Lorenz 1970), be covered up with new nest material, or be ignored. There is

Figure 6-14 Pronghorn antelope defecating. Courtesy T. Brakefield.

much variation even among individuals of the same species, and some nests may be very clean, whereas others are excessively fouled.

Eliminative behavior has evolved in many different ways, depending in part on the species' place and manner of living. Many (but not all) species that have relatively permanent dens, nests, or other living sites show a degree of voluntary control over when and where (but usually not how) they will defecate and urinate. The place usually is located away from normal living quarters. The selective advantages of this are obvious: indiscriminate defecation and urination could attract predators via the odor, could lead to disease and health problems, and could raise humidity to harmful levels: the waste products also could accumulate excessively. In species that normally move from place to place, such as ungulates, or that live above the substrate (as in arboreal species, including most primates, and aquatic species), the waste products simply drop away and pose little problem. There would be little selective advantage to voluntarily monitoring and controlling the process. Accordingly, the behavior remains under involuntary control. (This is probably the chief reason monkeys are not more popular as house pets.)

Vultures have an interesting behavior known as urohydrosis. They direct their defecation onto their own legs; the evaporation cools the legs and aids thermoregulation of the body during periods of high temperature.

Stretching and *yawning* are seen among osteichthyes, birds, mammals, and perhaps some of the other vertebrates. The function of stretching appears to be physiological but is not well understood.

Luttenberger (1975) showed in a study of tortoise *(Testudo hermanni)* yawning and stretching that yawning increases with muscular fatigue, an increase in CO_2, apparently hunger, and some temperature changes. He also showed that yawning was suppressed, but limb stretching was stimulated with a shortage of O_2. Yawning and stretching may affect the muscles directly or simply increase or improve circulation. Hadidian (1980) compared yawning in black apes *(Macaca nigra)* with numerous other vertebrates and showed that structurally it was highly stereotyped and clearly homologous. In addition to the simple maintenance function associated with fatigue, however, it was seen in other contexts, as described also by other researchers for other primates (reviewed by Hadidian). The other contexts include stress and threat or canine (teeth) displays, hence, communication. Yawning occurs most frequently in adult males; rates increase with age.

Yawning in fish, in which it would seem to have originated phylogenetically, seems homologous to yawning in terrestrial vertebrates but apparently is *not* analogous (Rasa 1971). In fish it is associated with increased activity rather than sleepiness or the relaxation of tension; it is not associated with breathing and O_2 or CO_2 discrepancies; and "it does not have the 'infectious' quality" as in other vertebrates.

Washing, bathing, and *grooming* are performed with either movements of the entire body or with sequences involving the head and/or various limbs. These movements are more noticeable than some others, such as simple intention movements, and are familiar to most people. Intensive studies of grooming behavior of hymenopteran insects permitted the interpretation of the phylogeny of the order (Farish 1972). Other detailed studies of grooming behavior in a wide variety of organisms are numerous. They range, for example, from grooming in flies (Dawkins and Dawkins 1976) and mantis (Zack 1978) to kangaroo rats (Randall 1981) and herring gulls (Rhijn 1977). Dust bathing, seen in many birds and some mammals, apparently functions to reduce parasitism. Many animals, particularly mammals, roll or wallow in dust or mud. Canids often rub their necks or roll in strong-smelling substances such as feces or rotten carcasses, much to the dismay of many dog owners. The function of this behavior is not known.

Preening is a very important and frequent behavior seen in birds and some insects such as flies. Avian preening helps spread oil obtained from the uropygial gland, near the base of the tail, onto the feathers for waterproofing. The movements also help reunite the hooks and hooklets of feather vanes that may have become separated. It seems amazing that some birds can preen, given the shapes of their beaks. A unique apparent grooming or preening-type behavior in many birds is called *anting*. They pick up ants and wipe them about on their feathers.

Scratching to remove surface irritations can be seen among all vertebrates, including amphibians and fishes. If the irritation is located where it cannot be reached with an appendage, such as in fishes, the animal may rub itself on rocks, the ground, or some other structure in the environment. Scratching also involves species-specific movements. For any particular part of the body, most individuals, even humans, will use the same appendage and manner of scratching.

Drying behavior, also remarkably species-consistent, may occur after deliberate or accidental wetting. Drying may be accomplished by shaking, rubbing, or extending various appendages to increase the surface area.

Thermoregulatory movements are shown by many if not most species. There is a large and somewhat confusing array of terms that pertain to temperature regulation in animals (Bligh and Johnson 1973, Ostrom 1980). Before focusing on the behavioral aspects, a few of these need to be defined and clarified. Many of the terms are more or less, but not exactly, synonymous.

Animals are classically divided into two main groups on the basis of temperature regulation: the so-called warm-blooded mammals, birds, and possibly dino-

Figure 6-15 Theoretical relationships between body and ambient temperatures in homeotherms and poikilotherms. In homeotherms the body temperature is largely independent of ambient temperature, within limits, whereas body temperature varies with ambient temperature in poikilotherms.

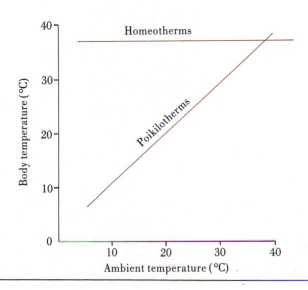

saurs (Thomas and Olson 1980) and all others, which are cold-blooded. Traditional technical names for the two groups are **homeothermic** and **poikilothermic,** respectively. The differences supposedly derive from the ability, or lack thereof, to generate body heat by carefully regulated, high rates of metabolism. The homeotherms would stay warm and maintain a relatively constant deep or core body temperature (peripheral and limb temperatures may fluctuate more) regardless of ambient (environmental) temperature, whereas the body temperatures of poikilotherms would fluctuate with ambient temperature (Figure 6-15).

The traditional view, however, is a bit oversimplified, and several other terms have been introduced to handle some of the differences. A term almost synonymous with poikilotherm is **temperature conformer.** In contrast, however, **temperature regulator** is not synonymous with homeotherm. Homeothermy technically refers to keeping the body temperature within relatively narrow limits. **Heterothermy** involves temperature regulation where body temperature is permitted to fluctuate more than in homeothermy, such as at night (e.g., in bats and hummingbirds) or during hibernation, but body temperature is nonetheless regulated. Furthermore, there are many ways of regulating body temperature. **Endothermy,** via controlled internal metabolic processes, is only one general way. **Ectothermy** refers to thermoregulation by acquiring body heat from external sources, usually via behavioral means. **Heliothermy** involves variation in exposure to solar radiation by behavioral means. The total heat budget involves heat absorption, generation, radiation, convection, and conduction.

In the real world of animals there is much diversity with little respect for taxonomic boundaries, and one can play many academic or even parlor games with the previous terminology. Mammals in general tend to be most homeothermic, except for bats and hibernators. Human body temperature, for example, is regulated quite precisely around 37° C. Body temperatures of birds generally are more variable, commonly in the range of 38° to 40° C. Sometimes it ranges from 35° to 41° C or more, depending on whether the bird is flying, active, excited,

Figure 6-16 Bumblebee body temperatures during flight (A) and warm-up (B) at different ambient temperatures. *Dark circles*, Thoracic temperature; *open circles*, abdominal temperature. From Heinrich, B. 1975. J. Comp. Physiol. 96:155-166.

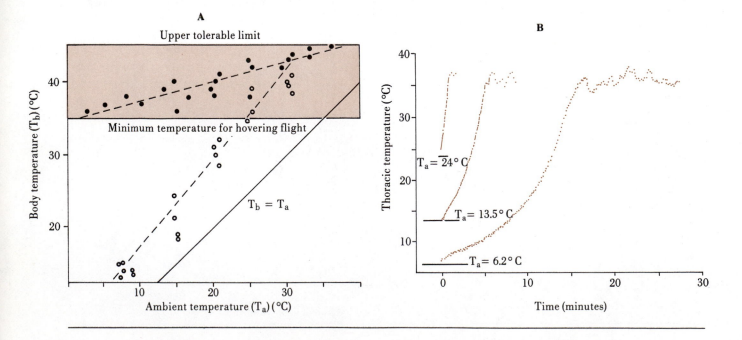

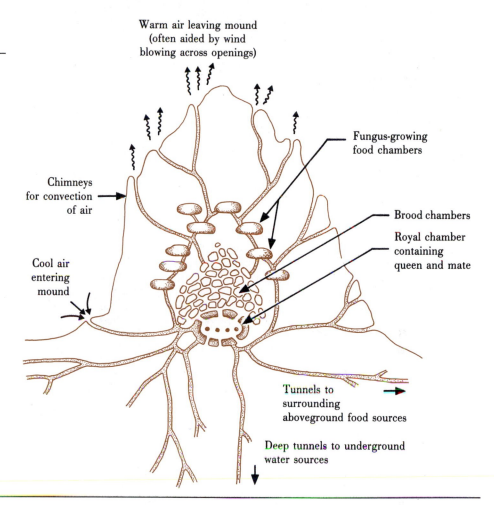

Figure 6-17 Temperature control and air conditioning in giant mounds of African termites (genus *Macrotermes*). Warm air rises in the mound, drawing cooler air from below. The pattern of air movements depends on the arrangement of passageways. Cooling also is achieved by evaporation of water droplets brought from underground tunnels and placed along the air passageways.

Warm air leaving mound (often aided by wind blowing across openings)

Fungus-growing food chambers

Chimneys for convection of air

Brood chambers

Royal chamber containing queen and mate

Cool air entering mound

Tunnels to surrounding aboveground food sources

Deep tunnels to underground water sources

calm, or sleeping. The body temperatures of hummingbirds can drop almost to ambient temperature at night.

Outside the groups classically considered homeotherms are many organisms, particularly among flying insects, that should be considered true thermoregulators and even a few homeotherms and good endotherms. Figure 6-16, *A*, illustrates, for example, almost classic homeothermy in the thoracic (but not abdominal) temperatures of a flying bumblebee. Bumblebees warm up (Figure 6-16, *B*), as do nocturnal moths, *some* katydids, beetles, flies, dragonflies, and other bees, by shivering and contracting flight muscles. Honeybees, some other hymenopterans, and many termites (Figure 6-17) achieve remarkable constant temperatures by heating and air conditioning their hives, nests, or mounds to within very narrow, precise temperature ranges.

Much thermoregulation is accomplished by basking and other behavioral means in virtually all temperature regulators, which brings one to the relevance of all this. Ectothermy and behavioral choice of the thermal environment are important to most organisms, including humans, regardless of the presence of endothermy. Ectothermy constitutes an important but often unrecognized component of the behavioral repertoire. Anyone who doubts that ectothermy is important even in humans has only to observe the species under the sun at the beach, on the lawn, or on rooftops, standing over radiators and hot-air registers, and in saunas.

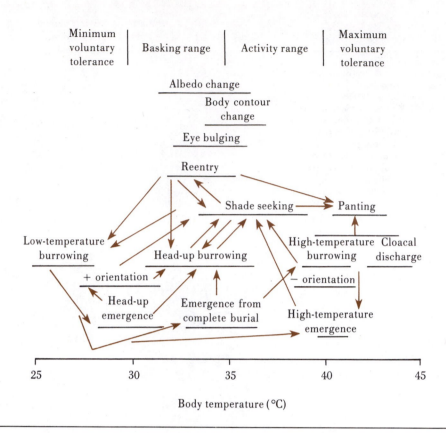

Figure 6-18 Behavioral temperature regulation in the horned lizard (*Phrynosoma* sp.) From Heath, J.E. 1965. U. Calif. Publ. Zoology 64:97-136. By permission of University of California Press.

When it comes to regulation of temperature in the home or work environment, Western man is almost in the same category as African mound termites.

Examples of behavioral thermoregulation are a cliché for reptiles (Greenberg 1976). A "thermoregulatory ethogram" for the horned lizard (*Phrynosoma* sp.) is shown in Figure 6-18. Numerous examples of ectothermal behavior also are well documented for many other organisms. Some arctic insects seek out the solar reflecting and focusing flowers of solar tracking plants (Kevan 1975). Many, if not most fish (Beitinger et al. 1975) have preferred temperatures and actively avoid temperature zones outside a particular range. Herring gulls will orient their bodies to the sun (often to reduce absorption) and rotate with the sun during the day (Lustick et al. 1978). The postures and locations of seals while out of the water have been shown to depend importantly on thermoregulatory factors (Gentry 1973). One can find numerous other published examples of behavioral thermoregulation.

A historically puzzling behavior, or group of behaviors, is spread-winged postures of perched or reclining birds of different species. The behavior has received much attention and generated some controversy (Kennedy 1969). Ohmart and Lasiewski (1971) proposed that roadrunners sunned to acquire heat. However, the researchers only looked at birds in sunning postures under light or in normal postures in the dark and not at birds using versus not using the postures under similar conditions of radiation—another case of the trained-cricket problem. I (1975) investigated spread-winged and nonspread-winged behavior of several species under natural and artificial radiation and found no thermal effects from the unique postures. Whether or not the birds' body temperatures or oxygen consumption changed depended on the presence or absence of radiation and not on the presence or absence of the spread-winged posture. This posture was influenced instead by the presence of water on the feathers in bright light, suggesting a drying function.

Shedding of epidermis (including hair and feathers) or exoskeleton, as in arthropods, can be seen in many species; the process is called molting or ecdysis and involves characteristic movements to accomplish. Some insects shed larval or nymphal exoskeletons, emerge as adults, and assume different habitats and ways of life. Shedding organisms may move to specific locations, and they usually engage in characteristic movements that remove the old skin or otherwise extricate the animal from it. There also may be characteristic postmolting behavior (Cloarec 1980). A somewhat analogous situation is seen in behavior associated with birth or hatching (Oppenheim 1972) where there are stereotyped patterns of movement.

Drinking is accomplished in most species with characteristic and familiar behavior patterns. Most birds scoop up a beakful of water, then tip back the head, letting the water run into the throat. A few birds, notably the Columbiformes (doves and pigeons) place the beak or head in the water and suck it up, as do most reptiles. (Columbiformes are considered relatively advanced, and, as so few other birds show sucking behavior, it may have been secondarily derived.) Many species obtain considerable water physiologically from their food. Drinking habits and frequency depend, in addition, on a number of other body characteristics (such as kidney, respiratory, and epidermal water losses) and several environmental factors, such as temperature and relative humidity. Many animals never drink during their entire lives.

There are a few drinking or other water-provision behaviors that are not as familiar. The male desert Namaqua sandgrouse *(Pterocles namagua)* in Africa, for example, soaks up water from pools in specially adapted abdominal feathers (Figure 6-19). It then flies as far as 80 km back to the young, which drink by

Figure 6-19 Namaqua sandgrouse soaking water into its feathers to provide water for the nestlings. From Cade, T.J., and G.L. Maclean. 1967. Condor 69:323-343.

Figure 6-20 Namib Desert Tenebrionid beetle (Lepidochora) collecting water from its fog-catching sand trench. From Seely, M.K., and W.J. Hamilton III. 1976. Science 195:485. Copyright 1976 by the American Association for the Advancement of Science. Photo courtesy Mary Seely.

pecking at and stripping the water from the feathers (Cade and Maclean 1967). A few other birds also have been reported to carry water back to the young on feathers or via the mouth.

Another unique technique for obtaining water is shown by three species of tenebrionid beetles that live in the Namib desert of Africa (Seely and Hamilton 1976). The beetles dig long trenches in dunes perpendicular to the wind during foggy periods. The trenches trap moisture, which is then consumed by the beetles (Figure 6-20). In one case where individuals from a population were sampled before and after drinking their fog water, body moisture increased nearly 14% during a single fog.

FEEDING AND OPTIMAL FORAGING

Feeding is a complex and important subject that has received considerable attention in behavior. It obviously overlaps with predator-prey considerations (Chapter 11) and also with investigations of central processing and control. An entire chapter (17) is devoted to the latter aspect. In addition, recall the notions of appetitive, consummatory, and quiescent stages of behavior such as feeding described in Chapter 2 (pp. 33-35). What will be considered in the present section concerns the visible manifestations of feeding along with the ecological and evolutionary implications. Just as surely as the function of sleep has been a puzzle, the function of eating is obvious. It is the means by which animals refuel, that is, replenish their bodies with chemical sources of energy and materials.

With the exception of some insects such as the mayfly, which never eats as an adult, having obtained all of its necessary food reserves from the larval stage, all animals confront the problems of finding adequate amounts of food and regulating intake to meet energy requirements and to provide protein and other nutrients essential for growth, maintenance, and reproduction.

In the broad sense the basic nutritional requirements of animals are similar. Yet these basically similar needs are satisfied by the consumption of an extraordinary variety of foods in accordance with the particular biochemical resources of the environments they inhabit. Every conceivable kind of food is eaten including such apparently unpalatable items as horn, carrion, poison ivy, and the crude tartar in the bottom of wine casks.

The meat-packing industry in the United States has always taken pride in its efficiency. ("Everything is used but the squeal.") The feeding behavior of animals is no less opportunistic or efficient. The various species display a wide range of physiological and behavioral adaptations enabling them to utilize a variety of sources as foods.

Feeding really is a large collection of functionally related behaviors. It involves several stages of search or locating, handling, and final ingestion. Some studies, often in conjunction with research on neural mechanisms (Chapter 17), have described the movements of feeding in very detailed manner (Zeigler 1976). Of present interest, however, are details on how animals accomplish the job of food input when faced with numerous choices and environmental challenges. The general energy budget formula for feeding requires that net energy gain (i.e., energy from food minus the energy spent searching, etc.) during a given period is greater than zero. This creates several biological problems that must be solved with appropriate behavior:

1. *What* to eat and how to recognize it
2. *Where* to search
3. How long to search before giving up, if unsuccessful
4. What to do about food unwilling to be eaten
5. *When* to eat (and when to *stop!*)

The fourth problem, what to do with animals unwilling to be eaten, involves ways of finding, catching, and handling prey. The ways, often termed **strategies** (see pp. 247 and 248 for a discussion of the semantic problems with the term), include variations on waiting and ambushing or trapping versus searching for, pursuing, and capturing prey and related behavior. These topics relate directly to predation; therefore the details will be covered in Chapter 11. The fifth problem, when to start and stop feeding, is largely a matter of internal control, the subject of Chapter 17.

Assuming that feeding or behavior leading up to feeding should commence or already has started, how does an animal deal with the first three problems listed? Also, aside from the specifics of predation in the common sense of the term, are there *general* problems and solutions for foraging behavior related to the fourth point? That is, can one generalize about food searching and handling behavior whether it is a lion searching for and attacking a zebra or a squirrel searching for and cracking a nut?

Description and Examples of Optimal Foraging

The questions previously listed are related to an important, recent, and rapidly growing body of theory referred to as **optimality theory**. The main development of this field has occurred with the topic of feeding (hence, its introduction here) under the name of **optimal foraging.** In essence it borrows from the subject of human economics and looks at feeding from a benefit/cost standpoint. Animals should feed and change their feeding in a manner that maximizes the benefits and minimizes the costs. The biological currency is fitness, that is, reproductive output. Reproduction usually is difficult to assess, however, and does not lend itself to short-term analysis. Thus energy in calories ingested minus calories spent during a given time period has been used instead as the standard measure.

The concept is natural and reasonable, so much so that it seems almost obvious. Furthermore, the basic idea has been inherent or implied in the theory of natural selection since the time of Darwin. This kind of thinking, although not necessarily identified as optimality theory, has played a large role in ecological, evolutionary, and population dynamic theory for many years (Cole 1954). Previously, however, attention focused mostly on ecological and evolutionary differences between species or viewed individual animals as single units that were adapted qualitatively as more or less fit than other animals.

Optimal foraging, in contrast, considers the behavior and choices of *individual animals*. Those individuals which can make optimal choices in their daily foraging should have a large selective advantage over those which cannot, and the ability would be expected to evolve. The subject of optimal foraging is considered to have started with the publication in 1966 of two important papers, one by MacArthur and Pianka and one by Emlen, which, incidentally, were published next to each other in the same journal.

These two papers were followed by a large number of others in several journals. Schoener (1971), among others, provided a review of the theory of feeding strategies through 1971. Two major reviews of optimal foraging were published subsequently by Pyke et al. (1977) and Krebs (1978). A symposium in 1978 was devoted to the topic (Kamil and Sargent 1981). Interest and publications since these reviews have not abated.

Figure 6-21 Optimal foraging among patches. A, Random distribution of two types of patches in a simplified illustrative model. A forager spends time in a given patch until the average rate of energy intake for that patch drops to or below the average rate that could be expected from the habitat as a whole; then the animal moves on to another patch. B, Time spent and energy intake in different patches should depend on the quality of different patch types. Modified from Charnov, E.L. 1976. Theor. Popul. Biol. 9:129-136.

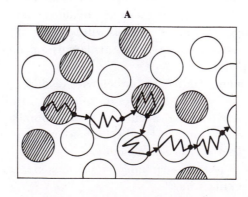

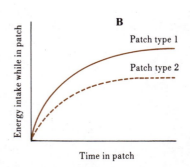

Optimal foraging, often termed *optimal searching*, has been concerned largely, although not exclusively, with four categories of choice: (1) what to eat (optimal diet or food type), (2) where to find it, particularly when food is not distributed uniformly in the habitat (optimal patch choice), (3) how long to spend (optimal allocation of time), and (4) optimal directions, patterns, and speed of movements (optimal search paths).

Much of the treatment of optimal foraging has been highly mathematical and is beyond the scope of this introductory book. Many of the findings, however, have confirmed rather remarkably some of the theoretical predictions, and a few examples will be given.

Optimal Patch Choice and Method

In the problems of choosing optimal patches, first introduced by MacArthur and Pianka (1966), the time spent searching includes the time spent in patches plus the time spent traveling between patches. The profitability of patches may change with consumption; that is, an animal depletes the food in a patch until there comes a point where it becomes more profitable to leave that patch and go to another. Furthermore, different patches may vary in quality. Animals should search within patches until it is no longer profitable to do so, then travel as quickly and in as straight a path as possible to the next patch.

Charnov (1976) described the basic mathematical aspects and provided some simple illustrations (Figure 6-21). The foraging patterns of most species investigated so far have conformed to the predictions. Bumblebees, for example, travel about and spend time in the most profitable places (Figure 6-22). Different hypotheses and the finer details for explaining how the animals determine profitability thresholds have been the subject of much recent research (Krebs et al. 1974, Hodges 1981).

Aside from the time spent, cost depends on the foraging activities themselves. As related to prey density, two different foraging methods may differ in their costs, as theoretically illustrated by Krebs (1978) (Figure 6-23). Ospreys *(Pandion haliaetus)* search for prey using several different methods: watching while perched, while gliding or flying forward, and while hovering in a stationary posi-

Figure 6-22 Diagram of actual foraging paths by an individual bumblebee worker *(Bombus fervidus)* in the same area on two different days *(solid* and *dashed lines)*. The bee spent much time within each patch, then traveled directly and quickly to the next patch. (Numerous movements within patches are not detailed in this illustration.) This bee primarily visited aster, as did some other individuals, but other individuals specialized in other patch types, such as jewelweed. Reprinted by permission of the publishers from *Bumblebee Economics,* by Bernd Heinrich, Cambridge, Mass.: Harvard University Press, Copyright © 1979 by the President and Fellows of Harvard College.

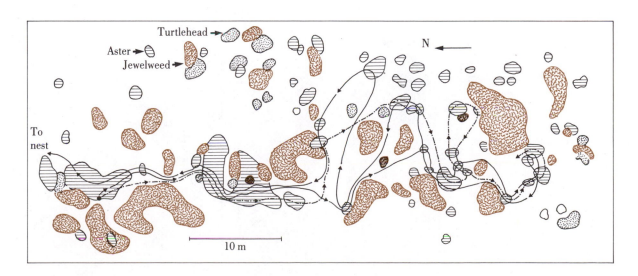

Turtlehead
Aster
Jewelweed
N
To nest
10 m

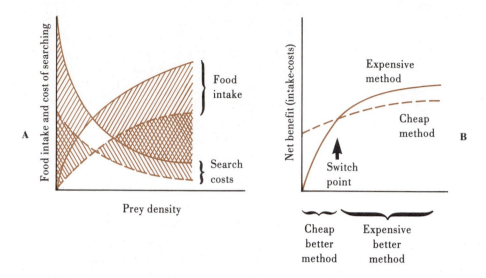

Figure 6-23 Methods of foraging depending on cost of method and density of prey. Theoretically it may be advantageous or better to use one method rather than another under different circumstances. The cross-hatched areas in (A), similar to hourglasses on their sides, represent the net benefit, or intake minus cost. To the left of the break-even point the animal is spending more energy than it is taking in (negative benefit). When net benefit is positive, there may be situations in which a cheap method, such as walking, is better than an expensive method such as running. The cheap method may result in lower intake, but because the cost is much lower, the *net* may be larger than for the expensive method (B). Depending on the shapes of the benefit curves, however, there may be a point where it is better to switch to the expensive method. Redrawn from Krebs, J.R. 1978. Optimal foraging: decision rules for predators. In Krebs, J.R., and N.B. Davies, editors. Behavioural ecology: an evolutionary approach. Blackwell Scientific Publications, Ltd., London.

tion in the air. Hovering is the most energetic, costly method but, overall, leads to successful catches 50% more often than do dives from other kinds of flight (Grubb 1977). The difference varies, however, depending on weather and visibility conditions so that under some conditions it would be more or nearly as profitable to hunt via the less costly methods. Pyke (1981a) has demonstrated expected differences in foraging methods by comparing the energetics and feeding of hummingbirds that hover and Australian honeyeaters (Meliphagidae), which perch.

The cost of foraging methods may vary not only among methods but also on distance traveled, particularly if the animal returns to a central place ofter obtaining food (Orians and Pearson 1979). Kramer and Nowell (1980) showed that chipmunks *(Tamias striatus)* fill their pouches with different amounts of food with the optimal load size depending on the travel time from the central location. Choice of feeding site also appeared to depend on travel time.

Figure 6-24 Specialized foraging behavior in the dung beetle *(Kheper aegyptiorum)*. In the sequence of behavior diagramed here, the beetle first cuts a ball of dung from a dropping (A), rolls it to a suitable location by pushing backward while walking on the front legs (B), buries it by digging a hole from under it (C), then lays an egg in it, covers it over, and leaves (D). From Ecology of the African dung beetle by Bernd Heinrich and George A. Bartholomew. Copyright © 1979 by Scientific American, Inc. All rights reserved.

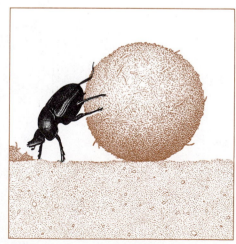

A   B

Optimal Food Type

The type of food that an animal chooses may be governed by optimality principles. Animals can be classed by diet on a broad continuum from generalists, such as earthworms, which eat almost anything they can ingest, to others that are specialists. Dung beetles, for example, are relatively specialized in foraging on dung, with the aid of specialized behavior (Figure 6-24). Oldsquaw ducks *(Chanqula hyemalis)* are specialized for feeding on invertebrates, primarily amphipods, after diving to depths of 46 m (150 feet) or more underwater (Peterson and Ellarson 1977). Some species are very highly specialized, feeding only on particular parts of a single species of plant or animal.

Within the range of adaptation of a species for feeding, animals may choose their particular food on an optimal basis. Theoretically, animals should choose the most profitable food items among those available, be more selective when profitable items are more common, and ignore unprofitable food outside the optimal range no matter how common it is (Krebs 1978). These predictions have been demonstrated now in numerous cases. Pinon jays *(Gymnorhinus cyanocephalus)*, for example, assess the quality of pinon seeds *(Pinus edulis)* and take only the good ones (Ligon and Martin 1974).

An excellent example of the subtleties in optimal choice of food items is provided by Northwestern crows *(Corvus caurinus)*, which feed on whelks—large, shelled marine molluscs (Zach 1978, 1979). To get at the animal the crows obtain them from the water's edge, fly inland (where shells do not bounce back into the water), and drop the shell onto a rock (Figure 6-25, *A* to *C*). The larger the whelk, the easier the shell breaks, but the more energy it requires to carry and drop from a height. If a shell does not break, the crow is faced with the problem of whether to try it again until it breaks or go get another (they repeatedly drop each until it finally breaks). In all aspects, including size and weight of whelk, choice of dropping site, height of drop for different-sized whelks, and numbers of drops, the crow behavior conforms to predictions of optimal foraging. In choice of size, for example, crows choose the larger whelks and ignore the smaller ones, although the smaller ones are more common and easier to carry (Figure 6-25, *D*).

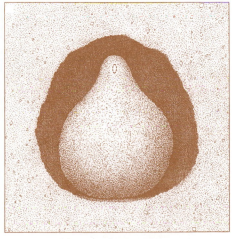

C   D

Figure 6-25 Whelk dropping behavior by northwestern crows. A, The crow flies from a perch to the water's edge where it searches for live whelks, then flies inland to make repeated drops until the shell breaks. After breaking a shell and eating the whelk, the crow either returns for another or goes back to its perch. **B,** Some crows drop the whelk from the top of the flight, which gives added height, but the crow cannot see where the shell drops. Others tip forward and lose height but can see better where they are dropping the shell. Several trade-offs are involved in this behavior. **C,** Examples of common types of breaks in dropped whelk shells. **D,** Differences in the frequencies of whelk sizes available versus The larger whelks are fewer and more difficult to find as well as heavier and require more energy to carry. But they break easier and hence provide a greater net benefit. **A** from Zach, R. 1979. Behaviour 68:106-117. **B** to **D** from Zach, R. 1978. Behaviour 67:134-148.

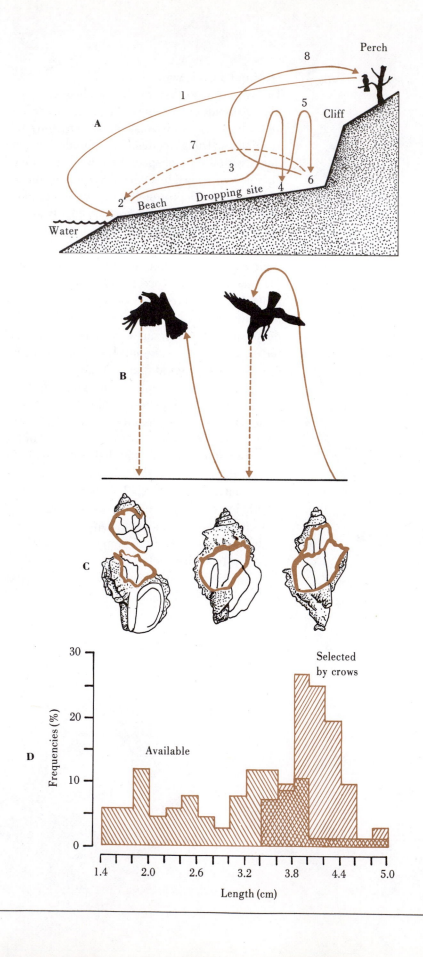

The smaller ones are more difficult to break and require more travel and more drops; the larger ones, hence, are more profitable.

Recognition of basic differences in adaptation among species is important to understanding optimal foraging. In a sense, basic evolutionary adaptations only roughly tune the organism to the environment. Fine tuning may be up to the individual. Ability to alter foraging behavior varies from species to species even among close relatives. In *Peromyscus* mice Drickamer (1972) showed that *P. leucopus* is more flexible than *P. maniculatus* and that the flexibility developed as the animals matured. Furthermore, feeding flexibility correlated with habitat conditions. *P. leucopus* live in a wide variety of habitat types. Partridge (1976b) came to a similar conclusion about the differences in optimal food habits of coal tits *(Parus ater)*, which live and feed in conifer habitat, as opposed to blue tits *(P. caerulus)*, which live in broad-leaved forest habitat. Fitzpatrick (1981) showed that differences in optimal foraging behavior in flycatchers (bird, Tyrannidae) varied with body size but otherwise were related to the various other optimal foraging factors involved in profitability. Moermond (1979) demonstrated foraging ranges or zones for several species of *Anolis* lizards. In addition to species differences, age differences also exist. Fox and Morrow (1981), during investigations of diet specialization in insects from an ecological viewpoint, concluded that herbivorous species that appear to be generalists over large geographical ranges may be specialists at the local level.

The Role of Learning in Optimal Foraging

Whether or not optimal foraging behavior is a learned or innate characteristic does not appear to be of much concern; it seems to occur regardless. The case of shell dropping by Northwestern crows, cited earlier, was interpreted by Zach (1978) as learned, whereas that in gulls is less flexible (but still presumably optimal) and perhaps innate (Tinbergen 1953). Optimal foraging in captive great tits *(Parus major)* improved in a manner interpreted as learning (Partridge 1976a). Through learning animals can discover some unique ways to increase their food supplies. Examples include crows that have learned to drop nuts in front of automobile tires for cracking, birds taking insects from spider webs and the radiator grills of cars, and woodchucks *(Marmota monax)* that learned to obtain aquatic vegetation by climbing trees that had fallen into a lake (Fraser 1979). Among insects, where behavior is commonly (perhaps mistakenly) viewed as largely innate, optimal foraging has been demonstrated, for example, in bumblebee foraging (Heinrich 1979), in whether army ants emigrate to new food locations (Topoff and Mirenda 1980), and in dietary choices of harvester ants (Davidson 1978).

Optimal Foraging and Search Image

This discussion of animal identification and choice of food items is closely related to the topic of "search image," first proposed by Uexküll (see Schiller 1957) and elaborated by L. Tinbergen (1960). According to this view, animals form a perception of what to look for, which improves chances of finding similar items in the future. Data have not supported the hypothesis in some cases (Smith and Dawkins 1971) or, if compatible with search image, have been interpreted in alternate ways, as discussed at length by Krebs (1973). Dawkins (1971) and Smith and

Dawkins (1971) point out that search image has been used in several different ways; hence it has become too imprecise to be useful. Others continue to use the concept of search image, at least in restricted ways (e.g., Alcock 1973, Pietrewicz and Kamil 1979). At present it appears that the search image concept is not dead but, nonetheless, has been supplanted largely by optimality views.

Optimal Behavior in General

The theory of optimal foraging can be generalized beyond feeding to a theory of *optimal behavior;* that is, choices do not have to be confined just to matters concerned solely with feeding. Many other second-by-second choices or decisions also must be made by living animals. These also have been shown to conform to optimality predictions. For example, backswimmer insects *(Notonecta hoffmanni)* balance foraging behavior when the advantages of concentrated prey are offset by risks of the backswimmers themselves being eaten (Sih 1980); mink *(Mustela vison)* balance between continuing to pursue potential prey underwater and interrupting the bout to go to the surface for oxygen (Dunstone and O'Connor 1979); juncos *(Junco hyemalis)* vary their foraging and watching-for-predator times depending on how many birds are in the flock and their position in the dominance hierarchy (Chapter 8); and great tits may either sing and engage in territorial activities or forage depending on whether the amount of light is sufficient to permit profitable foraging (Kacelnik 1979).

The theory of optimal behavior has been extended to many facets of behavior completely beyond feeding, including optimal choice of habitat, mates, and breeding conditions and as a description of choices between many cases of conflicting opportunities. The notion of optimal behavior will be encountered elsewhere in this book.

Problems with Optimality Theory

In spite of the promise that optimality theory offers, however, it has not been without its problems. Some critics have argued that the theory amounts to circular reasoning, that only examples which fit the predictions have been set forth, and that there are too many if's, and's, but's, and conditional statements. Examples of conditional statements can be found in many tests of optimal theory. Pyke (1981b), for example, notes a significant difference between observed and predicted behavior and suggests the "difference could have been due to the birds transient occupancy of the study area." In another paper Pyke (1981c) comments, "These movement patterns are consistent with the expectations of optimal foraging theory only if the hummingbirds cannot or do not determine the directions of possible inflorescence and if they cannot assess independently the sizes and distances of possible inflorescence." Kushlan (1978) demonstrated nonoptimal, "nonrigorous" foraging in great egrets *(Casmerodius albus)* wherein the birds used less profitable foraging methods over more profitable ones. Jaeger et al. (1981) described a number of conditions involved in whether or not red-backed salamanders *(Plethodon cinereus)* foraged optimally. Some tests of optimal diets have been clearly negative (Emlen and Emlen 1975).

Krebs et al. (1977) suggested that some differences between predicted and observed foraging depend on the need for animals to invest time in *sampling* for determining the quality and availability of food items. In his general discussion,

Krebs (1978) presents three reasons why the data may not fit the expected values: "The hypothesized cost function could be wrong, the premise of optimal behavior could be wrong, or both of these could be right, but the animal could have been tested in an environment to which it is not adapted." In considering whether digger wasps (*Sphex ichneumoneus*) commit what is called the "Concorde fallacy," whereby an animal continues to invest in doing something because of large expenditures already invested in it, Dawkins and Brockmann (1980) suggest that naive speculations of optimality pose real pitfalls. They suggest alternate interpretations such as evolutionarily stable strategies (ESS, Chapter 8).

What can be concluded from all this? Both reviews by Pyke et al. (1977) and Krebs (1978) are optimistic, even enthusiastic, and demonstrate reasonable examples of observations fitting predictions. The contemporary literature abounds with additional examples. The theory is reasonable, perhaps deceptively so. The objections to optimality theory are common to most scientific theories. Serious questions and alternate interpretations have been raised, however. Optimality theory should be considered as attractive and potentially valid, but, at the same time, it should not be taken too seriously yet. One should await further data, reviews, and discussions, all of which are accumulating rapidly.

Caching Behavior   Before leaving the subject of foraging behavior, the strategy of storing (hoarding, caching, etc.) food should be mentioned in passing. Many species of both invertebrates and vertebrates store food items. This behavior has come under comtemporary interpretations, including optimality. Numerous examples and interpretations of storing behavior exist in the literature, for example, Andersson and Krebs (1978) and Balda (1980).

SHELTER SEEKING AND CONSTRUCTION

A list of basic behaviors to meet minimum living requirements should include whatever animals do to obtain or provide protective shelter. Many species remain exposed in whichever environment surrounds them; they have no shelter or "home." At best there may be just a depression where they rest or give birth. At the other extreme, however, are species that construct elaborate structures. At the peak of such activity, excluding humans, are some species of termites and hymenopterans (bees, wasps, ants), some birds, and a few rodents such as the beaver. In between the two extremes are many tube-, den-, and hole-dwelling species. Many species construct the structure, including birds that build nests, many web- and cocoon-spinning arthropods, and miscellaneous shelter or shell builders such as caddis flies. Many of the most elaborate forms of animal engineering are social endeavors (Chapters 8 to 10), but much of the construction is on an individual basis. In almost all cases, except humans, learning from other animals appears to play little, if any, part of the specific patterns and choice of building materials used in construction behavior. However, animals may learn or improve with experience. The role of learning in animal construction has not received much research attention. Several examples of animal constructions are pictured in Figure 6-26.

Figure 6-26 A potpourri of animal construction. As demonstrated by studies of the construction behavior of naive animals and alterations caused by brain surgery (e.g., see Van der Kloot 1956), most of the construction behavior illustrated here is thought to result from innate motor programs in the brain. Illustrations are modified from several sources.

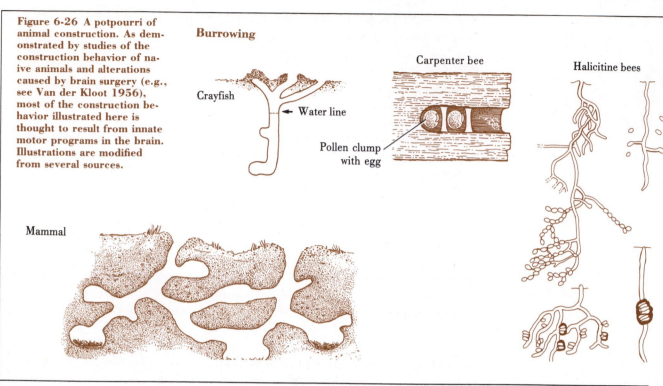

Burrowing

Crayfish

Water line

Carpenter bee

Pollen clump with egg

Halicitine bees

Mammal

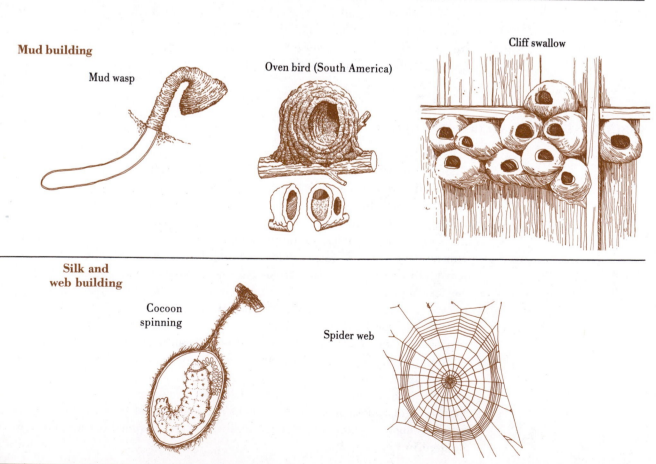

Mud building

Mud wasp

Oven bird (South America)

Cliff swallow

Silk and web building

Cocoon spinning

Spider web

Stick building

Caddis fly
larva

Beaver

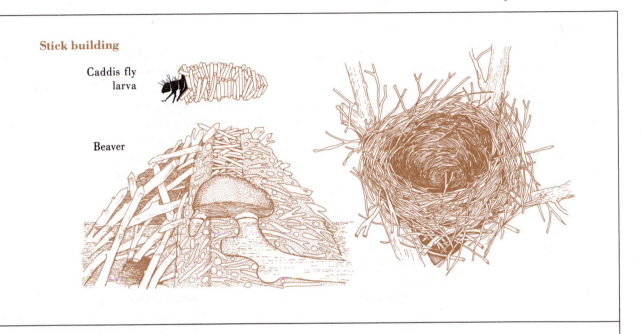

**Weaving and
rolling**

Weaver
ants

Leaf-rolling
beetle

Harvest mouse

Weaving birds

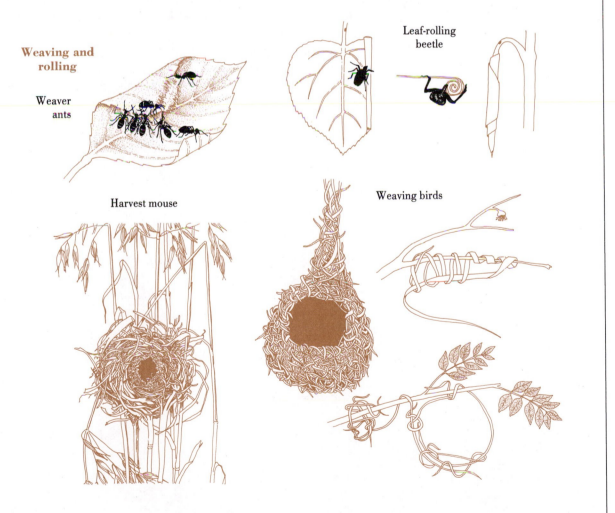

SUMMARY

Knowledge of the basic movements and individual behaviors of animals is important to an overall understanding of animal behavior for several reasons. These movements are perhaps the most basic to the individual animal's survival, generally form the greatest portion of the total repertoire, and consume most of the animal's time on a daily basis. Basic, relatively simple behaviors have served as useful subjects for studies of the internal control of behavior. These basic behaviors frequently are "conservative" in their original role; that is, they show little evolutionary change or lability, and they may be less subject to learning, individual variation, and voluntary control than many other forms of behavior. Finally, in new roles basic movements often form the substrate for evolutionary development of other behaviors such as communicative and reproductive displays, rituals, and highly species-specific behaviors.

Included among this survey of arbitrarily defined "basic" behaviors are general patterns of activity, sleep, and rhythms, movements of locomotion and intention movements, drinking, feeding and related behaviors, the construction of structures, and a wide variety of maintenance behaviors including respiration and related movements, elimination, stretching, grooming, scratching, washing and bathing, drying, thermoregulation, and the periodic shedding of protective body coverings.

Feeding has played a significant role in the development of the theory of optimal behavior, in which animals choose between alternate opportunities on the basis of processes that resemble benefit/cost analysis. Several examples and a discussion of the validity of the theory are presented.

**Recommended
Additional Reading**

Brady, J. 1979. Biological clocks. University Park Press, Baltimore.

Bruning, E. 1973. The physiological clock, 3rd ed. Springer-Verlag New York, Inc., New York.

Cohem, D.B. 1979. Sleep and dreaming: origins, nature and function. Pergamon Press, Inc., New York.

Collias, N.E., and E.C. Collias, editors. 1976. External construction by animals: benchmark papers in animal behavior, vol. 4. Dowden, Hutchinson, and Ross, Inc., Stroudsburg, Pa.

Emlen, J.M. 1966. The role of time and energy in food preference, Am. Nat. 100:611-617.

Enright, J.T. 1980. The timing of sleep and wakefulness. Springer-Verlag New York, Inc., New York.

Freemon, F.R. 1972. Sleep research. Charles C Thomas, Publisher, Springfield, Ill.

Frisch, K. von. 1974. Animal architecture. Harcourt Brace Jovanovich, Inc., New York.

Heinrich, B. 1979. Bumblebee economics. Harvard University Press, Cambridge, Mass.

Kamil, A.C., and T.D. Sargent, editors. 1981. Foraging behavior. Garland Publishing, Inc., New York.

Krebs, J.R., and N.B. Davies, editors. 1978. Behavioural ecology: an evolutionary approach. Blackwell Scientific Publications, Ltd., London.

MacArthur, R.H., and E.R. Pianka. 1966. On optimal use of a patchy environment. Am. Nat. 100:603-609.

Meddis, R. 1975. On the function of sleep. Anim. Behav. 23:676-691.

Menaken, M. 1971. Biochronometry. National Academy of Sciences, Washington, D.C.

Mills, J.N. 1973. Biological aspects of circadian rhythms. Plenum Press, New York.

Moore-Ede, M.C., F.M. Sulzman, and C.A. Fuller. 1982. The clocks that time us. Harvard University Press, Cambridge, Mass.

Pengelley, E.T. 1974. Circannual clocks. Academic Press, Inc., New York.

Pyke, G.H., H.R. Pulliam, and E.L. Charnov. 1977. Optimal foraging: a selective review of theory and tests. Q. Rev. Biol. 52:137-154.

Saunders, D.S. 1977. An introduction to biological rhythms. John Wiley & Sons, Inc., New York.

Schoener, T.W. 1971. Theory of feeding strategies. Ann. Rev. Ecology Systematics 2:369-404.

CHAPTER 7

John H. Gerard

Canada geese during migration.

FINDING A PLACE TO LIVE
HABITAT SELECTION, HOMING, AND MIGRATION

The ability to move about creates certain hazards for animals but also offers many opportunities. Unlike plants, most animals are not stuck where they initially find themselves. Animals accidentally can get themselves into unsuitable and dangerous locations; they can move to improved living places; and they can range far and wide among a variety of places. Many animals, although not all, establish various home bases or centers of activity from and to which they move away and return.

How do animals choose the best places to live or even know when they are in such a place? If they travel over long distances, beyond a distance over which they can maintain direct sensory contact with familiar landmarks, how do they keep from getting lost? How is something even as seemingly simple as familiar landmarks integrated in the brain?

Some species are capable of amazing feats of orientation and navigation. Even humans, although it is taken for granted, are able to go many places without a map, whether through forest, across desert or tundra, across a country to visit relatives, into a city to go shopping, or even through a maze of buildings and rooms on a university campus. The survival values for animals to move to favorable locations or remember important pathways seem fairly obvious. But how is it all accomplished?

This chapter begins with an example involving the monarch butterfly and the prodigious, long-distance feats of oriented travel of which animals are capable. Then the whole range of location or habitat-directed behavior from the simplest to the most complex will be considered. This will include very simple ways of finding suitable habitat, more complex forms of habitat selection, and finally longer-range homing and migration. The discussion of long-distance movements will be organized phylogenetically with special attention to insects, fish, birds, and mammals.

THE MONARCH BUTTERFLY: A CASE IN POINT

It seems incredible that fragile insects with big flappy wings such as monarch butterflies can fly at all. Yet they travel from Ontario, Canada, approximately 3000 km, to a small isolated location in the mountains of Mexico, then return the following spring! Others from elsewhere in North America join those from Ontario

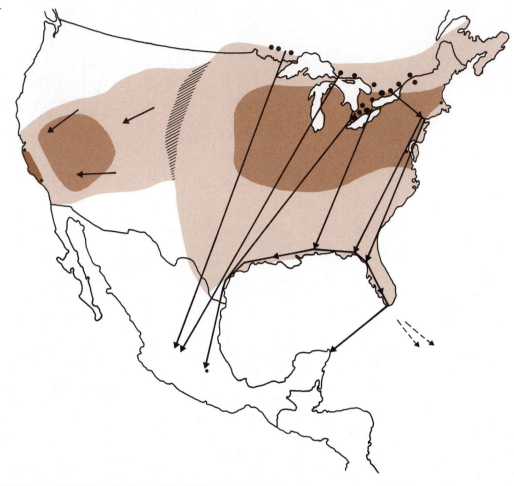

Figure 7-1 Major breeding ranges, migration routes, and wintering sites of the monarch butterfly. Dots and arrows from Canada indicate specific information based on wing-tagging studies. Arrows to the southeast from Florida indicate a possible route to the Antilles. Modified from Urquhart, F.A. 1976. Natl. Geogr. 150:160-173, and Urquhart, F.A., and N.R. Urquhart. 1979. Can. Field-Nat. 93:41-47.

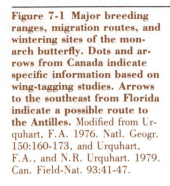

Monarch breeding range

Greatest concentration

Wintering areas

or travel to a second major location in California (Figure 7-1). The wing tagging and research that resulted in these findings were conducted over a 40-year period by the Urquharts (1976, 1979) and thousands of volunteers and fellow enthusiasts in the Insect Migration Association. Reports of butterfly locations came from many persons. In one case a monarch landed on a golf ball in California just as the golfer started into his swing. It was too late to stop the swing, but the remains of the monarch and its tag were recovered and mailed to the researchers. The wintering locations of monarchs in Mexico were discovered after several frustrating years. Newspaper advertisements eventually caught the attention of a dedicated couple, Ken and Cathy Brugger, from Mexico City.

The monarch butterflies breed in association with milkweed plants throughout much of the United States and in scattered locations along the southern edge of Canada (Figure 7-1). Two breeding distributions exist, one in the region of Utah and Nevada and the other in the northeast quarter of the United States and southeast part of Canada. The western population migrates to and winters primarily along the California coast between Monterey and Los Angeles. Butterflies from the eastern population were finally traced to a number of overwintering sites in the Neovolcanic Mountain region northwest of Mexico City.

In these mountain sites, at approximate elevations of 3000 m (10,000 feet) and temperatures that continually hover around freezing, the butterflies mass together by the millions in dense clusters that cover the trees, bushes, and ground (Figure 7-2). In one of the first areas discovered, the clusters of butterflies completely covered the branches and trunks of more than 1000 trees. The butterflies pass the winter at these high altitudes and low temperatures in a state of semidormancy. Then, apparently with the lengthening daylight of spring, clouds of butterflies take to the air and begin the return trip to the northeast.

The spring migration begins in late February and March. Mating occurs when the masses begin to disperse and continues along the northward journey. Few males travel beyond central Texas, but females may travel all the way back into Ontario and other northern areas, arriving throughout late May, June, and early July. They deposit eggs en route and next-generation adults, resulting from those eggs, begin appearing in June and early July. They are much brighter than the tattered, torn, and faded butterflies that traveled all the way to Mexico and back. First-generation adults from eggs layed in Canada show up from mid-July through August and yet another generation may appear through September, depending on the latitude. The adults from the previous year have all presumably died by this time. Then from mid-August throughout the fall the adults of the generations then alive begin migrating toward Mexico.

The butterflies form small clusters overnight at traditional trees and locations en route. The first butterfly at an overnight site flies around until a suitable location is found; then it lands and opens its brightly colored wings, which appears to attract others. The means by which they select the trees and why the same sites are chosen in subsequent years are not known. There is no evidence that a particular odor is involved. When trees are removed, the butterflies select

Figure 7-2 Overwintering roosting monarch butterflies in California. Photo from Animals, Animals. Copyright by M.A. Chappell.

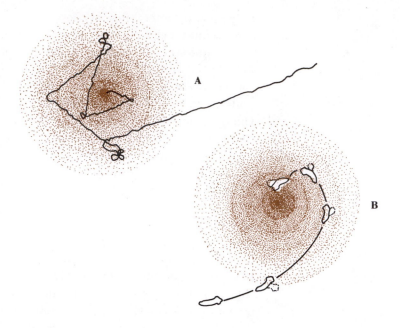

Figure 7-3 Kinesis in animal orientation. The animal moves until it encounters a favorable or unfavorable stimulus, at which time it changes direction or speed in a wandering, trial-and-error fashion. This eventually results in appropriate approach to or departure from the source of the stimulus. A, Track of a louse approaching a favorable diffuse chemical stimulus. B, Track of a planarian toward a favorable diffuse stimulus. The planarian movements of the whole body and then with direction testing via the head have been referred to as kinesis and taxis, respectively, or as a sequence of straight ("ortho") and turning ("klino") kinesis and taxis. Planarian orientation has received much attention since a study by Pearl in 1903; for a recent review see Mason 1975. A modified from Fraenkel, G.S., and D.L. Gunn. 1961. The orientation of animals. Dover Publications, Inc., New York. B modified from Carthy, J.D. 1956. Animal navigation. Charles Scribner's Sons, New York.

other trees nearby (Urquhart and Urquhart 1979). The mechanism or mechanisms by which the monarchs navigate to the small isolated wintering sites are completely unknown. It may involve some kind of a "species memory," fixed hereditarily in the nervous system, but it clearly does not involve individual memory because the migration each fall consists of new individuals. It seems more likely, because of the limited amount of neural tissue, that a fairly simple guidance system, perhaps based on the earth's magnetic field, exists but is not yet understood. It is tantalizing to consider the revolution that is likely to occur in human mechanical navigation systems, such as on aircraft, once it is determined how the monarchs and perhaps other species are accomplishing these feats.

Butterflies are not alone in their impressive achievements. Birds are perhaps more familiar in their migratory and navigational abilities and continue to attract attention in the department of amazing things. Considerably more research has been devoted to birds, and we may be closer to understanding the mechanisms they employ. This will be returned to later in the chapter.

SIMPLE FORMS

The two simplest categories of behavior that functionally move animals into suitable locations are **kinesis** and **taxis**. Kinesis is a *nondirected* movement in response to environmental factors such as light, temperature, moisture, or chemical cues. When in an unfavorable location or moving into an increasingly unfavorable place, the organism increases its speed, angle of movement, or amount of turning. These actions remove it from the present location and expose it to new conditions in new locations. When more suitable conditions are encountered, the animal simply moves less and remains in the new area. This process is haphazard and results in random movement into better locations (Figure 7-3).

Figure 7-4 Taxis in a crawling maggot. The maggot turns away from a light that has been directed toward its side. Technically this would be a negative-photo-klino-taxis ("away-from-light-turning-taxis"). Modified from Fraenkel, G.S., and D.L. Gunn. 1961. The orientation of animals. Dover Publications, Inc., New York. Based on Mast (1911).

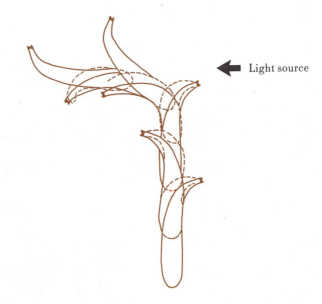

← Light source

Taxis, on the other hand, involves *directed* movements; that is, animals detect the source or differences in intensity of some factor, then move appropriately. Detection of concentration or direction may be accomplished by repeated testing of the environment. The best known examples of repeated testing involve planarians and fly maggots (Figures 7-3 and 7-4). An alternate method of sensing the source or intensity of something is shown in many species with directional sensory capabilities. Taxis is frequently identified with the environmental condition to which the organism is responding. For example, there are phototaxis, geotaxis, and hydrotaxis, to or from light, gravity, and water, respectively. Taxes, although still quite simple, do not seem as random as kineses.

HABITAT USE AND SELECTION

From the relatively simple level of kinesis and taxis, one can consider the problem of finding a place to live on a broader scale, involving organisms with more complex sensory and nervous systems and able to move considerable distances under their own power. Various categories of home finding are not necessarily exclusive or confined to certain groups of animals. Even the most mobile and complex animals may use taxes to some extent in guiding their movements. Chemical cues or odors are used not only in kinesis and taxis but also for some long-distance homing. Odor following is perhaps one of the most basic and general mechanisms for finding home and, although not particularly important among humans, can be found in almost all taxonomic groups from protozoa to molluscs, insects, fishes, mammals, and even birds (Grubb 1977).

The problem of how an animal ends up living in a certain area can be viewed in at least two ways: first, the *general* aspect of the *type* or *kind* of environment and, second, the *specific* location(s) within which a particular animal spends most of its

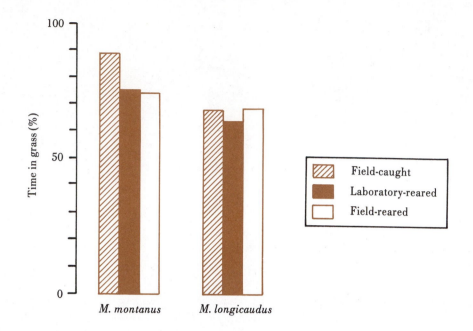

Figure 7-5 Habitat preferences of two sympatric species of *Microtus* mice in choice test outdoor enclosures. *M. montanus* is found normally in grass habitat and *M. longicaudus* in shrub habitat. The enclosures provided both types of habitat. Field-caught mice were live-trapped from the wild in their normal habitat as adults. Laboratory-reared were offspring reared in standard plastic laboratory containers with wood shavings. Field-reared mice were raised in *opposite* habitat types from which they normally are found; that is, *M. montanus* was reared in shrub and *M. longicaudus* in grass. Although field-caught *M. montanus* showed the greatest preference for grass, *all* mice showed a distinct preference for grass regardless of species or rearing. In interspecific encounters, *M. longicaudus* was observed to withdraw from *M. montanus*. Thus *M. montanus* is thought to occur naturally in grass habitat because of its preference for grass, whereas *M. longicaudus* probably occurs in shrub habitat because it is excluded from the grass by the more dominant *M. montanus*. Modified from Randall, J.A. 1978. Behav. Ecol. Sociobiol. 3:187-202.

life, including different locations at different seasons and the pathways between these places. The first topic usually is considered under the banner of **habitat use** or **selection** (not the same things) and the second under some name such as **home range** or **territory** (also not the same things). Habitat *selection* assumes that animals are choosing the area, whereas habitat *use* simply notes the presence of the animals and, hence, implies much less. Similarly, *territory* generally is considered to involve some degree of active defense and interaction between animals (Chapters 8 to 10), whereas *home range* just refers to where the animals spend their time, whether that place is actively defended or not. The different topics of habitat selection or use and home range will be discussed separately.

In distinguishing between habitat use and selection it must be emphasized that the places animals end up living are not always determined by the animals themselves. Many species, particularly some aquatic organisms and many of the flying insects such as locusts, are true drifters. They are carried about by the water or wind currents, disperse and occasionally become concentrated by eddies of movement, and generally end up simply where they find themselves. Some animals, such as many birds on migration, combine methods; they may drift with the wind or water, but they select which currents and when to ride so that there is a degree of active participation.

Another method by which animals may find themselves in particular locations without their active choice involves differential survival imposed by the environment, such as predation or weather factors. This may lead to the false impression that the animals chose their location. Imagine, for example, a large population of animals that disperses into the environment and settles randomly in all types of habitat. In some locations, however, the animals are more vulnerable to predators, perhaps because they are more conspicuous or there are fewer hiding places, and these individuals are eliminated. Others in more suitable habitat, however, survive and remain (Turner 1961). Their presence in just those habitats may give the impression that the animals chose the places when in fact it was determined by the absence of predation.

Competition also may influence where an animal lives. Studies of fish (Werner and Hall 1976) and mice (Randall 1978) (Figure 7-5) have shown that different species, when tested alone, may have similar preferences for habitat, but when together, one species may partially or completely exclude the other and "push" it into a less-preferred type of habitat. Ecologically this relates to the principle of **competitive exclusion**, by which one species may even completely eliminate another. Commonly, however, animals of one species may simply live in a habitat that they did not actively select. Even with increased density of a single species, some individuals may have to accept less-preferred places to live. Based on a study of reproductive output in herring gulls, Pierotti (1982) obtained results that supported a model proposed by Fretwell and Lucas (1970): "Increasing density in a preferred habitat can create a situation whereby fitness may actually be greater in less-preferred habitat."

In spite of these situations, however, it would seem advantageous for animals to actively choose where they will live, and there is at least circumstantial evidence that many, if not most, animals do choose within the range of options open to them. Any good bird-watcher, fisherman, or hunter knows that certain species will be found almost exclusively in certain habitats within a region, although the animals could easily move on their own to other places and different habitats. One of the first persons to demonstrate this quantitatively and employ the term "habitat selection" was Lack (1933). He demonstrated that different species groupings of birds were found in vegetation stands of different ages. MacArthur (1958), in a much cited classic study, showed that species may be segregated even in their use of individual trees (Figure 7-6).

One could state tautologically that the ultimate reason that particular species select particular habitats is because those are the places to which the animals are best adapted. But the semantics slide into the classic academic quagmire of the

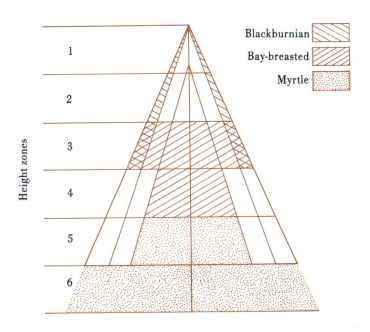

Figure 7-6 Preferred parts of trees for feeding habitat by different species of warblers. Diagram is based on percent of total seconds of observation for three (of several) species involved in the study. Modified from MacArthur, R.H. 1958. Ecology 39:599-619.

Blackburnian
Bay-breasted
Myrtle

Height zones

1
2
3
4
5
6

"niche." (For a good discussion of this problem, which is beyond the scope of this book, see Krebs 1978.) The problem of habitats to which animals are adapted is an important topic, however, not only for basic ecology but also for persons concerned with applied cases. The activities of humans, including habitat changes, have threatened the very existence of many species of animals, commonly classified as endangered species. To manage and correct the problems, one needs to better understand species habitat requirements and the environmental factors to which their senses are tuned. The factors which they need and those which they select may not always be the same; they may simply occur together. Furthermore, many persons are interested in maintaining various species, including domestic, under the best of conditions in captivity. Although the natural habitat contains many items and factors, only a few or a particular subset is really important to a given animal; there is no need to duplicate the entire set to satisfy the needs of the species. The trick is to find which factors are most important. It is the trained-cricket problem on a grand scale.

The ability to measure another species' habitat is very elusive. Habitat, rather than being a small, isolated object on which the animal and a researcher can focus their attentions, encompasses and surrounds the animal, with no well-defined boundaries. It generally is variable and under the best of conditions consists of many items and factors. The animal may respond to different factors at different times and seasons and under different circumstances. Habitat is a fairly vague, complex, and abstract entity.

In the case of simple organisms and their simple orientation movements, the animals often respond to specific, readily identifiable factors such as light or temperature. Habitat selection probably could be considered as being synonymous with orientation movement. But as the animals and their sensory systems become more complex, so does the sophistication of their habitat selection.

Habitat selection, as with any other kind of selection, clearly has to be based on sensory cues. But, except for a few cases, little work has been done on which cues are used by which animals under which circumstances. Furthermore, there has not been much research on the heritability of specific habitat selection. The identity of the preferred habitat in some species appears to be learned in part (via a form of imprinting, discussed in Chapter 19). It is clear that species are variable in their specificity. Many species are known as "specialists," and others are known as "generalists" in their habitat, as with diet.

The apparent choosing or selecting of habitat by many animals can be observed outright. Some rotifers, for example, become sessile, that is, fixed to one location, but initially they are able to move about. The habitats that some species of these microscopic animals choose may be as restricted as the left side of certain types of snail shells. Wallace (1980) described how larvae settle once they have reached that period of their life: they swim to various surfaces, pause, and contact the surface with the corona (the wheellike ciliated organ around the mouth from which they get their name); then depending on whether it is the appropriate surface, they temporarily attach or swim on and try a new location. In suitable habitat they may detach and reattach several times in different spots before becoming permanently fixed. Each time of initial attachment they twist around and bend over, touching the substrate with the corona as if trying out the site for preference.

Similar descriptive studies of habitat use have been published for many organisms, often showing much variability among different individuals, different seasons, and for different activities. Gottfried and Franks (1975), for example, showed differential use of habitat by two different flocks of juncos *(Junco hyemalis)* and for one of the flocks before and after a month-long period of heavy snow accumulation during which the flock disbanded (Figure 7-7). Hunter (1980) described "microhabitat selection" in great tits *(Parus major)* in which the birds chose different habitat for different activities (Figure 7-8). The most hidden, protected habitat was used when the birds were singing and might attract predators. MacArthur's study (1958) showing use of different parts of trees by warblers for foraging has been mentioned already.

Laboratory experiments to demonstrate that individuals select particular habitat and investigate the role of prior experience have been conducted with several types of animals. The basic approach has been to provide captive animals with a choice of habitats. Depending on the particular study, individuals have been captured from the wild or reared in the laboratory. Laboratory or hand-reared animals may have come from wild parents or from several generations of captive parents. They have been reared in natural or unnatural types of habitat or in very restricted conditions with no exposure to anything resembling a normal habitat. Klopfer (1963), for example, hand raised chipping sparrows in cloth-covered aviaries. He called them *Kaspar Hauser* birds.

The results of such experiments generally have demonstrated definite habitat preferences by individual animals and both species and experience effects with differences among closely related species. In a classic study of *Peromyscus* mice, Wecker (1963, 1964) tested a prairie subspecies, *P. maniculatus bairdi,* in a large observation pen that half enclosed a field and half enclosed the edge of a

Figure 7-7 Habitat used by two different flocks, including two different times for one flock, of one species of junco. The percentages of nonroosting time spent in the different habitats show considerable variability. Redrawn from Gottfried, B.M., and E.C. Franks. 1975. Wilson Bull. 87:374-383.

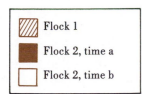

Flock 1

Flock 2, time a

Flock 2, time b

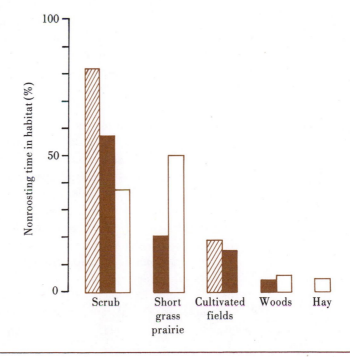

Figure 7-8 Microhabitat selection for different activities by great tits. From Hunter, M.L., Jr. 1980. Anim. Behav. 28:468-475.

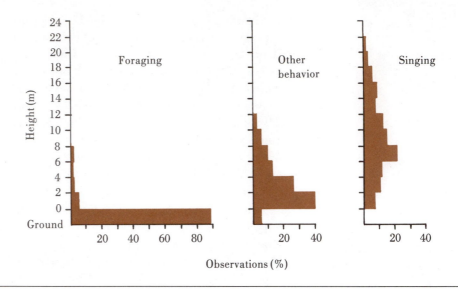

woodlot. The subspecies normally is found in fields. Wild-trapped mice as well as first-generation laboratory-raised offspring all showed a preference for the field side of the enclosure. After 12 to 20 generations of laboratory rearing, however, offspring showed *no preference* for either field or woods *unless* they were raised first in one or the other. In that case, field-reared offspring preferred the field and woods-reared offspring showed no preference, indicating that an inherent bias for the field still remained.

Similar results have been obtained among birds. Klopfer (1963) found that, when given choices between pine and oak branches on opposite sides of an observation chamber, white-throated sparrows showed a slight, but not strong, preference for the oak, whereas all chipping sparrows, whether wild-trapped, hand-reared in pine, hand-reared in oak, or *Kaspar Hauser*, preferred the pine. Those reared in oak, however, showed less of a preference for pine. In studies of blue tits *(Parus caeruleus)*, which normally are found in broad-leaved trees, and coal tits *(P. ater)*, which live in conifers, both wild-trapped and hand-reared birds tested in choice conditions preferred the habitat as predicted, but hand-reared birds showed less of a preference (Gibb 1957, Partridge 1974).

Sale (1971) experimented with coral habitat selection in reef fish. Thirteen groups of 10 fish each, all from the same species, were given choices of three types of coral. Seven of the groups were fish originally obtained from one type of coral, and six of the groups were from another type. None had been collected from the third coral type. All fish consistently, although not exclusively, chose the type of coral from which they had been collected. Considering that the fish were all from the same species, Sale suggested that the differences in choice could result either from genetic polymorphism for preference or as a result of experience. He believed that early experience was the cause, but he was unable to hand raise any of the species from eggs to find out.

Studies of animals under natural conditions, aside from the basic descriptive studies just mentioned, have been more indirect and often difficult to interpret.

The usual plan has been to observe a sample of individuals from a species where they are found and measure as many aspects of their surroundings as possible or at least as many variables as are deemed reasonable by the investigator. Comparable measures are made also for other species in the general area or for a number of randomly chosen points. Then the mass of data is subjected to multivariate analyses to boil it into something more comprehensible. The approach is an extension of studies in which distributions of birds are plotted along continuums of plant communities (Bond 1957).

Two of the pioneering applications of multivariate analyses for animal habitat use were by Cody (1968) and James (1971). In James' study, for example, 15 vegetation variables were measured for 46 species of birds from 18 counties in Arkansas. From her results she was able to determine the positions of the different species in an abstract "habitat space" (as was illustrated for woodpeckers in Figure 3-23). She also identified the characteristic habitat appearance, the *niche-gestalt*, from the viewpoint of the species. The characteristic appearance depended on the visible elements of the vegetational structure that were present consistently in the surroundings of animals for each species. Examples of James' *niche-gestalts* for four species of warblers are shown in Figure 7-9.

Subsequent multivariate analyses of animal habitat use proliferated, and various techniques were used to better illustrate differences among species while still allowing for variability (Figure 7-10). Numerous problems and questions arose, however. A symposium on the topic was held in April, 1980 (Capen 1981), during which additional examples were presented and valuable discussion and refinement of thinking occurred.

The basic problem with the multivariate approach to habitat use is that it is correlational at best. One cannot determine which, if any, factors the animals themselves are actually choosing, and it is not clear that the most relevant variables have been measured. Furthermore, outcomes of analyses of "preference" may shift and yield different interpretations, depending on which particular variables are included or omitted. Johnson (1980) discussed the problem of shifts of inferred preferences and proposed a solution whereby ranks rather than actual measurements be used for different elements. The problem of not including all variables was clearly recognized by James (1971). She stated, "This space also contains gradients in types of food, nest-sites, microclimate, etc. Although these variables are undefined in the present study, they would have to be included in a thorough analysis of the ecology of adaptation."

But even from among the variables measured one cannot be sure which are most meaningful to the animals themselves. Johnson (1981) noted this problem by saying that "a bird might find James' (1971) outline drawings of *niche-gestalts* to be meaningful, and would be willing to select its habitat based on those drawings. But the bird would be hard-pressed to plug the values of 15 or 20 variables into a number of linear combinations, compare the calculated values to one another, and select a habitat with value closest to its liking."

What really is needed to answer whether habitat selection is occurring and which factors are being chosen is to experimentally vary the components of animal habitat and measure the response of the animals. That, however, is a large order of business. The "test-tube" or experimental unit for such studies might be several hectares or even square kilometers in size; sufficient natural habitat may

Yellowthroat

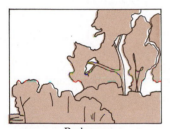
Redstart

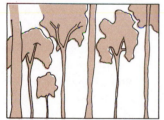
Black-and-white warbler

Hooded warbler

Figure 7-9 Four examples of F.C. James' *niche-gestalts* **for different species of birds. These outline drawings represent features of vegetation consistently measured where the birds were found.** Redrawn from James, F.C. 1971. Wilson Bull. 83:215-236.

Figure 7-10 Example of multivariate plotting by the first two principal components of the habitat, which indicates both species differences and species variability to a limited extent. The areas plotted are 1% confidence ellipses. These contrast with plotting of single point (0% confidence) at one extreme or large areas (95% or 99% confidence) at the other extreme. Points and 95% confidence areas are shown in Figure 3-20. Differences in the plotting techniques depend not on differences in the data (all plots can be derived from the same set of data) but rather on how much one wishes to illustrate the variability within the data. From White, D.H., and D. James. 1978. Wilson Bull. 90:99-111.

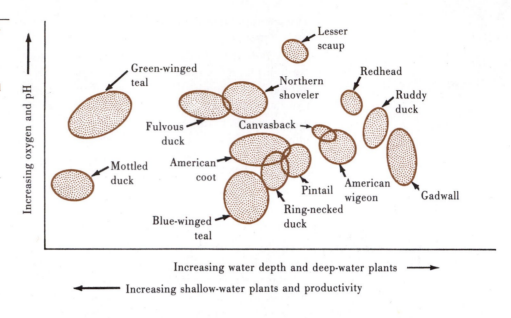

no longer exist for experimental manipulation; and one may not have the foggiest idea how to identify those components to be manipulated. Jacobs et al. (1978), for example, considered color in the nest selection by female village weaverbirds *(Ploceus cucullatus)*. They painted nests brown or green and found that females showed only a slight, nonsignificant preference in whether they accepted or rejected nests. Males, however, destroyed brown nests much more often than green nests after particular nests had been rejected by females. These results are difficult to interpret but suggest that color is *not* the only or major element in the choice of something as discrete as a nest. That is, the correct component or components have not yet been identified. It is most likely that several factors are important together in habitat selection. But, unlike the ease with which cardboard models of animals can be varied for different factors (as in Figure 2-10), habitat is not as easy to experiment with.

In a few instances the important factors in habitat selection may involve something as simple as presence or absence of nest holes in trees, cover (of almost any kind) from predators or inclement weather, or specific types of food. But more often than not, the situation is more complex, and the multivariate approach, used with caution and the best subjective insight from knowledge about the species, is the best tool available. At the same time one must be careful not to infer too much from the results of such studies.

Until it is shown beyond a reasonable doubt through experimental confirmation that an animal indeed is actively selecting its habitat, it must be considered that the animal may be where it is found by default. Default can result easily from predation and other sources of mortality or from competition with other animals, including those of the same species. Not all humans live in the dwellings or geographical areas of first choice, and it is unlikely that many other animals do either. Often there is a multitude of contributing factors. Furthermore, as is amply clear from a small sample of experimental studies of mice and birds, the factors can be quite different even for closely related species.

**Home Range and
Familiar Area**

Regardless of the reason that an animal lives where it does, most individuals tend to remain within a relatively confined area. Animals of some species do wander or roam more than others, and males commonly cover a larger area than females (Schoener and Schoener 1982). But even those individuals generally tend to return to places they have been before and stay more or less within a given area.

The ultimate advantages of being able to return to a familiar place, such as home, should be clear. Such places may afford protection, hence increasing the chances of survival, increased resources, and perhaps even stored resources. Simply being in a familiar area may aid survival. Mice allowed to become familiar with an area, for example, are less vulnerable to predation by owls than mice newly introduced to the area (Metzgar 1967). Any biological mechanisms that permit an animal to nonrandomly return to a previous location should have high selective value.

The area within which an animal stays is called its **home range.** Home ranges usually are dynamic; that is, they tend to change over time. Several techniques exist for estimating the size and boundaries of home ranges, and there are a number of associated terms and concepts. Lehner (1979) includes a survey and discussion of the older techniques. For recent methods, comparisons of alternatives, and general reviews of the subject, see Schoener (1981) and Anderson (1982).

The most common apparent method by which animals maintain their spatial orientation and ability to stay within a given area is what can be called simply the *familiar area* or *familiar path.* J. von Uexküll (1934) was one of the first persons to draw attention to the familiar path. The familiar area for an animal requires two important things: (1) the ability to *learn* and *remember* the area and (2) some form of at least minimal *sensory contact* with consistent features or **landmarks. Spatial memory,** whereby an animal forms some kind of mental image or map of its familiar area, may be one of the most basic and universal forms of memory and learning (Chapter 19). It is not known what the internal representation consists of, and it may be quite different in other species, perhaps consisting of series of directions more like a computer program rather than the apparent three-dimensional, external image which humans seem to experience.

The system, which is so familiar to humans that most of us take it for granted, operates by the individual starting from a given point and becoming acquainted with the features of the area or path during travels away from that point. If contact with the familiar area is broken, the individual is said to be lost, a familiar term.

Frequently the path's marks are made by the animals itself, such as when odor trails or signposts of some kind are left along the way. In many cases, such as with humans, a fairly complete image of the area may become established with experience. Rarely, however, is the image absolutely complete under the best of situations, and the sensory contacts with the environment can be quite minimal and often tenuous, particularly in new areas. Few people, for example, know all of the streets, insides of buildings, and nooks and crannies of the areas where they live. But most people know incredibly many places and can go to particular locations, including drawers, cupboards, and hiding locations, with extreme accuracy.

The details of learning or becoming familiar with the area or path vary among different species. Bees and spiders, for example, learn only during narrow and highly specific times (Chapter 19). Many vertebrates appear to learn with the first travel over a path and become gradually more and more familiar with it on subsequent travels. In all cases two attributes of the system are clear: (1) new familiar areas can be learned when the animal moves to or is moved to a new location, and (2) once established, sensory contact with and attention to details of the area may be reduced to the minimum needed to remain oriented. The latter point has been shown strikingly, for example, in bats returning to their caves. The bats will collide with obstacles not previously present although they easily could have detected them (reviewed in Griffin 1958). Griffin later (1976a) referred to this as the *Andrea Doria* effect, from a ship that struck an iceberg and sunk in spite of radar capability to avoid the obstacle. Apparently the bats form a three-dimensional image of the areas over which they routinely travel; then they rely on that familiar memory rather than constant echolocation.

The sensory channels that animals use in their familiar area vary widely among species. Senses in general are considered in Chapter 14. Bats orient spatially with echolocation and memory, as already indicated, but also with naturally occurring sounds. They have been shown, for example, to travel toward distant marshes (presumably rich in insect prey) by using sounds from frog choruses. They also will approach artificial playbacks of recordings of such sounds. Similarly, it has been proposed that birds flying at night may be able to orient toward certain sounds from below. Humans orient primarily on the basis of vision but also are capable of using sound and touch. Numerous other species also orient in their familiar area with vision and sound, whereas many species use odors, tactile cues, and a number of less familiar senses. Orientation on the basis of odors and air movements has been referred to as **anemotaxis.** It has been suggested that some fish identify their familiar area on the basis of electrical conductivities and resulting changes in the electrical field from different objects in their surroundings (Lissman 1958).

Regardless of which sense or combination of senses an animal uses, the familiar area probably is the most universal mechanism of spatial orientation found in the animal kingdom. As will be seen later, several species have supplemented their familiar area abilities with a host of additional orientation and navigational mechanisms. In many cases an animal possesses several redundant, backup mechanisms, often with a hierarchy of priorities for their use; that is, some mechanisms are employed under some conditions and others under other conditions. The process by which one mechanism or another is used often is not based on insight or intelligent choice. An animal that is relying on a mechanism that has been experimentally tricked, for example, may not switch to an alternate method of which it is capable under other conditions. Instead, the animal will become and remain lost and confused.

COMPLEX SYSTEMS OF ORIENTATION AND NAVIGATION

Ultimate Factors in Migration

The ultimate factors involved in long-distance migration can only be speculated on. The different locations at the two ends of the trip apparently are better suited to survival or reproduction at different times. The differential advantages, furthermore, have to be sufficient to outweigh the high risks of getting lost, running out of fuel, or getting caught by predators in unfamiliar territory during the long journey.

How long-distance migration originates in the first place is not clear. It would seem to be derived, as in other evolved characteristics, from the natural variation in the movements of the animals. Those individuals which move from areas that become inhospitable would, according to standard Darwinian reasoning, have better chances of survival than those which remain behind. Of those which do move away, some would then have to later return and have greater survival and reproduction than those which do not return.

Explanations for the origins of long-distance migration frequently invoke the roles of past glaciation and perhaps even continental drift, in addition to the pressures of present seasonal differences at different locations. Competition from other species already present in various locations also may be important. Otherwise it would not seem necessary, for example, for arctic terns to go all the way to antarctic regions or for other birds to go to South America just to escape the northern winters.

For many species of birds with a broad latitudinal range, those living at lower latitudes may show less tendency to migrate than others of the same species at higher latitudes; they even may be permanent residents. Birds from further north may fly on beyond the middle latitudes, in somewhat of a leap-frog fashion, and winter further south than the nonmigrants. In many species of birds there is a northward movement in the late summer and early fall, prior to the major migration to the south. And in some species, such as bald eagles, birds that nest far to the south, such as in Florida, may briefly leave and travel north after the nesting season.

Speculating on the origins of the monarch butterfly migration, Urquhart (1976) notes that most of the hundred species of milkweed in North America are native to Mexico. He considers that perhaps the monarch originated in Mexico:

Now, in returning there each winter, the butterfly is "going home," after straying, perhaps over eras of a warming trend, farther and ever farther north. Anyway, I'm convinced that the monarch's selection of the Sierra Madre for overwintering is no random choice.

Proximate Factors in Migration and Homing

The proximate factors that are involved in local and long-distance orientation are many and varied. The biological mechanisms by which different organisms find their ways about, some of the environmental factors that appear important, and the few generalizations that are now emerging will be considered. Discussion of the internal, physiological integrating mechanisms will be postponed to a later section of the book.

Before proceeding, it is necessary to clarify a few terms and concepts. The remainder of this chapter is primarily concerned with an animal's ability to *move to specific locations that cannot be sensed directly;* that is, the animal moves to a particular place beyond its familiar area with the aid of additional external cues. The ability to move to only a particular *kind* of place, without emphasis on the specific location, perhaps is viewed better as habitat selection. Simple movement *from* a place (i.e., dispersal or emigration) without a return or directed move to another specific location will not be considered here. **Migration** generally is

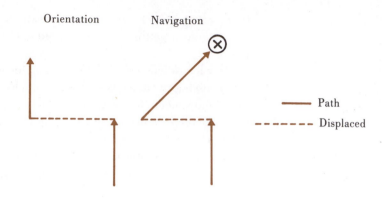

Figure 7-11 Difference between long-distance orientation and navigation, as indicated by experimentally displacing *(dashed line)* **the animal from its path** *(solid line)***. An animal that is merely orienting in a particular direction will continue in the same direction but not toward a specific goal or target. An animal that is navigating, however, will compensate for the displacement and travel toward the specific end point. The term** *orientation* **is used commonly in different contexts. Here it is used in the context of long-distance travel without possible direct sensory contact with the end point. Once the end point is in sight (or hearing, smell, etc.), then the term navigation is no longer appropriate as an alternative, and** *orientation* **is used to cover all situations.** Copyright © 1978 by R. Robin Baker. The evolutionary ecology of animal migration. Reprinted with permission from Holmes & Meier Publishers, Inc., New York.

considered to involve a move from one location to another with a subsequent return to the first. Orientation and navigation are terms that sometimes are used interchangeably. There is, however, a distinction: **orientation** refers to the ability to obtain or maintain a *direction* relative to an external cue, such as a compass direction, flow of current, or a particular stimulus such as the sun. **Navigation,** on the other hand, focuses on the goal or end point, the specific location toward which the animal is moving. In navigation the animal uses one or more mechanisms and is capable of adjusting for changes or displacements (Figure 7-11). A wealth of jargon has proliferated as a result of attempts to further separate and classify different categories of orientation, navigation, and migration (Baker 1978).

A Brief Phylogenetic Survey of Migration, Orientation, and Navigation

This survey necessarily will be brief, superficial, and sufficient only to permit some important generalizations. Persons wanting further depth can refer to the literature listed at the end of the chapter.

Aquatic Plankton
Although frequently referred to as "diurnal migration," the vertical movements of many small aquatic and marine organisms from deep water during the daylight hours to shallower depths during darkness probably should be considered as taxis rather than migration because the movements are in response to light.

Molluscs and Crustaceans
Movements of many of the larger crustaceans appear to involve a return to familiar or "home" locations. Limpets, snails, and some others have a home range and will return to known locations if artificially displaced. The mechanism is believed to involve a *familiar area*, based either on a spatial memory or a memory for olfactory, tactile, or other possible cues. Snails have been shown to follow mucous trails (Wells and Buckley 1972).

Arachnids
Although the evidence is not strong, some data suggest that wolf spiders form a home range and seasonally return to specific area. If true, such homing ability is hypothesized to be a result of a familiar area (Baker 1978).

Insects

A great many categories of homing and navigational abilities, or lack thereof, have been demonstrated in insects, perhaps reflecting the general amount of research devoted to this taxonomic group. Many species are simple drifters, moving about and cuing to appropriate habitat wherever it is, thus removing them from this discussion. Some disperse over long distances, often referred to as "migrations." In the African locust movement patterns exist in two forms, depending on the degree of crowding. But there is no evidence of return to the initial location; they largely drift with the wind.

Many species of insects, however, show homing ability. The large-scale population migration of monarch butterflies is a clear case. The proximate causes of monarch migration are not understood, and it is not known how they select the environmental conditions with which to fly, that is, whether they drift with prevailing winds or actively select the best winds.

Other insects that travel considerable distance (generally much less than the monarchs) and return use a familiar area, commonly with odor trails for the

Figure 7-12 Landmarks in the location of nest holes by the digger wasp *(Philanthus trangulum)*. A, This wasp first digs a hole; then, upon emerging from it, it spends a few seconds flying around the area learning the landmarks. The landmark in this instance is a ring of pinecones placed around the nest entrance while the wasp was inside. **B,** If the cones are moved to a new location while the wasp is away, it will fly to the new location with prey for the nest and be unable to find the actual nest entrance. The characteristics of landmarks used by these wasps were investigated further by using different shapes of surrounding objects, objects other than pinecones, incomplete perimeters, and combinations of objects, such as cubes and twigs; see references in Tinbergen 1951. Redrawn from Tinbergen, N. 1951. The study of instinct. Oxford University Press, Oxford.

Nest

familiar path. A classic ethological example demonstrating the use of landmarks involves the digger wasp (Figure 7-12). Odor trails are well known in ants. Specific chemicals, such as from the Dufour's gland in the abdomen (see Figure 9-6), are left on the substrate by dragging the abdomen. They are then followed on the return trip or by other foragers. The presence of the substance is easily demonstrated by interrupting the trail or by other manipulations such as by creating artificial trails with rods dipped in glandular extract or by moving trails so they form a closed circle. In the latter case the ants may continue to follow around the trail, going in circles, for extended periods of time.

Desert ants, which live in an unstable, shifting environment of dunes, home very accurately through the use of polarized sunlight. They adjust for the apparent motion of the sun in a typical sun-compass fashion (described later). Their eyes are specialized for the reception of polarized light (further discussion in Chapter 14).

Perhaps the most thoroughly studied instances of homing ability in insects involve various species of honeybees. Bees may find food and other resources via one or more of several senses. They then can return repeatedly to the same location, returning to the hive between trips. The ability to navigate back and

Figure 7-13 Use of landmarks, within limits, in honeybee orientation. A, When bees are trained to a feeding station past a tree line, then moved to a new location with a tree line in a different direction, they change direction and follow the tree line landmarks. B, If a single tree is used, however, the bees ignore the tree and travel in the former direction after the hive is moved. This figure is generalized from results of numerous experiments. For details and references, see Lindauer (1971a) and von Frisch (1967).

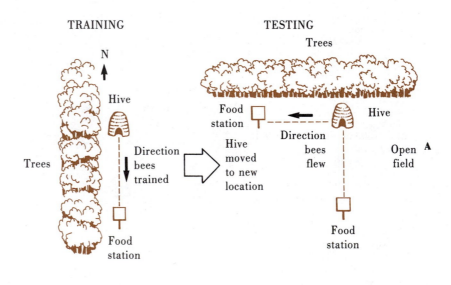

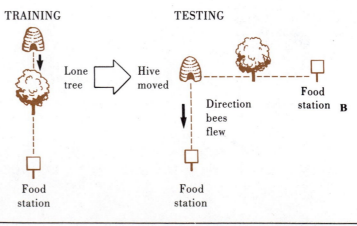

forth requires information on two aspects: (1) *distance* from the hive and (2) *direction*. (The communication aspect of this phenomenon, whereby individuals inform others of the location by the famous dance language, is discussed in detail in Chapter 9.)

The distance that the bees must travel is measured via the energy that they spend traveling out from the hive (Frisch 1967). Directional information can be obtained by at least two mechanisms. The first involves landmarks and familiar area. Bees can be trained, for example, to move parallel to a tree line at the edge of a field to a food source. If the hive is moved to a new location, the bees continue to follow a similar landmark form rather than the same compass direction (Figure 7-13, *A*). There are limits to the use of landmarks, however. If trained to a single tree, then moved to a new location, the bees move in the compass direction rather than relative to the tree (Figure 7-13, *B*).

The second mechanism used for local navigation by bees uses a **sun compass.** The position of the sun is used for orientation and is coupled with the internal circadian rhythm of the animal to compensate for the apparent motion of the sun across its daily arc. The sun compass can be used by the bees even under slightly overcast conditions with the aid of polarized light, somewhat as was described for the desert ants. The presence of the sun compass was suggested by observations on the changing orientation of the bees' dances inside their hives through the technique of **clock-shifting**, or altering an animal's internal diurnal rhythms relative to the external real time. Clock-shifting is accomplished by artificially changing the photoperiod. If the natural photoperiod is 12 hours of light from 6:00 AM to 6:00 PM, for example, it can be moved forward 3 hours by turning lights on at 3:00 AM and off at 3:00 PM and by preventing the animals from seeing any light after 3:00 PM (Figure 7-14). After a period of time the animal's internal rhythms change to the altered time period. Then if exposed to the external real photoperiod, the animal, in essence, subjectively perceives the real time now as being 3 hours later, as if the animal had been moved three time zones to the west. Shifts can be of any magnitude, either forward or backward. To help understand this, imagine being placed into these circumstances and imagine when and where the sun would appear at particular times.

Figure 7-14 Clock-shifting. The animal's subjective impression of time of day can be altered by modifying the times that the surroundings become light and dark. Technically the shifted *Zeitgebers* are said to entrain the diurnal rhythm to new times.

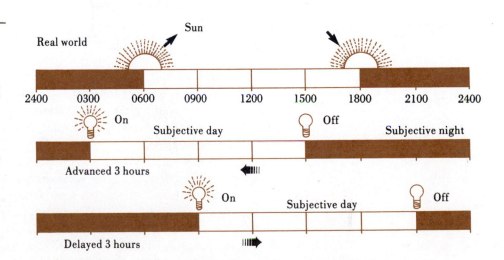

The use of sun compass by bees was confirmed by a number of experiments, including the movement of trained bees from Long Island to Davis, California (Renner 1960). This longitudinal shift of three time zones caused the bees to be incorrectly oriented in Davis by an angle of 45 degrees. This disorientation was as would be predicted if the bees were using the sun as a compass. (The sun's apparent motion changes 15 degrees per hour; it is located in the east at 6:00 AM, then moves 90 degrees to the south by noon and another 90 degrees to the west by 6:00 PM. When it is 12:00 noon at Long Island and the sun appears to the south, the time is 9:00 AM in Davis, and the sun appears in the southeast, a difference of 45 degrees.) Further experiments showed that the setting of the compass is learned and can be *reset* over a period of time. The internal circadian rhythm has about a 24-hour period (pp. 150-155) and will drift over time if not repeatedly synchronized with consistent external events (such as the sun), the *Zeitgeber*.

Resetting of the internal rhythm is possible unless the bees are transported across the equator, in which case the sun, in its apparent arc below the zenith, would appear to be going in the opposite direction. Offspring of bees transported across the equator do compensate for the sun, suggesting that they initially must learn the direction of travel. Further work has shown that the accuracy of the bee's compass resolution is about 3 degrees, the same angle that would be permitted by the angular limits of the ommatidia in the bee's compound eyes.

Fish

Movements and migrations of fishes have attracted almost as much interest as those of birds, partially because of the economic importance of fish. However, migration research is more difficult with fish than with birds. There are fewer recovery encounters between fish and humans than between birds and humans, which has resulted in a smaller return of fish data. Consequently less is understood about fish movements and migrations. Nonetheless, enough is known to demonstrate a considerable diversity among fish homing abilities and mechanisms. A few species probably do not home significantly but simply drift, as in the case of many species of insects. A large number of species probably return via the basic familiar-area methods. At least some, however, show remarkable migrational and navigational abilities. The details will be discussed for two cases: salmon and eels.

Salmon are **anadromous** fish, moving upstream to spawn. They spend most of their lives, often several years, in the ocean or other large bodies of water; then they return to freshwater rivers and streams to breed. Their migrational and homing abilities are legendary. They may travel 2000 to 4000 km out to sea; then when they return, they go back to the exact stream from which they hatched. Large numbers of tagging returns, for example, are found at the home stream and not at other nearby streams.

Understanding this homing behavior is important for economic as well as scientific reasons. Salmon provide a large commercial and sport fishing resource. At the same time numerous dams have been and continue to be constructed across streams, inadvertently blocking the salmons' return. Also, water pollution in the rivers and streams may make the home streams unsuitable for the fish or otherwise prevent their continued use. "Fish ladders" to aid the movement of returning fish around dams have been partially successful, but large numbers of fish continue to be lost.

A long series of studies, associated primarily with A.D. Hasler and a number of his students (Hasler et al. 1978) have established convincingly that the salmon home on the basis of stream odor. Different streams are believed to possess unique odors from the soil and vegetative characteristics of the drainage basin. Young salmon are believed to imprint (Chapter 19) to and memorize the odors of the streams where they hatch, then seek out those same odors at maturity.

These conclusions are based on a diversity of experiments that initially involved such manipulations as conditioning (Chapter 19) fish of various species to different odors using food rewards and shock punishment. Other work included cutting nerves from the nose to prevent any use of smell. Early objections were raised that interference with smell, for example, was interfering with other things, such as the ability to feed, and that one could not conclude that smell was responsible for homing (Peters 1971), another case of trained cricket.

Results of several different approaches to the subject, however, leave little room for further doubt. In an important laboratory experiment, fish were tested in a four-armed maze with water cascading down a series of steps into the center, from which it drained. Different odors, with which fish had or had not been raised, were introduced to different arms. The fish accepted or rejected odor-treated water and went up the appropriate arms as predicted.

In field tests researchers captured returning salmon and plugged the noses of half of them. Fish with plugged noses went to different streams at random, whereas normal fish with unplugged noses returned to their proper streams. In something of a hybrid laboratory-field experiment, juvenile salmon were exposed to artificial odors (morpholine or phenylethyl alcohol) or plain water (the control group) and then released into Lake Michigan. Eighteen months later when returning fish would be expected to move into streams along the lake, 19 streams were monitored. Two of these were artificially scented with either morpholine or phenylethyl alcohol. Of the treated fish that were recovered, 94% to 98% of the morpholine-treated individuals and 91% to 93% of the alcohol-treated fish were recovered in their respectively scented streams. Controls were recovered in a number of different streams. To further study the response of these fish, the researchers captured some fish and fitted them with telemetry transmitters. Fish homed to scented streams as would be expected if they were using the artificial odors. All of these results, particularly when taken together, indicate clearly that the fish were homing to odors to which they initially had been imprinted.

The next case of fish migration involves eels of the genus *Anguilla*. They are **catadromous,** which means they live in rivers and breed in the ocean, the opposite of salmon. Their movements are only partially understood after nearly a century of investigation and recent debate. Worldwide there are thought to be 14 to 18 species of eels, mostly in the western Pacific. The one or two species in the north Atlantic, however, have been the subject of a continuing controversy.

From what is known of the species, adult eels migrate to deep (400 to 700 m) warm ocean waters, which overlie much deeper cold water. The breeding area for the Atlantic is thought to be the Sargasso Sea region (Figure 7-15), in or near the so-called Bermuda Triangle. The eggs are shed into the open, deep waters where they drift with the currents and hatch into flat, leaflike larvae called leptocephali

Figure 7-15 Eel spawning area in the Sargasso Sea region of the Atlantic and known or hypothesized movements to and from the area. For explanation, see text.

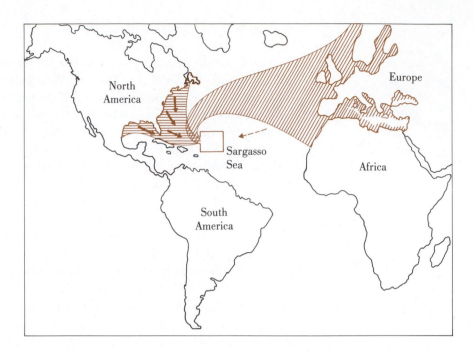

("small-headed"). They drift for 1 to 3 years or more, slowly growing, until they reach coastal regions, at which time they metamorphose into small eels called elvers or glass eels. They swim up into rivers where they spend several years growing and maturing and are called yellow eels, from their color. Then they take on a silver color and acquire the characteristics of a deep-sea fish, including enlarged eyes and different retinal pigments. They move back out to the ocean and are believed to return to their deep-water spawning areas.

Eels are found along the coastal rivers of both eastern North America and Europe. Early studies established that eels in the two areas have different growth rates and different numbers of vertebrae; therefore it was thought that they represented two different species. Initially it was not known where they spawned. Interest in the eels and their movements arose both from curiosity and from practical, economic concerns. Eels provide a considerable fishing resource. The catch along coasts of European nations in 1965 alone, for example, amounted to 17,000 metric tons.

A Danish biologist, J. Schmidt, began studying eels around the turn of the century, first with a tiny research vessel named Thor, then by hitching rides with commercial fishing ships. After being recognized as an outstanding authority on eels, he eventually obtained ships outfitted for systematic research. Part of the motivation for the work was concern that the species might be overharvested. Schmidt's research spanned more than 25 years. By following catches of progressively smaller and younger leptocephali, he tracked them backward to the Sargasso Sea. He also discovered that eels from both America and Europe were coming from the Sargasso Sea; apparently both species were spawning at the same time.

No adult eels were ever caught between the continental shelves and the Sargasso Sea, but based on the fairly complete picture from the leptocephali,

Schmidt hypothesized that there were two species, one North American and one European, which returned to the Sargasso to spawn. Schmidt published at least 21 papers on his research (cited in Harden Jones 1968).

Research on development of fish of other species, including trout and a blenny, however, have shown that growth rates and numbers of vertebrae may depend on water temperature at a critical period of embryonic development; that is, differences can be environmental rather than genetic. This led Tucker (1959) to propose that there was only one species of eel, with spawning stock coming from North America. He proposed that the eels found in Europe were simply from leptocephali that drifted there, taking longer and growing more slowly and with more vertebrae, but originating from American eels. Under this hypothesis the European adults did not return to the Sargasso Sea to breed.

A third possibility is that the two forms of Atlantic eels do represent a single species, as Tucker proposed, but adults from both sides of the Atlantic do return to the Sargasso. European adults could migrate directly across the Atlantic at great depths, backtrack over the route they had drifted earlier, or perhaps take some other indirect route. Eels are long-lived, and many things would seem possible.

Adult eels and eggs, however, have never been taken in the open ocean so the issue remains unsettled. A few other anomalies in the details of the story, reviewed in Harden Jones (1968), suggest that perhaps the Sargasso Sea is not the only location where Atlantic eels spawn. From the comfort of one's chair, in one's restricted surroundings, and faced with impressive amounts of material published by other people, it is easy to lose or not obtain in the first place a proper perspective on how large the world is. It can be very difficult to find things out there. Ships and airplanes are not the only objects that get lost in the Bermuda Triangle or anywhere else for that matter.

Amphibians

Several species of amphibians move seasonally between breeding and feeding areas. They are believed to rely primarily on familiar-area methods, although the sensory cues are not well known. The cues may be olfactory; however, much remains to be studied concerning the proximate factors in amphibian navigation. Female anurans may orient to and travel toward the breeding choruses of males.

Reptiles

As the diversity of reptiles waned after the Mesozoic Era and most groups and species became extinct, so did the story of their movements. It is unlikely that the extent or mechanisms of travel of such organisms as the pterosaurs, mesosaurs, mosasaurs, and dinosaurs, all animals capable of moving long distances, will ever be known.

A considerable body of knowledge does exist, however, for living reptiles. Among turtles, for example, there is some variability. The green turtle *(Chelonia mydas)* can home over 2000 km to Ascension Island, less than 10 km wide, in the middle of the Atlantic Ocean between South America and Africa. It is not known how they accomplish the feat, but it has been hypothesized (Carr 1965) that they use a combination of chemical cues in the ocean currents and visual cues from the sun's position. Other green turtles, as well as those of other species, home at egg-laying time to sites at warm coasts throughout the world and from which they themselves hatched.

The ability of young, newly hatched aquatic turtles hatched on land to find water is well known. Cues emanating from large bodies of water or patches of sky over such bodies have been hypothesized for guiding distant movements. Olfaction may play an important role. The familiar area is undoubtedly important on the local level.

Terrestrial reptiles do not move as far as marine reptiles, but some distances are still considerable. Snakes, for example, often travel to dens in the winter and then disperse to summer areas, with a subsequent return in fall. The basic mechanisms of terrestrial navigation in reptiles are likely to be based largely on the familiar area. The proximate, sensory factors are not well understood.

Birds

Many birds migrate between continents and return, often to the exact same nest site. The pied flycatcher, for example, breeds in northern Europe and winters in tropical Africa. Of 829 banding recoveries for that species during the breeding season, over half have been within 1 km of the nest where the birds were hatched (Berndt and Sternberg 1969). The tendency to return to the same area, "faithfulness to a place" or "site tenacity," frequently is referred to by the German term *Ortstreue.*

The best distance records are for arctic terns, which nest in the arctic and winter in the antarctic. An arctic tern banded in Russia was recovered 14,400 km (9000 miles) away. These birds are believed to make an annual round trip of over 35,000 km (22,000 miles). Not only the total length of trips but even the length of single nonstop flights by birds is impressive. A radio-equipped thrush was tracked 560 km (350 miles) in one night (Cochran et al. 1967). Golden plovers may leave Labrador, Canada, and fly 3800 km (2400 miles) nonstop to South America—on 2 ounces of fuel! The path of a flock of blue geese was observed and recorded by various commercial aircraft in 1952; the geese made a *nonstop* trip from James Bay, Canada, to Louisiana. They traveled 2700 km (1700 miles) in 60 hours (average 30 MPH) at an altitude of 900 to 2400 m (3000 to 8000 feet).

The altitude at which migrants fly is generally between 300 and 900 m (1000 and 3000 feet) but commonly goes up to 3300 m (11,000 feet), and large birds have been recorded up to 6100 m (20,000 feet) on overseas flights or over mountain passes. A human at those heights would have difficulty *walking,* even with added oxygen.

In addition to showing natural return behavior during migration, many birds can be displaced artificially, often by several hundred kilometers, and they successfully return. Homing pigeons are well known for these abilities, but similar displacements have been conducted with many other species as well. Present information on bird migration consists of a somewhat confusing and varied assortment of facts and hypotheses. One of the few conclusions that seems safe is that not all species navigate the same way.

Before plunging forward into the variety of information, however, some of the techniques used to study bird migration will be mentioned briefly. Because of their sizes, numbers, and differences in sensory and movement abilities relative to those of humans, birds are not particularly easy to follow. Early indications of their travels came from the familiar observations of the seasonal disappearances and returns of various species over many parts of the earth. Long-distance human travelers noted that certain species were seen at some times only in one region and at other times only in other, often far removed, regions. Banding (called ringing in England and Europe), color-marking, and other forms of individually identifying birds have produced a wealth of information concerning end points and occasionally intermediate points of journeys but not much information on actual paths, times of travel, and possible mechanisms. Solutions to some of these more difficult facets of the problem have been ingenious.

1. The direction and a few other aspects of migration have been studied with birds of some species held in circular cages. During the times when these birds would normally be flying, they show heightened activity or "migratory restlessness" (German, *Zugunruhe*) in the cages. Interestingly, the direction that they jump and attempt to escape from the cage often correlates with same direction they would otherwise be migrating. Because of this, external sensory cues, the birds internal daily rhythms, and several other factors may be experimentally manipulated and the subsequent effects on orientation recorded. The directions that the birds jump have been recorded, for later statistical analysis, visually, by treadles and electronic counters, and by the ink-pad technique (pp. 77-79).

2. Many facets of navigation have been investigated by transporting or displacing birds and then releasing them and determining the direction toward which they depart. Departure points have been plotted visually from the ground, with the aid of radio telemetry, which permits the birds to be followed beyond the horizon or nearby obstacles, or by tracking them via airplane. The data have generally been plotted on circular graphs (see Figure 7-18) for subsequent analysis and presentation.

3. Night migrants have been tracked with the aid of lights directed at the sky (ceilometers) and surveillance and tracking radar. Demong and Emlen (1978) prompted a proportion of experimental birds to initiate migratory flights by lifting them to the desired altitude by balloon; then he followed them with the tracking radar at Wallops Island (pp. 66 and 67). Birds have also been followed at night with radio telemetry and tracking from ground vehicles or aircraft.

4. The birds' senses have been impaired experimentally by cutting the appropriate sensory nerves, by placing the birds in artificial sensory environments, and by physically blocking the sensory organs at or near the surface. Nostrils and ears have been blocked, and the eyes have been covered with contact lenses in various experiments. Homing pigeons with frosted contact eye lenses, for example, have been able to return remarkably close to their lofts. Attempts to impair potential magnetic senses have included fitting the birds with magnetic bars and Helmholtz coils (discussed later).

5. One of the major techniques used in conjunction with the various methods listed here has been clock-shifting, as described earlier.

These and other techniques have been conducted under a variety of environmental conditions (such as night versus day, clear versus overcast skies, different seasons), with a wide variety of species, ages, and experiences of individuals, and with variable experimental designs and sample sizes. The variety of conditions, techniques, and factors has produced a volume of "data" that, until fairly recently, seemed rather bewildering. Patience, keen insight, and new experiments, however, have clarified much of the picture.

The genetic contribution for navigation to specific locations has not been well quantified in the analysis of variance sense, but it is clear that there is an important inherited component in many species. In some species the adults leave before the departure of the young of the year. Yet the young travel to the same area, a trip that obviously could not have been learned individually or made by traveling with experienced individuals. In European storks the direction that the birds take to travel around the Mediterranean Sea appears to be inherited. Young storks migrate independently from their parents. Storks in western Europe travel to the southwest and cross the Mediterranean Sea at the Strait of Gibraltar, whereas those further east move toward the southeast and cross on the eastern side. Young from eastern eggs moved to the west traveled in a southeasterly direction (Schüz 1971).

In many facets of avian migration and navigation there are also important roles for experience and learning. Learning commonly is involved in (1) the development of specific locations that are occupied at different times of the year and (2) how to navigate. In probably the majority of species the location of the initial home area is learned around the time when the young first begin flying. If they subsequently migrate or are artificially displaced, this is the area to which they will later return. If young are moved before they are capable of flight, as in the translocations and reintroductions of endangered species, the birds will later return to that new area. If the birds leave their natal area naturally on migration, they may fly directly to locations determined by inheritance (as in the arctic terns), accompany the adults (e.g., cranes), move in a general direction (e.g., storks), or more or less simply wander (e.g., bald eagles). After finding suitable wintering locations and even some particular temporary sites en route, these places apparently are remembered, and the birds then return at those times of the year in future years, thus establishing learned locations for different seasons.

Starlings in some parts of Europe have a tendency to migrate west in the fall and return to specific wintering locations. The difference between the inherited tendency to move west and the importance of learning a specific winter location was demonstrated by Perdeck (1958) by trapping and displacing birds of different ages during their fall movements. Immature birds simply continued westward from the new release sites, whereas adults corrected for the displacement and navigated to their normal specific wintering locations (Figure 7-16). (The tendency of European starlings to move west may, incidentally, be a partial reason for their successful introduction to the North American continent. Several small-scale introductions of starlings into North America were not successful. There were two known large-scale introductions, one in Portland, Oregon, in 1889, which was not successful, and another in New York City in 1890, which eventually led to the permanent establishment of the species in North America [Kessel 1953].)

We now have arrived at the most intriguing part of the whole problem of avian migration and navigation: regardless of the extent of learning involved, *how* do they manage to navigate so accurately? I will skim over many years and dollars worth of painstaking research and simply list the current information and conclusions. Several excellent reviews of the subject are available (Keeton 1974b).

Where possible, many species, such as gannets, appear to use landmarks and the familiar area (Figure 7-17). However, many species head straight home, often

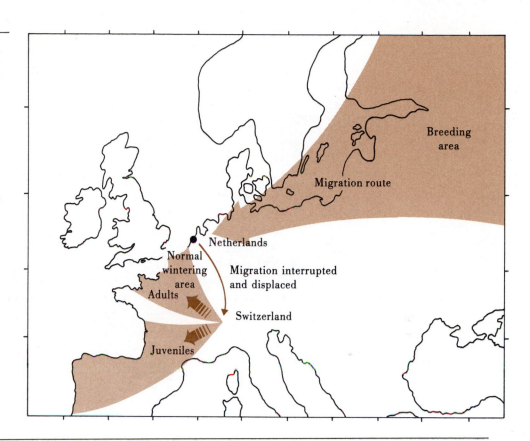

Figure 7-16 Interaction between innate orientation tendencies and experience as shown by displaced migrating starlings. Starlings from northern and eastern Europe migrate in a general southwesterly direction during the fall. Birds were captured in the Netherlands and transported to Switzerland where they were released. Banding recoveries depended on whether the birds were adults or juveniles: adults traveled northwest and returned to their normal wintering areas, whereas juveniles, including those which had been released in the company of adults, continued toward the southwest, including areas outside normal wintering range. Based on information and figures in Perdeck, A.C. 1958. Ardea 46:1-37.

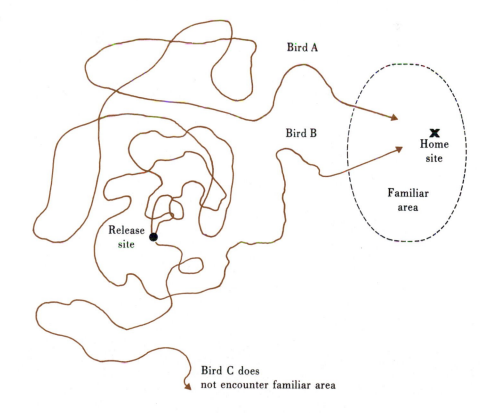

Figure 7-17 Wandering courses over increasingly larger area until familiar areas are encountered. This interpretation was suggested by following displaced gannets (*Morus bassanus*) with the aid of light aircraft at a distance. Based on information and figures from Griffin, D.R., and R.J. Hock. 1949. Ecology 30:176-198, and Griffin, D.R. 1955. Bird navigation. In Wolfson, A., editor. Recent studies in avian biology. University of Illinois Press, Urbana.

without visual landmarks. In many cases, particularly after experimental manipulation, birds appear to ignore landmarks and what should be familiar area.

Part of the story for many species involves one or more of various navigational *compasses*. Diurnal migrants and homing species (e.g., starlings, mallards, pigeons) can use the *sun* as a compass. That they use the sun at least partially has been demonstrated through the use of (1) clock-shifting, in which case the orientation is altered by the predicted angle, (2) by altering the apparent position of the sun with mirrors, with the birds orienting to the new mirrored position as would be predicted if they were using the sun, and (3) by keeping the birds under an artificial light source in a constant position, in which case the birds change their orientation by 15 degrees per hour, as if correcting for the normal movement of the sun.

Some species of nocturnal migrants have been shown to use *stars* in a compass manner. These species include European warblers, indigo buntings, and mallards. Evidence for mallards includes observations that birds released at night under a clear sky depart in a consistent, predicted direction, whereas those which are released under overcast skies depart more randomly (Figure 7-18) (Bellrose 1958).

Emlen (1975) has conducted perhaps the most thorough analysis of star-compass navigation using indigo buntings in conjunction with the *Zugunruhe* and stamp-pad technique discussed previously (Figure 3-18). By using naive young and artificial skies in a planetarium, Emlen showed that the star patterns (constellations) are important and that the young birds learn them at night as nestlings. The moon may cause disorientation, but the orientation to the stars is *not* affected by clock-shifting. The orientation is based on learning the position of the pole star among the other stars by the extent of rotation during the night. (Polaris is the current pole star in the northern hemisphere. The position of the pole star gradually changes over a 26,000-year period with the precession of the earth's rotation.) The general orientation of the birds, which depends on their

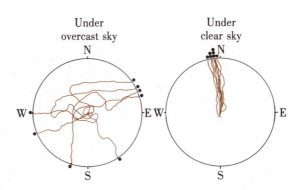

Figure 7-18 Paths of mallards released at night. Birds were tracked visually with the aid of small flashlights attached to their feet. Release point is at the center of the circle. Dots at the edge of the graph indicate the direction from the release point of each departing duck. Under overcast sky the ducks wandered in several directions and appeared disoriented. Under clear sky, however, all individuals departed directly and in a consistent direction. These results suggest that the birds needed and were using celestial cues. Some other species of birds are *not* disoriented under overcast sky. Modified from Bellrose, F.C. 1958. Bird Banding 29:75-90.

circannual rhythm and hormonal conditions, is *away* from the pole star, that is, south during the fall and *toward* it or north during the spring movements.

Without being able to see the stars, mallards and indigo buntings are confused, but European robins, swans, and many others are *not!* Also, pigeons trained to fly under such adverse conditions may home correctly under heavy overcast, with frosted contact eyeglasses, and at night. Some birds, at least under some combinations of conditions, are clearly using something more for navigation. The earth's magnetic field has long been suspected, but the topic has been beset with controversy, reminiscent of the Spallanzani bat-sonar story described in Chapter 1.

Evidence for the use of magnetism by birds for navigation includes the following:

1. Alternatives have been eliminated. Many birds orient correctly after all other apparent cues have been exhausted.
2. Caged birds of some species correctly orient without any visual cues.
3. Caged birds become disoriented in certain, imposed magnetic fields (Figure 7-19).
4. Experimental free-flying homing pigeons (without other cues such as the sun or landmarks) are disoriented with magnetic bars attached to them, whereas control birds under similar conditions but with brass bars attached are not disoriented (Figure 7-20).
5. Released pigeons that depart correctly with Helmholtz coil caps and a given battery polarity will depart in the *opposite* direction if the polarity of the battery connections is *switched* (Figure 7-21).

The next intriguing questions at this point are, *where* are the sensory detectors in the birds' bodies and *what* are they? Such organs were never previously identified or recognized, in spite of years of anatomical study of birds. There seems to be a similar state of knowledge as existed prior to Griffin's discovery of bat sonar.

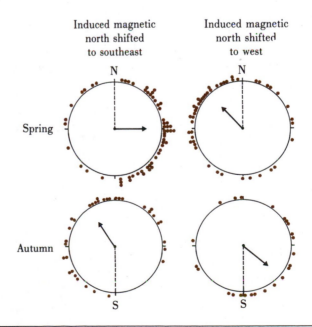

Figure 7-19 Example of induced, altered magnetic field effects on direction of migratory restlessness in caged birds. In this case European robins (Erithacus rubecula) were subjected to experimentally modified magnetic fields as indicated. Dashed lines indicate direction of normal migratory movements; solid arrows are mean direction of activity in altered fields. The birds' directions changed in response to the altered fields appropriately to the season (toward magnetic north in spring, away from magnetic north in autumn). Modified from Wiltschko, W. 1972. NASA Spec. Publ. 262:569-578.

Induced magnetic north shifted to southeast

Induced magnetic north shifted to west

Spring

Autumn

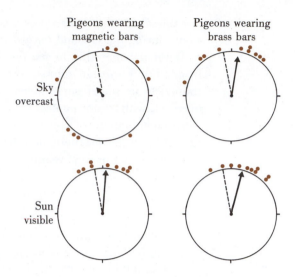

Figure 7-20 Effects of magnetic bars on homing pigeons. Arrows indicate mean direction of departure of pigeons from an unfamiliar release site. The length of the arrow indicates the consistency of departure direction among different birds. The dashed line represents the direction of the birds' loft from the release site. (Birds were released in several different locations; thus direction toward home loft is not any single compass heading.) Pigeons were disoriented and vanished in several directions if they wore a magnetic bar and were released under an overcast sky. Control birds with nonmagnetic brass bars or those released under a sunny sky (in which case the sun could be used for orientation) all departed in a homeward direction. Modified from Keeton, W.T. 1974. Sci. Am. 231(6):96-107.

Gould (1980) reported microscopic spicules of magnetite in a small region between the brain and skull of pigeons. Additional important findings on this topic should be forthcoming.

Some of the previous confusion in understanding bird navigation now appears to have been caused not only by the variability already discussed (among species, techniques, studies, etc.) but also by the fact that many species, particularly those with the most sophisticated abilities, possess much **redundancy** in their navigational systems. Many have more than one system, with the various systems serving as backups in case the others cannot be used. The presence of combinations and hierarchies of navigational cues has been demonstrated best in homing pigeons (Keeton 1974a,b). Experienced pigeons use the sun first if it is

Figure 7-21 Effects of electromagnetic coil (Helmholtz coil) caps worn on the heads of homing pigeons. Magnetic fields were induced by connecting the coils by wires to small mercury batteries worn on the pigeons' backs. The effect depended not only on whether or not the birds could see the sun (as in Figure 7-20) but also on the *direction of current flow and magnetic field in the coils*. Direction was reversed simply by reversing the connections to the battery. Details of the figure, such as arrow length, are as in Figure 7-20. Modified from Keeton, W.T. 1974. Sci. Am. 231(6):96-107. Based on material originally in Walcott, C., and R.P. Green. 1974. Science 184:180-182.

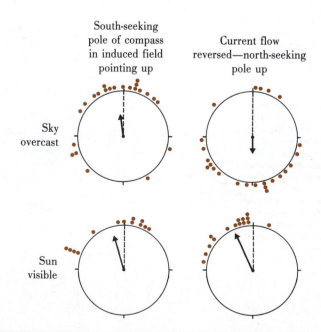

visible; clock-shifted birds will be off in the orientation under clear daytime skies. If the sun is not visible, however, then the experienced pigeons use the magnetic field, and clock-shifting will not alter the direction of their orientation under such conditions. Inexperienced homing pigeons need both the sun and magnetic fields for proper orientation. Experience permits the birds to use different systems alone or, in a slightly different interpretation, to get by with less information. Very inexperienced pigeons are unable to home at all.

This is still not the whole story: a compass by itself will not get one home. One must first know which direction home is; then the compass can be used to find that direction. Something else, such as a map or some kind coordinate system, is needed to determine the direction and distance between where one is and where one wants to go.

Matthews (1951a,b, 1955) proposed a ***sun-arc hypothesis*** for pigeon homing, which states that the coordinate information could be obtained from the path of the sun across the sky, resulting from the earth's rotation and inclination. This hypothesis has been discounted, but the means by which it was tested provides a valuable lesson in behavioral investigation. Therefore the details will be considered briefly. According to the sun-arc hypothesis (Figure 7-22), if one knew the time of day and the path of the sun at home or the location where one was headed, then one could figure where one was from the differences in the sun's path. For

Figure 7-22 Sun-arc hypothesis proposed by Matthews to explain bird homing. The bird is hypothesized to learn the path of the sun across the sky; then, if displaced as in homing experiments, it could compare the predicted path with an observed path and travel back home to correct the disparity. The concept is shown from two different angles to aid understanding. A, The view to an external observer, as originally diagramed by Matthews. B, View as might be seen by the bird. Open circle is the observed sun, and the solid circle is the predicted position. The observed sun, for the path it is taking, is too low and too far along in its arc. Therefore the bird must be north (for the sun to be lower) and east (for the sun to be ahead in its position) of home. To get home it should travel south and west. A from Matthews, G.V.T. 1955. Bird navigation. Cambridge University Press, Cambridge. B modified from Keeton, W.T. 1974. Sci. Am. 231(6):96-107.

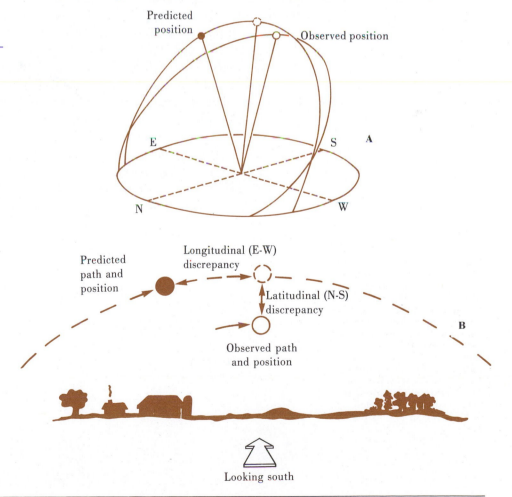

example, if one were in the northern hemisphere and one were located southwest of where one wanted to be, the sun would appear too high (i.e., in declination) because of one's being south and not far enough along its path (i.e., the azimuth would be too far east for that time) because one was west. To correct the situation and position oneself so the sun appeared in the right place, one would have to travel north and east the appropriate distances, hence moving to home or the desired location. This hypothesis suffered from two serious problems, however.

1. The apparent path of the sun changes only very slowly and would have to be traced for a period of time to determine its declination and azimuth. Released birds did not wait sufficiently long to have traced the path of the sun and obtain the necessary information before setting off in the proper direction.
2. Clock-shifting of birds to test the hypothesis resulted in an unexpected surprise: pigeons disappeared in the *opposite* direction to that predicted by the sun-arc hypothesis, as follows.

A hypothesis of Kramer's (1953) proposes that the sun is used only for compass information and that something else is serving as a map or coordinate system (the **map-compass hypothesis**). For example, consider an experiment in which a bird is delayed (i.e., clock-shifted back) by 6 hours and then moved south 100 km into an unfamiliar location and released at noon. Because its internal clock has been delayed 6 hours, the bird perceives its real home time as being 6:00 AM rather than noon. If the sun-arc hypothesis were correct, the bird would recognize that the sun is at the height of its path, a condition that would occur only much farther east. To get home it must travel six time zones to the west, a fourth of the distance around the earth! It also would notice that the sun is slightly too high because of the southward displacement, and it should go 100 km toward the north. Thus the pigeon would depart in a primarily western orientation. If, on the other hand, Kramer's map-compass hypothesis were correct, then the unknown map or coordinate system would somehow tell the pigeon that it was 100 km south of home and that it needs to travel north. The clock-shifted pigeon, on the altered basis that it is 6:00 AM, would determine north to be 90 degrees to the left of the sun. (The sun at "6:00 AM" should be in the east, and north would be at a 90-degree

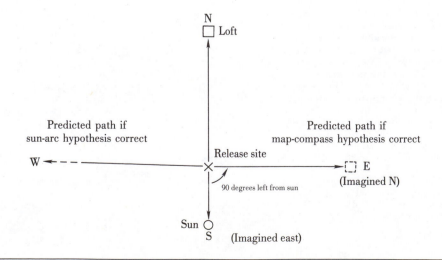

Figure 7-23 The critical test of the sun-arc versus the map-compass hypothesis. The birds were clock-shifted back 6 hours, as if they were living six time zones west, then released to the south of their loft at noon. Predicted paths under the two hypotheses are indicated. The clock-shifted birds actually traveled east. For further explanation, see text. Modified from Keeton, W.T. 1974. Adv. Study Behav. 5:47-132.

Figure 7-24 A few of the saw-whet owls netted during a migration study to determine proximate weather factors. These birds were being retained for banding, after which they were released back to the wild. Courtesy Lance Parthe and David Evans.

angle to the left.) Thus the bird should depart at a right angle to the left of sun. But the sun, rather than being in the east at 6:00 AM, is really in the south because it is really noon. The bird, in traveling at a right angle from the sun, would head east.

When the experiments were conducted, the pigeons headed east, as predicted by the map-compass hypothesis, instead of west, as was expected from the sun-arc hypothesis (Figure 7-23). For the moment, until better data or interpretations are obtained, the map-compass theory appears to be the best interpretation. However, the map-compass hypothesis has not been explained fully, and the physical basis of the map has not been demonstrated. Even this hypothesis should be held only tentatively. The real explanation may prove to be something not yet even hypothesized.

The proximate conditions under which birds initiate or continue migration probably have been investigated more thoroughly than in any other group of organisms. Because of the numerous factors involved in weather, multivariate analyses are preferred (Able 1974). In one such analysis of 1401 saw-whet owls netted in a banding operation (Figure 7-24) at Duluth, Minnesota, Evans (1980) showed that changes in migration volume correlated with windiness, barometric pressure, and temperature. In an extensive review of the literature Richardson (1978) concluded: "Causative and coincidental relationships remain difficult to separate, and at least a few birds migrate in almost any weather conditions. However, maximum numbers migrate with fair weather, with tailwinds and with temperature, pressure and humidity conditions that accompany tailwinds." Beyond this, specific conditions vary among different groups of birds such as waterfowl, shorebirds, and hawks.

Mammals

NOTE: Mammals are being considered last on the basis of our own mammal chauvanism; from a phylogenetic and geological standpoint, ancestral mammals arose before ancestral birds.

Most mammals probably home and navigate on the basis of familiar area, including those which move long distances, such as bighorn sheep, caribou, bison, and wildebeest. These animals usually travel in large herds, the young accompany the adults, and the familiar area probably involves a spatial memory that is passed by experience from one generation to the next as they travel together. The only species for which this is well documented, however, is bighorn sheep; if traditional routes are interrupted or stopped or young are translocated artificially, the long-distance travels cease.

Other mechanisms may exist in marine mammals such as cetaceans, pinnipeds, and polar bears; the mechanisms of their long-distance travels have not been researched to the extent of similar travels in birds. Natural navigational methods have not received much attention even for humans (excluding consideration of instruments). Experiments modeled after pigeon-homing studies have been conducted on humans (Baker 1980b, Gould and Able 1981) with mixed results. Those studies with the most controls and witnesses have not shown humans able to navigate without vision and familiar area (or instruments).

The lack of positive evidence for humans, however, may result in part from using subjects without extensive experience under conditions lacking visual cues. I would like to relate a rather remarkable personal anecdote. During 1970 while engaged in a peregrine falcon population research project with the Canadian Wildlife Service, a friend and I were traveling along the southwest coast of Baffin Island in a boat guided by an Eskimo. We came to a large bay, with Baffin Island on one side and the open Hudson straits on the other. We were returning to camp and, rather than waste time following the shoreline closely around the bay, we decided to save a few kilometers of travel by cutting across from one side of the bay to the other. As we were part way across, however, a very heavy fog developed. The fog became so dense that we barely could see the other end of the boat, much less where we were going. The sun could not be seen; there was no wind blowing for direction; and no waves could be heard breaking on the distant shoreline. We had several kilometers to travel, and inertial sense of direction seemed too inaccurate.

The Eskimo, however, proceeded forward at full speed. Two or three times we encountered seals coming to the surface for air, at which time the Eskimo would stop the motor and allow the boat to drift, hoping for a shot at the animal for food. He would light up a cigarette and perhaps shoot at the seal a time or two when it appeared. In the meantime we drifted. My friend and I could not tell how much or in what direction, but our movements in the boat were enough to cause it to change direction slightly. At one point the Eskimo shot one of the seals, paddled over to retrieve it, and hauled it over the side, all of which moved the boat considerably.

After each stop the eskimo would start the motor again, swing the boat around as if he knew exactly where we were and head off at full speed. After a long time the fog began to lift slightly, and the rocky point toward which we had been

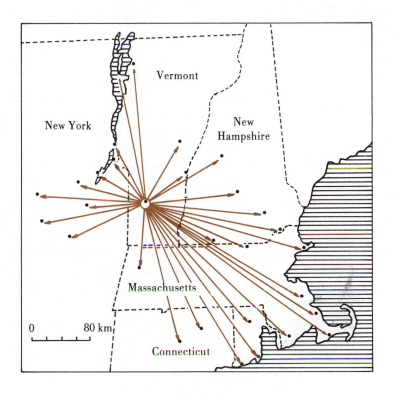

Figure 7-25 A few of the distant recoveries of little brown bats banded at a large hibernaculum in a mountain cave in Vermont. Bats were banded in the winter and recovered at other times of the year. In subsequent winters bats from this site would return to it. Note that to get to this location, bats from the southeast would have to travel *north* to spend the winter. In spite of being further north, however, the cave provided advantageous protection for hibernation. From Griffin, D.R. 1970. Migrations and homing of bats. In Wimsatt, W.A. 1970. Biology of bats. Academic Press, Inc. Based in part on results of Davis and Hitchcock (1965), who banded over 70,000 of these bats.

headed suddenly appeared before us! We were right on course. It did not occur to me to inquire at the time, so I never found out how that Eskimo knew where we were. But good navigation obviously is essential and adaptive in an environment frequently obscured by fog or snow and with wide expanses of featureless landscape or open ocean.

The most likely candidates for sophisticated navigational systems in mammals would be those capable of strong flight, the bats. They have been shown to home after artificial displacement up to a distance of around 500 km (Griffin 1970). A few species migrate in a north-south direction up to a distance of around 1200 km. But most travel only shorter distances, between seasons, to traditional protected hibernating sites *(hibernacula)*. An example of banding recoveries for little brown bats in the northeastern United States is shown in Figure 7-25. Bats may be affected by the earth's magnetic field in that they seem attracted to and hit radio towers under certain conditions of overcast and fog, similar to collisions by birds with towers. Bat navigation over long distances, however, has not yet received the extensive study given to birds, and the evidence is circumstantial at best. Further work is needed with individually tracked bats and using the diversity of approaches that have been applied to birds. The whole phenomenon of collisions with radio towers by both bats and birds needs further investigation. Are there differences in numbers of collisions when radio broadcasting is on or off? Are collisions caused by electromagnetic effects from large, grounded steel objects?

SUMMARY

First, many organisms do not migrate or navigate to specific locations. Many do not even have a specific home range or site to which they specifically return, unless by chance. Many species simply drift about and move only more or less at random within the limits of certain habitat types. A large proportion (majority?), however, do return and travel to specific locations. The most basic and probably most primitive mechanism of home range and travel to specific locations involves the familiar area, using some form of sensory contact and spatial memory.

A variety of forces or fields appears to be used by different organisms in the sense of a compass. Such means of orientation include celestrial cues (sun and stars), the geomagnetic forces, and water or wind currents. The compass senses probably operate in conjunction with some form of a "map" or "coordinate" sense, but the physical basis of the phenonemon is not yet understood.

Two potential cues have not been shown to be of much importance in animal orientation and navigation, except in a few isolated cases. The moon, which changes its position daily and would provide an "unreliable" predictor without further complicated adjustments, probably affects animals only indirectly, via the tides or by interfering with other systems such as star orientation. Inertial navigation has not been demonstrated to be of importance for long-distance travel in any organism.

Animals that routinely navigate or home over long distances, particularly some birds, have been shown to possess redundancy in their navigation. The different backup systems possessed by such animals generally operate in a hierarchy. Depending on the conditions and distance, parts of the hierarchy in homing pigeons are (1) familiar area, (2) sun compass, and (3) magnetic compass.

The topic of long-distance animal navigation is fascinating, still poorly understood, and receiving intense investigation. It is a field ripe for discoveries, and many should be forthcoming in the near future.

Recommended Additional Reading

Baker, R.R. 1978. The evolutionary ecology of animal migration. Holmes & Meier Publishers, Inc., New York.

Baker, R.R. 1980. The mystery of migration. The Viking Press, New York.

Carthy, J.D. 1956. Animal navigation. Charles Scribner's Sons, New York.

Galler, S.R., K. Schmidt-Koenig, G.J. Jacobs, and R.E. Belleville, editors. 1972. Animal orientation and navigation. Science and Technology Information Office, NASA Publication 262.

Gauthreaux, S.A., Jr., editor. 1980. Animal migration, orientation, and navigation. Academic Press, Inc., New York.

Harden Jones, F.R. 1968. Fish migration. Edward Arnold, Ltd., London.

Klopfer, P.H. 1969. Habitats and territories. Basic Books, Inc., Publishers, New York.

Klopfer, P.H., and J.P. Hailman. 1965. Habitat selection in birds. Adv. Study Behav. 1:279-303.

Lockley, R.M. 1967. Animal navigation. Hart Publishing Co., Inc., New York.

Matthews, G.V.T. 1955. Bird navigation. Cambridge University Press, Cambridge, England.

Schmidt-Koenig, K. 1975. Migration and homing in animals. Springer-Verlag New York, Inc., New York.

Schmidt-Koenig, K., and W.T. Keeton. 1978. Animal migration, navigation, and homing. Springer-Verlag New York, Inc., New York.

Storm, R.M. 1967. Animal orientation and navigation. Oregon State University Press, Corvallis, Ore.

Street, P. 1976. Animal migration and navigation. Charles Scribner's Sons, New York.

CHAPTER 8

Robert Gillmor/Bruce Coleman, Ltd.

A large rookery of gannets.

INTRODUCTION TO SOCIOBIOLOGY THE PHENOMENON OF GROUPING AND WHO'S WHO

Much of an animal's behavior involves interactions with other animals, particularly others of the same species. This brings us to a major subdiscipline in behavioral biology, *sociobiology*, which has been defined as the "systematic study of the biological basis of social behavior and the organization of societies in all kinds of organisms, including human beings" (Wilson 1980). The term *sociobiology* was used by C.F. Hockett in 1948, and a conference on sociobiology was held by the New York Academy of Sciences and New York Zoological Society in November of 1948. Sociobiology as a word was in occasional to common use by biologists during the 1950s and 1960s. The genetic concepts involved in sociobiology can be traced back at least to the 1930s. A number of papers and books dealing with the subject of social behavior were published: particularly important onés were by Hamilton (1964) and Williams (1966). But before 1975 there was no special significance or coherence attached to the topic per se. It was just one more topic in the field of behavioral biology.

The publication of *Sociobiology, The New Synthesis* by E.O. Wilson (1975), however, precipitated a surge of interest in social behavior. It fired the imagination of many behavioral scientists and quickly became a popular topic for graduate seminars and research projects. The topic began consuming many pages of journals, spilled into the public area, met some resistance among anthropologists and social scientists, created some controversy, and dismayed even some traditional biologists and ethologists. All the fuss, while attracting attention when it occurred, is not particularly unique in science (see Hull 1980). Sociobiology appears to have settled into a steady pace of research and output. All the controversy is by no means settled, but the public and the casually interested seem to have turned to other, more recent interests while the opponents in the sociobiological debates have dug in to hammer away at the topic. At first there seemed to be much confusion in the topic because of the infancy of the field, because evolutionary mechanisms (such as group and kin selection) were proposed that threatened the traditional concept of individual selection, and because it quickly became obvious that animal societies are *so diverse* that they defy classification. It appears that social behavior evolved independently in different groups of organisms and possibly for different reasons. Hence, social behavior is difficult to generalize. But the air has cleared somewhat on the subject, and it now is possible to obtain a reasonable, although still partial, picture of sociobiology.

ELEMENTS OF SOCIAL BEHAVIOR AND WHO'S WHO

Wilson's book was not the only synthesis of social behavior to occur in 1975. J.L. Brown, in the process of publishing a more general textbook, also provided an independent excellent review of the topic. Wilson and Brown presented somewhat different schemes of classification, and Wilson's scheme was incomplete; he presented different classifications for different taxonomic groups such as insects and mammals. In this book their schemes are merged and resynthesized with additional input and updating from other sources.

To begin with, *coincidental, passive aggregations* will not be considered as social behavior per se. Extrinsic factors, such as a concentrated food source or migration point, that bring animals together for the short term but without any lasting, evolutionary consequence may be interesting in other contexts but do not seem of much relevance to genuine social considerations. Such factors are included in the classifications of some persons (e.g., Emlen 1980), but they have been ignored here. The following, on the other hand, are partial indicators of a degree of true, intraspecific social behavior. One can measure how much sociality exists in the species by the prominence of each factor and by the combined number of the following elements that can be found in the organism's behavior:

1. The first and most obvious characteristic of social behavior involves the number of animals of the same species that actively come together or remain together in a group (Figure 8-1). The larger the number of animals in the group, the more social

Figure 8-1 Variation in numbers of animals involved in social groups. A, Badger, a relatively solitary mammal. B, Pair of bald eagles. During most of the year the birds are dispersed and can be found generally as pairs or with young of the year. During the winter, however, several bald eagles may be found congregated at feeding and night-roost locations. C, Moderate-sized social grouping of animals: a pack of African hunting dogs. D, Large group of animals: a flock of several thousand red-winged blackbirds. A courtesy Tom Brakefield. B by Irene Vandermolen, Leonard Rue Enterprises. C by Leonard Lee Rue III. D courtesy Ed Bry, North Dakota Game and Fish Department.

A

B

the species from the standpoint of this criterion. The minimum level of sociality would be two, such as in interactions between a male and female for mating purposes. This criterion, however, can be somewhat misleading and cannot be used alone. Many solitary species are nonetheless highly social in the sense that they have well-developed communication behavior when individuals do meet. Even "solitariness" may be a result of social behavior, such as territorial interactions, whereby the group is really an actively dispersed group. Some persons view active dispersion as social; others consider it antisocial, but it is largely a semantic matter. Some species are truly solitary and nonsocial in that individuals go about their business irrespective of what other individuals are doing. The number of animals in a group is an insufficient criterion by itself, and the following criteria also must be considered.

2. Social behavior depends in part on the *length of time or part of the life cycle that the group remains together*. For many species groups form only for brief, specialized functions, frequently involving reproduction. In other species the animals may remain in groups for longer periods, entailing several facets of their lives.

3. As opposed to the length of time that the members are simply in a group, that is, in physical proximity to each other, this element concerns the *amount of time or energy actually spent in social behavior*. Many animals, although physically grouped together, nonetheless may spend much of their time or energy in behavior that would be more properly classed as individual rather than as social behavior. Such behavior might include preening oneself or other toilet activities, eating, sleeping, etc. But in many highly social species even these types of behavior may acquire social overtones. Grooming (Figure 8-2) frequently serves a social, bond-

C

 D

Figure 8-2 Social grooming in primates. Courtesy Jean-Paul Ferrero.

ing function. In some species, particularly the social insects, even feeding involves communication and other social functions in the passing of food back and forth between individuals. A larger and larger proportion of the animal's time and energy becomes involved in truly social activities, and in such cases this has been considered as being more social.

4. *Reciprocal communication* is generally considered necessary as a mechanism for attracting and keeping the members of a group together. Thus a population differs from a society in that specific populations are bounded by a zone of sharply reduced gene flow, whereas particular social groups are bounded by sharply reduced communication. The majority of day-to-day communication occurs within the group, with much less communication between groups. But members may leave particular groups and enter other groups, commonly with subsequent mating, hence genetic exchange, so a given population may consist of several different social groups. The next chapter is devoted entirely to the topic of communication so further discussion will be postponed for the present.

5. Much social behavior is marked by a division of labor and *social structure or what is frequently referred to as roles*. The structure is extremely complex in some cases. In most systems, particularly mammals, the basis for the group is maternal, that is, the female and her young, but such structure is by no means universal.

The means of recognition of the structure varies. Recognition of role or one's place among invertebrates is generally based on *castes*. Animals recognize each other on the basis of their caste, frequently signaled via chemical releasers, that is, pheromones (Chapter 9). Social invertebrates may recognize their own colony or even their own kin (e.g., Buckle and Greenberg 1981). *Individuals* have not been thought to be recognized as such among invertebrates, although recent evidence indicates individual recognition indeed exists among some invertebrates. Caldwell (1979), for example, demonstrated individual recognition experimentally and clearly in mantis shrimps *(Gonodactylus festai)* by letting two individuals interact and establish dominance. Subordinates later were provided with and entered cavities containing clean water or water from aquariums with strange shrimps, but they refused to enter cavities with water from an animal that had defeated them. Johnson (1977) obtained similar results in another shrimp *(Stenopus hispidus)* by using mated pairs.

Recognition among vertebrates is more complex than in invertebrates. Vertebrates may respond to general stereotyped roles based on such factors as age, sex, possession of territories or resources, and body size. In addition, individual relationships and the ability to recognize individuals and previous experience with others are common and may become very important in vertebrate social interactions. The features used for individual recognition vary among species and groups. Humans recognize individuals primarily on the basis of facial features (with a specialized region of the brain devoted just to that function, Chapter 15), voice characteristics, and subtle behavioral mannerisms. Many mammals rely predominantly on olfactory or auditory cues. Birds that show individual recognition of mates, offspring, parents, or neighbors appear to depend largely on vocalizations and, to a lesser degree, visual appearances. Individual recognition in many vertebrates is based on olfaction. Combinations of visual, auditory, and olfactory cues may also be used. Both Johnson (1977) and Caldwell (1979) provide numerous references demonstrating individual recognition in vertebrates.

6. A feature of social behavior in many species is an *overlap of generations;* that is, families or parts of families may stay together. In the absence of this element, fertilized eggs are simply shed into the environment; the adult leaves or dies; and the offspring enter the world and have to fend for themselves. They never even meet or sense their parents. In other cases, with this element of sociality present, parents may remain with and in some cases defend the eggs or offspring for a part or all of the offspring's life cycle. Among the most highly developed situations, the group may include three or more generations that remain together throughout the entire year and for the animals' entire lives.

7. This last point represents the least common and, in the opinion of many biologists, the highest level of social behavior: *altruistic or aid-giving behavior where there is a cost to the altruistic individual.* In the most extreme forms it includes even the sacrifice of one's life and reproduction. The sacrifice of reproduction has caused the greatest theoretical problems and a reevaluation of the concept of individual selection. The obvious problem in sacrificing reproduction is that it would seem to prevent the spread of the genetic material. If there is a heritable component in the behavior, such behavior would seem to minimize or even eliminate itself by reducing the ability to reproduce. This topic has received much recent attention; it will be treated in detail later.

Using these characteristics as various measures of social behavior, the major taxonomic groups can be surveyed briefly for the incidence and extent of sociality. Beyond the basic interactions of mate and parent-offspring interactions, social behavior is relatively rare. Large groups of animals, such as a large wheeling flock of birds, may catch a person's attention, and the impression may be formed that such behavior is prominent and common. But on the whole, in frequency of species showing such behavior relative to all of the species that exist, the more advanced forms of social behavior are not common and certainly not prevalent. Wilson (1975b) identifies four "pinnacles" of social behavior. The information in Part III of his book is summarized in Figure 8-3 to illustrate

Figure 8-3 Who's who in the animal social world. A qualitative representation, not strictly to numerical scale, of proportions of animal species in different social conditions. (Invertebrates to vertebrates particularly are not represented to scale; vertebrates comprise only a small fraction of the world's total number of species.) Coincidental aggregations are excluded. **For explanation of details, see text.** Data from Wilson, E.O. 1975. Sociobiology, the new synthesis. Belknap/Harvard University Press, Cambridge, Mass.

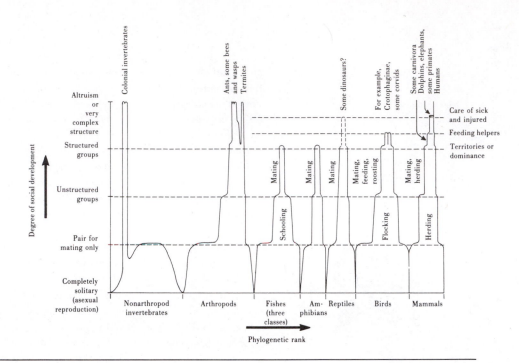

graphically the occurrence of social behavior among animal species. Wilson's four pinnacles all include at least some degree of altruism and very complex social structure. Following are the four pinnacles:

1. *Colonial lower invertebrates.* This is really a double pinnacle consisting of two groups, the bryozoans and, in the coelenterates, the siphonophores (Siphonophora). The siphonophores (such as the Portuguese man-of-war) are colonies of animals that appear to be single individuals composed of differentiated tissues. But instead of differentiated tissues, the parts are collections of highly specialized individual organisms that all exist together in one tightly knit colony. The individuals originate from budding or separate zygotes, then come together to form the colony. This is in contrast to the scyphozoans ("true jellyfish"), for example, which are other coelenterates in which the parts have a true embryonic origin. The difference can be seen, for example, in the coordination system. Scyphozoans have a nerve-net nervous system composed of individual cells arising from a common zygote. But in the siphonophores the coordination involves communication via sensory cells among the individual organisms.

   Although the distinction is real (if one does not press the word "real" too far) and the difference is one of organism versus superorganism, the boundary still seems a bit hazy conceptually. Budding of new individuals from a previous single individual and our lack of understanding about the earliest origins of differentiated tissues in complex organisms cause the lines to blur. Colonial invertebrates may have become social superorganisms to the point where the classification should drop them back to the point of being considered as individuals with an atypical origin. Take your choice.

2. *Colonial insects.* This pinnacle also has two major peaks, with one of the peaks consisting of several minor peaks. The peaks are the hymenoptera (ants, some bees and wasps) and the isoptera (termites). At least the major peaks probably evolved

independently, apparently involving different mechanisms, and some of the minor peaks among the hymenoptera also probably evolved independently but may have been spurred or permitted by a common underlying factor (haplodiploidy, to be discussed later).

3. *Nonhuman mammals.* This group includes dolphins, elephants, and some primates. Among mammals, particularly primates, carnivores, and ungulates, the most social in any particular taxonomic group are often (1) those with the largest body size or (2) forms that inhabit open grassland.

4. *Humans.* There seems little reason to separate humans from the other mammals unless the colonial lower invertebrates and colonial insects also are subdivided. The viewpoint that puts hymenoptera and isoptera on one pinnacle and humans and other mammals on separate pinnacles may simply be another expression of our basic *Homo*centric nature. But it does not make much difference; when constructing abstract entities, one is free to categorize as one chooses. Wilson did a big service by drawing attention to the most advanced forms of social behavior.

Placing the taxonomic groups in one comparison as in Figure 8-3 emphasizes a few other major points. Both unstructured and structured groupings are more common among vertebrates than among invertebrates. Birds have almost reached pinnacle status in a few instances. Examples of advanced forms of social behavior in birds include the Crotophaginae (a subfamily of cuckoos), with communal nesting and several females contributing to a clutch of eggs, and some of the Corvidae (crow family) in which other birds, usually relatives, help parents with the feeding of offspring. Some of the dinosaurs may have climbed quite high on the social ladder, although this probably will never be known for sure.

Structured groups, whether among invertebrates or vertebrates, are primarily concerned with reproduction. A particularly interesting form of structure for mating groups is arena or lek behavior, in which several males group together and display in one area for the attention of the females. (Arena behavior is discussed later in more detail.)

The literature is rich with examples of various degrees of social behavior among animals. For further details among the phylogenetic groups, see Wilson (1975b).

CAUSES OF GROUPING

Several major hypotheses have been proposed concerning the ultimate, ecological driving forces that lead to grouping and, subsequently, to social behavior. The temporary, short-term factors such as a brief concentration of food or a migratory point at which many individuals pass simultaneously are generally not of much consequence and, as already stated, will be largely ignored. With some species, such as vultures and some predators, however, concentrated food resources are the rule rather than the exception and may influence the species' social behavior. These situations will be included in the discussion that follows.

The hypothetical nature of these causes needs to be emphasized. Each of the hypotheses, alone or in combination with others, may or may not be true, and different factors are involved to different extents in the formation of groups in different organisms or even at different times in the same organisms. To reiterate,

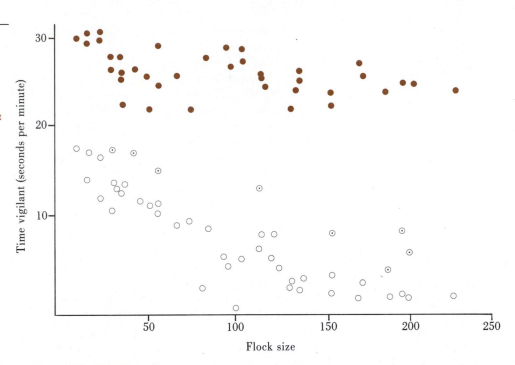

Figure 8-4 Vigilance time in relation to flock size and position in flock by starlings. Birds in each flock were categorized in three positions: central (○), midway (◉), or peripheral (●). Each point represents the mean time that birds in that position of a flock had their heads up and were watching for danger. From Jennings, T., and S.M. Evans. 1980. Anim. Behav. 28:634-635.

the tendency to group and the associated social behaviors most likely have evolved independently several times and for a variety of reasons. Empirical data and tests of these hypotheses are just now starting to accumulate. But, because many of these hypotheses overlap, many of the factors may be present together, as in the trained cricket. Furthermore, because of the usual difficulties in testing evolutionary hypotheses, firm conclusions are not likely for some time, and some of the competing ideas may never be resolved. The multivariate statistical techniques of principal component analyses and discriminant analyses may prove to be of some help in separating factors. The major present hypotheses are as follows:

1. *Antipredation:* safety in numbers. This aspect really is not a single hypothesis but a collection of closely related hypotheses.
   a. Improved detection of predators per se. With more eyes, ears, or whatever is being used to sense enemies, there is an increased chance that one or more individuals will detect a predator before the others and be able to warn the rest of the group.
   b. Detection by others permits more time for eating and other activities. Most animals have to remain alert to the possibility of predators and periodically interrupt other activities to survey their surroundings. Feeding birds, for example, can be seen frequently raising their heads and looking around in different directions. In the presence of a group with several watching, the frequency of surveying the environment by any one individual can be reduced. The advantages may vary depending on one's position in the group and vulnerability to predation (Figure 8-4). In the most advanced form, actual guard behavior is shown; one or a few animals assume the role of watching for the entire group, thus freeing the others almost completely.
   c. Cooperative defense against the predator. Although any one individual would be

Figure 8-5 Musk oxen defense ring against predators. Photo by L.L. Rue III.

helpless against a predator, several together may be capable of fending off an attack. Common examples of group defense include crows and many ungulates, with musk oxen defense (Figure 8-5) being particularly well known. Small passerine birds frequently "mob" a predator and even drive it from the vicinity. In advanced forms of social evolution, part of the structure includes individuals that are specialized for fighting and protecting the group. The most extreme cases are found in many of the highly social insects and in some human groups, with varying degrees of heritability and acquisition. Even in the "genetically programed" insects, whether or not an individual becomes a soldier is triggered importantly by its environment, largely via the diet.

d. Protection of young. This point is closely related to the previous one but differs in that the young only are defenseless by themselves, whereas for a variety of reasons such as increased body size, the adults are not normal prey or as easy prey and can protect the young. This hypothesis is reasonable as one of the more important factors in the evolution of parental care and a cause of family groupings.

e. Confusion of the predator. Many groups of animals may appear to be something different as a group from how they look as individuals. Or they may engage in a group behavior that creates a striking effect. Geese, for example, will clump into a tight flock in the presence of an attacking eagle and stir up the water, throwing it into a spray, and wave their wings simultaneously. The result is very effective; most eagles will break off the attack and leave. A group of animals may also confuse the predator by presenting so many choices that the predator cannot decide which one to take first and, in the meantime, the potential prey animals escape. Instead of choosing a particular prey item, some predators often just dash into the group. But even that frequently is not successful. The group may move out of the way in a coordinated fashion (Figure 8-6).

f. The "you-first" principle. If there are others around, there is less chance that any one individual will get taken by a predator; that is, chances are better that a neighbor will be captured first. In essence, one takes cover in the crowd. Hamilton (1971) introduced and discussed this concept as "an antithesis to the view that gregarious behavior is evolved through benefits to the population or species."

Figure 8-6 Movements of a coordinated group of animals in response to attack by a predator (at *dark arrow*). Similar responses are shown in many groups of animals under attack, including schools of fish, flocks of birds, and herds of ungulate mammals. The spreading of the group as the predator approaches has been described variously as "flash expansion" or "fountain effect." The fleeing animals then regroup and continue their escape as a tight school, flock, or herd. Modified from Partridge, B.L. 1982. Sci. Am. 246(6):114-123.

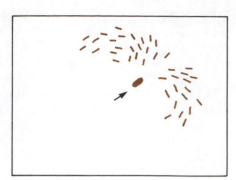
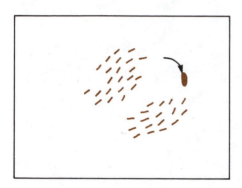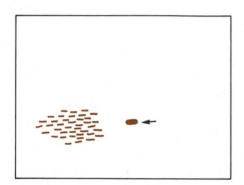

g. An extreme case of the previous point has been called the "cicada" principle: there are far too many to be consumed at one time. Cicadas are insects that spend most of their lives in the larval form underground, out of reach of most predators. When they emerge, the individuals are more or less synchronized and come above ground at about the same time. Although predators may get some of them, there are too many at once; the predators become satiated; and most of the cicadas survive. Many other species may group and show synchronized behavior for the same reason.

These factors cannot always be considered advantageous. Although grouping may thwart a predator in one case, it may actually encourage predation in another. Animals in a group may be more noticeable and easily detected or, because of the increased density, may encourage the predator to search more intensely for others. Under these conditions, dispersal rather than grouping would be advantageous from a predation standpoint. Although this may sound like confusing double-talk, it simply means that the outcome depends on the participants and the particular circumstances. There are no simple, universal principles that apply equally to all cases.

2. *Feeding efficiency and information sharing.* As with predation, this factor is not a single hypothesis but two.

a. Cooperative foraging. This in turn can be subdivided: (1) teamwork may help in the capture of prey, as in wolves, and (2) by meeting periodically it may be possible to exchange or obtain information of changing food sources, hence permitting moving and concentrated resources to be tracked through the environment. Ward (1965) proposed that night roosts of birds, in some cases, permit individuals to find where the food is located; birds that are unsuccessful in finding food one day may follow others to their food sources the next day.

Figure 8-7 Model of food and nest distribution proposed by H.S. Horn. *Dark squares,* **Food locations;** *circles,* **nests;** *lines,* **examples of a few of the paths to and from the nests for food;** *d,* **average distance.** Modified from Horn, H.S. 1968. Ecology 49:682-694.

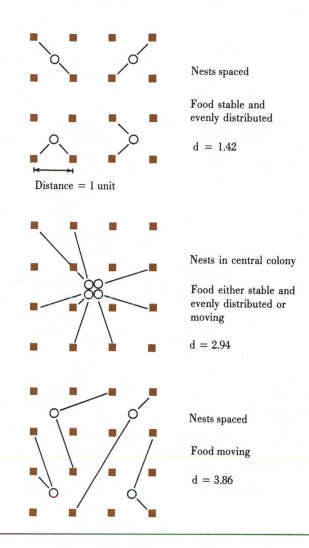

Nests spaced

Food stable and evenly distributed

d = 1.42

Distance = 1 unit

Nests in central colony

Food either stable and evenly distributed or moving

d = 2.94

Nests spaced

Food moving

d = 3.86

b. Efficiency and optimal foraging resulting from a centralized location. Horn (1968) proposed a model that central, grouped nesting is most efficient when food is irregular and moving, whereas spaced nesting is more efficient when the food is stable and dispersed more uniformly. Briefly, consider a simple case with 4 nests and 16 food units in 16 food locations. The locations are on a square grid and spaced 1 unit distance apart (Figure 8-7). Each nest requires 4 units of food. The average round-trip distance that must be made to obtain food under different conditions is shown. As can be easily seen, the distance is least for stable, equally distributed food if the nests are spaced. The animals make one trip to each of the four nearest locations. If, however, the food is all concentrated and moves from location to location so that the animals may have to move to any of the 16 locations to obtain their 4 units, then it is most efficient for all animals to be clumped together in the center.

3. *Facilitation of reproduction.* This point also consists of two separate considerations.

   a. It may be easier to find each other if individuals are not spread too far apart. In the extreme case the facilitation of reproduction may lead, theoretically, to very tight grouping as, perhaps, in the case of leks.

b. The presence of others reproducing may stimulate reproduction in a given pair. There is no reason a priori, however, for this to occur, and it is not obvious for the initial stages of its evolution or acquisition whether it would be a cause or simply a consequence. That is, the effect of sensing other pairs in reproductive activity could just as likely be inhibitory or completely neutral and without effect.

4. *Increased competitive ability*. A group of animals may be capable of overwhelming and driving others away from food, nest sites, or other resources. Packs of wild dogs in Africa, for example, are able to keep hyenas away from food.

5. *Increased probability of survival via population stability*. This is an elusive hypothesis that borders on group selection (Chapter 5). It proposes, in essence, that age-sex-role structure and such characteristics as territorial behavior may dampen population fluctuations. Excessive population fluctuation may easily lead to extinctions; therefore those groups which do not fluctuate so wildly are more likely to survive. It would be selectively advantageous for individuals or their genotypes to belong to stable groups.

6. *Division of labor and new solutions to environmental problems*. This hypothesis proposes that grouping makes possible the division of labor, which can lead to improved efficiency in all aspects of life. Rather than all individuals being equipped to do all things, specialization permits different individuals to become better at particular jobs

Figure 8-8 A living bridge formed of the bodies and legs of the weaver ant. From Hölldobler, B., and E.O. Wilson. 1977. Science 195:900-902. Copyright 1977 by American Association for the Advancement of Science.

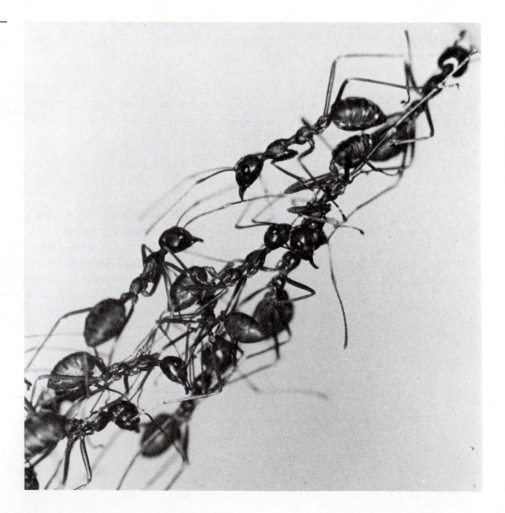

and then to share the outcome of their efforts. This is the familiar argument borrowed from division of labor in human societies. This facet of social behavior also permits entirely new outcomes, which could never be achieved by individuals working alone. Termite nests in some species, for example, are massive and microclimatically regulated for temperature, humidity, and CO_2 content (Figure 6-17). One nest is the product of millions of individuals working together in specialized capacities.

7. *Energetic efficiency.* The role of energy in animal groupings can be seen in two quite different examples.

   a. Thermal and other forms of conservation and efficiency. Voles in the arctic, honeybees, and others can reduce their surface to mass ratio by grouping together, thereby contributing to a common heat pool during cold periods. Slugs may group together to conserve moisture. In one of the more unusual forms of group efficiency, ants may form living bridges leading to shorter paths between locations (Figure 8-8).

   b. Efficiency in movement through a fluid. Because of the physics of moving through a viscous medium, a group may be able to move, particularly in some patterns of formation, with less total energy expenditure than if the same number were moving individually. This has been proposed for large birds, such as geese and other waterfowl, flying in V formation and for some fish swimming in schools. In the case of fish in schools, there may be a need for all individuals to be roughly the same size (although predation may contribute to uniform size in schools). There is some evidence that fish in schools can move faster partly because of heat conservation.

8. *Combinations of the preceding.* The various factors discussed previously are not always mutually exclusive. They probably combine in many cases to contribute to the overall formation of grouping; that is, when grouping may initially evolve and become established in a population for one reason or another, advantages also may accrue from other factors. This causes a form of positive feedback and strengthens the social bonding and behavior, but it also complicates attempts to understand the picture.

MAINTENANCE OF SOCIAL STRUCTURE AND SPACING

Recognition of one's place in the group, as well as that of others, appears to be fairly automatic and mechanical in invertebrates, based on various sensory features as discussed earlier. With vertebrates (and a few isolated cases in invertebrates), however, social position and spacing are determined and maintained largely through **agonistic behavior.** Because it is agonistic and may prevent further clumping and cooperation, this behavior is considered by many persons to be *antisocial* and disruptive rather than maintaining of social order; the matter is academic and can be debated either way.

Agonistic behavior includes aggression on the part of one (or more) animal(s) and the response on the part of another (or others) (Figure 8-9). The response generally is either to fight back or submit. Aggression may be actual and mortal or, more frequently, sublethal and ritualized. Sublethal aggression presents interesting theoretical problems such as what is the advantage of submissive behavior or why does the winner not kill the loser and avoid the risk of losing a later battle with the same individual? These problems will be returned to shortly. Agonistic behavior usually is expressed in one (or a combination) of two basic forms: territorial behavior and dominance hierarchies ("peck order").

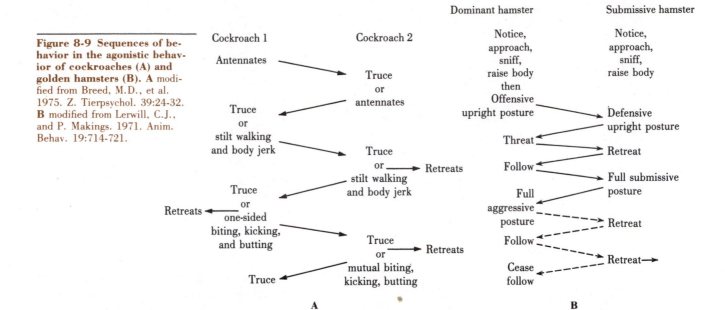

Figure 8-9 Sequences of behavior in the agonistic behavior of cockroaches (A) and golden hamsters (B). A modified from Breed, M.D., et al. 1975. Z. Tierpsychol. 39:24-32. **B** modified from Lerwill, C.J., and P. Makings. 1971. Anim. Behav. 19:714-721.

Territorial behavior is defined as aggressive behavior that permits individuals to maintain an area from which they exclude certain other individuals. There were several early ideas concerning the function of territories, depending in part on the type of territory involved (mating, feeding, roosting, etc.). Proposed functions included the following:

1. Assist pair formation
2. Maintain the pair bond and/or enable recognition of the mate by association with the territory
3. Reduce interference with copulation by others
4. Reduce the amount of time devoted to aggression
5. Permit improved defense of nest and young
6. Prevent epidemics
7. Limit the population density
8. Guarantee a food supply
9. Reduce losses to predators by establishing a familiar area (Metzgar 1967)

It now may be possible to generalize most (but not all) cases of territorial behavior to ecological situations with two characteristics:

1. There must be *limited resources* involved, such as food, nest sites, or mates.
2. These resources must be *defensible*. Spatially stable food, for example, can be enclosed within a fixed boundary, whereas food that is moving in water, such as a school of small fish, cannot be associated with a given, defended area.

Territories often are viewed (and measured) in a manner similar to home range, but as discussed in Chapter 7, the two are different. Territory involves active exclusion of some other animals. Evidence and measurement of territory require data on boundary agonistic behavior or boundary marking. Territories have been demonstrated in a large number of species (Figure 8-10).

Territories frequently pertain only to certain age and sex classes, particularly adult males, within groups. Territorial males, for example, may be confined to

Figure 8-10 Examples of territories in a variety of species and environmental conditions. A, Boundaries of communal group ranges of the bushy-crested jay in an area of rain forest and coffee plantations in Nicaragua. Area shown is approximately 700 by 900 m. B, Areas defended by four territorial fish, dwarf cichilds, in a 470 L aquarium. Markings at edge were used to identify locations. C, Hexagonal territories, the boundaries of which are visible in the sand substrate, in a tightly packed nesting group of *Tilapia* fish. A from Hardy, J.W. 1976. Wilson Bull. 88:96-120. B from Black, C.H., and R.H. Wiley. 1977. Z. Tierpsychol. 45:288-297. C from Barlow, G.W. 1974. Anim. Behav. 22:876-878. Photo courtesy George W. Barlow.

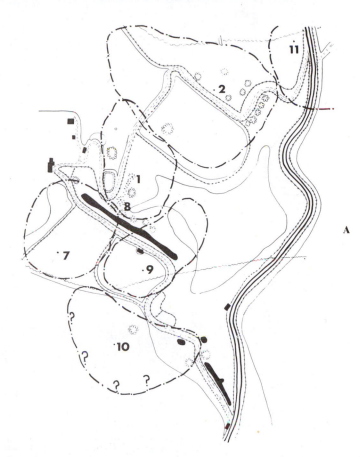

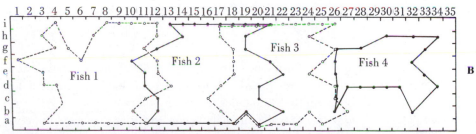

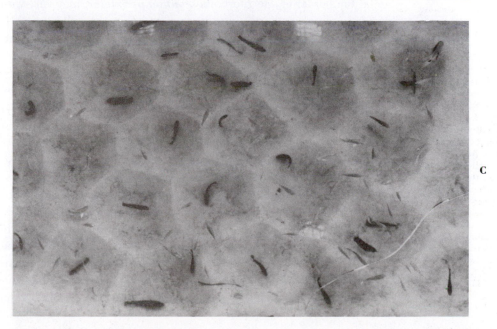

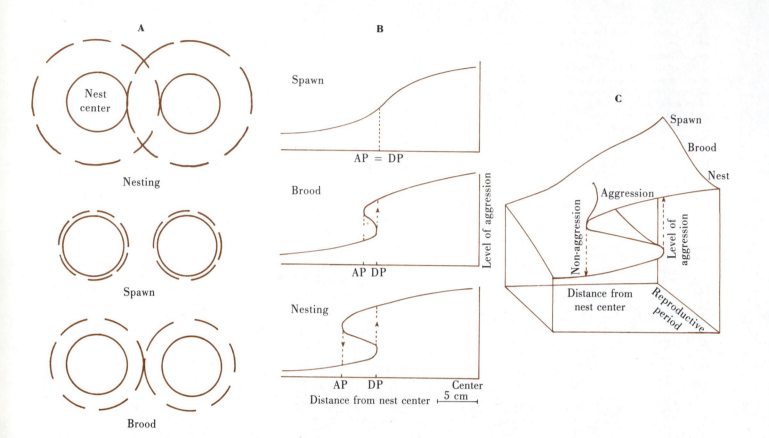

Figure 8-11 Changes in size of fighting contact perimeters and associated behavior in sunfish nesting territories. A, Perimeter and behavior depend on stage of reproduction and whether the intruder is approaching or withdrawing. The inner boundary marks the point at which a fish will fight an attacking intruder and the outer, the attack perimeter, is the distance to which a territory owner will fight a withdrawing intruder. During the nesting period the fish will not fight within the stippled "neutral" zone immediately between two adjoining territories, but they will between the two perimeters elsewhere. **B,** Changes in aggressive behavior with distance from nest and time of reproductive season. **C,** Mathematical catastrophe cusp model of the sunfish territorial behavior. From Colgan, P.W., et al. 1981. Anim. Behav. 29:433-442.

their own territories or surrounding "neutral" areas; nonterritorial adult males may be highly restricted, often to marginal habitat; and females and juveniles may be permitted to wander freely anywhere.

The points at which animals become willing to engage in actual physical contact at the border of a territory, the **contact perimeter,** may vary with season and also depend on whether an intruder is approaching or withdrawing (Figure 8-11). In the sunfish example shown in this figure, individuals may fight at a farther distance with a withdrawing intruder than one that is approaching. The shift between fighting and nonfighting behavior may occur suddenly and can be modeled, as shown here, with mathematical "catastrophe" theory. (For a simple, general introduction to catastrophe theory, see Woodcock and Davis 1978.)

Dominance hierarchies differ from territories in that the group is located together, with all members sharing the same physical area, but one or more individuals are behaviorally dominant over the other members. Resources (food, water, mates, etc.) go first to those at the top then, depending on the amount left over, on down to those beneath. If resources are insufficient for all, those group members at the bottom go without. There are several possible patterns of dominance relationships (Figure 8-12). Linear relationships are by far the most common. Measurement and evidence of a dominance hierarchy in damselfish are shown in Figure 8-13. The position of an animal in a linear dominance hierarchy often is identified by letters from the Greek alphabet. The top, most dominant individual, for example, is alpha; second is beta; third is gamma, and so forth.

Figure 8-12 Examples of dominance hierarchies. The linear form is most common and appears most stable. Other relationships frequently are temporary and may depend on the manner in which they form. In triads, for example, one individual may dominate two others, two may dominate one, or there may be circular arrangements in which *A* dominates *B* dominates *C* dominates *A*. Most of these lead eventually to more or less linear hierarchies.

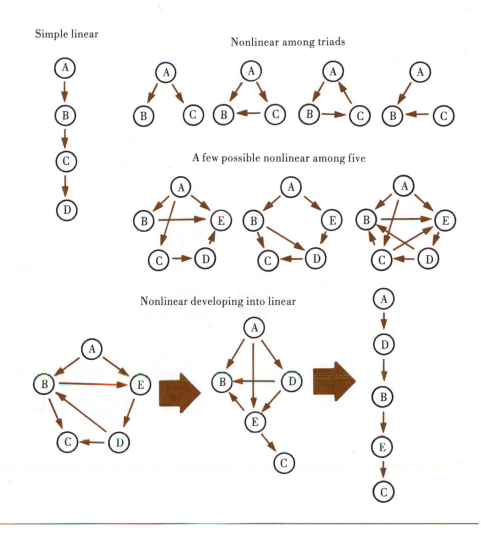

The two systems, territorial and dominance, occasionally merge in various ways. Some of the social behavior in cats, for example, involves dominance with a "spatial bias"; that is, in different cats' preferred resting areas, the cat preferring that area becomes dominant. In some situations of increasing population density a territorial system may break down and change to dominance hierarchies, and dominance systems may become more complex, with subgroups or complicated relationships among individuals.

The evolution of agonistic behavior, leading to territorial or dominance behavior, is not completely straightforward. The advantage of aggression, of having a good territory, or of being dominant would seem "obvious." The dominant individual should have first access to the resources, as has been shown to be the case in most studies. It is not clear, however, why the winners should not kill and simply eliminate the losers. In human social upheavals (where the extent of heritability is not known), winners frequently kill the losers and attempt to completely eliminate the chance of future trouble from those individuals.

What is the selective advantage of being submissive (Figure 8-14)? Several possible advantages have been proposed. If there is not enough of the resource to go around, the loser is going to lose either way, and there is no advantage to

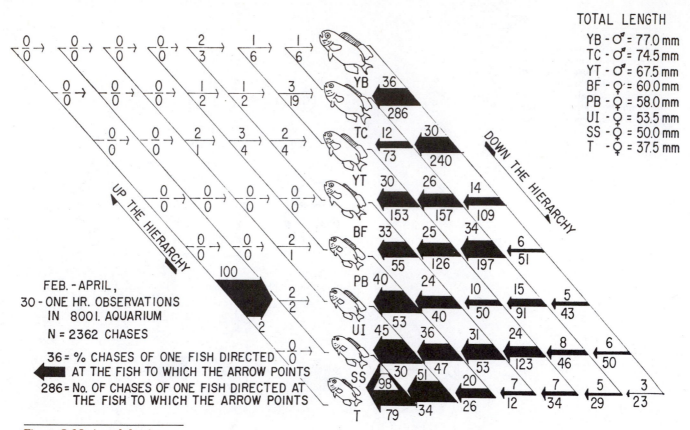

Figure 8-13 Actual dominance structure in a colony of bicolor damselfish. The linear sequence is shown along with an indication of who chases whom. In this case the hierarchy correlates strictly with body size, but this is not always the situation in other species and under other circumstances. From Myrberg, A.A., Jr. 1972. Behaviour 41:207-231.

wasting time, energy, and possibly life in useless fighting. It would be advantageous instead to conserve energy by not fighting. If there is enough, then the loser probably will get a turn later. The loser may be able to "sneak" in and get a bit of the resource, even after losing an encounter, or the loser may become dominant later. Both of these latter possibilities would be advantageous to the submissive individual, although seemingly disadvantageous to the dominant or territorial individual, as described earlier. Also, by being submissive and able to remain in the group, the individual may be able to obtain other advantages, such as protection from predators.

Figure 8-14 Normal versus submissive or frightened postures in the killfish. The lowered head and tail, arched-back submissive posture is common in many species of vertebrates. Also see Figure 9-14. Modified from Ewing, A.W., and V. Evans. 1973. Behaviour 46:264-278.

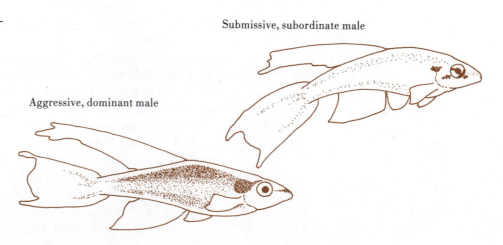

Submissive, subordinate male

Aggressive, dominant male

KIN SELECTION AND THE EVOLUTION OF SOCIAL BEHAVIOR

The discussion of individual advantages in agonistic behavior provides a good transition into the broader topic of the evolution of social behavior in general, including altruistic behavior. Many of the hypothesized functions and advantages of aggregations have already been discussed. But how do the finer details of interactions between individuals within the group come about, especially when the selective advantages would not seem to be equally distributed to all? First one must look at two terms pertaining to genetic "fitness," topics that were briefly introduced in Chapter 5.

1. *Individual fitness* refers to the reproductive ability of an individual (i.e., the ability to produce more individuals).
2. *Inclusive fitness*, on the other hand, places the emphasis on the genotype itself. This is the total fitness of not just the individual but also of other relatives that carry some of the same genes. To paraphrase an old saying ("A chicken is only an egg's way of making more eggs"), an organism is only DNAs way of making more DNA. It is the *genotype* that counts, not the individual, in this view. Hence, it may be possible under some circumstances that the individual is sacrificed for the sake of the total genotype.

During interactions between individuals in which resources may be gained or lost, with possible consequence to the genotype(s), there are four basic outcomes (see box below).

The theory of *kin selection* has been proposed to explain the possible evolution of these outcomes in terms of increases in inclusive rather than individual fitness. The amount of relationship between relatives can be measured by the *coefficient of relationship (r)* (not to be confused with other things that also are symbolized by the lowercase r). For siblings, on the average, $r = \frac{1}{2}$. For parents and offspring, $r = \frac{1}{2}$; for individuals and uncles or aunts, $r = \frac{1}{4}$; first cousins, $r = \frac{1}{8}$; etc. (Figure 8-15). If a particular act results in a gain for the genotype in one individual at the cost of an equal loss in the same genotype in a relative, the net advantage is zero, and the act should be of no evolutionary significance. If the overall ratio of gain to loss in inclusive fitness is negative, then the act should be selected against. But if the ratio (referred to as k) is positive and the gain outweighs the loss, then the act should possess a selective advantage (assuming it has a genetic basis). Except for identical twins, armadillo quadruplets, and similar cases, even the closest of relatives are not perfectly related, as indicated by r, and the ratio of gains to losses must compensate accordingly. In other words, a loss to an individual, if leading to a gain for a full sibling, must produce more than

---

Outcomes of Social Interactions Involving Resources

		Originator of an act	
		Gains	Loses
Recipient	Gains	Cooperation	Altruism
	Loses	Selfishness	Spite

Figure 8-15 Coefficient of relationship. Study the examples in this illustration to understand how the coefficient is calculated.

Father-son
(or other parent-child)
$L = 1, r = \frac{1}{2}$

Grandfather-grandson
(or other grandparent-grandchild)
$L = 2, r = (\frac{1}{2})^2 = \frac{1}{4}$

Half siblings
$L = 2, r = (\frac{1}{2})^2 = \frac{1}{4}$

Full siblings,
(e.g., brother-sister)
$L = 2$ twice, $r = (\frac{1}{2})^2 + (\frac{1}{2})^2 = \frac{1}{2}$

$r = \Sigma(\frac{1}{2})^L$ where L = number of generation links in the line of interest

● Individuals being considered

○ Other individuals involved in relationship

—— Links in the line of interest

--- Contributing links but not in the line of interest

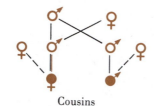

Uncle-nephew
$L = 2$ twice $+ 1$
$r = [(\frac{1}{2})^2 + (\frac{1}{2})^2](\frac{1}{2}) = \frac{1}{4}$

Cousins
$L = 2$ twice $+ 2$
$r = [(\frac{1}{2})^2 + (\frac{1}{2})^2] (\frac{1}{2})(\frac{1}{2}) = \frac{1}{8}$

twice as much of a gain to compensate for their sharing (on the average) only half of their genes. In simple, general mathematical terms:

$$k > 1/r$$

The loss in fitness to the originator of the act must be compensated by a gain of more than the reciprocal of the relationship in kin. Note that the gain in fitness must be among kin, but that kin do not have to be the direct recipients of the act—or the evolution of "spiteful" behavior cannot be accounted for and would require further explanation. For a simple, diagrammatic illustration of the principles, see Figure 8-16.

Kin selection conceivably could operate in any biological system. The more closely related individuals are, the smaller the reciprocal and the easier it would be to obtain gains in inclusive fitness. Kin selection in organisms that reproduce asexually would occur almost by definition, but asexual reproduction has certain other disadvantages such as relative (no pun intended) inflexibility (discussed in Chapter 10). The other extreme would be represented by normal, sexually reproducing organisms. An intermediate case, however, is found in hymenoptera (ants, bees, wasps), and this has been proposed as an explanation of the prevalence of highly developed social behavior in that group. The reproductive system of this group of insects involves *haplodiploidy*. The males are haploid; they develop asexually from unfertilized eggs (i.e., via parthenogenesis). Females arise from fertilized eggs and are thus diploid. Because the males, hence the fathers, are haploid, they have only one possible set of chromosomes to give, and the daughters all receive the same, identical genotype from their father. The contribution from a mother to the daughter is typical; either of the two diploid chromosomes from each pair can be passed on.

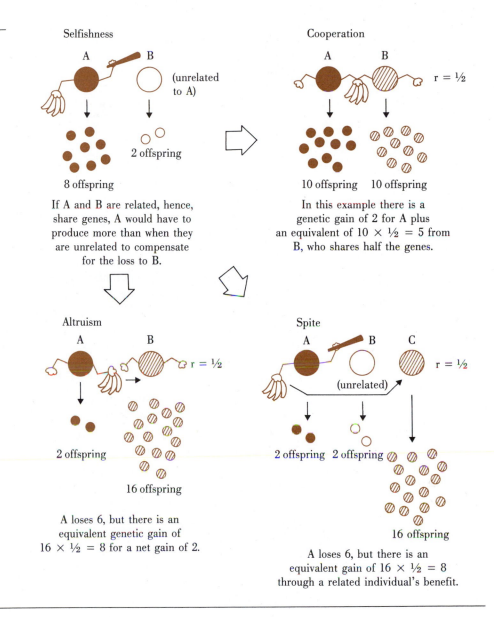

Figure 8-16 Hypothesized connection, via changes in gene frequencies, among various social relationships and kin selection. Shared genes are represented by degree of shading. Numbers of offspring indicate inclusive fitness. Bananas represent resources; sticks represent acts that reduce the fitness of another individual. Figure concept modified from Wilson, E.O. 1975. Sociobiology, the new synthesis. Belknap/Harvard University Press, Cambridge, Mass.

Selfishness

A   B
          (unrelated to A)

8 offspring   2 offspring

If A and B are related, hence, share genes, A would have to produce more than when they are unrelated to compensate for the loss to B.

Cooperation

A   B   r = ½

10 offspring   10 offspring

In this example there is a genetic gain of 2 for A plus an equivalent of 10 × ½ = 5 from B, who shares half the genes.

Altruism

A   B   r = ½

2 offspring

16 offspring

A loses 6, but there is an equivalent genetic gain of 16 × ½ = 8 for a net gain of 2.

Spite

A   B   C   r = ½
      (unrelated)

2 offspring 2 offspring

16 offspring

A loses 6, but there is an equivalent gain of 16 × ½ = 8 through a related individual's benefit.

The constancy of the genotype received from the father is what is believed to tip the scales in favor of kin selection. Siblings under normal diploidy are related to their parents by a coefficient of ½, the same as the coefficient with each other. But in the haplodiploid situation, assuming a single father, r between siblings (sisters) is ¾ on the average (see box on p. 246).

Thus consider two unrelated females striking off into the world on their own. One goes the normal route of reproduction and produces 100 of her own daughters, which, because of the mother-daughter r, results in what will be called (for the sake of simplicity) 50 new units of inclusive fitness. But the other female received a gene from her mother that leads to a form of reproductive behavior in which she does not mate and produce her own offspring but instead helps her mother produce 100 more young than the mother otherwise would have produced.

Possible Relationships at One Genetic Locus Between Sisters That Received Allele A from the Father and Either B or C from the Mother, as a Simple Illustration of the Principle*

	AB	AC
AB	1	$\frac{1}{2}$
AC	$\frac{1}{2}$	1

Average $= (1 + \frac{1}{2} + \frac{1}{2} + 1)/4 = \frac{3}{4} = 0.75$

*For a large number of chromosomes and loci the same principle applies. Hence, haplodiploid sisters have a coefficient of relationship of 0.75.

Because the average r for sisters under haplodiploidy $= 0.75$, the helping behavior of the sister produces 75 units of inclusive fitness. Both cases result in 100 offspring each, but in the helper case the genotype increases more than in the other. Thus the "help-mother-raise-your-sisters-rather-than-reproduce-yourself" genotype is selectively more fit and should increase in frequency, although some individuals are sacrificing individual fitness.

There are some nagging questions and a host of misunderstandings under the surface of all this. One problem is that the theory, even in its highly mathematical development, is overly simple by focusing on a few unique alleles that are shared by close relatives but not others in the population. In reality, however, *the total number of genes in an individual probably amounts to hundreds of thousands or millions, and a large proportion of these are shared widely among the population and entire species*, including all of those individuals that are not otherwise closely related kin. This would seem to pose potential problems for the theory of kin selection as currently envisioned. This and several other aspects of kin selection are discussed by Dawkins (1979).

Another question is partly semantic but also poses significant problems to the theory and the numbers that are plugged into particular variables: what is an individual? In a tightly knit, highly social group should all individuals be counted separately or should the whole group (i.e., superorganism) be regarded as the functional individual? Do nonreproductive individuals count the same as reproductive individuals, or, in the case of the helpers, are they merely contributing to the success, hence, fitness, of the one or few reproductives? Individual versus inclusive or gene fitness is not quite as clear cut as it may seem. One is dealing with a continuum from single-celled solitary animals to multicelled, highly colonial organisms (e.g., honeybees, army ants). There are many situations in between such as volvox (a collection of single cells), the Portuguese man-of-war (an organismic collection of different cells), and multicellular animals of various degrees of sociality. At all points on this continuum some units (whether individuals or colonies) survive and reproduce better than others. All that can be said for certain is that there are colonial hymenoptera with many different degrees of coloniality and there are noncolonial hymenoptera. Haplodiploidy may be the reason for the preponderance of sociality in hymenoptera. But at this point it is only one attractive hypothesis.

Haplodiploidy may help, but it certainly is not necessary. Kin selection could occur anytime that the gain in inclusive fitness is large enough to more than compensate for the fact that kin share only a fraction of their genotypes.

In addition to the theory of kin selection, there are alternate theories that attempt to explain the evolution of social behavior. Group selection, excluding the discounted Wynne-Edwards view, is a viable alternative, as briefly introduced in Chapter 5. A third theory is known as *reciprocity selection*, in which all participants gain from cooperative behavior, although they are not closely related. This amounts to something of "you scratch my back and I'll scratch yours."

Reciprocity theory has several pros and cons, but most of the arguments have been highly mathematical and are beyond the scope of this book. It is easy to see, heuristically, how reciprocal selection could favor cooperative behavior but not altruism or spite, where one or more individuals lose by definition.

Persons wanting detailed treatment of all three theories (kin selection, group selection, and reciprocal selection) should see Boorman and Levitt (1980). They suggest, incidentally, that *all three* selective mechanisms may be working in combination in many cases of sociality. In the hymenoptera situation it is obvious that helping behavior cannot be completely exclusive because at least one female has to produce her own offspring or else there would be nothing for the others to help! This raises the next major topic in the evolution of social behavior.

EVOLUTIONARILY STABLE STRATEGY

In many social groups one finds variation in the social behavior expressed by different individuals, such as between dominant and submissive behaviors. Many of these, such as dominance and submissiveness, may occur in the same individual at different times, ages, or under different circumstances.

The topic of optimality theory was introduced previously (p. 172). An animal searching for food in patches, for example, is expected to stay in a particular patch until the rate of finding food drops to the average success rate for the entire habitat. If the animal is alone or there is abundant food, optimality theory may work fine. If other animals are present, however, they also may be depleting the food supply, and the matter becomes more complicated. These and a number of other social considerations have led to a recent important subject in sociobiology: the **evolutionarily stable strategy** or **ESS**. By definition, an ESS is a behavior that, if more than a critical proportion of a population adopts it, continues to occur at the same frequency against alternate behaviors; that is, the frequency of its expression is not invaded by alternate strategies.

The phrase ESS is confusing or misleading on at least two counts. First, the word strategy has been borrowed, inappropriately in the opinions of some biologists (e.g., Louw 1979), from common human usage where it refers to conscious and deliberate planning and choice of actions. Animal behavior, however, is thought to only rarely if ever involve future planning and conscious choice, as discussed earlier. Rather we are talking about evolutionarily, ultimately determined options with perhaps proximate factors affecting the final expression. The behavior depends largely on the past and present and not on future planning; therefore the word "strategy" is a poor choice.

The second confusing aspect of the phrase is the meaning of stable. It is the single strategy that is stable, but not by itself—it is stable in a context of one or more alternatives. As Dawkins (1980) stated for a category of mixed ESSs (mixed is defined below), "The thing that is stable is the proportionate mix of strategies in the population as a whole." A given strategy, unless it has a monopoly in the population, achieves its stableness in a set of strategies.

The topic of the ESS will be further discussed below. Persons wishing to understand ESS better, however, should study (1) original introductions to the concept (e.g., Maynard Smith 1972), (2) discussions of proposed real-life examples (e.g., dung flies waiting for mates on cow pats, Parker 1978; digger wasps, Brockmann, Grafen, and Dawkins 1979), and (3) attempts to further explain ESS (e.g., Dawkins 1980). Many of the original papers have been collected together as readings; see the references at end of this chapter. Suffice it to say that there are many ways to misinterpret or disagree on ESSs, partly because of confusing terminology.

The basic theory, which can be traced back to Fisher (1930), is similar to the reason that one expects a 50:50 sex ratio. (For a good explanation of equal sex ratios, see Daly and Wilson [1978].) Different sexes or social behaviors may be expected to occur at particular frequencies. As soon as the frequencies change, the selective advantages and disadvantages change, and the subsequent frequencies are driven back to their stable levels.

A simple illustration of the principles of the ESS involves *hawk* versus *dove* behavior in a population. These labels were taken from human political terminology but were unfortunate choices because they also imply birds of the same name, and many students (to whom the simple illustration was directed in the first place) think in terms of two different species and become confused. We are talking about different behaviors present among members of the *same species* and, in fact, within the same population. Hence, different names for the behaviors, *aggressive* and *shy*, will be used here.

Maynard Smith (1976) introduces the basic concept by using the children's game "rock-scissors-paper." The game is played by two persons who each move a closed fist up and down three times in unison. On the third time they simultaneously make one of three figures with their hands. A closed fist is a rock; two fingers extended is scissors; and a flat hand represents paper. The winner is determined by rock blunting scissors, scissors cutting paper, and paper wrapping rock. If both players choose the same strategy, neither wins. For each win the loser forfeits one point to the winner. The payoff matrix (with payoffs to the player on the left) is:

	Rock	Scissors	Paper
Rock	0	+1	−1
Scissors	−1	0	+1
Paper	+1	−1	0

The contest is said to be symmetrical because each person has an equal chance of winning. A player could choose a *pure* strategy, that is, always play the same way. This would not last for long, however, because the other player would soon figure out what is going on and play the strategy that always wins. Thus one would expect a *mixed* strategy, where different strategies are played at different times and unpredictably. The contest is "zero-sum" because the three strategies have equal chances of winning when the other player's strategy cannot be predicted in advance, and the overall, long-term payoff should be zero.

The game can now be made a little more real, although still kept simple, and a population of social animals that are interacting with each other can be considered. To start there will be only two characteristics: aggressive and shy. Aggressive individuals fight and continue to do so until seriously injured or until the opponent withdraws. Shy individuals display until they encounter aggression, at which time they withdraw. When an aggressive individual (A) meets a shy individual (S), A always wins. When an A meets another A, they fight until one wins and the other loses, with each winning half the encounters. When an S meets another S, they display for a long time until one gives up, thus conceding the win to the other; again, in such encounters, each S has an equal chance and can be expected to win about half the encounters.

The payoffs now, however, are a little more complicated. One can arbitrarily assign 50 points for a win, 0 for a loss, -10 for wasting time over a long period of displaying, and -100 for getting injured. For two A's in a battle, the winner gets $+50$ for winning and the loser -100 for getting wounded; the average of these two scores is -25. This is the average payoff between two A's—they can be expected to lose in the long run. In the case of two S's, they both lose 10 points for the wasted time. Thus the winner gets $50 - 10 = 40$, and the loser gets $0 - 10 = -10$; the average of 40 and -10 is $30/2$, or 15. Under this system of symmetrical abilities and the assigned system of points, the shy individuals should stand to gain on the average, and a pure strategy of S where everyone is S should be a winning system and, hence, exist.

The problem is that in an encounter between A and S, A always wins. A gets 50 points, and S gets 0. The presence of one A in a population of S's gives the A a tremendous advantage. All of its encounters are with S's; it wins 50 points each time; and it stacks up points rapidly. With the presence of more A's, however, the advantage decreases. At the extreme, when all are A's as mentioned earlier, individuals will lose on the average, and the system should not last for long. One S in a population of A's will have the advantage because, although it always loses encounters and thus gets 0 points, this is better than being an A and having an average score close to -25. But somewhere between the two extremes, between 1 A in a population of S's and 1 S in a population of A's, there should be a point of equilibrium where the advantages of being A or S are equal. In the case of the arbitrary points used in the example earlier, the equilibrium comes when there are about 58% A's and 42% S's.

This equilibrium of mixed strategies should be stable. Any departure from it will confer a greater advantage to the other side, and the system should return to the equilibrium. If there is a genetic component in the two behaviors and the gene frequencies are susceptible to natural selection and evolution, then the "points" can be regarded as genetic fitness. If the payoff values are changed, the equilibrium points may shift to different proportions of strategies.

The example just described is, of course, rather oversimplified. There may be several other strategies that enter the picture. For example, there could be a *retaliator* that acts similar to an S unless it encounters an A, at which point it retaliates and fights back. Or there can be *bullies* that act similar to A's until pressed, at which point they withdraw. And there are possibilities for *cheating* and *deceit*, including self-deceit (e.g., always acting as an A although never winning). Furthermore, there are several possible asymmetries that must be considered. Asymmetries occur when the costs or benefits are not the same between different individuals. Some individuals may be stronger and thus more likely to win a fight that the opponent. Discrepancies in ability may arise from differences in body size, condition, age, sex, position in the social hierarchy, or possession of territory. Likewise, the benefits may vary. The reproductive value of winning a fight might not be nearly as much for an old individual that has already bred repeatedly as it would be for a younger animal trying to breed for the first time. All these things greatly complicate the calculations, to the point of needing a computer or a lot of paper and time. But the underlying principles remain the same.

The ESS concepts are briefly summarized and clarified as follows. A **pure** ESS is when all expressed behavior is the same, such as all "aggressive" or all "shy." **Mixed** ESSs occur when different strategies, such as aggressive and shy, occur at particular proportions in a population of animals. This may occur either when individual animals are always either aggressive or shy and the types of animals exist in particular frequencies or when particular animals show either strategy particular proportions of the time; that is, any one animal may be capable of displaying more than one strategy. In a mixed situation, however, the strategies occur at random and cannot be predicted in advance; that is, they are said to be probabilistic or stochastic. This is in contrast to **conditional** ESSs in which the outcome or expressed strategy may *depend* on something else, similar to the conditional "if-then" branch statement in a computer program. If an animal is large or on its own territory, for example, it should be aggressive, whereas if it is small or on someone else's territory, it should be shy. Strategies may be discrete, such as aggressive or shy, or they may involve continuums, such as waiting times. An animal can act immediately, for example, or wait for another animal to act first. The waiting time can be variable, and the strategy involves "how long to wait before acting."

Not all ESSs involve intraspecific aggression. The aggressive versus shy case is only one simple example. Other examples include different strategies of seeking mates (such as in good habitat where there is much competition versus in poor habitat where there is less competition), accomplishing copulation (e.g., holding territories as opposed to sneaking copulations off territories or on someone else's territory), obtaining food (catching one's own or pirating it from someone else), and obtaining nest sites (building one's own versus borrowing or stealing someone else's). Although all of these considerations do not necessarily involve aggression, all or most do revolve around competition for resources of some type. When competition for resources is not involved, the ESS concept may still apply (Dawkins 1980) but may not be necessary, and one can fall back on the concept of optimality theory instead.

The ESS concept has proven heuristically fruitful not only in several areas of intraspecific social behavior but also in interspecific competition and predation. ESS also appears to be of general use and has been applied in a variety of ecological and evolutionary contexts. The ESS might even underlie much of the long-term equilibria that can be seen in species that exist, apparently unchanged, for very long periods of geological time; that is, the ESS could be involved in the situation of punctuated equilibria (p. 125). If so, the ESS could represent the genuine "balance of nature." Dawkins (1980) has generalized the stable strategy concept even beyond evolutionary considerations to include developmentally and culturally stable strategies (DDSs and CSSs, respectively).

One ESS that has been tested experimentally is a conditional strategy: be aggressive if you already possess something and shy if someone else owns it (assuming no other asymmetry). It can be shown theoretically (e.g., Maynard Smith 1976) that this strategy is stable against any ESS that ignores ownership. It has been tested both in hamadryas baboons *(Papio hamadryas)* (Kummer 1971) and swallowtail butterflies *(Papilio zelicaon)* (Gilbert, personal communication in Maynard Smith 1976). In the baboon case ownership was over a female, and in the butterfly case the property was a territory at the top of a hill, the best location for obtaining females for mating. In both cases, prior owners are rarely challenged by other males. It could be that the owner is perceived as being stronger and that it would thus not be worth the fight for a challenger or, alternately, that the latecomer accepts the other's prior ownership. In the experiments one male or another was withheld (the baboons in cages and the butterflies in the dark) to allow the other to establish ownership. Subsequently, whichever male did not have ownership conceded to the one that did, indicating that the acceptance was not simply based on strength; that is, the direction of the ownership could be experimentally forced either way by manipulating ownership. It did not depend on any particular individual. And, as a further test, if circumstances were manipulated so that both males viewed themselves as the owner of the female or the hilltop, then when released together, a long, damaging battle ensued between the two.

In a spectacular, although inadvertent, case similar to the previous one, a trained female golden eagle with nest and young established ownership of a territory while a wild pair were gone from the area (Durden 1972). The problem developed despite deliberate attempts by the trainers to avoid such a situation. When the wild pair, previous owners of the territory, returned, a long drawn-out aerial battle ensued. The birds remained in the air for several hours and were followed in an airplane by the trainer, who tried to break up the fight. The fighting birds flew at times to a height of over 3 km (10,000 feet measured on the airplane altimeter). The same principle also is involved unknowingly in the inhumane sports practiced in some parts of the world in which two males of a species (chickens, dogs, horses, or others) are pitted against each other.

A note of caution, however, is necessary for these examples. There might be other interpretations that have not been explored. Also, there obviously are many species, such as impala *(Aepyceros melampus)*, where other males constantly challenge the territory owners. The ESS is a valuable concept, but it easily can become a tautology, a useless explanation of everything, if one only finds the right numbers so the data always fit the curve.

THE SOCIOBIOLOGICAL CONTROVERSY

Although sociobiology has seemed to some persons like a breath of fresh wind in biology, it has not been without its detractors. Some traditional biologists have grumbled because they did not see it as all that new; it is basically just good old-fashioned Darwinian biology dressed up in new clothes; and in some instances a few of the overenthusiastic proponents of sociobiology seemed to be overdressing. Some of the academic arguments have been over such things as group selection versus kin selection. But, by far, the loudest shouting has been over the application of sociobiology to human behavior.

Much of the problem over the acceptance of human evolution, starting with Darwin, has been associated with religious conflicts (see brief discussion in

Figure 8-17 Edward O. Wilson, author of *Sociobiology, The New Synthesis* (1975), which created a surge of interest in the biology of social behavior. Photo by Lilian Kemp.

Chapter 5). But there is more to the controversy than religious differences, including such things as nature-nurture, biological determinism, racial differences, intelligence testing, and predisposition to criminal behavior. One of the first notable flare-ups after Darwin was over Spencer's social Darwinism, or genetic determinism (Spencer 1892). Others, such as Allee (1943), considered the role of evolution in social behavior, ethics, and religion. This has continued to the present. Tiger and Fox (1971), anthropologists, asserted that in humans "identifiable propensities for behavior are in the wiring." But the publication of Wilson's book, *Sociobiology, the New Synthesis,* somehow triggered the latest eruption, and Wilson became the target of much hostility. The exchanges and emotions heated to the point where water was thrown at Wilson during a lecture before the American Association for the Advancement of Science. The academic rustling in the bushes became so loud that it was noticed even by passersby.

The most noted exchanges between the opponents of sociobiology and Wilson (Figure 8-17) in self-defense were in the *New York Review of Books* (Allen et al. 1975, Wilson 1975) and in *BioScience* (Allen et al. 1976, Wilson 1976). An excellent compilation of these and related documents is available in Caplan (1978), and an expanded, objective overview by a disinterested third-party philosopher of science is presented by Ruse (1979). For a more thorough level of discourse and debate, see Barlow and Silverberg (1980). Such exchanges are not unknown in science, and the situation in sociobiology is thus not unique (Hull 1980).

To begin with, in considering humans Wilson (1975) notes the considerable increase in human cranial capacity and the variability in human social behavior, and he clearly identifies the speculative and hypothetical nature of his statements (although his subsequent style of presentation sometimes seems less hypothetical). Several of these points are excerpted as follows, letting Wilson speak for himself. Highlighted emphasis throughout is mine:

At a distance a perceptive Martian zoologist would regard the globular head as a most significant clue to human biology. The cerebrum of *Homo* was expanded enormously during a relatively short span of evolutionary time [see Figure 8-18].

We have leaped forward in mental evolution in a way that continues to defy self-analysis. *The mental hypertrophy has distorted even the most basic primate social qualities into nearly unrecognizable forms.* Individual species of Old World monkeys and apes have notably plastic social organizations; man has extended the trend into a protean ethnicity. Monkeys and apes utilize behavioral scaling to adjust aggressive and sexual interactions; *in man the scales have become multidimensional, **culturally adjustable**, and almost endlessly subtle.* Bonding and the practices of reciprocal altruism are rudimentary in other primates; man has expanded them into great networks where individuals consciously alter roles from hour to hour as if changing masks.

It is the task of comparative sociobiology to trace these and other human qualities as closely as possible back through time. Besides adding perspective and perhaps offering some sense of philosophical ease, the exercise will help to identify the behaviors and rules by which individual human beings increase their Darwinian fitness through the manipulation of society. In a phrase, we are searching for the human biogram (Count 1958, Tiger and Fox 1971). One of the key questions, never far from the thinking of anthropologists and biologists who pursue real theory, is ***to what extent** the biogram represents an adaptation to*

Figure 8-18 Increase in brain volume of hominid species over evolutionary time. Reprinted with permission of Macmillan Publishing Co. from *The Ascent of Man* by D. Pilbeam, 1972, copyright © 1972 by David Pilbeam.

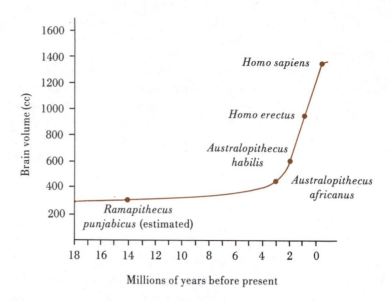

modern cultural life and to what extent it is a phylogenetic vestige. Our civilizations were jerrybuilt around the biogram. How have they been influenced by it? Conversely, how much flexibility is there in the biogram, and in which parameters particularly?

The first and most easily verifiable diagnostic trait is statistical in nature . . . [variability in social behavior]. [They] vary far more among human populations than among those of any other primate species. The variation exceeds even that occurring between the remaining primate species. Some increase in plasticity is to be expected. It represents the extrapolation of a trend toward variability already apparent in the baboons, chimpanzees, and other cercopithecoids. What is truly surprising, however, is the extreme to which it has been carried.

Why are human societies this flexible? Part of the reason is that the *members themselves vary so much* in behavior and achievement. . . . The hypothesis to consider, then, is that *genes promoting flexibility* in social behavior are strongly selected at the individual level. But note that variation in social organization is only a possible, not a necessary consequence of this process.*

Elsewhere in the book Wilson says that "the genes have given away most of their sovereignty," and later he stated that genes may account for perhaps 10% of social behavior. Wilson later (1980) clarified his views and provided further information and references on the genetic foundation of human social behavior.

Wilson's critics, however, in various degrees of emphasis accused him of advocating genetic determinism in human behavior. They have said, in so many words, that this justifies racism, sexism, imperialism, war, etc., and leads to acceptance of these characteristics as inevitable and even right. They accused

*Reprinted by permission of the publishers from *Sociobiology: the New Synthesis*, by Edward O. Wilson, Cambridge, Mass.: The Belknap Press of Harvard University Press, Copyright © 1975 by the President and Fellows of Harvard College.

Wilson of being politically motivated. Wade (1976), in assessing the situation, evaluated Wilson's response as follows:

The Wilson-Lewontin debate has every outward appearance of an illuminating battle between titans. Unfortunately the main issue is never joined, because Wilson denies that he says what the Sociobiology Study Group claims he says. The group has "utterly misrepresented the spirit and content" of the book, Wilson charges. They "cite piece by piece incorrectly, or out of context, and then add their own commentary to furnish me with a political attitude I do not have and the book with a general conclusion that is not there."

If one tries to cut through all the rhetoric, what are some of the more reasoned criticisms of sociobiology? Silverberg (1980), an anthropologist with extensive background in human behavior, correctly cautions that the whole picture of human behavior is far more complex than most biologists are prone to accept. He lists and discusses a large number of common pitfalls, a few of which include: the misuse of adaptation and selection as explanatory concepts, a careless view of phylogenetic relationships, minimizing intraspecific behavioral variation (including the downplay of human behavioral diversity), overgeneralization, taking what is familiar and widespread as universal, taking ideal behavior to be actual behavior, inferring biological significance from cultural trait distributions, an early Pleistocene behavioral fixation, using overly general and simple labels, and using evaluative labels (where behavior is evaluated against the observer's own behavior). As an example of minimizing variation and overgeneralization, a strong correlation often is drawn between polygyny and sexual dimorphism (both discussed in Chapter 10). But as outlined by Silverberg (1980:52), there is much variability and a very poor correlation between polygyny and dimorphism among the various great apes. Persons wanting further discussion should refer to Silverberg (1980).

Ruse (1979) takes a philosopher of science's tack to the subject by addressing the criticisms that sociobiology is incoherent, that it is unfalsifiable, and that it is false. (He ignores the charges that sociobiology is racist or sexist.) After treating each of these in detail, he concludes that "although [sociobiology's] advocates have made a start, much of their work remains in the realm of fascinating, unsupported hypothesis. It is, as the Scots say, 'not proven.' "

Criticisms that amplify Ruse's point and come from within the camp of evolutionary biologists (and from within the group of Boston critics, the "Sociobiology Study Group of Science for the People") have been ably expressed by Gould. Gould (1980b) has outlined the conceptual problems both in general and as they apply to humans:

When we examine the history of favored stories for any particular adaptation, we do not trace a tale of increasing truth as one story replaces the last, but rather a chronicle of shifting fads and fashions. When Newtonian mechanical explanations were riding high, G.G. Simpson wrote: "The problem of the pelycosaur dorsal fin . . . seems essentially solved by Romer's demonstration that the regression relationship of fin area to body volume is appropriate to the functioning of the fin as a temperature regulating mechanism." Simpson's firmness seems almost amusing since now—a mere 15 years later with behavioral stories in vogue— most paleontologists feel equally sure that the sail was primarily a device for sexual display. (Yes, I know the litany: It might have performed both functions. But this too is a story.)

Most work in sociobiology has been done in the mode of adaptive storytelling based upon the optimizing character and pervasive power of natural selection. As such, its weaknesses of methodology are those that have plagued so much of evolutionary theory for more than a century.

Sociobiologists have broadened their range of selective stories by invoking concepts of inclusive fitness and kin selection to solve (successfully I think) the vexatious problem of altruism—previously the greatest stumbling block to a Darwinian theory of social behavior.

Thus, kin selection has broadened the range of permissible stories, but it has not alleviated any methodological difficulties in the process of storytelling itself. . . . Behavior is generally more plastic and more difficult to specify and homologize than morphology. Sociobiologists are still telling speculative stories, still hitching without evidence to one potential star among many, still using mere consistency with natural selection as a criterion of acceptance.

David Barash (1976), for example, tells the following story about mountain bluebirds. (It is, by the way, a perfectly plausible story that may well be true. I only wish to criticize its assertion without evidence or test, using consistency with natural selection as the sole criterion for useful speculation.) Barash reasoned that a male bird might be more sensitive to intrusion of other males before eggs are laid than after (when he can be certain that his genes are inside). So Barash studied two nests, making three observations at 10-day intervals, the first before the eggs were laid, the last two after. For each period of observation, he mounted a stuffed male near the nest while the male occupant was out foraging. When the male returned he counted aggressive encounters with both model and female. At time one, males in both nests were aggressive toward the model and less, but still substantially, aggressive toward the female as well. At time two, after eggs had been laid, males were less aggressive to models and scarcely aggressive to females at all. At time three, males were still less aggressive toward models, and not aggressive at all toward females. Barash concludes that he has established consistency with natural selection and need do no more.

Consistent, yes. But what about the obvious alternative, dismissed without test in a line by Barash: male returns at times two and three, approaches the model a few times, encounters no reaction, mutters to himself the avian equivalent of "it's that damned stuffed bird again," and ceases to bother. And why not the evident test: expose a male to the model for the *first* time *after* the eggs are laid. [Even that test would not be conclusive; there are still alternatives. Another interpretation involves changes in hormones and behavior. See Chapter 16.]

We have been deluged in recent years with sociobiological stories. Some, like Barash's, are plausible, if unsupported. For many others, I can only confess my intuition of extreme unlikeliness, to say the least.*

For Gould's continuation of this argument (although presented chronologically before the above account) as applied to human behavior and noting his concerns about other, mostly nonscientists to misunderstanding and misusing sociobiological information, the following is a segment from Gould:

Human social behavior is riddled with altruism: it is also clearly adaptive. Is this not a prima facie argument for direct genetic control? My answer is definitely

*From Gould, S.J. 1980. Sociobiology and the theory of natural selection. In G.W. Barlow and J. Silverberg, editors. Sociobiology: beyond nature/nurture? Westview Press, Inc., Boulder, Colo. AAAS Pub. No. 35. Copyright 1980 by the American Association for the Advancement of Science.

"no," and I can best illustrate my claim by reporting an argument I had with a colleague, an eminent anthropologist.

My colleague insisted that the classic story of Eskimo on ice floes provides adequate proof for the existence of specific altruist genes maintained by kin selection. Apparently, among some Eskimo peoples, social units are arranged as family groups. If food resources dwindle and the family must move to survive, aged grandparents willingly remain behind (to die) rather than endanger the survival of the entire family by slowing an arduous and dangerous migration. Family groups with no altruist genes have succumbed to natural selection as migrations hindered by the old and sick lead to the death of entire families. Grandparents with altruist genes increase their own fitness by their sacrifice, for they insure the survival of close relatives sharing their genes.

The explanation by my colleague is plausible, to be sure, but scarcely conclusive since an eminently simple, nongenetic explanation also exists: there are no altruist genes at all, in fact, no important genetic differences among Eskimo families whatsoever. The sacrifice of grandparents is an adaptive, but nongenetic, cultural trait. Families with no tradition for sacrifice do not survive for many generations. In other families sacrifice is celebrated in song and story; aged grandparents who stay behind become the greatest heroes of the clan. Children are socialized from their earliest memories to the glory and honor of such sacrifice.

I cannot prove my scenario, any more than my colleague can demonstrate his. But in the current context of no evidence, they are at least equally plausible.

But why is this academic issue so delicate and explosive? There is no hard evidence for either position, and what difference does it make, for example, whether we conform because conformer genes have been selected or because our general genetic makeup permits conformity as one strategy among many?

The protracted and intense debate surrounding biological determinism has arisen as a function of its social and political message. . . . Biological determinism has always been used to defend existing social arrangements as biologically inevitable—from "for ye have the poor always with you" to nineteenth-century imperialism to modern sexism. Why else would a set of ideas so devoid of factual support gain such a consistently good press from established media throughout the centuries? This usage is quite out of the control of individual scientists who propose deterministic theories for a host of reasons, often benevolent.

I make no attribution of motive in Wilson's or anyone else's case. Neither do I reject determinism because I dislike its political usage. Scientific truth, as we understand it, must be our primary criterion. We live with several unpleasant biological truths, death being the most undeniable and ineluctable. If genetic determinism is true, we will learn to live with it as well. But I reiterate my statement that no evidence exists to support it, that the crude versions of past centuries have been conclusively disproved, and that its continued popularity is a function of social prejudice among those who benefit most from the status quo.*

In short summary of this section, the genetic role of altruistic behavior, with the concept of inclusive fitness providing the basic mechanism, seems not only plausible but almost without alternative for some of the invertebrate organisms with relatively simple nervous systems. Some genetic component (in the total

*From Gould, S.J. Biological potential vs biological determinism. Natural History, Vol. 85, No. 5. Copyright by the American Museum of Natural History, 1976.

variance, Chapter 4) also seems highly likely in higher vertebrate and human behavior. But evolutionary arguments can be tricky even for the simplest organisms and, in view of the extreme plasticity of human behavior and lack of data that permit the critical distinctions, the role of genetic input to human social behavior must still be considered an open question. It is not likely, based on general experience in science, that either of the extremes is correct. The human tendency toward war, etc., is not completely fixed by our genes, and one's fate is not sealed; nor is behavior 100% plastic, leaving one free to choose way(s) of life as oneself or some authority wills. Humans undoubtedly have *some* evolutionary baggage dragging us down. But just how much and what it means remain to be discovered.

SUMMARY

Sociobiology is concerned with the biological study of social behavior in animals. Criteria of socialness include the absolute numbers of animals in a group, the length of time they remain grouped, the amount of time or energy actually spent on social interactions, the extent of reciprocal communication, the presence of social structure within groups, overlap of generations in the group members, and the presence of altruistic behavior whereby an individual sacrifices some of its individual fitness for the benefit of another.

Social behavior, although quite noticeable in extreme form, is relatively undeveloped or absent among most species. The highest forms of social behavior are seen in a few colonial lower invertebrates, colonial insects in two orders (Hymenoptera and Isoptera), and a number of mammals, with some birds and perhaps some dinosaurs approaching similar peaks in social development. Several hypothetical causes of grouping are listed and briefly reviewed. Basic categories of hypotheses, with several subdivisions, include antipredation, improved efficiency in feeding and other activities, information exchange, facilitation of reproduction, competitive advantages, and combinations of these.

Territorial and dominance behaviors help maintain (or oppose, depending on one's point of view) social structure in some animals, such as many of the vertebrates.

Three theorized types of selection have been proposed for the evolution of social behavior: kin, group, and reciprocal. Kin selection is discussed most extensively. Haplodiploidy, where males are haploid and females are diploid, has been suggested as a special case of kin selection. Another important evolutionary topic relevant to social behavior is the evolutionarily stable strategy (ESS).

The subject of sociobiology has generated a considerable amount of controversy, which is centered primarily on the issue of the extent of genetic determinism in human social behavior. The controversy has quieted down but has not been resolved.

Recommended Additional Reading

Alexander, R.D., and D.W. Tinkle. 1981. Natural selection and social behavior: recent research and new theory. Chiron Press, New York.

Barlow, G.W., and J. Silverberg, editors. 1980. Sociobiology: beyond nature/nurture? American Association for the Advancement of Science Symposium 35. Westview Press, Inc., Boulder, Colo.

Boorman, S.A., and P.R. Levitt. 1980. The genetics of altruism. Academic Press, Inc., New York.

Brown, J.L. 1975. The evolution of behavior. W.W. Norton & Co., Inc., New York.

Caplan, A.L., editor. 1978. The sociobiology debate. Harper & Row Publishers, Inc., New York.

Clutton-Brock, T.H., and P.H. Harvey, editors. 1978. Readings in sociobiology. W.H. Freeman & Co., Publishers, San Francisco.

Dawkins, R. 1976. The selfish gene. Oxford University Press, New York.

Hunt, J.H., editor. 1980. Selected readings in sociobiology. McGraw-Hill Book Co., New York.

Krebs, J.R., and N.B. Davies, editors. 1978. Behavioural ecology, an evolutionary approach. Blackwell Scientific Publications, Ltd., London.

Ruse, M. 1979. Sociobiology: sense or nonsense. D. Reidel Publishing Co., Boston.

Wilson, E.O. 1975. Sociobiology, the new synthesis. Belknap/Harvard University Press, Cambridge, Mass.

Wittenberger, J.F. 1981. Animal social behavior. Duxbury Press, Boston.

CHAPTER 9

Ed Bry, North Dakota Game and Fish Department

Male ruffed grouse *(Bonasa umbellus)* drumming.

COMMUNICATION MODES, MECHANISMS, AND ECOLOGICAL CONSIDERATIONS

WHAT IS COMMUNICATION?

Communication is the passage of information (a representation of something) from one animal to another through messages or signals. Specific types of communication occur in various contexts, generally of a social nature and frequently in matters of or related to reproduction. Many signals are quite simple and direct, whereas others are extremely complex, involving complex interactions among animals. Visual signals, known as *displays*, often include specific movements and postures. Variations in the meanings of displays may be composed of movements that are quite striking and noticeable or very subtle (to a human observer). Basic behavior and, frequently, conflict behavior patterns are believed to be the major sources of movements that have evolved a communicative function.

The opening definition is overly simple, only partially satisfactory, and requires some discussion. This chapter will consider the general characteristics of communication, the problem of precise definition, methods of studying communication, transmission and sensory modes, and the evolution of communication.

A precise definition of communication is difficult and has received much discussion in the technical literature. Burghardt (1977), for example, stated:

Communication is usually treated as synonymous with social behavior. If we accept "communication" as equivalent to "social behavior," then we do not need the term "communication." However, "communication" is used to refer not so much to the behavior itself as to its signal function, its information content, and its reception and interpretation by other organisms. Most examples of communication are, however, inferred from intraspecific social interactions. . . . I have reviewed definitions of communication and rejected most of them.

He suggested an improved definition, which we will consider after discussing its background. Candland (1979) said of communication, "Some inquiry profits from exact definition, but communication exists in such a state of perplexity that premature definition is constraining."

Much of the problem results from the remarkable diversity and complexity of interactions among individual animals and groups of animals. As a result, there has been lack of consensus among biologists as to the *extents* that signals must be *intentional* (witting, conscious, etc.), *modulated*, *adaptive*, *verbal*, and so on before they qualify as communication. The main lines of disagreement have concerned, in addition to the basic problem of definition, (1) the functions of communication and (2) whether communication in other animals is substantially different from that in humans. Is the difference between humans and other animals qualitative or only a matter of degree (i.e., quantitative)?

The basic general characteristics of the common concept of communication include a **signal** (coded information or "message"), a **sender**, and a **receiver**. Other attributes include:

1. Both sender and receiver usually, but not always, belong to the same species.
2. The process is in *some way* adaptive to either or both sender and receiver.
3. The sender and receiver must possess the appropriate structures to respectively send and receive the message. This does not imply that the sending or receiving is conscious, voluntary, or even neural. Even in humans, where communication is least disputed, there may be facial or other expressions that are more or less involuntary, are universally understood—regardless of oral language—and may speak louder than accompanying words.

A case in point involves the messages carried in the appearance of some butterflies. For example, the butterfly may be a male monarch, a species that is poisonous and tastes bitter if eaten. The scales on the wings reflect light in different patterns at different wavelengths (colors). Thus the butterflies are equipped to send the messages. There may be several separate messages. One message in the range of light that is visible to vertebrates is a warning to potential predators; it "says" (which is short for "the meaning of the information carried in the message") that the butterfly tastes bad. There are different messages in ultraviolet wavelengths sensed by other butterflies concerning species and sexual identification.

The butterfly itself is unlikely to be aware of the messages in any case. But both types of receivers, vertebrates and other butterflies, can see their respective messages and act accordingly. The messages clearly are adaptive to the butterfly, the primary recipient of benefits in this case.

The message to the potential predator increases the butterfly's chances of survival. The message to other butterflies increases the sender's chances of reproduction. On the receiver's end, the message saves the predator from a bad meal and possible sickness. Or if the receiver is another butterfly, the meaning and value of the message depend on the species and sex of the receiver. The signals may enhance the chance of reproduction for a female monarch or reduce the chances of unnecessary interactions and wastes of time and energy for males or for butterflies of a different species. The monarch's messages are adaptive for all concerned, but the adaptive value varies: the predator warning is much more adaptive for the sender than for the receiver. For the sender his whole life is at stake; for the predator it is but one meal or perhaps just a bite of potential food versus a minor case of sickness. In other words, the messages evolved in the monarch primarily as a result of reduced predation and enhanced reproduction and not "to save the predator from getting ill" or "to save other species from wasting time attempting to reproduce." If the latter functions exist, they are adaptive to and evolved at the receiver's end. In other words, both sending and receiving species have *coevolved*.

	Observed Frequency of Behavioral Acts		
	Behavior 1	**Behavior 2**	**Behavior 3**
In absence of signal	300	500	2
After signal	100	500	200

Table 9-1 Alteration of behavior frequency in one individual as a result of a signal from another individual*

*Simplified hypothetical cases are given for illustration.

A commonly accepted operational definition of communication (e.g., Wilson 1975) is: *"Biological communication is the action on the part of one organism that alters the probability pattern of behavior in another organism in a fashion adaptive to either one or both of the participants."* In general mathematical formulation it is:

$P(x_2|x_1) \neq P(x_2)$ where

Individual	A	B
Behavior	x_1	x_2
Probability of act occurring	$P(x_1)$	$P(x_2)$

In words, the equation means that the probability of an act, x_2, occurring in individual B given that act x_1 has occurred in individual A is different from the probability of x_2 occurring in the absence of x_1. A would be the sender, B the receiver, x_1 the message, and x_2 the response.

For a simple hypothetical example, consider three different behaviors. Their frequencies of occurrence during a given time period are given in Table 9-1. Behaviors 1 and 3 appear to have been altered by the signal. Behavior 1 was decreased, and 3 was enhanced, whereas the frequency of behavior 2 was unchanged.

Before this definition could be applied, one would need a complete list of acts and frequencies or, more properly, good estimates of frequencies (Chapter 3). But a good quantitative approach has been accomplished in only a few instances. Two examples include mantis shrimps and Steller's jays (Tables 9-2 and 9-3).

It does not require much imagination to appreciate the magnitude of the job if one were to attempt to properly satisfy this definition for a more complete catalog of communication. Even those cases involving good quantification, as in Tables 9-2 and 9-3, are confined to only a portion of the intraspecific interactions. To pursue all the possible intraspecific messages and then move into those with other species and to do this for many individual species would entail an exponential, prohibitive increase in effort. Thus this definition may be operational, but it is not practical from a quantitative standpoint.

There are qualitative problems with this definition as well. For example, it is not clear how far one should extend into *interspecific interactions*. Many cases involve symbiotic or other relationships with plants, organisms that do not even have nervous systems. Many flowers, for example, have an appearance (message?) that attracts pollinating insects. Some plants have "messages" that repel insects (Figure 9-1).

Perhaps the most extensive considerations of interspecific interactions involve chemical signals and the resulting effects. A fairly complete classification for

Table 9-2 Comparison of frequencies of acts in one animal following acts by another animal: staged encounters between captive mantis shrimps*

Initial Act	Approach	Meral Spread	Lunge	Strike	Chase	Grasp	Coil	Avoid	Does Nothing	Total
					Following Act					
Meet	0(0.16)	3(1.9)	0(0.28)	5(1.8)	0(1.6)	0(0.08)	4(2.7)	2(3.5)	0(2.0)	14
Approach	0(0.46)	**15(5.7)**	0(0.81)	5(5.2)	1(4.7)	1(0.23)	10(7.9)	8(10)	1(5.9)	41
Meral spread	1(0.66)	10(8.1)	2(1.2)	3(7.5)	**0(6.8)**	0(0.33)	12(11)	**28(15)**	**3(8.5)**	59
Lunge	1(1.36)	2(4.4)	2(0.63)	0(4.1)	0(3.7)	0(0.18)	**17(6.1)**	9(7.9)	1(4.6)	32
Strike	1(0.73)	4(9.0)	0(1.3)	14(8.2)	**1(7.5)**	1(0.37)	13(12)	**28(16)**	**3(9.3)**	65
Chase	0	0	0	5	0	0	0	0	0	5
Grasp	0	1	0	−2	0	0	1	2	0	6
Coil	0(0.32)	6(3.9)	2(0.55)	6(3.5)	0(3.2)	0(0.16)	6(5.4)	8(6.9)	0(4.0)	28
Uncoil	0(0.24)	8(2.9)	1(0.41)	3(2.7)	1(2.4)	0(0.12)	0(4.0)	1(5.2)	7(3.0)	21
Avoid	1(0.95)	**0(12)**	0(1.7)	**2(11)**	**38(9.7)**	0(0.47)	**5(16)**	**2(21)**	**36(12)**	84
TOTAL	4	49	7	45	41	2	68	88	51	355
Expected	(1.1%)	(13.8%)	(2.0%)	(12.7%)	(11.5%)	(0.6%)	(19.2%)	(24.8%)	(14.4%)	

From Dingle, H.A. 1969. Anim. Behav. 17:561-575.
*Observed frequencies are shown to the left with expected frequencies in parentheses. This set of data is more complex than the hypothetical illustration in Table 9-1 in that it contains more behaviors, and the potential "signals" have not been identified beforehand. Thus all acts are considered for both participants, and expected frequencies are based on overall frequency of occurrence (marginal totals as a proportion of grand total). Acts in which following behavior was significantly altered (tested by chi-square) are in italics, and their significant data are in bold; these acts could be considered as "communication signals." Whether they increased or decreased can be determined by comparing the observed with the expected values.

Table 9-3 Inferred signals: comparison of behavioral acts of Steller's jays with the contexts in which they occur*

Context	Rattle	Musical	Growl	Wah	Too-leet	Shook	Total
		Number of Observations					
By supplanter	25	29	4	8	5	16	87
By supplantee	2	—	—	—	—	—	2
At individual but no supplanting	12	19	5	5	1	22	64
Aggressive sidling	—	2	6	2	7	32	49
In fight	—	—	—	1	—	2	3
After fight	1	—	—	1	—	—	2
Mobbing	1	—	—	31	—	—	32
Just alighted	—	—	—	26	—	2	28
On picnic table	—	—	—	28	—	—	28
Appeasement	—	—	—	1	—	—	1
Answering at a distance	—	—	—	—	—	2	2
Courtship	8	2	—	—	—	—	10
TOTAL	49	52	15	103	13	76	308
Context unspecified	78	88	2	131	17	77	393

From Brown, J.L. 1964. U. Calif. Publ. Zool. 60:223-328.
*This example lists the contexts in which Steller's jays gave various calls. Several of the vocalizations occur in several contexts with much overlap and variability. Thus only general inferences are possible. The rattle and musical calls are given mostly by one individual supplanting another or in courtship; the wah call is used mostly to maintain social contact, and the shook call is used by aggressive individuals toward others.

Figure 9-1 Coevolution of plant "messages" and butterflies. *Heliconius* butterfly larvae feed on the leaves of *Passiflora* vines. The larvae also are cannibalistic toward each other, which has led to female butterflies avoiding leaves with eggs already present for oviposition sites. Some *Passiflora* vine species have evolved leaves with fake egg spots, which results in butterflies avoiding these leaves. Should such "signals" of the plant be considered as "communication?" Modified from Gilbert, L.E. 1982. Sci. Am. 247(2):110-121.

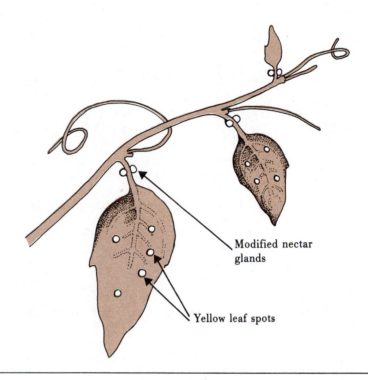

Modified nectar glands

Yellow leaf spots

chemical interactions has been proposed by Whittaker and Feeney (1971) (p. 266). Some of these go beyond what many consider to be within the domain of communication. One is faced with a continuum, and it is difficult to know where to draw the line.

The definition also includes other interactions that most persons would be reluctant to include with communication. For example, the sudden appearance of an attack by a predator will surely alter the behavior of its potential victim, but attacking behavior hardly qualifies as "communication." Similarly, consider the case of a resting animal that turns its head to watch another animal walk by. Its behavior has been slightly altered; it turned its head when it otherwise might not have. The passing animal might be of potential advantage or threat to the observer, or the observing may be satisfying a curious or exploratory state of the observer and, hence, the altered behavior could be viewed as adaptive. But the simple passing of another animal would not seem to be communication. It would be a form of incipient communication at best.

On the other hand, the definition may exclude some things that one would want to include as valid components of communication. What qualifies as an "act" or a "behavior"? The pattern on the butterfly's wing, even if the butterfly is dead, is clearly a message. Yet that and many other messages are morphological and not actions. Smith (1977) distinguishes between behavioral and nonbehavioral messages. Another problem is that the response on the part of the receiver may be extremely subtle. Many messages, for example, in human communication, can be received and recognized with no apparent change in outward appearance. Or there may be tremendously long time lags between the time the message is

Classification of Chemical Interactions Between Organisms*

Intraspecific Chemicals

Pheromones
  Releasers (affect behavior directly)
  Primers (affect physiology such as growth and development and may affect behavior
    indirectly)
Autoinhibitors (prevent others from settling or living too close)
Autotoxins and wastes (may be toxic, inhibitory, or without selective effect)

Interspecific Chemicals (Allelochemicals)

Allomones (favor emitter against receiver)
  Repellents
  Escape substances (as with ink clouds used by cephalopods)
  Suppressants
  Venoms
  Inductants (like primers among pheromones)
  Counteractants (neutralize chemicals from other)
  Attractants (baits or lures)
Depressants and wastes (selectively neutral to emitter but disadvantageous to receiver)
Kairomones (favor receiver, with or without advantage to emitter)
  Attractants (to food, to nest sites, for pollination, etc.)
  Inductants
  Alarm or warning signals
  Stimulants

Modified from Whittaker, R.H., and P.P. Feeney. 1971. Science 171:757-770.
*Whether or not these are considered as communication depends on one's viewpoint. If considered as messages or "information-carrying" molecules, they are known technically as semiochemicals. Names of categories most commonly accepted as communication are in italics.

received and the time the behavior is altered. Completely valid signals could be extremely difficult to detect quantitatively.

In addition to these considerations, one's definition of communication depends in part on the underlying hypothesis or perspective of the general function of communication in the first place. One view (Marler 1959, Smith 1977) emphasizes the **informational** aspects of communication and generally includes (depending on the situation) a degree of cooperation and mutual advantage between sender and receiver. According to this view, natural selection should improve accuracy and reduce ambiguity of the signals (e.g., Wilson 1975).

Animal signals, however, characteristically possess several attributes that do not fit well if one simply considers the mechanical transfer of information from one animal to another. Signals are not always directed, for example, at a particular receiver or, in many cases, toward any receivers of which the sender may be aware. There often is much redundancy and repetition of displays and calls, as anyone knows who has tried to sleep amidst the constant racket of songbirds in the spring or singing insects in the summer. Furthermore, some messages are considered (in some interpretations) to be "deceptive"; animals may withhold or exaggerate information. Animals may send false information, such as on the strength or intentions of the sender.

To accommodate these observations, the concepts and principles of game theory were incorporated (as in ESS, pp. 247-252). This view of communication places less stress on information transfer and considers communication to be more as **advertising,** whereby animals attempt through their "messages" to **manipulate** other animals to the sender's advantage (Dawkins and Krebs 1978). Whether or not there is any advantage to the receiver is irrelevant or secondary to the sender. If one were to define communication based on this emphasis, it would be: *Communication is the means by which one animal manipulates another with transmitted stimuli or messages.* This is the **manipulative** interpretation.

In consideration of the vast diversity of animal signals and their contexts, communication cannot be defined in the sole domain of either informational or manipulative views. Furthermore, they are not mutually exclusive. Hinde (1981) stressed that the traditional ethological view of animal communication is compatible with the game-theory view so that, although the identification with game theory might be new, the basic interpretation is not. Hamilton (1973) distinguished between *coordination displays,* which coordinate activities between individuals to the advantage of both sender and receiver, and *persuasion displays,* which are primarily to the advantage, interest, and benefit of the sender.

Discussions and quantitative analyses on whether emphasis should be placed on informational or manipulative interpretations (Caryl 1979, Hinde 1981) have revealed that communication may involve more than is embodied in either of the two simple viewpoints. Animals may not always be signaling clear, unambiguous information or intentions. Rather, they might indicate probabilistic, conditional behavior that subsequently depends on how the other animal reacts; that is, the signaler may not know what it will do itself but says, in essence, "I may do this or that if you do such and such." Thus one could construct a definition (or modify existing definitions) of communication to incorporate these **conditional** properties of the process. Even in the communication of nonhuman animals there appear to be elements of indecision, potential compromise, and possible diplomacy.

Is there a definition of communication, then, that accounts for all of the preceding considerations? Probably not. Perhaps the closest was given by Burghardt (1970b) as follows: "Communication . . . occurs when one organism emits a stimulus that, when responded to by another organism, confers some advantage (or the statistical probability of it) to the signaler or its group." This definition places the emphasis on the sender and neither requires nor excludes any advantage to the receiver. Organisms do not have to belong to the same species. The "response" does not have to be an observable movement on the part of the receiver; the message from a noxious potential prey animal, for example, might be "leave me alone," "do not move," or "go on about your other business." A major problem with this definition, however, is the difficulty of measuring "advantage" without simply inferring it in a circular manner.

One attribute of communication that has not been embodied in the definition, except as implied by "stimulus" or implied in the manipulative viewpoint, is that signals may not involve much force or energy themselves but can trigger or result in much expenditure of energy by the receiver. With the right signal one can move whole herds of elephants. It is the difference between pushing an elephant and telling it to move itself. In fact, this attribute of communication is considered to be one of its primary, general advantages; one can save vast amounts of energy or accomplish things that otherwise would be impossible by stimulating another to do it.

METHODS OF STUDYING COMMUNICATION

Most of the methods that have been used to study animal communication fall into one of two major approaches. In the first, basically exploratory approach, the animals are observed in natural or seminatural surroundings under a variety of interactions and conditions; events and behaviors occur naturally. Examples of this approach include the studies of mantis shrimps and Steller's jays cited earlier. From such information one can estimate frequencies of signals and behaviors, estimate probabilities of particular responses, and hypothesize about the identities and functions of particular signals.

In the second approach the researcher deliberately manipulates or experiments with the situation. The context, receiver, or the signal itself may be modified artificially. This approach generally is used to test hypotheses formed from the observations under more natural conditions. How the context or signal is modified depends on the modality, the species and situation, and the signal in question. The following list includes a variety of examples and illustrations of this general approach. All of these require imagination, ingenuity, and, when working with modalities such as chemical, auditory, or electrical fields, the technical ability to measure or modify the signals. Often the sensory systems of the animals themselves are used to measure the stimulus; one simply taps into the nerves coming from the animal's sense organ. The results of using an insect's antenna to monitor responses to various chemicals, for example, are known as "antennagrams."

- Isolate chemical signal extracts from the animals or their deposits to determine molecular composition or for presenting the chemicals in artificial, altered, or other controlled contexts.
- Create synthetic molecules for chemical signals. The molecules can be modified at different points, then presented to the animals to determine potency and which aspects of the molecule are most important.
- Use speakers for sound signals along with actual, artificial, or variously modified recordings. Emlem (1972), for example, modified the vocalizations of indigo buntings by cutting tape recordings and splicing the segments into new patterns. By presenting these modified songs to birds, he was able to isolate the critical aspects of the song.
- Use physical models for visual signals. Such models frequently have been remarkably simple, often consisting of little more than a piece of wire and yarn, a painted stick, or cut pieces of cardboard. The common use of simple models in early ethological research and the ready responses to them by many animals led to many of the early concepts of releasers (simple cues as the important triggers of behavior) and innate releasing mechanisms (discussed and illustrated in Chapter 2).
- Modify the appearance of the animal itself. In common flickers, for example, the male has a facial "mustache" marking that females do not have. Noble (1936) painted mustaches on females and covered the mustaches on males. This reversed the responses

from other flickers so that males were treated as females and vice versa. The mustache can be inferred to be a signal that identifies sex. Similarly, in numerous other instances, prominent markings of animals have been added, eliminated, exaggerated, or otherwise modified.

- Block the signal in either the sender or receiver. Covering or otherwise eliminating markings was just mentioned. At the receiving end, vision can be blocked with lenses, eye covers, or opaque screens. For chemical signals the openings or surfaces of glands can be shellacked or otherwise blocked; the receiver's sense organs can be blocked; or appropriate nerves in the receiver can be cut. Receivers can be deafened to block auditory signals. With appropriate combinations of techniques, some modalities can be blocked selectively. Animals can be placed, for example, in soundproof chambers with windows for vision.

MODES AND MECHANISMS OF COMMUNICATION

Signal Characteristics

The nature of the message depends on the physical source of energy and the manners in which it is produced or modulated. Animal communication may involve molecules, mechanical waves (sound), actual contact (i.e., touch or tactile), light, or electrical fields. What is called "information" or news about something is carried by a system through variation or changes of its state; that is, it is *modulated*. The simplest form of modulation is a simple binary, digital, or two-state condition such as on-off, yes-no, go-no go, presence-absence. A light, for example, can blink on and off. Examples of digital, or discrete, signals are shown in Figures 9-2 and 9-3. But this simple means of coding information is only one of

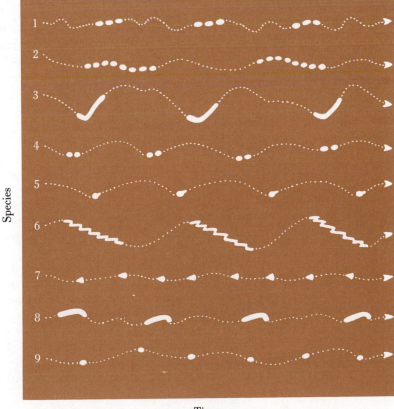

Figure 9-2 Discrete signals: firefly flash patterns in nine *Photinus* species. Some predatory fireflies, incidentally, mimic the signals of other species to attract and capture them (Chapter 11). Modified from Lloyd, J.E. 1966. Misc. Publ. Museum Zool. U. Mich. 130:1-95.

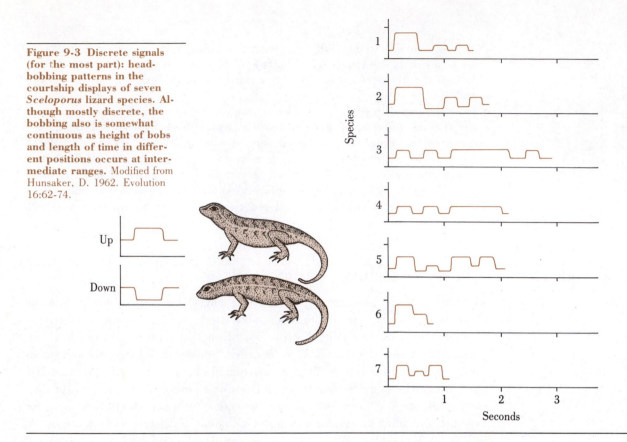

Figure 9-3 Discrete signals (for the most part): head-bobbing patterns in the courtship displays of seven *Sceloporus* lizard species. Although mostly discrete, the bobbing also is somewhat continuous as height of bobs and length of time in different positions occurs at intermediate ranges. Modified from Hunsaker, D. 1962. Evolution 16:62-74.

many ways by which signals may vary. The following is a list of the major ways by which signals, regardless of modality, are formed, modulated, and received in different ways by various animals:

1. *Qualitative differences.* These include different molecular configurations for chemicals, different colors for light, and different pitches for sound. These characteristics are affected by ecological factors and must account for "background noise" that could reduce signal effectiveness. Different fireflies, for example, emit yellow or green light, depending on whether the species is active at dusk or dark, respectively (Lall et al. 1980).

2. *Quantitative or intensity differences.* These include concentrations of chemicals, brightness of light, or loudness of sounds. The intensity of a message can occur in *discrete, digital* jumps as just discussed or by *gradual, graded, or analog modulation* (Figure 9-4). Some messages consist of combinations of digital and analog modulation (discussed further on pp. 276-279 and see Figures 9-11 and 9-12).

3. *Directional information.* Depending on the message, the situation, and the sensory capability of the receiver, the receiver may be able to determine the location or source of the signal.

4. *Patterns.* Patterns can be spatial, as in familiar complex arrangements of form and colors (e.g., the appearance of the butterfly wing) or a complex combination of chemicals or sound pitches. Or the pattern can be temporal with time sequences. Examples of temporal patterns include familiar sequences of sound as in songs or in human language. Visual examples include firefly flash patterns (Figure 9-2) or the head-boboing sequences performed by lizards (Figure 9-3). Pattern recognition by the receiver requires both advanced sensory capability to detect variation or changes and a well-developed nervous system to store and interpret the information.

Figure 9-4 Continuous signals: crest raising in Steller's jay. The crest can be raised or lowered over the entire range of movement. Meaning of the signal at different angles is indicated in the diagram. Modified from Brown, J.L. 1964. U. Calif. Publ. Zool. 60:223-328.

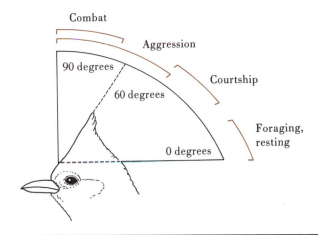

Messages may be modified not only by different, simultaneous channels from the sender but also by the environmental context. The same message can mean different things under different circumstances. The molecular "queen substance" in bees, for example, can lead to the development of new queen larvae if no queen is present, keep workers from rearing new queens if a queen is present, keep workers sterile, or serve as a sex attractant. The songs of male birds may mean one thing to a female and something quite different to another male. The same songs may have different meanings at different seasons. The same vocalizations that lead to territorial behavior and aggression and repel male red-winged blackbirds during the nesting season, for example, serve to *attract* males together outside the breeding season (Brenowitz 1981).

Comparison of Major Communication Modes and Channels

Chemical Communication

Molecular messages generally fall into two main classes: ***releasers***, which affect another's *behavior* (such as alarm and trail substances), and ***primers***, which affect another's *physiology* (such as queen substance in bees and odors in the urine of male mice that cause abortion of fetuses in females impregnated by other males—the Bruce effect).

Chemical messages, depending on how far one wishes to extend the definition of communication (p. 266), generally are called *pheromones*. They may well have been the first signals used by primitive organisms. They may be ancestors or descendent substances of internal hormones—at least they seem very closely related in modes of production and action and differ mostly in whether they carry the information internally or from animal to animal. Chemicals probably are the most universal of communication signals. Even animals that rely heavily on other modes also may use molecular messages.

The advantages of chemical communication are that it transmits through darkness and around obstacles; it has the greatest potential range (at least in air—often carrying for several kilometers); and it may be stable and last a long time. In the latter case, chemical communication transmits into the future, and its information does not depend on the continued physical presence of the sender.

The disadvantages of molecules, depending on the circumstances, are that they are slow to transmit (the larger the molecule, the slower it diffuses) and they may either fade too rapidly and are lost or do not fade quickly enough and interfere with subsequent messages.

Other considerations (not necessarily advantages or disadvantages) for chemical messages are that different molecules (hence, signals) require different glands for their production and release. Black-tailed deer, for example, have six different pheromone-producing glands (Figure 9-5). Many social insects are literally walking batteries of glands (Figure 9-6). Many of the pheromones are used in territorial boundary marking, as in deer, rabbits, and wolves (Figure 9-7). Graded aspects of the messages depend on the concentrations of the chemicals. The concentrations may depend not only on how much was produced or released but also on the fading with time (Figure 9-8), distance, and wind conditions (Figure 9-9). With time the molecules either may diffuse into the environment, or the molecular configuration may change. The time that a signal lasts, the distance it travels, and the ability to localize or pinpoint it depend, in part, on what is called the Q/K ratio, where Q is the emission rate and K is the threshold concentration at which animals respond. Sexual attractants have high Q/K ratios, whereas alarm and trail substances have low ratios.

Pheromone characteristics also depend on their molecular composition. The molecular alphabet consists largely of three letters, carbon, oxygen, and hydrogen, the most common compounds of virtually all biochemicals. The constraints on the size and amount of molecular specificity are:

1. Amount of information to be coded
2. Stableness, in terms of length of time and gradient
3. Metabolic cost of synthesis, storage, and transport (including volatility)

Insect pheromones, for example, usually involve 5 to 20 carbon atoms. The smaller ones include alarm substances that have molecular weights of 100 to 200. They should diffuse or disperse rapidly (to spread the message), and they do not have to be very specific or carry much information—just a simple message such as "Danger!" Therefore they can (because of low information content) and should

Figure 9-5 Pheromone-producing glands in black-tailed deer. Glands include metatarsal, anal, and interdigital. Scents from the metatarsal are rubbed onto other regions of the body, such as the forehead, which in turn, is rubbed on twigs. Odors also are left on the ground and diffuse into the air. Modif from Müller-Schwarze, D. Anim. Behav. 19:141-1!

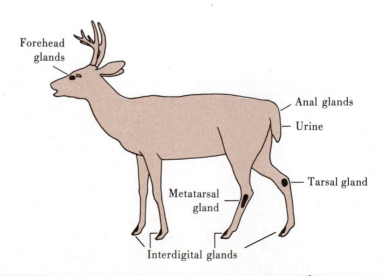

Forehead glands

Anal glands

Urine

Tarsal gland

Metatarsal gland

Interdigital glands

**Figure 9-6 Pheromone-pro-
ducing glands in insects: honeybee and ant. Many of the
glands are believed homologous among species, and
some are unique to different
species. Dufour's gland is important in trail marking.** Modified from Wilson, E.O. 1965.
Science 149:1064-1071.

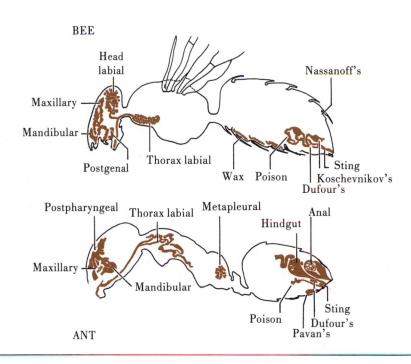

**Figure 9-7 Travel routes
(lines) and raised-leg-urina-
tion scent mark positions
(dots) of a wolf pack and six
neighboring pack territories.
The different packs are represented by different symbols
(combination of shape and
whether open or dark). Territory boundaries are clearly
visible in this diagram.** Modified from Peters, R.P., and L.D.
Mech. 1975. Am. Sci. 63:628-
637.

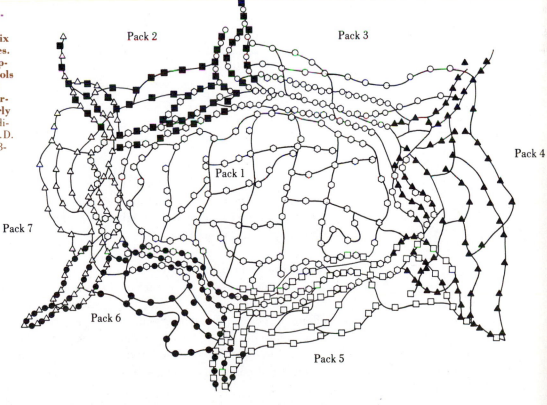

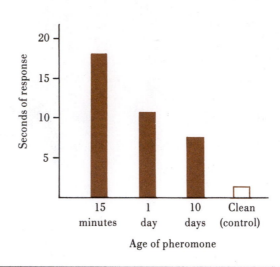

Figure 9-8 Signal fading with time as indicated by male hamster attention (sniffing time) toward defensive ("flank marking") pheromone artificially deposited in corners of their cages. Males ceased responding to male pheromones that were between 40 and 50 days old but continued to respond to female vaginal odor marks 100 days old. Modified from Johnston, R.E., and T. Schmidt. 1979. Behav. Neural Biol. 26:64-75.

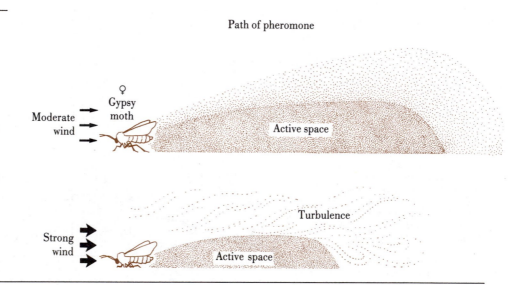

Figure 9-9 Active space, where the Q/K ratio is sufficient to produce a response, of a pheromone such as the gypsy moth sex attractant. Although the odor covers a larger area, it is sufficiently dense to produce a response only within a particular region. A breeze may carry the active space far downwind, but turbulence from strong wind may rapidly reduce the concentration and reduce the active space. Modified from Wilson, E.O., and W.H. Bossert. 1963. Rec. Prog. Hormone Res. 19:673-716.

(for fast diffusion) be small molecules. Insect sex attractants, on the other hand, need to carry more information and be highly specific, but speed of diffusion is not an important factor. Accordingly, those molecules are larger, having molecular weights of 200 to 300.

Auditory Communication

The advantages of using sound for communication are many. It can go around obstacles, such as in a forest, or be used in the dark; it is much faster than chemical communication. (Speeds of sound are discussed in Chapter 1.) It is much more flexible (is easily modulated and a variety of sounds can be produced by a single organ such as the larynx), and it allows much complexity, permitting lengthy and complex messages. Sound waves can carry a great deal of information.

The disadvantages of sound are relatively few and minor. Sound is subject to interference (constructive and destructive), and it distorts with distance. Thus only the simplest of messages can be sent over long distances in the air. (Low-frequency, complex signals are claimed to travel for miles in the ocean). Interference may arise not only from echoes and previous messages but also from other sounds and vibrations (background noise) in the environment. Another disadvantage of auditory communication is that it requires energy to produce signals each time.

The flexibility and complexity permitted by sound are familiar in human speech as well as in the vocalizations of cetaceans (whales, etc.) and birds. Auditory communication via patterned sound waves also is used by other mammals, reptiles, amphibians, many fish, several species of insects, and a number of marine arthropods.

The ability to produce and interpret large amounts of information carried in sound requires advanced sound-producing and sound-sensing organs and also a well-developed nervous system. Most organisms, aside from primates, cetaceans, and birds, apparently do not have the neural capabilities or else have not invested (in the evolutionary sense) much of their nervous system in the analysis of complex auditory discrimination.

Among pairs of animals of several species of birds and some mammals, the male and female sing in duet, that is, together. Crane duetting is illustrated in Chapter 5. The duets in some species are meshed so perfectly that the sound seems to come from a single individual.

Sounds may be transmitted by the air or water or via the substrate. Many spiders, for example, communicate by vibrations of the web or other substrate.

Tactile Communication

This form of communication generally is found in situations of close bodily contact between individuals, as might be expected. Tactile communication often occurs in conjunction with olfactory, visual, or auditory signals. Three important general functions that involve tactile signals are:

1. *Reciprocal feeding.* This involves passing food from one individual to another, usually from mouth to mouth. It occurs among many social insects, some birds, and a few mammals. The request for such feeding may be made by contact signals. For example, some insects tap their mouthparts together, and in many species of birds, such as gulls, chicks peck at the beaks of their parents.
2. *Grooming and initiation of grooming.*
3. *Initiation of physical transport.* Some species, particularly among social insects, may pull on another's mandible or bite at the neck of another to initiate carrying of one by the other.

The advantages and disadvantages of tactile communication are difficult to compare with those of the other sensory modalities. The major constraint obviously is that direct contact is required. Tactile communication might by viewed as a vibration sense, something like an auditory sense at zero distance. Assuming that direct contact is involved, tactile communication would require less-specialized structures for sending and receiving messages than hearing. Almost any external part of the body could be used for touching or bumping another animal, and most animals have numerous touch receptors on the surface of their bodies. All that is required is the generation or evolution of message-carrying function for these contacts; the signals must acquire meaning or, in other words, be interpretable.

Visual Communication

The advantages of visual communication include (1) it is transmitted instantaneously; (2) it may carry a large amount of information, assuming the receiver's eyes and brain are capable of processing it all; (3) it is highly directional, permitting the source to be located; and (4) some aspects, such as body coloration, are permanent, involving only an initial expenditure of energy and needing to be produced only once (although the subsequent *uses* of some structures entail much movement and energy expenditure).

Visual displays often involve movements of extensions of the body, such as the head, ears, legs and wings, and tail. Elephants make much use of their trunks, ears, and head for communication. Tail movements, sometimes in association with rump attention, serve communication in many species. Kiley-Worthington (1976), in an extensive review of mammalian tail movements, reported that tail (and head) elevations were associated with muscle tonus related to movement. In general, upright postures are correlated with preparation to move, alertness, and warning. Lowered tails are correlated with relaxation, fear, nonaggression, and submission. Lateral, wagging movements are generally associated with locomotion, intervals between other behaviors (Hailman and Dzelzkalns 1974), and such things as frustration and inhibition. There are numerous species-specific differences, and both ritualization and exaggeration are evident in many cases. In particular cases there have been a variety of hypothesized functions such as deer tail-flagging for warning (Alvarez et al. 1976), follow-contact signals, or as means of displaying dominance.

The chief disadvantages of visual communication are that (1) it cannot be used over too great a distance, depending on the visual acuity of the receiver; (2) it is blocked easily by obstacles so that the sender and receiver must be in direct line of sight; and (3) it requires light or the production of light.

Vision is perhaps the easiest sensory channel dispensed with for communication (although it still may be vital for flying, predation or avoiding predators, and living in trees). Almost no species relies on vision wholly for communication.

Electrical Communication

A number of electrical field–producing and sensing fishes, such as in the African family Mormyridae, communicate with each other via electrical signals in a manner analogous to the use of sound by other animals (e.g., Hopkins 1974, Kramer 1978).

Communication by Multiple Channels and Antithesis

Complete messages may depend on *combinations* of component signals. Simple signals, that is, with a single meaning, may be composed of several components. A defensive threat, for example, may involve sound, exaggerated movements, and perhaps even offensive odors simultaneously. Another way in which combi-

Figure 9-10 Greylag goose postures in a variety of situations. Modified from Tinbergen, N. 1965. Animal behavior. Time-Life Books, Inc., New York.

Normal, relaxed

Alert

Mild aggression (with hiss) or greeting (with cackle)

Strong aggression or greeting (depending on accompanying vocalization)

Mixture of mild aggression and fear

Mixture of strong aggression and fear

Defense posture

Submission

Submissive approach

Courtship approach

Figure 9-11 Signals formed by composite discrete and continuous messages. A, Zebra messages where threat (ears back) versus greeting (ears up) is indicated by ear position, and magnitude of the signal is graded by how open the mouth is held. B, Ant messages where odor trails guide following behavior, and head movements recruit to food (lateral movements) or nest site (backward-forward movements). A modified from Trumler, E. 1959. Z. Tierpsychol. 16:478-488, and Wilson, E.O. 1975. Sociobiology, the new synthesis. Belknap/Harvard University Press, Cambridge, Mass. B from Hölldobler, B. 1971. Z. Vergl. Physiol. 75:123-142.

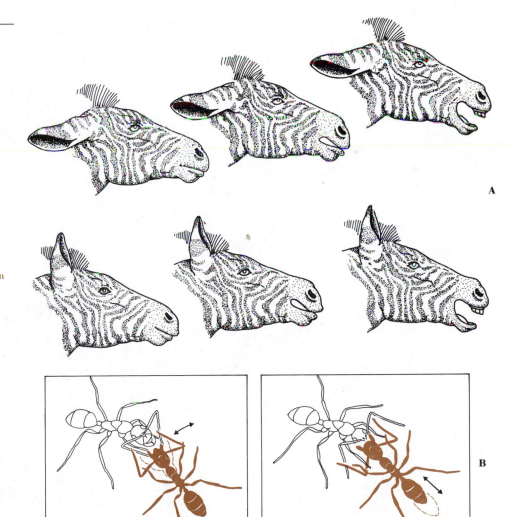

A

B

nations of signals are used is to modify the meaning of one signal by information carried in other signals. The lowered and outstretched neck of a goose, for example, may mean a threat if accompanied by a hiss or a welcome if accompanied by a cackle (Figure 9-10). Other good examples can be seen in zebras and ants (Figure 9-11). Note that in these examples, one of the channels is discrete, and the other is graded. In the zebra's message, for example, the position of the ear is either forward or back, hence, discrete, whereas the mouth can be open to a variable, graded extent.

Some of these complex, multiple-channel messages where the meaning is modified by other signals or context are referred to as **metacommunication**. Metacommunication is a two-stage communication wherein the first signal modifies or alters the meaning of subsequent messages. In essence, it creates a context. The two main categories of communication that have been dubbed as metacommunication are:

1. Invitation to play. The first message is that what follows is only "in jest" or "for the fun of it." This is seen in cats, dogs, humans, and others (Figure 9-12).
2. An indication of social status. The message sender indicates by its overall posture, carriage, and actions where it is on the social ladder. Examples for wolves are shown in Figure 9-13.

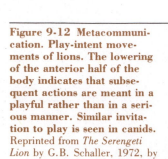

Figure 9-12 Metacommunication. Play-intent movements of lions. The lowering of the anterior half of the body indicates that subsequent actions are meant in a playful rather than in a serious manner. Similar invitation to play is seen in canids. Reprinted from *The Serengeti Lion* by G.B. Schaller, 1972, by permission of The University of Chicago Press.

Figure 9-13 Postural indications of status in wolves. The middle wolf in the background is a female in a normal nonrank-displaying posture. The three in the foreground are, left to right and forward, top- to lowest-ranking males, respectively. Rank is indicated by position of tail, head, ear, leg, and fur plus associated movements. From Zimen, E. 1976. Z. Tierpsychol. 40:300-341.

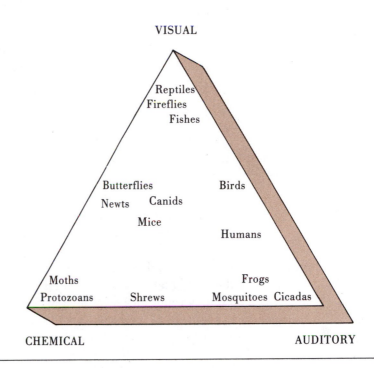

Figure 9-14 Role of different sensory channels, on a proportional basis, in the total communication repertoire of various types of animals. Modified from Wilson, E.O. 1975. Sociobiology, the new synthesis. Belknap/Harvard University Press, Cambridge, Mass.

The total communication repertoire for any given species may involve more than one modality. And within a particular taxonomic group, such as a family, order, or class, different species may vary considerably in the proportion of their total communication repertoire that is transmitted via one modality or another. As might be expected, the modalities used by any one species depend on environmental and life-history considerations. The proportions of their signals that fall into different sensory modalities can be placed on a series of continua (Figure 9-14).

A common property of many animal signals is known as **antithesis.** This means that messages with opposite meanings are frequently expressed by postures and behavior patterns that appear opposite. A classic example, first noted by Darwin (1859), can be seen in the communication of dogs (Figure 9-15, *B*). Another good example is provided by gulls (Figure 9-15, *A*). Although this principle seems intuitively clear, opposite meanings and signals are not always as obvious on closer inspection. What is really opposite of fear: anger, calmness, joy, confi-

Figure 9-15 Classic examples of antithesis in communication: gulls (A) and dogs (B). Essentially opposite postures are used in the expression of "opposite" conditions. A modified from Tinbergen, N. 1959. Behaviour 15:1-70. **B** modified from Darwin, C. 1873. Expression of the emotions in man and animals. D. Appleton & Co., New York.

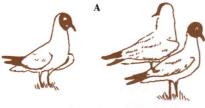

A

Aggressive upright
(facing opponent)

Facing away
(as in courtship)

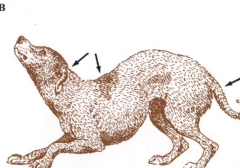

B

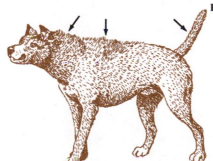

Dominant, aggressive

Subordinate, submissive

dence, or just what? These may be real emotional states with physiological bases, such as sympathetic or parasympathetic components of the nervous system. The subtleties, however, are less clear and may easily become entangled in semantic or anthropomorphic problems.

EVOLUTION OF COMMUNICATION

Specific Advantages and Functions

Thoughts on the general advantages and functions of communication were discussed in the introduction to this chapter. Whether the basic general function is to coordinate, persuade, save energy, or whatever, the specific functions seem more obvious. These can be inferred, or at least categorized, on the basis of *contexts* in which messages are given. The list is more or less arbitrary (see box below). It is taken largely from information in Brown (1975) but has been modified to account for recent information. This list can be used both for reference and, alternately, as a checklist to guide one's consideration of communication in any particular species; that is, it can serve as a means of organizing further studies and information, although it should not restrict new views on the subject. (Smith [1977:70] and others, incidentally, have objected to the functional classification of messages and have provided an alternate list. For a review of Smith's treatment of communication, see Hailman [1978c].)

The difference between alarm and distress calls is the difference between alerting to *possible* trouble and signaling that one is *in* trouble. Distress calls may be calls for help, or one might argue from inclusive fitness that they could signal, "I'm in trouble; save yourself!" These calls are very similar in many species of

Commonly Observed Contexts (Inferred Functions) of Animal Communication*

Survival (group)
    Assembly and recruitment
    Leadership and following
    Contact signals
    Alarm and distress
Social spacing
    Individual and/or caste
        recognition—status symbols
    Aggressive threat and inhibition
    Defensive threat
    Submission

Reproduction
    Courtship and pair formation
    Pair maintenance
    Stimulation (e.g., of ovulation)
    Precopulation
    Nest relief and exchange
    Synchronized hatching
    Parent-young
Miscellaneous and general
    Incitement to hunt and forage
    Grooming
    Greeting
    Initiation of physical transport
Metacommunication
    Invitation to play
    Status signaling (also listed above
        under social spacing)

*There may be several specific signals in any category for any species.

Figure 9-16 Similarity of alarm calls given by different species of birds. The long drawn-out horizontal call is given under conditions of likely danger. The gradual starting and stopping of the call create a ventrilquial effect, which makes it difficult to locate the source of the call. The sharp vertical calls, on the other hand, are given during mobbing or scolding circumstances where the bird may even be attacking the predator and its position is already known. Modified from Marler, P. 1959. In Bell, P.R., editor. Darwin's biological work. John Wiley & Sons, Inc., New York.

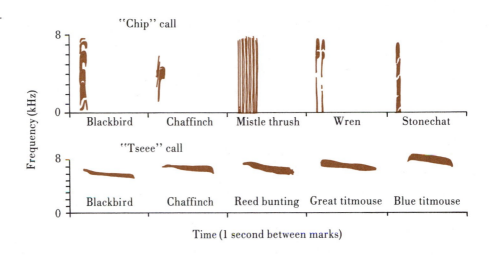

birds (Figure 9-16), thus are generalized, and may be understood by other species. Some convergence may exist even among widely different organisms. The distress screams of chickens, woodpeckers, rabbits, some ungulates, and humans, just to name a few examples, are remarkably similar to a casual listener. (Convergence is treated further below.) The alarm calls are thought to be of a sound quality that makes them difficult to locate, although there have been conflicting observations and opinions.

Aggressive and defensive threats differ in their intent and, often, in their appearance. Aggressive threat may use intention movements and incomplete motions to attack. Defensive threats, on the other hand, frequently use more symbolic signals that are less apparent to an outside and uninformed observer. In cats, for example, the defensive threat involves a sideways, arched-back posture, raising of the hair, backing movement, a grimacing facial expression, and hissing—the classic halloween-cat posture. Although initially uninformed, most animals, such as a dog, that ignore the message and press forward usually learn quickly what it means. Within the species the messages are more likely to be understood innately. Defensive threat postures commonly involve an exaggeration of body size as can be seen not only in the cat but also in, for example, many owls and some insects that spread their wings when confronted. Frilled lizards expand a collar of skin around the neck, raise their bodies, and hiss.

Reproduction signals, such as in courtship and pair formation, may be very specific (as will be discussed in Chapter 10) and often are easy to locate; they may be highly directional. Most of the communication of other animals that people are commonly familiar with or likely to encounter is directly or indirectly related to reproduction.

Under the "miscellaneous" category there are several less familiar signals. Incitement to hunt or forage occurs with bees and can be seen in the ritualized prehunting behavior of African wild dogs. Some insects, such as ants, may request or initiate physical transport (e.g., to new nests) by tapping their antennae. The one to be carried folds up in the pupal position and is then picked up and carried by the other. Functionally similar, or analogous, messages also can be observed in other species, particularly young mammals, including humans. Social grooming is quite common in mammals and some other groups; it obviously is derived from cleansing behavior. This type of grooming appears important in social bonding and is performed even when the individuals are clean and possess no parasites.

Evolutionary Origins of Displays

Most displays are thought to have their evolutionary precursors in the basic and maintenance movements of animals as described in Chapters 5 and 6. It is chiefly in displays that one finds the behavioral, evolutionary characteristics of ritualization and exaggeration. Simple movements are thought to initially acquire a display function, then become ritualized through subsequent evolution. Conflict behavior also is thought to be important in many cases. Aggressive and threat displays, for example, may originate when an animal is stimulated to both flee and fight at the same time. The animal may hesitate briefly, perhaps in an intention movement, or engage in a redirected or displacement activity (Chapter 2) in the presence of its opponent. If the result is selectively advantageous to the actor, even by only a slight amount, the behavior may become incorporated into the species' repertoire. During many subsequent generations the signal may change and become more prominent and effective. Some displays may become so altered that they no longer resemble the original, nonsignal behavior.

Signals in other contexts may develop from other original movements or borrow from signals that already are established. Huxley (1914), who first introduced the idea of ritualization, described, for example, European great-crested grebe courtship display, which involves mutual headshaking. The headshaking is believed to have derived from an appeasement display whereby two birds move their beaks away from each other, which in turn initially may have arisen from turning to flee. The mutual headshaking is then incorporated into a different display, the "penguin dance," which includes diving and the presenting of water plants by the two birds to each other. The part of the display involving the water plants is thought to have arisen from displacement behavior where the birds pick up vegetation during a conflict between hostility and sexual attraction.

Phylogenetic Comparisons of Communication

Humans possess the most highly advanced communication ability of any living species. This is one of the few safe categorical statements one can make. The capacity of human language is familiar to all; language is a major and everyday part of human life. What is not agreed on, however, is the nature of the uniqueness: is human language different from the communications of other animals in a significant, qualitative way, or is it simply a matter of degree and capacity? Do humans have a complete monopoly on language, and are humans the *only* species out of many to use language? Or do humans just have a whole lot more of something that other animals may possess to a lesser degree, with a huge gap between humans and other animals? This central question has aroused enough controversy that it will be returned to after other evolutionary considerations.

The differences between humans and nonhumans may be largely a matter of degree, and there must have been some evolutionary development and continuity. The gap may have occurred from rapid evolution of early *Homo* brains and language, with the extinction and loss of intermediate forms. Several different species of early man, living during the same time period on earth, may have possessed spoken language, but only one, the direct ancestors of humans, survived. Sagan (1977) speculated that the surviving species may have actually exterminated the others, stopping only when it reached the level of primates that lacked speech and were sufficiently different (e.g., chimpanzees) so as not to be in direct competition for resources. But whether one wishes to include humans or not, it still is possible to make phylogenetic comparisons of communication among other species of animals.

Perhaps the first question, which is partly stimulated by viewing humans at a peak, is whether there is a clear trend from simple to advanced in the phylogeny of communication. One cannot answer for sure, but it appears to be "no."

The communication of most invertebrates has received little attention. Thus much information is lacking, and of that which exists, there have been few, if any, reviews for organisms below the molluscs. If and when there is communication in the simpler animals, chemical and tactile signals appear to dominate, as would be expected from the limited sensory capabilities of these organisms. Depending on how one views the nature of the social interactions in colonial coelenterates, the degree of social behavior also is limited in the lower invertebrates. Thus one would not expect much communication in the simple animals in the first place. The simple interactions that do occur among these animals can be and probably are accommodated by the simplest of signals. Such signals may be highly species specific, but a given species would not have to possess much of a repertoire.

Above the simpler invertebrate forms one encounters a wide diversity of more advanced organisms. However, they cannot be placed together on a single ladder or path of related, increasing advancement. It is true that many forms resemble earlier stages in the phylogeny of other forms. But they are not all on one path, as is so commonly misunderstood, leading up to, say, humans. Rather, the relationships *branch* (Figure 5-4), and only a few organisms are related to our direct line of ancestry. The others have gone their own separate ways.

Among the more advanced organisms, there are the two major lines, protostomes and deuterostomes. Within the protostomes there are two major groups with much behavioral diversity and further advancement: the cephalopod molluscs and arthropods. Vertebrates form a big, behaviorally diverse group on the deuterostome branch. It is in these three most advanced groups, cephalopods, arthropods, and vertebrates, that communication has received the most research and comparative review (e.g., Sebeok 1977).

Within these three groups it appears, from what few generalizations can be made, that the most advanced communication (in terms of repertoire size, diversity, complexity, and specificity) occurs in species with the most sensory and social development and where the surrounding biological community is most complex. It may be disadvantageous to organisms if their communication signals are used by other species (such as predators or competitors) or are confused with signals of closely related but different species. Thus, it is thought that interactions with other species as well as within species may lead to more advanced forms of communication. In addition, there are the physical factors of the habitat, such as trees and other obstacles or the darkness of night, that make visual signals more difficult than in open environments. These are all essentially *envi-*

ronmental factors, and it appears that they may be far more important in the evolution of communication than a species' phylogenetic position. Beyond the broad (and almost useless) generalization that the simplest animals have the simplest communication and the advanced animals have more complex communication, phylogenetic trends do not seem to exist. One must look instead to sensory capabilities, environmental factors (physical, social, interspecific), and convergence.

Numerous examples of convergence in communication exist. Convergence in alarm signals was already mentioned (pp. 280 and 281). Cichlid fishes (Figure 9-17) use visual signals, changes in body patterning and color, that are similar to those seen in cephalopods (discussed further later), but fishes and cephalopods are not at all closely related. Cephalopods are much different from their other molluscan cousins, and cichlid communication is much different from that found in some other fish even in the same order. The communication of fish in complex marine reef communities may rival and resemble the communication of insects and birds in complex tropical rain forests far more than it resembles the simpler communication of fish in an arctic lake.

Collias (1960) classified bird and mammal calls in five groups: food calls, warning calls, sexual calls, parent-young calls, and group-movement calls. He noted the apparent convergence among species within these categories. Kiley (1972) discussed convergence in 14 context categories among 50 different ungulate species for which information was available.

Figure 9-17 Visual, poster coloration in a cichlid fish (*Aequidens paraguayensis*). From Timms, A.M., and M.H.A. Keenleyside. 1975. Z. Tierpsychol. 39:8-23.

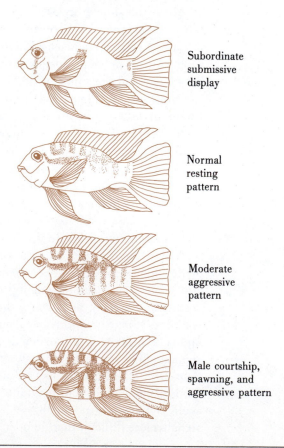

Subordinate submissive display

Normal resting pattern

Moderate aggressive pattern

Male courtship, spawning, and aggressive pattern

Figure 9-18 Generalized "growl," "bark," and "whine" vocalizations of birds and mammals. The frequency and time are variable, depending on species. For explanation, see text.

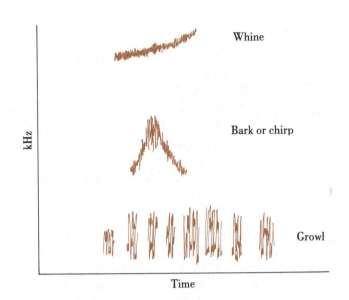

Eugene Morton (1977, Hopson 1980) suggests that there are even some basic underlying structures in bird and mammal (including human) vocalizations that correlate with emotion or mood. In particular, he has identified three common categories: the growl, bark, and whine (Figure 9-18). The growl is a harsh, low-frequency vocalization associated with anger. The higher-pitched, often rising-in-pitch whine is given by immature individuals and others when fearful, friendly, or appeasing. In between are the barks, which are chevron shaped, rising then falling in pitch, which indicate interest or curiosity. The characteristic chevron pattern shows up even in the chirp of a Carolina wren; when a recording of the chirp is slowed down, it sounds remarkably like the bark of a small dog. The same patterns also underlie the emotional intonations of human speech. For example, "I love you" and "please" are normally said in the rising and higher-pitched whine pattern. "Wow" produces the chevron pattern, and "I'll kill you," if meant, comes out in a typical low growl pattern. In fact, it is difficult to say such phrases convincingly without using the proper intonations.

If it is true that both birds and mammals do employ these common patterns, it would seem to be a good case of broadly convergent communication phenomena. It seems reasonable to assume that the stem reptiles, from which mammals and birds (and several other branches of reptiles) arose separately, did not use sophisticated vocalization. One can infer (although not be certain) that they did not from the limited use of vocalization in turtles, snakes, and lizards. The high- versus low-pitched sounds may be another example of antithesis. The low-pitched growls may be the vocal equivalent of puffing up and exaggerating one's size, as lower-pitched sounds are otherwise associated with bigger individuals. Phenomena such as shouting also may fall into this category (Zahavi 1979).

The comparative study of communication, in spite of an already vast literature on the subject, still is in its infancy. Most information is concentrated and biased toward only a few taxonomic groups. Many issues, definitions, and approaches still must be clarified before meaningful comparisons can be made. Tantalizing questions and suggestions abound, but we still are not quite sure what to do with them. Brown pelicans, for example, are highly social and live in colonies near other colonial species, surrounded by potential predators and competitors. Yet, according to one of the most exhaustive studies of brown pelicans (Schreiber 1977), the pelicans form pairs and reproduce with a repertoire of only five signals. Other birds (hence, close phylogenetic relatives) that also are colonial and live alongside the pelicans (under similar environmental constraints) have much larger and more complex repertoires of signals. Why are there such marked differences? Unfortunately Schreiber did not discuss the problem (review by Mock 1977).

As another example of an intriguing evolutionary question or suggestion, Gould (1975: footnote 37) proposed that honeybees are second only to humans and ahead of chimpanzees in the complexity of their language. Gould performed an admirable and seemingly objective analysis, using perhaps the best information available. His suggestions are most interesting if true. But, if true, what do they mean? If not true, how can the comparisons be improved? Gould based his calculations in part on the number of odors that bees apparently can distinguish. But the dance proper is known to code only distance and direction; odor information comes directly from odors in the hive or on the dancing bee's body; other types of information (rules) come from context (such as whether the dancing surface is horizontal and whether the sun or sky is visible [Brines and Gould 1979]). Thus the calculations for the bee dance may not be the best estimate.

Wilson (1972) simply counted the number of signal categories or displays known in honeybees and other social insects and found them to range from 10 to 20 per species. Moynihan (1970) compared 30 species of vertebrates and found the number of displays to range from 10 to 37 per species. There appears to be more variability among species within a class, such as within fish or mammals, than between classes. The average number of displays was slightly greater in mammals than in birds and in birds compared with fish, but the mammals with the fewest (deer mouse *[Peromyscus maniculatus]* and a monkey *[Aotus trivirgatus]*, each with 16 displays) had smaller repertoires than the fish with the most (*Badis badis*, which had 26 displays). Of the mammals studied, the rhesus monkey had the largest repertoire with 37 displays. A major problem in comparing animals on a basis even as simple as just counting displays is that there are no universal, standard ways of categorizing behavior (Chapter 3), and different observers may get quite different results even with the same species. Different techniques can produce not only relatively minor, quantitative differences in results but also major qualitative differences, as will be seen shortly when the controversy over honeybee communication is discussed. Comparisons of communication between different species are difficult at best, and there remains much work to be done in this field.

Thus without implying any phylogenetic significance, examples of what might be considered as relatively advanced forms of animal communication, throughout the three advanced groups, cephalopods, arthropods, and vertebrates, will be discussed. Communication is highly developed in many of the social arthropods. Among vertebrates examples can be found in all classes, except perhaps the cyclostomes (lampreys).

**Advanced Forms of
Communication**

This section partly reflects reality and partly reflects a combination of ignorance and human bias toward those modes of communication with which we are most familiar (e.g., visual and auditory forms). It may be discovered that the communication systems and capabilities of many animals are much more advanced than presently believed. They just may not be known about, or perhaps humans have not been able to detect the subtle variations of chemical, tactile, or other modes that are foreign to us. But in the following cases it is known without a doubt that the transmission and use of messages is complex and highly sophisticated. This includes many of the modes, such as olfaction and touch, that are not well developed in humans.

Cephalopods

The visual communication capabilities of cephalopods is thought to rival or exceed those of many vertebrates, with upward of 35 different communication patterns. Communication in this group of molluscs, which is only partially understood, has been reviewed by Moynihan and Rodaniche (1977). Cephalopods are relatively large and have quite flexible bodies, good vision, well-developed brains, and the ability to change their appearance rapidly through the use of chromatophores and other cells at the surface of the body. They are able to spread and withdraw, raise, lower and curl, and even change the surface texture of various body parts, change colors, darken and lighten, and form a variety of patterns on the body surface. A few of the varied appearances are illustrated in Figure 9-19. Cephalopods are able to change appearance even more than their vertebrate counterparts, fishes (compare with Figure 9-17). Fishes have much less flexible bodies and fewer appendages with which to work.

Many of the cephalopods, particularly the squids, are quite gregarious; virtually all use complex patterns of courtship; and most are social to some degree. In addition to their social characteristics, cephalopods are both predators and favorite prey. Thus communication signals may be combined in various ways with considerations of camouflage and mimicry. Some of the differences in signals appear to depend on context. In the squid *(Sepioteuthis sepioidea)* there are several alarm patterns. Two of the most common, transverse bars and longitudinal streaks, may occur when most other factors (hunger, courtship condition, etc.) are similar but differ depending on whether the animals are in open water near the surface or toward the bottom near cover. When a group is in an intermediate situation, some individuals will show the bars and some the streaks under threatening conditions, after which the whole group rapidly swims off together. The signal flexibility in cephalopods is so great, according to Moynihan and Rodan-

Figure 9-19 Coloration and body shape displays in octopus. Individual cephalopods are capable of expressing pigmented patterns from light to dark in a wide array. Pigment changes can be accompanied by changes in body shape. Apparently homologous patterns and displays are expressed among various species of both octopuses and squids. Modified from Moynihan, M.H., and A.F. Rodaniche. 1977. In Sebeok, T.A., editor. How animals communicate. Indiana University Press, Bloomington.

Longitudinal streak pattern   Flamboyant display   Dymantic display   Zebra stripe display (aggressive)

iche (1977), that an individual can send more than one signal at a time: "A squid in the midst of a group, for instance, can transmit at least three or four different 'messages' absolutely simultaneously, to completely different individuals and in different directions by assuming different color patterns on different parts of the body."

In addition to the visual patterns employed by cephalopods, some—particularly those living in deep water or those which possess shells *(Nautilus)*—may use touch and olfaction for communication. Some of the deeper water cephalopods also appear to have developed light-producing organs for communication. The only sense that does not appear to be used for communication by cephalopods is hearing. These animals lack hearing organs, do not possess air bladders as in some fishes, and most do not possess hard parts that could be used for sensing, transmitting, or producing sounds. Thus their lack of use of hearing may simply be an evolutionary constraint; that is, their bodies were not predisposed toward hearing. Moynihan and Rodaniche (1977) also suggest that sound communication by cephalopods might be risky and thus disadvantageous. With the animals being highly sought prey, sounds might unnecessarily expose them to other predators.

Honeybees

The dance language of the honeybee is quite familiar even to the general public now, although most persons do not know the details. Equally famous is the man, Karl von Frisch, who spent a lifetime studying honeybee behavior. His studies were partly responsible for his receiving the Nobel prize, along with Konrad Lorenz and Niko Tinbergen, in 1973. Von Frisch investigated the bees for many years on his own and then was joined by several others, including Martin Lindauer, Harold Esch, and Warwick E. Kerr, who have become nearly as well known themselves. The stories of the research and findings are published in several places and at different levels of detail and technicality (von Frisch 1967, Esch 1967, Lindauer 1971, and references contained therein).

In brief, foraging honeybees *(Apis mellifera)* perform a dance that symbolically codes information on the distance and direction of food from the hive. This dance recruits other workers who then are able to travel to the distant food site. There are two basic dances, the **round dance,** in which the dancing bee circles to the left and then to the right to indicate nearby food sources (Figure 9-20), and the **waggle dance** for distant food sources. The waggle dance consists of a figure-eight

Figure 9-20 Round and waggle dances of the honeybee. For explanation, see text. Reprinted by permission of the publishers from *The Dance Language and Orientation of Bees* by Karl von Frisch, Cambridge, Mass.: Harvard University Press, copyright © 1967 by the President and Fellows of Harvard College.

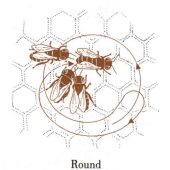
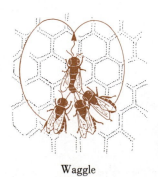

Round   Waggle

looping dance with a *straight run* performed in the middle of the loop. The waggle, in which the abdomen is wagged rapidly back and forth, occurs during the straight run. While the forager performs this dance, other recruit bees gather alongside and behind it and follow it through the dance, keeping their antennae in contact with it.

In addition to waggling the abdomen, the dancing bee emits sounds, thought to be produced by the wings, which serve to incite the following bees to forage (Wenner 1964, Esch 1967). The waggles and sounds have been studied with many ingenious techniques, including the placing of small magnets on the bees' abdomens and recording induced electromagnetic currents (Esch 1967) and by painting white spots on the abdomens and following the dance photographically with series of flashes (Wenner 1964). Sounds have been studied with the aid of small microphones, speakers, natural, modified, and artificial playbacks and by using both airborne and substrate-borne sounds. It appears that the bees are not receptive to airborne sounds but rather sense the vibrations through their feet or by direct contact with the antennae. During silent dances (i.e., lacking the sounds) other bees follow the dancer but subsequently do not leave the hive and forage. During the dance the recruits also are believed to sense the odor of the specific food source, which permits them to locate it exactly once they have used dance information to arrive at the general vicinity.

Distance to the food source from the hive is indicated in the dance by the number of waggles per straight run or the characteristics of the sound that is emitted during the run. Direction is coded inside a dark hive by the angle the straight run departs from straight up. This angle represents the horizontal angle outside the hive that the food is from the azimuth (a vertical line drawn to the horizon) of the sun (Figure 9-21). In other words, cues from gravity, sort of a negative geotaxis, on the vertical surface inside the hive are substituted for cues from the sun in a horizontal plane outside the hive.

This general pattern of dance communication is used by all bees of the genus *Apis*, but different species and races vary in the distance coding, that is, when they switch from the round to the waggle dance and how many waggles they make per straight run (Figure 9-22). In addition, there are two other variations on the theme. In one, the Italian honeybee *(A. mellifera ligustica)* incorporates a dif-

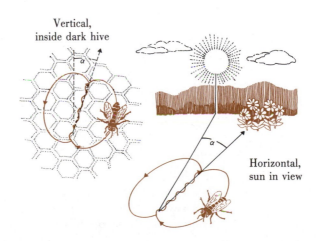

Figure 9-21 Horizontal orientation of the waggle dance to the sun outside the hive and vertical orientation with respect to gravity inside the hive. The angle from up or from the sun indicates the direction of the resource. Modified from Frisch, K. von. 1967. The dance language and orientation of bees. Belknap/Harvard University Press, Cambridge, Mass.

Vertical, inside dark hive

Horizontal, sun in view

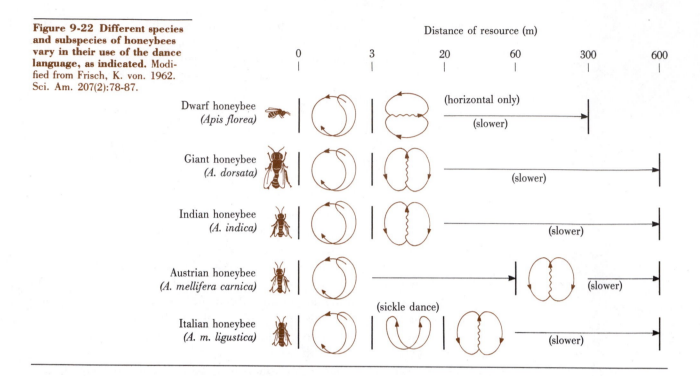

Figure 9-22 Different species and subspecies of honeybees vary in their use of the dance language, as indicated. Modified from Frisch, K. von. 1962. Sci. Am. 207(2):78-87.

ferent dance, the ***sickle dance,*** between the round dance and the waggle dance. In the other variation, the dwarf honeybee *(A. florea)* dances on a horizontal surface and in view of the sun or open sky. It orients the straight run directly toward the food and relative to the sun itself. Lindauer tried to force dwarf bees to dance on a vertical surface by tipping their hive and eliminating horizontal surfaces, but they would not dance on a vertical surface and apparently were unable to transpose from light to gravity.

The evolution of honeybee dance language has been elucidated by comparative study of many other genera of bees as well as ants and other flying insects. The family Apidae, to which honeybees belong, consists of three tribes: the Apinae or honeybees, Meliponinae or stingless bees, and Bombinae or bumblebees. Bombini are not known to use the dance language at all. Among various species of stingless bees, however, there are a variety of recruitment techniques that suggest steps by which the advanced honeybee language evolved. Foragers of some species of stingless bees return to the hive and excitedly dash around, bumping into other bees, passing on bits of food and spreading the odor, then *lead* the recruits all the way back to the food source. In some species of stingless bees, the returning foragers stop every few meters and leave strong scent marks, which are secreted by the mandibular gland, on rocks or vegetation. They subsequently use this scent trail in going back from the hive to the food site. This behavior is very similar to the use of scent trails by ants, which are in the same order (Hymenoptera). In other species of stingless bees, the foragers do not take the recruits all the way to the food source but only lead them part way, after which the recruits continue in the same direction to the food while the initial forager returns to the hive. Many of the species of stingless bees use sound to indicate distance to the food source.

The initial excitement or waggle movements may have been derived from a general tendency of flying insects, including moths, to rock back and forth after landing. The amount of rocking varies with the length of time the insect has been flying. The function of rocking outside bee language is not known; it may serve some physiological function related to respiration or muscle physiology.

The role of light versus gravity has been investigated by shining bright lights or using mirrors to reflect the sun into the hive. The result is that when light is available, the honeybees switch to that rather than transposing to gravity. For example, imagine that a food source is directly in line with the sun. Inside the dark hive the dancers perform the straight run of the waggle dance straight up. If, however, a mirror is placed under the hive and the sunlight is directed up from below, the dancers will point the straight run down, toward the light (Esch 1967). When the mirror is removed, the dancers reorient the dance up again. If a normally vertical hive is tipped so the bees have only horizontal surfaces, they will not dance in the dark, but they will dance horizontally if given a view of the sun or a bright light.

While focusing on the recruitment dance language of bees, one must not overlook the other communication capabilities of honeybees. In addition to their dance, they possess a large number of pheromones and also use sound in several other contexts. Honeybees in the hive make at least 10 distinct, different sounds, some of which are merely by-products of their activities and some of which are signals. Some of the sounds are types of pipping or beeping, which, depending on the particular sound, may either alert or soothe the bees in the hive.

During the replacement of a queen in a hive, after the original has been lost, the candidate queens will emit sounds known as "tooting" and "quacking." In the absence of a queen, several eggs or developing larvae are moved to larger queen cells and fed royal jelly, causing them to develop into new queens. The first one to emerge proceeds to go around and locate others, which she attempts to uncap and kill by stinging. The workers, however, may block her from the other queens, at which time she begins tooting. The toots become quite loud and can be heard by a human as far as 2 or 3 m from the hive. Meanwhile, as the other developing queens attempt to emerge, the workers keep pushing them back and resealing them in their cells. They begin to quack. The messages apparently announce to the workers that there is one free queen and one or more others attempting to get out. The workers release the quackers one at a time to fight with the free one. Eventually only one remains. She then flies from the hive for the nuptial flight with the drones. Following mating, the queen returns to the hive and begins egg production. Included in the pheromones are at least two inhibitors, known as queen substance, which are produced by the queen and which prevent any new queens from being developed as long as she is present.

During the 1960s a *honeybee language controversy* developed. The importance of bee and food odor to the lives and behavior of honeybees was well known; in fact, von Frisch's early work had concentrated on odor, and it was only several years later that he stumbled across the dance. There was no doubt that the dance contained information on the distance and direction of food (or, it should be mentioned, other important resources such as a new hive site). A human observer can easily decipher the distance and direction (this makes a good laboratory exercise if one has access to an observation hive). What came to be at issue, however, was *whether the bees themselves actually use the information contained in the dance.* The question was raised by Wenner and his associates beginning in 1967 (Wenner 1971). They believed, from good evidence, that honeybees were recruiting others and that the others were able to locate the food based on cues from the odor alone. The criticisms, charges, countercharges, and debate continued for nearly a decade, until the matter finally was settled by Gould (1975).

It might be possible to test whether other bees use the information by providing artificial dancing bees. But attempts to recruit bees artificially with model dancing bees largely failed. Esch (1967), for example, constructed an animated dummy bee, driven by a motor and tape loop. The model successfully stimulated other bees to follow the dance, but they would not leave the hive to search at the place indicated by the dance. Esch concluded that there was more to it than just the wagging dance. Gould's solution (1975) was involved and complex, but the gist of it was to use the fact that the bees prefer a light cue rather than gravity if given the choice. The light has to be at a particular threshhold brightness or the bees retain their orientation to gravity. Furthermore, the bees' sensitivity to brightness is determined not by any or all of its eyes but by the three simple eyes, or ocelli, located between the compound eyes. If these are covered, bees are still capable of seeing for flying and foraging, but they are six times less sensitive to light in the hive. The experiments consisted of covering the ocelli of some bees and not others and using light of proper brightness in the hive so that some bees (with covered ocelli) oriented toward gravity and the others (with ocelli functional) oriented toward the light.

Bees were provided food at various locations, the angle of the light in the hive was controlled, and possible feeding locations for recruits were provided at numerous positions in a circle around the hive. All procedures were rigorously and exhaustively controlled and monitored with automatic recording devices that excluded foreign objects such as ants from interfering and which permitted accounting for alien bees, flies, etc.

The results were clear and unambiguous: recruits went to the locations that were predicted, and they could be tricked or misled by combinations of covered ocelli and artificial lights. For example, foragers with covered ocelli would dance with respect to gravity in the presence of an artificial light. But recruits without covered ocelli interpreted the dance as oriented to the light. These recruits then traveled to the stations indicated by orientation to the light rather than to the station where the forager had actually fed. These results established without doubt that the recruits *can and, at least part of the time, do* use the symbolic information coded in the dance.

Why, then, were there discrepancies between the results of the two groups of researchers, von Frisch's group on one side and Wenner's on the other? Having confirmed that the bees *can* use the dance language, Gould experimented further, incorporating the techniques of both groups. In addition to a number of minor differences in method, there was one major difference: Wenner trained bees with concentrated sucrose solutions and the same scents that were used subsequently in experimental tests. The other workers used more dilute solutions and different scents, if any scent at all, than scents that would be used subsequently. By experimenting with combinations of conditions, Gould (1975) discovered that *both groups of researchers were correct!* He concluded, "Depending upon conditions, honey bee recruits use either the dance language *and* odor information, or odors alone" (emphasis Gould's). He stated further, "By their very different training techniques, von Frisch and Wenner may have been sampling two stages of the same process: exploitation of an abundant food source. Von Frisch's experiments could be seen as examining the early phase, while Wenner's would be exploring a later phase."

Several lessons can be drawn from the controversy over honeybee language, lessons that have been encountered previously (e.g., with bats and echolocation). First, different techniques of research may produce different results during investigations of the same phenomenon. Second, one often (in fact, usually) does not have the full story. Third, the intricacies of behavior, such as food recruitment in social bees, may be much more complex and subtle than supposed.

Birds

Bird vocalizations have received much attention in behavioral studies, partly because of interest in their own right and partly because of their importance in developmental studies. Much of the interest in bird song involves the ontogeny of bird vocalizations, a subject that has played a major role in understanding of learning. Bird song thus is discussed in detail again in Chapter 18. Bird vocalizations fall in two general categories: *calls* and *songs*. Calls usually are short and simple. They are used for most of the signal functions in birds. Songs, on the other hand, generally are more elaborate and complex. They are given by males, except in a few rare instances, and mostly during the breeding season, although they also may be given sometimes during the fall (by temperate latitude songbirds) when there may be a slight resurgence of gonadal hormones (Chapter 16).

It is a common question, even among the general public, to wonder why birds sing. On a proximate level it is because of hormonal influences on the appropriate parts of the brain and the presence of relevant external stimuli. From the ultimate standpoint, it is believed that the primary function in the evolution of song is the identification of the singer. The complexity of song probably has been influenced also by the very strong forces of sexual selection (discussed in Chapter 10). Song is thought to serve territorial identification to other males and sexual attraction to females.

Within the broad framework of identification, however, and depending on the species and the complexity of the song, songs may carry potential identity information of several different types. (This does not mean that other birds use all the information or that it is all necessarily functional.)

1. *Species recognition.* The basic song pattern is thought generally to be used by the birds within a species to recognize their own kind. It differs among species but is relatively constant within the species. The basic pattern tends to be more complex in areas with more species.
2. *Local dialects.* These are similar to dialects in human speech and may be either functional or merely consequences of slight geographical separation between pop-

ulations and the drift in expression that occurs over generations of copying the song through learning.

3. *Individual recognition.* Some characteristics of the song vary within the species but are constant for any given male. Many birds appear able to recognize individuals by their vocalizations, as is familiar in human speech.

4. *Motivational variation.* There also may be much variability among songs given by any one individual. Characteristics that vary in this manner include loudness, length, or frequency of repetition. These variations are believed to relay information on the internal state of the bird.

Further discussion of bird song, aside from a brief mention of mimicry, will be postponed until the discussion of song development in Chapter 18. A wealth of detailed information exists for bird song with numerous studies of particular species and groups of birds. Interested persons should consult an ornithology text or bird and behavior journals.

TAPPING THE LINES

One of the more interesting aspects of communication is that the channels of communication, regardless of mode, can be tapped and exploited. During the course of evolution there have been many instances in which one species has developed the ability to mimic the messages of another species. Examples include similar, often almost identical, appearances, vocalizations, chemical messages, or tactile cues. Often the complete and precise combinations of cues are mimicked.

Mimics may become social parasites or predators or enter into symbiotic relationships with the species they mimic. Such exploitation is, perhaps, most common among arthropods, particularly social insects. However, the phenomenon also occurs among vertebrates, the best known cases involving avian nest parasitism by cuckoos: the young of some species have food-begging behavior and color patterns inside the gaping mouth that closely resemble the food-begging signals of the host species' own young. Mimics, such as among birds, may be able to acquire and take advantage of other species' alarm and territorial calls, thus reducing interspecific competition and using other benefits. Further discussion of this topic will be postponed until Chapter 11, which deals with behavioral interactions among species.

A MAJOR CONTROVERSY: WHO HAS LANGUAGE AND WHO DOES NOT?

A lively debate developed over the past few years about whether human language, whatever it is, is unique or not. The controversy has centered on recent claims that great apes, particularly chimpanzees and gorillas, are capable of rudimentary forms of humanlike communication except that they lack vocal ability. On one side are those who say that the apes can. They include Beatrice and Allen Gardner, who initially worked with a champanzee named Washoe and who made the original claims. On the other side are persons such as Herbert Terrace and colleagues, who worked with a chimpanzee named Nim, maintain that apes cannot. In between are others, including several working with chimpanzees, who are less dogmatic. The controversy has spread to include symbolic capabilities of other animals as well including horses, dogs, pigs, goats, parrots, pigeons, and even people.

To the immediate participants (i.e., the Gardners and Terrace) the debate appears to be quite serious and even has become emotional. Some outside, third-party observers, however, have been watching and reporting almost with an attitude of amusement and levity. Articles have included catchy titles such as "The Great Ape Debate" (Benderly 1980) and "Does Man Alone Have Language? Apes Reply in Riddles, and a Horse Says Neigh" (Wade 1980). Several articles have been preceded by short poems or literary quotes; the style of writing has been almost tongue in cheek in many cases; and endings have often been amusing or philosophical, such as, "If only Koko could write." Wade (1980) began one paragraph, "A chimpanzee who asked not to be quoted by name told *Science* . . . 'Those who live in the academic jungle shouldn't ape the law of the jungle.' "

Thus in line with the tone already set by others, the present status of the debate will be introduced with the following story. A cat lover once claimed that she could speak and understand cat language. Several of her friends did not believe the story and decided to put her to the test. So they brought her a cat, waited for it to meow, then asked her what it had said. She replied she did not know because it was Persian.

In brief, there have been several ape-language projects and two major approaches. The initial findings and, to a lesser degree the subsequent controversy, have received wide publicity so that many of the names involved are likely to be somewhat familiar.

The Gardners, working with Washoe and then other chimpanzees, attempted to teach the apes to use American Sign Language (ASL) of the Deaf and compared their results to language acquisition in children (Gardner and Gardner 1969-1978). This *signing* approach (or variations of it) was subsequently used also by Roger Fouts at the University of Oklahoma (Fouts 1972, Fouts and Mellgren 1976), by Penney Patterson (1978), who worked with a gorilla named Koko, and by Herbert Terrace and colleagues (e.g., Terrace et al. 1979), who used Nim (Figure 9-23). Nim, incidentally, is short for Neam Chimpsky, after a famous linguist, Noam Chomsky. The apes were raised, to various extents, in a homelike setting where they interacted with humans in a variety of ways, including attempts to use ASL. Logistical problems with Nim, however, included shortage of time and volunteers who had difficulty coping with the chimpanzee, scratches, and torn clothes. This led to Nim being taught by a large number of persons, which may have complicated the results. On the other hand, Terrace used video recordings of Nim's progress, which permitted subsequent reanalysis.

The second major approach uses external objects, or *lexigrams*, rather than signs. David Premack used a variety of plastic chips that were associated with English words (Premack 1972). Rumbaugh and Savage-Rumbaugh used a computer keyboard with a variety of symbols (Rumbaugh 1977, Savage-Rumbaugh et al. 1980). Rumbaugh and associates first worked with a chimpanzee known as Lana and then two subsequent subjects, Sherman and Austin; the work was conducted at the Yerkes Regional Primate Center in Atlanta, Georgia, and the symbolic computer language was dubbed Yerkish.

Figure 9-23 Symbolic communication in apes. Nim answering *"Joyce"* in response to her question, *"Who?"* From Animals, Animals. Copyright by Dr. H.S. Terrace.

Figure 9-24 Symbolic communication between two pigeons. Pigeons were trained to spontaneously communicate with each other using color-coded keys on a sustained basis and without continuing human intervention. A, Jack *(left)* asks Jill *(right)* for a color name by depressing the WHAT COLOR? key. B, Jill looks through the curtain at the hidden (from Jack) color. C, Jill selects the symbolic name for the color while Jack watches. D, Jack rewards Jill with food by depressing the THANK YOU key. E, Jack selects the correct color as Jill moves toward her reward. F, Jack is rewarded with food. From Epstein, R., et al. 1980. Science 207:543-545. Copyright 1980 by the American Association for the Advancement of Science. Photos courtesy R. Epstein and R.P. Lanza.

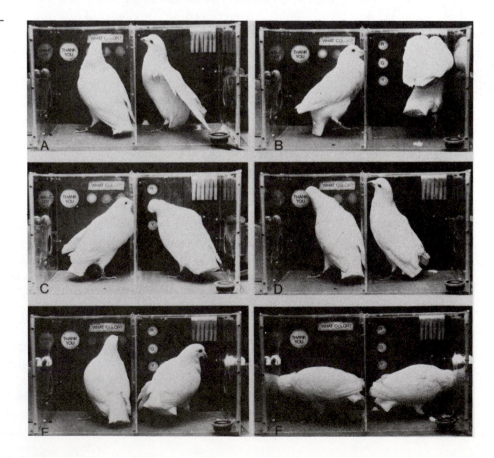

Those using the signing approach, except Terrace's group, have been adamant that they have solid evidence that the apes understand the symbols and can form rudimentary sentences. Terrace and colleagues have been adamant that they cannot and that "In sum, evidence that apes create sentences can, in each case, be explained by reference to simpler nonlinguistic processes" (Terrace et al. 1979). Those using the lexigraphic approach have generally taken somewhat of a middle ground. Savage-Rumbaugh et al. (1980), for example, concluded:

Symbols have merely served to replace or accompany non-verbal gestures the chimpanzee would otherwise employ. . . . Thus, it appears that chimpanzees, even with intensive linguistic training, have remained at the level of communication they are endowed with naturally—the ability to indicate, in a general fashion, that they desire another to perform an action upon or for them when there exists a single unambiguous referent.

But even the Rumbaughs have been entangled in the controversy, including a heated exchange with Thomas A. Sebeok during 1980 at a conference under the auspices of the New York Academy of Sciences. (The conference was described as "a celebration of deception in all its varieties.")

Other researchers have used other species. Pepperberg (1981), for example, worked with an African grey parrot and obtained results similar to the chimpanzee work. Epstein et al., including B.F. Skinner (1980), taught two pigeons, Jack and Jill, to communicate with each other with symbols (Figure 9-24). Epstein et al. (1980) concluded:

We have thus demonstrated that pigeons can learn to engage in a sustained and natural conversation without human intervention, and that one pigeon can transmit information to another entirely through the use of symbols. It has not escaped our notice that an alternative account of this exchange may be given in terms of the prevailing contingencies of reinforcement. . . . The performances were established through standard fading, shaping, chaining, and discrimination procedures [see Chapter 19]. A similar account may be given of the Rumbaugh procedure, as well as of comparable human language.

Thompson and Church (1980), who maintained that Lana's behavior (in particular) could be explained by paired associative learning and conditional discrimination learning (Chapter 19), went one step further and simulated the entire process and results in a computer. Their relatively simple program (available on request, Thompson and Church 1980: footnote 4) was written in BASIC.

Charges and countercharges have been hurled back and forth between those who maintain apes (and other animals) have language and those who believe they do not. The highlights of the arguments, as reviewed in part by Marx (1980), are as follow. Much of the debate is philosophical, concerning whether one views humans as unique from the start. A second major part of the controversy is semantic. Apparently no one is able to define what *language is*, or at least different workers do not accept the others' definitions. This essentially is the same problem previously discussed in defining *communication*. One leader in the study of sign language, Stokoe (cited in Benderly 1980), remarked on the issue of chimpanzee signing, "When language is reduced to that level, then to me it isn't language. . . . [The problem is that] everyone knows what languages are, but no one knows what 'language' is."

The third major source of disagreement concerns methodology. Terrace, for example, believes that what appears to be sentence construction by signing apes may simply be a result of the apes cuing to subtle, unconscious responses on the part of the researchers, a phenomenon known as the *Clever Hans effect*.

Clever Hans was a horse supposedly able to perform arithmetic calculations around the turn of the century. When given a mathematical problem, Hans would tap the correct answer with his hoof. The Clever Hans phenomenon was explained by a psychologist, Oskar Pfungst. It was discovered that, rather than calculating the answer himself, Hans was simply tapping until he perceived subtle cues from his owner that he had tapped enough. The owner's cues were unintentional, and the owner was not aware that he was giving them. By 1937 at least 70 other cases of "thinking" trained animals were known including dogs, cats, a goat, and even a "mind-reading" pig. A French horse, named Clever Bertrand, seemed even more psychic than Hans; he was blind. (Bertrand probably used auditory cues.) The Clever Hans interpretation was applied to claims during the 1950s that dolphins could be taught to communicate with humans. Terrace and several others, including experts on zoo animals, animal trainers, skilled magicians, and several zoologists working with animal communication in general, believe that the Clever Hans interpretation applies to ape language as well.

From the other side, there have been several complaints that Terrace used too many and incompetent volunteers in the training of Nim and that other aspects of Nim's environment were not comparable to those of the other sign-using researchers. Fouts complained further that Terrace, a student of Skinner, was simply using Skinner's techniques of operant conditioning (Chapter 19), which would be *expected* to produce a passive, imitative subject. They maintain that their chimpanzees were different, but they will not permit Terrace access to their data and have threatened to sue copyright infringement for Terrace using films of their work that were produced for television. Rumbaugh and Savage-Rumbaugh (reviewed in Wade 1980) have said that Sebeok does not understand their methods and that the criticisms simply reveal what they consider to be their critic's incompetence.

A major problem underlying the signing technique is that ASL is a whole separate language, having originated in France and developed linguistically on its own. ASL signers in the United States who do not know French can easily communicate in sign language with French signers who cannot speak English. ASL has its own structure and properties, which cross the boundaries of other spoken and written languages. ASL has its own grammar, which is not well understood (by linguists) and which clearly is not the same as in oral languages. All that is known for sure is that ASL works as well for those who use it routinely as any other human language. People who understand ASL as their original language, primarily the deaf or children of deaf persons, can communicate as completely and subtly in ASL as other persons can with spoken language. People who have not acquired ASL naturally (i.e., by growing up with it) rarely master the language. This applies, unfortunately, to the ape researchers, and as noted by Benderly, what they are using is not ASL but a form of pidgin sign English: "This rather impoverished form is really not a language, but a manual code for English."

One aspect of the problem is that no one yet has a method for knowing what another animal knows, the basic problem described in Chapter 5. A clear example of how this applies to language was given by Benderly (1980):

An American diplomat newly arrived in South America once watched as his 10-year-old son Tommy, who knew no Spanish, played soccer with some neighborhood boys. Tommy spotted a hole in the opposing line. "Aqui! Aqui!" he shouted, and a teammate kicked the ball to him. Later, while complimenting Tommy on picking up the language so quickly, the father asked what "aqui" meant. "I don't know, Dad," Tommy said, "but it sure gets me the ball in a hurry." . . . Is this all that the apes have learned to do, or have they truly entered into the cognitive world of human users of language? In Tommy's case we can ask; in Washoe's we cannot.

There are several other points of contention among those involved in the controversy, but we have covered enough. Do apes and other nonhuman animals possess humanlike language capabilities, even if only in limited degree? In light of remarks of Epstein et al. (1980) one might even ask whether *humans* possess language!

Regardless of which way one views all of this, however, there is still a substantial difference between humans and all of the others, even if the difference is only quantitative. Other species use communication largely or entirely during direct interactions, and the signals carry information about the *animals themselves* in most (although not all) cases. The number of such signals in the repertoire of a species varies from species to species, but depending on how signals are identified and counted, the maximum number per species apparently does not exceed 40 or so (e.g., Moynihan 1970). Human communication is at least related to that of other animals in that humans also possess a certain repertoire of nonverbal, species-typical signals. The underlying structure of emotional intonation in human speech may share characteristics with vocalizations in other vertebrate species, as suggested by Morton.

In addition, human communication has much more. In our language humans make extensive distinctions and references to other things, places, events, and abstract concepts that may be external and far removed from ourselves in time, space, and even reality. Human language is such an important part of existence that it undoubtedly has shaped much of what the species is and may very well be the most important, unique hallmark that distinguishes humans from all other animals.

Deese (1978) proposed even that *consciousness* is directly related to elaborate human use of language: "One of the chief linguistic functions of consciousness is to enable us to monitor our own speech. We listen to ourselves and we can tell, most of the time, when we have made a mistake, and we correct it."

Honeybees with their dances and even ants with their odor trails, as well as many species with alarm signals, refer to other things (e.g., food and enemies) external to themselves. But even here, the signals are limited, and the external items referred to are not very distant conceptually. The difficulty in teaching apes, which involves much time, effort, and research money, as well as the whole controversy over the results, simply indicates that even our closest (living) relatives are severely limited compared with humans. Even if the difference is only quantitative, it is so great as to almost qualify as qualitative. Whether or not it really is qualitative may be little more than an academic issue. Philosophically, one might ask whether it is a profitable research strategy in the first place to try to teach other animals a communication system, such as human language, to which they are not adapted. Instead perhaps humans should learn more about the communication systems that animals actually use.

SUMMARY

Social and interspecific interactions among animals are so diverse and complex that a universally acceptable definition of communication is not available. General attributes of communication that occur in some, but not all, instances include (1) a signaler, message or signal, and receiver, (2) adaptiveness to one or both (through coevolution) of the participants, (3) alteration of behavior as a result of the message, (4) function of carrying information, manipulating another's behavior, or both, and (5) low energy cost with much energy return or benefit.

Signals, which can be studied under natural conditions or through a variety of experimental manipulations, may occur in one or more of several modes: visual, auditory, olfactory, tactile, and, in some fish, electrical. Each means of conveying information has its advantages and disadvantages. Combinations of signals may be used to convey complex information or to alter the contexts and meanings of other signals (sometimes known as metacommunication).

Several types of communication, their evolutionary origins, and phylogenetic comparisons are discussed. Two major controversies in communication also are described: (1) whether or not honeybees actually use the information in their dance language (they do), and (2) whether human language is qualitatively or only quantitatively different from communication and potentials for communication in other species such as chimpanzees (issue not resolved).

**Recommended
Additional Reading**

Brown, J.L. 1975. The evolution of behavior. W.W. Norton & Co., Inc., New York.

Dawkins, R., and J.R. Krebs. 1978. Animal signals: information or manipulation? In J.R. Krebs and N.B. Davies. Behavioural ecology, an evolutionary approach. Blackwell Scientific Publications, Ltd., London.

Frisch, K. von. 1967. The dance language and orientation of bees. Belknap/Harvard University Press, Cambridge, Mass.

Frisch, K. von. 1971. Bees, their vision, chemical senses, and language. Cornell University Press, Ithaca, N.Y.

Gottlieb, G. 1971. Development of species identification in birds. University of Chicago Press, Chicago.

Hinde, R.A. 1981. Animal signals: ethological and games-theory approaches are not incompatible. Anim. Behav. 29:535-542.

Lindauer, M. 1971. Communication among social bees. Harvard University Press, Cambridge, Mass.

Marler, P., and W.J. Hamilton III. 1966. Mechanisms of animal behavior. John Wiley & Sons, Inc., New York.

Marler, P., and J.G. Vandenbergh, editors. 1979. Social behavior and communication: handbook of behavioral neurobiology, vol. 3. Plenum Press, New York.

Müller-Schwarze, D., and R.M. Silverstein, editors. 1980. Chemical signals, vertebrates and aquatic invertebrates. Plenum Press, New York.

Sales, G., and D. Pye. 1974. Ultrasonic communication by animals. Chapman & Hall, Ltd., London.

Sebeok, T.A., editor. 1977. How animals communicate. Indiana University Press, Bloomington, Ind.

Stoddart, D.M. 1976. Mammalian odours and pheromones. Camelot Press, Ltd., London.

Wenner, A.M. 1971. The bee language controversy. Educational Programs Improvement Corp., Boulder, Colo.

Wilson, E.O. 1975. Sociobiology, the new synthesis. Belknap/Harvard University Press, Cambridge, Mass.

CHAPTER 10

Male sage grouse *(Centrocercus urophasianus)* **displaying on its lek.**

BEHAVIOR AND REPRODUCTION
COURTSHIP, PARENTAL CARE, AND OTHER CONSIDERATIONS

Reproductive behaviors may be defined broadly as those activities directly involved with the production of new individuals. Such behaviors sometimes are quite simple. In many species, however, reproductive behavior may be most complex and involve spectacular displays and movements. This category of behavior, because of the sometimes remarkable displays, is frequently popularized and also has received considerable attention from biologists. Animal behavior in the minds of the public is generally associated with either some aspect of reproduction or predation, the other popular topic in behavior. Reproduction already has received attention in this book under the topics of evolution, sociobiology, and communication. It is, in fact, difficult to discuss any of these topics without also involving the others. Reproduction is a central issue in the subject of sociobiology.

Although reproduction is reviewed in Chapters 4 and 5 and the related subjects of kin selection and ESS are treated in Chapter 8, I would like to expand briefly on reproduction as a phenomenon, particularly one type of reproduction—sexual. Sexual reproduction is enormously complicated. At the proximate level it involves much cellular and biological machinery involved in the production, protection, and fertilization of haploid gametes (ova and sperm), and male and female must be coordinated, which may involve complex interactions.

Complexity at the proximate level, however, appears simple compared with complexity at the ultimate level. It is probably safe to say that no one, including those who think they do, really understands the evolution of sexual versus asexual modes of reproduction. Why go to all the bother of having two sexes, all of the anatomical and physiological machinery, and accompanying behavior? If a bird can lay a fertilized egg or a female mammal develop an embryo from a zygote in the uterus, why not simply reproduce from an unfertilized egg? Why even bother with an ovary, and why not just use some other cell, for example, one directly from the lining of the uterus? Many of the mechanics might seem not only unnecessary but even disadvantageous to life. Most women if given the choice, for example, probably would prefer to do without menstruation. Persons familiar with the present-day mechanics of differentiation and development will have ready answers. But one can look beyond these (minor?) complications and still ask, "Why?"

In one of the most clearly written and stimulating books on the subject, Williams (1975) opens the preface with, "This book is written from a conviction that the prevalence of sexual reproduction in higher plants and animals is inconsistent with current evolutionary theory." Then, after devoting an entire book to the problem, he closes with, "I am sure that many readers have already concluded that I really do not understand the role of sex in either organic or biotic evolution. At least I can claim, on the basis of the conflicting views in the recent literature, the consolation of abundant company."

The basic problem is that sexual reproduction raises a serious dilemma: it is so universal among organisms that one can infer that it is highly advantageous and offers a good solution to something. But by its very nature of dividing the genome during meiosis, it reduces, in all but a few cases (e.g., male hymenoptera), the genetic contribution of an individual to its offspring by 50%. If it is so good at producing answers, why does it toss half of the answer away each time?

Williams (1975) compares sexual reproduction to Sisyphus, a man who repeatedly pushed a boulder up a steep slope until nearly reaching the peak, at which time the boulder would roll out of control and down the slope so that he had to start over again. Sexual reproduction and recombination amount to continuously redoing the search for the best combination of genes. During recombination one is as, or perhaps more, likely to get bad combinations as good ones. And when a good combination is finally arrived at, it is cut in half again during the next generation. In another analogy, Williams suggests that sexually versus asexually produced offspring may be like lottery tickets: most persons in a lottery would prefer to have several different tickets rather than several copies of the same ticket. But what if the winnings depended on how many of the same ticket one holds? The problem refuses to go away. Asexual reproduction would seem, on the surface, to be the most advantageous, and when related species reproduce one way or the other, the asexual ones often do better than the sexual ones. Species groups of some weevils, for example, often have several parthenogenetic forms and one or a few sexual forms; the parthenogenetic ones are more widespread and numerous (Suomalainen 1969, cited in Williams).

The usual answer for the value of sexual reproduction is known as the *mullerian theory* (from H.J. Muller 1932) and has been popular in one form or another from around the time of Fisher (1930) up to the present. It has been accepted by Mayr (1963), and most other evolutionary theorists. It is used or assumed in Chapters 4 and 5 on genetics and evolution in this book. The theory is that genotypic variability can be generated without having to wait for mutations, which may increase the rate of evolutionary potential and change. Recombination in sexual reproduction theoretically produces genotypes that otherwise would not arise.

There seem to be problems, however, with this commonly accepted theory. Under artificial selection (i.e., selective breeding) evolution can be pushed to occur perhaps thousands of times faster than is occurring in nature. Williams (1975) notes that organisms can evolve faster than they do. In the last few million years, for example, *Drosophila* have not evolved as rapidly as humans in spite of their ability to evolve faster because of shorter generation length. Williams concludes that rather than focusing on factors that produce genetic change, one ought to explain "how populations can so effectively resist change over such amazingly long periods."

An alternate explanation for the presence of sexual reproduction is that it *facilitates adaptation to changing and new environments*. Some insight to this solution is gained from species that alternate between sexual and asexual modes of reproduction during their life cycles, such as in many invertebrate parasites; the sexual stage generally occurs just before dispersal into new and unpredictable environments. Most of the new offspring do not survive; the one or few who do, those who presumably have the best genetic answers, capitalize on their combinations by returning to asexual reproduction in their new, stable environments. But if this biological strategy works, why have most species dispensed with asexual reproduction? Williams argues that with increased, high fecundity (number of offspring per individual) and wide dispersal into newer and widely different environments, selective pressure increases, and with higher pressure the balance tips toward sexual reproduction, outweighing the costs of meiosis and recombination so that sexual reproduction becomes the exclusive mode. After further evolutionary change the phylogenetic descendants may lose the adaptations, or abilities, for asexual reproduction and become obligate sexual reproducers, even if they subsequently have lower fecundity and lower selective pressures. Thus Williams suggests that "genotypic variety provides a margin of safety against environmental uncertainty, and that . . . may be more important to population survival than is precise adaptation to current conditions. Sexual reproduction facilitates evolution indirectly by making extinction less likely, not by making phylogenetic change more rapid" (1975:154). Elsewhere, in reference to higher organisms, he says, "In these forms sexuality is a maladaptive feature, dating from a piscine or even protochordate ancestor (for vertebrates), for which they lack the preadaptations for ridding themselves" (1975:103).

This includes humans and most of the animals we are interested in from a behavioral standpoint. Which of the preceding theories, if any, or which parts are correct remains to be seen. It seems obvious that there are meiotic and recombinational costs to sexual reproduction, not to mention the proximate expenses associated with mating. Whether or not it is all worth it (i.e., that sex provides an advantage that outweighs all of this) or is simply something that was inherited and humans are stuck with cannot be stated for certain at the present time. Daly and Wilson (1978) reviewed the situation and commented: "Whatever its explanation, sex is with us. We seem to be stuck with sex, and only a disenchanted few regret it."

What difference does it make to behaviorists if there is sex in the first place? First, from the standpoint of the real world, without sex there would be no sexual behavior. Second, in a theoretical sense, general biological understanding is supported by an underlying basic framework of theory: when the basic framework is challenged and may be undergoing change, one needs to stay alert to possible ramifications in other, perhaps more immediate, aspects of one's understanding. Third, from a cautionary standpoint, uncertainty in fundamental theory should only cause one to be wary of virtually all theory. Understanding of reproductive behavior is *riddled* with theory that is untested to various extents. Thus, although it is important to learn and understand the hypotheses and theories, it is equally important that one does not place too much confidence in any of them but maintains an open and questioning attitude.

What appears safe to say is that asexual reproduction is not very prominent in higher organisms, although it is relatively common in the lower, simpler animals.

Figure 10-1 Diagrammatic courtship sequences: three-spined stickleback (A) and smooth newt *(Triturus vulgaris)* **(B). These examples illustrate the postures and sequences of behavioral interactions. A** modified from Tinbergen, N. 1965. Animal behavior. Time-Life Books, Inc., New York. **B** from Halliday, T.R. 1975. Anim. Behav. 23:291-322. Figure courtesy T.R. Halliday.

A

Male displays red belly   Female appears egg plump

1   2

In higher organisms asexual reproduction, when it does occur, usually is in the form of parthenogenesis, that is, development from unfertilized eggs. Parthenogenesis apparently is unknown in molluscs; it occurs surprisingly frequently in insects (e.g., the weevils mentioned earlier), and it is the source of male hymenoptera. It crops up occasionally among vertebrates in fish, amphibians, and reptiles but is not known to occur naturally in birds or mammals. White (1970) presents an excellent review of parthenogenesis. In the remainder of this chapter only sexual reproduction will be discussed.

THE KEY CLUB: WHO IS ALLOWED TO MATE AND WHY

Sequences of Behavior and Basic Biological Considerations

The complexities of reproductive activities arise from a multitude of different, interacting factors that must be satisfied if reproduction is to be successful. At the minimum, except for a few aquatic vertebrates that merely shed their gametes into the water, appropriate behavior must (1) bring the male and female into physical proximity and (2) lead to actual fertilization or copulation. (Copulation is the physical coupling of the male and female when fertilization is internal.) In addition, depending on the species and environmental conditions, there may be other behavior that (1) keeps the pair together, (2) results in incubation of the eggs and other care, maintenance, and defense of the nest, and (3) provides parental care of the resulting offspring.

The primary ultimate consideration for bringing the pair together involves species isolation. Individuals that are too unrelated (i.e., belong to "different species") have noncomplementary genetic makeups. They may possess different numbers of chromosomes, have vastly different loci or loci that are spaced differently on the chromosome, or otherwise be unable to combine gametes successfully if they copulate. Selection should favor individuals that do not waste gametes. Behavior is one of the observed mechanisms that confines fertilization within species.

Other ultimate factors may be observed in predatory species. Predators have the potential of killing and eating other members of their own species, even their own mates or potential mates. Behavior that kills the reproductive partner before copulation obviously would be unlikely to be passed on. Any mechanisms that prevent such cannibalism should be highly advantageous.

A few instances do occur in which reproduction is accomplished in spite of cannibalism between the partners; the best-known cases are preying mantises and black widow spiders. In the black widow fertilization usually is accomplished before the female eats the male. However, there is variation; in some cases the

Male zigzag dances and
brushes female
3

Female displays head-up posture
4

Male leads to nest
5

Female follows
6

Male turns on side and
points to nest
7

Female enters nest
8

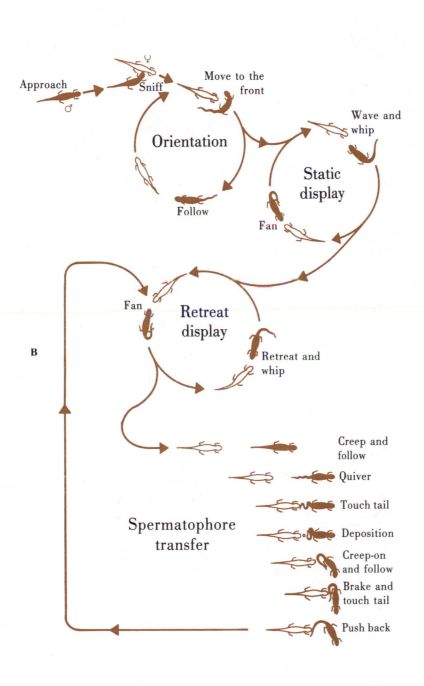

Approach

Sniff

Move to the
front

Orientation

Follow

Wave and
whip

**Static
display**

Fan

Fan

**Retreat
display**

Retreat and
whip

B

Creep and
follow

Quiver

Touch tail

Deposition

Creep-on
and follow

Brake and
touch tail

Push back

**Spermatophore
transfer**

Male prods female tail
9

Female spawns
10

Male enters and fertilizes eggs
11

male gets eaten first and hence does not reproduce. Sometimes the male gets away afterward.

Fertilization is accomplished despite cannibalism in preying mantises as a result of the senses and internal programing of the copulatory movements being located below the head of the male. The male's head contains neurons that inhibit copulation. When the head is removed, copulatory movements are disinhibited. While the female proceeds to eat the male's head, the remainder of the decapitated body moves into the correct position and copulates with the female. When females do not eat the male, the inhibiting neurons in the male's head apparently are disinhibited by other means. (The functioning of neurons and nervous systems is discussed in Part Three.) One could argue that cannibalism after copulation might benefit the male's offspring by providing the female with an additional meal. For most predators, however, cannibalism between the partners does not occur, as a result of the evolution of specific behaviors.

Some of the best-known examples of cannibalism-avoiding behaviors can be seen among the spiders:

1. Male orb weavers pluck the edge of the female web to test whether the female will run out as if for food. He proceeds into courtship only if she does not. Alternately, this could be interpreted as the male communicating to the female that he is a male and not food.
2. Male tarantulas have hooks on their forelegs to prop open the fangs of the females while copulating.

Figure 10-2 Examples of simplified flowcharts of courtship sequences in various species. A, Queen butterfly. B, Smooth newt. See also Figure 10-1, *B*. C, Cockroaches (genus *Periplaneta*). D, Domestic chicken. A modified from Brower, L.P., et al. 1965. Zoologica 50:1-39. **B** modified from Halliday, T.R. 1976. Anim. Behav. 24:398-414. **C** modified from Simon, D., and R.H. Barth. 1977. Z. Tierpsychol. 44:80-107. **D** modified from Fischer, G.J. 1975. In E.S.E. Hafez, editor. Behaviour of domestic animals. The Williams & Wilkins Co., Baltimore.

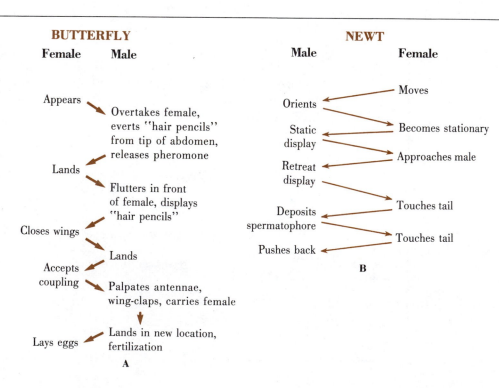

3. Males of another genus have hooks on the jaws to hold the females' jaws open.
4. Males of another species present "gifts" of prey to the female, then copulate while the female is eating (such as empid flies, p. 141).
5. In yet another species the male ties the front end of the female to the substrate with silk before proceeding to copulate.

In addition there are many more ways that mates among predatory species avoid cannibalism. To paraphrase the well-known Murphy's law: If something can happen in the course of evolution, it will!

In addition to accomplishing species isolation and preventing or sidestepping cannibalism, the reproductive behavior also must *synchronize the pair*; that is, they must be ready for fertilization at the same time. Not only must they be brought together in physical proximity (i.e., spatially), but temporal considerations must be met. Very few species are in a state of readiness for fertilization all the time. Much reproductive behavior may involve either testing or stimulating the readiness of the partner.

In many species the reproductive behavior is simple and straightforward. However, in many others there are very complex *sequences* of behavior and signaling. The sequences can be viewed somewhat like a series of locks and keys in which each lock must be opened before proceeding to the next one, with fertilization or copulation waiting as the final prize, in somewhat anthropomorphic terms. Examples of some sequences and general patterns of mating are illustrated in Figures 10-1 to 10-3. Complex sequences can be observed in species as diverse as fish

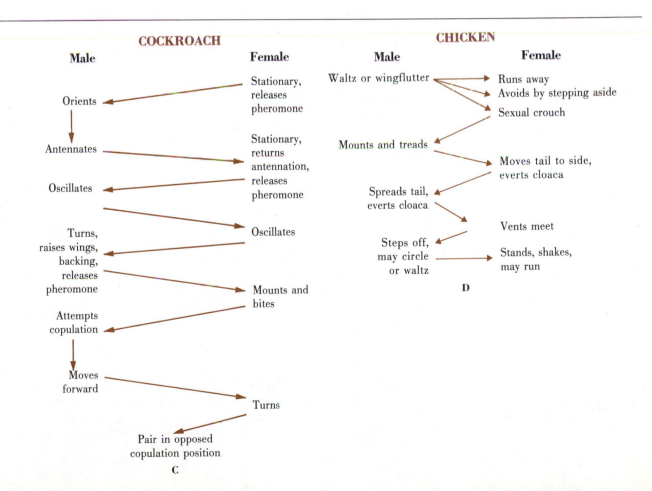

A

B

Figure 10-3 Courtship and copulation among animals. A, Butterflies mating. B, Female horseshoe crab laying eggs; 10 males are fertilizing them. C, Dwarf American toad. Females are larger. Eggs are fertilized outside the body. They are laid in strings encased in gelatinous tubes in water. D, Olive baboon mating, with young watching. A from Animals, Animals. Copyright 1970 by Robert Maier. B by Leonard Lee Rue III. C by John H. Gerard. D from Animals, Animals. Copyright by S.D. Halperin.

and butterflies. (One person reading an early draft of this book, incidentally, commented that she had never thought about butterflies copulating; it seemed out of character for them.)

A common but poorly understood act in the courtship or mating activities of many mammal species is called *Flehmen* (from German). After sniffing a female, the male lifts his head and raises his upper lip (Figure 10-4). It is thought to aid perception of the female's odor and perhaps detection of whether or not she is in estrus.

If two members of a pair are out of synchrony or otherwise are incompatible or if there is natural variation in the expression of the behavior, a relationship between two animals or a particular sequence may break off and go back to a previous stage (often with a new individual). What actually is observed in reproductive behavior often includes much variation (Figure 10-5).

Environmental Considerations and Timing

In addition to the basic biological considerations discussed earlier, reproductive behavior is influenced both proximally and ultimately by major environmental factors. The extreme variability that may occur in the environment generally results in certain periods of time being more suitable and favorable for reproduction than other times. Certain aspects of the environment, such as length of day or the appearance of heavy rainfall, may be correlated with future favorable times and, hence, may serve as cues to trigger or initiate reproductive behavior. As a result, one may observe and categorize several seasonal patterns of reproduction (Figure 10-6):

1. *Highly seasonal with a very specific time.* The palolo worms (a polychaete annelid) of the Pacific Ocean, for example, almost all mate during a single day at the end of November. The many species of temperate and upper-latitude vertebrates that breed only once a year, commonly in the spring, also fall into this category, although not quite so precisely.

D

C

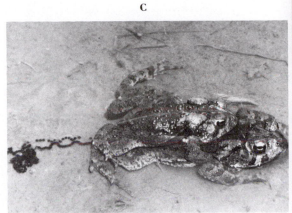

Figure 10-4 *Flehmen* as shown by an American bison bull. The mouth is partially opened, upper lip raised, and tongue is fluttered up and down in the mouth while the head is tipped upward. From Lott, D. 1981. Z. Tierpsychol. 56:97-114. Photo courtesy Dale Lott.

2. *More or less seasonal throughout favorable times.* Species in this category continue to reproduce over extended periods of time as long as conditions are not extremely hostile. Examples would include houseflies and earthworms in northern latitude, where they begin breeding in the spring, continue throughout the summer, and do not stop until going into diapause or until harsh winter weather makes further breeding impossible.

3. *Most of the time* (i.e., there is little unfavorable time). Many species, such as flies and earthworms in the subtropical and tropical latitudes where moisture remains adequate, may breed continually.

4. *Less-than-year cycles.* Army ants, birds with more than one brood per season, and many other species cycle through periods of reproduction that are not synchronized with obvious annual changes. The length of the cycle usually is less than a year and may go through three cycles in 2 years or seven cycles in 2 years. A few larger species have reproductive cycles that require much longer than a year for intervals between one productive period and the next. These very slow-breeding species include Andean condors, elephants, and many of the primates, including most humans.

5. *Irregular but at specific times.* Animals that are best known for this behavior pattern include many desert species such as insects, amphibians, and tropical and subtropical birds that breed only during or following periods of sufficient rainfall.

The "favorable" times usually are most important concerning survival of the offspring. If species with long periods of gestation or incubation did not begin breeding activities and fertilization until conditions were favorable, they would run out of time before the young were born or hatched and independent. Thus many species can be observed to engage in reproductive behavior when the conditions would otherwise seem to be unfavorable. For contrasting examples, consider the nesting of northern latitude birds as opposed to many large mammals at those same latitudes. The birds may not begin courtship and copulation until spring, which is followed by nesting, incubation, hatching, and rearing of young. Deer in the same region, however, typically engage in courtship and copulation

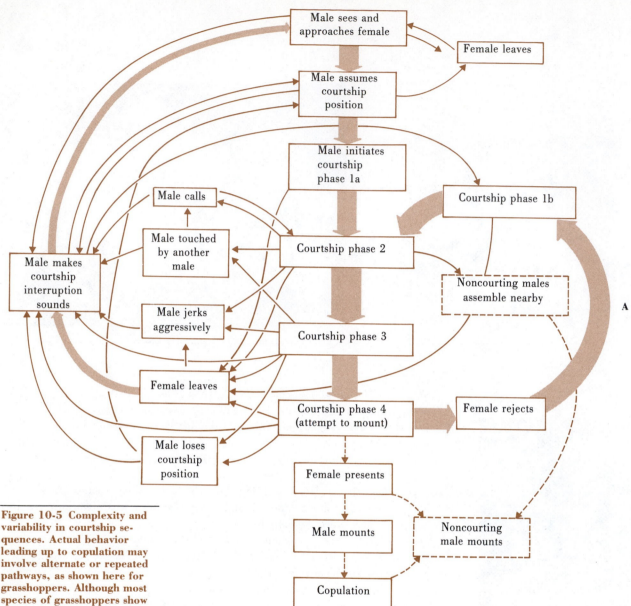

Figure 10-5 Complexity and variability in courtship sequences. Actual behavior leading up to copulation may involve alternate or repeated pathways, as shown here for grasshoppers. Although most species of grasshoppers show relatively simple sequences, some such as these are quite complex. A, Courtship in grasshoppers of the genus _Syrbula_. Frequencies of transitions from one behavior to the next are indicated by thickness of lines. B, Possible interactions between male and female Louisiana herons during the courtship period. Frequency of interactions is not quantified. A from Otte, D. 1972. Behaviour 42:291-322. B redrawn from Rodgers, J.A., Jr. 1978. Wilson Bull. 90(1):49.

during the fall; the embryos develop during the winter in the female's body; and birth and rearing of young occur during the spring and summer at about the same time as in the birds.

Predation also may be a significant environmental factor. This is thought to lead to the highly synchronized periods of reproduction commonly observed in many species. Many offspring produced simultaneously would overwhelm the feeding capacity of predators. The probability of an individual losing young would be less than if reproduction were spread over a longer period, which could provide the predators more opportunity to take the young. This hypothesis is known as the _Fraser Darling effect_ after the person who first suggested its importance (Darling 1938). (Also see the cicada principle, p. 234.)

The ability to synchronize reproductive behavior with environmental conditions thus may carry tremendous selective advantages. The physiological mechanisms by which all of this is accomplished are discussed in Chapter 16.

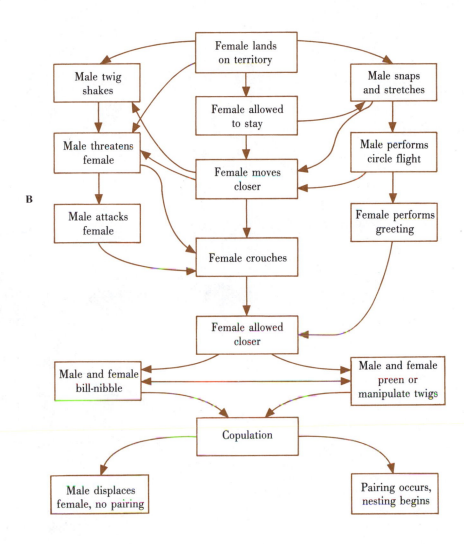

MATING PATTERNS IN REPRODUCTION

Several types of mating systems, that is, the patterns by which individuals come together in courtship and copulation, have been observed among various species. There is some overlap among categories and terminology, and many species show more than one type of pattern. Humans, for example, show virtually all patterns. The topic of mating patterns has been an active field of research, and much of its success and development has hinged on the ability to identify individual animals, such as through visible markers or radiotelemetry.

Before mating patterns and related topics are discussed in this chapter, it is necessary to insert a disclaimer concerning terminology. Many of the terms describing reproductive behavior in nonhuman species have been borrowed, rightly or wrongly, from human usage. Many biologists and anthropologists object, almost to the point of shock. They believe that such use of terms is anthropomorphic and that the words are loaded with meaning and implications

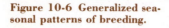

Figure 10-6 Generalized seasonal patterns of breeding.

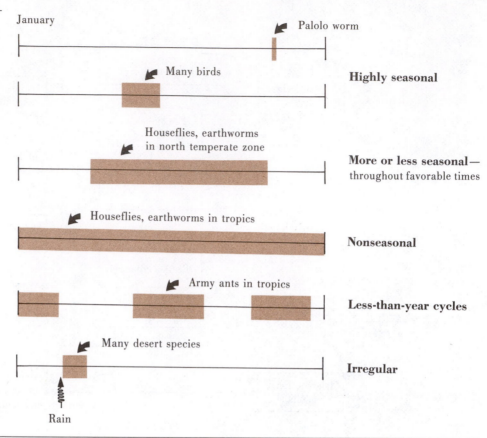

that are not appropriate in nonhuman contexts. Unfortunately, however, the use of the terms already is well established at all levels of discussion, including the professional journals. These terms will be used here also but with the understanding that they may be somewhat inappropriate. They should be treated with a minimum of human connotation. At the same time and in partial defense of these terms, however, humans are biological, reproducing organisms that should be subject to the same forces of nature as other species. Hence, terms that describe human reproductive behavior may not be entirely inappropriate.

In a few species mating is accomplished in some rather unusual (to humans) ways. These mating systems are *simultaneous or protogamous hermaphroditism.* In simultaneous hermaphroditism, a term derived from the names of the Greek god Hermes and goddess Aphrodite, an individual possesses the gonads for both sexes. Individuals usually exchange gametes with each other unless the opportunities are not available, in which case they may fertilize themselves. Protogamous hermaphrodites alternate between sexes; the individual is first a female or male (protogynous or protandrous, respectively), then the other at a later time. Occasionally it reverts again to the first. Secondary sexual behavior (and in some cases even primary sexual output—gametes) is hormonally determined in vertebrates, including birds and mammals, as opposed to strict genetic determinism in some other groups such as insects.

Figure 10-7 Maps of a distribution of protoandrous reef fishes and, to the right, the anemone colonies they occupy. Dark circles represent breeding individuals. From Fricke, H.W. 1979. Z. Tierpsychol. 50:313-326.

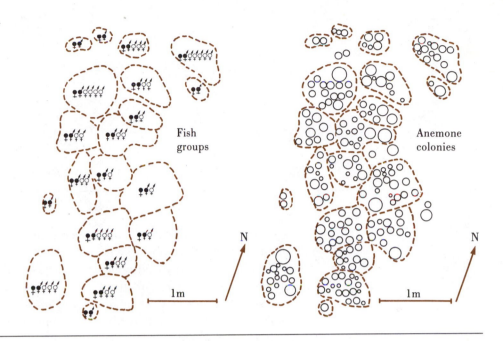

Fish groups

Anemone colonies

Natural instances of sex reversal include, for example, a type of marine cleaner fish in which the normal group involves a harem with a dominance hierarchy:

Male → Female 1 → Female 2 → Female 3 → Female 4 . . .

If the male is killed or lost from the group, the dominant female becomes the new male:

X → Male → Female 2 → Female 3 → Female 4 . . .

In another case involving anemonefish *(Amphiprion* sp.*)*, individuals are male first and then female. They live in groups among anemone colonies, but only the female and one of the males breed (Figure 10-7). If the female dies, the dominant male becomes female, and the next lower ranked male becomes the breeding male. Most mating patterns, however, are of the more familiar bisexual mating systems.

The basic patterns (Figure 10-8) that usually are considered are *promiscuity* or *polybrachygamy* ("many brief matings") (Selander 1972), *polygamy*, and *monogamy*. In promiscuity any male may mate with any female, usually more or less at random. Polygamy involves one individual of one sex mating with several individuals of the opposite sex. If it is one male with several females (the usual case), the specific term is *polygyny*. If one female mates with several males, the pattern is called *polyandry*. With monogamy there is one male with one female. Occasionally there may be a small stable group of two or more males and two or more females that mate with each other, a pattern called *polygynandry* (Daly and Wilson 1978). This form of group marriage occurs formally in northern India among people known as the Pahari and informally (not legally accepted) in Western culture. Polygynandry, which involves more stable groups than promiscuity,

**Figure 10-8 Mating patterns.
See text for further explanation.**

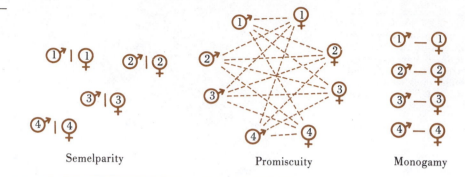

Semelparity   Promiscuity   Monogamy

has received little attention among species other than humans. It is partly a semantic issue; if the terms were to be clarified and distinguished more rigorously than at present, polygynandry might be discovered to be more common among nonhuman species than is realized.

These categories may be subdivided or mixed, and other patterns also have been identified recently. Polygynandry is closely related to a similar category, simultaneous polygamy, in which individuals may mate with more than one other individual with overlap between different polygamous relationships. Relationships in which one individual mates with only one other individual at a time but with new individuals over successive times have been called both *serial polygamy* and *serial monogamy*. If a given pair bond lasts through several breedings during one season, for example, one summer, but not into new seasons, the pattern is termed *seasonal monogamy*. Pair bonds lasting for the life of the individuals are called *perennial monogamy*. Wittenberger (1981:Table 11-1) classifies mating systems by both temporal and spatial categories. In monogamy, for example, temporally there are serial and permanent classes, and spatially there are territorial monogamy, female-defense monogamy, and dominance-based monogamy. (For explanation of differences and other cases, see Wittenberger.) Finally, individuals of some species mate only once during their lifetime, a situation known as *semelparity* or "big-bang" reproduction. (Repeated matings are classed collectively as *iteroparity*.)

What are the ultimate factors that lead to the observed diversity in mating patterns? In addition to the general biological and environmental factors already implicated in reproductive behavior (species isolation, prevention of cannibalism, synchrony of the pair, and synchrony with the environment), one now needs to consider two more factors to understand mating patterns: differential investment by the sexes and spatial (versus temporal) distribution of resources (see also Chapters 4 and 8). Spatial distribution of the resources may affect social behavior, such as territoriality and grouping in the first place, as well as the mating patterns per se.

**Differential Investment
by the Sexes**

The primitive condition of sexual reproduction is believed to involve essentially equal-sized, undifferentiated gametes *(isogamy)* and undifferentiation of the sexual parents. However, from this condition one gamete, the sperm, became much more mobile and smaller. It was stripped down to the bare essentials of its pack-

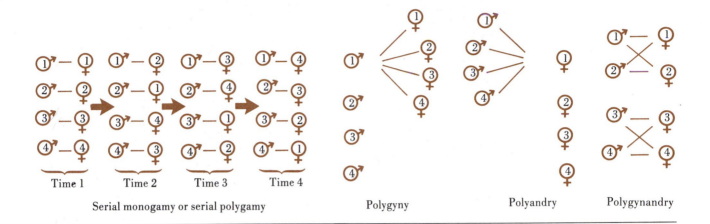

Serial monogamy or serial polygamy   Polygyny   Polyandry   Polygynandry

Figure 10-9 Examples of mating patterns and accompanying sexual dimorphism (differences between male and female) among birds, ranging from monogamy and little or no dimorphism to polygamy and dimorphism. The dimorphism is thought to have evolved in response to sexual selection. A, Pair of cliff swallows, not dimorphic, gathering mud for building the nest. B, Robin which is only slightly dimorphic; males have a darker head and redder breast than females. C and D, Male and female redwinged blackbirds in which the appearances of the sexes are markedly different. Photos courtesy Ed Bry, North Dakota Game and Fish Department.

A

B

C

D

age of DNA, an energy supply, a flagellum for movement, and a protein coat to contain these. Development of the fertilized zygote, however, requires additional energy and material, which became contained in the other gamete, the ovum. The condition of unequal gametes is called *anisogamy*.

In many species the amount of energy and material invested in the ovum may be considerable and highly significant in the life of the female producing it. The ova of reptiles, birds, and many other groups are composed largely of stored materials. In birds *one* egg may be 15% to 30% of the female's weight, whereas a male will not expend over 5% of his total weight in sperm and carrier fluids during an entire reproductive season. Mammalian ova do not contain much stored mate-

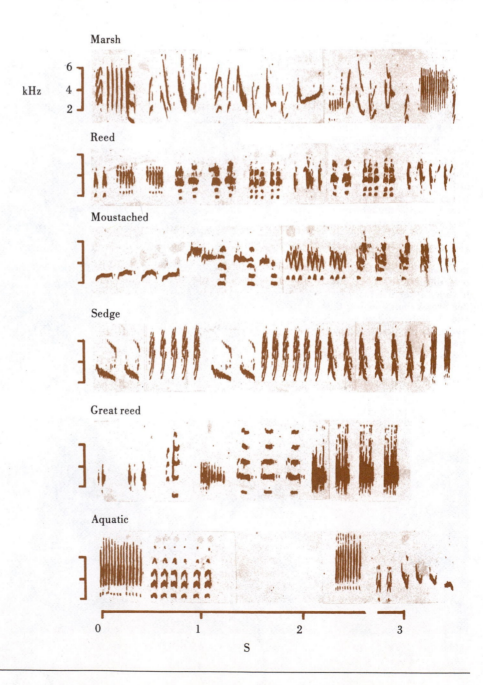

Figure 10-10 Complex songs of six species of European warblers, interpreted as auditory ornamentation resulting from female-choice sexual selection. From Catchpole, C.K. 1980. Behaviour 74:149-165.

Figure 10-11 Fighting between males in a variety of species. A, A midair clash between two fighting male rabbits. B, Male Masai giraffes fighting. C, Bighorn sheep fighting during the breeding season. All of these examples are mammals, but male fighting is common among some species of all vertebrate classes, arthropods, and others. See also Figures 10-12 to 10-16 and 10-19. A from Mykytowycz, R., and E.R. Hesterman. 1975. Behaviour 52:104-123. Photo courtesy R. Mykytowycz and E.R. Hesterman. B by Leonard Lee Rue III. C by Len Rue, Jr.

A

B

C

rial (yolk), but the female nonetheless invests considerably in pregnancy. Thus the bulk of the reproductive investment and responsibility in most species goes to the female, whereas the male mainly provides genetic information.

This differential investment by the sexes raises important implications for the mating of the individuals and, in the view of at least some sociobiologists (Wilson 1975:Chapter 15), introduces conflicts between the sexes, actually opposes cooperation, and forces what would otherwise promote social behavior. In particular it is hypothesized that the female should select males carefully and the males, on the other hand, would be expected to attempt to mate more frequently, other factors being equal. The females that are most successful (in terms of survival and reproduction) are those which pick the "best" mate (strongest, most likely to gather food if necessary, most likely to survive, etc.) for genes to combine with hers. The female stands to lose much, perhaps an entire breeding season, if she does not obtain the best mate. Males, however, do not have much to lose if they mate with unproductive females. All of this has led to what Darwin (1871) described as *sexual selection*. Sexual selection is a specialized category of natural selection, with emphasis on differential reproduction (rather than on survival). Such selection occurs in one or both of two basic forms: (1) female choice (also called *inter*sexual or *epigamic* selection) and (2) male fighting and dominance (or *intra*sexual selection). Huxley (1938) first used the terms epigamic and intrasexual to distinguish the two forms. Epigamic selection increasingly has been referred to as intersexual to contrast it with intrasexual selection.

In the first case, female choice, the males compete to *attract* the females. The "best" male is thus determined by the greatest ability to display, perhaps implying (in the evolutionary sense) that a male that has the time and energy for such display also has the time and energy for other aspects of survival and reproduction. This leads to the evolution of (1) elaborate "attractive ornaments" and displays in the males (Figure 10-9) and (2) "coyness" and caution on the part of the female. ("Courtship" might satirically be described under this circumstance as a male chasing the female until she catches him.) Elaborate ornamentation may occur not only in visual displays, but it may also play a role in complex vocalizations or songs (Figure 10-10).

In the second form, intrasexual sexual selection, the best male is determined by winning in fighting. The females do not "choose" but rather are just mated by the winner. This leads to the evolution of (1) *male aggression and fighting* (Figure 10-11), often with structures such as antlers and horns (Figures 10-12 and 10-13), (2) an increase in male mortality (and consequently an altered sex ratio), (3) success by dominance over other males, and (4) in many cases harems, which are herds or flocks of females under the jurisdiction of a single male (Figure 10-14). Occasionally males struggle with each other over the female right up to and including fertilization or copulation (Figure 10-15). Patterns in the evolution of

Figure 10-12 Examples of head structures evolved under intrasexual selection. A, Moose skull illustrating size and development of antlers. This skull weighs 17 kg (37.5 pounds), most of which is in the antlers. B, Male Hercules beetle from South America. C, Lucanid beetles *(Chiasognathus granti)*; male is in front, female behind. Photos by James W. Grier.

A

B

C

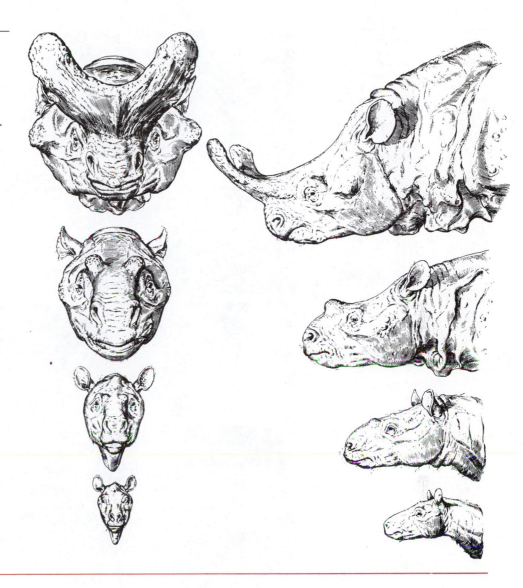

Figure 10-13 Head structures in males of closely related species. Reconstructions from a fossil series of extinct titanotheres (mammal). Similar series have been constructed for numerous organisms including dinosaurs, horned and antlered mammals, lizards, and several groups of horned insects. From Stanley, S.M. 1974. Evolution 28:447-457.

male fighting have been proposed for some groups of organisms such as ungulates (Figure 10-16).

Both forms of sexual selection (female choice and male fighting) may lead to the following:

1. Greater variance in the reproductive success of males than among females. Most females mate, but only a few males do during any breeding season. Those few males are responsible for most of the reproduction. It is difficult, however, to follow all individual males over the course of their entire lives. Some males that do not get the opportunity to mate one season may mate in subsequent seasons. Thus it seems likely that there is much less variation in matings per lifetime for males than occurs during a single breeding season.
2. Weak pair bonds.
3. Sexual dimorphism—physical differences between the sexes.
4. Delayed male maturity until the male is older and stronger.

Figure 10-14 Harems: the prize for winners of male fights. A, Red deer. B, Fur seal bull (in center) and harem of cows. Note dimorphism between the sexes. A from Animals, Animals. Copyright by Robert Maier. B by Leonard Lee Rue III.

A

B

Spatial Distribution of Resources

Uneven distribution of resources may pose important selective pressures on reproductive behavior. If males stake a claim to territories containing a great amount of such resources, it may be advantageous for a female to mate with him even if he already has another female. In fact, it has been postulated that under certain situations the presence of an already-mated male could serve as a signal of that male's capability to provide for additional females.

In other situations, however, particularly those involving predatory species, resources may be so dispersed and limiting that a male can provide for the offspring of only one female at a time, and both sexes may be required to care for the young. This may hinge, furthermore, on whether the young are altricial, that is, helpless and need to stay in one place while the parents provide for them, or

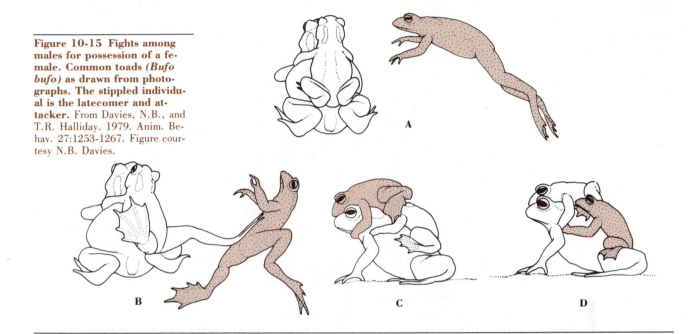

Figure 10-15 Fights among males for possession of a female. Common toads (*Bufo bufo*) as drawn from photographs. The stippled individual is the latecomer and attacker. From Davies, N.B., and T.R. Halliday. 1979. Anim. Behav. 27:1253-1267. Figure courtesy N.B. Davies.

precocial, that is, capable of moving about soon after birth or hatching and able to feed themselves. Altricial animals include, among vertebrates, rabbits, carnivores, many raptorial birds, and others. Examples of precocial animals include hares, many ungulates, waterfowl, shorebirds, and galliform birds. The presence of the male may be more beneficial, if not necessary, among altricial species. In precocial species, where young can feed themselves and require less protection, the male not only may be unnecessary but, in some cases, even disadvantageous by attracting predators.

Figure 10-16 Hypothesized stages in the evolution of cranial combat and displays in ungulate mammals. A comparative study of species suggests the sequences as described here. From Molnar, R.E. 1977. Evol. Theory 3:165-190. Based on Geist, V. 1972. Z. Säugetierkunde 37:1-15.

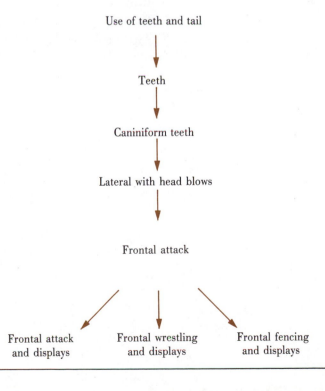

The Evolution of Mating Patterns

By sorting through the preceding factors, one now is in a position to consider the possible evolution of different mating patterns. The following model is a synthesis of the diverse models proposed by several others (e.g., see Emlen and Oring 1977), based primarily on studies of birds. The model is summarized in Figure 10-17. The most primitive and most frequently observed (including among mammals) condition is promiscuity or, more commonly among many invertebrates, semelparity. Any social order within this system usually is based on female feeding and care of the young. Such groups are named matriarchal (or *matrifocal*) (Wilson 1975).

Unevenly distributed food or other habitat resources may serve as a cause of competition. Several females may group together at the best location, and males may fight over the location. Whether over resources or females, competition may lead to the establishment of breeding or multipurpose territories. In other cases the link to food or other environmental resources is less direct or not defensible. The females themselves may be the object of male fighting, or several females may choose the same male. *In all of the these situations, sexual selection is the general factor that leads to polygyny.*

Monogamy is hypothesized to be derived from polygyny where the driving force involves the male's ability and need to help feed and care for the young, as in most birds and carnivorous mammals. Monogamy is the pattern found among approximately 90% of the bird species. The investment of males in parental care may cancel out any previous unequal investment by the sexes. If so, the male should become cautious of which offspring he cares for. A female is always certain that the offspring are hers, assuming no mix-ups after birth. But a male cannot know whether a female's offspring are from him or another male unless there are additional safeguards. This problem will be discussed further later.

Polygyny may be *secondarily* derived from monogamy in altricial birds with unevenly distributed resources, where one male may be able to provide adequate resources for more than one female.

Polyandry is rare. It is clear in only a few cases, such as jacanas and tinamous (types of birds). It may have evolved from monogamy through intermediate stages, whereby a female leaves a male while he is relieving her of incubation duties; then she does not return but goes on to another male and mates with him. This or

Figure 10-17 Proposed model for the general evolution of mating patterns.

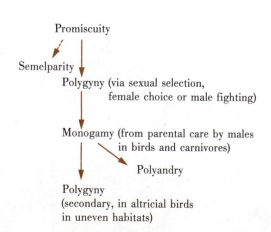

Figure 10-18 Mating patterns of grouse as found in different habitat types. From de Vos, G.J. 1979. Behaviour 68:277-314. Figure courtesy G.J. de Vos.

Monogamy

Tundra species:

Rock ptarmigan *(Lagopus mutus)*
Whitetailed ptarmigan *(Lagopus leucurus)*
Willow grouse + Red grouse *(Lagopus lagopus)*

Forest species:

Hazel gouse *(Bonasa bonasia)*

Polygyny, displaying males dispersed

Forest species:

Ruffed grouse *(Bonasa umbellus)*
Blue grouse *(Dendragapus obscurus)*
Spruce grouse *(Dendragapus canadensis)*
Sharpwinged grouse *(Dendragapus falcipennis)*
Capercaillie *(Tetrao urogallus)*

Leks

Steppe species:

Black grouse *(Tetrao tetrix)*
Sharptailed grouse *(Tympanuchus phasianellus)*
Prairie chicken *(Tympanuchus cupido)*
Sage grouse *(Centrocercus urophasianus)*

other sequences may result in a form of role reversal whereby the male ends up with most or all parental duties and investment of time and resources. A prime example is seen in seahorses and pipefish, where the female lays her eggs in a pouch on the males' abdomen. The male is left with the job of pregnancy. In such cases sexual selection may be reversed so that the male does the choosing, and the females do the displaying.

In other instances (e.g., some human systems in Tibet) polyandry may involve extreme forms of resource limitation where the joint efforts of two or more males, usually brothers, provide the care for a single female's young (Goldstein 1971). Polyandry would seem, however, to run against the basic forces—that a female should choose the best, not several, and a male should not waste his time taking care of offspring that could be from another male.

When all factors are considered together, including resource distribution, predation, and restraints over care of young, it appears that various environments may involve sets of factors that can lead to similarities in mating patterns among species living in those environments. De Vos (1979), for example, compared mating patterns among several species of grouse (Figure 10-18). De Vos' primary interest in the review was arena behavior, the subject of the the next selection.

Communal (Arena or Lek) Displays

An extreme form of intersexual selection, possibly in conjunction with ecological considerations (where it may be advantageous to remove the noisy and highly visible display area from the area occupied by hens and young), female choice may lead to communal displays, or leks, whereby all the males display in one common area (Figure 10-19).

These communal or arena displays represent a form of polygyny. Most of the females mate with only a small number of males, usually those in the center of the lek. The function or functions of communal displays are believed to be enhanced signaling and perhaps collective stimulation of the females. Groups of males are believed to be more attractive to a female than a solitary male, and thus the

Figure 10-19 Bird leks as seen in two closely related grouse. A to C, Prairie chicken *(Tympanuchus cupido).* D to F, Sharp-tailed grouse *(Pediocetes phasianellus).* In each series of three pictures the first shows several males on the lek; the second is a close-up of a displaying male; the third shows two males fighting at the boundary of their territories. Photos courtesy Ed Bry, North Dakota Game and Fish Department.

A

D

advantage to an individual male is that he is more likely to encounter a receptive female if he is in a group.

Communal displays are found among both invertebrates and vertebrates. Some fireflies in Asia display communally, with thousands in a single tree and synchronous flashing. The light is so intense and the trees so consistently used that these trees are reported to be used by humans in the area as navigational aids. Some fruit flies in Hawaii display communally on tree ferns. Other examples include singing by 13- and 17-year-old cicadas, nuptial flights of many insects, and dancing swarms of mayflies. The communal displays of many species of small insects may involve so many insects that they appear to be clouds of smoke from a

distance. The loud sounds of cicadas, incidentally, may serve not only as courtship signals but also, because of their high intensity and the associated discomfort for vertebrate ears, to repel potential predators.

Among birds communal display appears to have evolved independently in at least 10 different families (e.g., ruff, grouse, manakins, bowerbirds, and others). Arena displays may be observed also in several mammals. In some open-country antelopes (e.g., the Uganda kob) lek behavior is shown under certain environmental conditions. An African fruit-eating bat displays communally in trees with fast calls and wind beating.

Nonmodal Forms of Reproductive Behavior

A number of forms of reproductive behavior that are not considered acceptable in Western human society supposedly have been observed in other animals. The behaviors include those which, at least superficially, resemble rape, cuckoldry, and prostitution in humans. The terms are anthropomorphic and not considered useful by many behaviorists (see disclaimer, pp. 313 and 314). In addition, the interpretations of such behavior, regardless of which terms are used, have generated considerable discussion and lack of agreement in some cases. Differing viewpoints on rape or forced copulations, for example, are presented by Barash (1977a,b, 1978), Hailman (1978b), and several others. Barash considers rape as a possible adaptive male strategy from a sociobiological viewpoint. Hailman, on the other hand, noting that rape often occurs prior to development of spermatogenesis and ability to fertilize the female, interprets the behavior as a possible developmental stage of courtship.

Cuckoldry is a term derived from the parasitic nest behavior of cuckoos (as in the American cowbird) in which females leave their eggs in the nests of other birds to incubate, hatch, and raise. Although it is a strained term, as a variant of mating patterns cuckoldry involves females that get males to help raise young that were fathered by other males. A number of anticuckoldry behaviors have been postulated.

Little attention has been paid to possible "prostitution" in nonhuman animals, but a seemingly clear case has been presented for the purple-throated carib hummingbird *(Eulampis jugularis)* (Wolf 1975). Male hummingbirds defend flowering trees against all others of their species, both males and females. Males may permit females to feed at those trees, however, after they have copulated, including times outside the nesting season. The males copulate, permit the females to feed for a given period of time, then drive them off.

PHYLOGENETIC COMPARISONS OF REPRODUCTIVE BEHAVIOR

Much of the discussion about the phylogeny of communication simply can be transposed to this chapter. In short, there are few phylogenetic trends except within particular groups as was discussed, for example, for mating patterns in birds. Instead, the evolution of reproductive behavior may be viewed more appropriately relative to environmental factors and much convergence. Leks, for example, occur among insects, fish, birds, and mammals—all groups that are not closely related to each other. Similarly, monogamy, when found in birds and some carnivorous mammals, probably is convergent. Monogamy appears to be an adaptive response to the spatial distribution of resources and ability of the male to help care for the offspring. The most important roles of the male are in defense against predators and helping to provide food.

There are, however, some broad comparisons that can be made among major taxonomic groups. Between birds and mammals, for example, there is a major difference in the aid that males can provide in the care of offspring. The female is constrained to a much larger maternal role in mammals, partially precluding or reducing the male's role. In mammals the female carries the developing embryo within her body, then provides the major portion of food via the milk, a distinctly mammalian feature (except for a few isolated analogous instances, such as in doves and pigeons). In birds, on the other hand, the embryos develop in eggs outside the female's body. Feeding in altricial birds consists of finding food and taking it to the young. The male bird is able to help in both of these major tasks, incubation and initial feeding, whereas the male mammal is not unless it is indirect (e.g., protecting and feeding the female).

This major difference probably accounts for the broad differences between reproductive patterns in birds and mammals, for example, the prevalence of monogamy in birds and its rarity in mammals. For a further comparison, see Figure 10-20. Among precocial birds, those with young which are up and running and able to feed themselves soon after hatching, the mating systems are more similar to those seen in mammals. Most waterfowl and shorebirds, for example, are not monogamous or are only seasonally monogamous. Some of the larger species (e.g., geese and swans), however, are monogamous.

The role of the female in mammalian reproduction also may have predisposed mammals to the matriarchal social structure that is so common when social groups remain together beyond the immediate period of copulation and reproduction. The matriarchal groups consist of one or more older females, females of intermediate ages, one or only a few breeding males, and young immature males. Related females from several generations may remain together, but the male offspring are forced to leave after a minimum period of dependency. Only one or a

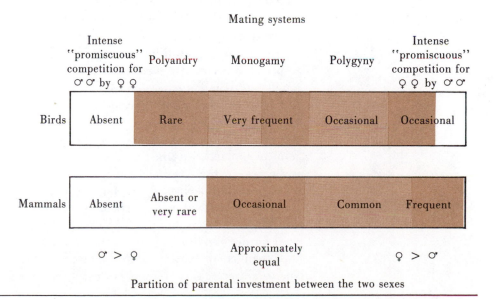

Figure 10-20 Mating patterns versus parental investment in birds and mammals. From Fig. 5-8 in M. Daly and M. Wilson, *Sex, Evolution, and Behavior*, © 1978 by Wadsworth Publishing Co., Inc. Used by permission of Willard Grant Press, a division of Woodsworth, Inc.

Mating systems

	Intense "promiscuous" competition for ♂ ♂ by ♀ ♀	Polyandry	Monogamy	Polygyny	Intense "promiscuous" competition for ♀ ♀ by ♂ ♂
Birds	Absent	Rare	Very frequent	Occasional	Occasional
Mammals	Absent	Absent or very rare	Occasional	Common	Frequent

♂ > ♀   Approximately equal   ♀ > ♂

Partition of parental investment between the two sexes

few adult males, initially unrelated to the females, are permitted within the group. (If the dominant male remains dominant long enough, he will begin mating with daughters and grandaughters, all of which stay with the group after attaining maturity.) Other males must fend for themselves or perhaps join bachelor herds. This matriarchal system is seen in elephants, some carnivores, a number of primate species, several ungulates, and a few other species of mammals.

Beyond the birds and mammals there are few instances of long-term relationships. Some examples include cichlid fish and a few other fish that maintain territories or occupy complex environments, such as tropical reefs. For the most part, however, reproductive relationships are simple and short term.

Among those invertebrates which have complex reproductive social relationships, such as many of the hymenoptera, the basis of the grouping is almost always mother, daughters, and sisters. In hymenoptera the queen mates only once. Termite queens must mate repeatedly, but only one male is involved; hence it is a monogamous system. Except for termites and the one-time mating in many others, lasting sexual relationships such as perennial monogamy are not known among any of the invertebrates. Furthermore, individual recognition is thought to be relatively rare among the invertebrates, as discussed in Chapter 8. Recognition and identification of social partners in insects are believed to be on the basis of caste rather than individual recognition (e.g., Wilson 1975).

PARENTAL CARE

The earlier discussion on evolution and the differences between birds and mammals in reproductive behavior raises an important topic for reproduction in general: parental care. In many species the attention given to reproduction stops once the eggs have been fertilized. The eggs are shed into the environment or otherwise simply abandoned. From this complete lack of parental care there ranges a continuum of increasing parental attention that reaches its peaks in species at the "pinnacles" of social development (Figure 8-3 and pp. 230 and 231). The individuals of many species of social insects cannot survive if separated from their extended family, which usually is synonymous with the colony. Humans are, perhaps, not quite that dependent on the family, and the nature of our social and family characteristics differs markedly from those of the social insects in many respects. But virtually everyone must be aware of the importance of the human family and parental care or substitute parental care.

In between the two extremes of zero to almost complete dependence on parental care are many intermediate levels (Figure 10-21). The next level above zero amounts to selecting a site for depositing eggs where there is either or both some form of protection against predators and the elements or a proximity to food resources for the emerging young. Eggs may be deposited, often by means of specialized morphological structures, ovipositors, in soil or sand, under vegetation, in crevices, or perhaps stuck to vegetation or other material that will later serve as food. The adults leave, and the young must hatch, enter the world, and fend for themselves.

Next in complexity of parental care are species that build definite structures, such as nests or some kind of egg case, that protect the eggs, but then abandon

Figure 10-21 Examples of parental care among species. A, Egg brooding by *male* water bug *(Abedus herberti)* with newly hatched nymph nearby. B, Adult urchin *(Strongylocentrotus franciscanus)* with juveniles under its spine canopy. Juveniles of a closely related species in the same genus are not commonly found under the adults. C, Lycosid spider carrying young via specialized abdominal hairs. D, Oral birth from gastric brooding by the frog *Rheobatrachus silus*. A from Smith, R.L. 1979. Science 205:1029. **B** from Tegner, M.J., and P.K. Dayton. 1977. Science 196:325. Figure courtesy Mia Tegner and Paul Dayton. **C** from Rovner, J.S., et al. 1973. Science 182:1153. Photo courtesy Jerome Rovner. Copyright 1973, 1977, and 1979 by the American Association for the Advancement of Science. **D** from Tyler, M.J., and D.B. Carter. 1981. Anim. Behav. 29:280-282. Photo courtesy M.J. Tyler.

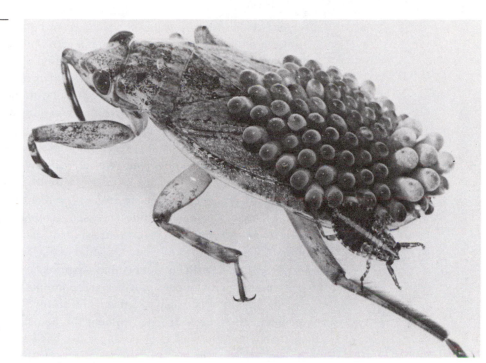

A

B

C

D

them. Next are those species in which one or both parents remain with the eggs until the point of hatching. Some remain with the young for varying periods after hatching and continue to provide some degree of protection or food. For those which stay with the young, the adults may either remain at the location where the eggs are, or they may carry the eggs or young with them, either in or on the body.

Finally, one encounters extended parental care, which, in the extreme cases, forms a family unit that continues for as long as the individuals live.

With increasing levels of parental and family care there are increasing numbers of social interactions and associated behavior patterns, such as between parents and offspring, among siblings, with other relatives or members of the group, and between mates concerning their offspring. The increase in parental care accounts for much of the increased diversity of communication and also permits more complex social relationships to develop in some species than in others. Species that totally lack parental care, for example, do not have signals for interactions between parents and offspring.

Extended parental care provides an opportunity for acquisition of experience and modification of behavior, that is, learning (Chapter 19) far more than in species with young that hatch without the benefit of parental care. Young that must fend for themselves generally must have the answers to life's problems already at hand, figuratively speaking.

The role of the parent in learning in most species is not to teach the offspring directly but rather to serve as a means of protection while the young learn by trial and error. Direct teaching by parents and other group members does occur in many mammals and some birds. Migratory routes in bighorn sheep and cranes, for example, are learned in this manner. Cultural learning is discussed further in Chapter 19. Parental care does not occur only in the context of opportunities for learning. The protection of parents may function simply to increase survival through the period when offspring are smaller, more fragile, less developed, and otherwise more vulnerable.

Differences in level and complexity of parental care cut across phylogenetic boundaries, and, as is encountered frequently in considering evolution of behavior, there are few phylogenetic trends. There are a few constraints. Birds, for example, are homeothermic and must provide heat for their developing eggs and newly hatched young (Figure 10-22), although they are not required to do it from

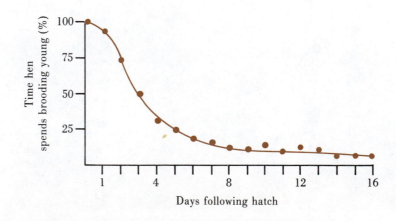

Figure 10-22 Decline of parental brooding of young in red junglefowl (*Gallus gallus*) as chicks get older and develop thermoregulatory ability. Modified from Sherry, D.F. 1981. Behaviour 76:250-279.

their own bodies. Megapodes, or brush turkeys, for example, lay their eggs in piles of decomposing vegetation from which the eggs get the necessary heat. The eggs of some birds, particularly the more primitive ones such as the loons, may withstand a considerable amount of periodic cooling. In a similar manner, female mammals bear their young alive (viviparity), and newborn mammals are dependent on milk from their mothers. This constrains certain types of parental care. Mammals are the only animals, for example, in which there are no known cases (excluding modern humans with milk supplements) of parental care in which the male may be left with total responsibility for the offspring.

Although factors such as thermoregulation are constraints in some species, they may play a role even in less constrained species. Many reptiles incubate their eggs (boas) or provide nests that produce warmth through solar heating (many species) or rotting vegetation (crocodilians). Thermoregulation of developing eggs and young is provided also by many colonial insects such as honeybees.

In some vertebrates, including birds, and arthropods there are species in which males assume complete responsibility once the eggs are deposited. The seahorses have already been mentioned, and there are several species of fish and a few birds where the males are left to guard the nest or carry the young about. But beyond these few constraints, examples at most levels of complexity can be found throughout the higher groups, particularly the arthropods and vertebrates. Remaining with the eggs, for example, occurs in octopuses, several insects, crustaceans, arachnids (e.g., some of the spiders), many fish such as the sticklebacks, a few amphibians, boas among snakes, crocodilians, and, of course, birds and mammals. Not all mammals retain the developing eggs in their bodies (e.g., platypus and spiny echidna), but even these latter forms stay with the eggs and nourish their young from mammary glands.

As should be increasingly apparent, environmental factors may be more important in the evolution of behavior than who one's phylogenetic relatives are. This clearly is the case for parental care. Wilson (1975) lists several factors that are thought to be conducive to the development of parental care. They include:

1. *Stable, structured environments.* The argument is that, in such environments, populations increase in size until competition becomes a very important factor. Organisms that help their offspring may obtain the competitive edge over those which do not. In unstable, constantly changing environments, on the other hand, conditions change and populations may be reduced so that the environment is never saturated and competition is less important. The best reproductive strategy under these circumstances is to produce a large number of diverse young, some of which may be adapted to the new conditions and may survive.

All of this often is discussed in terms of *r-versus-K selection,* a topic that has been deliberately avoided until now. Those species which have lower rates of reproduction, frequently more parental care, larger body size, and a host of associated characteristics, supposedly are K selected (Pianka 1970). The other extreme, with higher rates of reproduction and other characteristics, have been labeled as r selected. An r-versus-K explanation for parental behavior (as well as some other behaviors) is commonly used in the literature, including in Wilson (1975). Hence, the topic needs to be acknowledged here.

Many biologists, including myself, however, object to the concept of r-versus-K selection. Unfortunately, there is not space or time here to develop the necessary description and discussion of population dynamics and population genetics. In brief, r and K are abstract, derived mathematical constructs, similar to a quotient that is derived from a dividend and a divisor. There are many different combinations of dividend and divisor that produce the same quotient, and the same can be said of r and K. The quantities, r and K, are derived from simple mathematical models and are *consequences* of numerous interacting biological factors. Instead of selecting for r and K, natural selection more likely is having its effects on the components that go into r and K. There are numerous alternate hypotheses, for example, that better explain the evolution of larger body sizes or particular litter or clutch sizes (Emmel 1976) than selection for r or K. The phenomena (e.g., differences in reproductive output among different species) are very real. The hypothesis that stable, structured environments may lead to the observed phenomena may be valid (although not yet tested). The problem is the misleading r-and-K selection *label*. In the case of different numbers of offspring and associated parental care, the terms *high or low fecundity* (numbers of offspring produced) describe the situation adequately. Williams (1975), for example, uses the terms high and low fecundity without even mentioning r and K selection.

2. *Difficult environments*. This factor might seem to be a contradiction of the previous one, and indeed it may be. But different factors frequently may lead to similar outcomes, such as parental care. Some aspects of life, whether stable and predictable or not, may nonetheless be difficult and hazardous. Mature, reproductive organisms may travel into new environments where they survive because of larger body size or whatever, but any unprotected offspring would not. One or more factors in the environment may be adverse to life of smaller, more delicate young animals. As a familiar example, brooding birds can protect their chicks from precipitation and extreme temperatures by covering them when cold and shading them when under direct, hot sun. Many young animals can remain protected in burrows, dens, or nests while the parents go out to face a hostile environment to obtain food. Parental care permits the young to reach a stage where they can then deal with the environment on their own.

3. *Specialized diets*. This might be considered somewhat as a variation on the previous point; the young are not able to find, obtain, or handle the normal food of the species until they are larger, stronger, or otherwise more skillful or better equipped. This category applies particularly to predators that normally tackle difficult quarry or where prey is difficult to find. Many species of predators, with or without parental care, typically start with easier prey and then switch to more difficult types as they become older. Young kestrels *(Falco sparverius)*, for example, apparently prefer insects when first fledged, then later develop a preference for larger animals such as mice (Mueller 1974).

4. *Predator pressure*. There may be a very significant selective advantage to providing protection of young from predators. This might be viewed as another variation of the difficult-environment hypothesis, where something in the environment (predation) is posing a specific threat to the welfare of the offspring.

One or more of these factors may lead to the development of parental care and increased parent-offspring communication. Trivers (1972, 1974), in line with other sociobiological developments, published two important papers that stimulated much attention on the subject of parental care and behavior.

In his 1972 paper Trivers discussed the role of parental care in sexual selection. He defined *"parental investment as any investment by the parent in an individual offspring that increases the offspring's chance of surviving (and hence reproductive success) at the cost of the parent's ability to invest in other offspring."* He continued (1972:139-140):

One assumes that natural selection has favored the total parental investment that leads to maximum net reproductive success. Dividing the total parental investment by the number of individuals produced by the parent gives the typical parental investment by an individual per offspring. . . . Since the total number of offspring produced by one sex of a sexually reproducing species must equal the total number produced by the other (and assuming the sexes differ in no other way than in their typical parental investment per offspring) then the sex whose typical parental investment is greater than that of the opposite sex will become a limiting resource for that sex. Individuals of the sex investing less will compete among themselves to breed with members of the sex investing more, since an individual of the former can increase its reproductive success by investing successively in the offspring of several members of the limiting sex.

With this argument, Trivers essentially extended the anisogamy considerations and formally added parental care considerations to the list of items that could lead to sexual selection (Trivers 1972:168).

The differences in parental investment between sexes and at different times opened the door to a new theoretical framework. In particular, it permitted one to view the roles of the two sexes in their own selfish, competing contexts, rather than necessarily as cooperating on a mutually beneficial venture. This view fit other, contemporary views of selfish genes (as discussed on pp. 119-121). This lead to theoretical considerations of such things as desertion and cuckoldry, differential mortality of the sexes, and the evolution of a host of characteristics related to mating patterns and differences in behavior between the sexes.

Included in these considerations are one sex cooperating with or taking advantage of the other, honesty or deceit and cheating, and hypotheses as to when parental care should be biparental, that is, involve both parents as opposed to just one. Females have some options for extracting parental investment from the males; some patterns preadapt the pair for biparental behavior, such as the male guarding the female against rivals, and, depending on the situation, males also may have at least *some* investment worth protecting in the offspring. For details on these many aspects, see any of several recent books on social and reproductive behavior, for example, Wittenberger (1981).

Parent-Offspring Conflicts

Focusing on the selfish aspects also led to other hypotheses concerning parental care and behavior. In 1974 Trivers published a second important paper, in which he proposed that parent-offspring conflict resulted from the conflicts of interest between parents and offspring over continued parental care. In particular, there comes a point when the benefit of continued care diminishes from the parent's viewpoint, while the cost remains high or may even increase. It is to the adult's advantage to send the potentially independent offspring out on its own and begin

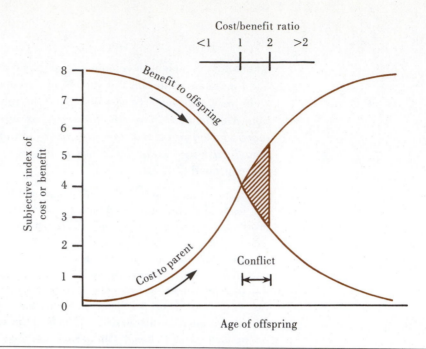

Figure 10-23 Parent-offspring conflict with changes in ratio of cost to parent/benefit to offspring, as proposed by Trivers (1974). The subjective index is merely to provide scale to help illustrate the ratio. Shapes of the curves shown here also are arbitrary. Modified from Trivers, R.L. 1974. Am. Zool. 14:249-264.

producing new offspring. But it still may be in the interest of the offspring to remain dependent, with the free protection and food that such has to offer. Eventually, as the offspring becomes stronger and more capable, the benefit diminishes so that the young adult is better off leaving and the conflict disappears. But there may be a period of time, depending on the species and situation, where there is a conflict of interest (Figure 10-23). The adult tries to push the offspring out, while the offspring tries to remain. This explanation has been provided for typical weaning behavior in mammals (Figure 10-24) and parent conflicts in other taxonomic groups.

That there frequently is serious conflict between parents and offspring at the stage before independence is quite clear. Some readers might even have experienced it in their own lives. But it is not yet clear that Trivers' explanation (1974) is necessarily the best. Earlier ideas included views that the conflict was somehow a necessary part of rupturing the relationship or a means of forcing the independence that the new adult would need in its life ahead (discussed in Wilson 1975). These interpretations viewed the conflict as the parent just doing what was necessary for the good of the young—a view that has not enjoyed widespread acceptance among sociobiologists.

But there are problems also in Trivers' explanation, as pointed out by Alexander (1974). In particular, the parent-offspring relationship is quite different from that between mates and sexes. Mates normally are not related closely, and there may be substantial genetic differences between the sexes. Parents and offspring, on the other hand, are directly related and carry some of the same genes. If inclusive fitness is a meaningful interpretation, then the interests of both parent and offspring must be taken into account. The genes are passed from parent to offspring, and those offspring someday may become parents. The successful sets of genes are those which work in both situations.

Alexander proceeds to reason that particular offspring are more expendable than the parents (which potentially could redistribute their reproductive investment) so the parents have control. If the offspring were to resist in a damaging manner, neither parent nor offspring (or, perhaps more importantly, the genes)

Figure 10-24 Weaning conflict. A sequence of drawings from photographs indicating a weaner's unsuccessful attempt to steal milk resulting in it being chased out of the harem. In the last illustration the weaner is seen resting outside the harem with a group of other weaned pups. From Reiter, J., et al. 1978. Behav. Ecology Sociobiology 3:337-367.

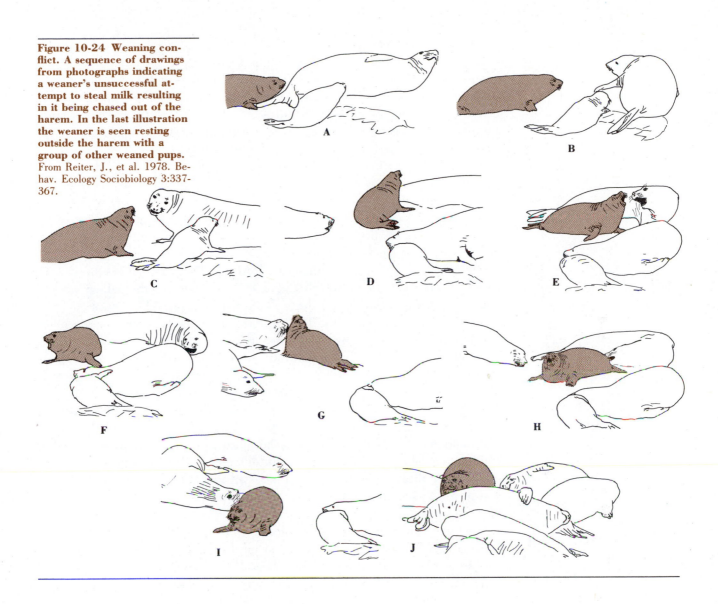

would win. Thus, according to Alexander, the end result is that *parents manipulate the offspring*. It is not yet clear which, if any, of these theories is correct. Perhaps new hypotheses will enter the arena, or new syntheses of several ideas will be forthcoming. The topic promises to be worth following in the future.

OTHER COOPERATION IN REPRODUCTION AND CARE OF YOUNG

The job of reproduction and parental care in many species is not confined strictly to the immediate parents. Additional helpers in rearing young are much more widespread than formerly supposed. Care of young by other than the parents is known as **alloparental care.** Depending on the sex of the helper, it can be *allomaternal* or *allopaternal*. Previous observers (e.g., Rowell et al. 1964) have used terms such as "aunt" or "uncle." But because these words carry connotations of genetic relationships, they have been dropped in favor of the more neutral alloparental or, simply, *helpers*.

The origins of alloparental helping are not well understood. The roots may lie in young from previous broods or litters that never fully leave, more or less hang around, or return, perhaps out of their own selfish desires for further protection or food. However it originated, it can be found in several of the most social species. It is the essence of insect sociality. It also occurs in at least 60 species of birds (with more cases being uncovered every year), porpoises, elephants, and carnivores, and it is particularly common among primates.

The benefits of alloparental care are easily hypothesized, even if the participants are not closely related. A major possible benefit for the helper is known as the *learning-to-mother* hypothesis; that is, the helper may learn to be a parent itself under the guidance of experienced supervision. Parenting is considered by some (references and discussion in Wilson 1975) to be sufficiently complex that it must involve a major learning component.

From the standpoint of the parent, the presence of helpers reduces the costs of caring for offspring. Provided the true parents are alert and the helpers not too incompetent, the presence of helpers would not be expected to expose the offspring to excessive, new risks. The presence of helpers also may permit other advantages of group living by perhaps increasing the division of labor and general level of efficiency, as well as by providing new social interactions and relationships.

Once extensive care and helping have become incorporated into a species' repertoire, it is a small evolutionary step to maintain long-term, continuous, and complex social groupings. In fact, at this point the role of reproductive behavior blurs with the whole of the life history. The entire life cycle then may be viewed as leading to enhanced, thereby successful, differential reproduction. Many biologists believe that reproduction is what life is all about anyway. It certainly seems reasonable, as outlined earlier, to postulate a most important role for reproduction in the development of the highest forms of social behavior.

Adoption, whereby an animal assumes responsibility and takes over parental care for an individual that is not or may not be closely related to it, is thought to be comparatively rare under natural situations. It is possible to force or trick adoption artificially, such as with animals in captivity or by switching young in species that identify their young by location (e.g., a nest) rather than individually. Except for a few recorded instances (Sade 1965, van Lawick-Goodall 1968), however, adoption is rare in the wild. In fact, in many species adults aggressively reject young that are not their own. This is as would be predicted by the selfish gene theory; one should not waste resources on anything that does not enhance inclusive fitness. Effort should be expended, instead, on producing one's own offspring. This principle, furthermore, may have played an important role in the ability of parents and young in some species being able to identify each other individually. When there are natural opportunities of offspring becoming mixed up with those from other parents, the ability to identify one's own would be advantageous (Burtt 1977).

Adoption has been observed, at least in the laboratory, among ants during territorial conflicts. Wilson (1975) describes the outcome of such raids: the losing adults may be killed or driven away, at which time the winners pick up the developing broods of the losers and take them back to their own nests. This behavior would seem to contradict the predictions of the selfish gene theory, and it is not clear whether it is merely accidental, offers selective advantages, or just what. Opportunities for such adoption would seem likely to occur frequently under natural conditions.

Adoption resembles and may be related to domestication (slave-making) and other types of behavior among some of the social insects. This leads to relationships and interactions among species (i.e., *inter*specific behaviors), the subject of the next chapter.

SUMMARY

Reproductive behavior may be quite simple and direct, but it frequently involves some of the most spectacular and complex behavior patterns known. The advantages of sexual reproduction, involving fertilization by two gametes and subsequent development, comprise an enigmatic topic. Regardless of why it occurs, from a behavioral standpoint it requires the successful coordination and bringing together of the two sexes. This is accomplished in a diversity of mating patterns described in this chapter.

Major evolutionary and proximate factors thought to be involved in the expression of reproductive behavior include species isolation, avoidance of mate cannibalism (particularly in predatory species), pair synchrony, temporal and spatial variability in the environment, predator avoidance, and differential investment by the sexes. These factors, in various combinations, have led not only to diversity and complexity in mating behavior but also to a particular category of natural selection referred to as sexual selection. This takes one of two forms of expression: male fighting and dominance (intrasexual selection) or female choice (epigamic or intersexual selection). A list of types of mating patterns, including several forms of monogamy and polygamy, is provided. An extreme form of display, seen among many vertebrates and invertebrates, is lek or arena display. In this type of courtship, several males display communally in one location. They are visited by the females, which appear to pick the male of their choice. Several nonmodal forms of reproductive behavior include rape or forced copulation, cuckoldry, and prostitution, where terms have been borrowed—perhaps wrongly—from human behavior.

Related types of behavior discussed in conjunction with reproductive behavior include parental care and care by helpers. Although some aspects of reproductive behavior may be constrained by phylogenetic considerations, such as viviparity and mammary glands in mammals, there are few broad phylogenetic generalizations that can be made. Instead, reproductive behavior, whether primary, parental, or auxiliary, appears to have evolved mostly in response to environmental constraints.

Recommended Additional Reading

Blum, M.S., and N.A. Blum, editors. 1979. Sexual selection and reproductive competition in insects. Academic Press, Inc., New York.

Campbell, B., editor. 1972. Sexual selection and the descent of man. Aldine Publishing Co., Hawthorne, N.Y.

Daly, M., and M. Wilson. 1978. Sex, evolution, and behavior. Duxbury Press, Boston.

Morris, D., editor. 1970. Patterns of reproductive behavior. McGraw-Hill Book Co., New York.

Williams, G.C. 1975. Sex and evolution. Princeton University Press, Princeton, N.J.

See also recommended additional readings in Chapters 8 and 9.

CHAPTER 11

A. Kerstitch

Starfish killing shrimp (*Hymenocera picta*).

INTERSPECIFIC BEHAVIORS SYMBIOSES AND PREDATION

Interactions among animals are not confined just to individuals belonging to the same species (i.e., intraspecific behavior). There are also numerous *inter*specific interactions and associated behaviors. Perhaps the most familiar general interaction between species is predation, whereby an animal of one species captures, kills, and (usually but not always) eats an animal of another species.

Studies of predation have focused on many different aspects of this phenomenon. Much of predation can be viewed simply as a special case of feeding, which is covered in Chapter 6. Predator or antipredator behaviors have figured importantly in considerations of ontogeny and learning (Chapters 18 and 19), ecology (Chapter 4), and the evolution of much of behavior (Chapters 5 and 8). Even the basic attributes that one associates with many species, particularly speed, strength, keen senses, advanced nervous systems, alertness and caution, and such things as various hard, protective body coverings, probably are direct evolutionary products of predation pressure. Species such as skunks that have other means of protection often lack much of the speed, agility, and alertness shown by other potential prey, such as rabbits. In another example, toads, which rely on a heavier defense of poison and bad taste than frogs, are much easier to capture than frogs.

After these aspects of predation are removed, what remain for the human observer and student of behavior are the chase, contest for survival, and all of the associated action (Figure 11-1). The general plot is the same over and over: the predator tries to find, capture, and kill the prey, and the prey tries to avoid those events. The variations on the theme—power, strength, speed, tactics, and uncertainty over the outcome in any particular encounter—make predation popular among the general public and professional biologists alike. The diversities of both predator and antipredator tactics rival or exceed the diversities in reproductive behavior. Before considering these, however, a broader perspective is needed, and one must look at other interspecific interactions. Predation is only one of many ways that animals belonging to different species interact.

Some interactions between different species bring the animals into close relationships throughout much or all of their lives. Such relationships are called *symbioses*, that is, living together. Wilson (1971) reviewed social symbioses among insects, the taxonomic group in which such relationships are most numerous, complex, and highly evolved. He subsequently (1975) expanded the treatment to other, noninsect organisms including vertebrates. Much of this chapter is based on Wilson's reviews. The social emphasis is reduced, however, to include single participants from one species or another; some categories are combined and relabled; and the subject is generalized to include predation. As will become

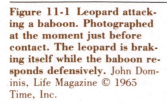

Figure 11-1 Leopard attacking a baboon. Photographed at the moment just before contact. The leopard is braking itself while the baboon responds defensively. John Dominis, Life Magazine © 1965 Time, Inc.

obvious, there is such diversity among interspecific interactions and graduations from one species and situation to another that various types of interaction merge. Attempts to classify them are necessarily arbitrary. Given the above qualifications, there are five major categories of interspecific interactions (see box, p. 343).

The first category, *simple aggregations* or *chance encounters* and *direct competitive interactions*, was not included by Wilson. This category is included here, however, because it often merges with and may lead to the evolution of other types of encounters. Furthermore, there may be direct behavioral interactions, particularly if competition over a resource results. Two scavengers, such as a badger and a cougar, for example, may happen across the same carcass at the same time. A definite fight, complete with sets of typical behavior patterns, may ensue.

In the second category, *commensalism*, one species benefits and the other is unaffected. *Mutualism*, the third category, differs from commensalism in that both species benefit. When one species benefits at the direct expense or loss of another, the interaction may be either *parasitism*, the fourth category, or *predation*, the last category. Predation generally involves a briefer encounter, the death of the victim, often some violence, and the consumption of most or all of the victim. Parasitism, on the other hand, may involve one of several forms of loss (not just consumption), and the interaction lasts longer—days, weeks, months, or years. Parasitism usually is not a primary cause of death, although it may contribute as a secondary cause.

The boundaries between predation and parasitism are not always clear cut, however, and some cases are not easily classified. The differences often are traditional and historical, or one may have to consider a combination of characteristics. Nestlings of some species of nest-parasitic birds may kill the young of the host species, but they do not eat them. Wolves, as predators of caribou, may follow the caribou herds and live in close association with them. Some social parasites in the nests of many insect species are also predators in that they kill and consume eggs, young, or other individuals of the host species. This kind of hybrid between a predator and parasite has been termed a *parasitoid*.

Most of the major categories of interspecific behavior just listed can be subdivided into other categories, as listed on p. 343. The remainder of this chapter is devoted to a survey of the various types of interaction, including examples and brief considerations of evolutionary origins.

Classification of Interspecific Interactions

I. Chance encounters
   A. Noncompetitive
   B. Competitive
II. Commensalism
   A. Plesiobiosis
   B. Interspecific groups
      (mixed flocks, etc.)
   C. Trophic commensals
III. Mutualism
   A. Interspecific groups .
   B. Trophobiosis
   C. Nontrophic symbioses
   D. Parabiosis

IV. Parasitism
   A. Trophic parasitism
   B. Xenobiosis
   C. Temporary social parasitism
   D. Dulosis
   E. Inquilinism
V. Predation

AGGREGATIONS/CHANCE ENCOUNTERS

Animals of many species meet; their paths cross with little or no significance; and the animals may show little notice of each other. They may meet coincidentally at a migration point or a nonlimiting resource such as a watering hole. Or they may, by chance, simply end up at the same place at the same time. In other instances, however, particularly over limited resources, animals of different species may compete and interact directly. Clashes over food are particularly frequent. The badger-cougar example was listed earlier. Many large mammalian and avian predators and scavengers are known to interact over prey. Wintering bald and golden eagles may fight over carcasses, as do several species of vultures and mammals in Africa. Various species of birds as well as birds and squirrels may interact at winter bird feeders. Similar encounters can be listed from many other taxonomic groups.

During competition over food many species display typical behavior toward animals of different species, and some are winners more often than others. At bird feeders, for example, some birds such as house sparrows or grosbeaks commonly establish a dominance over other species. Interactions among several different species of birds and mammals at a seed bait station are shown in Table 11-1. It seems most likely that such competition has led or at least contributed to the evolution of associated behavior such as removal of food from the original site, taking food to secluded or hidden areas for consumption, and caching or covering excess food. Many species bury extra food; leopards carry theirs up into tree branches; and some species urinate, defecate, or otherwise mark their food and prevent others from consuming it.

Aside from food, the resources next most likely to be limited and generate behavioral competition between species are holes or cavities for nesting or protection. Many hole-nesting birds and small mammals will fight for occupancy, and, again, some such as the European starling are commonly more successful than others.

Table 11-1 Results of encounters among antelope squirrels of known rank, quail (single or in groups), and single individual cottontails, scrub jays, and wood rats*

Opponents		Supplantations by	
A	B	A	B
Alpha, beta, or gamma squirrel	Gambel's quail (1 to 3 individuals)	19	12
Alpha, beta, or gamma squirrel	Gambel's quail flock	0	30
Alpha, beta, or gamma squirrel	Cottontail	5	11
Alpha, beta, or gamma squirrel	Scrub jay	13	2
Lower-ranking (delta or below) squirrel	Gambel's quail (1 to 3 individuals)	7	63
Lower-ranking (delta or below) squirrel	Gambel's quail flock	0	42
Lower-ranking (delta or below) squirrel	Cottontail	0	17
Lower-ranking (delta or below) squirrel	Scrub jay	4	10
Alpha squirrel	Wood rat	9	0
Lower-ranking (beta or below) squirrel	Wood rat	1	17
Gambel's quail flock	Cottontail	17	3
Gambel's quail (single or pair)	Cottontail	0	15

From Fisler, G.F. 1977. Anim. Behav. 25:240-244.
*Squirrel data are the combined results of encounters between single squirrels and each opponent (single or group).

COMMENSALISM

Plesiobiotic relationships are those in which two species commonly live in close relationship with each other but where the benefit to one or the other is slight or not understood. Some species of social insects, for example, live very close together, even in compound nests, but they keep their broods separate and otherwise live their own lives. Such relationships have not been well studied, and with further information many instances may turn out to belong to another category. Even if there is no benefit to either participant or if the relationship is one sided, plesiobiosis may preadapt the species for the evolution of more complicated relationships.

Interspecific groups (including flocks, herds, and schools) are those where *groups* of two or more species associate with each other. This category of commensalism, commonly referred to as *mixed groups*, is very similar to the previous category except the associations tend to be more transient in time and place. Most of these associations tend to be loose and not very consistent, but some can be quite stable. Mixed-species groupings are seen in marine and some freshwater fish, foraging and migrating birds, roosting bats, some mixed troops of primates, some dolphins, and open-plains ungulates. A few mixings extend even beyond class boundaries, such as between groupings of some monkeys and birds. There has been some attempt to classify the importance of different species to groupings based on their roles (e.g., Moynihan 1962): *nuclear species* are those primarily responsible for the attraction and cohesion of the mixed group; *attendant species* are the regular joiners; and *accidental species* are those which occasionally, but not commonly, become involved.

Which benefits occur in mixed-species groups and to whom they occur are not well understood. The benefits may be simply to the individuals. In many cases it seems almost as though species' distinctions are not relevant in these situations. This notion may be unsettling to many biologists, particularly those accustomed to "good of the species" thinking (pp. 127 and 128). The benefits may go primarily to individuals of one species (i.e., commensal) or to all (i.e., mutual). Because benefits can be to one or all, the category has been listed under both commensalism and mutualism (p. 343) (Wilson 1975). The major hypothesized benefits for mixed-species groupings are similar to those of intraspecific groupings: improved foraging efficiency, reduced predation, and epideictic functions. Epideictic behavior already has been discussed and discounted (pp. 117 and 118).

It is not clear just how mixed flocks can improve foraging efficiency, but circumstantial evidence supports the notion. Mixed foraging flocks in birds seem to occur during periods when conditions are harshest and food supplies are most limited, as during winter or in marginal habitat. Less successful individuals, regardless of species, that otherwise might fail on their own, may be able to take advantage of the experience and success of others. More eyes watching may make it more likely that food, which can be used by all, will be discovered. Among mixed herds of ungulates, such as those which formerly existed on the plains of North America, Europe, and Asia and those which still live in Africa, individuals of different species may facilitate each other's feeding by feeding on different parts of the vegetation. Animals of one species may remove those parts of plants that are obstacles or are indigestible to animals of other species.

Antipredation advantages would amount to extensions of those proposed for grouping within a species, namely the "you-first" principle, the confusion principle, and increased awareness to possible danger. Moynihan (1962) suggested that antipredation is the main advantage of mixed-species flocking in the tropics. Numerous recent reports support the role of antipredator functions for grouping, whether the grouping is intraspecific or interspecific.

Trophic commensals are those which live essentially off the garbage of another species. Remoras, or "shark suckers," for example, attach themselves to sharks and ride along to pick up bits and scraps of food from feeding sharks. Whole armies of scavenging birds, mammals, and arthropods may trail along after predators to clean up the remains. Most nests, dens, and hives, whether of birds, mammals, or social insects, have an entourage of arthropod scavengers, including numerous isopods, nonparasitic mites, collembolans, beetles, flies, and others. These commensals may be accepted and ignored by those with whom they live, or they quickly get out of the way and avoid direct encounters. Some interact directly with their hosts, using the same communication signals and displays. Some silverfish (insect) commensals that associate with tropical army ants even follow along during raids. Occasionally the chain becomes more complex. Antbutterflies follow army ants and feed on the droppings of birds, which also follow the army ants. The birds eat insects flushed by the ants. The butterflies follow the birds following the ants, but the butterflies apparently use cues, possibly the trail odors, from the ants (Ray and Andrews 1980).

MUTUALISM

Most forms of mutualism, where both species benefit, involve a benefit of food (hence, the term *trophobiosis*) for one of the participants. To the other participant goes a variety of benefits, depending on the situation, including protection (from predators, weather, etc.) and even dental and body hygiene. Body cleaning (Figure 11-2) is best known among several species of fish in which some clean others, but it also is seen in some reptiles and a number of large mammals over which certain birds forage and pick off parasites. Some species, such as cleaner fish of the genus *Labroides* are *obligate* or full-time cleaners (e.g., Potts 1973), whereas many others are only *facultative* or occasional cleaners (e.g., Sulak 1975, Brockmann and Hailman 1976). Specific, stereotyped behaviors often are involved in the interaction between the cleaner and its host.

A few species literally tend to and husband other species. Some species of hermit crabs, for example, attach sea anemones to their shells and, when they change shells, take the anemones along with them. Several species of ants and some stingless bees milk aphids and a few other types of insects for honeydew, a sugar-rich fluid that is excreted by many insects that feed on plant juices. The plant feeders extract only part of the nutrients and excrete the rest, something of an affluence in the presence of abundant resources. Some ants and bees simply follow the plant feeders and consume the honeydew, whereas many other species have evolved a mutualistic relationship, including signals from ants, which stroke or tap aphids with their antennae when it is time to feed. In the most advanced cases the aphids have lost all of their normal defense mechanisms, and protection is afforded by their ant caretakers. Some aphids have morphological adaptations that aid the ants feeding on the honeydew. In addition, the life cycles of the two

Figure 11-2 Cleaners removing parasites from the bodies of hosts. A, Cleaner shrimp cleaning a zebra moray. B, Neon goby cleaning a green moray. This cleaner is an obligate cleaner. Many species of fish, including some familiar ones, are facultative, or occasional, cleaners. Some fish species mimic cleaners and have become parasitic, nipping bites of flesh from the host instead of cleaning them. From Animals Animals. Copyright by Zig Leszczynski.

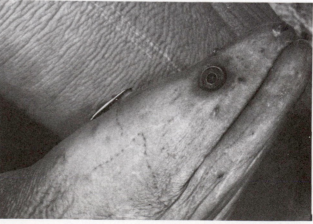

A   B

species are synchronized, and the ants are well coordinated in their care of the aphids. They take them underground into their own nests in the winter, transport them to new plants, and move developing aphids to appropriate parts of plants during the nymphal development. Such advanced levels of mutualism are not found among other species, particularly among vertebrates, except for the relationships among humans and their domestic animals.

Not all mutualistic relationships involve feeding. In one case involving a type of goby fish and a shrimp, for example, the shrimp dig holes that they both live in (Figure 11-3, A). The goby provides warning of danger. The relationship includes evolved tactile signals (Table 11-2) (Figure 11-3, B). The shrimp maintains contact with its goby through its antennae.

Figure 11-3 Mutualism between a shrimp and a fish. A, Shrimp *(Alpheus purpurilenticularis)* maintaining antennal contact with its goby fish *(Cryptocentrus steinitzi).* **B,** Flowchart of behavioral interactions between goby and its shrimp. Note that the goby's behavior depends on whether or not it is in tactile contact with the shrimp.

A from Karplus, I. 1979. Z. Tierpsychol. 49:173-196. Courtesy Ilan Karplus. **B** based on information in Karplus, I., et al. 1979. Z. Tierpsychol. 49:337-351.

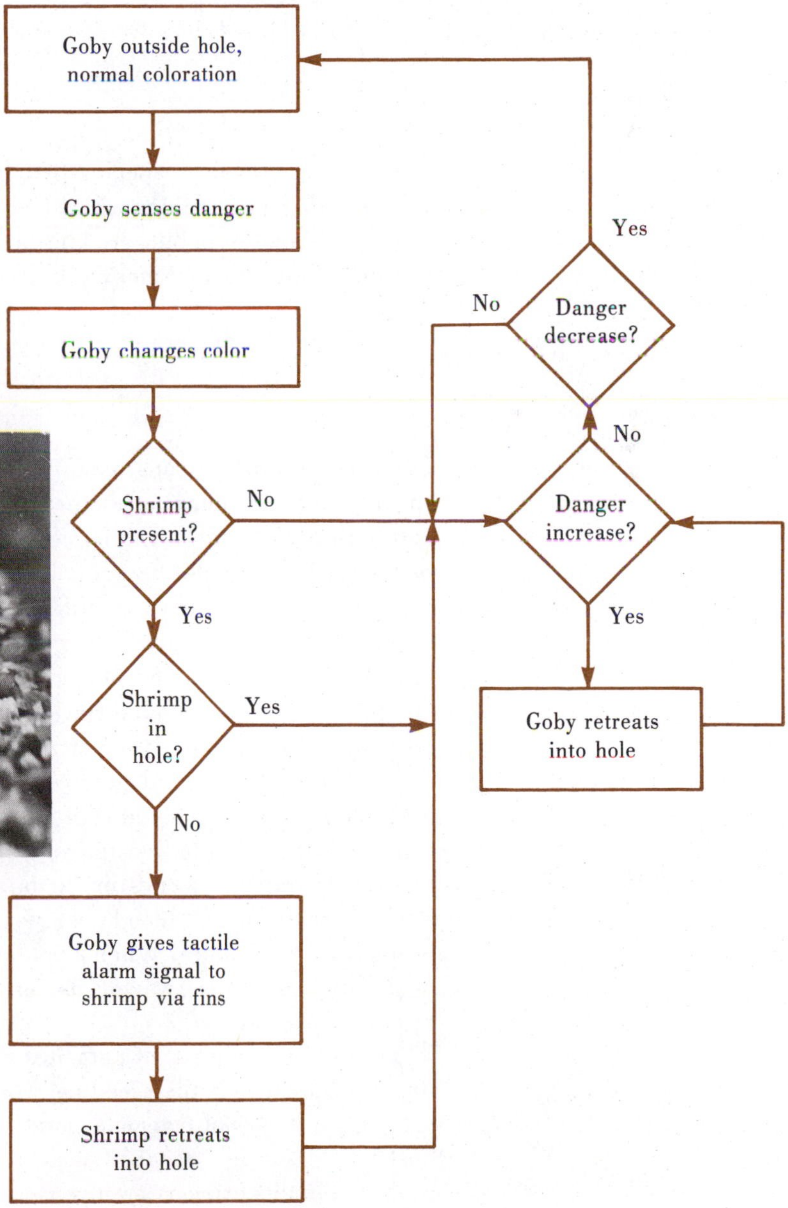

Table 11-2 Interindividual two-act sequences in interactions between gobies and *A. rapax*

	***A. rapax* Following Acts**						
Goby Initial Acts	**Flee**	**In Burrow**	**Manipulate Objects**	**No Change**	**Plough**	**Sit**	**Withdraw**
Dorsal erect	0	4	1	3	3	0	1
Flee	9	13	0	0	0	0	0
Guard	14	934	109	19	939	52	691
Move away	0	20	4	22	21	1	22
Nip sand	5	16	2	14	12	2	13
Pectorals wave	0	9	2	3	8	0	5
Sit away	3	8	0	0	7	1	4
Tail beat	3	0	1	1	0	0	1
Tail flick	119	17	3	134	22	82	65
Tail wave	0	22	3	25	15	1	38
Withdraw	1	15	2	3	10	2	6

From Preston, J.L. 1978. Anim. Behav. 26:791-802.

Parabiosis involves species that live so closely together that they share in mutual defense, foraging, actual feeding, and all activities except the rearing of offspring. Good examples are known at the present only for a few species of ants in Central and South America; however, more cases may be discovered.

PARASITISM

Contrary to most persons' understanding, there are many more ways that one species takes advantage of another as a parasite than just living in or on an animal and eating parts of it. The familiar forms of parasitism involve highly evolved behavior that permits the animals to be successful parasites. Fleas are classic escape artists, and they are difficult to kill if they happen to get caught. Many *ectoparasites* (those which live on the external surface of the host) hang on tight, have specialized, often flat shapes, and are otherwise well adapted for their style of life. *Endoparasites* often have reduced, simplified body structures and are specialized for simply eating or absorbing food and reproducing.

Most ectoparasites and endoparasites have complex life cycles for getting from host to host and, associated with these shifts of living, generally have specialized senses and behavior for locating and entering their hosts. They usually are guided by specific chemical cues or, among parasites of warm-blooded hosts, highly efficient heat sensors. Wood ticks, for example, crawl to the tips of grass and wait for passing vertebrates. When a potential host comes close enough, as sensed via heat, the tick begins waving its legs and will quickly attach to any host that brushes near enough.

These familiar types of parasites can be classed as *trophic parasites;* that is, they are parasites in their feeding. There are many other species that fall into this category but do not consume parts of their hosts directly. They may steal the host's food supply. Another name for this type of robbery is *kleptoparasitism.* Included in this category are hyenas, which steal food from African wild dogs and other large predators, many birds that rob food from other species and occasionally from certain mammals, and some fish and a number of arthropods that take food from others.

Probably the most numerous cases of food stealing are found in association with the social insects. Many species parasitize various ants, bees, and termites, and some steal from each other or enter nests and consume eggs or larvae. Many, if not most, species will rob from another if given the opportunity, but some have become obligatory parasites, where this is the only, hence, "required," way that they can survive. Some wait along odor trails and ambush ants returning with food supplies. One subgenus of ants that nest next to the nests of larger ants and parasitize their broods are known collectively as "thief ants." There are some species of termites that live in and consume the walls of other termite nests. (Technically, because they are not stealing *food*, they are not really kleptoparasites in the usual sense.) As Wilson (1975) put it, "Some termites have termites in their houses!"

Most of the trophic parasites are treated with hostility if detected and captured by the hosts. In *xenobiosis*, however, the robbers may become tolerated guests and even establish communication signals with the hosts, although the relationship is still parasitic and the benefits are in one direction only. The best examples come, again, from a number of different ant species.

From this point one encounters more advanced forms of parasitism—forms that generally involve more than mere feeding. *Temporary social parasitism* involves part of the life spent as a parasite and part spent in free life, during which the animals live on their own capabilities and resources. This form of parasitism has been identified among a number of social insects and in nest (egg and brood) parasitism among birds.

In the insects temporary parasitism occurs when a queen of one species invades the colony of another species, removes the resident queen, and then appropriates the resident workers for her own uses. The new parasitic queen lays her eggs, and the former workers help care for them. The new workers that result increase in number and eventually replace the old ones of the parasitized species. The former workers are not replaced by their own species because their queen is gone. When all the former workers are gone, the colony functions on its own.

Temporary social parasitism is found in a number of ants, some social wasps, and a few bumblebees. There are several ways, depending on the species, by which the parasitic queens gain entrance to the parasitized colony and kill the resident queen. Some force their way in; some more or less sneak in and associate with the workers until they have acquired the odors of the colony; and some use the normal communication signals of the parasitized species. In some cases (e.g., in the mound-building ant, *Formica exsecta*, which occasionally parasitizes *F. fusca*) the queen may lie down and tuck in the legs, as in the pupal posture, when approached by a worker of the other species. The worker then picks up the queen and carries her down into the nest. Different species also use a variety of ways to kill the resident queen. These usually involve some sort of strangling or biting of the neck with the mandibles after using different techniques to get the opponent in a vulnerable position.

Some temporary social parasites go to extremes as *hyper*parasites. Some ants of the subgenus *Dendrolasius* parasitize the nests of free living *Chthonolasius*, which had been acquired by parasitism in the first place by *Chthonolasius* taking over nests of the subgenus *Lasius*.

The evolution of social parasitism in insects has been suggested through comparative studies, particularly of the social wasps (references in Wilson 1975). It is thought to begin by occasional (or facultative) parasites *within* the same species, whereby queens attack other colonies of their own species. This eventually may lead to occasional attacks (in subsequent generations) on colonies belonging to *other* species. Further evolution may then lead to obligate relationships where one species becomes totally dependent on the other species.

Temporary social parasitism of a roughly similar nature is seen also among a few birds. Nest parasitism, in which the eggs of one species are laid in the nests of another species, is practiced by about 80 different species in 7 families or subfamilies (reviewed by Lack 1968, Meyerriecks 1972, and cited in Wilson 1975). Best-known nest parasites are the European cuckoos and the cowbirds, but there also are many others, including the black-headed duck *(Heteronetta atricapilla)* and a few species that are only occasional nest parasites. The parasites' eggs are incubated by the host, and the young are subsequently brooded and fed or otherwise cared for.

The complexity of the avian nest parasite relationship depends on the species. In some it is fairly simple, with the female simply laying eggs in other birds' nests and leaving; the hosts do not discriminate the presence of an odd egg or chick but simply accept them and raise them along with their own.

In other cases, however, there are various degrees of evolutionary advancement and coevolution, whereby changes occur in both species. In most of the obligate nest parasites, the females of the parasitic species possess a number of physiological and behavioral traits that improve their ability to parasitize. They may survey the neighborhood for suitable nests, be able to sneak into nests while the owners are away temporarily, or otherwise intimidate the owner; they may destroy nests that are too far advanced so that the owner has to start over; and they can usually lay eggs quickly, literally on the run.

The parasite's eggs themselves may be thicker shelled, can withstand dropping, and usually develop slightly faster than the host's, so that the parasite's eggs are the first to hatch. Then, depending on the species, when the parasitic chicks hatch, the chicks may maneuver the host's eggs or chicks on its back and dump them out of the nest (e.g., European cuckoos) or kill them with a hooked beak (e.g., the honeyguide, *Indicator indicator*).

Among many species the hosts recognize odd eggs and young and either destroy or abandon them. This apparently has led to a high level of mimicry in several species of parasites. In European cuckoos the eggs of the cuckoos resemble the eggs of the hosts with different females producing eggs that mimic those of particular host species. Females parasitizing a particular host species and laying the appropriate type of egg belong to a given *gens* (plural, *gentes*). This has posed an interesting genetic problem because females that produce eggs and lay them in the nests of different species all belong to the same species and even the same populations. Furthermore, a single male cuckoo may fertilize several females of the different gentes. The genes controlling which gens a female belongs to may be sex linked, as in baldness and hemophilia in human males. (In birds the female has the unmatched pair of sex chromosomes, which is opposite of the situation found in mammals.)

Mimicry in some cases extends to the begging signals that the chicks give to the adults. The young of the combasson and widow birds of Africa have similar markings on the inside of their gaping mouths, which are recognized by the adults. The species are closely related, however, and the markings were probably present in both, a preadaptation for the parasites and not a matter of convergence. (The egg mimicry of the cuckoos clearly is a case of convergence.) In some cases the parasitic species' offspring may learn the song of the host species and use that to ensure that male and female from the same genes mate, as with the Viduine finches.

The evolutionary sequence leading to nest parasitism in birds, as revealed by comparative studies among closely related species, is thought to begin with birds using abandoned nests of other birds or by appropriating the nests of others—but still incubating and rearing their own eggs and young. Next they occasionally may abandon their eggs in some cases, a few of which get incubated by other birds, but continue to take care of their own in other cases. In several species that lay large numbers of eggs, particularly among Galliformes (e.g., pheasants) and some waterfowl, hens may have "dump nests," where they lay large numbers of eggs early in the season. Several hens and even different species may dump in the same nest. Egg dumping or other forms of abandoning eggs in others' nests may lead to *facultative parasitism*, whereby the species is occasionally parasitic but not dependent on that mode. Such a pattern can evolve, through many generations, into greater and greater dependence on parasitism until the species reaches the point of complete dependence, that is, obligatory nest parasitism.

In a well-studied complex situation investigated by Smith (1968), there appears to be a mixture of parasitism and mutualism between several birds and their parasite, the giant cowbird *(Scaphidura oryzivora)* of South and Central America. This cowbird has gentes, as in the European cuckoos: three produce eggs that mimic the eggs of different genera of oropendolas (birds), one mimics the eggs of a cacique (another bird), and one is not a mimic but a dumper that produces a generalized sort of cowbird egg. Females of the mimic gentes are shy and elusive, whereas those of the dumper gens are more aggressive and may barge right into the nests of the hosts.

The different behaviors and egg-producing types of this cowbird species are a form of polymorphism. But the hosts also are polymorphic; some discriminate strange eggs, and others do not. The cause of this apparent polymorphism was found to be a problem with a local botfly *(Philornis* sp.), which infects bird nests and kills many of the nestlings. Nestling cowbirds protect themselves (and, additionally, any hosts) by snapping at adult botflies and by preening nestmates, which removes botfly eggs and maggots.

There are yet more participants in this tangled story: large colonies of social wasps and stingless bees also live in the region. The wasps and bees somehow repel the botflies. The result (and presumed cause of the polymorphisms) is that some oropendolas and caciques nest near wasp or bee colonies, receive botfly protection from them, and discriminate against the cowbirds—treating them as parasites. Other oropendolas and caciques, however, do not nest near the wasps and bees but accept the presence of the cowbirds. They live with the cowbirds in a mutualistic relationship: the hosts provide incubation, brooding, and food, while the cowbird nestlings provide botfly protection.

The most advanced forms of parasitism, exemplified by several species of

Figure 11-4 True slaves (of the same species) in ants *(Myrmecocystus mimicus).* **Individuals of opposing colonies are displaying in tournament fights. Behavior shown here includes stilt-walking and head-on (A), lateral (B), and lateral with antennal drumming (C). The winning colony, after tournaments involving hundreds of individuals, raids and enslaves the losing colony.** From Hölldobler, B. 1976. Science 192:912. Copyright 1976 by the American Association for the Advancement of Science. Photos courtesy Bert Hölldobler.

A

B

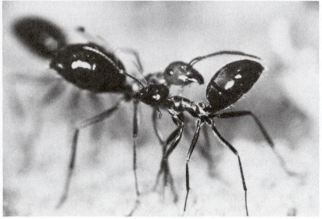

C

social insects, particularly among the ants, are *domestication or slave-making (dulosis)* and *inquilinism.* The difference between the two is that in domestication, ants of one species capture the pupae of another species and carry them back to the nest to become workers for the first species, and in inquilinism one species is totally dependent on and lives permanently with another species. (Although the term "slave-making" has been used in the literature, a better term borrowed from human use would be "domestication." In human society, slaves are from the same species, whereas domestication involves other species.) In inquilinism, one species becomes totally dependent on another species and lives totally within their hosts' nest and life cycle. At least one case of true intraspecific slavery has been reported in ants (Figure 11-4).

Ant domestication was first reported by Huber (1810). Since then the subject has received much attention. Darwin, in *The Origin of Species*, proposed that ant domestication began from ants of one species raiding another species for pupae that were used for food, but a few of the pupae survived and became workers. Wilson (1974), in an alternate explanation, suggested that the initial evolutionary steps involved a combination of territoriality and brood tolerance; that is, territorial ants fought with and killed other adult ants that nested too close, thus eliminating the adults, but they tolerated the presence of the pupae, which then emerged, and the workers began working for the winning colony.

Domesticator species usually conduct their raids with the aid of odor trails. Workers that discover a nearby colony lay a trail back to their own nest, which is then followed by others on the raid. The raiding ants have well-developed mandibles that are used for killing any resisting opponents. In some cases, such as *Formica subintegra*, the workers further disarm their opponents with *propaganda substances* (Regnier and Wilson 1971). The chemicals are produced in very large Dufour's glands. They cause alarm and dispersal in the defending workers but do not create adverse effects in the workers dispelling the substances.

Domestication and the raiding of other nests may, in a few instances, result in the raided queen also being captured and enslaved. When this happens, it may lead to a reduction in the number of capabilities of the raiding workers and the eventual evolution of a species that is permanently dependent on the other species (i.e., *inquilinism*). Wilson (1971, 1975) has proposed that this is one of three major evolutionary pathways that may lead to inquilinism. The others are plesiobiosis to xenobiosis to inquilinism and temporary parasitism to inquilinism.

Once a species has become inquilic, it rapidly (on an evolutionary scale) may degenerate to the point where all but the basic reproductive tasks are assumed by the host species. In the extreme case (e.g., the little ant *Teleutomyrmex schneideri*) the queens have concave undersides that fit the shape of their hosts; their legs are adapted for grasping and holding on; and they ride around on the backs of the host queens. The little ant queens receive nourishment from the host workers, and they somehow physiologically induce the host queen to produce only workers rather than any reproductively capable offspring.

The evolution of social parasitism in insects appears correlated with temperate zones in some manner. Richards (1927) and Hamilton (1972) proposed that it involves closely related species living in neighboring latitudes. Their ranges may merge subsequently. Differences in the timing of their life cycles may result in one emerging sooner than the other, with the first taking advantage of the other. Wilson (1971) proposed that cooling, whether seasonally, nightly, or altitudinally, and associated immobility may dull the responses of host species, making them more susceptible to the raiding species.

Tapping the Communication Lines

As briefly introduced in Chapter 9 and implied at several points in the discussion earlier, highly advanced parasites have evolved the ability to use the host's communication signals. This may occur not only between closely related species but may also involve convergence of distant species, such as between the cuckoos and their hosts. In insects the ability may extend even to other orders of insects.

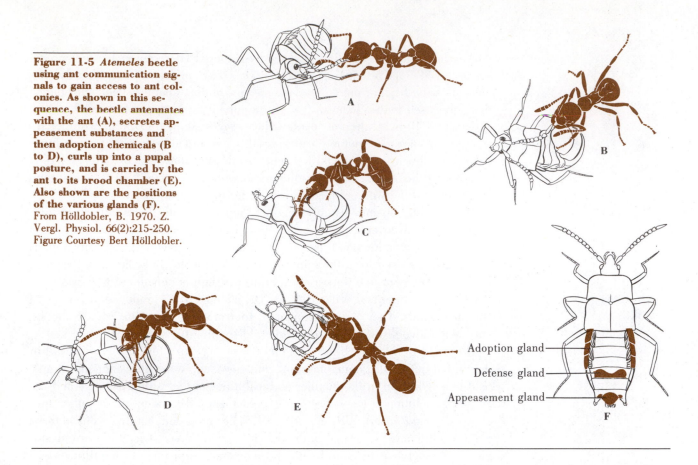

Figure 11-5 *Atemeles* beetle using ant communication signals to gain access to ant colonies. As shown in this sequence, the beetle antennates with the ant (A), secretes appeasement substances and then adoption chemicals (B to D), curls up into a pupal posture, and is carried by the ant to its brood chamber (E). Also shown are the positions of the various glands (F). From Hölldobler, B. 1970. Z. Vergl. Physiol. 66(2):215-250. Figure Courtesy Bert Hölldobler.

Figure 11-6 Drawing of a larval *Atemeles* beetle being fed by its *Formica* ant host. From Hölldobler, B. 1967. Z. Vergl. Physiol. 56:1-21. Figure courtesy Bert Hölldobler.

One of the best documented cases involves a beetle *(Atemeles pubicollis)*, which is a social parasite in colonies of ants of the genera *Myrmica* and *Formica* (Figures 11-5 and 11-6). This fascinating relationship was investigated thoroughly by Hölldobler (1970, 1971). The beetles locate the ants by the airborne odors of the colonies. They prefer to winter with *Myrmica*, which maintain larvae throughout the fall, winter, and spring; then they switch to *Formica*, which have larger colonies and more larvae during the summer. The beetles mingle with the ants by using the ants' communication signals. When the beetles first encounter a worker ant, they present an appeasement gland, at the tip of the abdomen, to the worker. This calms the worker, and the beetle then presents its adoption gland, which the ant licks. Subsequently the beetle is picked up and carried by the ant into the colony. If the appeasement and adoption signals fail and the worker ant becomes aggressive, the beetle has a system for bailing out of the tight situation: defensive glands that repel the ants. Once inside the ant colony, the beetles communicate with the ants by tactile signals and are fed by the ants in the same fashion that the ants normally feed each other. The larvae of the beetles have glands that cause the ants to pick them up and place them with the ants' own larvae, which the beetle larvae then proceed to eat.

In addition to the *Atemeles* genus, there are numerous other beetles and other arthropods that parasitize social insects by tapping the chemical and tactile messages of the hosts. The social insects are particularly susceptible to such intrusion, according to Wilson, because of (1) "the relative impersonality of insect societies" and (2) "the narrow sensory *Umwelt* of [the] hosts." (For a description

Figure 11-7 Tachinid fly (Euphasiopteryx ochracea) females approaching and attacking a cricket (Gryllus integer) using the cricket's song. In this series of photographs the cricket is dead but is mounted above a speaker through which the songs are being played. From Cade, W. 1975. Science 190:1312. Copyright 1975 by the American Association for the Advancement of Science. Photos courtesy William Cade.

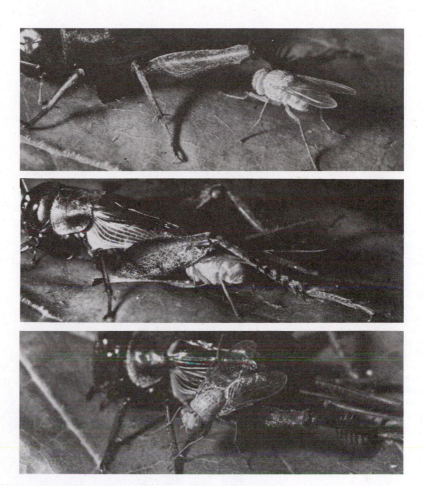

of *Umwelt*, see Chapters 1 and 14.) Wilson (1975) quotes a statement by Wheeler, who spent much of his life studying social parasitism, that compares the situation of code breaking and the subsequent effects to what would happen if humans were to "delight in keeping porcupines, alligators, lobsters, etc., in our homes, insist on their sitting down to table with us and feed them" to the neglect of our own children.

Exploiting the lines of communication of other species is practiced not only by parasites but also by predators (keeping in mind that the distinction between parasites and predators is not always very clear). Some predators locate their prey by the prey's intraspecific vocalizations and visual displays. Examples include tachinid flies (Figure 11-7) and frog-eating bats (Tuttle and Ryan 1981).

Fireflies belonging to the beetle family Lampyridae court members of the other sex by communicating with flashes of light. The males, which generally outnumber the females manyfold, fly about advertising themselves with species-specific flashes. Important characteristics of the flash are duration, frequency, and pattern of flashes, as well as repetition of sequences (Figure 9-2). The patterns of males are now known for over 130 species around the world. Different species using different flash patterns often live sympatrically. Females, which wait on the ground or perch on vegetation, answer males passing overhead with relatively simple flashes that depend partly on precise timing for species identification. When a male detects an appropriate female response, it hovers near the female. The two continue to flash to each other; then the male lands, walks to the female, and attempts to copulate.

In most species the behavior is basically as just described, without further complication or elaboration. The flash patterns are sufficiently distinct and stereotyped that species can be separated by the taxonomist. In some species, however, individual fireflies vary their signals (Carlson and Copeland 1978). A few species show variation to such an extent that they have caused taxonomic confusion. Some of the variation consists of males varying their signals under different contexts, such as for searching versus actual courting or variation used to more precisely locate the females.

Other variability, however, has been discovered to consist of aggressive mimicry on the part of the females, the **femme fatales,** in a few predatory species of

**Figure 11-8 *Femme fatales.*
A predatory female firefly
(Photuris versicolor) is eating
a male firefly of another species. When males advertise
themselves by flashing, the
predatory females attract
them by mimicking their own
females' response flashes.**
Photo courtesy James E. Lloyd.

**Figure 11-9 Luminescent sig-
nals of fireflies. Response
used by predator female is
shown below the female
answer she mimics. Vertical
bars at right indicate observed individual repertoires;
n is the number of females
exhibiting the repertoire.
Capture rates (percent) are
adjacent to prey species. The
flash rate of the *Photuris
congener* female is variable
and not well understood.**
From Lloyd, J.E. 1975. Science
187:452. Copyright 1975 by the
American Association for the
Advancement of Science. Figure
courtesy James Lloyd.

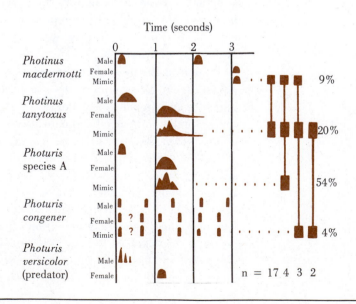

the genus *Photuris*. These females mimic the responses of females belonging to other species, attract and catch a proportion of their males, and eat them (Figure 11-8). Individual females may mimic the responses of three or more other species (Figure 11-9).

The *femme fatales* generally alternate between reproductive and predatory behavior. They first use their own species-specific patterns to attract correct males, with which they mate. Following copulation, they switch to calling in other males, which are eaten. Later they may revert to correct courting responses. The behavior has been studied extensively under both field and laboratory conditions (Lloyd 1975, 1981). Techniques have ranged from the use of simple stopwatches and penlights for simulating the flashes to complex photomultiplier and electronic recording systems. During field experiments where the researcher was artificially interacting with various *femme fatale* females, subjects occasionally would interrupt the experiment to answer passing wild males (Lloyd 1975).

To further complicate the *femme fatale* story, males of some of the predatory species also mimic the other species by using their male patterns (Lloyd 1980). These males normally use their own species-specific flashing, but, appropriate to the location and seasonal timing of other species that may be present, they occasionally mimic the other males. Over 15 hypotheses have been offered to explain the evolution of such mimicry on the part of males. After sorting through the pros and cons of the various explanations, Lloyd (1980, 1981) leans toward an explanation that basically consists of males tricking females that are attempting to trick other males. The males may mimic other males until they are called in, then switch to their own species-specific patterns and induce the female to change from predatory to reproductive behavior. Other explanations also may be partially valid. For example, there may be an element of "kamikaze-copulation" as in mantids and black widow spiders.

Humans use similar tactics, in a fashion, when artificial game calls, such as duck calls or elk bugling, are used during hunting or when insect pheromones are used to attract insects during control efforts.

The topic of exploitation of communication signals and the overlap between parasitism and predation leads to the next major category of interspecific behavior: predation. Predators and their prey come into close proximity and contact only occasionally and usually briefly, as indicated at the start of this chapter. The remainder of the time they live more or less separate lives and usually at some distance apart. Again, there are borderline cases, such as black-footed ferrets and rattlesnakes living within the tunnels and homes of their prairie dog prey. But, for the most part, the relationships between predator and prey are much different from the various symbioses discussed earlier.

PREDATOR AND ANTIPREDATOR BEHAVIOR

The term *predator* is derived from the Latin *praedari*, which means to plunder or take by force. Closely related terms are *raptorial* and *carnivorous*, which have roots in words meaning "to seize" and "flesh-eating," respectively. All of these, in their modern-day connotations, imply action and force. It is force and going against the actions of another individual that set predation apart from other forms of feeding behavior. This largely is a quantitative difference, or matter of degree, rather than a real, qualitative difference. Much of the recent scientific attention to predation concerns the topic of optimal foraging, which was introduced in Chapter 6.

Predation, like communication and many of the other concepts of behavior, is somewhat difficult to define. Many of the predator-prey interactions are commonly recognized, and few people would call them otherwise. Some persons, however, go as far as to extend predation to cover all of heterotrophy (eating other organisms), even herbivory (eating plants). They would say that herbivores are predators on plants. Curio (1976) offers a number of potential definitions: "A process by which an animal spends some effort to locate a live prey and, in addition, spends another effort to mutilate or kill it," and "Predation is best distinguished from other forms of foraging by only one of its consequences, in that it concludes with the mutilation or total destruction of an animal that offers some resistance against being discovered and/or being harmed."

The common denominators of predation and the contexts in which the word normally is used by biologists and lay public alike are the following:

1. Predation involves only *animal prey*, hence excluding plants. (Some plants such as Venus flytraps or pitcher plants eat animals, however, and might be considered as predators.)
2. The potential victims are alive, hence excluding scavenging (although many, if not most, predators also scavenge).
3. The intended victim is capable of resisting to some degree through the use of its own nervous and muscular or glandular systems.
4. The victim is killed, not just wounded or mutilated, if the predator is successful.
5. The interaction involves a relatively short period of time, normally only a few seconds, minutes, or, at most, hours.

The mutilation caused by parasites may range from insignificant to serious disfigurement, but individual parasites rarely kill their hosts. Even if the victim is killed, the interaction is often prolonged and may last for years. The intended victims of predation frequently are not killed, only wounded or perhaps just

Figure 11-10 A sampling of predators with prey. A, Jumping spider with fly. B, Sharp-shinned hawk with starling. C, Bobcat with rabbit. D, Mink with mouse. A courtesy Lyn Forster. See also Forster (1982). **B to D** courtesy Tom Brakefield.

A

B

frightened. But these instances would usually be classed as incomplete attempts or failures on the part of the predator.

Actual feeding is not necessary for an interaction to be considered as predation. Predators normally consume all or part of their prey, but not always. A pack of wolves trailing a herd of caribou throughout the year might seem little different from a parasite, but it is the interactions of individuals that compose the pack and herd with which we are concerned. Many insect parasites, such as mosquitoes, move in, take their meal, and get away fast; they do not interact for long periods of time—but they rarely kill the host. It is often the *combination* of killing (when successful) and brief periods of time that distinguish between predation and parasitism.

Evolution of Predatory Behavior

Predation as a way of obtaining food is so widespread among so many taxonomic groups and goes so far back into the pylogenetic past that it probably is not possible to trace the general development of predatory behavior. Predation undoubtedly has arisen independently numerous times and may have been the initial way of feeding. Herbivory and predation probably have been evolutionarily derived from each other numerous times over the eons in different taxonomic groups. In a particular lineage predation probably begins with scavenging or by feeding on smaller, easily handled animals, then eventually evolves to include the ability to handle larger and more difficult items. Some taxonomic groups of animals are exclusively or predominately predatory, including spiders, many of the extinct and living reptiles, many groups of fish such as the sharks, perch family, and particular orders or families of mammals and birds (e.g., Mustelidae and Falconiformes, respectively). A few examples of predators and prey are shown in Figure 11-10.

Within a particular taxonomic group, such as the mammals in general or Mustelidae in particular (Powell 1978), one can trace some predatory behavior similarities and infer the recent evolutionary origins. Eisenberg and Leyhausen (1972), for example, provide an extensive review of predatory behavior in 4 orders and 49 genera of mammals. Several aspects of hunting, capture, and

C
 D

killing behavior were considered. Some of the findings indicated, for example, that capture of prey in the mouth preceded the evolution of grasping with fore-paws. Vigorous shaking of prey evolved from primitive tossing of obnoxious prey, and whether or not the predators use precisely aimed bites at the back of the neck depends partly on the role of diurnal activity and use of vision by the species. It appears that several behaviors evolved in parallel or convergent fashion in different taxa under similar ecological conditions.

Predation or the avoidance of predation strongly influences the behavior of almost all animals. Predator behavior includes the capture and overcoming of the prey by using a variety of techniques, depending on the species involved, such as pursuit, ambushing, trapping, and various forms of tricking. Characteristic prey behavior may include means of avoiding or escaping from predators, hiding and camouflage, and various ways of being undesirable, hence, left alone. *Speed* and *alertness* are seen in both predator and prey and are believed to be direct evolutionary consequences of predation. These include high levels of muscular coordination, keen senses, and sophisticated nervous systems.

Sophisticated sensory, neural, and fighting systems as well as complex behavioral tactics appear to result from conflict and differential survival among all species of living organisms. If there were no bats, it is unlikely that moths would have bat detectors. Gradual improvements in the biological machinery and tactics are the very earmarks of evolution. Predation is responsible for a large proportion of these changes. Predators and their potential prey, in keeping up with or ahead of the other, are slowly but constantly improving in their abilities. Those which do not do not survive and reproduce.

The effects of predation on animal behavior and morphology are so pervasive that it has been theorized as one of the main forces of, for example, the development of vertebrates. Some of the chief expansions and radiations of new forms have come after new weaponry (such as jaws rather than round mouths) or new guidance and movement systems (such as balance organs, streamlined bodies, rayed fins that permit fine control of swimming) developed. The other chief factors in major evolutionary developments have been the entrance into different habitats (such as from aquatic to terrestrial or finches moving into the Galapagos Islands) and the direct or indirect effects of competition. Thus much of an animal's behavior and morphology (e.g., long legs, cryptic coloration, keen senses) can be interpreted from the evolution of predator-prey interactions.

Much of the literature on predation is descriptive and anecdotal. Controlled studies in the laboratory or deliberate manipulations of predation in the field are highly susceptible to experimental artifact. Predation is extremely difficult to study under natural conditions; it is relatively infrequent, and one usually cannot predict when and where it will occur. Nonetheless, there have been several excellent studies (Schaller 1972) and, apart from the optimal foraging aspects, at least one major review of the subject (Curio 1976).

Particular taxonomic groups of predators have received much attention and generally are familiar. They include the cats, canids (dogs, foxes, wolves), raptors or birds of prey (eagles, hawks, falcons, owls), spiders, snakes, a number of predatory fish, and others. The largest and most powerful predators include the large cats, wolves, grizzly and polar bears, eagles, the extinct meat-eating dinosaurs, large snakes and crocodilians, and a host of marine organisms including large sharks, killer whales, and the giant squid and octopus. Predators at the other, small-size extreme, however, are no less dangerous and awesome; only the scale is different. Shrews, predatory protozoans, and predatory insects are as dangerous to other animals within their own size range as any of the larger predators are to larger vertebrates (including humans).

Human opinions toward predators tend to be polarized. Many people, particularly biologists who appreciate the evolutionary, ecological, and behavioral implications, tend to view predators neutrally or favorably. Some persons associate predators with traits of strength, power, and courage; in fact, the national symbols of most countries tend to be predator species. Many others, however, have a strong dislike and negative view of predators. These views stem from the fact that many predators are potentially dangerous to humans (although attacks on humans have been remarkably infrequent during recorded history); some predators are in direct competition with humans for game and livestock; and many persons associate the behavior of predators with undesirable human behavior.

Proximate Factors of Predation

Predation forms an abstract continuum not only with other forms of foraging but also with nonpredatory behavior in the individual predator. A predator that is resting or otherwise not engaged in predation may suddenly encounter an opportunity to catch something and take advantage of it. Or a predator that is roaming about, perhaps returning home from a visit to a territorial boundary, may begin to search for something to catch and eat. Likewise, an act of predation may turn less serious or the individual may be interrupted or distracted by something else, such as its mate. Some female arthropods (e.g., some mantids and spiders) will mate and eat their partner at the same time. The variety of animals that are called predators range over such a wide spectrum of invertebrates and vertebrates that all of the possible behaviors involved in predation defy classification.

It is possible, however, to consider some of the proximate factors involved in predation. The most obvious is hunger on the part of the predator. Hunger is difficult to define and quantify because it involves internal neurophysiological mechanisms. For the present discussion, however, hunger will be used in the familiar sense. Operationally it generally is measured in terms of time since previous feeding, amount previously eaten, and an animal's normal or average body weight. This view of hunger is oversimplified (see Chapter 17), but it will do for now. Much, although not all, of what applies to feeding in well-studied species such as the fly and rat (Chapter 17) undoubtedly applies also to predators. There are probably some phylogenetic constraints and affinities in the control of feeding, such as between insects and mammals, but a number of feeding considerations are either unique or accentuated in the case of predators.

Perhaps one of the most common problems of predators, as opposed to herbivores or omnivores, is the feast-or-famine situation under natural conditions. When food is at hand (or paw), there may be much more than the animal needs for its immediate nutrition. Subsequently, the animal may be forced to go with little or no food for relatively long periods of time. Other species tend to have a more constant or stable food supply. But for predators near-starvation and actual starvation in many cases are routine facts of life and death. As a result, predators may gorge themselves much more than other animals when given the opportunity.

Spotted hyenas *(Crocuta crocuta)*, for example, may eat up to seven times as much (Kruuk 1972, cited in Curio 1976) and lions up to four or five times as much (Schaller 1972, cited in Curio) during a big meal as what they would normally eat during a subsistence diet. Some fish, snakes, and frogs will eat so much that they cannot get it all into the stomach at once, and the remainder has to wait in line, with tails or other parts of the food extending out of the mouth until digestion can make room for it.

At the other extreme, most large predators are able to go many days or weeks without eating. A hyena followed continuously by Kruuk (1972) went up to 5 days between meals, and Mech (1970) reported that wolves can go at least 17 days without eating. Eagles may go 2 weeks or more between meals in some cases. Predators, especially the young and inexperienced, frequently must go too long between meals and starve. Whereas predators impose a major source of mortality on their prey species, the major cause of natural mortality among predators is starvation. (During recent times, humans have become one of the major sources of mortality for large predators. In spite of legal protection, most of the reported cases of dead bald eagles in the United States, for example, are from gunshot.)

The ability to gorge when food is available, however, is tempered by considerations other than physical capacity, particularly in smaller predators. Animals that are too heavily loaded down may become more vulnerable to predation themselves. Excessive gorging is observed mainly among the larger predators, which have few, if any, predators above them.

Figure 11-11 Comparative, subjectively determined hunger thresholds at which predators engage in different behaviors. Note that mantids, which normally do not pursue their prey, will capture and eat at a lower threshold than is required for active pursuit, whereas the stickleback and cat, which normally do pursue prey, will pursue or stalk at lower thresholds than at which they eat. Also, some predators may capture prey but not eat it if below the hunger level at which they normally eat. Modified from Curio, E. 1976. The ethology of predation. Springer-Verlag New York, Inc., New York. Based in part on Holling, C.S. 1966. Mem. Entomol. Soc. Can. 48:1-86.

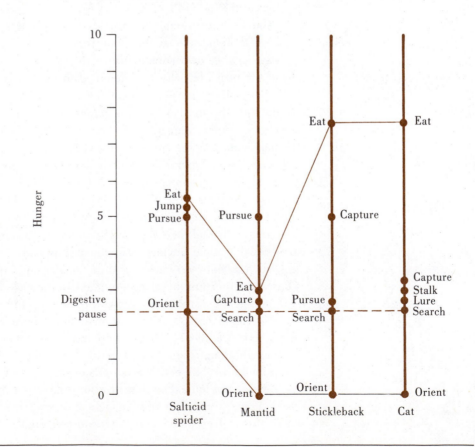

A complete sequence of a predatory act, from initial sensory contact with the prey to eating, can be broken into several component acts and movements. Some of these are associated with the level of hunger, and some are not. Furthermore, those which are associated with hunger may vary from species to species. For example, hunger may alter the thresholds at which predators chase or strike prey or the size of the animal they will tackle. Curio (1976) lists several traits that may be altered by hunger level in several different species. He modified and extended a comparison among a few representative species done by Holling (1966) (Figure 11-11). An ambushing predator such as a mantid may require a higher level of hunger before pursuing prey than predators that normally pursue their prey. Vertebrates appear to have a higher threshold for actual eating than other components so that they are more likely to capture without eating than arthropod predators. Furthermore, recovery after exhaustion of different components, such as searching, lurking, chasing, catching, killing, and eating, occur at different levels of hunger for each component (Leyhausen 1965). This suggests that an act of predation is not a single, unified behavior but consists of a collection of components.

Several components of predation may not be affected by hunger level, depending on the species of predator. Speeds of stalking and pursuit, for example, are relatively constant regardless of hunger level. Similarly, the time required to eat a given amount of food may be more or less constant in many species, but not in others. Once a mantid strike is released, it is fairly mechanical, and the success rate of mantids does not depend on hunger level (Holling 1966). In many vertebrates, however, the seriousness and persistence of an act of predation tend to increase with hunger, and the success rate may go up.

Several species of predators, particularly among spiders, some insects, birds, and mammals, store excess prey. Whether or not storing varies with hunger, however, appears quite variable and depends on the species. In some cases the handling of prey before storage involves different behavior from the handling of prey to be eaten immediately, which, again, suggests that separate mechanisms are involved internally. Curio (1976) provides several examples among birds and mammals.

In addition to hunger, there are numerous other factors that govern the proximate aspects of predation. Predation may be strongly affected by both diurnal and seasonal rhythms; predators will hunt and kill at some times but not at others, or they may change their tactics and prey preferences on a regular basis. Diurnal versus nocturnal patterns are quite obvious, with predator and prey patterns most likely affecting each other.

Competing species of predators also may influence each other. Short-eared owls on the Galapagos Islands, for example, hunt both night and day on islands where the Galapagos hawk *(Buteo galapagoensis)* is absent but only during the night where the hawk is present. Any owls that do fly during the day where hawks are present are attacked immediately by the hawks (de Vries 1973). Circannual rhythms, aside from those clearly associated with prey abundance, have received little attention regarding predatory behavior.

A major proximate factor in predation is the presence of dependents. A mate or offspring that require feeding will greatly increase the amount of predation and also may alter the type of prey and the manner in which it is handled. Norton-

Figure 11-12 Syrian wood-pecker foraging behaviors: self-feeding versus feeding young. Frequencies differ depending on the circumstances of feeding. Modified from Winkler, H. 1973. Oecologia 12:193-208.

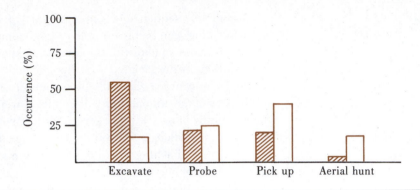

Griffiths (1969), after a thorough study of parental feeding in oyster catchers (birds), concluded that both self-feeding and the feeding of young shared common internal behavioral mechanisms. This does not appear to be the case, however, in numerous other species where the parent may or may not feed itself during the same time that it feeds young and where the specific behaviors may be quite different. Raptors, for example, will pluck and decapitate their prey before giving it to a mate or young in a manner not seen when the birds are simply feeding themselves. Likewise, whitethroats *(Sylvia communis)* remove the heads of flies before feeding their offspring but not before feeding themselves (Sauer 1954). Winkler (1973) showed marked differences in the foraging methods used by Syrian woodpeckers that were feeding themselves and those feeding their young (Figure 11-12).

One somewhat intangible factor in predation, not closely tied to hunger, has been called *readiness to hunt.* In birds of prey this sometimes is referred to as "yarak" or "sharp-set." It was proposed that prey species could recognize the appearance of raptors that were ready to hunt (e.g., Hamerstrom 1957). But such would seem disadvantageous for the predator and should be selected against; I (1971) was unable to find any evidence of hunting readiness in the postures and external appearance of hawks before being offered food, although they quickly showed readiness after food became visible. Although the visibility of hunting readiness in the absence of potential prey may vary (e.g., obvious in African hunting dogs and absent in raptors), most persons working with predators generally attest to its presence. In a manner that does not always seem to correlate well with body weight or other measures of hunger, some predators will be very ready, if not almost on a hair trigger, and others will completely ignore certain opportunities. In anthropomorphic terms, some predators at some times seem to anticipate and relish a good hunt. At other times or with other individuals in the same situation, predators may appear as if they could not care less about an opportunity. Lack of response may depend, in part, on experience and past failures and successes.

Readiness to hunt generally involves a higher level of excitement and alertness, occasionally to the point that a predator will perform predatory acts even in the absence of prey. These vacuum behaviors, as they are called (Chapter 2), may include the full sequence of chasing, catching, killing, and eating—all with an imaginary prey. Furthermore, some components of predatory behavior will be displayed at high levels of excitement when not appropriate, such as "killing" a prey object given to an animal when the prey already is dead. Similarly, predators deprived of certain components of their behavior may engage in it excessively when later given the opportunity. Birds of prey, for example, typically pluck some of the feathers and hair from bird or mammal prey before eating it. If fed meat only rather than whole animals for a period of time, raptors sometimes will ignore meat even if quite hungry when given a choice of an object that can be plucked. Furthermore, they may pluck it much more than normal.

Problems of Finding, Recognizing, and Selecting Prey

Different predators eat quite a variety of prey. Some predators, the specialists, have a narrow range of prey type, consisting of a single species of prey in the extreme case; more general predators may take quite different animals at different times. Because of this diversity among both predators and prey, it is difficult to further generalize beyond the discussion of optimal foraging in Chapter 6. Many of the problems and solutions are unique to particular cases of predation.

Many species of prey, for example, are small, camouflaged, hidden under or behind objects, scattered throughout the habitat, or all of these. The biggest problem facing the predators of such prey is how to find them in the first place. Capture and handling of the prey, once found, may be minor problems. On the other hand, some prey may be large and conspicuous. In this case the predator

Figure 11-13 Prey preferences of newborn garter snakes *(Thamonphis sirtalis)* from different geographical locations. Responses were measured as attack behavior shown toward water extracts from various prey species. Responses were inferred to be innate and appropriate for the particular locality. From Burghardt, G.M. 1970. Behaviour 36:246-257. Courtesy Gordon Burghardt.

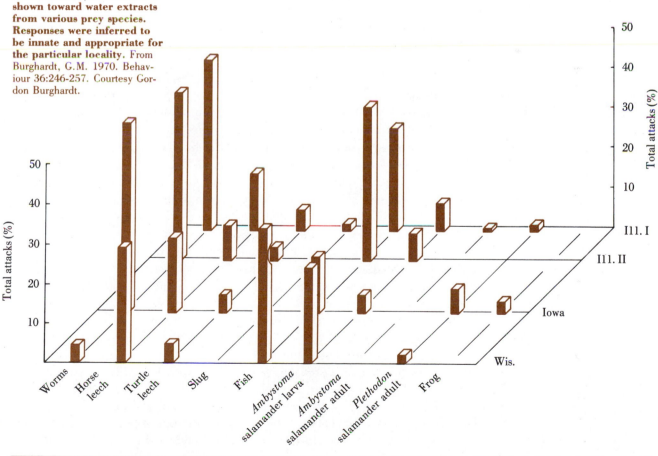

may have little difficulty finding them; the problem is how to catch and handle them once contact has been made. Some potential prey are quite capable of fighting back, are dangerous, or have other specialized defenses, such as porcupine quills, and require specialized tactics on the part of the predator. Specialized problems require specialized solutions. Fishers, for example, have solved the problem of capturing porcupines by attacking them in the face.

Because the different categories are only loosely related, different aspects of the problem of finding suitable prey simply will be briefly discussed.

Which Prey Species Are Taken?

As just mentioned, some predators have narrow diets, and some have wider ones. Examples of highly specialized predators include the Everglades kite, which eats one species of snail, the apple snail, and the hognose snake, which exclusively eats toads (*Bufo* sp.). There is clearly some kind of prey selection operating in these species, and the predators usually are selecting on the basis of specific cues. Many snakes, such as the hognose, find and identify their prey on the basis of prey odor. Some species show geographical variation in their apparently innate prey preferences (Figure 11-13).

Predators that eat a wider variety of prey frequently concentrate on particular species at particular times, although they may take many different species throughout the year. The predators often do not take animals in proportion to their availability; they may take animals of one species and seem to ignore or refuse animals of another species that is equally abundant. This has been discussed under optimal foraging but merits brief further discussion in the points that follow.

Prey Switching

Many predators will concentrate on one species of prey for a period of time, then suddenly *switch* to another type of prey. Many insectivorous birds and fish, for example, will change the species of insects they are eating and refuse those which they ate previously, although they still may be abundant. Hyenas (Kruuk 1972) and African wild dogs (Schaller 1972) will hunt different large prey, for example, wildebeest or gazelles, then switch to something else such as zebra. Furthermore, depending on which they are going after, they may start the hunt with different pack sizes and different prehunt rituals. After hunting wildebeest for several days and switching to zebra, the predators may leave the den and travel past or through large herds of wildebeest, ignoring them completely, on their way to hunt for zebra. In some cases, such as perhaps with insectivores, different prey may provide different nutritional items, and the switch might be explained by dietary needs. In others, however, such as those involving large mammal prey, the dietary differences would seem insignificant. Furthermore, it is not always a matter of simply taking easier or more generally preferred prey when it becomes available; the predators may subsequently switch back to the first prey type. Anthropomorphically, the predators sometimes simply seem to be bored and want a change.

Some of the switching has been interpreted in terms of different appetitive behaviors (Chapter 2); that is, there may be different specific "appetites" for different prey, and after one has been satisfied, its threshold is raised and another becomes more likely. To say that predators switch because they have different appetites, however, runs into the circular logic of many drive concepts when appetite is inferred from the switch. A better understanding of what occurs inside the animal is needed.

Pigmy owls in the laboratory exhibit different tactics for hunting and killing small birds than for mice (Scherzinger 1970). Birds are hunted by hiding in dense foliage and ambushing them or by dashing through vegetation. They are killed by simply grasping with the feet and talons. Mice, on the other hand, are hunted by watching from high, open perches. Mice are killed by biting the head and neck with the beak. There may be separate neural mechanisms for the two types of prey, and at different times one or the other may predominate.

Switching also has been studied in the drinking behavior of laboratory rats (Morrison 1974). When offered more than one solution to drink, the rats will concentrate on one, then switch to another, then switch back again. The observations are consistent with (but do not prove) the interpretation that palatability declines with consumption until it falls below the palatability of an alternate solution.

One cannot say, however, that switching is universal. It has already been noted that some predators stick to the same type of prey throughout their lives. In other species and on different occasions, some predators are very reluctant to switch to different prey, sometimes even showing obvious fear of different potential prey. It is clear that we are far from understanding the real causes, proximate or ultimate, of prey switching phenomena in predatory behavior. It should be a rich area for future investigation.

Search Image and Associated Notions

Von Uexküll (1934) originally proposed and L. Tinbergen (1960) revived the concept of "search image" (introduced in Chapter 6). This refers to a selective attention, or perceptual change, by which an animal searches for a particular object. The idea is supported by the apparent nonrandom choice of prey items (i.e., not in statistical proportion to their availability), the fact that animals may not take a particular prey item although available but then suddenly switch to it, and the further observations that prey is ignored or selected irrespective of its position in the habitat (e.g., part of a tree), its novelty (some new items may be taken readily), or its conspicuousness. In fact, it is the latter point where search image may be most important; it may permit a predator to easily and efficiently find items that otherwise are more difficult to detect. It would seem to be less important with obvious prey that require less searching. Humans searching for relevant items among other meaningless material, such as for particular words or phrases on a printed page, report that the relevant items stand out, whereas the other material is blurred (Neisser 1966). It is a common experience among hunters, bird watchers, nature photographers, and wildlife researchers that experience permits one to spot distant or hidden animals to a degree that can surprise even the observer.

The issue of selective perception raises the topic of internal neural control and integration (Chapters 14 to 19). The search image concept is attractive, but a number of problems with it were already mentioned in Chapter 6. Concerning predation, there are complicating factors and a host of associated terms and ideas. Krebs (1973) discussed search image relative to different types of learning: how useful the term is depends on the forms of learning with which it is associated. Phrases synonymous or nearly synonymous with search image include "learning to see," "shifts of attention," "sharpened peaks of generalization gradients," and others.

Holling (1959), in considering predation from an ecological rather than purely behavioral standpoint, introduced the concept of *functional responses* in predators (Figure 11-14). Functional responses refers to the number of prey taken per predator. When combined with numerical responses, that is, the number of predators present, one gets the total predatory response to a prey population. Only some responses, in the presence of alternate prey, are consistent with the notion of search image; that is, with an increase of prey density there is little increase in predator attention to that prey until suddenly there is a rapid increase in the numbers of that prey taken per predator. With further increases in prey density the numbers taken per predator level off as the predators become satiated. The net effect of acquiring the selection and then becoming satiated produces an S curve.

Figure 11-14 Predation responses to increases in numbers or density of prey animals. A, *Numerical responses* as shown by two genera of shrews *(Sorex* and *Blarina)* and deer mouse *(Peromyscus)* to changes in sawfly cocoon density. With increased numbers of cocoons there were increased numbers of *Sorex* and *Peromyscus*, up to a point, preying on them. *Blarina* showed little numerical response. B, *Functional responses.* These refer to the numbers of prey taken by *individual* predators. In this case, *Blarina* responded the most, up to a point, and *Sorex* the least, just the opposite of the numerical response. Modified from Holling, C.S. 1959. Can. Entomol. 91:293-320.

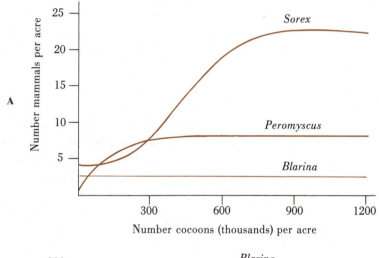

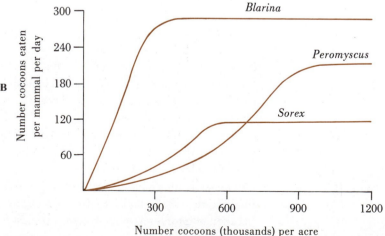

There are, however, several other variables that complicate or potentially detract from the concept of search image. First, there is the common phenomenon of *area-concentrated search* in predators. Many species of prey animals live somewhat concentrated and in clumps. Perhaps as an evolved or learned response to this, many predators, on encountering a prey item, will change their pattern or speed of traveling and searching so that they concentrate their efforts in the same area. The result is somewhat similar to location orientation by kinesis (p. 190 and Figure 7-3). The effect, which may resemble search image, is that the predator stays in the vicinity of a higher concentration of prey, encounters them at a higher probability, and takes more than would be expected from their average density over a larger area. (One might predict that the prey should respond in turn by living scattered. For further discussion see Curio [1976].)

A closely related concept to area-concentrated search is *profitability of hunting*. According to this idea, which is based on the principles of optimal foraging, a predator should concentrate only on those areas which produce optimal return of resources for the effort. When the return drops in a particular patch of habitat, the animal should go elsewhere, perhaps to other similar patches. This also could produce an effect similar to search image; that is, the predator is not searching the environment randomly but concentrating on those areas which are most productive and, as a result, may increase the probability of encountering some prey relative to others.

There are additional problems in interpreting search image. *Social facilitation*, whereby predators take what they see others taking, somewhat like a fad or everyone choosing something because it is popular, can lead to a nonrandom selection of prey items. There may be a *training bias* in which animals simply choose what they are used to, that is, are familiar with from previous experience. The predators might be able to detect and find all items equally well but stick with the familiar, which superficially could resemble searching based on a search image. This would amount to a resistance to switching, however, and would not be expected to produce an S-shaped functional response curve.

Fear of new items (neophobia) is another phenomenon that could resemble search image by restricting what an animal preys on. Many predators refuse to take some potential prey apparently because they literally are quite afraid of them. Eagles and many other large predators, unless extremely hungry, may show explicit signs of fear and distress, including vocalizations and attempts to escape or withdraw when exposed to unfamiliar, but otherwise suitable, potential prey species. Curio (1976 and references contained therein) cites numerous examples of predators, including owls and mustelids, that have been observed to give alarm calls and respond with panic at the sight of novel prey. In such cases, the predator gradually may have to become familiar with new prey by scavenging on parts of or whole carcasses of dead animals; even these may be approached with caution and hesitation. Other predators, such as some of the small *Accipiter* hawks, on the other hand, may tackle novel prey items seemingly with abandon and carelessness, including larger and potentially dangerous prey such as various gulls and rooster pheasants.

Finally, one must consider negative experiences that predators often encounter with some types of prey. Bad-tasting prey may cause taste aversion, or injury from dangerous prey may cause aversive conditioning (Chapter 19), both of which may lead to the avoidance of otherwise potential prey. This may shift the predator's concentration to other forms of available prey and, again, may produce an end result that resembles search image. As in the case of prey switching, the whole topic of search image and related concepts deserves much more study and clarification.

Prey Selection

In the preceding sections, the problem of *prey recognition*, that is, how predators find and recognize potential prey in the first place was discussed. The associated concepts of searching image, etc., imply that predators are *looking for* common cues or animals that look (or sound or smell) alike. Now a different aspect of predation is encountered—how *individual items* of a given prey type are chosen. It has been proposed that predators may select *odd* or *different* prey. In the case of *prey selection*, as opposed to prey recognition, one is dealing with predator-prey interactions in which the problem facing the predator is not one of finding the prey. Rather, the predator may not only have found the prey but may have found so many that now it is faced with the problem of deciding *which* to try to take.

Alternately, there may not be several at once from which to choose, but over a period of time the predator may encounter several potential prey of the same species. Different potential prey may vary in their ease of capture or likelihood of fighting back and inflicting injury. In this case it may be advantageous for the predator to somehow discriminate between those which are worth the effort and those which are not. Can and do they?

One of the stock answers in popular biology is that predators take only the sick and unfit, thereby preserving the quality of the prey and maintaining ye olde balance of nature. It is unlikely that predators are acting in the interests of the prey or some abstract general notion about nature, even de facto. But their own interests in not wasting effort or avoiding injury might produce similar effects. There is some evidence that, at least in some species and under some circumstances, predators definitely select odd or different prey (e.g., Mueller 1977). But there are numerous other instances in which predators appear to be taking from the prey population at random across the various age and sex classes

Figure 11-15 Anglerfish of the genus *Antennarium*, in luring posture, displays a bait that bears a remarkable resemblance to a small fish. When a prey animal is attracted close to the lure, the anglerfish jumps forward and snaps it up. From Pietsch, T.W., and D.B. Grobecker. 1978. Science, vol. 201, cover photo, July 28. Copyright 1978 by the American Association for the Advancement of Science. Photo by David B. Grobecker.

(Pearson 1966). Three-spined sticklebacks preferred odd-colored prey *(Daphnia)* depending of the density of prey. They chose odd prey at high densities of prey and common prey at low densities. Prey selection in many preadators is influenced by several factors, such as ease of capture, in addition to or instead of oddity per se. One cannot generalize that predators always take the sick, injured, and odd, only that sometimes they do.

One can now leave the more or less academic aspects of predation and just stand back to appreciate the evolutionary outcome: an incredibly rich diversity of hunting and antipredator behavior. The remaining sections of this chapter will largely list the potpourri of predator and antipredator tactics that can be observed among living animals. Most readers may be familiar with many of these; the purpose is to bring everything together in one place and remind one of the diversity to which millions of years of interactions can lead.

Hunting and Capture Methods

Following are the various hunting and capture methods used by predators:

1. *Groping and flushing.* Many predators, such as octopuses, many arthropods, and other animals using their feet or feelers, work their way through the environment groping and feeling for possible prey. One recalls the familiar picture of a raccoon feeling around under water for prey items. In some cases the prey is caught as quickly as it is encountered, and in other cases it is flushed and then chased. Many bird and mammal predators may crash through bushes or tall grass, flushing victims that are then chased and caught.

2. *Stalking and ambushing.* Mantids and true chameleons are familiar examples, as are many of the cats. Some predators use other tactics in conjunction with stalking, and various fish wiggle bait or items that attract the curiosity of unsuspecting prey (Figure 11-15). Some are *aggressive mimics* (i.e., "wolves" in "sheeps' " clothing). Some mantids, for example, resemble harmless leaves or even flowers. As an added complication in the ant-aphid relationship, in one case there is a predaceous lacewing larva that attaches aphid "wool" to its body so that it resembles the aphids (Figure 11-16). The larvae, masquerading as aphids, escape detection by the guarding ants and proceed to feed on the aphids.

3. *Chase and pursuit from a distance.*

4. *Interception of flight path.* This is a variation on outright chase and pursuit where a predator anticipates a flight path and moves at an angle or by a different route so that it meets the prey at a particular point.

5. *Exhaustion of prey.* Some predators will not outrun or outfly their prey but simply stay with them and cause them to continue moving or otherwise harass them until the victim is finally exhausted. This technique is used not only in cases where the prey may be dangerous but also where it has other tight defenses. Starfish overcome bivalve molluscs, for example, with the aid of their tireless hydraulic system of suction feet by which they grasp the mollusc's shells and pull until the mollusc eventually yields.

6. *Tool use to get prey.* The most explicit users of tools are, for example, woodpecker finches and chimpanzees, which use twigs or stems from plants to pry out certain prey items from crevices. Some vultures use rocks to break into eggs. But, in a sense, one might consider other structures external to an animal's own body as tools when they aid prey capture. Classic in this category would be the webs of spiders. A few other species also use webs or sticky substances, and some, such as the ant lion, build pits or other kinds of traps.

Figure 11-16 Wooly aphids and predaceous lacewing larvae—"wolf-in-sheep's-clothing." A, Typical habitat of wooly alder aphid *(Prociphilus tesselatus).* An aphid colony appears white on branch of alder bush in foreground. **B,** Close-up of part of an aphid colony with a larva of *Chrysopa slossonae* (arrow). **C,** *Chrysopa* larva in its normal, wax-covered (shielded) condition. **D,** Ant protecting aphids by biting an "attacking" finger. **E,** Ant imbibing a droplet of honeydew delivered by an aphid. **F,** Ant biting a shielded *Chrysopa* larva that was released in its vicinity. **G,** Ant biting a denuded larva that it has just detected. **H,** Denuded larva applying plucked wax to its rump with the head. From Eisner, T., et al. 1978. Science 199:790-794. Copyright 1978 by the American Association for the Advancement of Science. Photos courtesy Thomas Eisner.

7. *Communal hunting.* Many predators hunt socially and use their joint efforts to overpower, confuse, or exhaust prey, or to communally round up their victims. Familiar examples include wolves, several other (but not all) species of canids, killer whales, brown pelicans, some communal spiders, and many others. Communal hunting has been shown to increase the numerical success ratio of prey capture, and it is particularly advantageous in dealing with large victims that could not be handled by individual predators alone. Curio (1976, modified from Schaller 1972) tabulated the increase in success rate of lions with increased numbers of animals hunting together (Table 11-3). Other advantages of group hunting, as compiled by Curio, include more economical consumption of carcasses, consumption with less interference of other competitors, improved ability to feed young with large carcasses (often by carrying parts back), easier stalking of herds without stampeding, and sharing of food sources. Curio provides further discussion on all of these points.

8. *Combinations of tactics and different techniques at different times and places.* Many predators possess a diverse repertoire of predatory behaviors and employ different hunting methods, some of which may be very situation specific, at different times.

The improved success of communal hunting raises the general question of success rates in different tactics of hunting. Success is a difficult subject to quantify because it depends on many variables and subjective judgments on the part of the observer. The difficulty of observing predation under natural conditions was mentioned earlier. One is not always sure even when an act of predation

A

B

E

F

Table 11-3 The relationship of hunting success to the number of lions stalking or running

No. of Animals Hunting	Thomson's Gazelle		Wildebeest and Zebra		Other Prey		Total	
	No. Hunts	Success (%)	No. Hunts	Success (%)	No. Hunts	Success (%)	No. Hunts	Success (%)
1	185	15	33	15	31	19	249	15
2	78	31	17	35	11	9	106	29
3	42	33	16	12.5	5	20	63	27
4 to 5	42	31	16	37	4	25	62	32
6+	15	4.1	21	43	7	0	43	33
TOTAL	362		103		58		523	

Reprinted from *The Serengeti Lion* by G.B. Schaller, 1972, by permission of The University of Chicago Press. Cited in Curio 1976.

has been attempted for subsequent tally of whether or not it was successful. Nonetheless there are some crude estimates for different predators under different situations. For a comparison of success rates among different predators, see Table 11-4. Table 11-5 compares success rates depending on the type of prey. As can be seen, success rates span the entire range, from 0% to 100%, with some under 10%, a few over 90%, and most somewhere in between. The range is so wide that it provides little useful information, particularly in view of all the variables and subjectivity that enter into the tallies. All one can safely conclude is that the

C

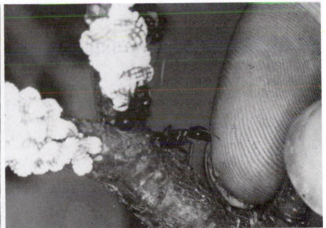 D

G

 H

**Table 11-4 Examples of
hunting success
of various predators***

Predator	Prey	No. Attempts	Successful (%)	Comments	Reference†
Woodruffia metabolica (Holotricha)	*Paramecium* sp.	Many	14	2.2%/encounter; attempt = dilate mouth to engulf prey	Salt (1967)
Busycon carica (Gastropoda)	*Venus mercenaria*	26	58		Carriker (1951)
Cuttlefish	Shrimp	?	ca.90		Messenger (1968)
Largemouth bass	Fish	85	94	Prey fishes without evasive movements	Nyberg (1971)
Forster's tern (*Sterna forsteri*)	Fish, ca. four species	1538	24		Salt and Willard (1971)
American kestrel	Rodents (insects?)	?	33	On familiar hunting ground	Sparrowe (1972)
Osprey	Fish	469	80 to 96	Both dives and snatches from surface	Lambert (1934)
Various raptors (*Falco columbarius, F. peregrinus, Accipiter nisus, Haliaetus albicilla*)	Birds	688 (60 to 260)	7.6 (4.5 to 10.8)	On migration	Rudebeck (1950, 1951)
Black bear	Salmon	1481	38.6	During 310 fishing sequences	Frame (1974)
Wolf	Moose	77	7.8	From all moose "tested"	Mech (1970)
Spotted hyena	Wildebeest calf	108	32	Similar for wildebeest adult, Thomson's gazelle, zebra	Kruuk (1972b)
Puma	Deer, elk	45	82	Excluding aborted hunts	Hornocker (1970)
Cheetah	Thomson's gazelle	87	70	Only fast chases tallied	Schaller (1972)
Chimpanzee	Mammals, six species	95	40	Including primates	Teleki (1973)
	Olive baboon (*Papio anubis*)	18	36	Adolescent victims <2 years	Teleki (1973)

From Curio, E. 1976. The ethology of predation. Springer-Verlag New York, Inc., New York.
*Additional examples are provided by Schaller 1972: Appendix B.
†See Curio (1976) for references.

success rate for predators is rarely very high and may be surprisingly low. Most of them miss their prey much of the time and must just keep trying over and over. Many fail too often and starve.

Handling and Killing Tactics

Once the predator has made physical contact with its prey and brought it down, the job often is not done; the victim has to be subdued and killed. There are four main categories of technique: (1) use of teeth, beak, or other mouthparts to bite or tear, (2) use of claws on the feet to puncture or tear into the animal, often in

Table 11-5 Hunting success in relation to type of prey

Predator	Prey	No. of Attempts	Successful (%)	Comments	Reference*
Hierodula crassa (Mantidae)	Fly walking	898	63.0	70.2% strikes/contact	Holling (1966)
	Fly flying	112	13.4	33.1% strikes/contact	
Herring larva, 35 to 42 days	*Artemia nauplii*	81	100		Rosenthal (1969b)
	Larger plankton, *A. metanauplii*	303	96.5		
Red fox	Rodents	ca.58†	25-100	Depending on snow conditions	Palm (1970)
	Larger prey‡	9	0		
Wild dog	Thomson's gazelle <2 months	22	95	Mean = 70%, including other prey except zebra	Schaller (1972); see also Estes and Goddard (1967), H. and J. van Lawick-Goodall (1970)
	Thomson's gazelle >2 months	47	49		
Cheetah	Thomson's gazelle fawns	31	100		Schaller (1970), Schaller and Lowther (1969)
	Thomson's gazelle adults	56	54		

From Curio, E. 1976. The ethology of predation. Springer-Verlag New York, Inc., New York.
*See Curio (1976) for references.
†162 km tracked.
‡Roe deer (*Capreolus capreolus*), European hare (*Lepus europaeus*), red squirrel, birds.

conjunction with squeezing, (3) use of constriction such as with body coils in snakes (Figure 11-17) (Greene and Burghardt 1978, Greenwald 1978) or otherwise smothering or suffocating the prey, which is perhaps just a variation of the preceding, and (4) use of poison or some form of stinging. The latter category includes coelenterates, stingrays, poisonous snakes of several types, spiders, cone snails, scorpions, shrews, and a few other animals with poisonous bites or stinging structures. A unique but relatively rare form of immobilizing or killing prey involves the use of electrical shock, primarily in a few chondrichthyes (e.g., electric and torpedo rays) and some electricity-producing osteichthyes (e.g., electric eels and electric catfish).

Antipredator Defense Tactics

The evolved diversity of predatory tactics is matched or perhaps even exceeded by the diversity of defense tactics and behaviors:

1. *Escape by fleeing and outrunning, outswimming, or outflying the predator.* This is one of the basic and more familiar tactics. It undoubtedly has led to much of the fleetness and agility seen in animals. In some species the intended victims literally may jet away, as in many of the cephalopod molluscs and scallops, or they may twist away or jump or fly into a different habitat, such as diving into water or the air or climbing a tree.

Figure 11-17 Corn snake killing a mouse. Constriction patterns vary among snakes. Some species coil tightly around the prey, whereas some use coils consisting of lateral twists and loops of the body. Different species use relatively stereotyped patterns of constriction. Photo courtesy Tom Brakefield.

2. *Use advance warning systems.* Many prey species have exceptionally acute sensory systems and are able to detect the predator before it comes dangerously close. The "bat detectors" of noctuid moths are discussed in Chapter 1.

3. *Use unpredictable movements.* These acts sometimes are called *protean displays*, as used by Curio (1976). (The term *protean* has not always been used consistently in the literature; also see item 6d below.) Fleeing animals may zigzag, jump, turn, change direction of flight, and otherwise move in an unpredictable fashion that is difficult for predators to successfully follow. An example of this tactic is given in Chapter 1 as one technique by which moths evade bats.

4. *Use hard armor.* Many prey species have shells or other forms of hard or tough outer covering or manage to block themselves behind a shield of some form.

5. *Use spines.* Examples of spiny animals include porcupines, hedgehogs, porcupine fish, the crown-of-thorns starfish, and numerous others.

6. *Employ offensive defenses.* This is a whole set of related tactics.

   a. *Fight back.* Many prey species are extremely aggressive themselves. Several of the insects and many rodents, for example, can inflict serious bites and may turn and attack their attackers. Some salamanders are able to repel even shrews (Brodie 1978). Prey that are as large or larger than their predators have equal or greater strength and can usually fight back in many cases.

   b. *Retaliate with shock or poisonous stings, bites, etc.*

   c. *Use borrowed poison.* Some molluscs, for example, consume coelenterates and then use their poison as protection against their own predators. Monarch butterflies, in a more familiar example, incorporate poisonous glycosides into their bodies from feeding on milkweed (*Ascepias* sp.).

   d. *Surprise the attacker.* Numerous species have different tricks of creating sudden visual displays, such as suddenly exposing eyespots on the wings, or loud noises such as claps, vocalizations, and hissing. Some animals will quickly change body posture, shape, and size or spread their wings or legs. These startle responses are called *protean displays*.

   e. *Use chemical defenses.* The classic chemical user is the skunk, but numerous other animals, particularly among the insects (Figure 11-18), may fight back with obnoxious sprays and odors or be poisonous, as in the case of the monarch butterfly.

f. *Use aposematism.* Many species, particularly those carrying chemical defenses, also carry conspicuous *warning displays*. These include the monarch butterfly's bright orange-and-black pattern, the skunk's black-and-white pattern, and numerous others.

g. *Use Mertensian mimicry.* The prey may reverse the tactics and resemble something dangerous, which they are not—sort of the sheep in wolf's clothing. Some snakes, for example, may shake their tails, hiss, flatten their necks, and engage in striking and more subtle movements that resemble other, dangerous species. Some flies resemble more dangerous bees. This category of mimicry is similar to Batesian mimicry (below) but does not involve bad taste or permit trial-and-error learning. The dangerous model may be so lethal that, if the real thing were encountered, the victim would not have a second chance. This form of mimicry receives its name from Mertens (1956), who participated in an ongoing controversy over mimicry in coral snakes and whether the most poisonous forms were the models or the mimics (for a summary discussion of that problem, see Wickler [1968] and Greene and McDiarmid [1981]).

7. *Camouflage.*

a. *Use camouflage patterns and disruptive coloration.* In a few extreme cases the animals actively cover themselves (Figure 11-19).

b. *Use protective mimicry.* This is a camouflage carried to the extreme. The animal may appear as a conspicuous object but be something other than what it really is. Some species resemble flowers, leaves, dead objects, and even bird droppings.

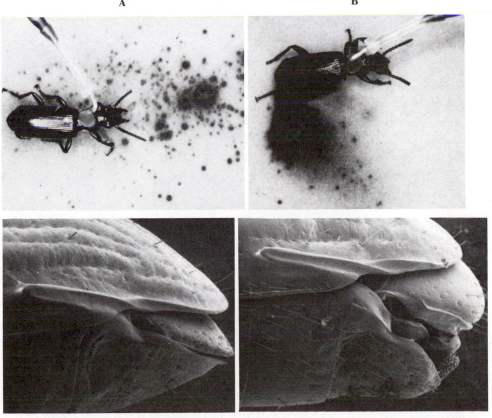

A   **B**

C   **D**

Figure 11-18 Bombardier beetle defense. Bombardier beetles, of the Carbidae subfamily Paussidae, spray a hot, quinone-containing secretion at potential predators. The secretion is formed by mixing component reactive chemicals from two-chambered glands. It is aimed and sprayed with the aid of flanges on the body surface, which direct the spray toward its target. Beetle discharging on chemical indicator paper in response to stimulation of left foreleg (A) and right hindleg (B). C, Tip of abdomen in lateral view of beetle immobilized while discharging toward a foreleg. D, Same of a beetle immobilized while discharging toward a hindleg. Movable "lip," lowered in (D) from its position in (C), is shown in center of photograph. From Eisner, T., and D.J. Aneshansley. 1982. Science 218:84. Copyright 1982 by the American Association for the Advancement of Science. Photos courtesy Thomas Eisner.

Figure 11-19 Camouflage by decorator crabs. Decoration behavior and hooked setae, which help hold materials, are shown for *Oregonia gracilis*. The crabs manipulate pieces of vegetation and debris with their mouthparts and chelae and then attach them to various parts of the body. From Decorator crabs by Mary K. Wicksten. Copyright © 1980 by Scientific American, Inc. All rights reserved.

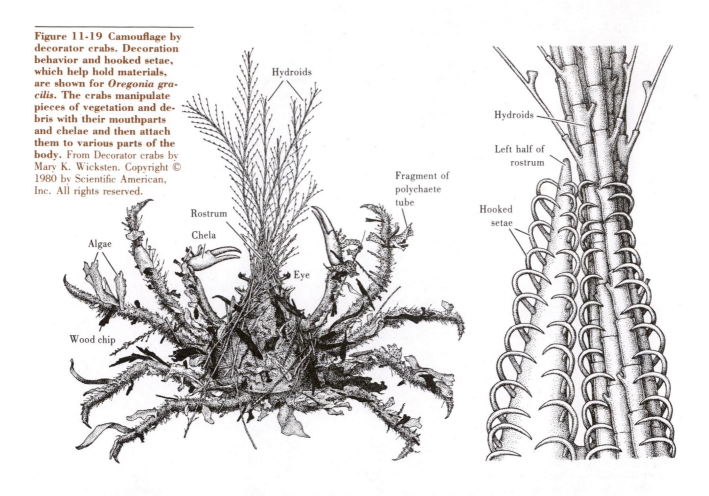

Figure 11-20 Owl facial mimicry of predators inhabiting their nesting areas. A, *Lynx lynx*. B, *Martes martes*. C, *Vulpes vulpes*. D, *Bubo bubo*. E, *Asio otus*. F, *Asio flammeus*. From Mysterud, I, and H. Dunker. 1979. Anim. Behav. 27:1.

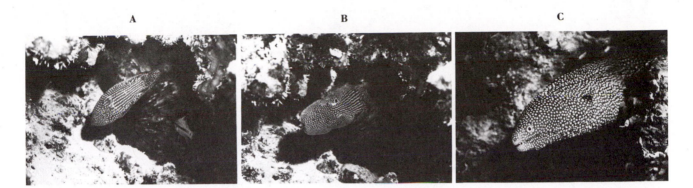

A   B   C

Figure 11-21 Pleisiopid fish (*Calloplesiops altivelis*) mimicking a noxious moray eel (*Gymnothorax meleagris*) when frightened by a predator. A, *Calloplesiops altivelis* (total length approximately 15 cm). Normal posture. Note the ocellus at the posterior base of the dorsal fin. B, Intimidation posture. The fish is now facing away from the camera. C, *Gymnothorax meleagris* (total length approximately 1 m, head length approximately 15 cm). Posture when confronted by a diver. From McCosker, J.E. 1977. Science 197:400-401. Copyright 1977 by the American Association for the Advancement of Science. Courtesy Steinhart Aquarium, San Francisco.

Mysterud and Dunker (1979) proposed that the "ear tufts" on owls (which are predators themselves) may serve to mimic the facial expressions of mammalian predators that they may encounter in their nesting environment (Figure 11-20). In other striking examples, juvenile lizards *(Eremias lugubris)* mimic the pattern and posture of noxious beetles (Huey and Pianka 1977), and a plesiopid reef fish when frightened resembles the head of a moray eel (Figure 11-21).

c. *Use false heads and other misleading directional cues.* Many prey species have false heads or prominent markings located in the least vulnerable parts of the body, such as at the tip of the tail, which may fool and misdirect the predator. In one extreme case, a butterfly has an inconspicuous body with an entire false body marked on the tip of the wings; even the shape of the wings points toward the false markings (Figure 11-22).

d. *Use Batesian and Müllerian mimicry.* Many potential prey animals are edible but mimic inedible ones by carrying the same aposematic warning colors and patterns. These are Batesian mimics. Some mimics are so close that they even have fooled entomologists and have been misclassified in insect collections. Some flies and beetles, for example, remarkably resemble various hymenopteran insects (Figure 11-23). Many of the aposematic tactics work on the basis of learned taste aversions; that is, the predator learns from killing or at least getting a bite of one animal to leave similar-appearing ones alone. The first is, in a sense, sacrificed for the sake of the others. This is a loose, uncritical interpretation, however, and one must be cautious not to fall into the "good of the species" trap of thinking (see Chapter 5).

Figure 11-22 False body markings on butterflies of the genus *Thecla*. The wing pattern displays a false head with striped pattern that draws attention away from the vulnerable true body and head. Modified from Wickler, W. 1968. Mimicry in plants and animals. McGraw-Hill Book Co., New York.

Wing stripes lead toward false head

False head and antennae

Figure 11-23 Examples of Batesian mimicry among insects. Many harmless insect species resemble dangerous or noxious species, as indicated. Modified from Wickler, W. 1968. Mimicry in plants and animals. McGraw-Hill Book Co., New York.

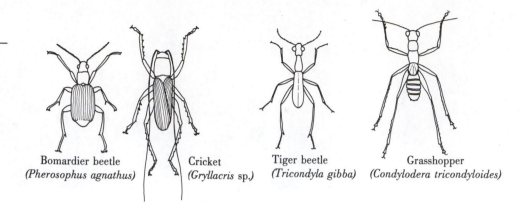

Bomardier beetle
(*Pherosophus agnathus*) Cricket
(*Gryllacris* sp.) Tiger beetle
(*Tricondyla gibba*) Grasshopper
(*Condylodera tricondyloides*)

If the first one encountered is an edible mimic, the predator does not learn and may continue killing animals that look like the same, perhaps including several more edible ones, until it encounters one or more poisonous forms. Thus for Batesian mimicry to be successful, predators should encounter a minimal proportion of the real thing, and in general, mimics cannot outnumber the inedible forms.

Several studies (e.g., reviewed in Morrell and Turner 1970) have shown that, at least among vertebrates, predators can learn to both *discriminate* and *generalize* concerning mimics. These are two different aspects of recognition or abstraction. Discrimination refers to the ability to distinguish between items, and generalization refers to perception of common characteristics (Figure 11-24). With further trial and error or increased hunger, predators may be able to discriminate better until only "perfect" mimics are avoided; that is, they can learn to distinguish only partial mimics (Figure 11-25).

Müllerian mimics are inedible themselves but share the same appearance and warning coloration with another inedible species. This forms a sort of double protection by which two or more carry the same warning flag.

8. *Create confusion.* A few species prevent attack by confusing the predator. Octopuses eject a cloud of black ink from which they escape while the predator cannot see. Geese in the presence of an attacking eagle may stay at the surface of the water and wildly splash water with their wings, which surprises and confuses the eagle.

9. *Hide or seek protective cover.* This is a well-known tactic that probably ranks with simple fleeing as a major means of protection. A large number of species will try to get under or behind cover during an attack by a predator, and many stay in cover during vulnerable periods, such as while sleeping or when young are present. Many species bury themselves in the substrate (Figure 11-26). A few animals do not go into cover but actually go into the open when under attack. A prime example is shown by caribou during the winter in forested lake country. When surprised or frightened, such as by a wolf pack, the caribou run from the woods out onto the ice and snow of the frozen lake.

10. *Seek safety in numbers.* One of the most common defense measures is to group together with other individuals. If any are taken, there is a good chance that it will be another—the "you-first" principle (p. 233). But in many cases the simple presence of the group will prevent further approach by the predator so that no individuals are lost. One of the best-known group-defense tactics is shown by musk-oxen, which form a tight circle with the young in the middle and their formidable, armored heads all facing out.

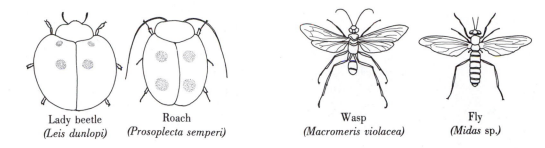

Lady beetle
(*Leis dunlopi*)

Roach
(*Prosoplecta semperi*)

Wasp
(*Macromeris violacea*)

Fly
(*Midas* sp.)

Figure 11-24 Generalization and discrimination of mimics by predators. Background patterns of different colors and shapes were used. Artificial prey—insect-shaped pastry either edible (soaked in water) or distasteful (soaked in quinine hydrochloride)— were placed on these backgrounds. Wild birds, mostly starlings, were permitted to take the food items in natural settings. Colors and patterns of backgrounds are shown along with mean number of items that were eaten. The noxious and perfect mimics were eaten least. Imperfect models were generalized and also partially avoided. The lack of complete avoidance, however, indicated that the birds were discriminating between perfect and imperfect mimics. Modified from Morrell, G.M., and J.R.G. Turner. 1970. Behaviour 36:116-130.

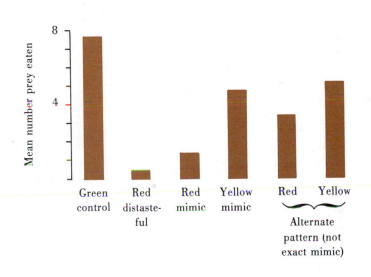

Figure 11-25 Sequence of bird choices of food items described in Figure 11-24. The noxious red model and perfect red mimic were not eaten by the wild birds after a few brief trials in which the birds experienced the bad taste. The birds generalized and partially avoided yellow mimics but learned to discriminate and eventually accept the nonperfect mimics. Control items were green. Modified from Morrell, G.M., and J.R.G. Turner. 1970. Behaviour 36:116-130.

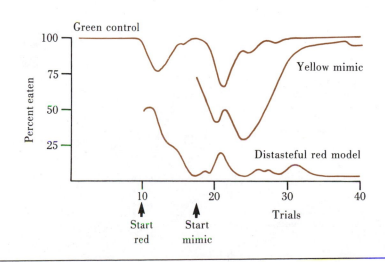

Figure 11-26 American toad hiding by burying itself under sand. Photo courtesy Tom Brakefield.

11. *Live in a neutral zone between predator territorial boundaries.* This tactic is exemplified by deer living between the territorial boundaries of neighboring wolf packs (Figure 11-27). The wolves avoid these buffer zones between territories to avoid encounters, often fatal, with neighboring packs. The deer use these zones as refuges (Mech 1977, Nelson and Mech 1981).

12. *Use combinations of tactics.* Most prey species have more than one tactic. Depending on the situation and predator, potential prey may use different defense techniques or several in combination. Geese using the wing-waving and splashing distraction mentioned earlier, for example, generally engage in that behavior while also tightly flocked together. The effect of several birds displaying simultaneously is very impressive and apparently is intimidating to a predator.

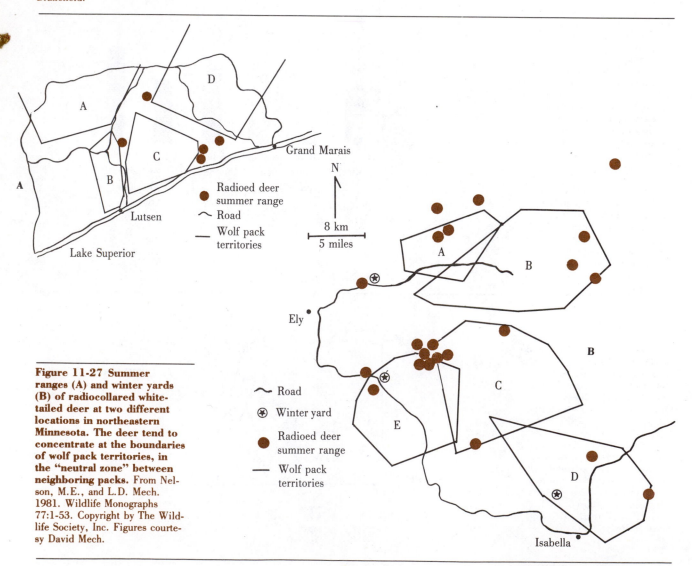

Figure 11-27 Summer ranges (A) and winter yards (B) of radiocollared white-tailed deer at two different locations in northeastern Minnesota. The deer tend to concentrate at the boundaries of wolf pack territories, in the "neutral zone" between neighboring packs. From Nelson, M.E., and L.D. Mech. 1981. Wildlife Monographs 77:1-53. Copyright by The Wildlife Society, Inc. Figures courtesy David Mech.

**Long-term
Predator-prey
Relationships**

Many predator-prey interactions have coevolved over long periods of time, with both sides showing highly advanced tactics for dealing with the sophisticated methods and defenses of the other. The relationship between fisher and porcupine, in which the fisher is behaviorally adapted to the porcupine's defenses, was mentioned earlier. Another fascinating relationship involves the octopus and moray eel. Morays hunt octopuses at night by odor and by investigating crevices where an octopus may be hiding. When the moray encounters an octopus, it immediately clamps onto the octopus with its viselike jaws. The octopus responds by clamping onto the moray's head with its tentacles. To this the moray responds by tying its body into a knot and slipping the knot down over its head, which forces off the octopus' tentacles (Cousteau 1973). The moray retains its hold on the octopus, which it then proceeds to kill and eat.

SUMMARY

Interspecific interactions and behaviors form a large category ranging from relatively neutral situations to various forms of direct competition, predation, and parasitism. In many cases different species live together (symbioses). Advantages of such living may be one sided (commensal) or accrue to both participants (mutualism).

Because of the diversity of interactions and behaviors that fall under the banner of interspecific, there is no underlying theory that pertains to all. This chapter thus consists primarily of a descriptive survey with theoretical comments where appropriate. Inquilinism in ants, for example, involves one species being dependent on work provided by another species. Three hypotheses have been proposed for the evolution of such life histories: (1) via intermediate levels of raiding and domestication (or slave-making) of other species, (2) by simply living in close association with other species through intermediate levels of evolution, and (3) by an evolutionary route involving parasitism. Numerous other categories of interspecific behavior are listed and discussed.

The diversity of interspecific behaviors, even within one category such as predator-antipredator interactions, illustrates the biological analog of the facetious Murphy's Law: If something can evolve, it will (or already has).

**Recommended
Additional Reading**

Cousteau, J. 1973. Attack and defense. World Publishing, New York.

Curio, E. 1976. The ethology of predation. Springer-Verlag New York, Inc., New York.

Eaton, R.L. 1974. The cheetah. Van Nostrand Reinhold Co., New York.

Eisenberg, J.F., and P. Leyhausen. 1972. The phylogenesis of predatory behavior in mammals. Z. Tierpsychol. 30:59-93.

Hölldobler, B. 1971. Communication between ants and their guests. Sci. Am. 224:86-93.

Mech, L.D. 1970. The wolf. Natural History Press, New York.

Schaller, G.B. 1967. The deer and the tiger. University of Chicago Press, Chicago.

Schaller, G.B. 1972. The Serengeti lion. University of Chicago Press, Chicago.

Wilson, E.O. 1975. Sociobiology, the new synthesis. Belknap Press, Cambridge, Mass.

CHAPTER 12

Ed Bry, North Dakota Game and Fish Department

Two red fox *(Vulpes vulpes)* **playing near their den.**

PLAY

*The only difference between the men and the boys is
  in the price of their toys.*

Anonymous

Play is an enigmatic topic. It is something that almost all persons have done. Yet authors have difficulty trying to "define" play behavior, depending on which kind of definition one is searching for (see Fagen 1981 for discussion). It also may be difficult to define an apple. Apples are easy to *describe*, however, and may be defined easily in that sense. Similarly, play can be defined more or less by description and by listing examples. Play has been recognized in other animals that act at times in a manner that seems similar to human play. But it is not really certain *why*, in the ultimate sense, animals play. It is the *function* of play that is not understood and cannot be included in the definition.

Play commonly is considered along with ontogeny and development (Chapter 18) because it frequently is associated with young, growing animals. But this association may be partly misleading. The ontogenetic function of play is an important hypothesis that still is being considered (discussed later), but it does not account for all behavior that is considered to be play. In some species play may continue to be expressed, albeit perhaps less frequently, in older individuals or in association with other behaviors, such as reproduction. Some eagles, for example, seem to become quite playful annually at a preliminary stage of reproduction, before their nest-building behavior has become more organized and functional (Grier 1973).

The role of play even in humans is highly conjectural and academic. Human play is influenced strongly by experience and culture. Some societies (e.g., the Hutterites) display little play after certain ages, whereas people in other cultures retain interest in play through old age. Although not well understood, there may be correlations between play, creativity, and art in humans.

Fagen (1981) interprets the basic problem in past scientific attempts to study play as being a lack of adequate theory. An underlying theory or understanding, if only rudimentary and even if wrong, guides the activities that are called science. If one does not know what he is looking for, it sometimes is difficult to see anything.

The problem stems from two underlying causes. First, play seems to have no *immediate or obvious* function, and, second, play is an extremely difficult subject with which to experiment. Because of traditional biological beliefs that attributes arise only under selective pressure, it is easy to think that play should have *some* advantages, effects, or "function" or else it should not have evolved. But, if so, the advantages have been difficult to see and associate with the actions of play.

The other major problem facing play research and understanding has been the difficulty of dealing with play experimentally. It is the classic problem of the trained cricket (p. 81). Trying to experiment with play, for example, by preventing an animal from doing it, may inadvertently affect *many* important aspects of the animal's life, biology, and development. Some of the things affected actually may be correlated with play, and some may be totally irrelevant but are inescapably caught up in the experiment. This has proved a nasty and persistent problem and has produced disagreement over opposing hypotheses.

DESCRIPTIONS OF PLAY

As observed in its basic form among animals, depending on the species, play consists of frolicsome leaps, running and quick turns, rolling about, climbing, and other movements in exaggerated or otherwise unconventional manners, plus knocking, kicking, throwing, dropping, and wrestling with other objects—including other organisms. Play may become quite elaborate, particularly in social situations.

Examples of Play Behavior

There are several types of play and a number of general characteristics that can be described for play. Fagen (1981) provides an excellent review of the lists formulated by other persons. Before looking at generalizations, however, some specific instances of play will be noted. Ideally actual play or films of it should be observed. Because this is not possible in a book, however, we will have to settle for a few pictures (Figure 12-1) and some verbal descriptions. Fagen sifted through the voluminous natural history literature, and several of these are accounts originally cited by Fagen.

Koford (1957) describes the play of young vicuna *(Vicugna vicugna)*, a South American camelid ungulate species. This play seems more or less typical of many ungulates:

Often a group of young run away from the adults and back, several times in succession, one then another in the lead. At 2 weeks of age the young rarely race more than 50 yds. from the family group, but at 1 month of age they run away as far as 100 yds, and at several months of age, twice as far. On occasions the galloping youngsters make high leaps over rocks, ditches, or other obstacles. Sometimes a juvenile acts as if chasing an invisible partner, running fast and recklessly, jumping high, and kicking his hindfeet back. After jumping down over a bank, one fell and turned a somersault. Another fell on a road and skidded along on its side.

Solo play similar to that just described is familiar in domestic species, including horses, cattle, and sheep. But ungulates certainly do not have a monopoly on solo play. Fagen (1981:99), for example, summarizes solo play as seen in a monkey, Lowe's guenon *(Cercopithecus campbelli)* and as reported by several persons (see Fagen for references):

Solo play, including arboreal acrobatics and object manipulation, occurs frequently. A young guenon may carry some item (branch, heavy rock, or the like) up a tree, drop it, then descend, and repeat the game several times.

Figure 12-1 Examples of play as shown by lion cubs (A), domestic dog pups (B), and juvenile rhesus monkeys (C). A from Animals, Animals. Copyright by Margot Conte. **B** from Animals, Animals. Copyright by Paula Wright. **C** from *Animal Play Behavior* by Robert Fagen. Copyright© 1981 by Oxford University Press, Inc. Reprinted by permission. Photo by John Bishop.

Play frequently involves other animals, particularly conspecifics. The following is a description of "aggressive" play in ferrets (European polecats, *Putorius putorius*) by Poole:

Aggressive play in the young polecat is initiated by an attacker which jumps on to the back of its opponent and bites it on the back of the neck. The opponent rolls over on to its back and makes a series of snapping bites at the muzzle and neck of its attacker, at the same time pushing it away with its paws. In response to this, the attacker either stands above its opponent and snaps its jaws in a playful attempt to bite its neck or alternatively rolls on to its back and the two animals exchange their roles of aggressor and defender. If the animals become very excited, the aggressor shakes its opponent vigorously by the neck and may drag it around the arena by the scruff of its neck. Sometimes both animals lie side by side on their backs, snapping their jaws at one another and waving their legs in the air. Occasionally polecats direct bites at one another's limbs and tails but this is not a common pattern in aggressive play.

If one animal is in a playful mood whilst the other is not, the playful animal teases its victim in an attempt to incite it to join in the play. Teasing takes several forms. A polecat may mount the other animal whilst holding its neck with the teeth until the opponent is incited to roll over and playfully retaliate or it may dance up to the other polecat, finally jumping on it, or it may chase it around, frequently biting it in the pelvic region.

A polecat which does not wish to indulge in play or has already had enough, threatens its opponent by hissing and baring the teeth; this results in the attacker desisting. If one of the animals is smaller or weaker than its opponent which is being too rough, it cries plaintively until it is released. Playful aggression frequently is accompanied in the older animals by vocalization which takes the form of an excited clucking.

The above patterns constitute, together with chasing and flight, the most frequent patterns of aggressive play.*

The next example concerns the most frequent patterns of play seen in wild African lions:

(1) After a posture indicating readiness, one cub rushed at another or at its mother, flung its body over the partner, pawed it and perhaps used its mouth to seize and then lick the partner's cheek or ear or neck while lying in a relaxed body contact. The intention posture was often taken up by the partner who would then in turn perform the rush; alternatively he would stand his ground expecting the other. In most cases, while lying in close contact or perhaps whilst rolling over each other, both partners used their paws and mouths as mentioned before.

(2) A cub suddenly paid attention to a stick, a little shrub or a piece of grass. It pawed the object, perhaps rolling on the ground, stretching out its forepaws to reach the object. Very often another cub joined the scene.

(3) Sometimes even the mother took part in a restrained manner: she moved her tail repeatedly, thus stimulating a cub to play with it. Alternatively she knocked a cub over very gently with a paw or a push of her nose, and while it pawed and struggled with all four legs, she licked it with repeated strokes of her tongue.

(4) From time to time a cub ventured a few yards from its mother and examined the surroundings—especially strange apparitions like the Landrover of the observer—with concentrated attention showing an ambivalent attitude between curiosity and shyness.†

Animal play may engage individuals of other species, as indicated in the next two examples. The first is an instance described by Stefansson (1944). It involves caribou *(Rangifer tarandus):*

Three large bulls were feeding quietly and near them were eight yearlings. . . . Several of the yearlings started on a sudden run. . . . When I looked at them through the binoculars I saw they were chasing a fox. Three or four of the yearlings chased it for two or three hundred yards and then returned to where the

---

*From Poole, T.B. 1966. Aggressive play in polecats. Symp. Zool. Soc. Lond. 18:23-44.
†From Schenkel, R. 1966. Play, exploration and territoriality in the wild lion. Symp. Zool. Soc. Lond. 18:11-22.

rest were feeding. The fox now waited a little, as if to see if the caribou had given up the game, and then ran in among them again and was chased by others. Sometimes the fox dodged in and out among them and when it had secured the interest of the yearlings and induced them to the chase, it would run around the three old bulls using them as a sort of protection from the yearlings. The old bulls paid not the slightest attention to either fox or yearlings and went on feeding quietly. I watched the game about half an hour, after which I mounted the ridge and approached till they saw me.

The next case describes an interaction between ravens and wolves as written by Mech (1966):

As the pack traveled across a harbor, a few wolves lingered to rest, and four or five accompanying ravens began to pester them. The birds would dive at a wolf's head or tail, and the wolf would duck and then leap at them. Sometimes the ravens chased the wolves, flying just above their heads, and once, a raven waddled to a resting wolf, pecked its tail, and jumped aside as the wolf snapped at it. When the wolf retaliated by stalking the raven, the bird allowed it within a foot before arising. Then it landed a few feet beyond the wolf and repeated the prank.

This case introduces an element of seeking thrills, playing pranks or "jokes," or "expecting the unexpected" that can be seen occasionally in the play of many animals, not just familiar human humor and pranks.

The next case presents further instances of social play and what might be described as "thrill" or surprise seeking. It is a continuation of Fagen's summary of guenon play:

Social play includes arboreal and terrestrial chasing and wrestling. A monkey may vault over another's back. One monkey may present itself as if to solicit mounting, then suddenly flee at the instant its partner attempts to mount. In an arboreal game two to five monkeys scramble up to a swaying tree branch and wrestle there. One monkey may abruptly fall off the swinging bough, which then springs upwards, knocking other monkeys off balance. Free falls are frequent in play. One monkey may drop down by stages through small trees or underlying bushes. Subadults appear to "seek thrills" . . ., breaking or tearing open the nests of stinging weaver ants (Oecophylla longinodis), jumping around frantically while the ants attack them, then returning immediately to the nest to repeat the process. Juvenile Lowe's guenons play interspecifically and reciprocally, chasing and being chased by wild squirrels (Heliosciurus gambianus) or fleeing from a charging (but tame) mongoose (Crossarchus obscurus), then returning to "tag" the mongoose and flee again. The monkeys also pursue other mammals and birds, apparently for the sole purpose of scaring them. . . . The habit of teasing hornbills (Tockus fasciatus) seems best developed. The hornbills roost in a large tree. Once they are settled, the juveniles leap toward them and shake the branches until the hornbills take flight. The monkeys then wait in the tree until the birds return. When the birds have again settled, the monkeys repeat their mock attack.*

Good descriptions of play abound in the literature. Fagen (1981) details many others throughout his book and provides a thorough taxonomic review of the

*From Animal Play Behavior by Robert Fagen © 1981 by Oxford University Press, Inc. Reprinted by permission.

published observations. One might be surprised to learn, for example, that complex play has been described even for bats (Chiroptera), with young chasing each other, wrestling, romping, slapping each other with their wings, and otherwise playing in typical mammalian fashion. Most people are familiar with the play of puppies and kittens and of various species commonly confined in zoos. These various cases, plus perhaps some insight from human play—which is remarkably similar in many ways to the play of other animals (Figure 12-2)—provide good examples from which one now can make several generalizations.

General Attributes of Play

Play lacks apparent, external goals. This point has already been mentioned, and it is, perhaps, one of the most obvious characteristics of play behavior. One or more objects, including other animals, may be involved (and may even be modified or damaged) in an animal's play, but no recognizable function is accomplished with the object. The animal simply "plays" with the object—it is hard to avoid the self-definition, circular trap.

Although play may be very demanding of energy and time and thus involve work in a physical sense, it generally seems to lack "seriousness" or "work" in terms of human connotations (i.e., something that needs to be done). It is not temporally associated with the "serious" aspects of life such as predation, escaping from predators, eating, or reproduction. And other, frequently older, members of a group who are engaged in less playful activities may literally frown on those individuals which are playing.

Not only does the play itself appear to lack a goal, but it also generally is engaged in only in the absence of other goal-directed situations. Other more obviously functional behaviors may inhibit or interrupt play. An animal may break off

Figure 12-2 Similarity of human play to that of other primates. A and B, Play-wrestling by lowland gorillas in an upright position. C, Play-wrestling by Sherpa boys, Nepal. A from Animals, Animals. Copyright by M. Austerman. B and C from *Animal Play Behavior* by Robert Fagen. Copyright © 1981 by Oxford University Press, Inc. Reprinted by permission. Photos by John Bishop.

A   B   C

a bout of play at the appearance of danger, food, a chance to mate, or other opportunities or needs.

The movements of play generally are borrowed from other behaviors. There may be unique motor patterns used as play-soliciting signals preceding or during social play, which are discussed further later. But beyond these, there are few, if any, movements seen in play that are not also seen in other contexts. The movements include those displayed in predatory behavior, social fighting, fleeing, reproduction, and eating. Loizos (1966), in a qualitative assessment of the use of various motions in play, suggested that these borrowed motions may be reordered in sequence with many possible permutations, exaggerated, repeated, fragmented and shown in incomplete sequences, or displayed in combinations of these (e.g., exaggerated and repeated or fragmented and repeated).

Bekoff (1976) stresses that these characteristics have not received adequate quantitative analysis yet. From the few data available (e.g., for black bear cubs, *Ursus americanus,* and New England coyote-canids, *Canis latrans*), however, there is little or no evidence that movements are exaggerated in play, and in some acts the opposite is true. Bekoff suggests that Markov models, information-theoretic analysis, and measures of conditional uncertainty provide possible means for quantitatively assessing the sequences of movements and for comparing the amount of order or randomness in play versus nonplay behaviors using similar movements. Limited information that is available (e.g., on vervet monkeys, *Cercopithecus aethiops,* black-tailed deer, infant canids, and young rats) supports the notion that play movements may be reordered in sequence but has not permitted a measure of whether play motions are more random than in nonplay.

Play generally is seen more frequently in young animals than in adults. Adults of some species, however, may play considerably, particularly when the other needs, such as feeding, have been met.

Play is most likely to be shown, at least in the few mammals studied, under relaxed and familiar conditions. The mammals studied include lions (Schenkel 1966), polecats (Poole 1966), and humans (Hutt 1966). Young mammals may require familiar surroundings, the presence of the mother, or a familiar object ("security blanket") before they will engage in play. In the absence of the mother, for example, young lions will behave cryptically and show no play or exploratory behavior.

Play is potentially dangerous and costly. The risks and costs of play have not been well quantified. The topic is just beginning to be subjected to optimality considerations. From the meager published information that Fagen was able to summarize, however, it is clear that play may not be cheap. Accidents that occur during play may be common and may lead to temporary injury, permanent disability, and even death. In a study of confined ibex *(Capra ibex),* for example, at least 5 of 14 kids sustained injuries during play that produced visible limps (Byers 1977). Most injuries related to play result from falls. Even among humans, about two thirds of nearly 100,000 annual injuries on playground equipment in the United States are the result of falls.

Animals expose themselves not only to accidents while playing but also may become more vulnerable to predation. Juvenile vervet monkeys *(Cercopithecus aethiops),* for example, were caught by baboons *(Papio cynocephalus)* most often when playing in groups away from the adults (Hausfater 1976). This may be a factor in the lack of play by some species, such as lions, unless they are in the presence of adults or in familiar surroundings. Playing animals may run risks while playing also from their own species when play escalates into serious, aggressive fighting or if it attracts punitive intervention by other individuals.

In a less direct sense, play also may be costly in terms of time and energy. Again, good quantitative data are skimpy, but available studies (reviewed in Fagen 1981), mostly from primates and other confined animals, show that young animals commonly spend 1% to 10% and sometimes 50% or more of their time in play. This may be physically demanding, is often exhausting, and consumes considerable metabolic energy. Time and energy spent on play are not available for other activities. Even in human terms, time, energy, and money spent on recreation, sports, and entertainment may cut significantly into one's budget and reduce the supply available for more "necessary" items.

All of this strongly suggests that play has some selective advantage that outweighs the potential costs and risks. Otherwise it would seem that play should be selected against. An intuitive, subjective understanding of at least the proximate benefits was expressed by a parent who was overheard to say, while watching her children climb about in a tall tree, that she would rather take the chance of her children suffering broken bones than broken spirits.

Play appears to involve feelings of pleasure. This is a subjective and difficult point. The interpretation is derived from the human perspective and, hence, may seem anthropomorphic. Also, some interpretations are redundant and circular: "Something is pleasurable because the animals feel pleasure." Such aspects of behavior have received considerable academic discussion. Nonetheless, the notion that other animals may experience pleasure seems reasonable. Physiological evidence exists for "pleasure centers" in the brain (Chapter 16). The presence of pleasure sensations in other species, whether or not associated with play, can be argued theoretically from two evolutionary standpoints:

1. If humans experience pleasure and humans evolved, then it is possible, if not likely, that our relatives also possess the sensation—the continuity of evolution.
2. Ultimate factors require proximate mechanisms. The physiological process that is subjectively called "pleasure" could be an important proximate mechanism; that is, pleasure simply may be the name that is given to the neurological process that motivates animals to eat, engage in sexual and grooming behavior, and, perhaps, play.

Play often contains elements of surprise and, apparently, the seeking of the unexpected or "thrills." This is familiar in human play, as seen in activities of daring, stunts, simple games such as peekaboo, and the punch lines in humor. Curiosity and interest in things where the outcome, or at least the timing of the outcome, is not known beforehand may underlie much of our daily activity. Consider, for example, interest in mysteries and novels, sports events where the winner is not known beforehand, or even preoccupation with news, information, and science. In human games and humor there often is the wide-mouthed period of anticipation and suspense followed by the point of release, often accompanied by laughter or other sense of exhilaration.

These elements of the unexpected are by no means confined to human play.

Perhaps the most familiar instances of anticipation and surprise can be seen in the ambushing and pouncing play of kittens. They crouch and hide behind objects, with the tips of the tails twitching, then suddenly jump or strike out with their paws. Dogs often will race circles around each other or their master, coming as close as possible, then run past as if daring to be tagged or caught. Tree squirrels and many other species often will run up and tag another animal and then race away. The young of many ungulate species seem to delight in chasing birds that are on the ground. They will chase after them until they fly, wait for them to land elsewhere, then chase and flush them again. Fagen (1981:325-328) cites a great many more instances of "thrill seeking."

Play is probably a category *of behavior, including many specific types, and appears to overlap in some cases with another category, exploration.* There are at least two kinds of exploration, *specific* and *diversive* (Hutt 1966). They may merge over time from one category to the other. In specific exploration the animal's senses and attention are directed toward the object, and, depending in part on the complexity of the object and number of senses it stimulates, the animal's attention decreases in time with repeated exposure to the object. This type of exploration might be viewed as learning rather than play.

Diversive exploration, however, seems to fall more into the class of what most persons would call play. In cases where the properties of the object are known, senses are directed more away from the object, and attention is given to doing things with the object. The behavior seems to shift from "What does *it do?*" to "What can *I do* with it?" This has led some persons (e.g., Hutt 1966) to distinguish between exploration as acquisition of information and play where learning is incidental. Exploration and play are difficult to distinguish in infants (human) but diverge and become more distinct as the individual becomes older. Exploration also appears to differ among different species and may be correlated with predatory and nonpredatory modes of life (Hutt 1966). Rodents, for example, more readily explore new environments than objects in familiar environments, whereas the opposite is true of predators, including humans.

Diversive exploration may be associated strongly with creativity in humans. Dissanayake (1974) has proposed that art originated from play. At least subjectively, there appear to be strong correlations among humor, play, creativity, and the development of thinking ability (see Adams 1974). The converse also appears to apply; persons who are more "serious" or socially inhibited (often described as "good" and "obedient" children) may be more reluctant to play and are less creative (Hutt 1966).

Social play, where the object being played with is a conspecific, generally includes clear and unambiguous signals. These signals, described previously as metacommunication, alter the meaning of the subsequent actions, rendering them to be "not serious." Actions that would be responded to as outright aggression without the play signals are treated differently by the respondent in the presence of the play signal.

Play seems to be restricted among various taxonomic groups. Few instances of play have been observed among invertebrates or poikilothermic vertebrates. Thus it appears to be confined primarily to birds and mammals and, among these, is most conspicuous in mammals. Play is seen most frequently among predators or species with the greatest abilities of vision and manipulation.

WHY DO ANIMALS PLAY?

Although one cannot be certain of the emotions of animals of other species, it is fairly clear that many, if not most or even all, animals play because it is "fun" (i.e., pleasurable). But this is, at best, simply the proximate cause of behavior.

The causal factors of most interest concern the ultimate, evolutionary causes (assuming that they exist) for behavior. There are many candidate hypotheses, all with their advocates and opponents. They have been discussed by several persons, for example, Smith (1977) and Caro (1981), with the most recent major review by Fagen (1981).

The best basic explanation of play appears to be the muscular and physical development of young animals in a context that is less serious and dangerous than is encountered by older animals. But this hypothesis has not been tested, and training clearly is not sufficient to explain all play behavior. Thus the practice theory along with several alternate ideas that have been proposed will be reviewed briefly. The various hypothesized functions of play are surplus energy, pleasure (only), arousal/stimulation, practice, exercise, and social functions. Social considerations merit their own section of text and will follow the review of basic explanations.

Surplus Energy Theory

Friedrich Schiller (the poet), Herbert Spencer, F. Alexander, and others have proposed that play results from the use of excess energy in animals. Attempts to deal with this theory, however, have met with little success. Müller-Schwarze (1968) found no significant increase in play actions of play-deprived black-tailed deer except for an increase in running speed of one subject. Chepko (1971) obtained inconclusive or insignificant differences between play-deprived and nonplay-deprived young Toggenberg goats *(Capra hircus)*. Again, it is a difficult hypothesis to test, although one approach might be to experiment with animals restricted on food (energy) intake rather than restricted on play.

On logical grounds alone, excess energy, that is, energy not needed for immediate or short-term needs, would seem more likely to be physiologically metabolized and stored as fat rather than being blown off and wasted. Furthermore, as considered by Bekoff (1976), "The attributing of play to an excess of energy does not lead to any further clarification of the characteristics of play."

Pleasure Theory

Play appears to give pleasure to the participants. The pleasure may even be shared by onlookers. Aside from the proximate aspects, pleasure might be selectively advantageous in itself. The pleasure theory, which can be traced in one form or another back at least to Pycraft (1912), suggests that play occurs for the immediate pleasure or fun of it only and that there is no other adaptive significance. However, this would be difficult, if not impossible, to test for. One may be able to discard the idea that play exists for pleasure only by elimination, that is, by adequately demonstrating that an alternate explanation exists and that there is a reasonable function or ultimate causal factor.

Arousal/Stimulation Theory

Based on earlier psychological notions of drives, Ellis (1973) suggests that play helps elevate an animal's level of arousal. In this view the behavior is involved in a generalized stimulus seeking that may expose it to other things needed in life. This harks to exploratory behavior and may be partially involved in some cases. But, again, it is a vague concept and would be difficult with which to experiment. Bekoff (1976) references and comments on semantic problems with the term "arousal." One possibility might be to measure levels of arousal in the reticular-activating system or from surface brain waves as during measures of sleep and then correlate them with play behavior. Until such measurements are obtained and carefully interpreted, there is not much more that can be said about a possible arousal function for play.

Practice Theory

Practice would seem one of the most reasonable, intuitive explanations of play, based on present understanding of learning, but evidence does not support the theory. According to the practice theory, through practice and repeated performance in a nonserious setting, an animal learns or neurally perfects movements and behavior needed in more serious contexts. This concept was suggested by Groos (1898) and has been maintained by subsequent authors, for example, Aldis (1975). One problem, again, is that it is difficult to deprive animals of play to see if this interferes with the performance of other behavior.

But in the few cases where naive or socially deprived animals have been studied, there is little evidence that subsequent behavior such as prey killing or copulation suffers. Vincent and Bekoff (1978) attempted to deal with the problem by correlating performance in play with performance of prey killing in infant coyotes. They found, however, no significant correlations involving such things as frequency of participation in play, participation in agonistic interactions, or frequency of most motor patterns. Only the frequency of pouncing in play was correlated with prey-killing ability.

In other cases the array and performance of motor patterns in play versus other behaviors are confusing, but they do not seem to support the practice theory. In the aggressive play of polecats (Poole 1966), for example, the patterns of play develop with age, but as they appear, various components of the patterns emerge in stereotyped and completed adult form with no subsequent modification. Thus many borrowed behaviors that show up in play appear in already complete form, without need for practice. On the other hand, some motor patterns that are very important in the serious context are either absent, for example, in the play of young spotted hyenas, *Crocuta crocuta* (Kruuk 1972) or are unimportant in the play behavior, such as stalking in lions, *Panthera leo* (Schaller 1972).

Thus there has been little evidence to support the practice theory of play. Nonetheless, the lack of evidence (as with the other theories) does not by itself discount the theory. Sample sizes in studies have been small, which commonly creates a statistical type II error problem (i.e., makes it difficult to reject the null hypothesis), and only a few species have been carefully studied. Even if practice is not involved in such things as the frequency of expression of certain movements, it still might affect the efficiency or finer aspects of the movement.

The role of practice and learning, at least in some species and involving some types of play, remains attractive and should receive further study.

Before proceeding to the next theory, it might help to stress the difference between two of the theories already mentioned: the arousal theory and the practice theory. Both concentrate on use of the nervous system in play. They differ in that arousal involves more general, nonspecific stimulation of the nervous system, whereas practice focuses on specific neural patterns affecting the performance of particular movements. The next theory concerns general, nonspecific stimulation of the *muscular* system.

Exercise Theory

According to the exercise theory, the use of muscles and associated (e.g., cardiovascular) systems in play stimulates the development and maintains the physiological condition of these systems in the absence of risky, serious contexts. Then in the presence of serious situations, which are encountered more frequently when an animal grows up and loses the care and protection of its parent(s), the animal is in physical condition and better able to cope with the demands of the serious situations than if it were not in good physical shape. This theory differs from the practice theory in that it emphasizes physical endurance systems rather than the neural coordination aspects. Persons suggesting general neural or sensory stimulation as a function of play (i.e., the arousal theory) also generally have pointed out the possibility of muscular and cardiovascular effects (references in Fagen 1976).

Although also skimpy on direct experimental data, the exercise theory of play is the best candidate for explaining play. The selective advantages of strength and endurance are obvious. Muscular and cardiovascular systems appear to require development and maintenance and suffer from disuse as much as or more than the nervous system.

Before leaving this review of basic explanations, however, it is important to emphasize that there is no necessity for a single explanation for all play. There are many different species, and they face a diversity of environmental challenges. Factors underlying the development of what is called play in some species might be quite different from those causing similar-appearing behavior in other species. Although the exercise theory appears to be the best general explanation at the moment, it does not have to be universal, and it may appear best only because competing ideas lack data and theoretical development.

PLAY IN A SOCIAL CONTEXT

Play with other animals, primarily belonging to the same species, represents play in its most elaborate, sophisticated form. The animals may play in a rough-and-tumble manner but not so rough, normally, as to seriously hurt or damage each other. Social play often shows a degree of cooperation and even what might be called "fairness." Larger and tougher animals may play by restraining or self-handicapping themselves (Figure 12-3). Cats generally keep their claws retracted during play; bears cuff and wrestle while using their paws in a manner that does not expose their sharp claws to scratching and tearing of the other; and horned animals either do not use their horns during play or else use them only gently. When using their mouths and teeth, playing animals usually only nip or mouth their play partners rather than truly biting in such a way that would break the skin or cause damage.

Figure 12-3 Self-handicapping among animals of unequal size and strength. When animals that are not matched in physical ability play, the larger and stronger often restrains itself and does not use its full strength in the interaction. From *Animal Play Behavior* by Robert Fagen. Copyright © 1981 by Oxford University Press, Inc. Reprinted by permission. Photo by John Bishop.

Much of the apparent fairness seen in play is accomplished by matching of animals of similar strength, ability, and social status. Discrepancies may be balanced by compensation in other aspects. Older animals commonly are stronger than younger, for example, and males may be stronger than females. Abilities can be balanced by an older female playing with a younger male. But even with age, sex, and most other things being equal, one would not always expect individual interests to coincide. Thus the presence of any degree of fairness or the fact that animals play together at all, let alone animals that are obviously not matched in strength or ability, raises several interesting questions.

Social play usually is accompanied by a set of relatively unambiguous communication signals, discussed previously under the topic of metacommunication. The relaxed, open-mouth ("smile" in primates) play-face is almost universal among mammals (Figure 12-4). A southeast Asian primate, the douc langur *(Pygathrix nemaeus)*, has a bright yellow face, white whiskers, blue eyelids, and brown and chestnut on the remainder of its head. During play, and only during play (or playful preliminaries before copulation), this langur closes the eyes and displays the blue eyelids toward other individuals.

Other visual signals include distinct postures, such as the familiar play-bow (probably derived from stretching) shown by dogs and lions and a particular position of the tail shown by bovids and equids. Some primates bend over and look between their legs. The signals may take the form of movements such as rolling around on the back or gamboling play-gaits seen in many species and perhaps derived from solo play.

Play signals are not confined to vision; various species may use signals in any of the sensory modalites. There are a number of auditory play signals plus many that use the tactile senses, with various body contacts, poking, tagging, and jostling. Olfaction has received the least attention by humans (who do not use olfaction to the degree that many other animals do), but there is some evidence that other animals use play-smells. Wilson (1973) applied ether extracts from the bodies of playing short-tailed voles *(Microtus agrestis)* to the bodies on nonplaying voles. Nonplaying voles with the extracts then elicited play from other voles, whereas nonplaying voles with ether alone did not.

The essence of the play signals, as judged by the reactions of the conspecifics, is the general message that "this is done in fun or jest and not to be taken seriously; I am only kidding or joking." However, most species have a number of different signals. The need for several is not clear. Such may be for redundancy or to ensure that the chances of misunderstanding are minimized, or there may be much more to the message than simple play versus nonplay. The various signals, or gradations of signals, may carry information on interest and motivation in play, strength, or social status of the players or other messages (perhaps even including deception) that have not yet been considered.

If social play and the associated signals and phenomena were found only in humans or a few other species, it might seem to be simply a cultural phenomenon. But the behavior patterns and signals are too widespread among many species; there are too many similarities in the behavior among different species; and the behavior develops in naive animals. All of this would not seem likely just on a cultural basis. It is reasonable to infer a large genetic contribution shaped by selection and evolution.

As a set of behaviors, most play seems to be very stable; that is, it appears to be a solid ESS. Cheating, bullying, refusing to play, eruptions of play-fights into genuine, serious fights, and other nonplay or antiplay actions exist and can be observed. But they do not occur frequently and do not seem to be overly disruptive on the whole.

If one were to consider the selection of play purely from the standpoint of the individual, one would expect many conflicts of interest between individuals of different sizes, strengths, sexes, and—particularly—different ages. If play were to function primarily for physical or neurological training, as has been argued earlier, then animals of different ages should have greatly differing needs and interests in play. Some of this is seen, but it is not the whole story, as stated by Fagen:

Older animals, especially certain individuals, appear to become increasingly conscious of status, increasingly unwilling to lose playfights, and increasingly unwilling to accept a subordinate position in play, whereas younger playmates appear tense or anxious when older individuals solicit play from them. . . . These changes may suggest that as physical ability improves through play, play must be more like true fighting in order to further improve ability, behavior of immatures becomes a better predictor of adult behavior (including competitive ability), animals begin to assess each other, to misinform each other, and to cheat in playfights, the cost of play increases and the benefits of play decrease, trust breaks down, and play decreases in frequency. In this sense, play may be said to contain the seeds of its own destruction. The facts are not quite this simple, however.*

Although cases of the ontogenetic transition from play to more serious living can be cited, there still are many counterexamples and enough variation, includ-

Figure 12-4 Mammalian play-face. The open-mouth grin is shown by many species during play. Shown here are mother and infant chimpanzees playing. Photo by Baron Hugo Van Lawick © 1965 National Geographic Society.

*From *Animal Play Behavior* by Robert Fagen © 1981 by Oxford University Press, Inc. Reprinted by permission.

ing older animals playing well with much younger individuals, to cloud the picture and let one know that more thought and research are needed.

One might expect the breakdown in play to involve similar-aged individuals competing for status, whereas younger animals would not pose a threat; hence, play would not be disrupted. Sociobiology, including kinship, inclusive fitness, and game theory, are proving most useful at suggesting resolutions for some of these problems. (It is perhaps appropriate that game theory be applied to play behavior.) It has been shown, for example, that playmates of mixed ages are generally siblings or that third-party intervention in play that is not going well is by a relative of the loser. Intervention that disrupts play, on the other hand, usually is by individuals that are not closely related to the players.

Results of modeling play with game theory, to consider various factors of age, relatedness, social rank, strength, and attributes such as generosity or stinginess, show some promise. But these results still are quite preliminary, theoretical, and untested. Persons wishing further discussion and references should refer to Fagen (1981:Chapter 7).

SUMMARY

Play behavior is fascinating and familiar, but the function or functions, if any, are not well understood. Play can be defined only descriptively at present. Several examples and general attributes are presented in this chapter. Part of the lack of understanding apparently results from scientific neglect of the subject and the resulting lack of a theoretical framework. Part of the problem results from the apparent lack of goal direction in play.

In spite of the difficulties of experimenting with play or seeing an obvious function, several explanations have been proposed for the evolution and presence of play behavior. They include the following play theories: surplus energy, pleasure, arousal, practice, and exercise. Arousal, practice, and exercise differ in that they emphasize general neural effects, specific neural (learning) effects, and muscular and endurance effects, respectively. Of the various theories, the exercise theory has the most support. The others, however, cannot be discounted yet, and different factors may be operating in different instances.

Social play appears to be advanced and perhaps derived from basic, individual forms of play. Optimality principles and ESS predictions are beginning to be applied and tested relative to play behavior.

Recommended Additional Reading

Bekoff, M. 1976. Animal play: problems and perspectives. In P.P.G. Bateson and P.H. Klopfer, editors. Perspectives in ethology, vol. 2. Plenum Press, New York.

Fagen, R.M. 1976. Exercise, play, and physical training in animals. In P.P.G. Bateson and P.H. Klopfer, editors. Perspectives in ethology, vol. 2. Plenum Press, New York.

Fagen, R.M. 1981. Animal play behavior. Oxford University Press, Inc., New York.

Hutt, C. 1966. Exploration and play in children. In P.A. Jewell and C. Loizos, editors. Play, exploration and territory in mammals. Symposium of Zoological Society of London, No. 18. Academic Press, Inc., London.

Jewell, P.A., and C. Loizos, editors 1966. Play, exploration and territory in mammals. Symposium of Zoological Society of London, No. 18. Academic Press, Inc., London.

Loizos, C. 1966. Play in mammals. In P.A. Jewell and C. Loizos, editors. Play, exploration and territory in mammals. Symposium of Zoological Society of London, No. 18. Academic Press, Inc., London.

PART THREE

INTERNAL
CONTROL
OF BEHAVIOR

PHYSIOLOGICAL
AND PSYCHOLOGICAL
ASPECTS

CHAPTER 13

Frederic A. Webster, K.D. Roeder, and A.E. Treat

Bat chasing a moth. A simultaneous record of the calls of a free-flying bat *(top trace)* and the neural output from the ears of a noctuid moth.

NERVOUS SYSTEMS, NEURONS, AND MUSCLES

The internal systems that mediate the proximate factors in behavior are the nervous and endocrine systems or, if one chooses to consider them together, the neuroendocrine system. The nervous system controls the actual movements and postures of the individual animal. Even something as simple as standing requires considerable neural and muscular coordination; a dead animal falls into a heap.

The actions or behaviors of animals obviously depend greatly on basic, gross morphology. Whether an animal runs, jumps, walks, flies, or swims is determined largely by the type of appendages it possesses. But the detailed pattern of movements and how the animal accomplishes these depend on a highly sophisticated (except in some invertebrates) nervous system. This system, in essence, takes information or input from outside the animal and other information from inside the animal, screens, processes, and integrates everything, and then directs the appropriate action or activity.

Thus the nervous system can be divided conceptually into three main parts: input, processing, and output, plus nerves that serve as the cables to connect the other parts. Input occurs via the senses. The output primarily is via the muscles, which contract in various combinations and sequences to create patterned movement.

This chapter provides a general overview and introduction to nervous systems, with emphasis on principles of how neurons operate. Understanding these principles is central to understanding many other aspects of the nervous system. This chapter also considers briefly the physiology of output. Subsequent chapters are devoted to senses, processing of information in the brain, the roles of hormones and other chemicals in behavior, and how processing may be modified by development and learning.

NERVOUS SYSTEMS

Different types of animals vary widely in the structure and organization of their nervous systems. Although some forms are unique, there are two basic patterns among animals with complex systems. The two patterns correspond to the two major evolutionary branches of the protostomes and deuterostomes (see Figure 5-4). The protostome pattern consists of solid, paired ventral nerve cords with periodic enlargements called *ganglia* or, for the larger ganglia, *brains*. Most of the groups of protostomes have a dorsal brain in the head region, above or surrounding the eosophagus, and connecting to the ventral nerve cords.

Deuterostomes, as exemplified by the vertebrates, have a dorsal hollow nerve cord, the spinal cord, the anterior end of which is enlarged to form the brain. Vertebrates also possess ganglia, collections of nerve cells outside the brain and spinal cord. Nerves connecting with sense organs, muscles, and glands enter and leave the brain and central nerve cord(s) at numerous and various points in all organisms.

The brain plus the major central nerve cord is referred to as the **central nervous system** or **CNS.** The remainder generally is called the **peripheral nervous system.** In the evolutionary development of nervous systems, it appears that the more recent, "advanced" organisms have *refined* and *added* parts to the old nervous system patterns, with few replacements of parts. Mammals, for example, apparently retain all or most of the reptilian components (and associated functions) of the brain and have added a few more. This has led to some interesting speculations by Paul MacLean and others that much of human behavior, such as aggression and territorial behavior, results from the ancient, ancestral traits that occasionally may be in conflict with more recently acquired behavior patterns (Holden 1979, on MacLean, also see Sagan 1977).

A great deal of evolutionary divergence has occurred in the nervous systems of unrelated groups of animals. In addition to the major ventral or dorsal nature of the CNS of invertebrates or vertebrates, respectively, there are numerous other differences. Many invertebrates have *identified* neurons in their central systems; that is, all individuals of a species have the same number and kind of neurons, and the individual neurons can be identified and named or numbered (Figure 13-1). Populations of vertebrate neurons appear less stereotyped and are more variable from individual to individual. Most invertebrates are smaller than most vertebrates, yet invertebrates commonly have larger neurons than vertebrates; thus it is clear that most invertebrates must have many fewer neurons, often by several orders of magnitude. But they still have to use these systems for the myriad of tasks required in daily living. Many species of invertebrates have existed longer and have larger populations than many species of vertebrates. The differences in success certainly are not always or even usually a result of nervous system factors, but occasionally they might be.

Figure 13-1 Identified neurons in the supraesophageal ganglion of a larval tobacco hornworm moth *(Manduca sexta).* **The large cells visible (in boxed areas) are medial 1a neurosecretory cells. Smaller, more numerous neurons, which would require staining to make them readily visible, also can be individually identified and named.** Photo courtesy Gerald G. Holt, U.S.D.A. Metabolism and Radiation Research Laboratory, Fargo, North Dakota.

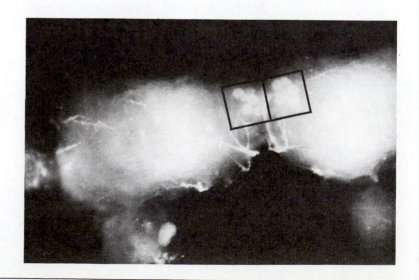

Roeder (1948) referred to the ability of invertebrates to get by with much smaller but highly efficient systems as **neural parsimony.** Not only do vertebrates have more neurons than invertebrates, they also appear to have a greater number of synapses per neuron on the average. The invertebrates apparently compensate through a number of other characteristics, such as fewer alternate nervous pathways and fewer connections to muscles.

Differences in nervous systems among different groups of animals are further elaborated in Figure 13-2 and in the following brief descriptions for some of the major taxonomic groups. To keep phylogenetic relationships in mind, consider this material in conjunction with Figure 5-4.

Protozoans are unique because they function as a single cell, unlike other animals that have whole systems composed of millions of specialized cells. Even on the organelle level no miniature nervous system–like elements have yet been discovered. It appears that the mechanisms of behavior in protozoans are basically chemical processes.

Sponges are not known to possess any nerve or nervelike cells or any other elements of a nervous system. Nor do they show any contractions or movement aside from ciliary action at the surface of cells, the choanocytes. Just as there are a few "animal-like" plants, such as the Venus flytrap, sundew, and nematode-trapping fungi, that show simple movements (which may include changes of electrical potential but which are nonetheless not considered to be neural or "behavioral"), sponges are sessile "plantlike" animals in the strictest sense of the word. Because of this they are of little interest from a neural or behavioral standpoint.

Coelenterates, according to Rushforth (1973), "have proved to be perplexing animals to the behavioral physiologist." Various studies of responses within tissue lacking known neural elements suggest that even epithelial cells are both sensitive to stimuli and capable of conduction. Among the truly neural elements, different types of cells and conduction pathways have been described, but they generally are diffuse (the so-called nerve net) and are not well organized centrally.

With the *Platyhelminthes,* or flatworms, one comes to the first group (among those studied) that shows more complex nervous systems and a much greater capacity for behavior.

Molluscs occupy an enigmatic position in the phylogenetic tree. Because of their lack of segmentation (for the most part) and based on a number of other developmental and morphological considerations, molluscs generally are placed below the annelids and arthropods on the main protostome-coelomate branch of the phylogenetic tree. There are six or seven (depending on the author) classes of molluscs, only three of which have been of much economic importance or are familiar to most persons. They are the gastropods (snails, slugs, etc.), bivalves (oysters, clams, etc.), and cephalopods (squids and octopuses). Organisms in all but one of the classes, the cephalopods, behave as might be expected from their phylogenetic position. The cephalopods represent outstanding misfits in the otherwise orderly scheme. They have highly advanced senses and nervous systems as well as separate appendages with which they move and manipulate objects.

Annelids all have a similar basic nervous system differentiated into a "brain" surrounding the eosophagus, ventral nerve cord, and ganglia in each body seg-

Text continued on p. 410.

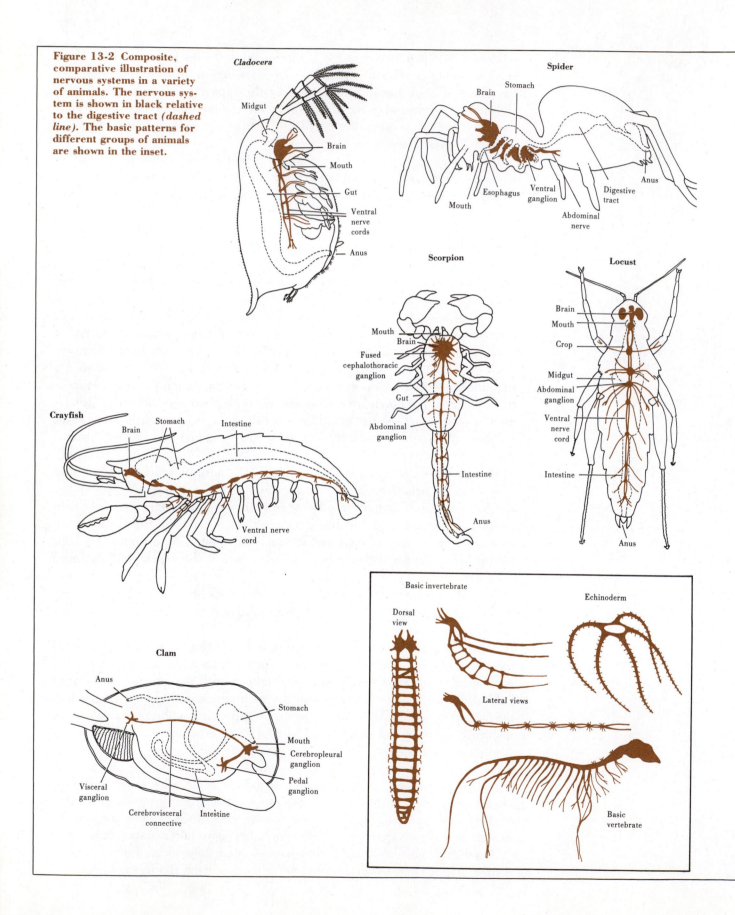

Figure 13-2 Composite, comparative illustration of nervous systems in a variety of animals. The nervous system is shown in black relative to the digestive tract (*dashed line*). The basic patterns for different groups of animals are shown in the inset.

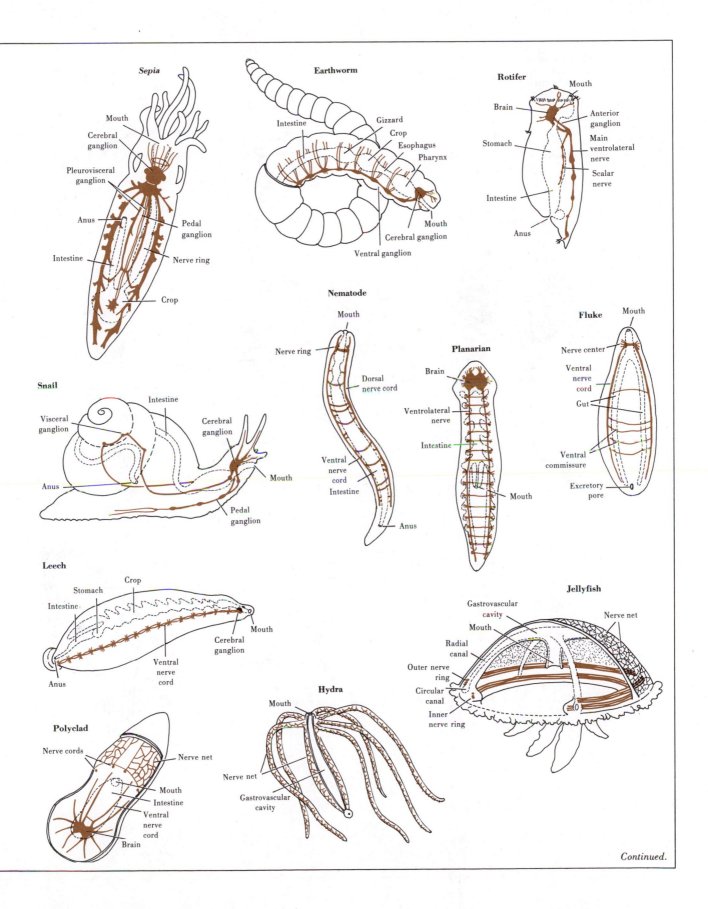

Continued.

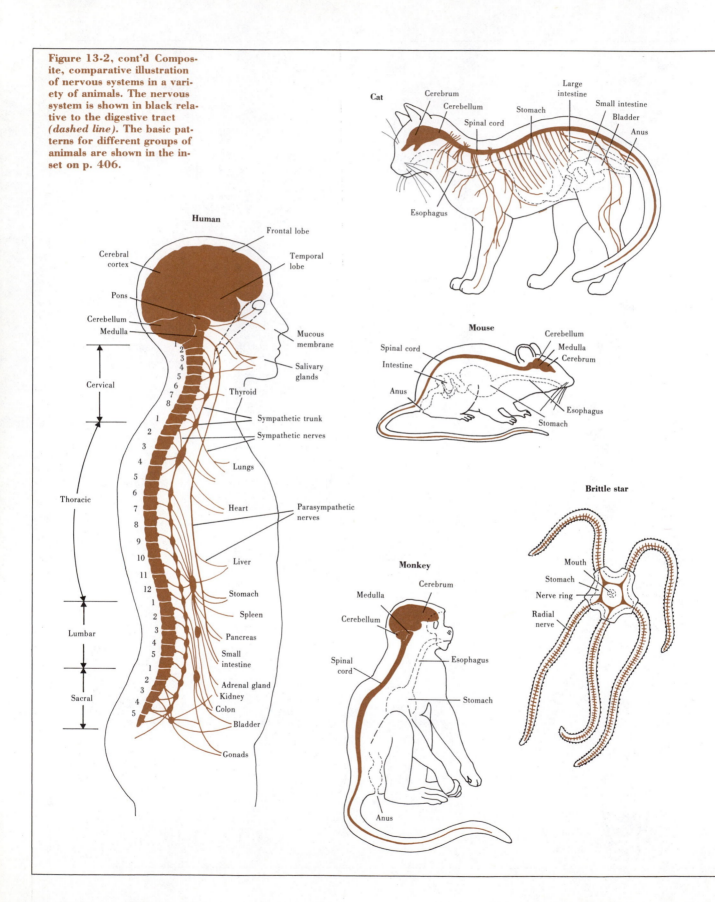

Figure 13-2, cont'd Composite, comparative illustration of nervous systems in a variety of animals. The nervous system is shown in black relative to the digestive tract (*dashed line*). The basic patterns for different groups of animals are shown in the inset on p. 406.

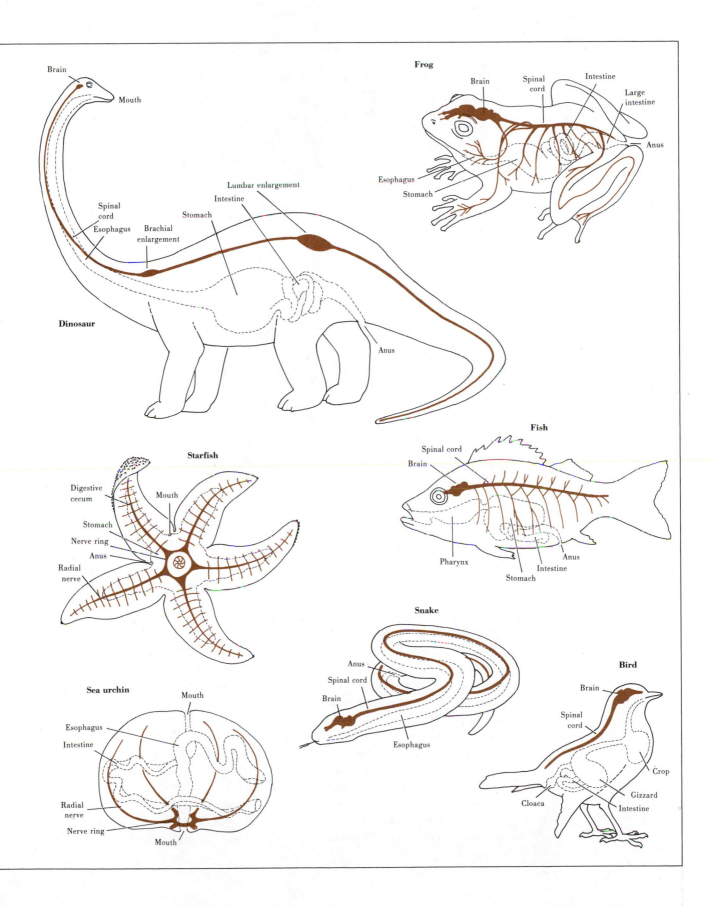

Figure 13-3 Insect (grass-hopper) brain illustrating the complexity found in arthropod central nervous systems. A, Lateral view. B, Frontal view. From Matthews, R.W., and J.R. Matthews. 1978. Insect behavior. John Wiley & Sons, Inc., New York. Based on Snodgrass, R.E. 1935. Principles of insect morphology. McGraw-Hill Book Co., New York.

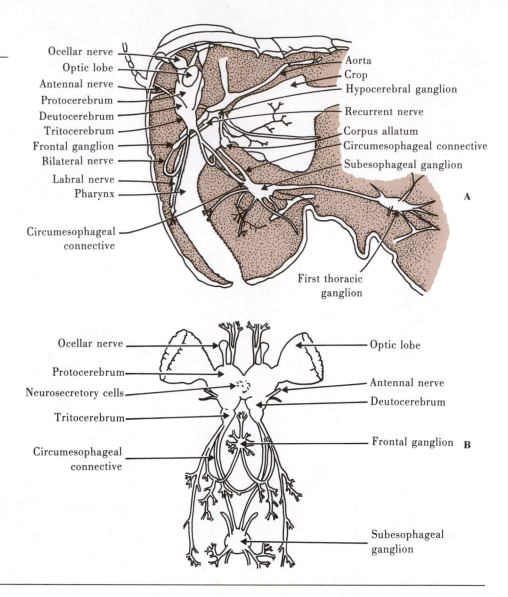

ment. Their sensory systems are primarily mechanical-tactile, surface photoreception (with much variability from simple photoreceptive cells to simple ocellar eyes and complex eyes with cuticular lenses), chemoreception, and, in a few marine tube-dwellers, there are statocysts for balance. The sensory capability varies greatly among species, with corresponding degrees of differentiation in the brain and nervous system so that the fine structure of the brains is fairly diverse among annelids. The polychaetes, in spite of being considered most primitive otherwise, show the most advanced sensory and neural development.

Arthropods are believed to have descended from segmented, annelid-like ancestors. They possess highly advanced sensory systems capable of sensing distant events, appendages capable of manipulating objects, and an accompanying advanced, highly differentiated nervous system. Arthropods perhaps best exemplify the popular picture of mechanical, robotlike animals. Their nervous systems are far simpler than those of vertebrates, but they nonetheless are quite complex (Figure 13-3).

Figure 13-4 Vertebrate brains. A, Basic form and general divisions. B, Comparative differences and evolutionary development among different classes of vertebrates.

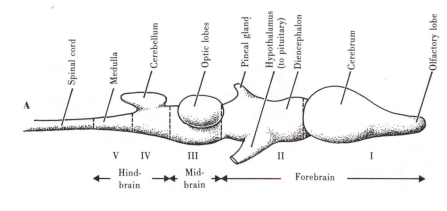

Spinal cord · Medulla · Cerebellum · Optic lobes · Pineal gland · Hypothalamus (to pituitary) · Diencephalon · Cerebrum · Olfactory lobe

A

V IV III II I

← Hind-brain → ← Mid-brain → ← Forebrain →

B

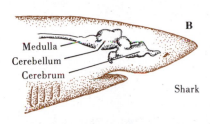

Medulla
Cerebellum
Cerebrum

Shark

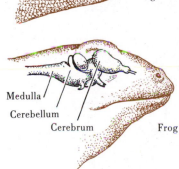

Medulla
Cerebellum
Cerebrum

Iguana

Medulla
Cerebellum
Cerebrum

Frog

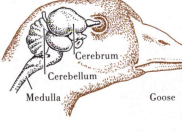

Cerebrum
Cerebellum
Medulla

Goose

Echinoderms are fascinating because of their uniqueness. As a group they seem to be out of place on this planet, as if they belonged somewhere else in the universe. They begin life as bilateral larvae, which change through a complex series of metamorphic stages to end up as radially symmetrical "things" that have no head; the nervous system is barely centralized and almost does not even resemble a nervous system. Furthermore, they have no excretory system and possess a unique hydraulic water-vascular system. Many of them eat by everting their stomachs out through the mouth and onto their food, to digest it in place. Some are predators, but none are parasitic, and few have any social life. They are both ancient, apparently surviving long past the time when they might have become extinct, and successful, being quite numerous in many parts of the ocean. They appear to be most closely related to chordates—including humans. Their earliest stages of embryological development are remarkably similar to those of chordates; only echinoderms and chordates have well-developed internal skeletons; and humans share a number of biochemical affinities with them.

Lower chordates do not show much more advanced nervous systems than echinoderms. Of the two living subphyla of lower chordates, larval urochordates have a slight but simple enlargement of the dorsal nerve cord, which has been called a "brain." At metamorphosis the urochordates degenerate into what looks like a cross between a sponge and a mollusc and are slightly less impressive than echinoderms. In the other subphylum, the cephalochordates (which include amphioxus) retain the elongated form throughout life, but the senses are limited,

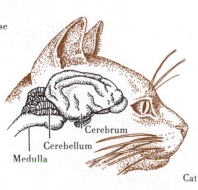

Cerebrum
Cerebellum
Medulla

Cat

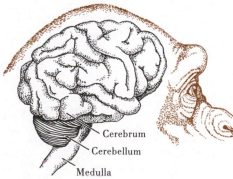

Cerebrum
Cerebellum
Medulla

Chimpanzee

and the nerve cord is essentially a simple, straight tube with little differentiation and nothing resembling a brain.

Vertebrates, a subphylum of chordates, are the organisms with which most persons are familiar. All have the single dorsal, hollow nerve cord and well-developed brain. The basic pattern of the vertebrate nervous system is similar among all classes, but the brains differ vastly in degree of development from group to group, as indicated in Figures 13-2 and 13-4.

NEURONS

Figure 13-5 Variety of neurons found among and within different animals.

The nervous system is composed of a variety of cells and tissue types. The basic functional cells that perform the actual processing of information are called *neurons.* Neurons differ greatly in size, shape, and physiology among different animal groups and different parts of the nervous system of any one animal (Figure 13-5).

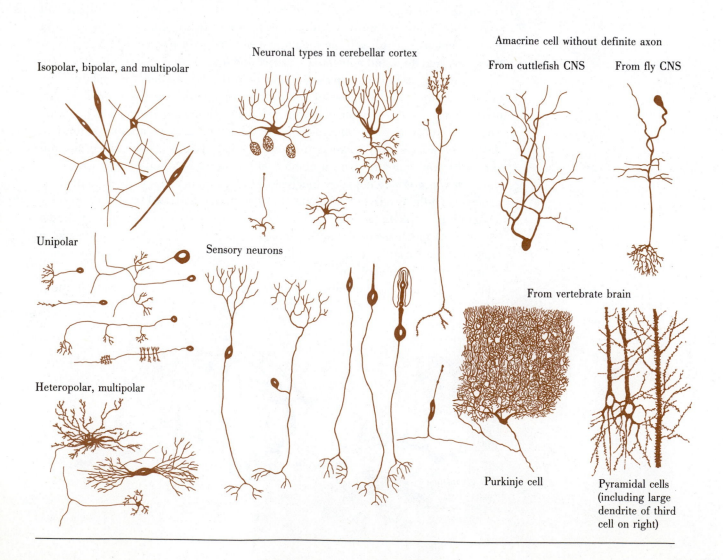

Isopolar, bipolar, and multipolar

Neuronal types in cerebellar cortex

Amacrine cell without definite axon

From cuttlefish CNS   From fly CNS

Unipolar

Sensory neurons

From vertebrate brain

Heteropolar, multipolar

Purkinje cell

Pyramidal cells (including large dendrite of third cell on right)

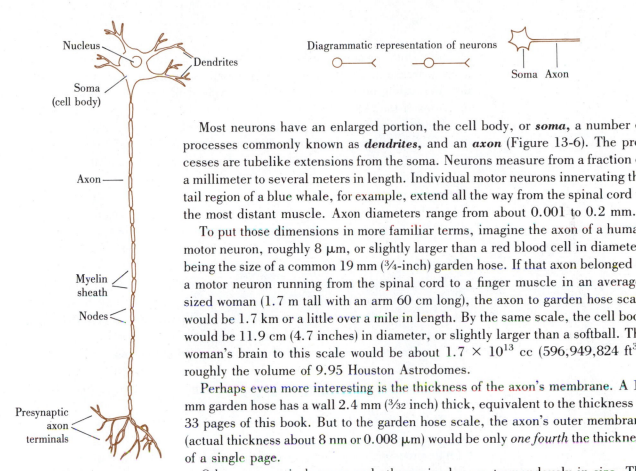

Diagrammatic representation of neurons

Soma   Axon

Figure 13-6 The vertebrate motor neuron as a representative illustration of neuronal structure.

Most neurons have an enlarged portion, the cell body, or **soma,** a number of processes commonly known as **dendrites,** and an **axon** (Figure 13-6). The processes are tubelike extensions from the soma. Neurons measure from a fraction of a millimeter to several meters in length. Individual motor neurons innervating the tail region of a blue whale, for example, extend all the way from the spinal cord to the most distant muscle. Axon diameters range from about 0.001 to 0.2 mm.

To put those dimensions in more familiar terms, imagine the axon of a human motor neuron, roughly 8 μm, or slightly larger than a red blood cell in diameter, being the size of a common 19 mm ($\frac{3}{4}$-inch) garden hose. If that axon belonged to a motor neuron running from the spinal cord to a finger muscle in an average-sized woman (1.7 m tall with an arm 60 cm long), the axon to garden hose scale would be 1.7 km or a little over a mile in length. By the same scale, the cell body would be 11.9 cm (4.7 inches) in diameter, or slightly larger than a softball. The woman's brain to this scale would be about 1.7×10^{13} cc (596,949,824 ft^3), roughly the volume of 9.95 Houston Astrodomes.

Perhaps even more interesting is the thickness of the axon's membrane. A 19 mm garden hose has a wall 2.4 mm ($\frac{3}{32}$ inch) thick, equivalent to the thickness of 33 pages of this book. But to the garden hose scale, the axon's outer membrane (actual thickness about 8 nm or 0.008 μm) would be only *one fourth* the thickness of a single page.

Other neurons in humans and other animals vary tremendously in size. The somas, for example, vary in diameter from around 2 μm to over 500 μm.

Neurons have two primary functions: (1) process and (2) transmit information. Processing sometimes is referred to as **discrimination,** a kind of simple yes-no decision making; the **transmission** of that decision is often called, somewhat confusingly, *communication*. The decision, to fire or not to fire, is not a conscious process on the part of the neuron. It occurs as a result of properties of the neuron in conjunction with its input. The whole system operates by the combined actions of millions of unique, individual neurons interacting with other neurons in a generally orderly fashion, including priorities and hierarchies that generally resolve conflicts.

The Structures of Neurons

There is so much diversity and specialization among neurons that a generalized, typical neuron does not really exist. But one can look at the general properties of most neurons and, at least on paper, consider a generalized neuron. The type of neuron that is usually chosen as "representative" is the vertebrate motor neuron. The internal and gross morphologies of such a neuron are shown in Figure 13-6.

The soma is that part of the neuron which contains the nucleus. It is generally larger in diameter than other parts. The soma also contains a relatively large amount of cytoplasm plus many of the standard organelles found in most cells, including nucleoli, mitochondria, endoplasmic reticulum, ribosomes, Golgi bodies, centrosomes, and lysosomes. This region of the neuron is primarily involved with the synthesis of many (but not all) of the important molecules of the neuron, such as transmitter substances. The membrane surface of the soma may or may not, depending on the neuron, also be involved in the information-processing activities of the cell.

In addition to the more or less standard organelles, neurons also contain some other important structures, particularly fibrillar structures and Nissl bodies. The fibrillar structures extend throughout the neuron, not just the soma, are unbranched, and vary in size. The smaller ones (6 to 10 nm) are called neurofilaments, and the larger ones (20 to 30 nm) are microtubules. Microtubules are believed to be important in the transport and recycling of various neural molecules. The function of neurofilaments is unclear. Nissl bodies appear to be composed, in part, of ribonucleoprotein, include large numbers of ribosomes, and are believed to be important in neural protein synthesis.

The branching cellular extensions or *processes* of neurons were formerly classified into two groups, *axons* and *dendrites*. The distinctions recently have become hazier. In general, one can talk about parts of the cell that receive information (traditionally called dendrites) and parts that send information (axons). The problems, however, are twofold: (1) dendrites also may pass information to neighboring neurons, and some axons can receive input directly from other neurons, and (2) many types of neurons, particularly but not exclusively in the invertebrates, are not as differentiated into distinct parts for receiving and sending. This has led to an array of new terms, built on old categories, for processes and parts of processes. Bullock et al. (1977) use **neurite** in the text as "a general term for processes, noncommital as to their type" and in the glossary as "the main or longest process of a nerve cell; usually equivalent to axon and axis cylinder." The terms axon and dendrite are firmly and historically established, however, and will continue to be used here.

The whole neuron can be classified on the basis of its processes. Depending on the number of processes attached to the soma, neurons are **unipolar, bipolar, or multipolar.** All are found throughout the animal phyla, but unipolar cells predominate in invertebrates, whereas bipolar and multipolar neurons are more common in vertebrates. Unipolar neurons in vertebrates are found mostly outside the CNS. Many neurons have their own specific names such as basket cells, pyramidal cells, and Purkinje cells.

At the distal ends of axons or axon branches are the **terminals.** Large prominent terminals are called **boutons.** These are the most familiar points of contact with other neurons. (Neurons also have several other important points of contact with each other, as will be discussed later.) The axon terminals usually make contact on dendrites or somas, but they also may connect at other axons. Connections on dendrites frequently occur at dendritic **spines,** important structures that appear to vary with development in some organisms and are thought to be significant in learning and memory.

Direct contacts between cells occasionally occur via fused membranes, such as between the paired giant ganglion cells in a leech (Eckert 1963), but generally the connections are across a very small gap (1 to 30 nm), the well-known **synapse.** A very large number and variety of types of synapses now are recognized (e.g., Bodian 1972). A few examples, along with their names, are illustrated in Figure 13-7. Bullock et al. (1977) provide an excellent review and further details. Brief-

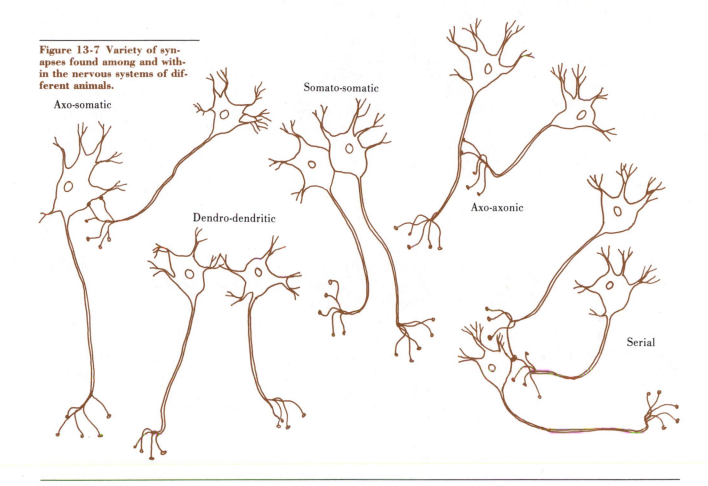

Figure 13-7 Variety of synapses found among and within the nervous systems of different animals.

Axo-somatic

Somato-somatic

Dendro-dendritic

Axo-axonic

Serial

ly, there may be simple synapses, classically from the axon of one neuron to the dendrite or soma of another, where the information is passed in one direction; there are several chains of synapses; there are complex nests or glomeruli of several synapses; and there are reciprocal junctions where information passes in both directions. Synapses can occur not only between axons and dendrites, as indicated earlier, but also between two axons, between two dendrites, between two somas, and in other combinations. Some synapses are at one point, and some involve numerous points of contact or a large area of membrane. These findings have led to a whole new field of neuroanatomy—study of the microarchitecture of neurons.

Inside the terminals and among the other organelles present are small membrane-enclosed structures called *synaptic vesicles*. These vesicles contain a chemical, the *transmitter*, which carries information across the cleft. There is only one kind of transmitter per synapse, but there are different kinds in different synapses.

Neurons can be classified not only by their morphologies but also by their *position* in the nervous system. Neurons that deal primarily with input are *afferent* or *sensory*; those dealing with output are *efferent* or *motor*; and those between are called *interneurons* (also called internuncial, connecting, or association neurons). Afferent is as in "affect," and efferent is as in "effect."

Interneurons that primarily connect forward or backward to other parts of the nervous system are **projection** neurons, whereas those which connect more or less laterally to corresponding regions on the other side, in bilateral organisms, are **commissural.** Branches of neurons that travel to the *opposite side* of the body are **contralateral,** and those to the *same side* are **ipsilateral** relative to their soma or point of origin. Neurons also may be named by the region where their somas reside, such as spinal, dorsal root ganglion, cerebral, etc. *Thus, depending on the topic of discussion, particular neurons may be named in many different ways.*

Collections of somas are called **centers, nuclei, masses, or sometimes ganglia** inside vertebrate CNSs or **ganglia** in invertebrates and in the periphery of vertebrate systems. Bundles of processes are called **tracts** inside the CNS and **nerves** if outside the CNS.

Before proceeding further one needs to consider the means by which neurons are studied. Whole animals or parts thereof are relatively easy to study. But as the dimensions of the subject become smaller, as with neurons, specialized techniques and equipment usually are required.

Neurophysiological Methods

Knowledge in neurophysiology has proliferated since the period of 1942 to 1952, but the roots of the understanding of neurons go back a surprisingly long way, to at least Malpighi in the seventeenth century. Although there were many early misunderstandings and misinterpretations, some progress was being made throughout the entire nineteenth and early twentieth centuries.

Visual Techniques

The development of magnifying lenses and microscopes with modest magnification was a great breakthrough for studying biological structures. Present-day light microscopes are capable of around $\times 1000$ magnification. The next major visual step was the development of the electron microscope, which permits magnification up to $\times 250,000$. The limits are not really an issue of magnification but rather resolution, the smallest dimensions that can be distinguished. The shortest visible wavelengths of light permit one to resolve objects to about 0.2 μm, whereas the electron microscope takes resolution down to a theoretical limit of about 4 or 5 Å, or 0.0004 to 0.0005 μm. (Recall that the motor neuron axon diameter in the example earlier was 8 μm and its outer membrane was 0.008 μm.)

Neurons are not normally pigmented. Thus to be seen, in addition to being magnified, they need to be fixed or preserved, stained, and occasionally sectioned. There are numerous techniques for staining neurons.

Recording of Bioelectric Events

The bioelectric events of neurons have been recorded using several different techniques. Voltage differences and current flow, described later, have been measured by inserting a recording electrode into the cell while keeping another outside. Electrodes are the points or terminals of an electrical source or recording device. Metallic wire is unsuited for small electrodes; they are made instead by carefully drawing out heated microcapillary glass tubing and filling them with salt solutions. The size of the tip of modern microelectrodes is from 0.1 to 0.5 μm. Some of the first recordings from the inside of neurons were accomplished by inserting the inner electrode through the cut end of a giant squid axon. This giant axon (300 to 800 μm) was used initially because of its large size. More recent techniques involve penetrating the neuron's membrane directly. These methods obviously demand much practice, patience, and skill. The minute size of the electrical events, in millivolts or $\frac{1}{1000}$ volt, requires specialized amplifiers that do not interfere with the events.

If the simple presence of a neuron's impulses, rather than their actual magnitude, is of interest, then both electrodes can be placed on the surface of the cell. When the two electrodes are placed on the same neuron at separate points, they measure the relative difference between the two points instead of the difference between the inside and outside of the neuron. It is also possible to place one electrode on the cell and the other in the external medium away from the neuron. If several axons are in contact with the electrode, the impulses of all are recorded together, and the recorded event is called a *compound action potential.* Probably the simplest technique of all is to use a large (relatively) wire electrode that is bent into a hook. With such an electrode an entire nerve or part of a nerve can be lifted from the surrounding tissue for simultaneously recording the activity of all of the neuronal processes that pass through it.

The amplified impulses can be traced visually with an oscilloscope on its cathode ray screen, marked on paper with pens, or detected acoustically via clicks on a speaker. The sequence can be tape-recorded for later replay, sent to electronic counters, or forwarded directly to a computer for analysis.

Relatively recent developments in neurophysiological techniques have shown that neuronal activity can also be recorded optically (Tobias 1952, Ross et al. 1974). These techniques may permit the activities of smaller neurons and larger numbers of neurons to be recorded simultaneously.

Stimulation, Ablation, and Lesioning

Much information and insight into the properties and pathways of neurons have been gained by stimulating one part of the system and recording subsequent activity elsewhere. For example, one can stimulate sensory endings in a variety of manners to discover the characteristics to which the neurons are sensitive. The resulting impulses, called *evoked potentials,* can be recorded directly from the sensory neurons or traced into other neurons further along in the system. Likewise, one can stimulate parts of the brain, spinal cord, or ganglia, depending on the organism, to discover which muscles will be caused to contract. Stimulation is accomplished by electrodes that supply electrical currents to the neurons. Instead of stimulating various regions, one can damage or interrupt various parts of the system and then determine the effects. Effects also can be followed by injecting or implanting chemicals such as radioactive-labeled precursors, hormones, and poisons.

The placement of electrodes, lesions, or introduced substances may be relatively easy in small systems or surface preparations. Neurons or regions of neurons that are deep in the brain usually are located with the aid of precision tools— stereotaxic instruments (Figure 13-8) that permit exact location in three-dimensional space. Stereotaxic maps and atlases of the brain (Figure 13-9) are now available for many organisms. After the affects of stimulation or implants have been observed in the live organisms, they are killed, the brain or other parts of the

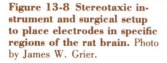

Figure 13-8 Stereotaxic instrument and surgical setup to place electrodes in specific regions of the rat brain. Photo by James W. Grier.

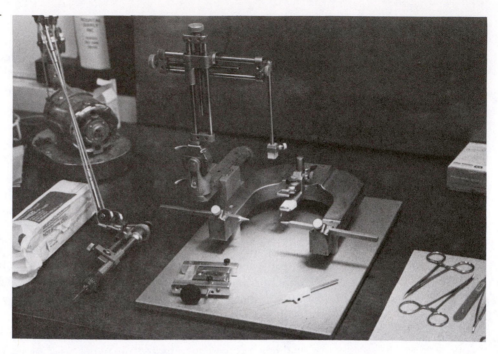

nervous systems removed, fixed, and then cut into thin sections of viewing under a microscope to either confirm or determine the exact locations of the electrodes or implants.

Experimental Manipulation and Analysis

Electrical and chemical processes and events by which neurons operate have been investigated through a variety of experimental techniques such as submersing isolated neurons in solutions of different chemical compositions. In addition to a large number of important ions, including Na^+, K^+, Cl^-, and HCO_3^-, there are many biochemicals involved in neural functioning, particularly at synapses.

One technique that has been important in conjunction with such work is the *voltage clamp*. Normal events in the neuron, such as changes in voltage, occur too rapidly to allow the measurement of all components of the event. The technique of voltage clamping holds voltages at experimentally controlled levels, permitting the determination of electrical and chemical processes occurring in the neuron at that voltage. Voltage clamping is accomplished by using the usual electrodes to record voltage plus additional electrodes to artificially feed current in or out of the cell, as determined through a feedback apparatus.

The Operations of Neurons

To understand how the nervous system codes and transfers information, we will take a brief close-up view of how neurons operate. Of several forces in nature (including, for example, gravitational and magnetic), the electrical forces between charged particles form the basis for neural mechanisms. Because of this neurons share certain properties with other electrical systems, and much of the terminology and notation has been borrowed from physics. A full explanation of the membrane and electrical properties of neurons requires a deeper background in mathematics, electrical principles, cable properties, physical chemistry, and physiology than can be assumed or developed here. Such detail, however, is not necessary for present purposes.

Electrical forces arise from the repelling or attracting interactions between charged particles. The particles of interest are negatively or positively charged electrons or ions—atoms or molecules that have either an excess or a deficiency of electrons. Like charges repel, and opposite charges attract.

When a substance contains an equal number of positively and negatively charged particles and when they are free to move about so that the charges are more or less equally distributed, the substance is electrically neutral, or at zero potential. But substances differ in the ease with which the charged particles can move through them, and it is possible to create an imbalance in the total charge.

The amount of imbalance, or **potential difference,** is measured in **volts (V).** A standard flashlight battery is 1.5 V, and most modern automobile batteries are 12

Figure 13-9 Plate from a stereotaxic atlas of the rat brain, illustrating coordinates of particular regions of the brain. From König, J.F.R., and R.A. Klippel. The rat brain. Copyright 1963, the Williams & Wilkins Co., Baltimore.

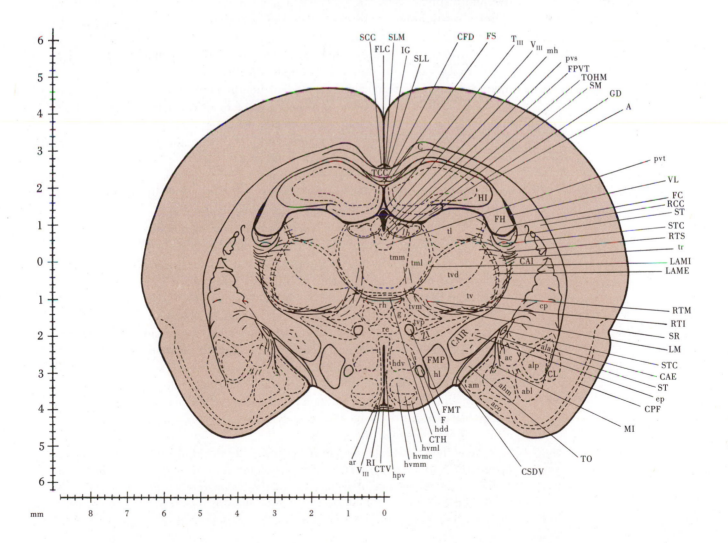

V. Neuron and muscle potentials are measured in *thousandths* of volts or **millivolts.** Potential difference usually is shortened just to potential.

When a charge is built up, like carrying water up a hill, then allowed to flow back, the resulting movement is called **current,** a flow of charge over time. With two different charges in electricity, particles can move in either direction, depending on the circumstances.

The charged particles that lead to the electrical properties of neurons are several kinds of ions, particularly sodium (Na^+), potassium (K^+), calcium (Ca^{++}), chloride (Cl^-), and a number of negatively charged organic ions. The extracellular and intracellular (cytoplasmic) fluids are relatively good conductors, although cytoplasm has a higher resistance than the extracellular fluid. The cellular membrane is extremely thin, as mentioned previously, but it is a good insulator; that is, it is relatively impermeable to the movement of the various ions.

The membrane of the neuron (as in virtually all living cells) actively transports various ions across itself. The processes are not yet completely understood, but they result in an imbalance of ions of different kinds. That the transport is active and involves metabolic processes has been shown by experiments in which cellular respiration is blocked and the ionic concentration gradients disappear.

The ionic pump is believed to be linked for Na^+ and K^+ so that Na^+ is pumped out of the cell as K^+ is pumped in (Figure 13-10). The charges of these two ions might seem otherwise to cancel each other, but the permeability of the membrane is greater for K^+ than for Na^+. As a consequence K^+ is constantly "leaking" back out, which leaves the outside with a greater positive charge. Other ions, particularly Cl^- and HCO_3^-, also leak across, and this contributes to the overall potential.

Differences in concentrations of the ions between the inside and the outside reach equilibrium levels. The voltage potential that is associated with these levels is called **equilibrium** or **resting potential,** in millivolts.

Thus the neuron is somewhat like a microminiature battery except that, instead of having just two contacts as in batteries, the difference in charge is between the two sides of the cell's membrane, and most or all of the surface of the cell can discharge or recharge. Furthermore, different parts of the cell may discharge while other parts of the same cell are simultaneously recharging.

Another way of describing the charge difference or potential is to say that the neuron is electrically **polarized** between the inside and the outside. If the polarity is reduced, the neuron is **depolarized,** and if increased it is **hyperpolarized.** The inside of the neuron has a negative charge relative to the outside. Thus depolarization means an increase in positive charge inside the cell so that the inside is more like the outside. Depolarization is, in a sense, a double negative in that it is a decrease of negative charge. Conversely, increased polarity, hyperpolarization, means a more negative charge inside.

In some less active neurons or if the cell is deprived of its normal sources of surrounding stimulation, that is, if it is isolated from those neurons which normally join it, resting potential remains relatively constant at around -50 to -70 mV (Figure 13-11). If the charge is slightly altered, depolarized or hyperpolarized, the potential will return, exponentially, to the "resting" value.

Figure 13-10 A, Sodium-potassium pump in the membrane of neurons. These ions are actively transported against their respective ion gradients. The membrane is partially permeable to the ions, and some "leak" back across. The gates and pump shown here are simply mechanical models to illustrate incompletely understood molecular mechanisms. **B,** During an action potential some of the ions flow passively but rapidly back across the membrane. See also Figure 13-13.

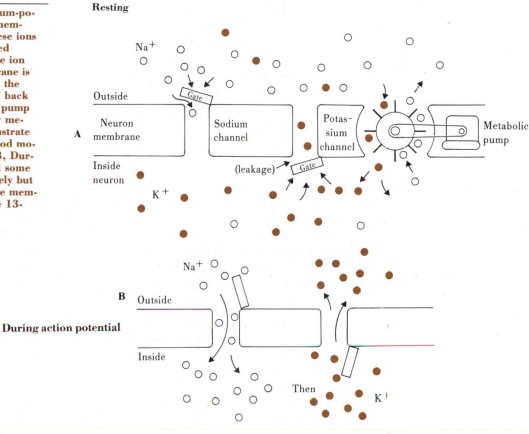

Resting

A

During action potential

B

Figure 13-11 Terminology and events associated with changes in neuronal membrane potentials. *1, Resting potential; 2,* more negative—*hyperpolarization*—as in *IPSP; 3,* less negative—*depolarization*—as in *EPSP; 4, summation; 5, action potential*—inward rush of Na⁺; *6,* return to negative state—outward rush of K⁺; *7, overshoot*—*refractory period; 8,* intrinsic rate of firing—spontaneous depolarization; *9a,* stimulus of EPSP; *9b, increase in rate* of firing; *10a,* inhibition—hyperpolarization by IPSP; *10b, delay* in firing. Also see text for further explanation.

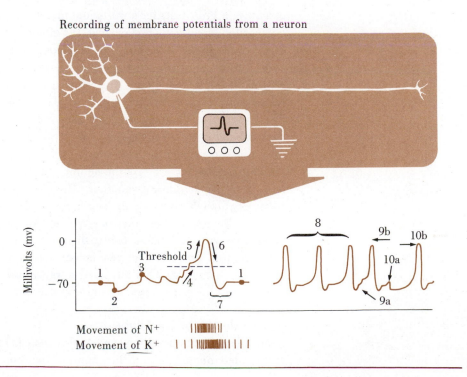

Recording of membrane potentials from a neuron

Discrimination by Neurons

Most neurons are not isolated or static. Under natural conditions the cell may have different charges at different locations. Likewise, if an electrode is kept at one location in the cell, the potential may be found to change over time. These changes are only local and do not affect the entire cell. They occur primarily in dendrites and the cell body with very complex local circuits between neighboring dendrites. Potentials in these dendritic interactions are called **graded (dendritic) potentials.** These changes in the membrane potential are caused by events in the immediate vicinity and at the surface membrane of the neuron.

The two main types of events that cause the graded changes are (1) **transduction** of sensory input from outside the nervous system and (2) **input from previous neurons,** usually after crossing a synapse. Transduction involves a change from one energy form to another, such as in the conversion of sound waves to electrical waves in a microphone. The graded potentials from sensory input are specifically named **generator potentials.** Potentials caused by synaptic input from previous neurons are aptly termed **postsynaptic potentials (PSPs).** Postsynaptic potentials, depending on the characteristics of the particular synapse, are either **excitatory (EPSP)** or **inhibitory (IPSP).** EPSPs lead to a slight amount of depolarization, and IPSPs cause hyperpolarization. Generator potentials are excitatory in any given sensory neuron, but that neuron may, in its turn, inhibit subsequent neurons.

Graded potentials, as already indicated, are local and transitory; that is, the effect of the potential or charge diminishes or decays exponentially with distance and time (Figure 13-12). The larger the change in potential, the greater the area affected and the longer it lasts. The spread of the potential involves the combined electrical-chemical properties of the neuronal membrane. It is **passive,** and it is termed **electrotonic.**

The effects of graded potentials are also **additive.** A given graded potential, for example, a PSP, lasts for a while and then disappears. If, however, another PSP comes along before the first is gone, the effect of the second adds to the remaining effect of the first. Additional PSPs can come either from repeated impulses (i.e., over time) in the same synapse, from neighboring synapses, or from a combination of the two. The effects are referred to respectively as **temporal** or **spatial summation** (Figure 13-12). Most neurons receive a tremendously large number of synapses (up to an estimated 80,000 in vertebrate cerebral basket cells) and are thus subject to a constant barrage of EPSPs and IPSPs, all of which are summed over time and space in the neuron.

Thus far the properties of the neuron resemble those of a metal cable such as a wire from one building to another. The electrical effects move from one region passively to the next within the conducting material whenever there is a change in voltage. But in neurons this passive, electrotonic effect does not go very far. Compared with a telephone cable, the cytoplasm has much higher resistance than the wire, and the cell membrane is leakier and not nearly as good an insulator as the cable wrapping. This affects the transmission of impulses both within the neuron and the conveyance of the impulse onto the neighboring neurons.

Transmission of Decisions

Transmission of information over short distances may be accomplished by passing the graded potentials onto neighboring neurons. This requires that the mem-

Figure 13-12 Local nature and summation of graded potentials. An individual postsynaptic potential diminishes with time and distance from its initial occurrence. Strength of the potential (depolarization) is indicated on the surface of the neuron by amount of shading. Each series of neurons represents sequence in time. A, Single EPSP, fading with time. B, Two EPSPs at the same site. C, Two EPSPs arriving at different locations. D, A combination of temporal and spatial summation.

Single EPSP

Single EPSP from incoming spike

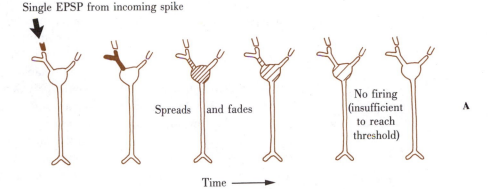

Spreads and fades

No firing (insufficient to reach threshold)

A

Time ⟶

Temporal summation

First EPSP

Second EPSP at same synapse

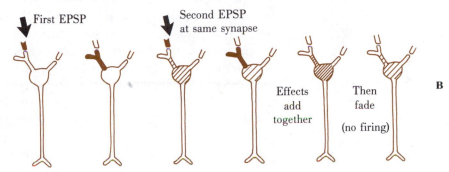

Effects add together

Then fade (no firing)

B

Spatial summation

EPSPs at two locations

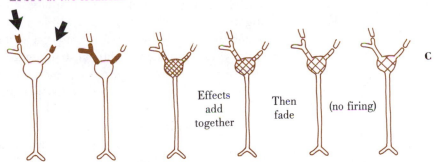

Effects add together

Then fade

(no firing)

C

Spatial and temporal summation

First EPSPs at two locations

Second EPSPs

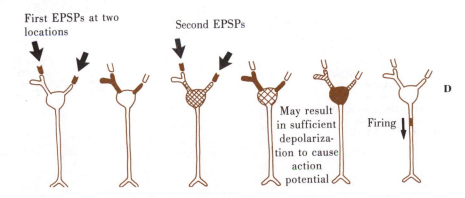

May result in sufficient depolarization to cause action potential

Firing

D

branes of the two cells be extremely close together and have certain properties. Some specialized junctions, known as **gap junctions** or **nexuses**, allow the current to flow in either direction or just in one direction, as in an electrical rectifier. The passage of current remains passive and thus is not easily blocked, and it occurs without the delays observed in chemical synapses (discussed later).

It is interesting historically that a controversy developed during the first half of the twentieth century over whether transmission from one neuron to another was electrical or chemical in nature, the so-called spark versus soup schools. It turned out that both were correct for different connections. Chemical synapses, however, are by far the more common of the two types. To understand them one must return to the normal processes occurring inside a given neuron and then consider how the information is passed on to the next neuron in line.

If depolarization becomes sufficiently large, a different phenomenon occurs; an **action potential** is initiated (Figures 13-11 and 13-12), and it affects the entire cell eventually. Action potentials also are commonly called *spikes*, *nerve impulses*, and *firings*. The present picture of the events occurring during an action potential is based primarily on a series of elaborate experiments, relying heavily on isolated giant squid axons and voltage clamping, which were done by A.L. Hodgkin, A.F. Huxley, and their students and colleagues.

During the action potential, at the point where it is occurring, the membrane's conductance properties suddenly change so that sodium ions rush in, making the inside positive (Figures 13-10 and 13-13). Starting about the same time but occurring more slowly, potassium rushes out, making the inside negative once again, and in fact even slightly hyperpolarized. Then the membrane again "closes down" and becomes relatively impermeable to K^+ and Na^+. The two ions are actively "pumped" back to the other sides (although K^+ continues to leak back slightly more than Na^+), and the resting potential is restored. The changes in the membrane have been modeled somewhat as the opening and closing of gates that guard different channels through the membrane. Na^+ and K^+ each have their own channels. The whole process may take 0.5 msec to several milliseconds ($\frac{1}{1000}$ of a second). Repeated action potentials can occur, depending on the neuron, up to 1000 times per second. For a brief period after firing (the *refractory* period), the neuron will not and cannot be caused to fire again.

The change in the membrane that accompanies an action potential at a particular point affects the surrounding membrane. This sets off a similar change in the neighboring regions, and by this means, the spike does not diminish, as in graded potentials, but is maintained or **propagated** in an active chain reaction. It travels throughout the neuron (Figure 13-13), normally from the cell body down

Figure 13-13 Ionic flow during an action potential. The event propagates to neighboring regions, causes similar ionic flows there, and thus travels over the entire surface and length of the axon. See also Figure 13-10.

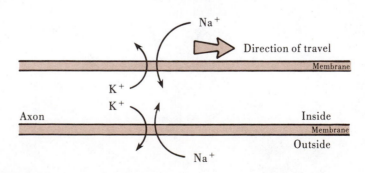

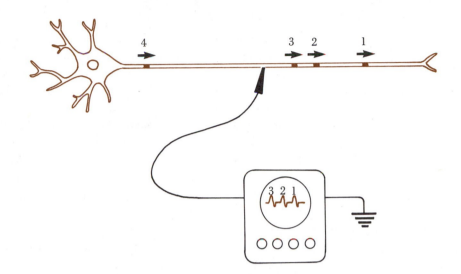

Figure 13-14 Repeated action potentials in a neuron. After one spike passes a particular point on the axon's surface and the membrane has recovered, another spike can follow soon thereafter. Within a given neuron, several spikes may be traveling down the axon simultaneously. The frequency of spikes over time is the means by which neurons code information.

the axon to the other end of the axon. The process of propagation is somewhat analogous to the burning of a firecracker fuse.

Imagine a neuron with spikes being propagated down it in rapid succession (Figure 13-14). The number of spikes per unit of time is the ***spike frequency***. *This rate or frequency of firing is the means by which neurons normally code information.* With increased stimulation one usually finds an increased rate of firing. The amplitude of a spike is not important for coding information; each spike is "all-or-none," that is, either present or absent.

Neurons do not (generally) code information by simply firing or not firing but rather by changing their rate of firing. Under natural conditions in the intact animal, depolarization may occur as a result of generator potentials, from postsynaptic potentials, or spontaneously; that is, many neurons do not really have a resting potential but rather a ***resting or spontaneous rate of firing***. A common role of generator and synaptic potentials is not to cause an action potential per se but to alter the neuron's spontaneous or endogenous firing rate. If a cell is depolarizing intrinsically, EPSPs speed up the rate of depolarizing via the normal summation process and cause the cell to fire sooner. If this process is repeated, it can greatly increase the frequency of spikes. Similarly, IPSPs can reduce the frequency. See Figure 13-11 (*9a, 9b, 10a,* and *10b*).

By looking at spike frequency, one can "break the code," as it were, tap the nervous system, and begin looking for eventual correlations with behavior. It should be stressed, however, that this is only a start. A complete picture will require information on the whole system. Without information from several points, it is similar to trying to understand the operations and behavior of a city by tapping only one phone line.

The speed that an action potential travels down the axon varies from a few centimeters per second to around 100 m/second (roughly 200 miles/hour). The speed of transmission is affected by several factors, such as the size or *diameter of the axon* (the larger, the faster), the *temperature* (the warmer, the faster), and whether or not *myelin* is present. Myelin is a fatty insulated wrapping around the axon.

Myelination permits the action potential to leap electrically *(saltatory conduction)* from point to point along the axon without being slowed down by the normal biological membrane processes that otherwise would occur in the intervening spaces. Periodically along the axon there are bare points of conduction—the **nodes of Ranvier.** The mechanism was elucidated by showing that poisoning or anesthetizing the axon membrane between the nodes does not interfere with the impulse, whereas at the nodes it does.

The important point of myelination for purposes here is that it permits a combination of small diameter and rapid transmission of action potentials in the axons. A myelinated frog axon 12 μm in diameter conducts at the same velocity (25 m/second) as an unmyelinated squid axon nearly 30 times the diameter. In a competitive and dangerous world where speed counts, the advantage of rapid transmission is obvious. Large diameters have some disadvantages that will be discussed shortly.

Returning to chemical synapses and the action potential in a given neuron, when the impulse reaches the terminal, it is translated into a chemical message or *transmitter.* PSPs were observed to occur in discrete, quantal jumps. This led to the hypothesis that certain quanta of transmitter were being released from the presynaptic terminals. Subsequent electron microscopy confirmed the existence of the small organelles previously mentioned, the synaptic vesicles, and other techniques have permitted the vesicles to be isolated and analyzed, showing them to be the discrete packages of transmitters. The picture still is not completely understood, but the present view is described below.

When an action potential arrives at the terminal, some of the synaptic vesicles move to the presynaptic membrane, join with the membrane, and expel their

Figure 13-15 Synaptic cleft and transmission of information via the spread of transmitter chemicals released from vesicles at the membrane surface into the cleft.

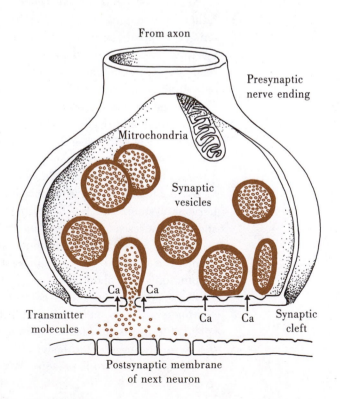

From axon

Presynaptic nerve ending

Mitrochondria

Synaptic vesicles

Ca   Ca

Transmitter molecules

Ca   Ca   Synaptic cleft

Postsynaptic membrane of next neuron

contents into the synaptic cleft or gap (Figure 13-15). This process is somehow coupled with the presence of calcium ions. Calcium ions normally enter the neuron along with sodium ions during action potentials, but the inward rush of calcium is particularly important at the terminal where it is involved in dumping vesicles.

The expelled transmitter moves across the gap and alters the ionic conductances of the postsynaptic membrane of the neighboring neuron. Receptor sites for the transmitter occur on the postsynaptic membrane and somehow bind the transmitter. The altered conductance leads to a change in the equilibrium and, hence, a changed potential or PSP. The gap between presynaptic and postsynaptic neurons is so small that the chemicals diffuse across and transmit the message in an estimated 0.5 msec. This is fast relative to our familiar time frame but slow relative to the speed of electrical conduction in wires or even spike conduction in axons. Each synapse slows down the total message, and the number of synapses can be estimated from the time it takes for a neural response to occur.

Several chemical substances have been identified as neurotransmitters. Two of the most familiar are *acetylcholine* and *norepinephrine*. Once they have done their job of transmitting the message across the synapse, they are about as useful as the news in old newspapers and can create problems if they accumulate. There are several means by which the transmitters are inactivated and recycled: by enzymatic destruction, active uptake into the axon terminal or glial cells, or diffusion into the intercellular spaces. Acetylcholine is metabolized by cholinesterase; the choline is actively reabsorbed back into the terminal; and new acetylcholine molecules are manufactured. Norepinephrine apparently is reabsorbed intact.

Synapses and their complex molecular and membrane interactions are highly susceptible to a variety of exogenous chemicals such as cocaine (which blocks the uptake of norepinephrine in certain synapses), caffeine, nicotine, diazepam (Valium), and many others. The introduction of such chemicals into the body can wreak havoc on an intricate and otherwise well-tuned system.

Maintenance of Neurons: Recycling and Recharging

Recharging the neuron after action potentials, via the sodium pump, obviously is important, but it is easily misunderstood. The simplified diagrams that illustrate ionic exchange (Figure 13-11) do not usually convey a good impression of the true relationship between the action potential and the ionic pump. In particular, the neuron does not have to stop to recharge after each action potential. The driving forces behind action potentials are the ionic concentration gradients. Once these gradients are established, an individual action potential diminishes the total gradient by only a very small amount; in larger axons the effect is almost negligible. The action potentials still can continue for some time even if the ionic pump is blocked or inhibited. The gradients eventually run down, and the impulses will stop, but, at least in the squid axon, this may take several minutes (Caldwell and Keynes 1957). Perhaps a good analogy would be the recharging of a battery in an automobile. As long as the car is running, the battery is being charged, but only as much as needed. But even with the engine turned off, the lights still can be operated by the battery for some length of time. Discharging and recharging are largely separate processes.

Figure 13-16 Compound action potentials from sensory neurons A_1 and A_2 and the B cell of a moth ear under a variety of conditions to illustrate principles of neuronal coding of information. For detailed explanation, refer to the text and study this illustration in conjunction with that information. Based on photographs and discussion in Roeder, K.D. 1967. Nerve cells and insect behavior. Harvard University Press, Cambridge, Mass.

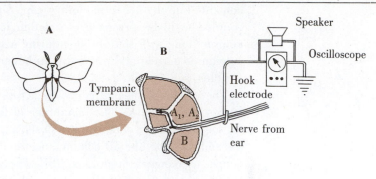

Recording from one ear

Nonneuronal Components of the Nervous System

In addition to the neurons, which are responsible for the actual processing of the information, there are many cells that support or contribute to the functioning of neurons in various ways. The general term for such cells is **neuroglia** or **glial** cells. There are several types.

Some examples that illustrate many of the preceding principles in this chapter follow, and all of this information is applied to behavior. These examples also will serve to introduce several new principles about neurons.

EXAMPLES: MOTH EARS AND COCKROACH KICKERS

The responses of moth ears to bat calls offer a classic introduction to the connections between nerve cells and behavior. Thus the example of moths and bats that was begun in Chapter 1 will be continued. This will carry us to the output side with a brief glimpse at the problems of integration. After that a few additional points gained from insight into the evasive behavior of cockroaches will be considered.

Neuroethology, as this general topic is called, is rapidly becoming an established discipline of its own, and well-studied cases are becoming numerous. Other well-known examples include cricket calling, insect catching by frogs and toads, and basic movements in chickens. For further details and a general introduction to the subject, see Ewert 1980.

Moth Ears

The moth ear has several advantages for neural research: (1) moths are relatively common and easy to obtain; (2) the ears are relatively easy to get at; (3) each ear has a total of only *two* acoustic sensory neurons (A_1 and A_2 cells) plus one proprioceptive sense cell (the B cell), which greatly simplifies the interpretations of results; and (4) the responses of these sensory cells are typical, hence representative, of sensory cells in many other organisms and more complex sensory structures. With the use of an extracellular electrode on the auditory nerve and a recorder (Figure 13-16, *A* and *B*), one can infer what a bat sounds like to a moth. Note that we are recording a compound action potential from 3 axons: A_1, A_2, and B. The following is based on photographs and extensive discussion in Roeder (1967).

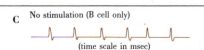

C   No stimulation (B cell only)

(time scale in msec)

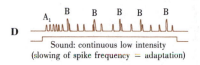

A_1 B   B   B   B   B

D

Sound: continuous low intensity
(slowing of spike frequency = adaptation)

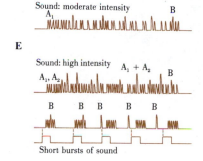

Sound: moderate intensity
A_1 B

E

Sound: high intensity $A_1 + A_2$

A_1, A_2 B

B   B   B   B   B

Short bursts of sound

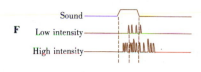

Sound

F   Low intensity

High intensity

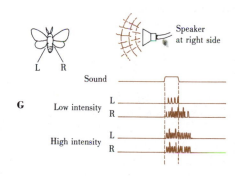

L   R

Speaker
at right side

Sound

G   Low intensity   L
                    R

High intensity   L
                 R

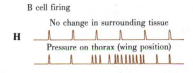

B cell firing

No change in surrounding tissue

H

Pressure on thorax (wing position)

A moth's ear is sensitive to sounds in the frequency range of 3 to 150 kHz, but it is most sensitive to the range of 50 to 70 kHz. Within this range, however, the moth is tone-deaf; it apparently cannot distinguish pitch but reacts in a similar manner to all. Recall that bat calls range in frequency from 20 to 100 kHz, in the same range as the moth's hearing. (Human hearing is in the range 0.04 to 20 kHz.)

A recording from one moth ear under quiet conditions is diagrammed in Figure 13-16, *C*. The periodic spikes are from the B cell. In the presence of a pure continuous tone of varying intensity, the responses are as shown in Figure 13-16, *D*. Several characteristics of the response can be seen. (1) There is a brief delay between the start of the sound and the onset of the action potentials at the site where they are recorded. (2) The frequency of spikes, that is, the number of spikes per unit of time, increases with an increase in sound intensity. (3) At low sound intensities only one cell, A_1, responds, and then at higher sound intensities A_2 joins in. In some instances A_1 and A_2 occur simultaneously and produce a double peak. (4) The spike frequency at all intensities begins to decrease with time (*sensory adaptation*), although the sound intensity stays the same. With short bursts of sound, rather than with continuous tones, the adaptation does not occur (Figure 13-16, *E*). With short bursts of increasing intensity (up to the point of saturation), another difference becomes apparent: a longer afterdischarge occurs, that is, the length of time that spikes continue after the stimulus has stopped (Figure 13-16, *F*). The B cell during these experiments shows no alteration in frequency.

The moth has no independent means of assessing the sound other than the series of spikes coming in from those few neurons. A low spike frequency could arise from several different situations such as a distant, low-intensity signal in the sensitive range of detection, a loud signal that is outside or on the edge of the ear's sensitivity range, or sensory adaptations and the decline in a cell's sensitivity. There would seem to be no way for the moth to distinguish between these and other alternatives. But the moth does have more than just these three neurons.

The results of recordings made simultaneously from *both* ears are shown in Figure 13-16, *G*. Unless both ears are saturated, the neural response starts sooner, has higher spike frequency, and lasts longer on the side from which the sound is coming. It is not very far from one side of a moth to the other, and the sound will hit both sides at practically the same time; thus the delay in arrival time is too small to be useful. But the ears are pointing in different directions and receive a given sound at different *intensities*. This permits a means of determining at least a directional component in the sound, via the combination of input from right and left sides. In addition to spike frequency in particular neurons, the nervous system has access to *combinations* of inputs, in this case from A_1 left, A_2 left, A_1 right, and A_2 right. These findings, incidentally, are what prompted the search for the *directional behavior*, which is described in Chapter 1.

A complete moth still has not been described. Recordings were made from both ears, but the wings were removed to get at the neurons, and the moth was fixed in a holding device. Ideally we would like to record from the neurons of a free-flying moth being chased by a real bat. But that would be very difficult and most likely be too uncontrolled for interpretable results. Some additional information came from the artificial application of pressure to the thoracic membranes

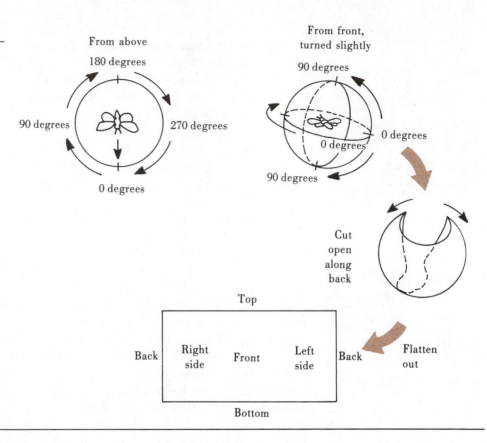

Figure 13-17 Technique for deriving a flat map of the three-dimensional sensitivity of moth hearing to surrounding bat calls. The map is formed by unwrapping and distorting a sphere from around the moth and laying it out flat, as illustrated.

near the ear, as might occur during movement by the wing muscles. This pressure caused a change in the B-cell spike frequency (Figure 13-16, *H*); thus it was suggested that such internal (proprioceptive) information might also be of use to the moth, giving it information on the position of its wings.

So the recording setup was further improved by leaving the wings on the moth in different positions and presenting the sound from all around the moth. There were several technical problems such as the prevention of echoes and the need for many standardized recordings in short periods of time. The whole system needed to be mechanized.

The solutions were ingenious. In brief, the researchers used a mechanical stage to hold the speaker in different positions and then varied the sound (automatically via negative feedback) to keep a constant nerve response. This permitted the sensitivity to be measured indirectly by the sound intensity required to trigger the response (e.g., if it needed a louder sound, the ear was less sensitive under the given conditions).

From these results the sensitivity was recorded as on a sphere surrounding the moth. Three-dimensional spherical structures are difficult to display in a flat two-dimensional manner and must involve some distortion, as in a map of the Earth. Imagine the spherical surface being rubber and stretchy with small holes at the poles. If the structure is stretched open at the holes to make it cylindrical, then cut open at the back, it can be laid out flat—a Mercator projection (Figure 13-17). Simplified Mercator projections of the moth's hearing sensitivities with the wings in different positions with a summary interpretation are illustrated in Figure 13-18.

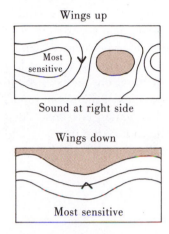

Wings up

Most
sensitive

Sound at right side

Wings down

Most sensitive

If bat is at:		
1		2
6	5	6
3		4

More sensitive		Least sensitive

+ = to sounds; 0 = average; − = to sounds

	Wings up		Wings down	
	R ear	L ear	R ear	L ear
1	+	−	−	−
2	−	+	−	−
3	+	−	+	+
4				
5	0	0	0	0
6				

Figure 13-18 Simplified map of moth hearing sensitivity. The most sensitive regions are indicated; the least sensitive are solid. Note that the pattern shifts depending on whether the wings are up or down. Inputs to the two ears, with wings up or down, for bats at different positions relative to the moth are indicated in the table. Parts of the table for positions *4* and *6* have deliberately been left blank for completion by the reader as an exercise to test understanding of the material. Modified from Roeder, K.D. 1967. Nerve cells and insect behavior. Harvard University Press, Cambridge, Mass.

The wings block or baffle the sound, depending on their positions. From this it can be seen that, from a total of only six sensory cells, two acoustic and one proprioceptive on each side of the body, the moth potentially can extract sufficient information to localize the source of the sound. Furthermore, with moving wings a "flicker" would result in the sound pattern from all locations except from directly behind the moth—the best position for a fleeing moth to keep the bat.

Further research (see Chapter 1 and Roeder 1967) confirmed that the information on sound intensities and location actually was useful to the moth. It was also determined that moths can detect the bats at a distance of about 30 to 43 m, whereas the bat cannot detect the moth until it is within about 2.5 to 4 m.

This still is not the whole story. Bat chirps have not yet been distinguished by the moth's nervous system from all the other sounds in the environment, such as cricket chirps. Once the input has been filtered and identified as "bat call," there remains the problem of translating this into the appropriate, directed, muscular-controlled flight response.

Penetrating the moth's nervous system further, via evoked potentials, one can briefly glimpse some of the different interneurons that help filter the input (Figure 13-19). There are "repeater" neurons that basically relay the initial A-cell input,

Figure 13-19 Output from moth interneurons during the process of stimulus filtering and identification of bat calls. Different neurons respond to similar input (from the sensory neurons) in different ways. The combination of these different outputs permits the system to abstract and extract information.

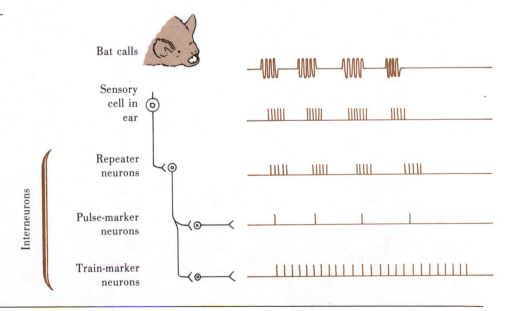

Bat calls

Sensory
cell in
ear

Interneurons

Repeater
neurons

Pulse-marker
neurons

Train-marker
neurons

relatively unchanged, to the nervous system. Other neurons respond differentially to different parameters of the input. "Pulse-marker" neurons respond only once per ultrasonic sound pulse and only if cocked by intervening periods of silence. "Train-marker" neurons maintain a continuous output, at their own spike frequency, during a series of sound pulses. These and other neurons are believed to sequentially and selectively filter through the characteristics of the initial input, somewhat like a combination lock, until the right combination is reached that identifies a "bat call." The moth example has thus shown how the intensity and identity of sensory input is coded by *spike frequency* and how the system begins to *sift and filter* the input to extract relevant information from it.

Cockroach Cerci

The next example involves another simple picture of neural integration. The subject is the detection and evasion of potential predators through the startle response of cockroaches. The account presented here is a brief summary of material reviewed by Roeder (1967).

Cockroaches run at the least provocation and are fast, as anyone who has ever tried to squash one can attest. Unlike the fairly specific bat response of moths, the evasive response of cockroaches is more generalized. The response is characterized by speed rather than specificity or information content; that is, the cockroaches do not take time to identify the source of the stimulus.

The sense organs that trigger the response are *cerci*, hairlike projections on the tip of the cockroach's abdomen. The cerci are sensitive to slight air movements, including low-frequency sounds. When the cerci are stimulated, the cockroach jumps and runs, scurrying off to a dark hiding place.

Total response time from a puff of air to the start of a jump was measured in cockroaches by a ingenious setup as shown in Figure 13-20. The cockroach was affixed to a support with a drop of hot wax; then the support was connected to a phonograph pickup, which in turn was connected to an amplifier and oscilloscope. Another pickup was connected to a flag that was placed near the cockroach's cerci to detect the stimulus. The cockroach was given a little ball to hold in its feet. When a puff of air was given to the cerci and the cockroach jumped, kicking away the little ball, the pickup would record the rebound of the cockroach.

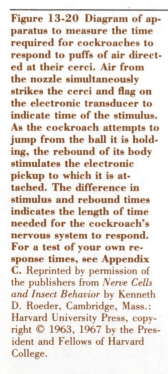

Figure 13-20 Diagram of apparatus to measure the time required for cockroaches to respond to puffs of air directed at their cerci. Air from the nozzle simultaneously strikes the cerci and flag on the electronic transducer to indicate time of the stimulus. As the cockroach attempts to jump from the ball it is holding, the rebound of its body stimulates the electronic pickup to which it is attached. The difference in stimulus and rebound times indicates the length of time needed for the cockroach's nervous system to respond. For a test of your own response times, see Appendix C. Reprinted by permission of the publishers from *Nerve Cells and Insect Behavior* by Kenneth D. Roeder, Cambridge, Mass.: Harvard University Press, copyright © 1963, 1967 by the President and Fellows of Harvard College.

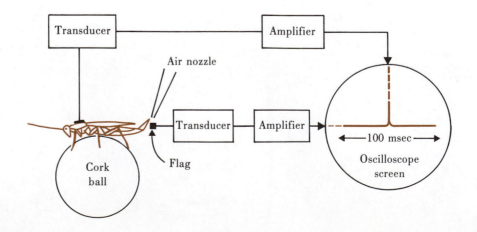

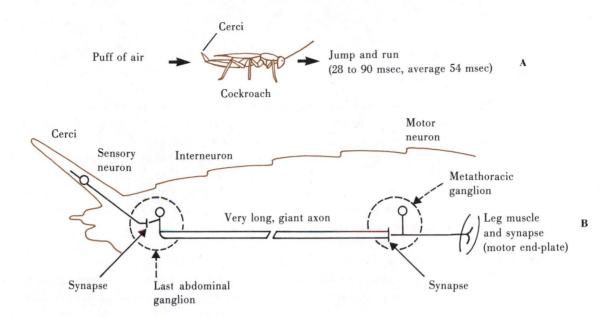

Figure 13-21 Cockroach startle response times, as measured in Figure 13-20, and times for neural events measured with electrodes placed at different locations within the cockroach nervous system. A, Sequence of events. B, Diagrammatic section of cockroach body. C, Times (in milliseconds) measured at different points. Based on information in Roeder, K.D. 1967. Nerve cells and insect behavior. Harvard University Press, Cambridge, Mass.

Times (in milliseconds) measured at different points:

Response time of cercal sensory neuron	0.5
Conduction time (short axon)	1.5
Synaptic delay	1.1
Conduction in long, giant axon	2.8
Synaptic delay	4.0
Conduction in motor axon	1.5
Neuromuscular synaptic delay	4.0
Development of muscular contraction	4.0
TOTAL (close to observed but lower)	19.4 msec.

The research was not without problems. Cockroaches often would not accept the setup quietly or readily but instead walked, cleaned themselves or otherwise vibrated the pickup, or even kept kicking the little balls away before the time of the desired stimulus. But good results finally were obtained. The measurements of 23 successful cockroach reactions averaged 54 msec (0.054 second) over a range of 28 to 90 msec. Persons not familiar with millisecond units of time can refer to Appendix C to obtain a useful point of reference with human response times.

Next a number of subsequent spike recordings and experiments were performed at various locations inside the cockroach. The parts of the nervous system and the minimum measured times for the messages to traverse different individual segments are shown in Figure 13-21. These times, plus other characteristics of the spikes, revealed several important characteristics of the neurons and the whole system. First, the discrimination or filtering of the input is accomplished by passage through neurons with different properties, as was seen in the moth. Furthermore, spatial and temporal summation are involved. The giant fibers do not fire unless they receive impulses from *several* of the cercal sensory axons more or less simultaneously, that is, via spatial summation; motor neurons do not fire until they receive two or more *successive* impulses from the giant fibers, that is, via temporal summation. The practical implications of this are that slight air currents do not trigger the escape response, but those from more substantial input, such as a predator or an approaching folded newspaper, do.

Each of the neurons in line has a unique response to the preceding neurons. Motor neurons, for example, maintain a significant afterdischarge of spikes that continue long after the giant fibers cease firing. This afterdischarge is believed to help *maintain* the response, that is, to help keep the cockroach running after the senses serve to *trigger* it. The sustained running carries the cockroach out of the range of danger. Although discrimination of the input is occurring in this example and although more sensory neurons are involved (150 cercal neurons versus the 6 moth ear neurons), the discrimination does not appear to be nearly as complex as in the moth. Translated very roughly to English, the message in the cockroach is "Danger!"; the message in the moth's system is more like "A distant cruising bat approaching from the upper left!"

The cercal and giant fiber axons could fire 200 to 300 impulses per second for extended periods, but the synapses would fail after just a few seconds at this frequency. The synapses were much more disrupted by various anesthetics and drugs than by the action potentials. For the distance involved, the synaptic transmission was much slower than for axonal transmission. Thus of the two main functions of neurons, discrimination at the synapses is much more time-consuming and subject to disruption than is the transmission of the impulse.

Finally, although the speed is least variable in the axons, the speed varies depending on characteristics of the neuron, particularly its diameter. The impulse travels the length of giant fiber in 2.8 msec, which is estimated to be nearly 10 times faster than in the other normal-sized axons. This speed is interpreted as beneficial: it may make a sufficient difference (even if only 1% to 2%) in the overall response time to mean the difference between life and death. Over long periods of evolutionary time the faster cockroaches, with the larger axons, evidently survived and reproduced in greater numbers than the slower ones.

But there is a cost; the large diameters take up space that could be used by a larger number of small axons. Recall that part of the information processing may occur through different *combinations* of signals. With more units the increase in number of possible messages is multiplicative, not additive. Fewer axons mean many fewer possible combinations. Nature rarely goes to one extreme or the other, however, but usually strikes a compromise, even in cockroaches. Thus the ventral nerve does not contain *only* giant fibers but also a number of smaller axons (Figure 13-22), which also must contribute to the overall processing, discrimination, and probably much of the variability among responses.

Compared with the moth, the cockroach takes sensory input from a larger number (150) of similar sensory neurons and boils it down to a relatively simple, rapidly transmitted message. The result is a response time of around 54 msec. The moth, on the other hand, uses fewer (6) but dissimilar sensory neurons and apparently extracts more information by more internal processing. But it takes more time; the turning responses of moths to bat calls (see references in Roeder 1967) range from 75 to 252 msec, averaging around 140 msec.

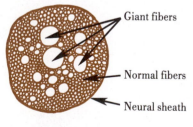

Giant fibers

Normal fibers

Neural sheath

0.1 mm

Figure 13-22 Cross section of cockroach ventral nerve cord showing both giant axons and axons of smaller, normal size *(smaller specks)*. This view is as if one were to cut across a telephone cable carrying a large number of individual wires. The different wires would represent the axons of individual neurons. Based on photographs in Roeder, K.D. 1967. Nerve cells and insect behavior. Harvard University Press, Cambridge, Mass.

Response time obviously is important in the evolution of behavior. But further evidence for the validity of these inferences would be comforting. One would predict, for example, that predators and prey should be closely matched in speed. In their long-term interactions each side would be constantly changing in the process of staying ahead of or even with the other side. Because of fast prey, faster predators survive and reproduce differentially better than slower predators. This tends to push up the average speed of the predators. In turn this increases the pressure on the prey, causing an upward shift in the speed of prey. But because increased speed costs something (such as a loss of ability to discriminate), the pace of the evolution would not be expected to be excessive nor would either side be very far ahead of the other. Large differences between the abilities of the participants should cost more than the advantages and thus not be predicted.

Unfortunately, one usually is stuck in biology with a picture composed of pieces that come from different puzzles and do not always fit together as one would like. Such is the case here. The match between moths and bats is hard to evaluate. Moths detect bats well in advance of the bat's detecting the moth, and there are measurements of the moth's response time, but bats fly faster, and all the figures for the bat's responses and ability to connect with the moth are not known. All things considered, the two should be closely matched. For cockroaches one has only their side of the picture. Information does exist, however, for another predator-prey interaction: that between mantids and flies (also reviewed in Roeder 1967). Mantid strikes from ambush take 50 to 70 msec to complete, and laboratory-raised mantids miss about 10% to 15% of their strikes. Measurements of fly response times show that it takes 45 to 65 msec for them to start escaping. The match in timing is remarkably close.

●　　●　　●

In summary, these examples illustrate how sensory input generates action potentials and how subsequent neurons process the information. The speed at which spikes move depends on several factors, one being the diameter of the axon. Large axons transmit spikes faster, but they take up space and reduce the number of other axons that can pass through the same area. Information in particular neurons is coded by spike frequency. Increased intensity of input, if excitatory, leads to an increased frequency of spike output. Different neurons vary from each other in the nature of their input and subsequent responses (i.e., output). Further processing of information is accomplished by filtering and sorting through combinations of several neurons. This eventually leads to muscular output that forms the appropriate behavior.

Nervous systems sometimes are compared heuristically with man-made computer systems. The comparison may be useful for suggesting general features or principles, the analogs of which can then be considered in biological systems. But, as can be seen from these descriptions of neurons, there are many major differences between neurons and the components of computers. Digital computers have only two states for each unit. Neurons also either fire or do not, but they also have many possible states with graded potentials and with variable spike frequencies and patterns. Action potentials and the quantal release of transmitters might be considered digital properites, but the remainder of the processing is analog. Computers have physical-electrical mechanisms, whereas biological systems have chemical-bioelectrical processes that are vastly different.

Next the chief output side of the nervous system—muscles—will be considered. Other forms of output of effectors include cilia, flagella, pseudopods, neural secretion, endocrine and exocrine secretion, electric organs, bioluminescence, and many other things, some of which are discussed elsewhere in this book.

The purpose of this chapter is to focus on the properties of individual neurons and provide a general overview and introduction to the whole system. A more detailed picture of how the brain functions deserves separate, additional treatment and is presented in Chapters 15 and 17. In the remainder of this chapter muscles will briefly be considered as important behavioral output mechanisms.

MUSCULAR OUTPUT

The final result of most neural input and processing is muscular contraction. Muscles and muscular contraction are extremely interesting subjects in their own rights. The present concern, however, is with the connections between the nervous and muscular systems. A particular movement, hence behavior or part of a behavior, depends on *which* muscles contract and the spatial-temporal patterning of the contractions. This is largely under the control of the nervous system, and finally mediated via the motor neurons.

The control that the nervous system exerts over the muscles can be easily demonstrated by a person's handwriting style. The writing style of most persons is unique. Writing normally employs a large number of muscles in the hand and forearm, with letters being formed by a very exact sequence of contractions. If a person writes in large letters on a wall or blackboard, however, completely different muscles, largely in the upper arm and shoulder, are involved. Yet the writing style normally shows little or no change (see Figure 15-22). This is because the nervous system is directing the writing; it is simply sending the

Figure 13-23 Motor endplates. The connections between a motor neuron and the muscle fibers it ennervates. Photo by James W. Grier.

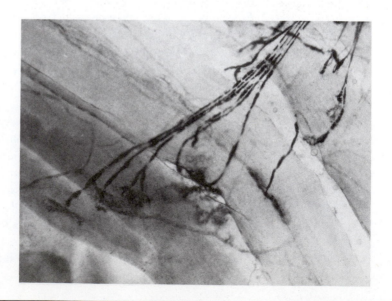

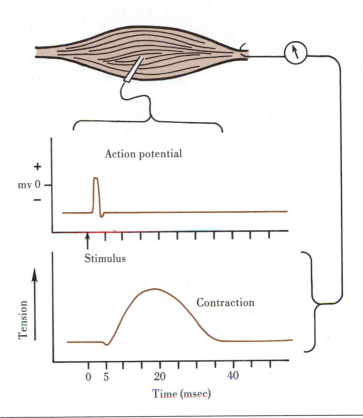

Figure 13-24 Time sequence of action potential and contraction in a muscle. The first event is an action potential, similar to that which occurs in a neuron. The contraction of the fiber then develops subsequently and more slowly.

instructions to different muscles. Compare your hand and arm writing with that of one or more friends.

On the final steps of its journey through the nervous system, information in the form of action potentials travels down the axon of a motor neuron. The last synapse in the system is a *motor end-plate,* the synapse between the motor axon and the surface of a muscle fiber (Figure 13-23). An action potential, similar to that seen in neurons, is initiated in the muscle fiber. This action potential triggers a sequence of chemical reactions that culminate in the observed contraction (Figure 13-24). A given fiber (in familiar twitch muscles) either contracts or does not contract (all-or-none). Individual muscles are composed of large numbers of fibers, and the overall graded strength of contraction in a muscle depends on which proportion of its fibers are contracting at any one time.

Although the preceding describes the general picture, nature has found much room for variation. Neural parsimony, the ability to function with relatively few neurons, in arthropods has already been discussed. At the output side parsimony is achieved in a variety of ways but, basically, by sending different information down the same axon rather than using different axons for different information. A given motor axon may innervate several muscle fibers in both vertebrates and invertebrates, but axons typically innervate a higher proportion in invertebrates. In many cases all the fibers in a whole muscle may be innervated by only one or a few axons' impulses, and recruitment of muscle fibers is accomplished by different patterns or frequencies of firing. If the pattern rather than just the frequencies of impulses is important, the nerve-muscle synapses are said to be *pattern sensitive*. In the crayfish claw opener muscle, which is innervated by a single axon, a train of impulses results in a particular strength of contraction. If the impulses are grouped in pairs, the muscle contracts more strongly. In vertebrates, on the other hand, recruitment is mostly accomplished by adding more motor neurons, which, in turn, fire more fibers.

The all-or-none principle is strictly true only for twitch muscles—fibers that contract quickly with a rapidly propagated action potential. There are more slowly contracting fibers and fibers in which only local regions of the fiber contract, which show graded contractions. This includes a number of arthropod muscles, some vertebrate smooth muscle, and even the "slow" fibers of vertebrate striated muscle. Muscle fibers vary tremendously in their properties and, as a result, are difficult to classify. Categories of fast (phasic) or slow (tonic), red and white, voluntary or autonomic, smooth or striated, etc., are only partially useful. Muscle fibers also vary in size (up to 1 mm diameter in a barnacle, *Balanus nubilis*) and length of time they can remain contracted. The shells of some molluscs are held closed by "catch" muscles that can remain contracted for days.

Some muscles are capable of speed and precise fine control, as in the eye and human fingers, whereas others are involved more in slower, power movements such as in lifting weights. Precision control muscles generally have a higher ratio of neuron-to-fiber innervation. In many muscles different muscle fiber types (e.g., slow and fast) are intermixed.

Another means of varying recruitment is via **polyneuronal innervation.** Different axons from different neurons may impinge on a given muscle fiber. Some are excitatory, and some are inhibitory, and a given axon may have endings that deliver fast (phasic) short bursts or impulses and other endings that deliver slower (tonic) trains of impulses. These in turn affect the *speed* of the ensuing contraction and thus affect the development of strength in the muscle. The summed effects of polyneuronal innervation are reminiscent of discrimination normally associated with the nervous system. Some observers have quipped that crabs can think in their legs.

Whether muscle contraction results strictly from a chain of input (including proprioceptive feedback) to integration to output (the peripheral control hypothesis) or strictly from the central nervous system alone (the central control hypothesis) has been a point of some debate. It is mostly a matter of degree, however. It is clear that *some* external input is almost always required, even if only to trigger or steer the movement. The control cannot be generalized simply—the importance of sensory input varies among different taxonomic groups and under different circumstances. Even in the most extreme cases, cardiac and some smooth muscles with their own pacemakers or other spontaneous rhythmicity and passive, electrotonic spread of impulses, there still may be a speeding or slowing of contractions resulting from sensory or contemplative input.

The frequency of rapid wing movements of some insects would seem to exceed the capacity of neuronal action to fire them. Indeed, some are not caused to contract at each beat by neural impulses. The flight muscles of some of the more primitive insects connect directly to the wings and are contracted in standard fashion. In more advanced forms, however, the flight mechanism is very complex and varies among species (Pringle 1957). In essence, the muscles are connected indirectly—not to the wing but to the elastic thorax. This snaps back and forth (the "click mechanism"), which moves the wings, via lever principle, and also stretches the flight muscle on the rebound. This stretch, rather than a neuronal action potential, causes the next contraction, and the movement continues. The role of the nervous system is only to trigger the initial contraction, periodically maintain the process, and apparently modify the frequency to achieve controlled movements. A similar system is involved in the sound-producing tymbal organ of cicadas.

One intriguing aspect of nerve-muscle connections that deserves mention concerns the *specificity of innervation,* at least in vertebrates, including mammals. Apparently because of some chemical means of identity or recognition, specific motor axons innervate specific, "correct" muscle fibers or types of fibers. This normally occurs during initial development of the animal. If cut, they will grow back to the correct fibers: if displaced, they will find circuitous routes back to the fibers; and if incorrect connections are forced, the correct axons may eventually reconnect and displace incorrect ones. If the incorrect ones remain connected, the correct ones somehow repress the synaptic influence of the incorrect ones. Incorrect connections usually result in inappropriate movements, which often cannot be compensated through experience; that is, relearning does not seem to occur very easily or at all. But after the axons reconnect appropriately, the movements are then correct once again. Thus at the final stages of neuronal control it appears that the correct muscular contractions depend on the specific connections within the motor neuron-muscle fiber units.

SUMMARY

The nervous system controls and coordinates the behavior of an animal with the rapidly changing, surrounding conditions of the environment. Sensory input is integrated with the current internal state of the organism and leads to the appropriate output—sequences of coordinated muscular action (or other output such as glandular secretions). The basic structural and functional units of the nervous system are the neurons.

Neurons vary greatly among different animals and different parts of any one animal. Accordingly, neurons are classified in many different ways. A large array of specialized, modern equipment and techniques has vastly improved the understanding of neurons.

Neurons are negatively charged internally as a result of active and passive events that lead to imbalances in the concentrations of several ions. Input from sensory endings or from other neurons produces local, graded changes in the membrane potential. Input from various impinging sources adds and subtracts, depending on whether it is excitatory or inhibitory; that is, it summates over time and space in the neuron. An action potential may be initiated if changes in the potential reach a sufficient magnitude. During an action potential the membrane permeability for several ions changes rapidly. Among the ions that cross the membrane, Na^+ rushes in and creates a temporary increase in the positive charge inside the neuron; then K^+ rushes out, which reduces the positive charge and helps restore the internal negative condition. Action potentials, or spikes, are propagated down the length of the axon until they reach the end, or synapse.

The synapse then affects the next neuron (or muscle or gland), either directly in gap junctions or indirectly, in chemical synapses, by expelling a transmitter into the synaptic cleft. Many different chemicals serve as transmitters, although only one kind is present in any one synapse. The transmitters are synthesized and recycled in the axon terminals where they are packaged in synaptic vesicles. The vesicles release the transmitter in quantal amounts. The transmitter diffuses across the very small space and affects the membrane permeability of the next neuron, causing a new local, graded potential.

Neurons code information through spike frequency. Different neurons vary, however, in the input they receive and the characteristic manner in which they respond to that input. Characteristics that differ include, for example, the length of delay before spikes occur, the number of spikes, and the amount of afterdischarge. Some neurons fire spontaneously and have a resting (or pacemaker) rate of firing. Input may alter this intrinsic rate. These differences permit different combinations, and patterns of information flow through the combined responses of many different neurons simultaneously.

Moth hearing is used as an example to illustrate some of the properties of neurons in general and sensory neurons in particular and to relate these principles to behavior as observed output. Recording and experimental techniques, spike frequencies, and the different responses among different neurons to different inputs are illustrated in this example. The moth's system, although extremely simple and limited, extracts a considerable amount of information from sound waves. The cockroach startle response is used to illustrate principles concerning synapses and the speed of processing and transmission. The differences between the moth and cockroach examples emphasizes the coding of information by several neurons working together, not just in isolation.

This chapter concentrates on neurons to lay the groundwork for further study of nervous systems and to indicate their importance to behavior. Little attention thus is given to the whole nervous system; treatment of this topic has been postponed to subsequent chapters. A few general comments are offered, however, to further place neurons in proper perspective. Whole nervous systems have become more differentiated and organized in more advanced organisms. Nervous systems might be compared somewhat to man-made computer systems, but the analogy has many limitations.

**Recommended
Additional Reading**

Bullock, T.H., R. Orkand, and A. Grinnell. 1977. Introduction to nervous systems. W.H. Freeman & Co., Publishers, San Francisco.

Cold Spring Harbor Symposia on Quantitative Biology. 1952. The neuron. Biological Laboratory, Long Island, N.Y.

Eccles, J.C. 1977. The understanding of the brain, 2nd ed. McGraw-Hill Book Co., New York.

Ewert, J.P. 1980. Neuroethology, an introduction to the neurophysiological fundamentals of behavior. Springer-Verlag New York, Inc., New York.

Fentress, J.C., editor. 1976. Simpler networks and behavior. Sinaurer Associates, Inc., Sunderland, Mass.

Kandel, E.R. 1976. Cellular basis of behavior. W.H. Freeman & Co., Publishers, San Francisco.

Mountcastle, V.B. 1980. Medical physiology, 14th ed., Vol. one. The C.V. Mosby Co., St. Louis.

Roeder, K.D. 1967. Nerve cells and insect behavior, revised ed. Harvard University Press, Cambridge, Mass.

Scientific American. 1979. The brain (special issue). Sci. Am. 241(3).

Stevens, C.F. 1966. Neurophysiology: a primer. John Wiley & Sons, Inc., New York.

CHAPTER 14

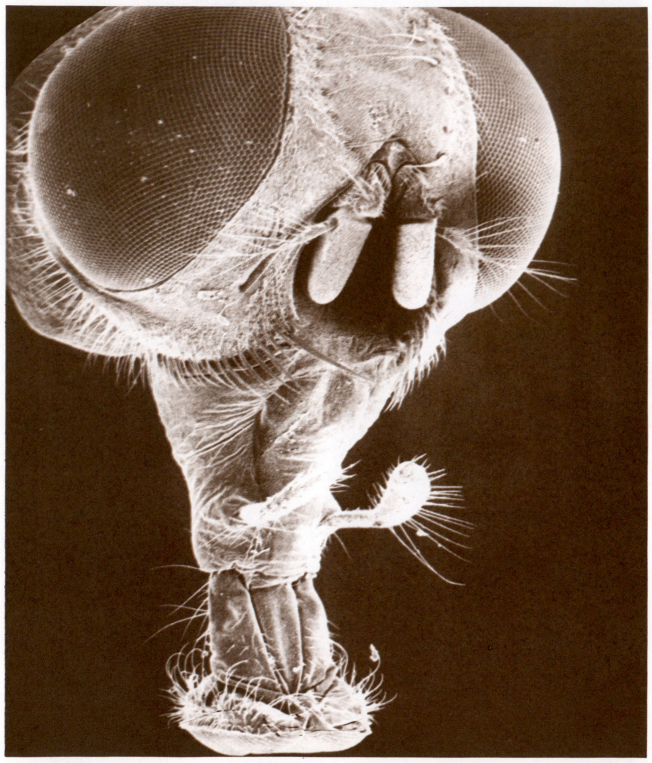

North Dakota State University Electron Microscope Laboratory

Scanning electron micrograph of a housefly *(Musca domestica)* **head.**

DIFFERENT SENSORY WORLDS

Differences among the senses of animals contribute importantly to the differences one observes in behavior of different species. At the other (output) end of the nervous system, there are far fewer differences in muscular systems. There certainly are various kinds of muscles; they connect in many ways to other parts of the body; and muscles are of much interest from a physiological viewpoint. But, from the standpoint of behavior, most of the observed differences in muscle contractions simply arise from different patterns of action potentials coming from the nervous system. Muscles, for the most part, just do as they are told; they contract on command. The senses, however, provide the first link of an animal with the outside world, and this greatly affects subsequent behavior.

The concept of the *Umwelt* (von Uexküll 1909) was introduced in Chapter 1: each animal lives in its own sensory and perceptual world. For example, the bats and moths flying about on a warm summer evening sense their environments in ways that are different from each other, and both are much different from humans. Technically, the *Umwelt* as a sensory and perceptual world is only a crude approximation of von Uexküll's original concept. Originally, the *Umwelt* meant much more, including also the behavioral, effector output of the animal, and involved several German terms, *Merkzeichen, Merkmal, Wirkzeichen,* and *Wirkmal,* which do not translate easily into English. For insight into the original concepts proposed by von Uexküll, see the translation by Schiller (1957), including her introductory note.

To expand the picture somewhat, imagine now a different scene: a young couple, for example, two college students, on a walk in the woods with a dog on a warm spring afternoon. Within several meters of them in that same woods are many other species of animals. Try to imagine how the world "appears" to each of them.

First consider the humans. They may not seem to be sensing very much; they are mostly paying attention to each other and largely are unaware of their surroundings. This probably is correct, but it gets one into other topics, perception and attention, that go deeper into the nervous system than needed at the moment. Much of the **stimulus filtering** that occurs in an animal is **central**; that is, it involves neural integration at higher levels within the central nervous system. We are concerned now more with **peripheral** stimulus filtering, that is, what the senses are capable of receiving if the individual is paying attention.

Thus if and when they pay attention to them, the couple can sense a large number of things in the environment with which most persons are familiar. There are the sights and colors of leaves, branches, and tree trunks, blue sky and clouds overhead, sounds of breezes, singing birds, and their own footsteps. There are a number of odors that they detect faintly, such as her perfume, his body odor (which she may detect but he himself may not), and the smells of the damp woods. You can fill in the rest of the picture yourself.

What about the other animals? The dog is lower to the ground, for one thing, and has a different perspective simply for that reason. In addition, the dog sees somewhat more poorly and lacks the visual resolution that humans have. But the dog receives a rich combination of smells much more strongly than the people and also hears higher pitches, such as those of nearby insect sounds. Scurrying under the leaves is a shrew that sees even more poorly than the dog but probably has keener senses yet for smell, perhaps hearing, and touch. The shrew's world is dominated by chemical senses, touch, and vibrations received from around its body, particularly various hairs and vibrissae about the head. It is visually aware of light and dark but not much more.

The shrew is hunting earthworms, which in turn are tuned primarily to odors and a world full of vibrations in and on the soil. Vision is practically nonexistent for the earthworm; it is aware only of the presence or absence of light. As the people walk by and their shadows fall across the earthworm, the worm's world suddenly becomes darker and is filled with a tremendous number of vibrations. The earthworm is aware of light in a similar manner that humans are aware of temperature (heat). Not too much further down the path is another animal, a copperhead snake, that "sees" very accurately via heat. It also constantly tests the air for molecules that it picks up with flicks of the tongue and which it places in the Jacobson organ in the roof of its mouth for chemical sensing.

As night falls and the woods become dark, the two people find it more difficult to find their way about. The light on which they depend has rapidly disappeared. The change to night will be less noticeable to the shrew and the dog. At the bottom of a nearby muddy stream is a small catfish that senses little change in light or temperature but is surrounded by a world of currents and water pressure changes, aquatic sounds (some coming in from the stream bed and bank), and a rich world of chemical cues, which may be detected at many places over the fish's body.

This scene is designed to *emphasize* the *Umwelt*. One cannot stress too greatly the importance of recognizing that other animals sense their surroundings in a multitude of vastly different ways. One does not have to travel to other worlds in outer space to find strange creatures unlike ourselves. They are all about us. Yet the senses of other animals are all too easily taken for granted, forgotten, or overlooked, whether by hunters, fishermen, photographers, or students of animal behavior. Begin to practice thinking in terms of how the world might "look" to other animals not just as humans see it.

The remainder of this chapter surveys briefly and systematically the sensory capabilities of various species and considers the mechanisms by which senses transduce different forms of energy into nervous impulses. First, some of the methods used to investigate the senses of other animals will be considered. Because they do not possess the same experiences, perceptions, thoughts, and forms of communications as humans, we cannot share their worlds directly. Humans are forced to use indirect means and *infer* what information is going into other animals' systems.

METHODS OF INVESTIGATING SENSES

Different techniques have been used in sensory research. They fall broadly into four major categories:

1. *Evoked potentials: neural activity recorded from sensory neurons.* Either by microelectrodes inserted directly into the axons of sensory neurons or by external electrodes placed nearby, one can measure changes in spike frequency associated with different stimuli impinging on the sense organ. This technique was used extensively in the example of the moth ear. In an analogous manner one can record the activity from other sensory neurons as well.

2. *Blocking, interruption, or removal of sense organs or neural connections.* This was the general method used by Spallanzani (e.g., plugging the bats' ears, covering the eyes, etc.) when he inferred that bats were using hearing to navigate. By interfering with the normal sensory processes, one can observe subsequent changes in the animal's behavior.

The pectines of scorpions (reviewed by Carthy 1968) provide a good example of the use of this method. Pectines are paired, small appendage-like structures that are located just behind the bases of the last legs of scorpions. On the ventral surface of the pectines are "teeth" and pegs (Figure 14-1). When a scorpion moves, it turns the pectines so that they brush along the substrate. The function of these unique organs has caused much discussion. Several functions were proposed, but comparisons between scorpions that had their pectines amputated or covered with varnish and others with intact pectines showed that the pectines were not used to sense airborne and substratum-borne vibrations, collect water, or create air movements over the openings of the book lungs. Anatomically the pectines were shown to carry sensory cells with axons to the CNS.

Other considerations, however, including recording from the sensory cells, have suggested that the pectines are used for tactile discrimination of the substrate. Different species choose sand grains of different size and, in particular, an area of sand that presumably provides the maximum stimulation to the pectines. The pectines differ in size and number of teeth, hence, in the physical characteristics to which they would be sensitive, among the different species of scorpions.

This method must be used with care and supplemental information to avoid incorrect interpretations, as in the trained-cricket problem.

3. *Learned responses in conjunction with sensory stimuli.* This is perhaps the most commonly used technique in the study of sensory discrimination. Learning may involve conditioned reflexes (e.g., Pavlovian or classical association, Chapter 19). If food is preceded by a signal, dogs become conditioned to salivate on

Figure 14-1 Diagram of a scorpion pectine organ. A pair of these is located near the base of the backmost legs. Their function was puzzling until research revealed them to be mechanoreceptors that sensed the size of sand grains. From Carthy, J.D. 1968. Symp. Zool. Soc. Lond. 23:251-261.

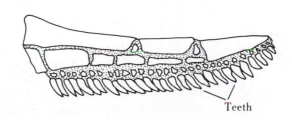

Teeth

Figure 14-2 Demonstration of negative afterimage. Color the circle red, such as with a red felt-tip pen. Gaze directly at the red spot for a minute or more under bright light and then look at the gray section of the diagram. You should see a green spot, the negative afterimage resulting from staring previously at the red spot.

hearing the signal, if they can perceive it. The signal can be changed, such as in intensity or pitch, until the dog no longer shows a response.

A succinct example of the use of this technique is shown in the determination of intensity thresholds of marine teleost fish to single-tone sounds (Tavolga and Wodinsky 1963, reprinted in Tavolga 1976). Fish were placed in a tank with two chambers separated by a shallow barrier, similar to avoidance conditioning shuttle-boxes used with rats. Sound of a particular tone was presented, followed 10 seconds later by a mild electric shock. Sounds were presented in a carefully controlled and monitored manner through underwater speakers. Most of the different species of fish learned to swim across the shallow barrier to avoid the shock 90% of the time after 5 to 7 days of 25 trials per day. After the avoidance behavior in the fish was established, the loudness of the sound at a particular frequency was successively decreased from test to test until the fish ceased to avoid it, at which time loudness in subsequent tests gradually was restored. This technique is known in sensory psychophysics as the *staircase method* (Cornsweet 1962). Different individual fish were tested several times, and the sensitivity threshold was calculated as the loudness that elicited avoidance behavior 50% of the time. The whole process was repeated at other frequencies of sound to provide a profile of the different species' hearing abilities.

4. *Natural responses in conjunction with experimental stimuli.* Rather than use artificial, conditioned responses, many workers have investigated senses via the animal's own responses, for example, the turning of flying moths in response to bat cries. Perhaps one of the best-known responses, used in tests of visual discrimination, involves movements of the eyes, head, or whole body of animals toward moving objects. This frequently is accomplished by placing the subjects inside a rotating cylinder with patterns placed on the inside of the cylinder. When the cylinder moves, the animals will turn to follow the pattern—if they can see it.

One of the most striking uses of a natural response was given by Moericke (1950), who investigated color vision in peach aphids. These aphids stab at a substrate if, and only if, the material is placed on a green background. He then exposed them for a length of time to a background of purple (the color complement of green). Subsequently they would stab at a neutral gray background—a behavior they otherwise would not display. This demonstrated a negative-color afterimage in the aphids. To experience a similar effect yourself, see Figure 14-2.

Another striking but more frequently used response for investigating senses uses animals' natural intraspecific responses, such as the movements of males toward the females and territorial interactions between males. A female's odor, for example, can be simulated and then modified to find the thresholds of the male's ability to discriminate.

SENSORY ATTRIBUTES

Any given physical entity, such as light or sound, has a great many different characteristics that potentially might be sensed. Different animals vary tremendously in regard to which or how many aspects they discriminate. One can consider the following general categories for all sensory modalities.

Sensitivity

This concerns whether the modality can be sensed at all, that is, whether the sense is present or absent, and, if present, the magnitude of change or difference that can be discriminated. Animals also vary in whether they can detect only relative changes or if they have an absolute sense for the particular modality.

Humans are not believed able to detect magnetic fields (see discussion in Chapter 7). But it now is clear that many other species can. Humans, however, have a very good sense of hearing, whereas many other species apparently are completely deaf. Noctuid moths are capable of detecting sounds within the range of bat cries, but they only detect the presence or absence of the sound— they are otherwise tone-deaf. With vision, many species have lens-containing eyes and can see detail at various intensities, but they are incapable of discriminating wavelengths, that is, color, whereas humans have very good color vision.

Humans have a general external temperature sense, but it is only relative, and it is not nearly as precise as in many other species. Mosquitoes have been reported (Herter 1962) to be sensitive to as small a temperature difference as 0.002° C and fishes to as small a difference as 0.02° C (reviewed by Hess 1973b). Many species can be shown to possess an absolute sense for temperature. Rodents, bees, and fish can be trained to choose a particular temperature that does not depend on a relative difference with a previous temperature. Honeybees maintain the temperature inside their hives, and various species of African termites control the interior temperatures of their mounds, as described in Chapter 6. Probably the most striking and best-known case of absolute temperature sense is possessed by the Australian brush turkey (Megapodiidae, *Leopoa ocellata*), a bird that maintains a nest temperature at 33° C through the use of decomposing leaf litter—not its own body heat. The nest activities are performed by the male while the female goes on to lay more eggs in the nests of other males. The bird frequently tests the temperature inside the nest mound by taking samples of the material in its mouth. If the mound has become too warm, the bird kicks material away from it to let it cool, or if it is too cool, it kicks more material onto the nest.

Human sense for external temperature is only relative, as mentioned. Thus it is possible for different parts of the body to sense the same temperature differently. Bathwater, for example, commonly feels hotter to the cold feet than to other, warmer parts of the body. Likewise, it is disconcerting to have one's hands adapted to different temperatures, such as by holding one in warm water and one in cold, then pick up an object that is of an intermediate temperature; it will feel cold to one hand and hot to the other.

Temperature has been chosen only as an example. Whether a sense is relative or absolute can be applied to other sensory modalities such as sound as well. A few people have "perfect pitch" for sound; that is, they can determine the absolute tone of a sound without an external reference. Most persons, on the other hand, can only tell whether one pitch is higher or lower than another or if, in a series of tones, one is "out of tune."

Sensitivity may decrease over short periods of time by the sensory receptors adapting or by neural pathways deeper in the nervous system temporarily habituating to the information. A familiar example is known as *olfactory fatigue*. When a person first encounters an odor, it may be quite noticeable. After a few minutes, however, it no longer can be perceived.

Spatial or Temporal Pattern Discrimination

In many cases a particular mode, or form of energy, cannot only be sensed, but the animals may be capable of distinguishing spatial or temporal variation. *Spatial* pattern is of obvious and great importance to vision. It is so familiar to humans that we are likely to take it for granted. We may infer that spatial variation and pattern also are useful in sensing the Earth's magnetic field, particularly for birds and perhaps bats. Likewise, the ability to sense variation from place to place in other forms of energy is important for various species.

The ability to discriminate *temporal* variation also can be observed. Because of the time component, however, such discrimination is largely a consequence of processing at higher levels within the nervous system. Memory and learning may or may not be involved in the recognition of time sequences. (Recall that the ability to identify the patterns of bat calls by moths is a consequence of processing by combinations of interneurons.) Examples of temporal pattern recognition include the flashing patterns given and received by different species of fireflies, head-bobbing patterns in lizards, and the many song and call patterns of insects, birds, mammals, and many others. Striking examples of call patterns were shown earlier for crickets (p. 97). Humans are particularly well equipped for sensing and recognizing sound patterns in conjunction with very sensitive pitch discrimination. Humans commonly recognize, even after long intervals of time and only by brief excerpts, the voice of a friend or public personality or the strains of a familiar tune.

Ability to Localize the Source

There is a great difference between being able to detect the presence, absence, or level of whatever is being sensed and being able to identify the location of its source. Pinpointing the source requires ability to discriminate *direction* and *distance*. This is most easily and most quickly done, particularly if the distance involved is considerable, with vision. Light waves are extremely directional and obviously fast; thus direction is (relatively) easily determined if the animal has proprioceptive information (see p. 449) on the direction it is pointing its eyes. Distance can be measured in vision in a variety of ways, such as through parallax or accommodation.

Parallax is used by detecting differences in the angle of direction to a source as seen from different positions. The different positions may be by two eyes located side by side and facing forward (i.e., binocular or stereoscopic vision). This form of vision is familiar among primates and many other mammals and birds plus some others, generally among tree dwellers, flyers, or predators—to all of whom distance information is vital. Parallax also may be achieved by movements of the entire head or body, as in owls. *Accommodation* involves the detection of changes required to focus the image, which may be achieved either by changing the curvature of the lens or by moving the lens different focal distances from the retina.

Localization in other sensory modalities may be somewhat analogous to that in vision, using focusing and intensity-measuring devices, such as ears for sound. Cats, bats, and owls, for example, have binocular (or more properly, binaural) hearing. Some snakes have organs, pits, that crudely focus heat.

Many organisms operating with other senses may localize the source by testing for small differences in intensity, such as the strength of an odor or the amount of heat, in different locations. This often requires concomitant information on the surrounding water or air currents. Mosquitoes, body lice, and bedbugs can locate prey from a distance of several meters. Some lice, incidentally, also have an absolute temperature sense and have been used in primitive societies as a thermometer, in a sense. As long as a person had the normal complement of lice, he or she was healthy, but if a fever developed or the body temperature would go too low, the lice would leave that person and move to others (cited in Milne and Milne 1964:90).

BRIEF SURVEY OF THE SENSES

Mechanoreception

Mechanoreception refers to senses that depend primarily on physical contact (touch, tactile) and associated pressure, bending, or stretching. This includes many of the internal (and often subconscious) *proprioceptive* senses. Pain is considered frequently in this category; some of the surface skin receptors are considered as pain receptors. Pain, however, may be largely a function of higher integrative processing that comes about with *excessive* stimulation of almost all senses. Too much light or sound, for example, can create painful sensations just as surely as exposed and overstimulated skin sensors (as during cuts or other wounds). Odors, at least as sensed in humans, are not directly related to pain, but indirectly they may be very objectionable, perhaps leading to nausea, which in turn can be painful.

A sense of pain, incidentally, appears to be present in humans and many other vertebrates, including fish, but not in all other species. Hess (1973b) mentions

Figure 14-3 Section of human skin showing numerous sensory endings and nerve pathways. The various types of sense cells are sensitive to different effects such as heat, pressure, and pain. From Mountcastle, V.B., editor. 1980. Medical physiology, 14th ed. The C.V. Mosby Co., St. Louis. Courtesy Dr. M.E. Jobaley.

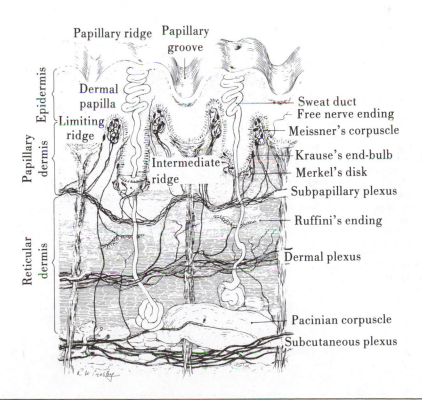

evidence for pain in invertebrates only in squid. Evidence that pain may not exist consists of observations that the animals *act* as vertebrates do when the pain senses have been blocked or otherwise anesthetized. Evidence that feelings of pain do exist in some, such as fish, includes the presence of substance P and endorphins, chemicals associated with pain and pleasure in vertebrates (Chapter 16).

A wide variety of touch, pressure, and other surface sensors is illustrated in Figure 14-3. The pectines have already been mentioned as examples of tactile senses. Touch also occurs with many devices that extend from the body surface such as in vibrissae, "whiskers" of many species (i.e., long, stiff hairlike structures about the mouth), antennae, and many other specialized projecting structures.

Mechanoreception is important not only at the surface but also internally. Most internal mechanoreceptors convey information on the positions and stresses of numerous body parts, such as the B cell in the noctuid moth ear. A few other examples include hinge organs in arthropods and muscle stretch receptors in vertebrates.

The two senses, inertial and vibrational, discussed in the next sections involve very advanced forms of mechanoreception. (All sensory receptors are advanced or specialized to some degree.) Inertial and vibrational sense organs could be considered as mechanoreceptors because they generally possess cilia that transduce the information through physical bending of the structures.

Inertial Senses

These senses usually operate in one of two general ways. The first involves a small mass, the statocyst, contained within the organ and which impinges on the cilia,

Figure 14-4 Semicircular canals and principles of fluid movement in different planes. A, Illustration of fluid movement with a jar of pickles. If the jar is tipped, the fluid moves with it. If the jar is rotated, however, the container turns, whereas the fluid stays in the same position. B, Depending on the plane of movement, fluid at a particular position (a, b, or c) will move with or away from a position on the canal surface (A, B, or C). Motion of the fluid relative to the canal creates a current or shearing force that is detected by ciliary sense cells located in the ampullae of the semicircular canals in the inner ear (see Figure 14-5).

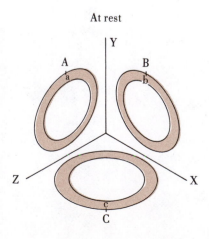

Rotation: container moves relative to fluid inside, creating a shearing motion inside

At rest

Tipping: contents move with container

A   B

depending on the pull of gravity or the forces of acceleration caused by movements. The second, often in conjunction with the first, operates through movements of fluid through circular tubes or canals, the semicircular canals found in vertebrates. These canals are placed in the three different dimensional planes. When movements are across the plane, the fluid moves with the canal, as in tipping a jar of pickles forward or backward. But when the movement is within the plane, the fluid remains stationary, as when the jar of pickles is rotated instead of tipped, and this creates a motion of the fluid against the walls of the container. Movements that are not exactly within any of the three primary planes are sensed from the vectored components of the contributing planes. The principle of this fluid movement is illustrated in Figure 14-4.

The effects of gravity and inertia also are detected via the weight of various body parts (e.g., through sensory hairs between the head and thorax in many insects) or even from the floating force of air bubbles trapped on the body of various small aquatic organisms.

Vibrational Senses

A number of senses that depend on airborne, aquatic, or other substrate vibrations, chiefly hearing, are also closely related to mechanoreception, as they depend on physical deformation of the sensory neurons for transduction and frequently use ciliary-type structures. We have already discussed hearing in the moth.

Hearing has been studied most extensively and is most familiar in vertebrates. In all vertebrate classes the sense of hearing depends on a portion of the inner ear, an outgrowth known as the lagena, or in its more highly developed forms, the cochlea, which is derived from and closely related to the system involving bal-

B, cont'd

ance. It includes fluid that transmits vibrations and hair cells, which detect the fluid movement and transduce the energy into nerve impulses. The inner ear, in both balance and hearing aspects, furthermore appears related to the lateral line system in fish in which clusters of hair cells are located in tubes or sunken channels in the skin of the fish. These sense movements and low-frequency vibrations in the water at the surface of the fish. In the most primitive fish the hair cells are distributed over the surface of the body rather than in the specialized lateral lines.

Sound vibrations may be picked up directly by the sensory cells in many or most fish (their bodies being similar in density to the surrounding medium, hence, relatively transparent to sound), or via an intervening system of mechanical detection, transmission, and modulation (usually amplification). Detection generally occurs at a membrane, such as the swim bladder in some fish or the tympanic membrane (eardrum) in terrestrial vertebrates. Transmission of vibrations to the inner ear and modulation are accomplished via one or more bones, the Weberian or auditory ossicles. These are derived from primitive gill arch or jawbone structures. In modern fish of the superorder Ostariophysi (including min-

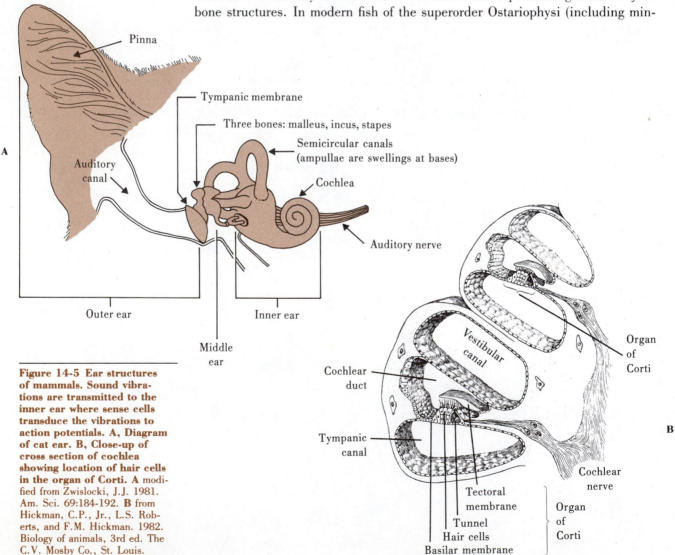

Figure 14-5 Ear structures of mammals. Sound vibrations are transmitted to the inner ear where sense cells transduce the vibrations to action potentials. A, Diagram of cat ear. B, Close-up of cross section of cochlea showing location of hair cells in the organ of Corti. A modified from Zwislocki, J.J. 1981. Am. Sci. 69:184-192. **B** from Hickman, C.P., Jr., L.S. Roberts, and F.M. Hickman. 1982. Biology of animals, 3rd ed. The C.V. Mosby Co., St. Louis.

nows, goldfish, carp, and others) this series of bones connects between the swim bladder and the inner ear.

In amphibians there is a transition during larval stages from one bone, the branchial columella, which connects between the inner ear and a part of the lung sac to the tympanic columella, which connects between the adult eardrum and the inner ear. The columellar structure in reptiles generally consists of two columellae connecting between a tympanic membrane and the inner ear. Snakes have only one columella, which is embedded in tissue and connects to the quadrate bone. Snakes lack an external ear, tympanic membrane, and middle ear. Birds have a single columella connected to the eardrum and an external ear, which is hidden under the feathers at the side of the head. Owls have asymmetrical ears, one being larger than the other, which aids their ability to precisely locate sound sources in the dark. Mammals have three auditory ossicles and perhaps the most advanced vertebrate hearing sense (Figure 14-5).

Bats are not the only mammals that can hear high, ultrasonic pitches. At least 23 species are known to be capable of hearing above the 20 kHz upper limit of humans (Hess 1973b). Several, including chimpanzees, hear up into the range around 30 kHz, and many small mammals, including bats, mice, and shrews, hear up to the range of 90 to 120 kHz. Porpoises and seals may produce and hear underwater sounds up to around 180 kHz, although seals apparently have an upper limit for airborne sounds around 22 kHz (Mohl 1968). Sound travels nearly five times faster in water than in air, and because of this and other differences, mammalian hearing may be different for water and air, and upper-frequency limits in the two media may not be strictly comparable.

Figure 14-6 Hearing organs of insects. Insects use a variety of structures for hearing including pressure chambers with a tympanic membrane, movement receivers such as hairs or antennae that extend from the body and vibrate with sound waves, or combinations of pressure chambers and projecting structures. A, Cricket with its ears on its legs just below the knee. In the cricket, incidentally, a tracheal tube connects between the ears on the two sides of the body, permitting sound to travel from one side to the other. B, Pressure-sensitive mouthpart palps for hearing in hawkmoths (Sphingidae). The moth's right palp *(left arrow)* is intact, whereas the other *(right arrow)* is broken open to show air chamber. Eyes are to sides, and tongue is in center. A heavy layer of facial scales (hairlike) had to be removed to view the palps. C, Vibration-sensitive antennae on the face of fruit fly. D, Projecting rocking club structure *(light object at point of arrow)* connected at its base to the tympanum of the ear on the thorax of a water boatman. NOTE: This structure is best seen on fresh specimens; it commonly is damaged during drying, as in collections. For further information on insect hearing, see Michelsen (1979). Photos by James W. Grier.

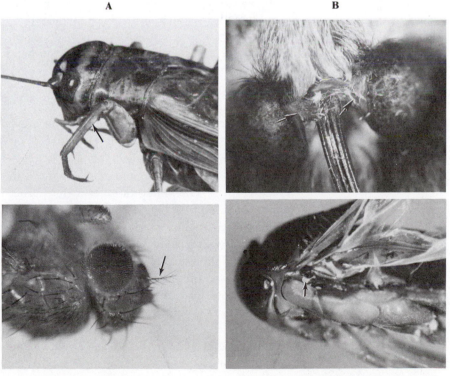

Persons wanting further details on the mechanics and interspecific differences of vertebrate hearing should refer elsewhere, such as to descriptions commonly found in introductory biology, zoology, anatomy, and physiology texts. Also, see Hess (1973b). Popper and Coombs (1980) provide an extensive review of hearing in fish.

Hearing in invertebrates is usually claimed only for arthropods, particularly among insects where several different mechanisms of pressure-sensitive chambers and vibration-sensitive hairs, antennae, and other body extensions are found (Figure 14-6). The ear of the noctuid moth is discussed in Chapter 1. Hearing is also likely in several crustaceans and spiders, some of which produce sounds believed to be used in communication.

From the point of advanced, specialized hearing structures, sensing of vibrations appears to merge with a more general sense of touch and vibration among both vertebrates and invertebrates. The more general sense is mediated usually through body mechanoreceptors but may include specialized detectors on the limbs, between body parts, and on various hairs, antennae, and vibrissae. Many organisms, including lower invertebrates, are sensitive to low-frequency vibrations in or at the surface of the ground or water. Such vibrations are frequently used for detecting prey, warning of danger, or in communication. Vibrations in the substrate are used, for example, by water striders, many spiders, coelenterates and molluscs, and probably most underground dwellers, including earthworms, snakes, and burrow-dwelling mammals.

Vision

Vision also depends on vibrational energy and might seem as though it should be considered under other vibrational senses. The vibrations have a much shorter wavelength, however, and the transductional process is very different. The physics largely are different; the detection is different; and we now are dealing with a substantially different sense.

Vision is based on light, a very narrow band in the electromagnetic spectrum (Figure 14-7). Light is one of only two general types of electromagnetic energy that reach the surface of Earth from outer space. The other form involves radio waves. All other wavelengths are filtered out by the Earth's atmosphere (Figure 14-7). Hence, it probably is not a coincidence that biological organisms tuned into light wavelengths during the course of evolution.

Light waves share many of the same properties with sound and other waves and can be described with some of the same general equations and mathematical terms (p. 9). The chief differences (for purposes here) are that the wavelengths are *much* shorter; hence, the frequency is much higher, and the speed is very much higher. The wavelengths are so small, less than a thousandth of a millimeter, that rather than interacting with and vibrating whole structures (as with sound), light interacts with matter at the molecular level. Biological light sensors thus must contain light-absorbing, light-sensitive pigments or photochemicals.

The speed varies slightly as it passes through or encounters various materials, which leads to refraction and reflection. The speed is so high that, for distances encountered on Earth, it may be considered as instantaneous for all practical purposes. Thus an animal using a light-sensing mode has access to information about distant events essentially as they happen. Information from other sources,

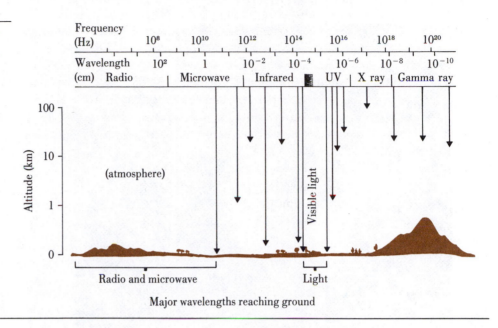

Figure 14-7 The electromagnetic spectrum, illustrating those wavelengths which penetrate Earth's atmosphere. Light is only a very narrow part of the total spectrum. Different colors represent different wavelengths of light within the narrow band of light.

such as sounds or, particularly, chemicals, may involve delays of time. Light also has the advantages of being easily focused (if the proper structures are available), highly directional, and in ready supply in most places during the daylight period. The chief disadvantages of light are that it is easily blocked (in fact most of the entire major source, the sun, is cut off every night), and it is more difficult for animals to produce their own supply than for other forms such as sounds and odors.

The sensitivity of different organisms to light and the types of structures they possess for sensing it vary considerably. Even all of the animals that use vision do not see the same thing. Most animals are sensitive to light in one manner or another. The protozoans possess photochemicals or light-sensitive organelles and have a diffuse light sense. *Euglena*, a flagellate protozoan, has 40 to 50 orange-red granules in the stigma or "eyespot" near the flagellum. When exposed to light, it swells and affects the direction of the flagellar beat. Coelenterates, annelids, and most of the invertebrates, in fact, have at least light-sensitive cells located somewhere on or in their bodies. Molluscs, arthropods, and vertebrates possess a variety of eye structures (Figures 14-8 and 14-9), some quite advanced and with image-focusing lenses.

Color Vision
The range of wavelengths that humans call light can be broken into arbitrary segments identified as colors. The range goes from infrared at one end to ultraviolet at the other. Depending on the structures and pigments they possess, different species are sensitive to different segments of the spectrum, just as animals vary in which frequencies of sound they can hear. Note that humans can see into the red end, although not into infrared, whereas many species, particularly nocturnal ones, cannot see into red. This is the basis for using red lights at zoos or in behavioral research when one wishes to view nocturnal animals without

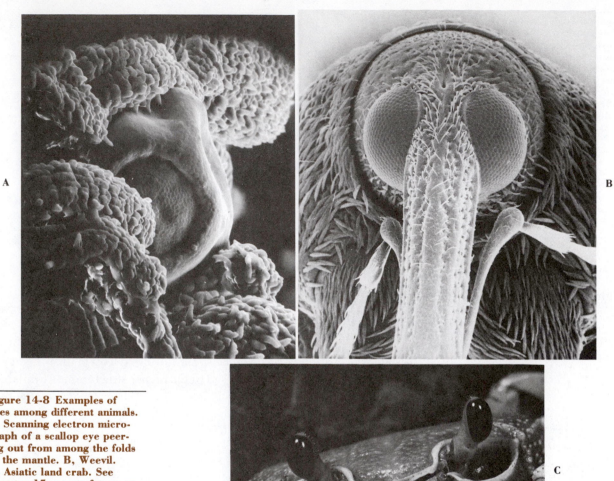

Figure 14-8 Examples of eyes among different animals. A, Scanning electron micrograph of a scallop eye peering out from among the folds of the mantle. B, Weevil. C, Asiatic land crab. See Chapter 15 opener for an example of vertebrate eye—the large raised eyes of bullfrog typical of amphibians. A by P.P.C. Graziadei. **B** courtesy North Dakota State University Electron Microscope Laboratory. **C** from Animals, Animals. Copyright by Zig Leszczynski.

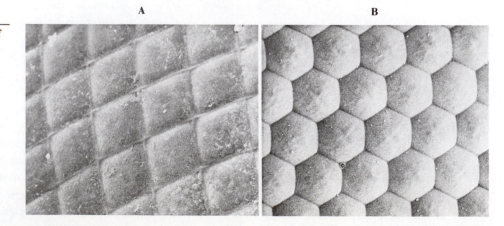

Figure 14-9 Close-up view of surface of arthropod compound eyes. A, Square ommatidia of crayfish eye. B, Hexagonal ommatidia of noctuid moth. Photos courtesy North Dakota State University Electron Microscope Laboratory.

exposing them to "light." Because many animals cannot see the red light, they would sense the surroundings as dark. This also is the reason that red or yellow "bug lights" do not attract night-flying insects.

Even human eyes, however, are not very sensitive to red light. At night, when one needs to see in dim light, it is useful to use red lights for such things as instrument panels, passage ways, and night charts. Shorter wavelengths, even at relatively low intensities, eliminate the dark adaptation and leave humans unable to see in the dark.

Figure 14-10 Flower appearances under visible (*left frames*) **and ultraviolet** (*right frames*) **light. These appearances are as they would appear to a vertebrate eye if one could see into the ultraviolet. Insects can see ultraviolet, but their eye structure does not permit the detailed vision that humans possess. For a suggestion of how the flowers might look to an insect, see Figure 14-11. The dark centers of the flowers in ultraviolet are thought to serve as targets or guides for foraging insects.** From Silberglied, R.E. 1979. Ann. Rev. Ecol. Syst. 10:373-398. Reproduced, with permission, from the Annual Review of Ecology and Systematics, Volume 10, © 1979 by Annual Reviews, Inc. Photos by Robert E. Silberglied, Annette Aiello, and Thomas Eisner.

As seen through vertebrate eye

As seen through insect compound eye

Figure 14-11 A graphic concept of how a flower might appear through the compound eye of an insect. This is only a crude, speculative representation, however, as humans do not yet know how the perception might be processed and "appear" in the insect's CNS.

Red apparently represents the limit for photochemical visual processes, or at least it is as far as evolution has carried vision in that direction, although sunlight and artificial incandescent lights carry much energy in infrared wavelengths. Species that detect infrared do so, for the most part, through other, thermal mechanisms rather than with visual pigments.

While human vision seems to lean toward the red end of the spectrum and does not extend very far into the blue, many species, such as some arthropods, are sensitive to ultraviolet. They have entirely different eyes from ours and do not focus and sense patterns in the same manner that we do; thus it is difficult to compare human vision with theirs. If human wavelength sensitivity were shifted toward the range of a bee's, flowers, for example, would appear to have different patterns than we normally see (Figure 14-10). Because of the different structure of compound insect eyes, using ommatidia, the pattern of the vision also would be quite different (Figure 14-11). In addition to these major differences, insect and vertebrate eyes also differ greatly in terms of acuity, as will be discussed further later.

The preceding comments on sensitivity to various wavelengths apply whether an animal is aware of *which* wavelength it is seeing or not. To be able to **discriminate** color is another matter—it requires the presence of *more than one* visual pigment and the neural ability to keep information from the multiple channels straight. Most species capable of sensing color have three, or occasionally more, visual pigments and associated sensory neurons. If different pigments with different sensitivities to particular wavelengths are present simultaneously, then this provides the animal with independent measures on the characteristics of the light; that is, different colors can be distinguished. Figure 14-12 illustrates the three-pigment visual sensitivities for the goldfish.

Whether or not other species can distinguish color has been a question of perpetual interest to the lay public and many biologists alike. Frisch (1914) was one of the first to apply training techniques to resolve the question with bees. Since that time, training methods, the optokinetic nystagmus response (inside the rotating cylinder), and even some evoked potential recordings have been applied to a wide variety of animals to determine whether they have color vision.

Color vision has been demonstrated in bees and a wide variety of other, but not all, insects, cephalopods, many fish, many amphibians, diurnal but not nocturnal reptiles and birds, and some mammals including primates, squirrels (but not many other rodents), many carnivores, and others that have been tested such as pigs and horses. Color vision as in other aspects of senses, however, varies, and it is not simply a question of possessing it or not possessing it. Color vision may be possessed or used to a variable degree. Several primates, including some individual humans, have weak or missing ability to discriminate red. Cats, although demonstrating an ability to discriminate colors, nonetheless show much better brightness and pattern discrimination. It is clear that in vision and not just color vision not even all mammals have the same view. Furthermore, it appears to vary even among individuals of the same species.

Night Vision

Although the sun's light may be largely cut off at night, some is reflected back onto the dark side of Earth by the moon. Some light also reaches Earth at night

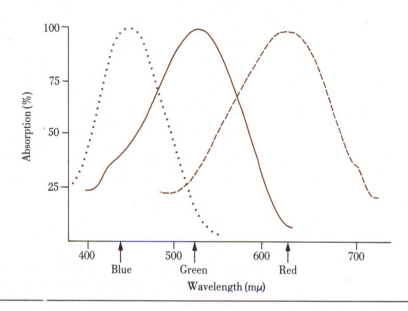

Figure 14-12 Three-point cone sensitivity in goldfish, a species believed capable of color vision. Color sensation is believed to result from *combinations* of input from these three sensors. Yellow, which is between green and red, for example, would be sensed by moderate stimulation of green cones plus moderate stimulation of red cones. Modified from MacNichol, E.F., Jr. 1964. Sci. Am. 211(6):48-56.

from stars other than the sun. During recent years the level of light at night has increased severalfold from artificial sources, particularly near metropolitan areas. The increases have been too recent, however, to have been taken advantage of by evolution, as far as is known. And even with artificial sources, the general level of illumination at night is several orders of magnitude less than during the day. But few places on Earth are totally dark, and there are many biological gimmicks to use the faint remaining light.

By essentially using two different sets of visual pigments and sensory neurons, according to the so-called duplicative theory, the eye can switch between high and low light levels. The basic cell types are *rods,* the most sensitive and used primarily at night, and *cones,* the least sensitive and used during the day and for color vision. Most invertebrates that rely heavily on vision apparently have only one set and thus are active at either day or night, but not both. Vertebrates, however, may have populations of both cell types and, in some instances, may be active at either or both times. Most vertebrates have both types, but the proportions vary; diurnal species have cones predominantly, whereas nocturnal species have a much higher proportion of rods. Persons interested in the chemistry of vision should consult a physiology text or the recent review by O'Brien (1982).

One of the most interesting techniques for detecting light at extremely low levels amounts to recycling it. After the light enters the eye and has passed through the sensory cells, a reflective surface at the back sends it back through the sensory cells for a second pass. This is the reason for the shine that can be observed in the eyes of many nocturnal animals, including deer, cats, and some moths. The reflective surface in the eyes of vertebrates that possess it is called the *tapetum lucidum.* Diurnal species, on the other hand, may be faced with the opposite problem—too much light in many circumstances. In these species a reflective surface would be unnecessary and even could create serious problems by flooding the inside of the eye with too much light. Accordingly, the back surface of the eyes of chiefly diurnal species have become black, which absorbs the light.

Polarized Light Detection

A few species, primarily among arthropods, are able to detect and use information in the plane of polarization of sunlight. Among vertebrates the ability apparently amounts not so much to the sensing as the screening of light via polarization, much as with polarized sunglasses. This has been suggested (cited in Hess 1973b without reference) for herons as a means of reducing glare from the surface of water. It has also been suggested, but not proven, as a possible aid for orientation in some long-distance movements by some vertebrates. Kreithen and Keeton (1974) demonstrated that pigeons are sensitive to polarized light.

In insects the ability to detect the plane of polarization appears to function in orientation. In the case of bees, it permits them to orient with respect to the sun when it is obscured by clouds, as long as at least one patch of cloudless sky remains visible. Bees can infer the position of the sun because the angle of polarization varies in a systematic fashion with the angle from the sun. Desert ants, which live among constantly shifting sands and which thus lack stable landmarks, also use polarized light for orientation, as was introduced in Chapter 7. It has been shown that polarized light is important only at a restricted segment of the ant's eyes. The reception of polarized light is permitted by the unique molecular configuration of the insect eye (Figure 14-21).

Acuity and Form Vision

Many species have only a diffuse sense of light detection, as discussed previously. With the ability to focus light and include a large number of sensory neurons in different positions somewhat like the dots that compose a television picture, it is possible to discriminate form, pattern, and other characteristics of the picture. Acuity or resolution refers to the smallest separation of points that can be distinguished. It is usually measured by the angle between the point of observation and two points being observed. A 25-cent piece held at arm's length, about 0.7 m, for example, provides an angle of over 2 degrees across its diameter. The diameter of the full moon as viewed from Earth has a visual angle of only 0.5 degree or 30 minutes. Acuity depends on the focusing quality of the eye, the number of receptors packed into the receptive field, and the ability of the nervous system to handle the large amount of information being received. The limit of human visual acuity (around 1 minute of angle) is very nearly the apparent angle between two readily located stars, Mizar and Alcor. Mizar is the bright star in the middle of the handle of the Big Dipper. Alcor is a faint star very close to it, and the two form an asterism (a pattern smaller than a constellation) known as the Horse-and-Rider. The ability to distinguish these two stars by naked eye on a clear night is a good test of one's vision.

The limit of a honeybee's visual acuity, in contrast to a human's, is only about 3 degrees. A bee could resolve neither a quarter at 0.7 m nor the appearance of the full moon. The insect eye is limited by the structure of the ommatidia, or individual elements of the eye.

If an animal has reasonable acuity and a good field of receptive neurons, it may be capable of discriminating shape, form, patterns, and movement within the field of vision. This demands not only a large number of sensory neurons but also a good neural processing capability. The retina in the vertebrate eye develops as an outpocket from the embryonic brain and amounts to an extension of the brain itself. It is a complex structure, although extremely thin and about as fragile as a spider's web. It consists of three general layers: sensory cells, bipolar neurons, and ganglion cells. Processing of visual information begins here and is continued at higher levels within the CNS (Chapter 15).

Infrared and Heat Detection

Heat and infrared sensing are similar to vision in that they are widespread among animals and vary from a generally dispersed, diffuse sense in many animals to an advanced, highly developed sense in others. Heat and light sensing differ, however, in two important aspects. First, advanced forms of sensing are quite common for light but are fairly uncommon for heat. Although most animals probably have some temperature sense, it is almost universally of the diffuse sort, like that of humans. Only a few specialized forms have the advanced capabilities and highly specialized organs. Species or groups that are so specialized include the megapodes, a number of arthropods that are parasitic on warm-blooded vertebrates, and two families of snakes (Crotalidae, pit vipers, and Boidae, boas and pythons). Night moths have infrared receptors by which the males apparently locate the females at close range, after first using olfactory modalities.

The second major difference between heat and light sensing involves the transduction process. Heat detection, rather than employing photochemical molecular changes, uses the thermal properties of molecular motion in some manner that is not well understood. Some sensors, such as in the night moths, snakes, and many of the arthropods, use radiated heat that is sensed from a distance. Others,

Figure 14-13 Infrared heat sensors in snakes. A, Rattlesnake showing the infrared detecting pits in front of the eyes and beneath the nostrils. B, Infrared "shadows" cast in the pits of a pit viper, depending on how the head is oriented toward the heat source. A by James W. Grier. B modified from Gamow, R.I., and J.F. Harris. 1973. Sci. Am. 288(5):94-100.

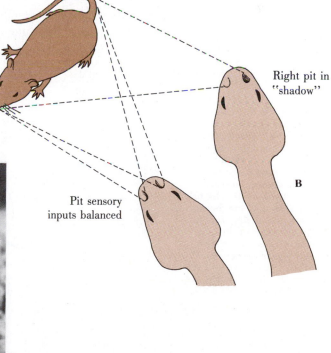

Right pit in "shadow"

Pit sensory inputs balanced

B

A

however, rely on conduction through contact and the associated heat exchange at the surface. The latter include fish and other aquatic organisms, humans, and probably most organisms with only a diffuse sense of temperature.

The means by which snake pits localize a heat source are shown in Figure 14-13. Direction is determined through the differences in maximum and minimum areas (shadows) of stimulation. Distance is sensed from parallax, as in binocular vision. For further information on infrared sensing in snakes, see Gamow and Harris (1973).

Electric Senses

One of the most fascinating senses involves the ability to sense electrical fields. Most organisms are unable to do so, except perhaps for the vague uncomfortable feeling that is detected indirectly by other sensors in a strong field. But a number of specialized electric fish, including many that commonly are sold in pet shops, are able to produce their own electrical fields and then sense changes that occur in the fields because of nearby objects. A few species, electric skates, rays, and eels, can generate enough electricity to use it in defense and for stunning prey. Most species, however, appear to use it merely for sensing their environment and for communication and interaction with others of the same species.

That these fish can detect differences in electrical fields and conductivity of objects has been demonstrated by two-choice tests in which the animals were trained to choose, for example, a conductor over a nonconductor. Figure 14-14 illustrates the difference in the electrical fields.

Magnetic Senses

Perhaps even more amazing than the ability to detect electrical fields is the sensing of Earth's magnetic field by many invertebrates, birds, and perhaps some other species. One of the reasons that humans find electric and magnetic senses so intriguing is that we do not, as far as is known, possess them. Humans are

Figure 14-14 Electric senses. Electrical field surrounding a fish, showing perturbations caused by conducting or insulative objects. A rigid body and undulating fins minimize the fish's disturbance of its own electrical field. The fish shown here is of the genus *Sternopygus* (order, Gymnotiformes). Several fishes, including sharks and skates, possess electrical senses. Modified from Lissman, H.W. 1963. Sci. Am 208(3):50-59, and Matsubara, J.A. 1981. Science 211:722-725.

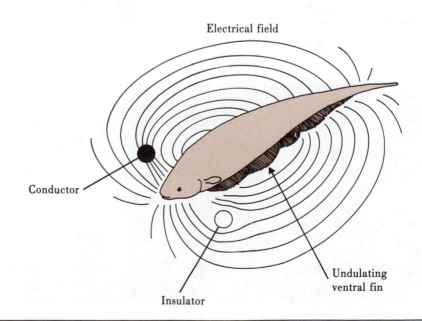

Electrical field

Conductor

Insulator

Undulating ventral fin

unfamiliar with magnetism except through artificial means such as compasses and electronic equipment. Because we lack these senses, it is easy to assume that other animals also lack them. After it was first suggested that other species might be sensing Earth's magnetic field, much time and some very convincing evidence were required before most biologists would accept the idea. But the evidence is now very strong, and, furthermore, there appears to be some progress in locating the possible sensors. Magnetic material has been discovered in the bodies of many organisms including bacteria (Frankel et al. 1979), bees (Gould et al. 1978), pigeons (Walcott et al. 1979), and dolphins (Zoeger et al. 1981). The usefulness of magnetic orientation in some organisms is not very clear. In some African termite mounds the orientation may function for a proper orientation of their nests to the sun for temperature regulation. For the birds, bees, dolphins, and others, however, the value of a magnetic sense is obvious—it appears to form or could form a very important component in their arsenal of orientation and navigational mechanisms (Chapter 7).

X-ray and Radio Wave Sensing

Many forms of energy, including the vast majority of the electromagnetic spectrum above and below the region of light and heat, appear *not* to be used by any Earth-based biological organisms. Several species commonly used in psychological laboratories, including rats, mice, cats, rhesus monkeys, and pigeons, have been shown to respond behaviorally to low levels of ionizing radiation (e.g., Haley and Snider 1964). But the mechanisms involved are not well understood and probably are indirect, related to nausea and taste aversion—like effects (discussed in Chapter 19) or some unknown effects related to sleep arousal in the brain.

Radio waves, while reaching the surface of Earth through our atmosphere, appear not to have been used naturally by any living organisms on Earth. The only, but substantial, use has involved the artificial production and reception by humans. The physics of radio waves, their wavelengths, energy needed for production and modulation, and rather complex circuitry needed for processing do not seem to make radio waves very accessible to biological systems.

Chemical Senses

The chemical senses, including taste and olfaction, are perhaps the most universal among animals. For many species almost their entire *Umwelt* is based on chemical senses. Various chemicals are vital to life itself, and it may well be that there is not a single animal that does not have at least some chemical sense. Even the simplest, one-celled protozoans must select the molecular food substances they take into their bodies and the microclimates of the chemical environments in which they exist. The sensing of various molecules is used in the detection and attraction to prey or other food and the avoidance of noxious environments and substances. Chemicals commonly are used in communication, for example, as sex attractants, territory markers, and as warnings or repellents.

As a biochemical instrument for the analysis and recognition of molecules, even the human nose is in many cases much better than the most sensitive and sophisticated of man-made instruments, although many chemicals are dangerous to health or may damage the nose. Yet our chemical sense appears to be as crude and insensitive compared with those of some species as their sense of vision or hearing is to ours. Our own poor chemical senses have hampered our views and research on the sensory capabilities of other species, and, until recent years, humans commonly assumed that other animals used chemical senses much like ourselves.

But many other animals differ significantly from humans in not only their sensitivities to chemicals in general but also in terms of the specialization of their sensory organs, location of the organs, and the evolved specificity as to which particular chemicals or classes of chemicals can be sensed. The simple four-class taste system (salt, bitter, sour, sweet) of humans, for example, certainly does not apply to all animals. Many aquatic organisms and some arthropods have receptors

Figure 14-15 Identification of chemical cues by combinations of sensory input. A, Firing rates of four olfactory receptors in the tobacco hornworm larva *(Manduca sexta)* in the presence of 11 different odors. **B,** Examples of simultaneous sensory input from the four sensors in response to different odors. Combinations of input from just these four permit the distinction of many unique odors based on their total sensory "complexion." Modified from Dethier, V.G. 1971. Am. Sci. 59:706-715.

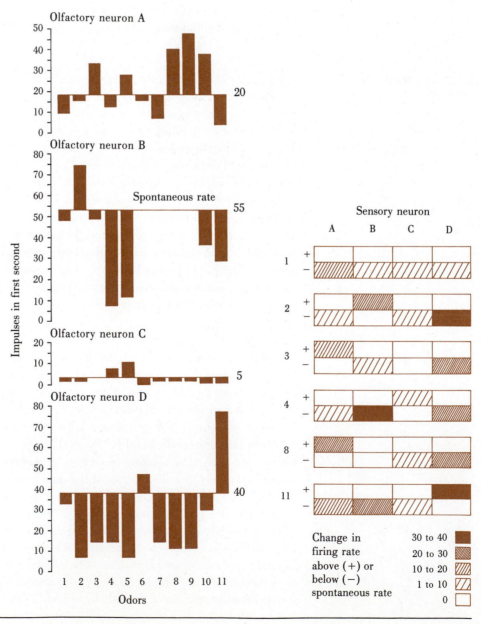

all over the body surface or on legs, antennae, and other structures away from the mouth and "nose."

The mechanism of odor and taste transduction is quite different from all other modalities discussed so far. It depends on molecular configuration and characteristics such as charge. Because of the chemical nature of the sense, the sensory organs must have a moist surface for solubility. The surface area also is extremely important. The operation of chemical senses involves molecular movement and the probability of the particular sensory neurons encountering molecules of the proper characteristics for stimulation. The greater the evolved match between the sensor and the molecule and the larger the surface area, or number of sensors, the greater will be the chance that the molecules will be detected. The chances also increase with greater numbers of molecules, that is, the concentration, that is present.

Patterns of smells, that is, specific odors that permit the identification of a large number of different substances or sources, such as the individual odors of different persons, are discerned in a manner similar to pattern discrimination in other modalities. Many different sensors are present; they operate simultaneously and report their results independently to the CNS. The total picture is composed of the interpreted composite obtained from *combinations of input*. Most smells consist of a complex of many different molecules, and their specific identity depends on the proportions of sensors of different types. Thus just as many different sentences can be composed of different combinations of only a few letters, many different smells can be discerned from the combinations of a few basic types of molecules (Figure 14-15).

One of the best-studied chemical sensors is that of the silkworm moth *Bombyx mori* (Schneider 1974). Bombykol is a sex-attractant pheromone that attracts males to females (Figure 14-16) solely over other sensory modalities. Males move to and attempt to copulate with any object emitting bombykol, regardless of sight, sound, or touch, and they ignore females from which they cannot detect bombykol. Bombykol is an unsaturated fatty alcohol, *trans*-10-*cis*-12-hexadecadien-1-*ol*. Furthermore, only the *trans-cis* isomer is very effective. Other geometric isomers of the same compound stimulate males only weakly. Unlike many olfactory systems where a variety of sensors in combination discriminate a variety of odors, the bombykol system uses large numbers of sensors to respond identically to only a single kind of molecule. It is a finely tuned system indeed. The

Figure 14-16 Male silkworm moth *(Bombyx mori)* responding to olfactory but not visual stimuli from a female. The female is releasing the sex-attractant pheromone bombykol. Modified from Schneider, D. 1974. Sci. Am. 231(1):28-35.

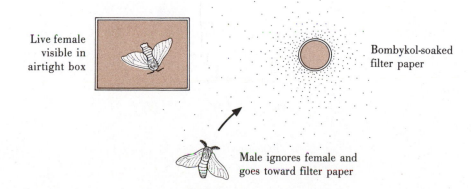

Live female visible in airtight box

Bombykol-soaked filter paper

Male ignores female and goes toward filter paper

sensory cells are located on the male antennae (and only on the male) and respond proportionally to increasing concentrations of the pheromone as shown in Figure 14-17.

EMITTED-ENERGY SENSES

The source of energy for the sensory systems of many species of animals is the animal itself. The individual emits the energy by which it then senses the environment or a component of the environment. Examples include the bats, which emit cries and then listen for the echo, and electric fish, which generate an electrical field, then detect changes resulting from objects within the field. Other examples of animals that use sonar include many of the marine mammals and the oilbird (Figure 14-18).

The use of sound underwater presents some interesting problems and opportunities. Sound does not reflect well until it encounters a medium that differs in density from the one in which it is traveling. Much biological tissue is largely composed of water and, hence, would transmit rather than reflect sound underwater; such tissue would be largely or partially "invisible" to sonar. The sonar echoes that reflect off many fish probably reflect from their internal air bladders and perhaps bones rather than from their outer body surface. This would give sonar-using animals such as dolphins something like "x-ray vision." They might even be able to "see" (hear the echoes of) such things as air bubbles in the digestive tracts of their young!

Some aquatic insects use water surface waves. Whirligig beetles (*Gyrinus* sp.) whirl rapidly and then detect changes in the returning wavelets. The sensory cells that are used by the beetles are in fine hairs at the second joint from the base of the antennae (reviewed in Griffin 1959). They may be able to detect waves as small as 4×10^{-7} mm, less than a millionth of a millimeter, and the beetles are unable to orient or avoid obstacles when these particular hairs are removed. The

Figure 14-17 Increased magnification of the chemical sensory organs on the antenna of the silkworm moth *(Bombyx mori)* and neural output in response to varying concentrations of the sexual attractant bombykol. Redrawn from The sex-attractant receptor of moths by Dietrich Schneider. Copyright © 1974 by Scientific American, Inc. All rights reserved. Recordings by E. Priesner.

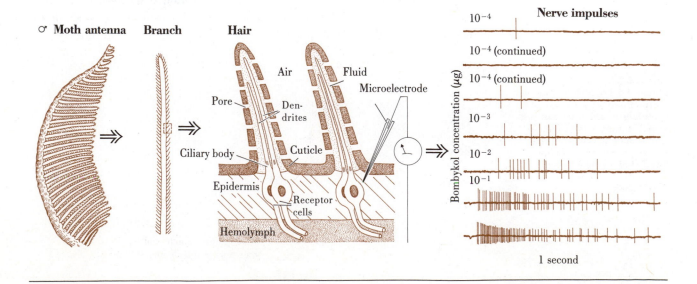

Figure 14-18 Oilbird *(Steatornis caripensis)* using echolocation to avoid an obstacle placed in its path. The bird was captured with this flash photograph just as it came to a momentary stop and was hovering in midair inside a dark cave. From Konishi, M., and E.I. Knudsen. 1979. Science 204:426. Copyright 1979 by the American Association for the Advancement of Science. Photo courtesy Eric Knudsen.

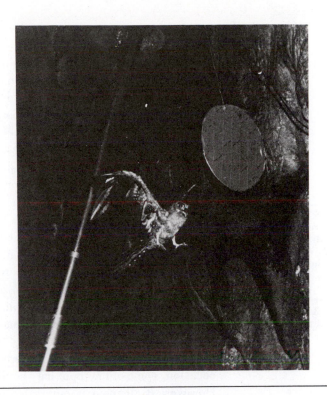

beetles whirl intermittently, apparently waiting for the return of waves echoing off obstacles, each other, and other objects, between periods of whirling.

A few deep-sea fish and perhaps some other marine organisms produce their own light by which they see (Figure 14-19). They may have exceptionally sensitive and specialized eyes.

A number of organisms produce light but, apparently, do not use it for their own sensory purposes. The function, if any, of the luminescence in many of the marine organisms is not well understood, although there are several hypotheses. In other cases, such as the fireflies and other deep-sea marine animals, the luminescence clearly serves a communication function (Chapter 10).

Figure 14-19 Use of emitted light by the flashlight fish *(Photoblepharon palpebratus).* It has been postulated that the light functions to obtain prey, deter predators, communicate intraspecifically, and aid general seeing. Fish has luminescent organ lid open. From Alex Kerstitch/ Tom Stack and Associates.

COMBINATIONS OF SENSORY INPUT

Very few species rely solely on a single sensory mode for any given behavior or function. Even as simple a situation as bodily orientation, knowing which end is up, usually involves input from more than one sensory source. Body orientation in most vertebrates uses the balance organ, or semicircular canals and associated structures, but only in part. Proprioceptive input from the weight of different body parts, the stretch of various muscles, and vision also play important parts. "Up" for fish, for example, usually involves the brightest part of their environment because of daylight entering the water from the surface. If, as in an aquarium, illumination is brightest at the side, then many fish will tip to orient their dorsal surfaces to the brightest light level. The tipping is rarely complete, however, but rather at an angle that vectors, or compromises, the input from vision and that from the semicircular canals (Figure 14-20). Likewise, body orientation in humans depends in part on the perceived horizon. If the horizon is gradually changed so that the change is not detected, as in so-called gravity houses at amusement parks, our sense of orientation may be altered and, for example, water may appear to run uphill. Combinations of sensory input are involved in most behaviors of sensory interest.

With combinations of input, there is a certain amount of redundancy, and on occasion, some of the input may be dispensed of without noticeable impairment. However, if all input is eliminated or if the system places high priority on an important sense that is lost, serious problems may develop. Confusion with balance, or dizziness, can leave one with the sensation of being stationary but the rest of the world is spinning even when it can be seen that it is not. In extreme cases, all sense of orientation is lost and *vertigo* develops, a very uncomfortable and alarming state in which to be. This may occur in flight when clouds, snow, or similar obstructions obscure the horizon and turning motions interfere with the inertial senses. Jet pilots on complex turns and maneuvers sometimes have to learn to live with vertigo and trust only their instruments for long periods of time. Vertigo also develops sometimes with damaged or diseased nerves from the inner ear, and it also may be triggered by extreme nausea or emotional stress.

All of this, again, gets into higher levels of neural integration. It should be emphasized at this point, however, that sensory input from different modalities may be weighted in importance. There generally is a hierarchy, as was seen in bird orientation mechanisms. If high-priority inputs are available, even if they are incorrect, lower-priority inputs may be ignored. Furthermore, the hierarchies of input may switch within the same individual under different circumstances. A sense that is used heavily in one behavior may be totally ignored during the course of a different behavioral situation.

Figure 14-20 Fish vertical orientation in respect to the combination of input from light (normally coming only from above) and gravity pulling from below. A, Fish in normal upright position with light from above. B, Fish tipped in a vectored response to light from the side and gravity from below.

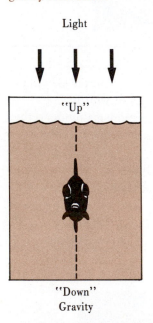

Light

"Up"

"Down"
Gravity

"Up"
Light

"Down"
Gravity

MECHANISMS OF TRANSDUCTION

The means by which external energy is changed into action potentials by sensory neurons or sense cells are not well understood for any of the sensory modalities. In the pacinian corpuscle, a pressure receptor in the skin of vertebrates, the transduction is primarily associated with a change in sodium permeability from mechanical changes of the membrane (Gray 1959, Loewenstein 1971). In cells

Figure 14-21 Structure of visual sense cells of different animals. Light is sensed via a molecular interaction with light waves occurring in highly specialized, pleated membranes within the sense cell. The molecular arrangement in some organisms, such as insects, but not other organisms permits the detection of polarized light. The sensitivity of cells to particular wavelengths, or colors, in different cones depends on the particular chemicals the cells possess. A, Section of photosensitive part of rod and cone. B and C, Visual cells of vertebrate and insect (*arrows indicate orientation of molecules*). D, Close-up cross section of vertebrate rod.

A redrawn from Brown, K.T. 1980. Physiology of the retina. In Mountcastle, V.B., editor. Medical physiology, 14th ed. The C.V. Mosby Co., St. Louis. **B** and **C** from Polarized-light navigation by insects by Rüdiger Wehner. Copyright © 1976 by Scientific American, Inc. All rights reserved. **D** reproduced from De Robertis, E., *The Journal of General Physiology*, 1960, vol. 43, pp. 1-13 by copyright permission of the Rockefeller University Press.

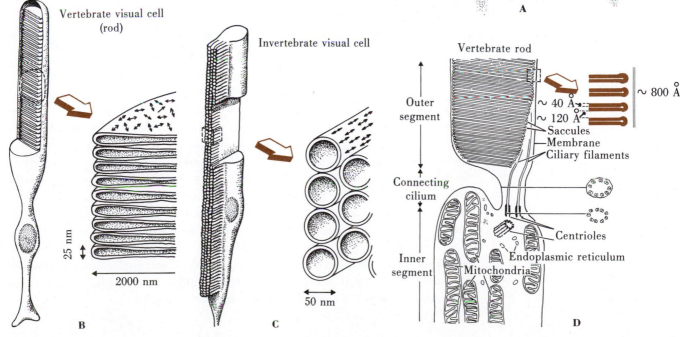

with cilia, electrical changes correlate with the beating of the cilia. Horridge (1965) has suggested that depolarization triggers the beating. Barber (1968) hypothesizes that the reverse might happen, that is, that the passive bending of the cilia might initiate depolarization through the mechanical deformation of the membrane in the surrounding area. A mechanical membrane effect might be expected in many of the mechanoreceptors and closely related cilia-bearing cells, including statocyst and auditory hair cells, touch, stretch receptors, and many relatively unspecialized cells with free or bare endings.

The structures and molecular physiology of vision have been studied intensively for many years. The vision sense cells of complex eyes generally contain complex folded molecular structures (Figure 14-21). In some organisms, such as insects, the structure may be sensitive to polarized light as discussed earlier. But just how all of this is converted into generator potentials at the last stage is not yet known. See O'Brien (1982) for the most recent review of current understanding.

In vision as well as the chemical senses, transduction, although it is not really known, is thought be be related to the molecular makeup of the cellular membrane. As mentioned previously, it also is not known how transmitter substances alter the membrane at the final stages of synaptic transmission, but the general picture of receptors and alteration of conductances for the various ions could easily apply also to external chemicals being sensed by the sensory neurons. The sense cells would have receptors that, when combined with other molecules of the appropriate characteristics and configuration, would alter the membrane conductances. If true, chemical sensing would differ from synaptic transmission primarily only in the specific characteristics of the membrane receptors and the source of the impinging chemicals. This, however, is purely speculative.

The picture presented above might also explain the sensory adaptation that occurs in the chemical senses. When the receptor sites have combined with molecules, those molecules would have to be destroyed or somehow discharged before that site could be activated again.

SUMMARY

Although it is easy to take our own senses for granted and assume that other animals sense the world in the same manner as humans do, it is a mistake to view the world as appearing the same to all species. In fact, it is wrong to think that the world even has a set appearance. What *we* see and think of as the world is only *as we see it*. Other animals, to the contrary, sense the world in many and greatly different ways. They live in sensory worlds that are totally alien to humans; we do not have to travel across the voids of outer space to distant planets to find strange life forms—there are many of them all around us sharing the planet Earth and even our own homes. We sense many things, in sights and sounds, that they are totally or mostly insensitive to, and likewise they may sense things, such as magnetic or electrical fields, high-pitched sounds, wavelengths of light, and molecules, of which we are largely or completely unaware.

Because animals cannot communicate their subjective perceptions of what they sense, humans must use a number of techniques to infer and translate between their senses and ours. The chief methods used during recent years include the recording of evoked potentials from the sense cells, disruption of the normal sensory process with observation of the subsequent effects, and the use of conditioned or natural responses to determine whether or not an animal can sense something.

For any given sense there are many general attributes that can be considered. In addition to the simple ability to detect the presence or absence of a particular energy form, there is much variation among animals in their sensitivities and whether only relative changes or absolute levels can be sensed. Different organisms also may possess highly advanced and specialized sense organs and processing nervous systems that permit the discrimination of spatial and temporal patterns and finer divisions of the modality, such as pitch in sound and color in vision.

Most of this chapter is devoted to a systematic brief survey of the variety of senses found in different animals. The survey includes mechanoreceptors and closely related vibration-sensing organs (such as statocysts and ears), proprioceptors, vision in its multitude of forms, the less familiar senses for infrared, electrical and magnetic fields, and the chemical senses. Many species use emitted energy; that is, they produce the signals or fields, then sense changes in what is reflected back.

Most behavioral situations, even as simple an action as bodily orientation, involve the combined input from several, often redundant, sensory systems. Finally, a few facts and speculations are presented concerning the possible mechanisms by which external, environmental information is transduced, that is, converted, into nervous impulses by the sensory cells.

Recommended Additional Reading

Boudreau, J.C., and C. Tsuchitani. 1973. Sensory neurophysiology. Van Nostrand Reinhold Co., New York.

Carthy, J.D., and G.E. Newell, editors. 1968. Invertebrate receptors. Academic Press, Inc., New York.

Gersuni, G.V. 1971. Sensory processes at the neuronal and behavioral levels. Academic Press, Inc., New York.

Hess, E.H. 1973. Comparative sensory processes. In Dewsbury, D.A., and D.A. Rethlingshafer, editors. Comparative psychology. McGraw-Hill Book Co., New York.

Lorenz. K.Z. 1952. King Solomon's ring. Thomas Y. Crowell, New York.

Milne, L., and M. Milne. 1964. The senses of animals and men. Atheneum Publishers, New York.

Tavolga, W.N., editor. 1976. Sound reception in fishes. Benchmark Papers in animal behavior, vol. 7. Dowden, Hutchinson, & Ross, Inc., Stroudsburg, Pa.

CHAPTER 15

John H. Gerard

Close-up of a frog *(Rana catesbeiana)* **eye, the start of a complex pathway through the nervous system.**

The centipede was happy quite
  Until a toad in fun
Said, "Pray, which leg goes after
  which?"
That worked her mind to such a pitch
She lay distracted in a ditch
  Considering how to run.

Nineteenth century, Anonymous

It frequently is claimed that the human brain contains around 10 billion neurons. However, based on a density of around $100,000/mm^2$ of surface area (given in Hubel and Wiesel 1979), there may be at least 10 billion in the cerebral cortex alone. The cortex is only part of the brain. Recent guesses for the number of neurons in the whole brain range as high as a trillion. But no one knows for sure, and it is anybody's guess; I will be conservative and use the traditional figure of 10 billion for discussion purposes. For other parts of the nervous system it is estimated (e.g., Nauta and Feirtag 1979) that in humans there are about 2 to 3 million sensory neurons and roughly an equivalent number of motor neurons. Four to 6 million sensory and motor neurons may seem like a lot, but in comparison with the estimated 10 billion total, that is less than 0.01 of 1%; 99.99% of the neurons are interneurons. And the number of synapses on any one neuron may be as high as 80,000.

Not only is one faced with large numbers of neurons and connections, but the structure and arrangement of the brain, although ordered, is very complex. Even the gross anatomy of the brain requires much time and effort to learn.

A rough but useful model of the human brain, incidentally, can be formed from one's hands (MacLean 1981). Close the fingers into two loose fists and bring them together with the fingertips touching, the thumbs together pointing up, and the two forearms together down to the elbows. The tips of the thumbs represent the prefrontal cortex of the cerebrum, and the heel of the hand would be the occipital lobe, where the primary visual area is located. The primary auditory area is over the middle knuckles. The middle fingers are the motor area of the cortex, and the ring fingers represent the sensory cortex for touch and other sensations throughout the body. The touching fingernails would be the corpus callosum, which connects the right and left sides of the cerebrum.

By separating the two fists, one has a half section of the brain. Under the folded fingers, in the palm of the hand are such things as the caudate nucleus, putamen, and globus pallidus. The brainstem would be in the region from about the base of the thumbs across the base of the hands to the wrists. Compare this model with Figure 15-1, which will be discussed later in the chapter.

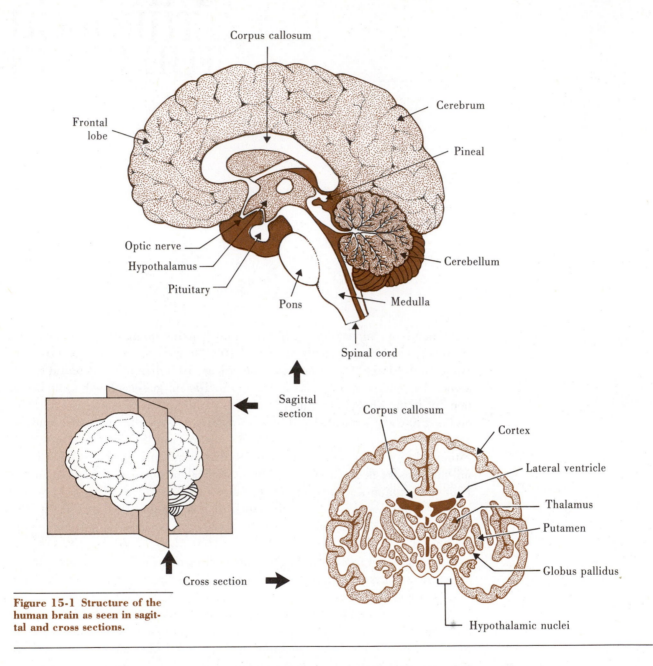

Figure 15-1 Structure of the human brain as seen in sagittal and cross sections.

This hand model, of course, is only approximate. It is about the right size but slightly smaller than the actual brain. The spinal cord would be represented by the two forearms, but it is actually much smaller—closer to the size of one's little finger. And, in the hand model, the cerebellum is missing; it belongs about in the position between the projecting wrist bones and the heel of the hand. To continue describing the human brain in familiar terms: the adult brain weighs about 1.4 kg (3 pounds) and has a texture about like soft avocado or banana.

The brains of humans and a few other mammals, such as other primates, carnivores, and cetaceans, represent the highest known stages of neural development on Earth. The other mammals are not far behind, mostly differing quantitatively, and the remainder of the vertebrates have very complex brains. The brains of many invertebrates also are complex and highly developed. The gross anatomy of a grasshopper brain, for example, is shown in Figure 13-3. Even the brains of some of the lowly molluscs, where every neuron can be counted and identified, are not what one really would call simple, except in the most relative sense.

In spite of all of the seemingly bewildering complexity and large numbers, however, neuroanatomy, neurophysiology, and behavior have made significant advances on the subject of neural integration in recent years. It is now possible to get at least a few glimpses into what is occurring within the central nervous system and obtain suggestions as to how behavior is integrated. (At the same time one must be careful not to overstate the level of understanding. Above the level of the simplest behaviors in invertebrates, we are still a long way from completely understanding the integration and controls of behavior, even of something as "simple" as feeding [Chapter 17].)

To comprehend the internal integration of behavior one needs to know something of the anatomy, arrangements, and relationships of the parts, something of the pathways and connections between one part and another, how information is passed and modified, and how the sum of the component processes results in the well-orchestrated, integrated output that is called behavior. Some of the basics of the individual units, neurons, were previously looked at in Chapter 13, and how different senses provide input to the system was discussed in Chapter 14. This chapter will trace a few of the pathways of information through the systems of different organisms then in Chapter 16 the role of chemicals in the nervous system will be considered. Chapter 17 will apply all of this to a category of behavior, feeding, and explore the limits of understanding of a behavior and its internal and external (proximate) controls. In Chapters 18 and 19 how all of this develops in an individual and how it might be modified by experience (i.e., learning) will be considered.

VERTEBRATE NEURAL PATHWAYS WITH EMPHASIS ON HUMANS

Reflexes

It may seem strange to start with the organisms that have the most complex systems. However, mammals, including humans, probably have received the most comprehensive studies; they are most familiar; and they provide the full spectrum from simplest to most complex pathways.

The simplest possible functional unit of behavioral integration is somewhat arbitrary. Among invertebrates there are some units of integration that do not involve even neurons. In some insect flight muscles, for example, the contraction is stimulated by the stretch of the same cells, without input from other cells (pp. 438 and 439). In the single-cell protozoans integration involves parts of the cell, and movement is achieved through cilia, flagella, or pseudopods and molecular mechanisms.

For purposes here, the simplest neural pathway is the **simple reflex**. This reflex path consists of two neurons: a sensory neuron connected directly to a motor neuron. It is analogous to a doorbell with a push-button switch for the stimulus and the buzzer for the response. Although simple, such reflexes are relatively uncommon. They are found primarily in the coelenterates (invertebrates) and again in the terrestrial vertebrates. In the terrestrial vertebrates they are mostly postural reflexes and probably evolved secondarily. Without the supporting medium of water, sudden slips of an animal from a tree, the edge of a cliff, and the like can be fatal. Such accidents probably are quite common for terrestrial organisms. The individuals that survive these slips either are small, with low mass and slight momentum so that "crashing" is not serious, or are those which can react quickly and catch the fall before it goes too far. The fewer the synapses, the faster the information is transmitted; some muscle stretch receptors appear to have evolved direct synapses on motor neurons in the spinal cord (Figure 15-2). The receptors are stimulated during a sudden stretch, as when a slip occurs, a limb moves, and the muscles are pulled unexpectedly. This is the basis, incidentally, for the knee-jerk reflex.

Even the simplest two-neuron system shows some of the central principles of neural pathways and integration, particularly the "filtering'" and modification of the signal. Not just any stimulus or any degree of stimulus will result in a muscular contraction. The stimulus has to involve a mechanical stretch, not light or sound waves impinging on the receptor, and it has to be a stretch of a certain minimum threshold before the muscle will contract. Greater stimulation may lead to greater contraction by enlisting greater numbers of muscle fibers.

The *selectivity* and *specificity* of response and differences between different neurons are central principles in the passage of information through different neural pathways. This was introduced in Chapter 13, will be encountered frequently, and should be kept firmly in mind. Neurons are not simple connectors, like wires, that merely pass the information from one point in the path to another.

Figure 15-2 Simple, two-neuron stretch reflex. An unexpected stretch, such as from an accidental slip, stimulates a proprioceptive stretch receptor in the muscle, which synapses directly to a motor neuron in the spinal cord, which, in turn, causes a reflex contraction of the muscle. The sequence of events can be traced by following the numbers.

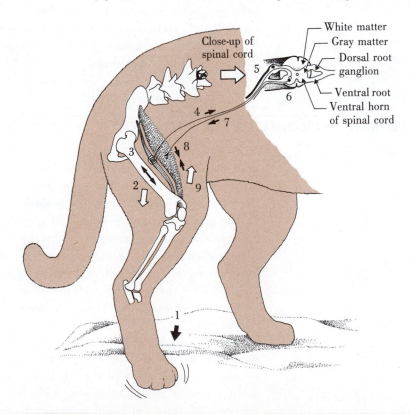

Recall the moth hearing example. A few of the interneurons, the repeaters, did more or less relay the signals with relatively little change. However, there were other interneurons such as pulse markers and train markers (pp. 431 and 432), the output action potentials of which were significantly different from the inputs and which were in response to specific patterns of input.

The pathways of neurons are not isolated even in the simple example involving just two neurons. In fact, one is not dealing with just one sensory neuron and one motor neuron. Many similar stretch and motor neurons are involved. Thus there are really a large number of these two-neuron paths. Furthermore, the sensory neurons contact interneurons through collateral branches. These interneurons may in turn contact the motor neurons or other neurons. Some of the axons cross to motor neurons on the other side of the spinal cord, and some travel up to the brain itself, which brings us to three more principles found in the operation of nervous systems (Figure 15-3):

Divergent paths

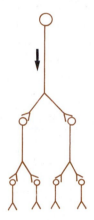

1. *Divergence.* The information is sent out to several other points simultaneously; that is, it spreads or diverges to other neurons.
2. *Convergence.* Signals from several neurons impinge on any given neuron, such as the motor neuron. This permits the temporal and spatial summation of postsynaptic potentials (PSPs) as discussed earlier (pp. 422 and 423).
3. *Parallel paths.* Several similar paths occur side by side.

Convergent paths

Most of the reflexes that are found in vertebrates involve not just two neurons, sensory and motor, in a given path but also contain one or numerous intervening interneurons. There may be much crossing from side to side in the spinal cord, and a lot of information passes to and from the brain. A cross section of the spinal cord (Figure 15-2) shows both gray and white matter. The gray matter consists of motor (in the ventral horn) and interneuron somas plus processes (dendrites and axons) traveling relatively short distances within the local neighborhood. For most of the distant travel, such as to and from the brain, the processes move into the outer white regions. The white appearance results from the white fatty material in the myelin. Thus, this white matter represents myelinated fibers carrying impulses rapidly over relatively long distances.

Parallel paths

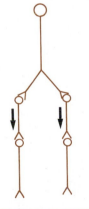

To put this into an example, imagine that you are walking through the kitchen and happen to bump an elbow on a hot stove (Figure 15-4, *A*). The first response, a reflex, is to jerk your arm away. The jerk, much of which was probably mediated and integrated through neurons in the spinal cord, also probably occurred a few milliseconds before you were consciously aware of it. The pathways from the temperature-sensing neurons to the motor neurons are shorter and faster than those going up to the brain. But the signals also went to the brain so you may have sensed the pain, became aware of the jerk, and perhaps responded by swearing.

Now imagine two different scenarios. In the first you are carrying a pan that is full of near-boiling grease when you bump the stove. But this time you do not jerk your arm away. Rather you move it slowly and tolerate the burn a little longer. An explanation would be that there is increased information coming down the spinal cord from the brain, caused by the awareness of the pan of hot grease and the consequences of spilling it, so that impulses from the higher, conscious levels in the cerebrum are *inhibiting* the action of the motor neurons that would otherwise

Figure 15-3 Principles of synaptic pathways. Passing of information among neurons is indicated.

lead to the jerk. The presence of these active IPSPs cancel out (in the summation process) the incoming EPSPs from the temperature sensors in the elbow. But the pathways back up to the brain and pain centers, etc., are not inhibited; you become aware of the burn; and the same swearing responses come out. Diagrams of these various routes are illustrated in Figure 15-4, *B*.

In the next scenario (Figure 15-4, *C*), which may be with or without the pan of hot grease, you are not alone but have a new friend from college standing by. This time, as the higher centers become aware of the burn and evaluate the responses, other centers inhibit the vocal swearing response. You may swear to yourself but also quickly consider that you do not wish to embarrass your friend and you must keep calm, with much of that processing going on subconsciously. The centers that would lead to the stimulation of motor units in the vocal apparatus are not

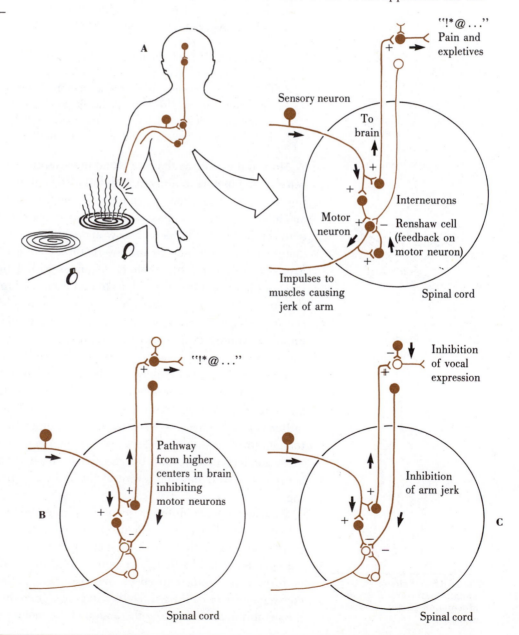

Figure 15-4 Diagram of a cross section of spinal cord, sectioned at the point where a spinal nerve is connected. Scenarios of reflex pathways illustrate basic paths of neural information to, from, and within the central nervous system. In this illustration darkened neurons indicate activated pathways; open neurons indicate inhibited or nonactivated pathways. A, Accidental bumping of a hot stove. B, As in A but with a pan of hot grease that should not be spilled. C, As in B but in the presence of company.

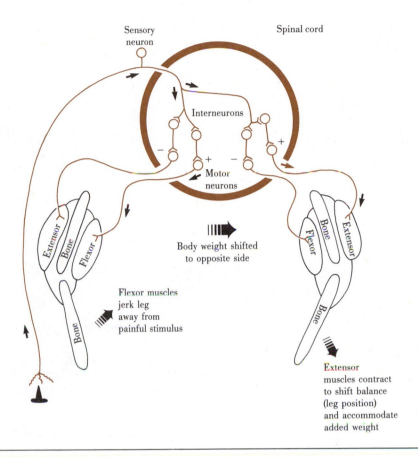

Figure 15-5 Collateral pathways and crossing of sensory information to other motor neurons.

Sensory neuron

Spinal cord

Interneurons

Motor neurons

Body weight shifted to opposite side

Flexor muscles jerk leg away from painful stimulus

Extensor muscles contract to shift balance (leg position) and accommodate added weight

Extensor   Bone   Flexor   Bone

Flexor   Bone   Extensor   Bone

stimulated. Or the swearing may have already begun, but other centers cut it short or modify the words to something less objectionable.

The preceding example involves arm movements, which do not seriously interfere with balance. Now consider, however, that instead of bumping your arm on the stove, you step barefooted on a sharp tack or start to step over the edge of a drop-off that you did not realize was there. This leads to a reflex that causes the foot and leg to be jerked rapidly back. But the remainder of the body's movement and weight is already going forward, on the premise that the jerked leg was going into its next position. If nothing else intervenes and the remaining movement continues, the balance will be lost with the changing center of gravity, and you will fall down. But falls do not always occur. Instead they may be caught by the simple reflex discussed at the first. The stretch receptors signal that stretch is occurring faster than it is supposed to. The appropriate motor units are excited to fire, causing contraction of various muscles that quickly reverse the motion of the body and restore the center of gravity to a new balance. Also, collateral branches cross the spinal cord (Figure 15-5) and synapse with appropriate motor units on the opposite side of the spinal cord to cause the appropiate corrective action.

Much work has been done in recent years, and many of the specific pathways through the spinal cord now are well mapped and understood. For illustration, see Figure 15-6. Persons wanting further details should consult a contemporary text on human anatomy and physiology.

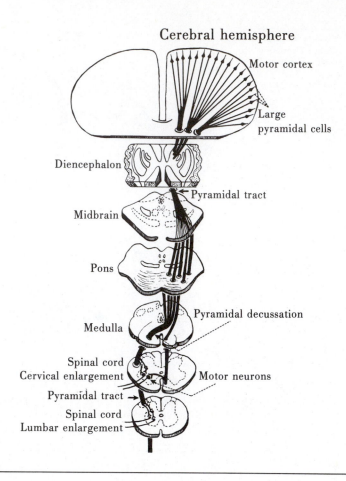

Figure 15-6 Example of a pathway of information from the brain down the spinal cord to motor neurons. Other pathways are used for different movements. Similar but different pathways are involved carrying sensory information. From Eccles, J.C. 1977. The understanding of the brain, 2nd ed. McGraw-Hill Book Co., New York.

Reflexes are believed to differ from more complex forms of behavior mostly in degree, that is, quantitatively. Reflexes are simply at one end of a continuum of coordinated movement. At the other end are longer sequences of more complex and sophisticated actions of the type that humans usually are interested in, such as predator-prey interactions, long-distance migration, courtship, and other social interactions. These higher levels of behavior are believed to simply (if simply has any meaning here) involve additional numbers and sets of interneurons, although no one yet has traced any of these complete specific pathways for complex behavior in vertebrates. In addition there are two other matters of extreme importance: (1) the problem of *conflicting* impulses and (2) the problem of *voluntary* action—behavior that somehow and to some degree originates within the brain itself.

Based on information from simpler cases, such as sensory input and reflexes, it seems likely that conflicts in general are handled by one of the major processes of neural integration: **inhibition.** That is, different stimuli are clamoring and competing for attention by higher levels within the nervous system (see abstraction, discussed later), but only one or a few actually get through. When a stimulus does make it through, in essence by shouting the loudest, it not only stimulates or excites certain pathways, but it *inhibits competing traffic.* This results in a neural filtering process whereby everything does not happen at once.

The problem of voluntary control of behavior is perhaps least understood and is one of the main targets of present research and interest. The intrinsic, spontaneous firing of some neurons, such as command cells, may stimulate some voluntary actions. Some spontaneous activity may not be able to get through the inhibition of other traffic except under certain circumstances or during lulls. Some activity, which appears voluntary, may simply be a result of delayed responses or be stimulated by subtle or internal (e.g., hormonal) unrecognized stimuli and, hence, not really be "voluntary." Voluntary behavior is of interest not only because it is so poorly understood but also because of its relevance to human initiative and creativity, "free will," and consciousness. Integration and action within various higher centers of interneurons may or may not be modified by the effects of experience (i.e., learning).

Some of the centers of neurons that control more complex behavior may reside mostly or completely within the spinal cord. If frogs receive a slight irritation on the back, for example, they will direct the hind leg to that point and scratch at it. A frog completely deprived of the brain and all the higher facilities still will correctly orient the leg and scratch at an irritation on the back. Scratching reflexes in other vertebrates similarly may be coordinated within the spinal cord. Many of the basic locomotion patterns of movement, such as general swimming, walking, running, and flying are thought to be controlled within the spinal cord. An example of behavioral control outside the brain in invertebrates was noted earlier (p. 308) in male praying mantids, which do not have a spinal cord but which have an analogous region in the central nervous system. If and when the female eats the male's head—including the brain—the remainder of the system still will direct the copulation behavior. Most of the complex patterns of behavior in advanced organisms, particularly the arthropods and vertebrates, however, are controlled by the brain.

Mapping the Brain

It has been known for over 100 years that certain parts of the brain are associated with particular neural functions. Localized injuries and damage to the human brain were known to consistently affect a person's sensory or motor abilities, depending on the specific location. Then electrodes placed at the surface of the brain in work done on other species, particularly cats and various primates, demonstrated similar topological properties in their brains. In particular, if mammals are exposed to a variety of stimuli, such as sounds, flashes of light in front of the eyes, or touch at different points on the body surface, effects can be detected at recording electrodes in certain positions. Or if electrical impulses are given at different locations on the surface of the brain through stimulating electrodes, muscular contractions can be elicited. A stimulus at one location, for example, will cause a leg to jerk, whereas at another position, a paw may be caused to twitch. During recent years, in the course of brain surgery, medical doctors have had the opportunity or need to record from or stimulate various regions of the human brain.

Results from all of these sources have demonstrated consistent locations of particular functions and have permitted the mapping of several important regions of the brain. Figure 15-7 shows the sensory and motor maps for the surface of the human cerebral cortex. Note that the amount of area occupied for different

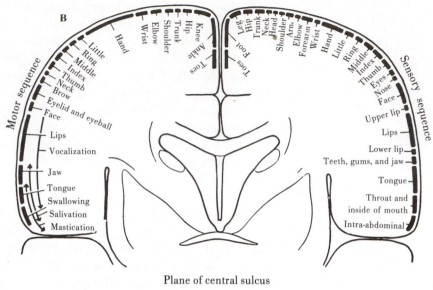

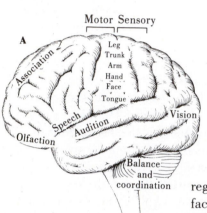

Figure 15-7 Graphic map of sensory and motor regions of the cerebral cortex in humans. A, External view. B, Cross-sectional view. A from Hickman, C.P., Jr., L.S. Roberts, and F.M. Hickman. 1982. Biology of animals, 3rd ed. The C.V. Mosby Co., St. Louis. **B** from Rasmussen, T., and W. Penfield. 1947. Fed. Proc. 6:452.

regions of body is not proportional to their actual surface area; some, such as the facial and hand regions, receive proportionately much greater representation than other areas, such as the back and legs. Not only are there differences from top to bottom and anterior to posterior but also between the right and left sides, an indication of asymmetry between the two sides of the cerebral cortex. Similar mappings have been accomplished in a great many other species, such as iguanas (Figure 15-8) and barn owls (Figure 15-9).

Recent radiographic techniques of positron-emission have permitted the mapping of brain activity from outside the head in normal, healthy subjects. Examples are shown in Figure 15-10.

Figure 15-8 Somatic sensory map of the optic tectum region of the iguana *(Iguana iguana)* brain. From Gaither, N.S., and B.E. Stein. 1979. Science 205:595. Copyright 1979 by the American Association for the Advancement of Science. Courtesy Barry E. Stein.

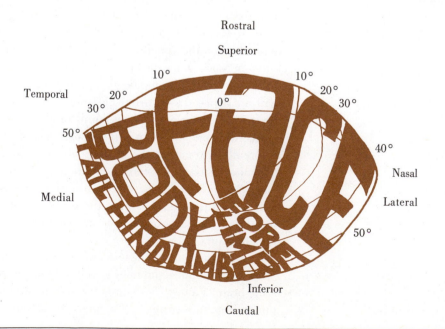

Figure 15-9 Neural map of auditory space in the midbrain auditory area *(MLD)* of the barn owl *(Tyto alba).* A, Photograph of a barn owl listening for prey. B, Diagram of hearing area regions of the brain corresponding to surrounding locations in the owl's hearing receptive fields. Particular locations of sounds around the owl map to particular locations in the brain.

A courtesy T. Brakefield. **B** from Knudsen, E.I., and M. Konishi. 1978. Science 200:795. Copyright 1978 by American Association for the Advancement of Science. Courtesy Eric Knudsen.

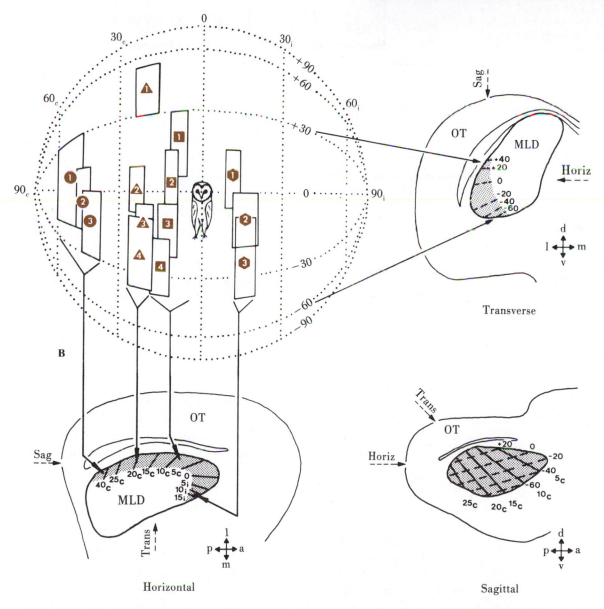

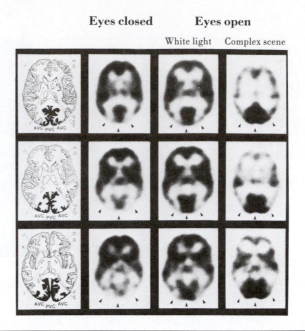

Figure 15-10 Positron-emission maps of the human brain under different visual experiences. These pictures result from radiographic emissions from 2-^{14}C deoxy-D-glucose being metabolized by active regions of the brain and recorded photographically from outside and above the head. Drawings on left represent cross sections of brain illustrating the primary *(PVC)* and associative *(AVC)* visual cortices. In the remaining columns, lighter areas represent regions of highest metabolic activity. From Phelps, M.E., et al. 1981. Science 211:1446. Copyright 1981 by the American Association for the Advancement of Science. Courtesy Michael Phelps.

These maps are important because they demonstrate that at least some functions of the brain are compartmentalized; that is, they occupy particular positions. And they also give one some points of contact with neural processing. Thus one now has information at the level of the senses, contractions of the muscles, and regions of sensory and motor processing at the surface of the cerebral cortex. But there still is a vast amount in between. To obtain further pieces of the puzzle, one must go deeper and consider additional principles of neural integration.

Abstraction and Pattern Recognition: Vision as a Case in Point

Although one can discuss simple tracing and mapping of pathways through the nervous system, remember that the pathways are not just open roads or relays; rather different interneurons are filtering and modifying the messages. This leads to a principle that can be referred to as ***abstraction.*** The verb *abstract* literally means to draw from. Different neurons in the pathway abstract or extract different bits of information from the incoming impulses, and together they may derive a generalized picture. The noctuid moth with only six sensory neurons in its ears, counting the stretch receptors, is able to identify a bat call from among the bewildering hodgepodge of other environmental sounds. This is because, in addition to the sensory neurons, there are different interneurons. Some relay the sensory information with little change; some fire only when the pulses fall within the range given by bats; and some mark the lengths of trains of such pulses. *Taken together* they identify and spell *bat!* This unlocks the pathway for a particular behavioral response much like the proper combination of numbers in a combination padlock. One can get an even better picture of this by looking at the processing of visual information in vertebrates.

To ensure understanding of the problem, consider what happens when an object, for example, in the form of the letter "A," is viewed. The image projected onto the retina is not simply passed onto the brain intact and unaltered, like a television picture to its screen. For a system such as that to work and permit abstraction on a very simple basis, the image would have to be more constrained. If either the eye or the object is moved, for example, if the object is moved closer or farther away so that its apparent size changes or if the object is rotated or shifted up and down or sideways, then the image would fall on different sensory cells. Somehow the system must abstract the qualities that make the object look like the letter "A" over a range (within limits) of sizes, rotation, and other variation in appearance. Also consider other letters such as "N's" that turn into "Z's" when rotated or "M's" that turn into "W's." For other species and situations, abstraction might take the form of recognizing, for example, "bug" or distinguishing edible bugs from inedible or harmful ones or visually identifying mates and predators. All might occur at different distances, at different positions in the visual field, against differing backgrounds, and so on. Abstraction involves much more than simple passing of a picture as occurs in a photographic camera, television system, or the transmission of a news Wirephoto.

The visual pathway in vertebrates begins at the retina of the eye (Figure 15-11, A). The retina actually is an extension of the brain, being derived embryologically

Figure 15-11 Initial retinal processing of visual information in the vertebrate retina. A, Diagram of the eye showing pathway of light and position of retina. B, Close-up of retinal structure showing variety of cells and the pathway of neural impulses away from the sense cells and toward the brain via the optic nerve. Modified from Polyak, S. 1941. The retina. University of Chicago Press, Chicago.

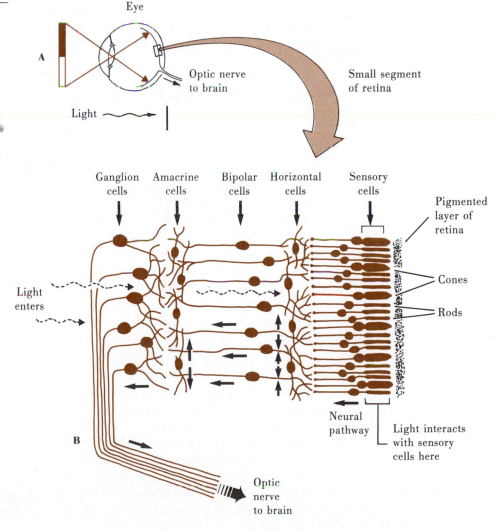

Figure 15-12 Demonstration of the blind spot and visual processing in the sensory field of the human retina. Close the right eye and focus the gaze of the left eye on the hunter riding the horse. Slowly move the book page back and forth at different distances from the eye. At one point the fox will become invisible because the image is falling on the point in the retina where the optic nerve leaves the eye and there are no sensory cells. If, however, one switches eyes, closes the left eye, and focuses the right eye on the fox, the horse and rider cannot be made to disappear completely. This is because the horse and rider occupy a larger area, extending beyond the blind spot, and the visual processing system "fills in" some of the missing information from retinal areas surrounding the blind spot. This visual processing is described in following figures and text. From Lorenz, K. 1952. King Solomon's ring. Thomas Y. Crowell Co., New York. (Figure originally used in a different context.)

from the neural tube. Accordingly, it is more complex than most other sensory structures. But, at the same time, it is removed from the rest of the brain and is thus simpler and easier to get at; it has provided relatively easy access for understanding the vertebrate brain.

The retina, which is about the thickness and consistency of a spiderweb, is composed of five general types of neurons, arranged in three basic layers (Figure 15-11, *B*). The cells are (1) the sensory cells (rods and cones of various types), (2) the horizontal cells, which carry information horizontally or across the retina between neighboring sensory cells, (3) the bipolar cells, which further integrate information and carry it to the next layer, (4) amacrine cells, which also integrate horizontally and may help detect directional movement, that is, fire when sensory information crosses the visual field in one direction but not the opposite, and (5) the ganglion cells, which synthesize qualities or abstractions from the previous cells and then forward them to the brain via the optic nerve; that is, the optic nerve is composed of axons from the ganglion cells. There are no sense cells, incidentally, at the point of the retina where the optic nerve leaves the eye, which creates a *blind spot* (Figure 15-12).

The visual cells do not operate quite as simply as the general pattern described in Chapter 13. They are short and release transmitter chemicals without the need for action potentials and long axons. Furthermore, when stimulated, they do not respond by becoming depolarized and releasing more transmitter but by the opposite—becoming hyperpolarized and releasing less transmitter than when "resting." Again, one is reminded that nature is not obliged to keep things simple, and there is much variety in the nervous system. Beyond the point of the sensory and horizontal cells, the others operate in the more familiar pattern of action potentials and code information by firing.

These few types of cells permit an amazing amount of initial abstraction to take place within the eye itself. Amacrine cells with different polarities, for example, allow movement in different directions to be recorded.

The extent to which different species take advantage of the preliminary processing depends on the species and is thought to have some relationship to their environment and life-style. Frogs show a high degree of early processing, with several general categories of information being synthesized by the ganglion cells and sent down the optic nerve (Figure 15-13). The overlapping area of sensory cells covered by a particular ganglion cell is called the *excitatory receptive field (ERF)*. Particular ganglion cells and their optic nerve outputs are classed according to the type of information (i.e., visual quality) of what is being detected. In addition to those illustrated in Figure 15-13, there are brightening detectors for any increase in brightness, particularly in blue wavelengths.

The outputs of these classes of ganglion cell fibers go to different regions of the brain (Figure 15-14). The first four go to different regions of the midbrain and the optic tectum, and the fifth goes via the dorsolateral geniculate region of the diencephalon and from there to the cerebrum. All of this is further integrated,

Figure 15-13 Output of different ganglion cells in the frog retina in response to different visual events. Based on Maturana, H.R., et al. 1960. J. Gen. Physiol. 43:129-175.

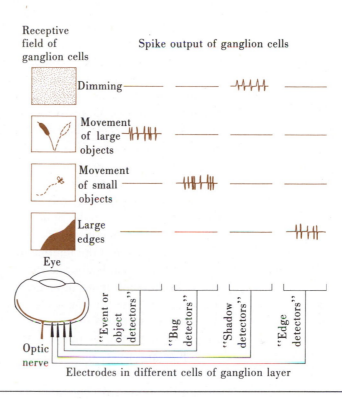

Receptive field of ganglion cells

Spike output of ganglion cells

Dimming

Movement of large objects

Movement of small objects

Large edges

Eye

"Event or object detectors"

"Bug detectors"

"Shadow detectors"

"Edge detectors"

Optic nerve

Electrodes in different cells of ganglion layer

including perhaps some minor modification from previous, learned experience, and the appropriate motor responses are chosen. Most of this is believed to be accomplished in the midbrain of the frog. Then, depending on whether the abstraction comes up with, in simple English, "bug," "solid object in the way," or "approaching large object (possible predator)," the frog will, respectively, aim and steer its tongue at the object, avoid the object while moving, or jump away and attempt to escape. Note that much of the visual abstraction occurs directly in the retina, and the synthesized information is relayed largely into the midbrain and onto the appropriate motor output. It all works quite well; frogs have been around for many millions of years; and they continue to populate Earth wherever suitable habitat remains.

Figure 15-14 Basic visual pathways from the retina to the brain: frogs versus cats. Modified from Tinbergen, N. 1965. Animal behavior. Time-Life Books, Inc., New York.

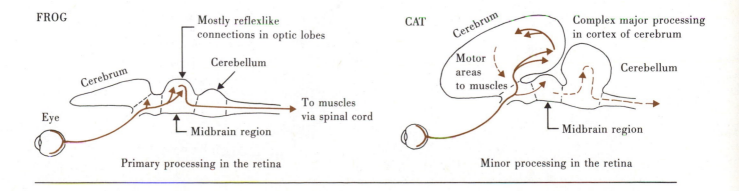

FROG

Mostly reflexlike connections in optic lobes

Cerebellum

Cerebrum

Eye

To muscles via spinal cord

Midbrain region

Primary processing in the retina

CAT

Cerebrum

Complex major processing in cortex of cerebrum

Motor areas to muscles

Cerebellum

Midbrain region

Minor processing in the retina

Other vertebrates, such as mammals, show similar visual systems with some modifications. The proportions of retinal ganglionic responses vary. Mammals tend to have more of the dimming detectors (called *off sensors*) and brightening detectors *(on detectors)*, with fewer ganglion cells responding to more complex features. In rabbits and some squirrels, for example, only about 34% of the optic nerve fibers carry complex responses, such as to movement in the ERFs. In the cat it is even less, only about 8% for complex classes and the remaining 92% for on and off detectors. (It was earlier thought that mammal eyes were entirely dominated by on or off detectors, but the more complex types have been documented and quantified.) In other words, cats (and to variable extents other mammals, including primates) rely less on retinal processing of the information and shift the job more into the brain itself, presumably where there is much greater capacity for abstraction.

In addition, *where* the information is processed in the brain has shifted in the course of evolution leading to the mammals. Most, although still not all, of the visual information from the mammalian eye goes to the dorsolateral geniculate region of the diencephalon and from there onto the visual cortex of the cerebrum (Figure 15-14) where extensive abstraction occurs. There is much less of the simple relay that takes place as in the frog midbrain.

So how and where does all of this take place? Among at least carnivores and primates with binocular vision, information from half of each retina goes to each side of the brain. (Among other mammals and other vertebrates that have been investigated, all the information from each eye simply crosses entirely to the opposite side of the brain.) The information goes first from the optic nerve to the

Figure 15-15 Visual pathway from the retina to the brain in primates (as seen from above the head). The right visual field from both eyes goes to the left side of the brain and vice versa for the right side. Note that it travels through the brainstem and lateral geniculate body, and some information goes to the superior colliculus, although the majority goes to visual areas of the cortex. Modified from Polyak, S. 1957. In Kluver, H., editor. The vertebrate visual system. University of Chicago Press, Chicago.

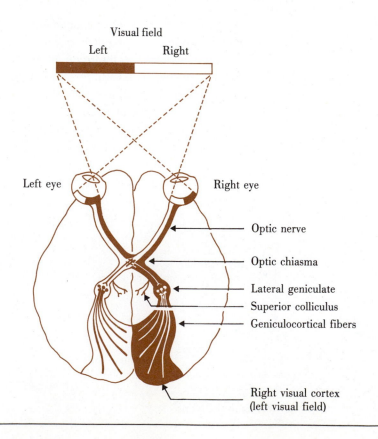

dorsolateral geniculate region, then to the visual cortex (Figure 15-15). The two basic categories of visual information, from the ganglion cells, are on-center, off-surround and off-center, on-surround. (Figure 15-16 shows patterns of action potentials depending on the patterns of light falling in the ERF.) Similar responses are shown at both the ganglion cell level and at the dorsolateral geniculate.

In the visual cortex the brain tissue is organized in columns and arranged in complex patterns, where there are "simple" cells, "complex" cells, "hypercomplex" cells, etc. All these sequentially abstract further qualities out of the information. This is all perhaps best explained with diagrams (Figure 15-17). These pictures may not be immediately obvious, but by reading the legend and spending a few moments to study the illustration, the reader should be able to understand what is occurring in that cortex.

Figure 15-16 On-off centers of visual processing in the mammalian retina showing ganglion cell firing under different stimulation of the receptive fields. Based on Hubel, D.H. 1963. Sci. Am. 290:54-62.

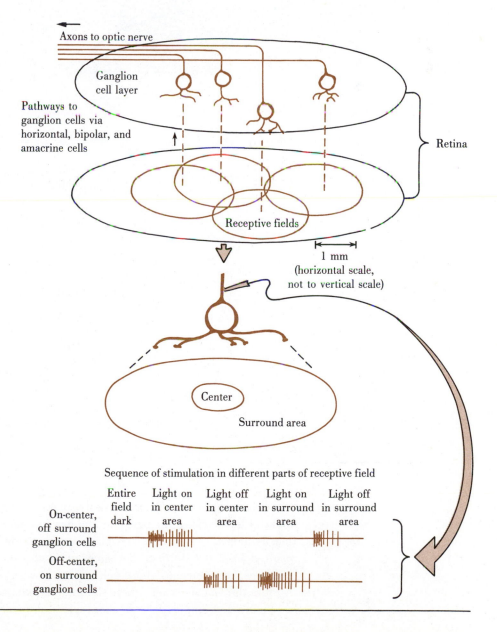

Figure 15-17 Higher-order visual processing in the mammalian visual cortex of the brain. Study these diagrams to understand the details. The broad outline of what is occurring is explained in the text. Images are inverted by the lens onto the retina. *LGN* **is the lateral geniculate nucleus. Different parts of the illustration refer to other parts as indicated by the arrows.** Based on Hubel, D.H., and T.N. Wiesel. 1979. Sci. Am. 241(3):150-162, and Bullock, T.H., et al. 1977. Introduction to nervous systems. W.H. Freeman & Co., Publishers, San Francisco.

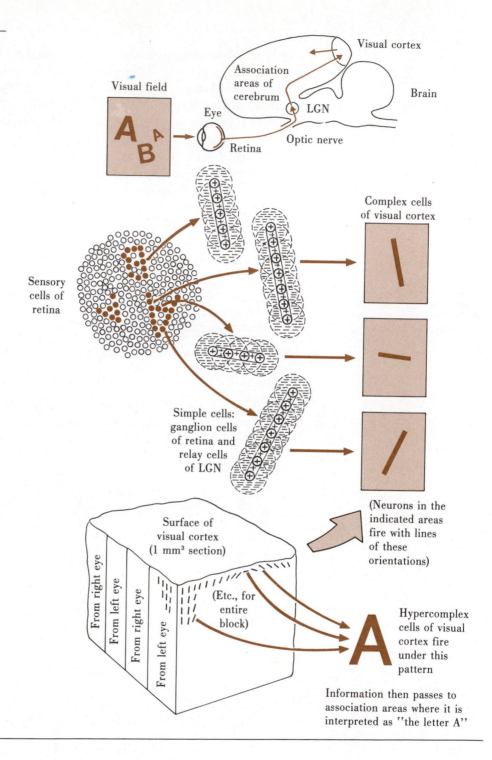

It is interesting to note, incidentally, that the human cortex is only about 2 mm thick, although it is quite irregular and has a surface area of about 200,000 mm². Just as amazing as what takes place in the spiderweb-like retina is what has resulted from that little 2 mm sheet of tissue: society, architecture and great buildings, art, science, and much more.

From the visual cortex, where the information (such as for the letter "A") has been extracted, impulses go to other regions of the cortex and brain, get combined

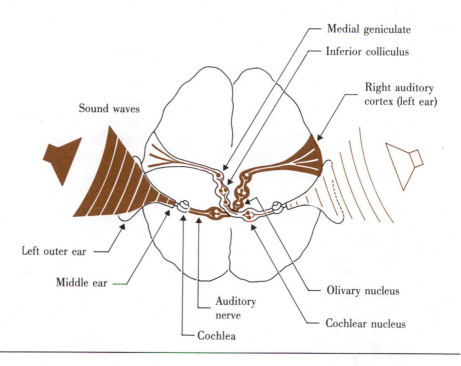

Figure 15-18 Auditory pathway in the human brain. Modified from Picton, T.W., et al. 1973. Electroenceph. Clin. Neurophysiol. 36:179-190.

Medial geniculate

Inferior colliculus

Right auditory cortex (left ear)

Sound waves

Left outer ear

Middle ear

Auditory nerve

Cochlea

Olivary nucleus

Cochlear nucleus

with other information, and lead eventually to muscular output and observable behavior. Other senses have their own pathways. Olfaction comes in at the anterior end and through the olfactory nerves, and audition arrives from cranial nerve VIII. Figure 15-18 illustrates the auditory pathway.

Motor Centers in the Brain

Just as the sensory side of the pathways does not involve simple passing of information but includes modification and abstraction at various steps, the motor side also involves sequences, but somewhat in reverse. At the motor side, as far as can be inferred, the information starts as somewhat of an abstract command, often in ***command neurons.*** Then it is broken down in an administrative fashion until it reaches the lowest levels of spinal motor neurons and, finally, the muscles themselves. One can use stimulating electrodes inserted into the nervous tissue and, depending on which level or pathway, cause simple muscle twitches or release entire patterns of behavior.

Holst and St. Paul (1960, 1963), for example, were able to insert stimulating electrodes in chicken brains and elicit whole, functional movements. When the electrodes were closer to particular centers, they could be stimulated with less current, and the response occurred sooner, that is, with a shorter latency period, than when the electrodes were farther from a particular area. With the electrode in a particular region, the outcome depended on the strength of the current.

At lower centers for simple behaviors, such as turning the head, standing up, or walking, two centers could be stimulated simultaneously, and the movements would add together or cancel out, depending on their mechanical compatibility. A chicken, for example, could be stimulated to stand up and turn its head to the right at the same time. But if the centers to turn the head to the right and turn the head to the left were stimulated simultaneously, the chicken would do nothing.

Higher up in the neural center hierarchy, centers for complex behaviors were found. These include sleep, which, if stimulated, led to a sequence of the chicken looking around, yawning, sitting down, fluffing the plumage, retracting the neck, and closing the eyes. The complex sequence depended in part on what the chicken was doing when the current was turned on, the remainder of the sensory environment, and its internal physiological condition. If the chicken was already sitting down, for example, the sequence was shorter. If it was eating, the chicken

Figure 15-19 Hypothesized stimulatory and inhibitory relationships for interruptions in the incubation behavior of the herring gull. *N,* Incubation system; *E,* escape system; *I,* inhibition input unit. For other symbols and detailed explanation, see Baerends (1976). Compare this contemporary interpretation of functional organization of behavior with the classic Tinbergen releaser model (Figure 2-15). From Baerends, G.P. 1976. Anim. Behav. 24:725-738. Courtesy Gerard P. Baerends.

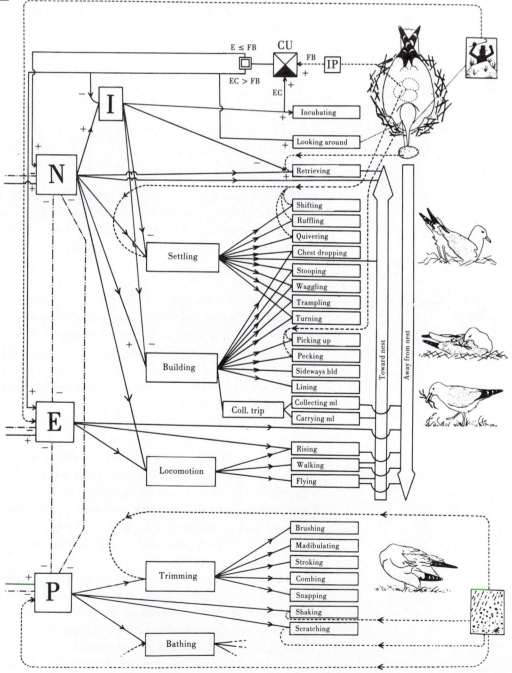

would stop eating and perhaps walk before starting to sleep. Unlike the simple behaviors, such as turning the head, the complex centers did not add together but rather inhibited each other so that only one would be active at a time. If one center was stimulated before another, the first was expressed and the second was inhibited. If two were activated at the same time, one would suppress the other, which either would not be expressed at all or would wait until after the first was completed. Sometimes they would alternate. Chickens stimulated to simultaneously reconnoiter and feed would alternately make pecking movements and raise its head and look around. In a few instances, opposing behaviors such as aggressive pecking and fleeing could be stimulated equally so that a different behavior was expressed. In the case of a hen stimulated to be aggressive and flee at the same time, she would run back and forth with her wings up and scream.

From this emerges a picture whereby there is a hierarchy of administrative levels on the motor side. Higher command centers do not direct the muscle-by-muscle movement but somehow direct lower centers, which direct still lower centers until the level of specific muscle movement is reached. The lower centers are employed in common by several of the higher centers. The running motions of a chicken, for example, can be incorporated into higher functional categories of running toward food, running toward a mate, or running away from an enemy. This interpretation is partly compatible with Tinbergen's synthesis of hierarchies of centers (pp. 42 and 43). It differs, however, in that it is not just a matter of releasing inhibited drives; the whole picture is more complex and may include stimulation or excitation as well. Contemporary, updated models of the functional organization of behavior (Figure 15-19) are approaching physiological mechanisms closer and closer. Inhibition is not just a matter of something needing to be released in the classic view but rather is a process that different pathways, particularly at the higher levels, do to each other. The neural message that gets through is the one that, in a sense, manages to shout the loudest and inhibit the others.

All behavior control is not quite this simple. Two or three aspects that are not understood yet are where learning fits into this scheme (Chapters 18 and 19), how "voluntary" processes arise in the first place, and how everything is integrated at the highest levels. One thing that has become fairly clear is that there is no single highest center, the so-called hidden observer or man at the top running the control panel. Rather, there are several highest centers. They may have connections and communicate, in the sense of inhibiting each other, but all parts otherwise are thought to operate independently of each other. The apparent overall control of behavior may be either just an illusion or simply a property of the whole.

Language Pathways

One can supplement this picture by tracing the language pathways, from input to output, in the human brain. It may seem strange that one can turn to perhaps the most complex neural control of all, human language, to help understand neural control in general. But the study of language control has several advantages: it is a subject of great importance and interest and hence has provided much motivation for persons to understand it; there is a long history and large body of information and knowledge pertaining to it; and by its very complex nature, it provides enough detail to be broken into components and show which parts of the language process are being handled in which parts of the nervous system. Much, if not most, of the understanding of language pathways has resulted from injuries and disease that affected language in different ways.

Most persons would be repulsed at the thought of doing any kind of damaging or harmful experiments or research on human brains. Accordingly there are, or should be, strict regulations against doing research with human subjects that might endanger or damage their health or mental facilities. Nature, however, is not so kind. Through the years there have been numerous accidents, diseases, birth defects, and other disorders that have led to various damage to the brains of humans. These problems, observations on their effects, and attempts to correct or otherwise deal with them have led to a wealth of information and insight into the probable workings of the brain. Four examples of such disorders are Broca's aphasia, Wernicke's aphasia, prosopagnosia, and Gilles de la Tourette's disease. Epilepsy and attempts to minimize its effects by cutting the corpus callosum, a band of tissue that connects the right and left hemispheres of the cerebrum, also have aided understanding of the human brain by providing the so-called split-brain experiments. Following the results of the split-brain work, researchers have found means of simulating the experiments in normal subjects by splitting and screening the sensory input rather than through surgery and dividing parts of the brain.

Broca's aphasia, named after Paul Broca who studied the problem beginning in the 1860s, is a speech disorder that results from damage to part of the cerebral cortex shown in Figure 15-20. The speech of a person with such damage is broken

Figure 15-20 Language and facial recognition areas of the human brain. The size of these areas differs between the right and left sides of the brain, occupying more space on the left side in many cultures.

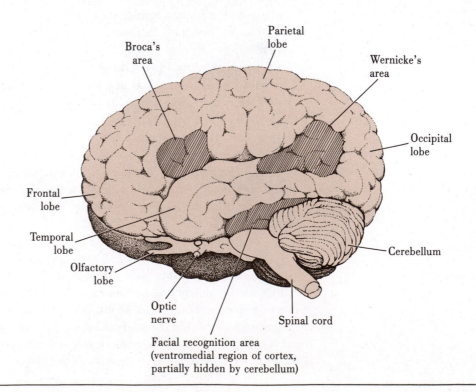

and difficult to follow. Words are not spoken smoothly. But they do make sense, and the person can be understood.

Problems associated with Wernicke's area (Figure 15-20), named after Wernicke, who studied patients in the 1870s, are different from those in persons with Broca's aphasia. In Wernicke's aphasia the speech is smooth and sounds normal superficially—but it generally comes out as nonsense. Even if one pays close attention, it is difficult or impossible to understand what a person with Wernicke's aphasia is trying to say. Wernicke's area is thus believed to be the region where language is interpreted, that is, takes on meaning.

A tract of nerve fibers, called the *arcuate fasciculus*, connects Broca's and Wernicke's areas of the cortex. Thus it is inferred that the language pathway is connected in the middle at this point, but what about the input and output? Visual and auditory input was discussed earlier. Language information can enter visually by reading something written or through the auditory system by hearing something spoken. In the case of blind persons, other pathways, such as using touch for braille, may be developed. The components are abstracted, as discussed earlier, then forwarded to association areas and onto Wernicke's area for comprehension. In the case of vision, the abstracted information goes from the primary visual area to the angular gyrus for association and then to Wernicke's area (Figure 15-21). From Wernicke's area, if something is to go to output, perhaps after further processing and other input from other parts of the cortex, it is given meaning and then sent to Broca's area for administrative cleaning and polishing. From there one can infer that it goes to the motor cortex.

Which part of the motor cortex the information is sent to depends on the specific form of output. If it is to be spoken, the chain of information goes to motor areas that direct word formation by the muscles in the larynx, throat, face, and jaw. If to be written by hand, the information goes to centers controlling the forearm and hand. If the information is to be written larger, such as on a blackboard, then the output goes to centers controlling muscles of the upper arm, shoulder, chest, and back. That all of this is under control of higher centers can be illustrated easily by comparing the writing styles of different persons. Whether one writes small and uses muscles of the hand and forearm or large and uses quite different muscles in the upper arm, etc., the writing style is the same (Figure 15-22).

Therefore one can see that Wernicke's and Broca's areas are at a common intersection between different paths of input and output. Language information can come in by one or more channels and go out by several channels.

There also is more between. Language is not just cold, logical chains of information. It may take on subjective and emotional flavors, which are contributed from other regions of the brain. Injuries or problems in another part of the cortex lead to an interesting (except to those directly involved) syndrome known as Gilles de la Tourette's disease. With this condition the speaker is a compulsive swearer and continuously speaks in a manner that is socially unacceptable. Such patients, or victims, are unable to inhibit swearing except with utmost effort. Thus it appears that there are regions in the cortex associated with language of an emotional content and with normal (to various degrees) social inhibition.

Damage to another region of the cortex (Figure 15-20) leads to a condition known as *prosopagnosia*. In this case, the unfortunate persons are unable to

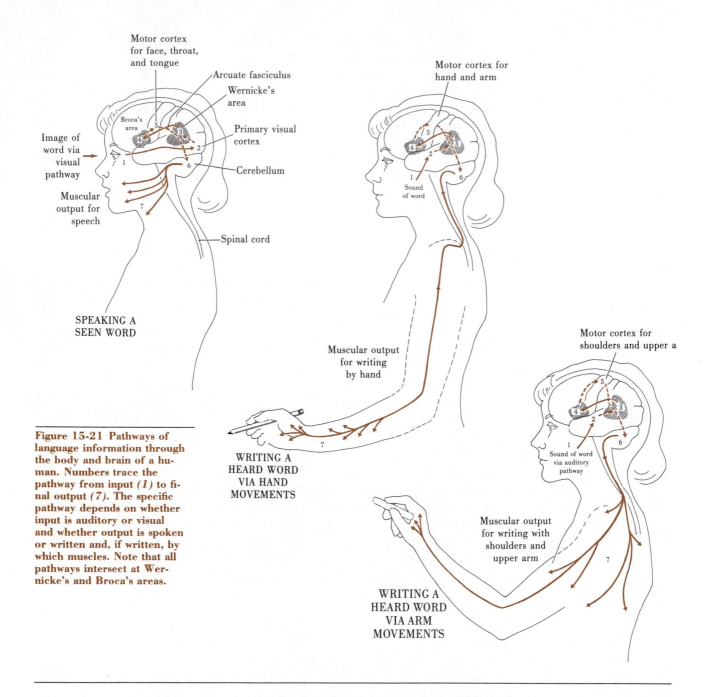

Figure 15-21 Pathways of language information through the body and brain of a human. Numbers trace the pathway from input (1) to final output (7). The specific pathway depends on whether input is auditory or visual and whether output is spoken or written and, if written, by which muscles. Note that all pathways intersect at Wernicke's and Broca's areas.

identify faces. If they are shown pictures of familiar faces or when they see the persons, they are unable to say who it is. The problem from this region of the cortex, however, is completely visual. If the person hears a familiar person speak, he or she can readily identify the voice. Thus there appears to be a region of the cortex devoted specifically to facial recognition and another area for speech recognition. It is likely that other species have similar specialized regions of the brain appropriate for individual recognition. Parents and offspring of penguins, for example, are able to identify each other vocally even in the midst of nesting colonies that may have millions of individuals.

Right versus Left Sides of the Cerebrum: Brain Lateralization

Other parts of the language (and thinking) picture involve differences between the two sides of the cerebrum. Broca was the first to detect these differences. Injury to the left side of the brain produced much more serious aphasia (inability to speak) than injury to the same parts of the right side. Present understanding of this increased greatly with recent research in primates and humans from split-brain work conducted by Sperry and his associates (Sperry 1970, 1974, Nebes and Sperry 1971).

There is a prominent band of neural tissue, containing about 200 million nerve fibers, that connects the right and left sides of the cerebrum, called the *corpus callosum*. Although it may be carrying as many as 4 billion or more impulses per second, no one was really sure of its function. Some persons thought it served just to hold the two sides of the brain together. A few unfortunate persons with severe and constant epilepsy, who could not be cured with medicine, were able to recover substantially by having the corpus callosum cut. The epilepsy involved neural electrical storms that traveled back and forth between the two sides and, apparently, built up intensity in the process. Superficially the patients appeared normal after the operation of cutting the callosum. Further testing, however, revealed interesting, although not normally debilitating, problems. The results perhaps are best illustrated (Figure 15-23).

Figure 15-22 Writing styles of two different persons writing with either the hand (A and C) or arm and shoulder (B and D). Styles may differ widely among different persons, but for any one person there is less variability depending on which set of muscles is used.

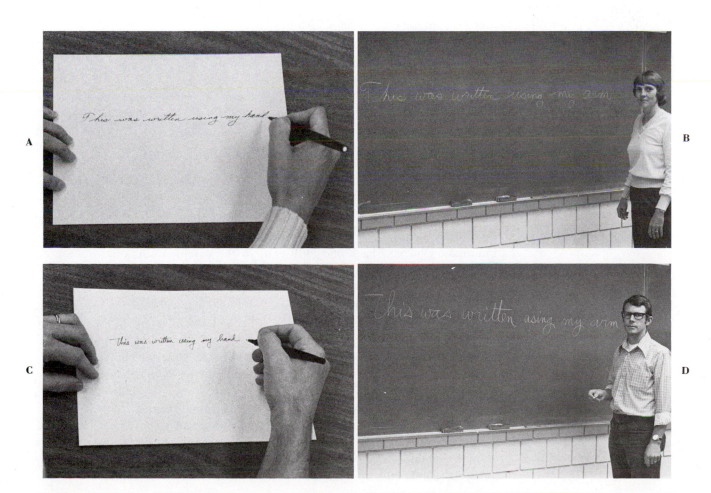

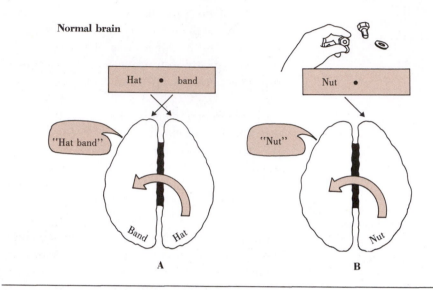

Figure 15-23 Examples of results from split-brain experiments. Subjects are asked to look at the dot in the center of a screen and report the words they see flashed briefly. Because of the visual pathways, as described earlier, words on the left go to the right hemisphere and vice versa. Information in the right hemisphere is normally transferred via the corpus callosum to the left hemisphere for synthesis with information already in the left side for assembly in the speech areas of the left hemisphere. A, Normal subject reads and reports "Hat band." B, Normal subject reads and reports "Nut" and also reaches behind a screen and correctly selects a nut by touch from among several objects not visible to the subject. C, Only information going directly to the left hemisphere can be verbally reported because the cut corpus callosum does not permit passage of information from the right hemisphere. D, With information going only to the right hemisphere, the subject is not able to report seeing any words, but the left hand is still directed by the right hemisphere, through neural pathways that cross to the other side below the corpus callosum, to select the correct object by touch. Modified from Sperry, R.W. 1970. Res. Publ. Assoc. Res. Nerv. Ment. Dis., Vol. 48, and Sidtis, J.J., et al. 1981. Science 212:344-346.

Following these results on patients who had the corpus callosum actually cut, techniques were developed for simulating the results on normal persons. The procedures involve presenting photographic images so quickly that there is not enough time for the information to travel across the corpus callosum. From this research it was discovered that the two sides of the cerebrum function essentially as two separate computers with two different functions. The left side, in most persons in Western culture, is involved with series or sequential and analytical processing, which includes language and such things as mathematical types of analysis. The right side, on the other hand, is involved with parallel processing, in a sense, and looking at things more holistically and synthetically. This side of the brain is concerned more with the abstract, artistic, and subjective side of thinking. The corpus callosum is the normal communication channel that permits information to pass back and forth and, in the complete thinking process, to combine the results of both analytical and synthetic forms of thinking.

The left side is normally the dominant side and, it has been proposed, leads to the prevalence of activity and analytical thinking commonly seen in humans. R. Ornstein (cited in Sagan 1977:177) has likened the situation to the difficulty humans have seeing stars in the sky during the daytime. All of the stars are still up there, but because the sun lights up the atmosphere and dominates the daytime, we cannot see them. We have to wait until night, when the sun is out of the way, before the stars become visible. Similarly, according to the analogy, humans find it difficult without practice and effort to think synthetically. It is much easier to be active and analytical because thinking is dominated by the left side. This may be why persons get most insights and solutions to problems while taking a shower, listening to music, lying in bed, or otherwise relaxing—the left hemisphere becomes idle and permits the right hemisphere to be expressed.

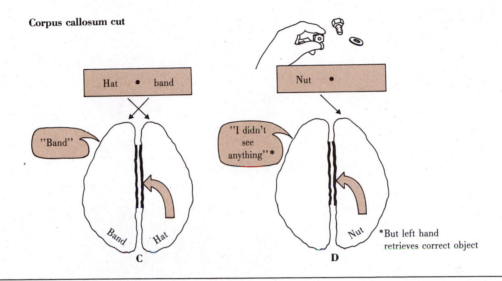

Corpus callosum cut

Much if not most of the properties of human awareness and consciousness are associated with the left hemisphere and may be tied to language. Deese (1978) proposed that consciousness developed as a process associated with language to monitor, detect, and correct routine errors in speaking. Persons with split brains or those allowed input only to one side seem unaware of things they cannot process in language, although the brain on the other side can still understand and perform. This has been shown by the ability to pick up an object that could not be identified verbally. The hand performed properly, but the person could not state what had been done. Connections to lower, emotional regions of the brain also still function normally. A female shown a nude photograph to the right but not the left hemisphere became embarrassed but did not know why.

The topic of language processing in humans dredges up earlier discussions in this book about the differences between language in humans and communication in other animals (pp. 294-300) and the problems of communication of consciousness, self-awareness, and anthropomorphism (pp. 125-127). Do other animals constantly live, even when awake, in a mental world that would seem like ours when we are asleep or anesthetized? It is an intriguing question but, unfortunately, still beyond our grasp. For further discussion of this fascinating subject, see Eccles (1970, 1977), Crick (1979), and others.

Before leaving the subject of differences between right and left sides of the brain, it should be noted that some of the differences may be developmental and depend on culture. The localization of right and left brain functions may be different in humans raised in Japanese or Polynesian cultures, regardless of genetic race (Tsunoda 1978, *The Japanese Brain: Brain Function and East-West Culture*, not available in English but described in Sibatani 1980). Persons of an Oriental race raised in Western culture have the Western right-left differences,

Figure 15-24 Sensory lateralization in hearing: Western versus Japanese differences in processing of sounds by the brain. Modified from Sibatani, A. 1980. Science 80 1(8):22-27.

Western pattern of lateralization

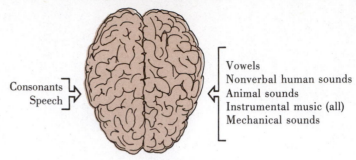

Consonants
Speech

Vowels
Nonverbal human sounds
Animal sounds
Instrumental music (all)
Mechanical sounds

Japanese pattern of lateralization

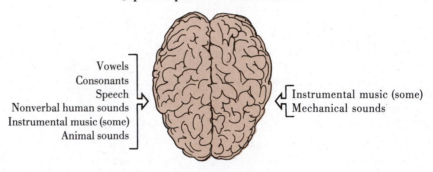

Vowels
Consonants
Speech
Nonverbal human sounds
Instrumental music (some)
Animal sounds

Instrumental music (some)
Mechanical sounds

whereas Caucasions, for example, raised in Japan have the "Oriental" differentiated brain. Figure 15-24 illustrates some of the reported differences for the hearing of language and other sounds. It is likely that some of the differences in organization of the brain are developmental and depend on which inputs are encountered first during compartmentalization processes.

Contributing Components and Other Pathways

Three other major topics are of interest to the preceding discussion of pathways in vertebrate brains: the role of the cerebellum, the reticular activating system, and the limbic system.

Based on clinical information from injuries and other problems in the cerebellum, it was inferred that the cerebellum was the major point of administrative handling of muscle control; that is, commands came down from higher centers in the cerebrum (of mammals) or elsewhere (e.g., midbrain in amphibians), and the cerebellum worked out the details. The cerebellum is second only to the cerebrum of humans in size, number of neurons, and complexity of neurons. The Purkinje or basket cells, those neurons with an incredible amount of dendritic branching (Figure 13-5), are found in the cerebellum. There are numerous known feedbacks between the cerebrum and cerebellum so that the administrative functions of the cerebellum are not only quite reasonable but highly likely. For a good introduction to the inferred processing of motor control by the cerebellum, see Eccles (1977). This is not the whole story, however, and there may be significant differences in other vertebrates. There is a rough correlation between proportional size of cerebellum and amount of activity and agility. (Birds and some tree-climbing mammals, for example, have very large cerebellums.) Within particular groups, such as some of the amphibians, however, some of the more active and agile species have smaller cerebellums than those of their sluggish relatives. There is much yet to be learned about the cerebellum and most other parts of the vertebrate brain.

The *reticular activating system (RAS)* is a loose net of neural connections largely in the brainstem but extending into the cortex. It appears to function importantly in alertness and attention. It does not connect specifically to any of the other pathways but, rather, with essentially all of them. It is somehow involved with monitoring, sorting out, and mediating all the pathways and maintaining some order in all the traffic. The RAS may be the closest thing to the "hidden observer." When the RAS is activated, the animal is alert and awake. Sleep, on the other hand, involves a reduction of activity in the RAS. The RAS seems to serve as something of a central switchboard operator or relay system.

The *limbic system* is a collection of regions, or nuclei, in the brain that borders on and overlaps with (depending on authority) the brainstem and the outer layers of the cortex. The brain is formed of general layers that are complex and intricately connected and show an array of homologies among different groups of vertebrates. The first region of the forebrain, the telencephalon, can be subdivided into three basic sections: the first or inner, concerned basically with the primitive sense of smell: the second, known as the striatum: and the third, outer or mantle layer, which is the pallium or cortex. These may be subdivided further into, for example, archistriatum, paleostriatum, and neostriatum and archipallium, paleopallium, and neopallium. The human cerebral cortex is largely composed of neopallium (neocortex).

During the course of vertebrate evolution the brain has gradually acquired new layers and parts, building on top of the old, and there has been some transfer of function from one region to another. Because of the constant remodeling, it is difficult to provide a simple picture. As a start, however, the limbic system can be conceived somewhat as a middle general layer between the older and newer parts of the brain. It is often thought of as the old "mammalian brain," although, because of the remodeling, it does not correspond entirely to any single part of the striatum or older parts of the pallium; traces of it can be found far below the mammals, certainly in the reptiles and even in fish brains.

MacLean (discussed extensively in Sagan 1977) simplified this picture even further to consider that the human brain is something of a triune brain: the inner and oldest part inherited from ancient ancestors, the so-called reptilian brain (although it certainly goes back to the fish) concerned with basic matters of survival, such as eating and sex; the middle, limbic system involved with emotion and motivation, such as anger, rage, and love; and the third, outer, modern part devoted to human intelligence, consciousness, and language. The triune brain is a vast oversimplification, but it is useful conceptually for emphasizing that humans have inherited a highly evolved organ with advanced abilities built on ancient characteristics and behaviors, with such things as human language being the most advanced of all. But even our most advanced neural processing still

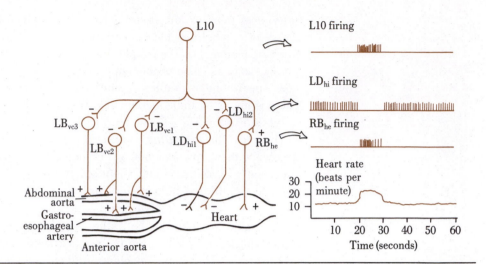

Figure 15-25 Control of cardiovascular output in *Aplysia:* effects from a single higher-order identified neuron, L10. Schematic diagram showing the relationship and connections of L10 to motor neurons and the heart system and the effects of activity in L10 on two other neurons and heart rate. The labels (e.g., *L10, LB_{vc3}*) identify individual neurons. Modified from Kandel, E.R. 1979. Sci. Am. 241(3):66-76.

retains connections with and input from other, older parts of the system. What this means for future human life, even continued existence, is a matter of ongoing discussion and debate that goes beyond the scope of this text.

Tracing the Actual Circuits: Pathways Through Invertebrate Brains

The preceding discussion has only scratched the surface of what may be going on inside the vertebrate brain. To trace neural integration to its finest level of operation one needs to account for the precise, neuron-to-neuron pathway over the entire route of the messages. This is still far from being accomplished in vertebrate brains and behavior. But in some of the simplest invertebrates that have nervous systems with limited and consistent (identified) numbers of neurons, a number of these pathways now have been traced from beginning to end, or at least have been elucidated sufficiently to permit one to understand the pathway. Elementary circuits have been described for a variety of invertebrate movements and behaviors including insect walking and flying, crayfish tail flexion, movements of the crustacean's gastric mill (by which they chew their food), mantid strikes, and many more. For an illustration, consider the neural control of the heart and simple gill reflex in a marine snail, *Aplysia*—molluscs that have received extensive neurobiological study (Kandel 1979a,b).

Some *Aplysia* neurons have been shown to control the heartbeat through specific channels of excitation and inhibition (Figure 15-25). Others have been shown to mediate the gill withdrawal reflex when the siphon or mantle has been stimulated (Figure 15-26). Note that virtually all the connections have been accounted for and there are both excitatory and inhibitory synapses. Not only are the specific neural connections now known, but progress has been made in understanding, at least for these systems, how the pathways may be modified (i.e., involving learning) for the gill withdrawal reflex. The subject of behavioral modification in the individual, however, will be postponed to Chapters 18 and 19. In the next chapters the role of chemicals in neural processing, then the problem of controlling food intake in animals, as an example of the overall understanding of behavioral integration, will be considered.

Figure 15-26 Neural control of the gill withdrawal reflex in *Aplysia*. A, A snail showing the position of the gill when extended in normal position and after withdrawal following stimulation of the siphon. **B,** Schematic diagram of neurons involved in the withdrawal reflex. **A** modified from Kandel, E.R. 1979. Behavioral biology of *Aplysia*. W.H. Freeman & Co., Publishers, San Francisco. **B** modified from Kandel, E.R. 1979b. Sci. Am. 241(3):66-76.

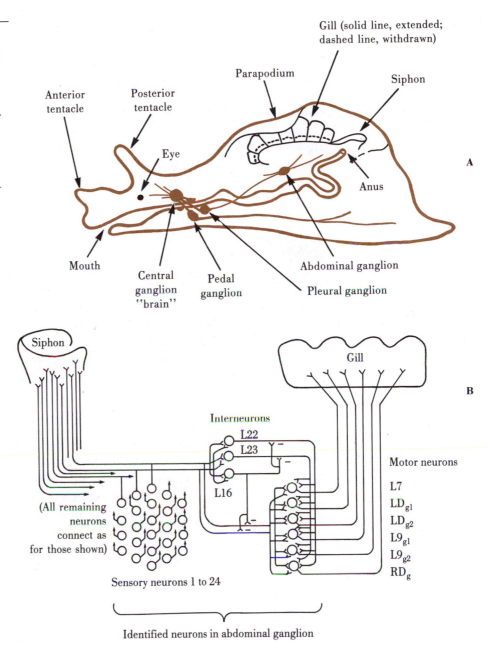

SUMMARY

Although we still are far from a complete detailed understanding of how the nervous system operates, astonishing progress has been made during recent years, and now at least some of the details can be filled in. For some of the simpler organisms with simpler nervous systems we are approaching a description of most or all of the connections and knowledge of which are involved in certain behaviors. In more complex organisms, particularly the arthropods and vertebrates, we still are tapping general lines and tracing broad pathways, plus gaining insight on particular details of bits and pieces of the system. Visual pathways are described as a model. It is clear, however, that there are several important principles that apply generally to all nervous systems.

First, neurons do not simply pass information from one point to another like so many wires or passive conductors. Rather, different neurons act differently on incoming information, depending on the summation of excitation and inhibition and internal properties. Different neurons respond differently to the same sorts of input and may have quite different patterns of output from their neighbors. They pass their output on to other neurons in a variety of paths that may run parallel to output from other neurons, diverge to several other neurons, or converge with input from several others. All of this sifting and sorting permits the process of information extraction or abstraction. In other words, various attributes or qualities of input can be separated and abstracted, for combination or comparison with input from elsewhere in the system, such as from adjoining regions of the visual field.

A second major principle of neural processing involves inhibition by some impulses over other, neighboring impulses. All of the surrounding groups of neurons are attempting to do the same thing, and it is the most strongly excited pathway that makes it through. This reduces the traffic and sharpens the effectiveness of information passage in different parts of the system, although a variety of other information may be processed simultaneously elsewhere in the nervous system. One can think, chew food, breathe, and tap the foot simultaneously, for example.

As sensory input comes into the nervous system, it is sequentially abstracted further and further as it progresses to higher levels of processing. In the converse, motor output is thought to start at higher, general command levels and then is administratively broken down for the final muscular (or other) output. For example, language may be commanded, that is formed and given meaning, in a higher level such as Wernicke's area of the cortex and then be polished up and prepared for final output in a lower area such as Broca's area of the cortex. Higher command centers may command that an animal run toward food, toward a mate, or away from an enemy; then lower centers take care of the actual job and ensure that the movements are smooth, coordinated, and corrected on the basis of changed, incoming sensory input.

At least heuristically, these various functions can be viewed somewhat as command, administrative, and secretarial. In a sense, the reticular activating system of the vertebrate brain is the telephone operator or central switchboard that answers incoming calls and directs them to the appropriate destinations. Furthermore, in this analogy, when that switchboard closes down, the animal goes to sleep.

Different parts of the nervous system are organized and localized for different specific functions, such as for particular sensory or motor jobs. Many of these have been specifically mapped in vertebrate brains. In addition to mapping the major processing and association areas and centers of particular activities, many of the specific pathways from one part of the nervous system to another have also been mapped.

Much of our understanding has come from studies of particular pathways and functions, such as vision and human language processing, both of which are described in this chapter in some detail. Present understanding, however, still is relatively superficial. The specific neuron-to-neuron pathways involved in particular behaviors are known for only a handful of simple behaviors in simple organisms. In addition, we have a poor understanding at best of the origins of voluntary or willful behavior, of what, if anything, serves as the overall process of integration, and of how learning enters the pathways. There are some general hypotheses of how learning works; these will be discussed in subsequent chapters.

Recommended Additional Reading

Bullock, T.H., R. Orkand, and A. Grinnell. 1977. Introduction to nervous systems. W.H. Freeman & Co., Publishers, San Francisco.

Eccles, J.C. 1977. The understanding of the brain, 2nd ed. McGraw-Hill Book Co., New York.

Ewert, J.P. 1980. Neuroethology, an introduction to the neurophysiological fundamentals of behavior. Springer-Verlag New York, Inc., New York.

Fentress, J.C., editor. 1976. Simpler networks and behavior. Sinauer Associates, Inc., Sunderland, Mass.

Kandel, E.R. 1976. Cellular basis of behavior. W.H. Freeman & Co., Publishers, San Francisco.

Sagan, C. 1977. The dragons of Eden. Ballantine Books, Inc., New York.

Scientific American. 1979. The brain (special issue). Sci. Am. 241(3).

CHAPTER 16

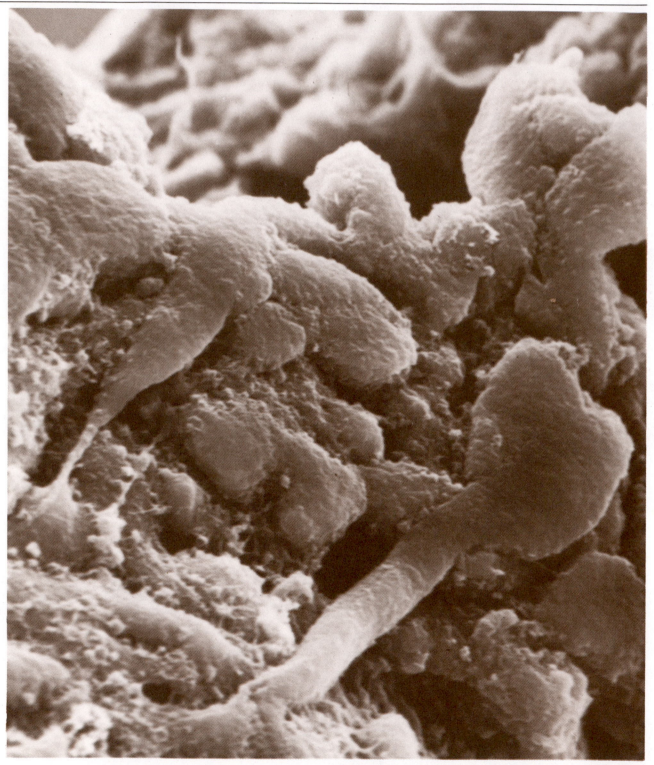

Neurosecretory cells in the brain of a larval tobacco hornworm moth.

HORMONES, NEUROCHEMICALS, AND BEHAVIOR

In the last chapter behavior was discussed in terms of neural pathways and connections. The internal integration of behavior, however, involves much more than just the neural, "electrical" aspects. Virtually all behavior also involves participation by chemicals in various ways. At the minimum in a behavioral sequence there are transmitter chemicals that move from one neuron to another. Even the electrical events can be viewed as chemical in a sense, with the biochemistry of cell membranes and the electrical charges being carried by ions. But chemicals also play a role in a much different sense. Many of them are quite complex and carry information themselves, often in close conjunction with the neurons and often on their own via the circulatory system. The point of this chapter is that hormones and other neurochemicals importantly influence behavior and vice versa (Figure 16-1).

The neural and endocrine systems are so closely intertwined that they often are viewed as one system, the neuroendocrine system. There is a continuum between transmitters that cross the short distances of the synapses from one neuron to another and hormones that are carried in the bloodstream with few, if any, interactions with the nervous system. Between these two extremes there are neurosecretory cells that are located in the brain and resemble neurons except that they release their chemicals into the bloodstream rather than into synapses, and there are endocrine glands that are stimulated by direct nerves from the central nervous system. Some chemicals, such as epinephrine, serve both as synaptic transmitters and glandular hormones.

At least conceptually, if not biologically, endocrine glands seem to merge with exocrine glands, those which secrete externally rather than internally. In the discussion of communication and social behavior, the powerful effects of pheromones in the lives of many animals have been seen. Pheromones may act as primers that affect the physiology of other individuals or releasers that stimulate specific behavior. Pheromones are, in effect, hormones that travel outside one body to have their effects on another body.

The conceptual relationships of all these components in behavior are shown in Figure 16-2. The positions of chemicals in this scheme are highlighted with the shaded boxes. Figure 16-2 obviously is abstract and oversimplified. There are, for example, many different glands, some of which are endocrine, some exocrine, some stimulated by nerves, some only by circulating substances, some by both nerves and substances, and there are many different senses, as described in Chapter 14. Pheromones are placed in parentheses because they are not universally important to all species and because they also may be viewed in a sense just as another form of behavioral output, such as from the muscles. But the principles as diagrammed here apply to all multicellular animals that possess nerves, muscles, and a central nervous system, that is, most living animals. To apply this to any particular organism or group of organisms, one only needs to fill in the details.

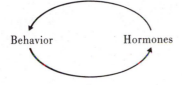

Figure 16-1 Basic dependence of behavior and hormones on each other.

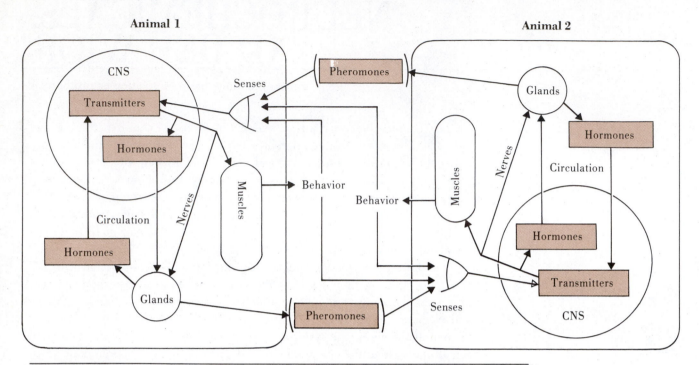

Figure 16-2 General relationships between the intraanimal and interanimal components of behavior and chemicals. The three major categories of behaviorally important chemicals are transmitters, hormones to and from the nervous system, and pheromones between animals. The behavior of an animal, resulting from patterned muscular output, may affect, via the senses, chemical interactions both in the behaving animal and in other animals.

Figure 16-3 Major glands and neural-endocrine integration in insects.

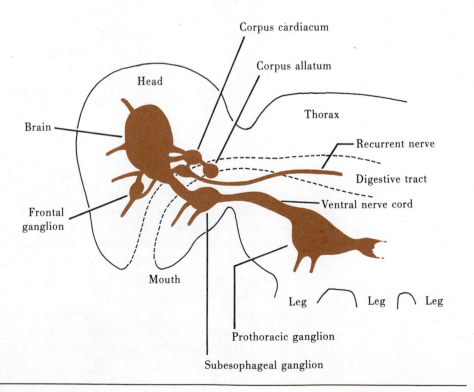

INSECT BEHAVIOR AND HORMONES

In insects, for example, there are two major endocrine glands closely associated with the brain, the corpora cardiaca and the corpora allata (Figure 16-3). Neurosecretory cells in the protocerebrum produce brain (activation or prothoracic-trophic) hormones that move from the brain, along the axons of these neurosecretory cells, to the glands. The numbers of neurosecretory cells range from 5 to several thousand, depending on the species involved. The hormonal outputs from the corpora cardiaca and allata produce various morphological/physiological effects and behavioral effects, including the movements associated with molting, finding hiding places to spend diapause (similar to hibernation), cocoon-spinning patterns, feeding, and many more. The diversity of other insect glands and behaviors involving pheromones, including those for reproduction, caste production and recognition, trails, alarms, etc., has already been indicated in earlier chapters.

VERTEBRATE BEHAVIOR AND HORMONES

The general behavior-hormone picture as applied to vertebrates is shown in Figure 16-4. This figure focuses on reproduction and related behavior, a central matter in vertebrate life. Glands other than the gonads also may affect behavior, such as the adrenals during emergencies. But they are less prominent; the principles are more or less similar; and they are represented indirectly in the diagram. Short-term emergencies and long-term stress, for example, and the result-

Figure 16-4 General picture of neural-endocrine integration in vertebrates.

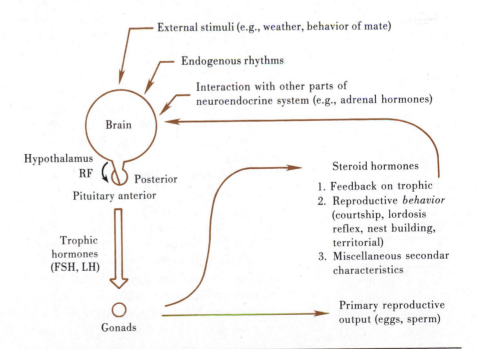

ing output from the adrenal glands may affect neural functioning in particular parts of the brain. In addition to interactions with hormones from glands other than the gonads, the neurons in various parts of the brain are affected by endogenous rhythms (e.g., diurnal) and—via the senses—external stimuli.

The external stimuli vary from species to species and include such things as environmental conditions of the physical habitat (e.g., effects of rainfall on desert species, photoperiod, temperature, water pH), vegetation (e.g., odors from plants, chemicals in plant food), and *the behavior of other animals, such as mates, offspring, and social peers or superiors.*

All of these are routed through the brain presumably in familiar neural fashion, involving particular pathways, regions of the brain, abstraction, and so forth. Some of the neural pathways lead to neurosecretory cells in the hypothalamic region at the ventral side of the diencephalon or thalamus. Some neurosecretory output extends into the posterior pituitary gland where it is released into the general circulation. Examples include oxytocin and prolactin. Other output is released more locally into the hypothalamic portal blood vessels and carried a short distance to the anterior pituitary. The chemicals released over this short path are called **releasing factors (RFs).** At the anterior pituitary they stimulate the release of further hormones known generally as **trophic ("feeding") hormones.** They are not related to food eating but are called trophic because they "feed," in the sense of stimulating, other glands and tissues in the body. The gonadotrophic hormones, in particular, are follicle-stimulating hormone (FSH) and luteinizing hormone (LH). The releasing factors from the hypothalamus that stimulate their release are FSHRF and LHRF, respectively. FSH and LH travel via the bloodstream throughout the body and are picked up by cell membrane receptors at appropriate tissues, particularly in the gonads.

The gonadotrophic hormones then cause glandular development in the gonads, depending on which sex is involved. This results in primary reproductive output, eggs or sperm, and another type of hormone, the sex steroids. The steroids travel throughout the body via the circulatory system and have several effects, including familiar morphological developments such as deeper voices and beards in men and changes in body shape in women. They also feed back on the nervous system in two important ways: they modify neural output in particular regions of the brain, which affects behavior directly, and they interact with the pathways that lead to the hypothalamic neurosecretory cells. The modified behavior of the animal may serve as a stimulus (e.g., a releaser) for the behavior, nervous system, hormones, and reproductive development of another individual that may be a mate or potential mate.

All of this can now be applied to specific examples. Consider what takes place during the sequence of reproduction, for example, in a pair of birds. The birds first engage in a period of courtship, then construct a nest, then lay eggs and incubate them, often with the male and female taking turns. Finally, when the chicks hatch, the parents feed and protect their offspring. There are many ways in which this sequence *could* be interpreted. Perhaps, in the anthropomorphic or action-by-reason view (pp. 127 and 128), the birds want to have some babies, understand what needs to be done to accomplish that, and set about doing the job, step by step. Alternately, maybe there is simply a chain of complex events that set up a series of stimuli such that, in the presence of the proper stimulus, the next step is triggered. But neither of these two interpretations appears correct. If birds are given stimuli, such as eggs in a nest or chicks that are begging food, they will not necessarily behave properly unless they are in the correct hormonal condition. Sometimes the birds will not incubate when there are eggs, and at other times they may incubate when there are no eggs.

If the stimuli are not appropriate for the particular hormonal state of the animal, the birds generally either will completely ignore the stimuli or else become aggressive, destroying the eggs or killing the chicks. These outcomes should not occur if either the adults "knew" what they were doing or simply were responding to particular stimuli. Instead, one must turn to an interpretation based on the neural-hormonal picture discussed earlier. The proper hormonal conditions *in conjunction with* certain stimuli are required before particular behaviors, such as nest building, incubation, or chick feeding, will be expressed.

One of the first examples of this in birds was demonstrated by a series of studies of ring doves by D.S. Lehrman (e.g., 1964). When a male and female ring dove are placed together, the male first begins a bow-and-coo display. If provided with nesting material and everything else is suitable (e.g., they are not frightened or stressed such that adrenal inputs block reproductive pathways), then the pair of doves will go through a sequence of stages whereby they pair, build a nest, lay eggs, incubate, hatch and raise the young, and then abandon the young and start all over again.

If one focuses on just one of the stages, for example, *incubation* behavior, the birds, stimuli involved, and internal hormones can be altered experimentally, as was done by Lehrman. A summary of his experiments is as follows:

1. If given a normal mate and nesting material, a female will begin incubating eggs (her own or others that are provided artificially) in 5 to 7 days, but not before. If given eggs in a nest before day 5, she will completely ignore them or perhaps take the nest material for building her own nest.
2. If given a normal mate but no nesting material to work with, it will be about 8 days before the female will incubate eggs that are provided to her.
3. If given a castrated male for a mate, the female will not incubate eggs at any time. Presumably, although this was not done by Lehrman, a castrated male given hormone injections would serve as a normal male and stimulate a female to eventually incubate.
4. If a lone female dove is given a nest and eggs, she will not incubate at any time.
5. If a lone female dove is given hormone injections, the results depend on the hormone:
   a. With *progesterone* she will begin incubating within *3 hours*.
   b. With *estrogen* she will incubate after 1 to 3 days.
   c. *Testosterone* injections have no effect on the female, and she will not begin incubating.

Incubation is only one of several behaviors involved in the whole reproductive sequence. In the case of feeding behavior, whereby the dove produces a "crop milk" from the lining of the esophagus and regurgitates it to feed the young, *prolactin* is the required hormone. The whole sequence of interacting factors—behaviors, hormones, and internal, physiological conditions—is shown in Table 16-1.

Table 16-1 Summary of events during reproduction in the ring dove

| | | Behavior | | | | Major |
	Day	Male	Female	External stimulus	Major hormones	physiological effects
1 week	0 (start)	Bow and coo		Mate	Testosterone in ♂	Ovary and oviduct development
	1	Nest call	Nest call		LH and FSH (pituitary)	
	2	Nest build (bring materials)	Nest build (build)	Nest material	Estrogen and progesterone (gonads)	
		Copulation	Accepts copulation			
	7		Egg laying			
2 weeks	~9					
		Incubation	Incubation	(Eggs in nest)		
	~23				Prolactin	Crop development
	(hatch)					
2½ to 4 weeks		Feeds young	Feeds young	Young begging		
	40	(Goes back to start)				

Birds are not the only vertebrates in which these interactions occur. The same general processes can be seen in virtually all vertebrates and usually involve the same or similar hormones. Anole lizards, for example, demonstrate the interaction between ovarian development in the female in response to male behavior very clearly (Figure 16-5). Normally the male anoles go through a period of aggression toward other males, after which they court the females. The females normally show ovarian development only after the courtship period begins. In different experimental combinations, the ovaries of females exposed only to courting males developed early and continued to show follicular growth. On the other hand, ovaries of females exposed only to aggressive males failed to develop at all. Females exposed to a reverse situation, that is, first to courting males and then to aggressive males, showed early ovarian development, followed by early regression (Crews 1975, 1979).

Details of the interaction between behavior and hormones vary depending on the extent and nature of parental care, whether eggs are layed in nests or developing embryos are carried internally (e.g., in most mammals), and the timing and patterns of ovulation and embryonic development. In mammals, for example, there are different patterns of ovulation in relation to courtship and copulation. In some mammals with induced ovulation, including rabbits and a few species from numerous other orders, ovulation is triggered a few hours after copulation. In most other species, however, ovulation occurs spontaneously at periodic times.

The complex pattern of reproductive hormones in the female human is discussed in most introductory biology and health classes. It will be reviewed briefly here, however, to place it in context with the rest of this discussion and to emphasize that humans fit into the general vertebrate scheme. Most persons are

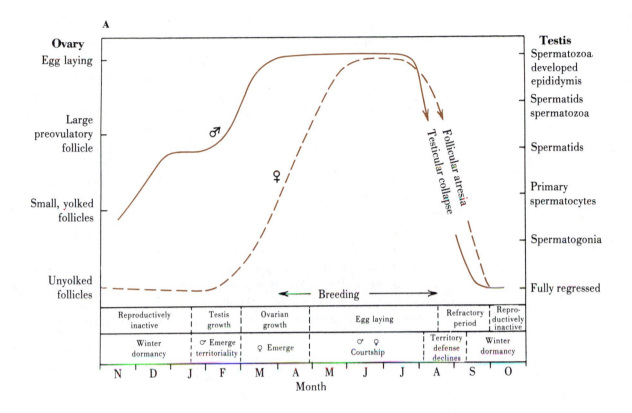

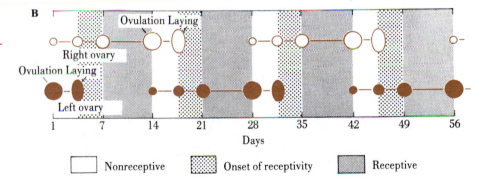

Figure 16-5 **Effects of male behavior and female hormones on anole lizards. A, Sequence of behavioral and physiological events in the annual reproductive cycle of** *Anolis carolinensis.* **Note that the males normally engage in a period of male-male aggression and territoriality before commencing courtship behavior with females. B, Normal sequence of female behavior patterns and ovarian development during the breeding season. C, Behavioral receptivity of females to males after the females' ovaries were removed and they received injections of various female hormones of different concentrations. A** and **B** from Crews, D. 1975. Science 189:1060. Copyright 1975 by the American Association for the Advancement of Science. Courtesy David Crews. **C** modified from Crews, D. 1975. Science 189:1060.

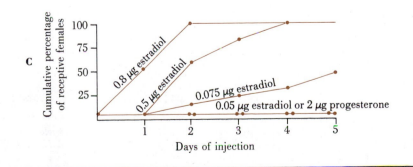

familiar with the changes in behavior and mood that occur with the monthly cycles in the human female; all persons are either female or have girl friends, wives, mothers, or daughters. The changes that occur during the menstrual period involve drops in the levels of circulating steroid hormones (Figure 16-6). This amounts, in essence, to a form of natural drug withdrawal. The human female shows some changes in behavioral response to a mate or potential mate at different points in the monthly cycle, but they are much less prominent than in other mammals. The human female is more or less responsive, depending on learned and cultural factors, at all times once she has entered puberty.

In other mammal species, however, the female may be receptive to males only during restricted, specific times, commonly known as "heat" or *estrus*. During these times they usually display certain behaviors, again in response to internal hormonal conditions. Pet owners usually are familiar with the behaviors and may find them to be a nuisance, such as in some female house cats. The rigid, downward-arched-back posture of females during estrus in many species of mammals is known as the *lordosis reflex*. It is prompted by high levels of estradiol in the animal's bloodstream (Figure 16-7). At other times the females will not be receptive to the males and may even be aggressive toward them and actively reject any

Figure 16-6 Hormonal and anatomical changes during the menstrual cycle of human females. This complex set of interactions can be understood most easily by slowly and carefully following the numbered sequence. For additional explanation, refer to a supplementary text. Note that scales, such as blood concentrations of hormones, are not indicated. They can be researched from the literature as an exercise.

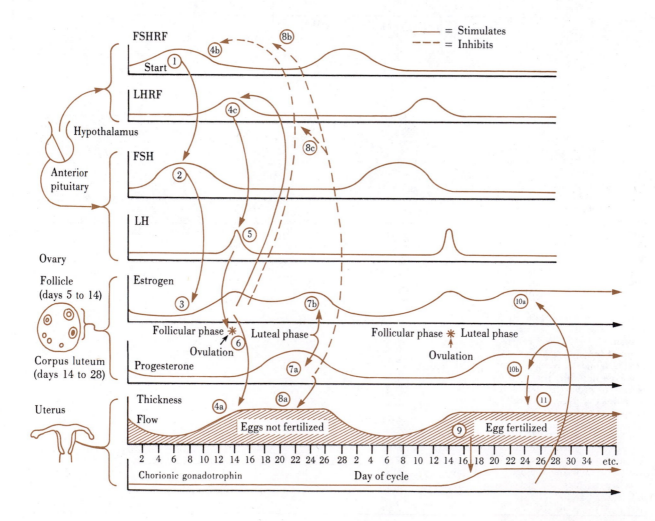

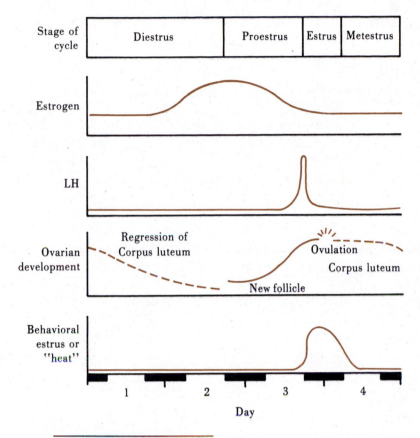

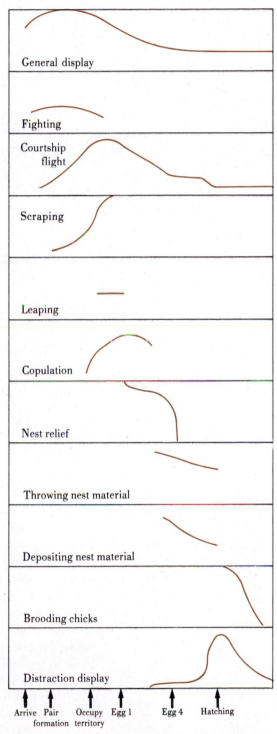

Figure 16-7 Normal estrous cycle and behavior of the female white rat. *Black bars*, **Periods of nighttime darkness.** Modified from McEwen, B.S. 1976. Sci. Am. 48-58.

Figure 16-8 Several behavioral components involved in the reproductive cycle of the ringed plover. Frequency of each behavior is shown by the height of the line. Modified from Laven, H. 1940. J. Ornithol. 88:183-288.

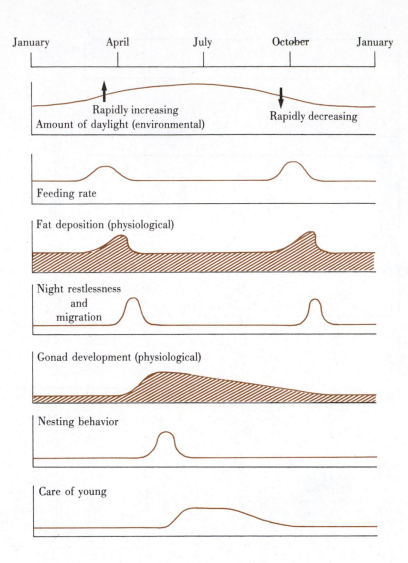

Figure 16-9 General annual sequence of hormonally dependent behavioral and anatomical events in north temperate migratory songbirds.

attempts to copulate. In most species the two sexes are more or less synchronized, mediated by hormones on both sides, and males may not show much or any behavioral interest in courtship and copulation either except at appropriate times. For additional information on reproduction, see van Tienhoven (1968).

So far only a few segments of an animal's behavior as affected by hormones have been discussed. These segments need to be placed in perspective of the individual's entire life history and life span. Examples of the whole complex of behavior for reproduction and overlaps among components are illustrated in Figure 16-8. Placing this into an annual cycle is illustrated not only for reproduction but also for migration and other behavior for the north temperate migratory songbirds in general (Figure 16-9).

Similar examples abound in the published literature and can be found for nearly all multicellular animals, including noninsect invertebrates. Thus behavior, via the nervous system, affects hormones, which in turn affect behavior.

PATHWAYS AND MECHANISMS OF HORMONE ACTION

Where and how do hormones have their effects on neurons and the nervous system? In general, there are specific *target neurons located in particular regions of the brain.* These neurons possess the proper cellular receptors to interact with specific hormones. Many, although not all, of these receptors are located in the brainstem and limbic systems (recall the triune brain of the previous chapter). Some of the receptor areas have been located with the aid of radioactive-labeled hormones. Tritium-labeled estradiol has been injected into the body of female rats, for example. It was given time to be picked up by the brain; then the rats were killed and the labeled areas of the brain identified (McEwen 1976). In another example tritium-labeled testosterone demonstrated hormone-concentrating regions in auditory and vocal centers in male frog brains (Figure 16-10).

As might be expected, however, there are *many* kinds of hormones and neurochemicals that exist and affect the nervous system in a variety of ways. The list of hormones and transmitter substances includes nearly 50 or more substances (box, p. 518). The list continues to grow as more neuroendocrine substances are identified. It has been suggested that only a small fraction of the actual number of naturally occurring neural chemicals has been discovered and that the total could be around 200 or more (Barchas et al. 1978).

Depending on their general structure and how they interact with neurons, there are two broad classes of chemicals: *proteinaceous or peptide-type hormones and transmitters* and *steroid hormones.* Examples of the molecular structures for a few of them are shown in Figure 16-11. The two classes interact with cells in quite different ways.

The steroid hormones are believed to enter the cells, such as specific neurons in the brain, and combine with receptor molecules inside the cytoplasm. The hormone-receptor complex moves to an acceptor site on the DNA in the nucleus and affects subsequent expression of the genes via the messenger RNA and protein synthesis (Figure 16-12). This, in turn, may alter the functional properties of those particular neurons.

The peptide-related hormones, on the other hand, affect the cells at specific molecular receptor sites at the cell's surface. Through a complex sequence of molecular interactions, including a secondary internal messenger, cyclic AMP (cAMP), there are both short-term changes affecting the membrane properties of the cell and long-term effects on protein synthesis (Figure 16-12).

Figure 16-10 Uptake of androgen by auditory and vocal centers in the brain of clawed frogs *(Xenopus laevis).* Dots indicate locations of uptake as revealed by autoradiographic studies. After the animal has been injected with radioactive hormone, it is killed and the brain is removed. Sections of the brain are placed against photographic plates to locate areas of radioactivity. Similar results have been obtained for many other vertebrates including anoles (Crews 1975) and rats (McEwen 1976). From Kelley, D.B. 1980. Science 207:553-555. Copyright 1980 by the American Association for the Advancement of Science.

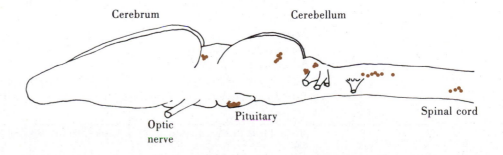

Cerebrum   Cerebellum

Optic
nerve   Pituitary   Spinal cord

Amine or peptide-related hormones

Figure 16-11 Molecular structure of various peptide-related and steroid hormones. LH, FSH, prolactin, and other proteinaceous hormones are large long-chain sequences of around 140 to 200 amino acids and cannot be easily illustrated because of their sizes.

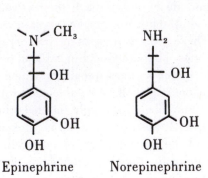
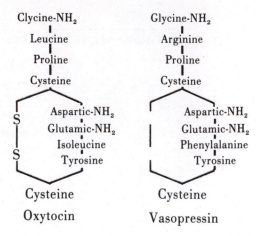

Epinephrine   Norepinephrine   Oxytocin   Vasopressin

Known and Hypothesized Chemicals Involved in Regulation of Animal Behavior

Dopamine	Prostaglandins
Norepinephrine	Corticosteroids
Epinephrine	Estrogens
Tyramine	Testosterone
Octopamine	Thyroid hormone
Phenylethylamine	Follicle-stimulating hormone (FSH)
Phenylethanolamine	Luteinizing hormone (LH)
Dimethoxyphenylethylamine (DMPEA)	
Tetrahydroisoquinolines	Enkephalins
	Beta-endorphin
Acetylcholine	Substance P
Histamine	Bradykinin
Gamma-aminobutyric acid (GABA)	Neurotensin
Gamma-hydroxybutyrate (GHB)	Somatostatin
Glycine	Angiotensin
Taurine	Bombesin
Purine	Cholecystokinin
Aspartate	
Glutamate	Follicle-stimulating hormone-releasing factor (FSHRF)
Serotonin (5-hydroxytryptamine)	Luteinizing hormone–releasing factor (LHRF)
Melatonin	
Tryptamine	Vasoactive intestinal polypeptide (VIP)
Dimethyltryptamine (DMT)	Adrenocorticotrophic hormone (ACTH)
5-Methoxytryptamine	Thyroid-releasing factor (TRF)
5-Methoxydimethyltryptamine	Sleep factor delta
5-Hydroxydimethyltryptamine (bufotenin)	Sleep factor S
Tryptolines	

Modified from Barchas, J.D., et al. 1978. Science 200:964-973, and Gurin, J. 1980. Science 80 1(1):28-33.

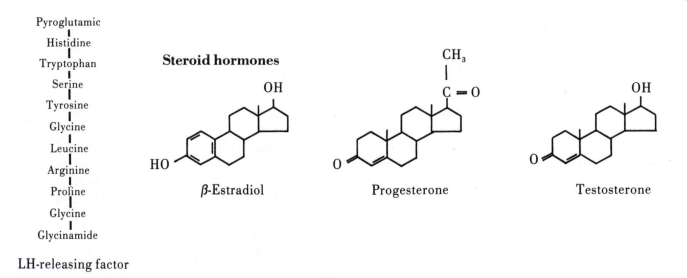

Pyroglutamic
Histidine
Tryptophan
Serine
Tyrosine
Glycine
Leucine
Arginine
Proline
Glycine
Glycinamide

LH-releasing factor

Steroid hormones

β-Estradiol   Progesterone   Testosterone

In both cases, steroids and peptides, it is easy to see how, by interacting with the machinery of gene expression and protein synthesis, hormones exert effects that span hours, days, and months and make even permanent changes. In some cases hormones can affect the very organization and functioning of the brain itself. A particularly important example involves sexual differentiation of the brain in mammals, including humans.

In mammals, the female and male nervous systems are different in relation to particular behaviors and the production of certain neurochemicals. LH production in females, for example, is cyclic with prominent peaks, such as just before ovulation, whereas LH is produced at more constant levels in males. The sexual differentiation of the mammal brain, in the few species that have been studied, occurs somewhere around the time of birth. In humans it is just before birth. In

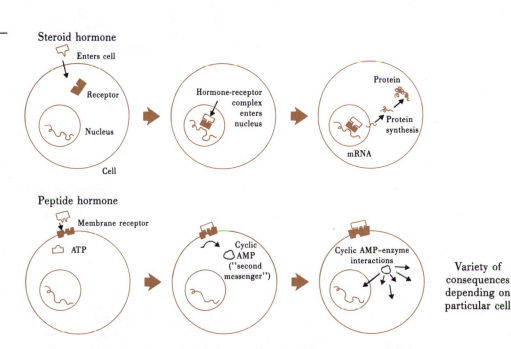

Figure 16-12 Modes of action for steroid versus peptide hormones.

Steroid hormone

Peptide hormone

rats it occurs a few days after birth, which has made the phenomenon easily accessible for research outside the mother rat's body.

The basic, sexually undeveloped rat brain is "female." Development of the "male" brain results in response to a sudden spurt of testosterone secreted by the young male's testes around the time of birth. If a male is castrated just before the event, its behavior will remain feminine for the rest of its life. It will not become aggressive toward other males and will show, for example, the lordosis response if given a shot of estradiol. A male that is castrated but then artificially given a shot of testosterone at the right time will behave as a male later if supplemented with the proper adult male hormones. A female, on the other hand, if given a shot of testosterone at the critical period, will develop a masculinized brain. The results of these experiments are illustrated in Figure 16-13. Partial effects sometimes occur from fetuses located next to each other in the uterus (Meisel and Ward 1981).

The masculinization of the brain, incidentally, is slightly more complex than just described, and testosterone is only the intermediate chemical in the process. Paradoxically, it is estradiol, a "female" sex hormone that actually causes the masculine development. This comes about by testosterone entering the neurons in the appropriate regions of the brain and being converted intracellularly into estradiol. Why does not the female's own estradiol, in young females, cause the

Figure 16-13 General pattern of sexual differentiation in the brains of neonatal mammals, as exemplified by experiments on rodents.

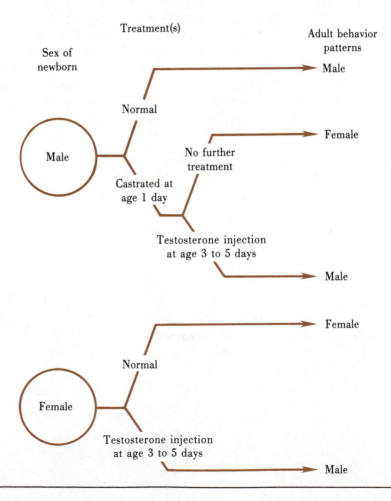

same effect? First, the levels of estradiol in young females are very low, and secondly, the liver of females produces an estrogen-binding protein, alpha-feto-protein, which binds up any endogenous estradiol and prevents it from causing male differentiation in the female brain (McEwen 1976). The enzyme, incidentally, is produced only for a short period of time, which occurs during this critical period of the brain's development.

PLEASURE, PAIN, AND EXOGENOUS MOOD-ALTERING CHEMICALS

Not only are particular categories of behavior, such as reproduction and migration, affected by chemicals, but there are also many other ways in which behavior is integrated chemically. An animal's general overall moods and levels of motivation, for example, may be affected significantly by particular chemicals. Of particular interest during recent years have been findings related to the phenomena of pleasure and pain. The subject has been of particular relevance to humans because of problems related to drug use in human culture, particularly in present-day Western society. Both legal and illegal drugs have assumed an increasing importance in human daily living and welfare.

Research has progressed along several lines. One of the first insights into internal mechanisms came in the 1950s when Olds and Milner (1954) discovered the so-called pleasure centers in the brain. Rats with electrodes inserted into these regions would press bars that resulted in self-stimulation of those areas with electrical impulses. No other reward was required; the rats pressed the bars simply to stimulate those regions of the brain.

Along another line of inquiry, it was asked why the brain should be so receptive to certain narcotic drugs such as opium and morphine. Why, for example, would evolution have produced receptors in the nervous system with the molecular specificity for these particular chemicals. It was hypothesized that it was not for external chemicals but rather for naturally occurring molecules of a similar nature that were already within the system. Searches for such molecules soon turned up several. They have been named *endorphins* (for endogenous morphine) and *enkephalins* (for "in the head"). There are several different molecules in this class, varying in size and shape.

These chemicals are just about as their names imply: internal, naturally occurring morphines or opiates. They only resemble the external opiates, however, and there are some important differences. For one, they break down more rapidly in the body, and the effects from particular molecules do not remain as long. For another, they are natural parts of the system and, in a sense, belong and fit better in the whole integration process. But because of this, they may be required for normal operation of the nervous system. In this sense, every individual vertebrate is "drug dependent."

Under normal conditions these naturally occurring chemicals are thought to function in two main areas: motivation and control of pain. They may be important in motivation by providing the mechanism, through the "pleasure centers," for stimulating an animal to do something. Certain behaviors are rewarding or pleasure producing. These may include some basic behaviors such as eating and sexual behavior.

Figure 16-14 Hypothesized pain pathway in the central nervous system of vertebrates. P substance carries pain information from sensory neurons to neurons in the pain pathway of the spinal cord. Release of P substance may be blocked, however, by presynaptic enkephalins released from neurons in a pathway from higher centers such as the "pleasure center." Externally introduced narcotics may simulate the enkephalins, and naloxone will occupy the enkephalin receptors but not block the release of substance P, hence, interfering with natural or external analgesic routes. This enkephalin–P substance pain pathway is only part of the total picture of pain, however; see text.

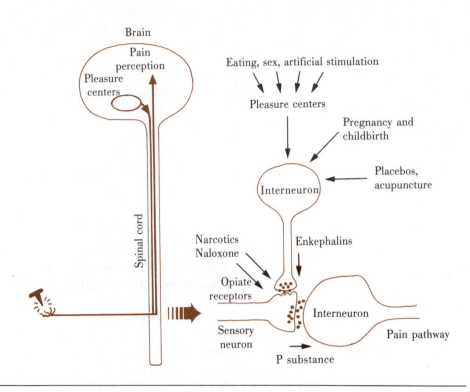

In the control of pain, some of the chemicals apparently serve as natural analgesics that reduce the perception of pain by inhibiting the sensory pathways of pain up to the brain. This mechanism, as currently interpreted, is illustrated in Figure 16-14. Pain, although it hurts or perhaps because it hurts, has a very important function and is highly advantageous. It warns that something is wrong and corrective behavior is required. When pain is dulled, one is only covering over the real problem, which may remain or even get worse.

Why should evolution have produced a natural pain suppressant? It has been proposed that in some instances, such as childbirth in mammals, the pain cannot be avoided and the signals should simply be suppressed (pregnancy and childbirth are important events associated with the release of natural opiates). In other situations, there may be critical emergencies where survival itself is at stake and the animal might be killed if it were to stop and pay attention to the pain; in this instance, it would be more advantageous to suppress the pain and wait until later to deal with the problem. In other words, the short-lived natural, internal opiates temporarily might get the animal through in a tight pinch.

The opiate receptors are located, among other places, on the axons of sensory neurons in the spinal cord. These receptors also bind to two other types of chemicals: exogenous narcotics and a most interesting chemical, *naloxone.*

Naloxone has proved most useful as the key to unraveling the story of the mechanism. Naloxone, although it fits the receptors, does not produce the narcotic effects; it simply plugs the sites and prevents other chemicals from occupying the receptors and producing their effects. Thus in the presence of naloxone, the pain pathways are not inhibited and the pain is felt.

The implications of all of this are many and may be, although speculative, significant. Not only may several drug addictions be related, but many other compulsive behaviors may have a related underlying mechanism. For example, compulsive behaviors such as overeating or compulsive sexual behavior may be indirectly related to such things as heroin addiction. The difference is that the opiates are being turned on internally via the pleasure centers rather than being received from outside through a needle under the skin. The problems can be nearly as compulsive and difficult to correct in either case. This is speculative, however, and we have a very incomplete understanding of the whole mechanism. Furthermore, the opiate pathways do not account for all pain suppression or analgesia. Some types of analgesia appear to involve pituitary hormones, with or without opiates, and some involve neither hormones nor opiates. Thus at least four categories and presumably mechanisms of natural analgesia exist (Watkins and Mayer 1982).

The complete picture, particularly in human behavior, appears to go much beyond a simple interpretation of pain and pleasure. Some people are addicts, and some are not. With the normal operation of pleasure and pain, some persons become compulsive and some do not. Why are there differences? In the case of alcoholism, for example, there are numerous factors that involve a complex of family and social interactions and developments during the life of the individual.

The opiate pleasure/pain-suppressant system may be involved in many, perhaps even most, "high-producing" phenomena, including many of the pleasures of daily living. This system may come into play in producing pleasure after long-term participation in otherwise uncomfortable activities such as jogging. Additionally, the endorphins are thought responsible for such things as *placebo effects* (where persons feel less pain because they *think* they have gotten a pain killer when they actually did not) and phenomena such as acupuncture, where pain is suppressed by placing pins in particular positions in the skin. Naloxone, incidentally, blocks both the placebo effects and acupuncture.

It has been proposed that drug addiction may result from exogenous narcotics replacing and suppressing the natural systems. The normal operation of the nervous system apparently depends on at least a minimum level of these types of chemicals. When the external sources are denied, the natural sources may have been suppressed and are not present or present at insufficient levels; the nervous system no longer operates properly; and the individual suffers withdrawal and even death in extreme cases.

It should be emphasized that not all mood-altering chemicals operate through the opiate system just described. Many exert their effects at other points in the complex set of interactions. Some resemble synaptic transmitters and stimulate postsynaptic receptors, as if there were increased firing of presynaptic neurons, and the circuits are hyperactivated. Mescaline, for example, resembles norepinephrine and dopamine. LSD and psilocybin resemble serotonin. LSD is particularly potent, although its exact mode of action is not known. The most widely used stimulant drugs, methylxanthines (e.g., caffeine and theophylline, which are found in coffee and tea), compete with adenosine in the brain and reduce adenosine's normal inhibiting effects on neurotransmitters, which, as a result, leads to increased neurotransmission and neuron firing. Several drugs, such as cocaine and a number of pharmaceutical antidepressants, operate by blocking the normal metabolism and uptake of certain transmitters at the synapses, which amplifies the effects of those transmitters.

The proper functioning of the neuroendocrine systems can be viewed as somewhat analogous to the operation of a finely tuned, high-performance automobile engine. Tampering with it by introducing exogenous chemicals, unless one is a skilled mechanic (competent physician), is similar to putting wrong chemicals in the gas tank, putting water in the oil, or upsetting the timing.

SUMMARY

Behavior is coordinated not only by the nervous system but also by the endocrine system and a large variety of chemicals. Chemical transmitters carry information from neuron to neuron within the nervous system. In addition, neurosecretory cells within the nervous system, after receiving input directly from other neurons, release hormones into the circulation for other parts of the body. External stimuli exert an influence on several facets of an animal's life, including growth and development, migration, reproduction, and social interactions, through this mechanism.

Similarly, hormones released by other parts of the body, such as the gonads, are received at target neurons within the central nervous system and affect behavior. The interactions both ways, brain to body and vice versa, as well as interactions with other animals, relevant environmental stimuli, and the present internal state of the animal, are numerous and complex. The expression of reproductive behavior in many animals, for example, depends on hormones from glands that were stimulated by hormones from the brain, which were, in turn, stimulated by sensing the appropriate behavior of a mate. Examples are provided for insects, doves, and lizards, but analogous hormonal interactions can be found in virtually all multicellular animals.

Behaviorally important chemicals within the body fall into two main classes: peptide-protein types and steroid hormones. The former operate through cell membrane receptors and an intermediate messenger chemical, cAMP, acting within the cell. Steroids apparently enter their target cells directly to exert their subsequent influence.

Chemicals exert their effects not only on immediate behavior, but some hormones impose long-term effects by organizing the brain during critical periods in early development of vertebrate animals.

Some chemicals, such as the endorphins, are involved normally in pleasure, which may serve the proximate function of motivation, and pain, which serves to warn of problems. Some exogenous, externally introduced chemicals, such as the opiates, resemble, simulate, and may interfere with some of these natural neurochemicals. Other mood-altering, psychoactive chemicals exert their effects in other manners.

**Recommended
Additional Reading**

Aideley, D.J. 1978. The physiology of excitable cells, 2nd ed. Cambridge University Press, New York.

Barker, J.L., and T.G. Smith, Jr., editors. 1980. The role of peptides in neuronal function. Marcel Dekker, Inc., New York.

Beyer, C. 1979. Endocrine control of sexual behavior. Raven Press, New York.

Bloom, F.E., editor. 1980. Peptides: integrators of cell and tissue function. Raven Press, New York.

Bloom, F.E. 1981. Neuropeptides. Sci. Am. 245:148-168.

Crews, D. 1975. Psychobiology of reptilian reproduction. Science 189:1059-1065.

Crews, D. 1979. The hormonal control of behavior in a lizard. Sci. Am. 241:180-187.

Iversen, L.L. 1979. The chemistry of the brain. Sci. Am. 241(3):134-149.

Kandel, E.R., and J.H. Schwartz. 1981. Principles of neural science. Elsevier/North-Holland, Inc., New York.

Lehrman, D.S. 1964. The reproductive behavior of the ring dove. Sci. Am. 211(5):48-54.

Leshner, A.I. 1978. An introduction to behavioural endocrinology. Oxford University Press, New York.

McEwen, B.S. 1976. Interactions between hormones and nerve tissue. Sci. Am. 235(1):48-58.

Snyder, S.H. 1980. Brain peptides as neurotransmitters. Science 209:976-983.

Watkins, L.R., and D.J. Mayer. 1982. Organization of endogenous opiate and nonopiate pain control systems. Science 216:1185-1192.

CHAPTER 17

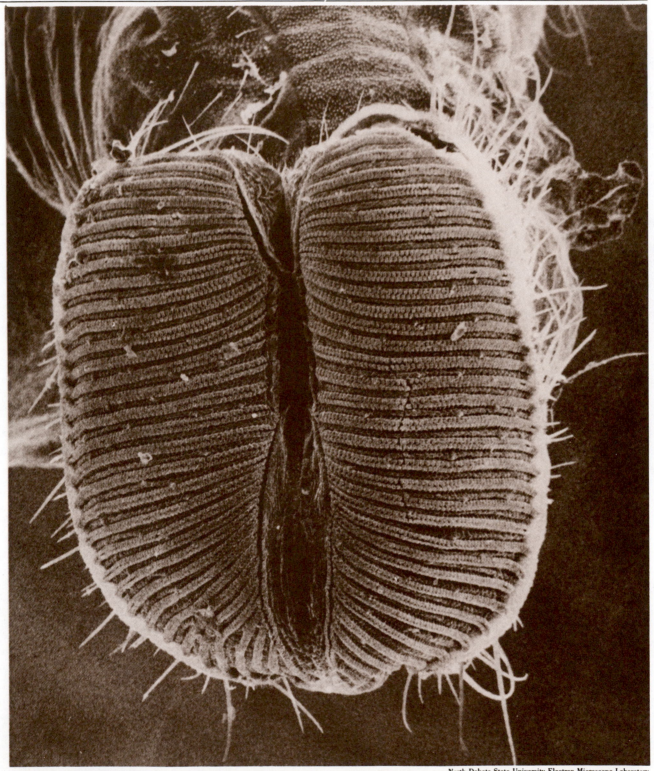

Opened oral surface and surrounding chemosensory hairs on the labellum of a housefly *(Musca domestica)* proboscis.

NEURAL MECHANISMS AND FEEDING A CASE IN POINT

William W. Beatty

The four previous chapters discussed some of the underlying neural and endocrine components of behavior, simple movements, simple animals, or small segments of the picture for more complex behavior in advanced animals. This chapter illustrates approaches and general problems of studying brain mechanisms of a larger, functional unit of behavior: the control of food intake and body weight. As will be seen, even something as seemingly simple as the initiation and cessation of eating may involve numerous, interacting factors and neural mechanisms.

We now are dealing in part, incidentally, with the related topics of *drive* and *motivation.* These topics have a somewhat embroiled, controversial history, as introduced in Chapter 2. Brief mention of a recent view of motivation relative to pleasure centers is presented in Chapter 16. Persons wanting further information on motivation per se should consult discussions elsewhere, such as Bolles (1975).

Feeding is an activity of enormous importance to all animals, as discussed in Chapter 6. There is much diversity in the kinds of foods eaten, the manners in which they are obtained, and the temporal pattern of feeding. Some "filter feeders" feed more or less continuously, extracting nutrients from a very dilute medium that typically contains only 4 to 5 mg of organic matter per liter. Other animals feed intermittently, alternating periods of feeding with rest and other behaviors. What, if anything, governs this behavior? Before we can consider possible mechanisms, we need to ask whether or not feeding is regulated in the first place.

IS FEEDING REGULATED INTERNALLY?

Most physiological psychologists and biologists have assumed that feeding is somehow regulated internally. This is a natural assumption based on the familiar observation that individuals of most species fall within a relatively narrow range of size and weight. Underweight animals usually can be found to be suffering from lack of food or some abnormal condition. When animals become overweight, the cause usually can be traced to unnatural, abnormal conditions. Thus it seems reasonable that regulation exists.

The evidence that a *particular* level of weight is maintained in particular animals is quite convincing. The adult rat, for example, has a remarkable ability to control its food intake to maintain reasonably constant energy reserves in the face of changing environmental conditions. Enforced exercise (usually accomplished by forcing the rat to run on a treadmill) greatly increases energy expenditure. The rat compensates for this drain on fuel reserves by a proportionate increase in caloric intake up to a certain point. If more than about 5 hours of running is demanded per day, body weight is not maintained; instead the animal controls weight at a somewhat lower-than-normal level. Rather precise caloric adjustments in intake also occur in response to variations in environmental temperature, which affects the energy expended to keep warm. Body heat is a significant factor for a small mammal such as the rat. Changes in temperature cause appropriate changes in feeding; rats eat more in cold environments, less in hot.

Variations in the caloric density of the diet also initiate appropriate alterations in feeding behavior. Precise caloric control of intake occurs when solid diets are diluted by as much as 25% with nonnutritive bulk (Figure 17-1). Increases in consumption occur up to 75% dilution. On liquid diets diluted with water, caloric control is even better.

But the rat's control system, as in many other animals including humans, has one very important "flaw." It is vulnerable to the sensory quality of the diet. Fed only the nutritive but (evidently) bland chow used in most laboratories, the rat controls its weight very nicely. However, if given an assortment of sweet and fatty foods, rats overeat and become obese (Sclafani and Springer 1976). The effective ingredients to cause dietary obesity in the rat are very familiar. They are available in abundance and variety at any supermarket. Conversely, the rat controls its body weight at a lower than normal level if forced to subsist on a foul-tasting diet (such as a standard laboratory diet mixed with bitter quinine). The animals totally refuse such diets for a time and then, with considerable reluctance, consume them in quantities that are adequate to maintain life, albeit at a lower-than-normal level of body weight (Peck 1978).

Figure 17-1 Effect of dilution of the diet with nonnutritive kaolin in normal male white rats. The animals maintained relatively constant nutrient intake and body weight for dilutions up to 50% by increasing the total food intake. Modified from Kennedy, G.C. 1950. Proc. R. Soc. Lond. 137:535-548.

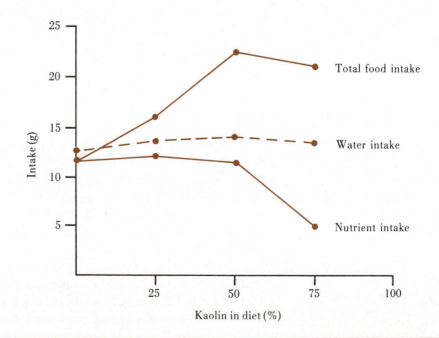

The fact that body weight is maintained at a relatively constant level has lead many researchers to postulate that there is some sort of internal *set-point* mechanism that achieves this end. Because fat is the major energy reserve in the body, it has been proposed that somewhere in the body there is a *lipostat*, a mechanism that measures the amount of fat, and presumably initiates appropriate behavioral and physiological responses to correct deviations from the body fat set point.

As has been seen, there is fairly good evidence that food intake is altered to compensate for changes in body weight above or below the set point. In adult mammals weight and fat are well correlated. Direct support for the lipostatic theory is sparse, although there are a few reports indicating that removal of the body fat organs initiates increases in food intake and increases in the amount of fat in remaining fat organs (Mrosovosky and Powley 1977).

Although little progress has been made in understanding the signals that the hypothetical lipostat detects, it has become increasingly clear that bodily energy reserves are vigorously defended. In addition to alterations in feeding that were already described, changes in metabolism are involved in conserving energy stores. If food is scarce, basal metabolism declines. Conversely, if the organism overeats, the body initially stores the excess as fat. But if overeating persists, weight does not increase further. Apparently the body increases its metabolic rate and burns up the extra calories (Figure 17-2).

Figure 17-2 Weight gain in a human subject after prolonged period of greatly increased caloric intake. Despite the fact that the subject nearly doubled his caloric intake, he never gained more than 14% above his initial weight. After prolonged overfeeding even a 40% increase in intake above his initial consumption did not sustain his modest weight gain. Modified from Sims, E.A.H. 1976. Clin. Endocrinol. Metab. 5:377-395.

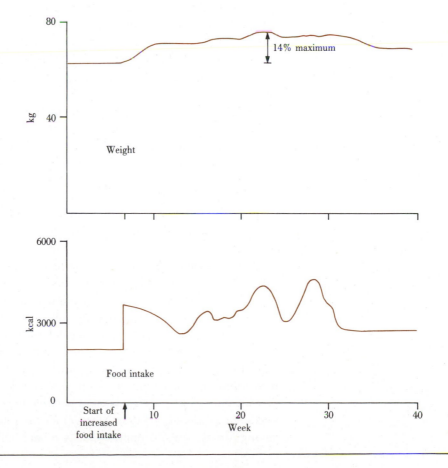

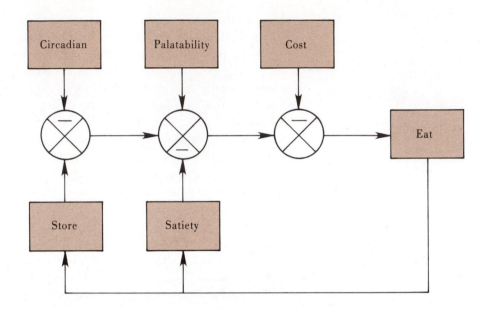

Bolles (1980), however, challenged the view that regulatory mechanisms exist. In this view of feeding, any controls, particularly satiety, evolved because they protected the organisms from hazards in the environment. One of the big hazards is the occasional presence of too much food, which could lead to bursting. Flies deprived of that safety mechanism by cutting the particular nerves, as discussed later, will overeat until they literally burst and kill themselves. Likewise, other mechanisms protect the individual from ingesting poisonous substances. Mechanisms involving learning with respect to feeding can provide the animal with a means for evaluating whether or not something is worth the effort, based on previous experience.

Bolles proposed that, instead of being "regulated," an animal's food intake and body weight simply represent an equilibrium or balance of evolved species-specific characteristics in a particular environmental context. The amount of food eaten, in this view, is the summation of several factors (see also Figure 17-3):

$$\text{Eating} = A(\text{Palatability}) + B(\text{Satiety}) + C(\text{Storage}) + D(\text{Cost}) + E(\text{Circadian})$$

Thus as palatability, for example, is modified, so is food intake. With increased palatability the animal eats more and gains weight to a new equilibrium. On poorer-tasting diets the animal loses weight. Or, for another factor, as cost of obtaining and processing the meals goes up, food intake goes down. If a rat, for example, has to work harder for its meals, it eats less. In an illustration provided by Bolles, consuming a zebra is a major and dangerous undertaking for a lion. The job is not over once the zebra is captured and killed; zebras come wrapped in leather. If lions have to catch their own zebras, they remain relatively sleek. If provided with easier meals, lions become fat and heavy.

In this sense feeding might be viewed somewhat analogously to human economy where the amount of money in the bank is a balance between expenditures and income. Most persons desire a positive balance with a certain level of safety and savings (analogous to fat storage) for the future. Some individuals will go on saving almost without limit, whereas others will increase their expenditures or stop working so hard once there are adequate savings. Part of the economy equation, like palatability, depends on how easy or acceptable it is to obtain the money by a given method. There is no set amount of money in the bank toward which everyone carefully regulates. The amount in the account simply is a balance between inflow and outflow and varies from individual account to individual account and, for any one account, from time to time. This viewpoint already has been discussed from the ultimate standpoint in Chapter 6, under the topic of optimal foraging. What is of concern now is the proximate means by which animals achieve some balance in their feeding behavior.

For a simple analogy consider a bucket that holds only so much water, for example, 10 L. If it is filled to overflowing and any excess spills out, it still holds only 10 L. If, however, a valve is placed in the water line leading into the bucket, the water flow can be turned off when the bucket fills up and before it overflows. Now the bucket holds 10 L, but for a different reason from in the first case: a mechanism is present for controlling the input.

To refine the picture further, imagine that the valve is used to turn water on when the water drops to 5 L and off when it reaches a certain level, for example, 8 L. But it can be used to keep the level more precise, between 7.5 and 7.7 L, or right at 7.6 L. One hand can be used to turn the valve on and off, or different hands can be used, one for on and one for off. The amount of water, the 7.6 L, might be arbitrary and by chance, or perhaps it was determined by an order from the city water department.

In this analogy, what does "regulation" mean: the fact that the bucket contains only so much water, the presence of a mechanism to control input, the control of amount within any limits or only within precise limits, or that something (e.g., the city water department) specifies the particular level? The problem emphasizes the ambiguity of a common word such as regulation and exposes some of the semantic problems with which one is faced. It also, however, breaks the topic of interest into a set of more specific questions that can be applied to feeding behavior: are there control mechanisms; if so, how precise are they; how do they compare among different species; and what, if anything, determines the particular level at which weight and feeding are expressed?

It is on the latter aspect that Bolles' view differs from the commonly held notion. Bolles says that the expressed level is simply a balance of several factors, with different factors possibly involving mechanisms. One can consider the separate mechanisms, but it is a waste of time to look for a single level-determining mechanism, commonly referred to as a set point.

The opposing view states that the level is determined (although it is not known how) by some master factor to which the contributing mechanisms respond. In this view one can investigate both the contributing mechanisms (analogous to the valves and hands that turn them) as well as the master, set-point mechanism (analogous to the city water department). This view does not require the set point to be at the same level in all individuals—only that each animal has a set point or master mechanism.

In the remainder of this chapter we will consider the component control mechanisms, which are known to exist, and remain alert for the presence of a master control, which may or may not exist. Feeding in two distant but well-studied organisms, a fly and a rat, will be described and compared.

CONTROL OF FEEDING IN AN INSECT: THE BLOWFLY

Blowflies are relatively large flies (as flies go), and as is common among insects, there are several species. Largely through the pioneering efforts of Vincent Dethier and his students a vast amount of information on the feeding behavior of one species of blowfly, the black blowfly *(Phormia regina)*, is available (Dethier 1976).

A typical blowfly *(Phormia)* weighs 25 mg when full grown and has a life span of 42 to 60 days under optimal conditions. Although the fly usually eats complex meals that contain carbohydrates, fats, and proteins, it can maintain itself entirely on a carbohydrate diet to supply energy. Females cannot reproduce, however, unless they eat a protein meal at the appropriate time in the reproductive cycle.

The fly's major method of getting around in the world, flying, consumes enormous amounts of energy. Energy to power flight is derived almost exclusively from the aerobic oxidation of glycogen (a polymer of glucose) stored in muscle and from oxidation of the fly's blood sugar, trehalose (a disaccharide of glucose). Blood trehalose probably is the most important fuel. Flies flown to exhaustion will resume flying almost immediately after consuming a carbohydrate meal or after injection of sugar into the blood.

Feeding Behavior and Patterns

A completely satiated blowfly is relatively inactive. As deprivation increases, the fly's activity also increases (Figure 17-4) until the fly finds a suitable meal or dies of starvation (about 2½ days in the blowfly). Although it may seem paradoxical that the fly should expend more and more of its dwindling energy supplies, a

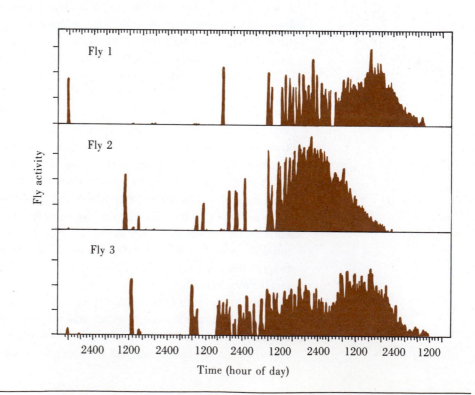

Figure 17-4 Activity levels of three individual flies over a 5-day period in the absence of food, under constant light conditions. Although faced with decreasing energy supplies, they became more and more active until they died of starvation. In other words, they did not become inactive and conserve energy. Reprinted from *Journal of Insect Physiology*, vol. 10, G.W. Green, The control of spontaneous locomotor activity in *Phormia regina* Meigen—I, copyright 1964, Pergamon Press, Ltd.

Fly 1

Fly 2

Fly activity

Fly 3

2400   1200   2400   1200   2400   1200   2400   1200   2400   1200

Time (hour of day)

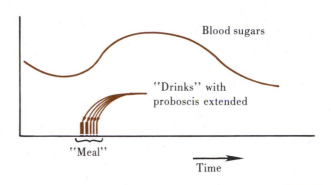

Figure 17-5 Typical pattern of feeding and blood sugar in a fly.

positive relationship between activity and deprivation is generally (although not invariably) observed in animals. Presumably the increased activity increases the probability that the fly will discover a suitable meal to replenish its energy deficit before the increased drain on energy leads to death. The mechanism, although risky, seems to work fairly well judging by the world fly populations.

During daylight, when blowflies are generally active, the fly alternates between periods of random flight and rest. While airborne, the fly's behavior is governed by visual stimuli, wind currents, and olfactory stimuli. Flies and other insects orient in flight toward the direction of the wind. This tends to place the fly in optimal position to detect food odors, although it still is a matter of debate whether wind currents or odors are most important in guiding flight. Both mechanisms may contribute.

Once on the ground the fly starts to walk. Its behavior is then controlled by olfactory stimuli (if they are present). Unacceptable odors lead to an abrupt cessation of walking, followed by flight, turning away, or frantic grooming. Acceptable odors lead to a variety of behaviors including stopping, extending the proboscis (the fly's mouth), grooming, and approach. If the surface does not contain a volatile (and hence odorous) chemical, the fly walks randomly until it encounters a liquid with its legs (Dethier 1976).

Suppose that a hungry fly encounters a source rich in carbohydrates. In the wild this might be a nectar-filled flower, an open wound on an herbivore, or a bit of decaying material. (Very little is known about the feeding habits of wild blowflies.) In the laboratory more artificial but better-controlled food sources, such as solutions of known sucrose or glucose concentrations, ordinarily are used. If the fly lands in the sugar-water solution, it immediately extends its proboscis and begins to feed (actually drink). Feeding continues without pause for perhaps 60 seconds if the sugar solution is fairly concentrated. During the next minute the fly may feed for 20 seconds or so. Then it pauses for a few minutes, drinks again for about 10 seconds, pauses for a few minutes, and drinks again for 10 seconds or so (Figure 17-5). Feeding then ceases for 30 minutes to 2 hours, even if the fly remains standing in the sugar-water solution.

Control of Feeding

Initiation of Feeding
Feeding (and drinking of water as well) is initiated by sensory stimulation. As already mentioned, olfactory stimuli sensed by receptors in the antennae determine the fly's approach to potential sources of food and water at a distance. After the fly locates a meal, olfactory inputs are relatively unimportant.

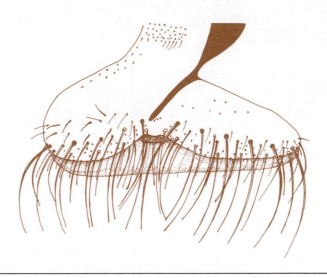

Figure 17-6 Sensory receptors on the labellum (lips) of the fly *(Phormia regina)* proboscis. Diagram of side of labellum showing the numerous, long, chemosensory hairs. (See Figure 17-8 for basic anatomy.) See also chapter opener for scanning electron microscope photograph of the surface of the labellum showing the fringe of chemosensory hairs of a housefly. From Wilczek. M. 1967. J. Morphol. 122:175-201.

Ingestion is controlled by three sets of chemoreceptors located on hairs (Figure 17-6). One set is located on the tarsi of the fly's legs, the second on the lobes of the labellae at the distal end of the proboscis, and the third set is within the lobes of the labellae in the interpseudotracheal papillae.

Each hair contains five receptors, one of which is a mechanoreceptor. The other four cells are chemoreceptors, the dendrites of which extend down the shaft of the hair. The tip of the hair has a small opening or pore permitting liquids to make contact with the chemoreceptor dendrites (Figure 17-7).

Electrophysiological recordings from individual hairs suggest that each of the four chemoreceptors responds to a different class of chemicals. One receptor responds to sugars, primarily simple pentoses and hexoses. A second responds to water. The third responds to salts, especially sodium chloride, and the last chemoreceptor remains something of a mystery. It has been called the anion cell. It seems to respond to sodium salt anions of long-chain fatty acids, but its exact chemical specificity remains unclear.

What role do these chemoreceptors play in the control of the fly's feeding behavior? Receptors in the tarsi make first contact with liquids that might satisfy the fly's hunger. If the solutions stimulate the sugar receptor, the fly extends its proboscis. Varying degrees of extension are possible, depending on the strength of stimulation of tarsal sugar receptors (and also the strength of inhibitory factors). But suppose the fly extends its proboscis fully. This brings the hairs on the surface of the labellar lobes into contact with the sugar-water solution. These hairs also contain sugar receptors. When the sugar receptors of the labellar hairs are stimulated, the labellar lobes are opened, and the fly begins to suck. Opening of the labellar lobes exposes the receptors in the papillae, which perform a final evaluation of the solution before it is ingested.

Thus the initiation of feeding is controlled by four sets of chemical receptors. Olfactory receptors make the initial determination. Then taste receptors on the legs, labellae, and papillae are exposed in sequence. The potential meal is screened for suitability, and as might be expected, the papillae are the most demanding. Certain substances that are acceptable to the tarsal and labellar receptors are rejected by the papillae. Both stimulation and rejection are

Figure 17-7 Close-up of a chemosensory hair (minus the long middle section) and associated sensory neurons. Reprinted by permission of the publishers from *The Hungry Fly* by V.G. Dethier, Cambridge, Mass.: Harvard University Press, copyright © 1976 by the President and Fellows of Harvard College.

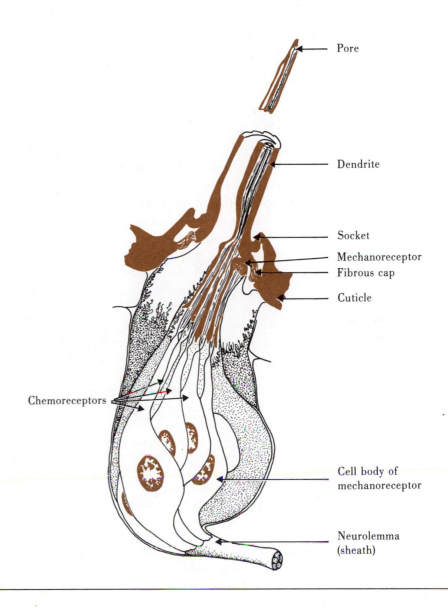

Pore

Dendrite

Socket

Mechanoreceptor

Fibrous cap

Cuticle

Chemoreceptors

Cell body of mechanoreceptor

Neurolemma (sheath)

extremely sensitive. Sucking can be initiated, for example, by stimulating a single labellar hair (i.e., one sugar receptor).

Cessation of Feeding and Satiety

The salt and anion receptors obviously are important in protecting the fly from ingesting unpalatable food, but they are not involved in controlling feeding in the presence of palatable substances such as sugar-water solution. What causes the cessation of feeding, and what maintains the state of satiety?

Electrophysiological studies indicate that the chemoreceptors emit a burst of activity on first contacting a solution and then quickly reduce their firing to a lower, relatively steady level. Both the initial burst and the steady state evidently convey information to the central nervous system, but the maintenance of ingestion requires sensory input from the sugar receptors above a certain minimum level. Sugar receptors cease firing within milliseconds after chemical stimulation is removed; sucking may continue for 10 to 20 msec longer; then it also ceases. Thus continuous sensory input to the sugar receptor is required to maintain ingestion, but it is not enough.

In the short run, that is, over a period of a few seconds or minutes, two mechanisms seem to determine whether or not input from the sugar receptors will be adequate to maintain feeding. First the receptors themselves exhibit adaptation, a reduction in response, beginning within a second or less. The receptors also exhibit rapid disadaptation—a return toward the initial high level of response. Under normal circumstances when the fly's behavior is under control of many different receptors located on many different hairs, different receptors are adapting and disadapting at different times.

A second mechanism, habituation, a longer-lasting depression of the sensory response, is also important. Habituation is presumed to arise at synapses deep inside the fly's nervous system. Again, in natural feeding there are dynamic changes in the state of central synapses supplied by peripheral receptors. Together adaptation and disadaptation of receptors and habituation and dishabituation of central synapses determine whether the total amount of sensory input exceeds the threshold necessary to maintain feeding.

A number of other factors also are important:

1. *Deprivation.* This generally reduces the acceptance threshold for palatable substances (i.e., the minimum concentration that will be ingested). Greater deprivation also seems to increase the range of acceptable substances, probably by overriding inhibitory mechanisms.
2. *Concentration of the sugar.* The amount ingested is an increasing function of concentration, at least up to the point where the viscosity of the substances limits the rate of sucking.
3. *Palatability of the substance.* This is *not* the same as the nutritional value of the food.

Perhaps the most convincing demonstration of the last point comes from studies in which flies are given a choice between a highly stimulating but completely nonnutritive substance such as the sugar fructose and a nutritive but relatively nonstimulating substance such as mannose or sorbitol. Flies invariably choose the nonnutritive but stimulating substance and starve to death. Under natural conditions the substances that stimulate the receptors normally occur together with, and hence serve as reliable indicators of, the nutritional substances.

Adaptation and habituation interact with the fly's state of deprivation and the nature of the carbohydrate to determine the fly's feeding behavior during the first few minutes after it encounters a suitable meal. But adaptation and habituation are neuronal processes that typically have a time course of seconds or minutes. What accounts for the fact that the fly ceases to feed for a period of hours although food is readily available?

To answer this question a brief description of the digestive system of the blowfly is necessary. The digestive system is divided into three parts: the foregut, consisting of the pharynx, esophagus, and crop, the midgut, and the hindgut (Figure 17-8). Fluid is sucked into the pharynx and passed into the esophagus. Waves of peristalsis drive the fluid initially into the midgut (some goes into the crop as well). When the midgut is full, valves close and shunt fluids into the crop.

Figure 17-8 Internal anatomy of *Phormia* showing relationships of digestive system, associated nerves, and central nervous system. Modified from Dethier, V.G. 1976. The hungry fly. Harvard University Press, Cambridge, Mass.

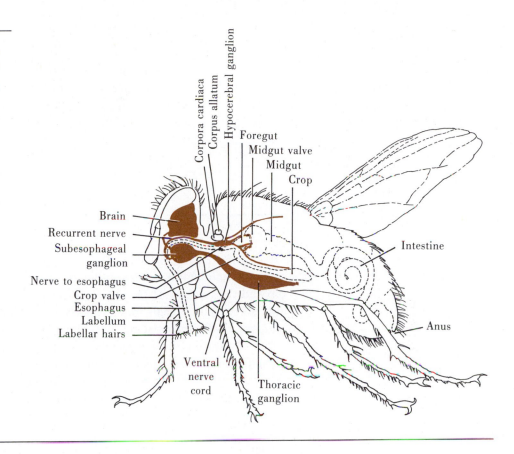

As the midgut empties, fluid is regurgitated from the crop into the esophagus and back into the midgut.

Any number of mechanisms involving various portions of the gut or the blood might be invoked to explain the long-term inhibition of feeding. Bood sugar level or some other humoral factor that changed as a result of nutrient absorption might be involved, but experimental data rule them out.

To summarize a great deal of painstaking research rather quickly, injections of glucose throughout the gut and into the blood do not affect feeding. This is rather remarkable because such injections supply energy and will restore active flight in a fly flown to exhaustion. Similarly, other possible humoral factors seem to be quite unimportant. Flies whose "bloodstreams" are connected (parabiosed, Figure 17-9) feed as individuals. Consider, for example, that two members of a parabiotic pair of flies are starved, and one is fed to repletion. The fed partner fails to feed when offered a suitable carbohydrate meal. But the starved partner exhibits normal acceptance thresholds and feeds normally, despite being exposed to any potential humoral satiety factors.

These and other experiments suggest that satiety in the blowfly is not related to the nutritional consequences of the ingested meal. Why then does feeding remain suppressed long after sensory adaptation and habituation have dissipated?

The answer is that there are two sets of inhibitory inputs, both arising from abdominal receptors and both of which can override excitation from the sugar

Figure 17-9 Parabiotic blowflies. These flies have been surgically joined at their thoraxes to permit interchange of blood. They are allowed to exercise on rolled screen wheels suspended from rods and onto which they hold and walk. Modified from Dethier, V.G. 1976. The hungry fly. Harvard University Press, Cambridge, Mass., and Belzer, W.R. 1970. Ph.D. Thesis. University of Pennsylvania, Philadelphia.

Wire screen exercise wheel

Rod supporting screen wheel

receptors. One inhibitor is a set of stretch receptors located in the region between the crop and the midgut. These receptors monitor the passage of fluid from the foregut to the midgut. Signals from the receptors inhibit feeding via the recurrent nerve (Figure 17-8). The existence of such a mechanism is supported strongly by the results of severing the recurrent nerve. Flies so prepared overeat (become hyperphagic) and in some instances may ingest so much that they burst. The duration of the first drink is perfectly normal; that is, the fly stops drinking after a minute or so, presumably because of sensory adaptation and habituation. But unlike the normal fly, once adaptation and habituation have dissipated, it resumes feeding to excess.

A second system of feeding inhibition involves stretch receptors located in the body wall. These receptors are distended as the crop fills, although they are not actually located in the crop. These abdominal receptors send inhibitory inputs via the ventral nerve cord. If this nerve is cut, hyperphagia also occurs, but the pattern is different from that after recurrent nerve severing. Flies with the ventral nerve cord severed are reported to drink one long meal that terminates when the fly bursts.

Thus the basic control of feeding (at least feeding on carbohydrates) in the blowfly turns out to be appealingly simple. Sensory stimulation initiates and maintains feeding until adaptation and habituation combine to terminate a drink. Disadaptation and dishabituation result in recovery of feeding, which is short lived. Eventually stretch receptors monitoring distension in the crop and at the junction of the foregut and midgut provide longer-term inhibitory control. It is even possible to draw a flowchart for this set of interactions (Figure 17-10), as if one were to write a computer program to operate the fly's feeding behavior.

As elegant as this picture is, there are some disquieting questions. Neither the excitatory nor the inhibitory mechanisms involve monitoring of any factor directly related to available energy or potentially available energy. The sugar receptor responds very nicely to substances that the fly cannot metabolize—so nicely that the fly may starve to death in the presence of other adequate but "unattractive" energy sources. Similarly, the major long-term inhibitory mechanisms also do not monitor energy. Perhaps this is not a real problem. Natural stimuli that activate the sugar receptor usually will be nutritious also or occur with nutritious substances. The fly's feeding system obviously works.

But there is one disconcerting fact in this otherwise simple story. Deprivation exerts powerful effects on feeding. This occurs partially because the empty crop does not release fluids to stimulate foregut stretch receptors inhibiting feeding. Then there are the effects on acceptance thresholds for both palatable and unpalatable substances, mentioned earlier. But eating and activity rates continue to increase after the crop is completely empty. In spite of the negative results from nutrient injections and the parabiotic experiments, this suggests that somewhere nutrient reserves may be monitored. Where and how (and, in fact, whether they are) are not known.

Figure 17-10 Flowchart of the control of feeding in the blowfly. For explanation, see text.

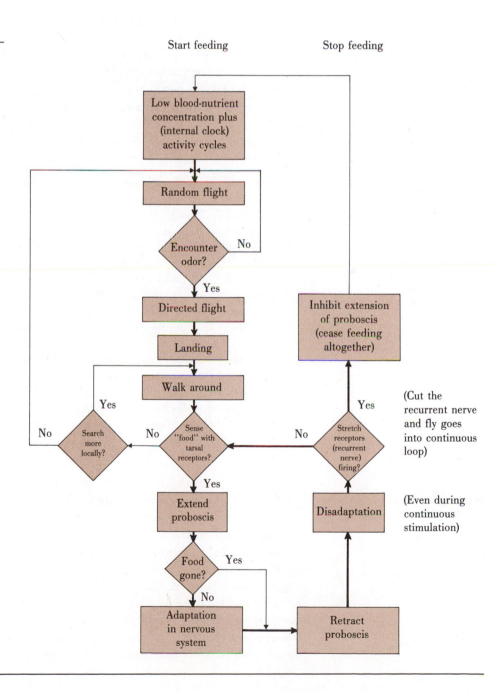

In contrast to the wealth of truly elegant work on peripheral mechanisms of blowfly feeding, almost nothing is known about what goes on deeper in the central nervous system. How, for example, do inputs from the recurrent nerve and the ventral nerve cord inhibit feeding? These and other questions await empirical answers.

CONTROL OF FEEDING IN A MAMMAL: THE RAT

The control of feeding in mammals is a great deal more complex than in the blowfly. Although the control of food intake and body weight has been studied in many mammals, the most complete information exists on the rat, particularly laboratory strains derived from *Rattus norvegicus*. Fortunately enough data exist from other species to suggest that the major features of control of food intake and body weight are fairly similar among various mammals, including humans.

From a biological perspective rats are enormously successful creatures. No doubt a portion of the rat's biological success results from the fact that they are omnivores; like humans, they eat a variety of plant and animal food. Nonetheless, rat feeding behavior is quite discriminating. Following are several aspects to the rat's food selection habits worthy of note:

1. Rats are quite suspicious of unfamiliar foods. When first exposed to a strange food, rats eat very little of it initially. This aspect of rat feeding behavior is termed **neophobia** (literally, fear of the new) and may in part explain why rats notoriously are difficult to poison.
2. Rats have well-established and perhaps unlearned preferences with regard to foods. Although omnivores, they prefer foods that taste sweet to humans. They also like mildly salty foods. On the other hand, they avoid sour and bitter-tasting substances, eating them only as a last resort. This aspect of rat feeding behavior also may have adaptive value, since sweet-tasting substances generally are rich in substances that can provide energy, whereas bitter substances are often poisonous.
3. The rat's nutritional requirements are similar (although not identical) to those of humans. As a consequence rats long have been used by nutritionists to assess commercially prepared foods.
4. Rats (and mice) display obesity syndromes caused by a single recessive gene, fa/fa for fatty (ob/ob for obese in the mouse). Genetically obese animals seriously overeat, produce elevated levels of insulin, and gain weight, nearly all of which is fat.

Feeding Behavior and Patterns

The fact that the rat can adjust its intake rather precisely to compensate for changes in caloric density of the diet and variations in energy expenditure might lead one to suppose that feeding is initiated when some sensor detects that energy reserves have fallen below a certain threshold and continues until reserves are replenished. Studies of patterns of meal taking in freely fed rats, however, do not support this expectation. There appears to be no relationship between the size of a particular meal and the length of time since the preceding meal. The experiments were performed under conditions of constant diet formulas and reasonably constant energy expenditure by the animal so the state of energy reserves should be related directly to the length of time since the last meal (LeMagnen 1971). The

.duration of feeding during a given meal, however, seems to depend more on the taste properties of the meal than on the depletion of the animal's energy reserves. It is similar to humans stuffing themselves on a big Thanksgiving meal but still having "room" for dessert, even if it hurts.

After the rat has eaten a large meal, however, it is likely to *wait longer* before taking another meal than if it has eaten a small meal. Thus, apparently, rats (and presumably many other mammals) achieve their long-term control of food intake by varying the interval *after* a meal, depending on how much they ate, rather than by varying the amount to be eaten depending on how long it has been since the previous meal. In other words, if a rat eats a lot, it waits longer until eating again. If it eats only a small amount, it will try to eat again sooner.

All of this may seem confusing, but the outcome makes sense if one considers that the laboratory rat's ancestors lived in a world where food resources may have been scarce, depending on environmental conditions and competition. In such a world it would not be adaptive for the animal to pass up a good meal that it happened upon, even if it had fed recently. On the other hand, having gorged itself, the rat has no reason to go out seeking food again soon.

The foregoing interpretation of rat meal patterns is speculative. Taste quality complicates the picture. At present there are few very detailed data on meal patterns in rats under natural conditions or good comparative studies of feeding patterns and state of nutrition in many other mammals. Furthermore, there appear to be no studies that compare the effects of free-running feeding patterns, as discussed earlier, with effects of *imposed regular* feeding schedules, such as with regular daily meals in many human cultures or in the food provided to domestic animals on a regular basis.

It should be clear, however, that feeding patterns in rats are not as simple as in flies. Some of the differences may be only superficial. Perhaps because flies have smaller bodies and are flying machines, they carry less fuel and run out sooner and, hence, have more predictable feeding patterns. Or because flies and rats are so unrelated, there may be major differences in underlying mechanisms. What is known about mechanisms controlling feeding in the rat?

To begin with, the factors cannot be easily divided into those which initiate and those which stop feeding as was the case with the blowfly. Instead, there are numerous interacting factors, and it is not known yet just how they achieve the final control of feeding. The important factors can be divided into two groups: peripheral (sensory, digestive tract, liver, en route from the viscera, hormonal, and genetic) and central nervous system (hypothalamic, centers, and pathways). In the following pages there is space only for a brief review of a few of the highlights from extensive research. Persons interested in more detail should see Carlson (1980).

Peripheral Controls

Much as with the fly, olfactory, taste, and digestive system sensory factors are also present in the rat. Taste, olfaction, and other factors include some that have not been well examined. As strange as it may seem, we really do not have any idea what importance (if any) the *sight* of food is to the rat. Not much more is known about the role of *olfaction* in controlling feeding. Most knowledge is based on studies aimed at determining the importance of *taste*.

Taste stimuli exert powerful control over the rat's behavior, as indicated earlier and by studies that eliminated the sense of taste through self-administering feeding catheters implanted in the stomach. In the most revealing arrangement (Snowdon 1969), food was delivered to the stomach only as long as the rat depressed a lever. Rats feeding in this way maintained original body weight, but they did not grow as normal rats do, and there were several other differences. In a maze-learning situation, for example, oral rewards promoted much more rapid learning than intragastric rewards (Miller and Kessen 1952). Taste appears to be responsible mainly for the rewarding effects of food or, in less stodgy terms, the joy of eating.

Theories of hunger historically (e.g., as proposed by the famous American physiologist Walter Cannon) placed great emphasis on stomach sensations. Stomach contractions were assumed to stimulate receptors in the stomach, which in turn gave rise to the sensation of hunger (or "hunger pangs"). Most persons would swear that they "feel" hunger in their stomachs, and, in fact, many studies have confirmed that gastric motility is fairly well correlated with subjective sensations of hunger in humans.

However, many lines of evidence make it clear that neither the stomach nor sensory inputs from it are essential to hunger or satiety. Surgical removal of the stomach or section of its nerve supply, for example, does not greatly disturb control of feeding or appetite. Human patients whose stomachs have been removed because of disease or injury continue to experience hunger. Although they must eat smaller meals more frequently, their feeding behavior seems otherwise normal. This does not mean, however, that the stomach is unimportant under more normal circumstances.

The possible role of stretch receptors in satiety was considered early. Both humans and rats were induced to swallow balloons that could be filled with air or water, thereby distending the stomach. As might be expected when the balloons were pumped full, the subjects lost interest in eating. It was quickly noted that the experience of having one's stomach filled with air was aversive to humans (and also seemed aversive to rats) so the experiments were hardly definitive. However, rats will tolerate a fair amount of gastric distension that accompanies eating diets diluted with nonnutritive bulk, so most workers concluded that gastric distension was not very important, except under extreme conditions.

For many years it has been known that intubation of hypertonic fluids into the stomach or the duodenum suppressed feeding in hungry rats (Schwartzbaum and Ward 1958). The effects are seen even when nonnutritive substances (such as NaCl) are used, so the inhibition of feeding most likely occurs either because of the activation of osmoreceptors or because of distension produced by entry of water into the gut across the osmotic gradient. It is not clear which of these mechanisms is involved, but the evidence tends to favor distension (e.g., Collins and Davis 1978). The duodenum is more sensitive to these inhibitory effects than is the stomach (Snowdon 1975).

There also is evidence that chemoreceptor mechanisms important to satiety are localized in the gut. Gastric loads of nutritive substances are generally more effective in inhibiting feeding that nonnutritive loads (Snowdon 1975). Recent research indicates that the stomach may contain specific nutrient detectors (Deutsch et al. 1978).

Perhaps the most exciting development in recent years is the discovery that duodenal hormones in part may be responsible for satiety. The presence of hormonal and humoral (carried in the blood) factors contrasts sharply with the lack of such controls in the blowfly. Three substances have been implicated: enterogastrone (Schally et al. 1967), cholecystokinin (CCK) (e.g., Antin et al. 1975, Kraly et al. 1978), and bombesin (Gibbs et al. 1978).

Major control over the absorptive and fasting phases of metabolism is exerted by hormones produced by the pancreas, the adrenal medulla, the adrenal cortex, and the pituitary gland. Branches of the autonomic nervous system in turn regulate the release of these hormone products and exert direct controls over metabolism.

The major hormonal control over the absorptive phase of metabolism is exerted by insulin, a protein hormone produced by islet cells in the pancreas. Insulin is secreted in response to a rise in blood levels of either (1) glucose or (2) amino acids. Insulin promotes the storage of glucose and other nutrients as fat.

Several other important hormones that affect energy and food metabolism, hence (directly or indirectly) feeding behavior, include glucagon, epinephrine, growth hormone (somatotrophic hormone), and certain hormones of the adrenal cortex (called glucocorticoids, of which corticosterone is most important in the rat). In general, these hormones promote the release of energy from bodily stores.

The hormones of the gonads also influence feeding, growth, and body weight regulation. Testosterone stimulates feeding and body weight gain and reduces the proportion of body fat, apparently by facilitating growth of muscle. Estradiol reduces food intake, body weight, and body fat. Both testosterone and estradiol can cause these changes when injected directly into the brain in very small quantities so they may regulate feeding by influencing brain control mechanisms. Alternatively, the effects of the gonadal hormones may result largely from influences on peripheral metabolism.

Investigations of other peripheral controls have included a large number of experimental manipulations such as self-administered feeding via intravenous catheters, cutting and exteriorizing the esophagus so that the mouth is separated from the gut (sham feeding), infusion of a variety of nutrient and nonnutrient solutions into the hepatic portal vein (between the intestine and liver), cutting the vagus nerve, which supplies the viscera, and perfusion or blood-mixing experiments as in the parabiosed flies. The results of these various studies indicate that the mouth, stomach, duodenum, liver, vagus nerve, and hormonal and chemical factors all play important roles in the control of feeding. When any one or a few of these at a time are eliminated, the rat still controls its food intake but, in all cases, imperfectly.

The Hypothalamus and Central Mechanisms

Unlike the paucity of information concerning central nervous system mechanisms in controlling fly feeding, there is much information for possible central mechanisms in the rat. Unfortunately, much of it is conflicting or ambiguous. The history of the story provides a revealing journey through the perils and problems of trying to understand neural mechanisms in complex brains such as those possessed by mammals.

Early workers in the physiology of behavior believed, or perhaps more accurately hoped, that the brain would be compartmentalized simply and discretely into functional units that fit with ways of categorizing behavior. This view of brain function can be traced historically to the ideas of the phrenologists, Gall and Spurzheim, in the early 1800s and is, in fact, observed in the central projections of the major sensory systems (Chapter 15). This view, however, does not apply to the control of feeding, as revealed by extensive recent research.

For many years is appeared that, in its most basic outline, the control of feeding in mammals such as the rat might be quite similar to the control of feeding in the blowfly. Recall that in the blowfly feeding is excited by the detection of sugars by chemoreceptors. Feeding ceases when the chemoreceptors adapt and when an inhibitory state of satiety is produced by foregut and body wall receptors that monitor distension and connect to the brain via the recurrent nerve and ventral nerve cord. The dual center model of feeding in mammals, first proposed by Stellar (1954), basically is a similar mechanism. It postulates the existence of an excitatory feeding center in the lateral hypothalamus (LH) and an inhibitory (satiety) center in the ventromedial hypothalamus (VMH).

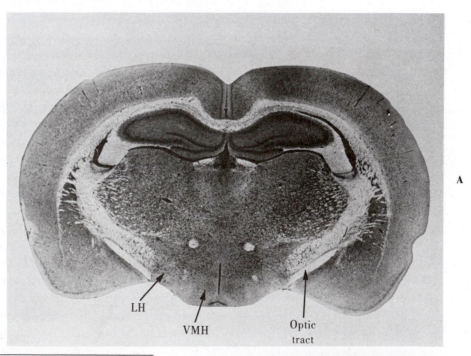

LH

VMH

Optic tract

A

Figure 17-11 Photomicrographs of cross sections of the rat brain at the level of the hypothalamus illustrating various regions discussed in the text. A, Normal brain indicating critical areas and pathways for feeding behavior. B, A bilateral lesion of the ventromedial hypothalamic region (VMH lesion).

A from Konig, J.F.R., and R.A. Klippel. 1963. The rat brain: a stereotaxic atlas of the forebrain and lower parts of the brain stem, © 1963 the Williams & Wilkins Co., Baltimore. B from Ferguson, N.B.L., and R.E. Keesey. J. Comp. Physiol. Psychol. 89:478, © 1975 by the American Psychological Association. Reprinted by permission of the authors.

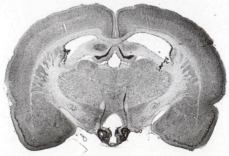

B

Clinical data dating back at least a century have implicated the hypothalamus in the control of food intake and body weight. The hypothalamus is located on the bottom surface of the brain just above the pituitary gland (Figure 17-11, *A*; see also Figures 13-9 and 15-1). It receives sensory input from several parts of the body and influences both parts of the autonomic nervous system. The hypothalamus has been implicated in many functions in addition to feeding, including the control of body temperature, drinking and water balance, sexual behavior, control of the pituitary gland, and many physiological and behavioral rhythms. Early accounts described hyperphagia (overeating) and obesity from tumors that involved the basal and medial parts of the hypothalamus, but opinion remained hotly divided as to whether the hypothalamus or the underlying anterior pituitary (which was typically also affected by the tumors) was responsible for the altered feeding.

The issue was not resolved until the early 1940s when Hetherington and Ranson (1940, 1942) and others showed that appropriate lesions (Figure 17-11, *B*) in the hypothalamus led to both hyperphagia and obesity. Destruction of the pituitary caused neither effect unless the hypothalamus also was injured. Hypophysectomy (removal of the pituitary) did not prevent the development of either hyperphagia or obesity from VMH lesions.

If given unrestricted access to reasonably palatable food in the home cage, a rat with VMH lesions begins to overeat almost immediately. The hyperphagia (which may result in a doubling or even tripling of daily food intake) persists for 3 to 12 weeks after surgery. During this **dynamic stage** body weight increases rapidly and may reach twice the preoperative level or that of an intact control. Nearly all of the excess weight is fat; lean body mass is not increased and may in some instances be reduced. Thereafter the rat enters the **static stage,** and food intake returns to nearly normal levels. Body weight is maintained indefinitely at the elevated level (the static obese plateau).

If a static obese rat with VMH lesions is starved and thus caused to lose weight, another dynamic stage of hyperphagia and rapid weight gain occurs as soon as food is again made freely available. This experiment can be repeated indefinitely with identical results; each time the VMH rat overeats voraciously until it regains its static obese body weight level, indicating that the effect of the lesion is permanent (Mook and Blass 1968).

The fact that hyperphagia persists only until a certain higher-than-normal weight level is reached suggests that the VMH lesion has elevated the level at which body weight (or body fat) is being controlled. Several other lines of evidence support this conclusion (Figure 17-12) (e.g., Hoebel and Teitelbaum 1966). In addition, the level of feeding in lesioned rats is affected by taste quality and energy demands such as from exercise or cold.

Even if palatable food is available, however, hyperphagia and obesity do not develop after VMH lesions if the animal has to exert much energy to obtain the food. There are a variety of ways of manipulating effort; all of them fortunately point to the same conclusion.

Information about the other implicated region of the hypothalamus, the LH just lateral to the VMH, began appearing in the early 1950s. Anand and Brobeck (1951) reported that bilateral lesions in the LH resulted in aphagia (no eating) and death. Within a few years other workers confirmed these observations but also

Figure 17-12 Body weight and food intake of a rat that sustained a bilateral VMH lesion at day 0. Note the initial period of overeating (hyperphagia) and rapid weight gain followed by return toward normal intake as the rat's weight reached a static plateau about 100 days after surgery. After subsequent force-feeding, which drove the rat's weight above its plateau level, its food intake was reduced until it lost weight. After a period of forced starvation food intake increased until the lost weight was regained. Modified from Hoebel, B.G., and P. Teitelbaum. 1966. J. Comp. Physiol. Psychol. 61:189-193.

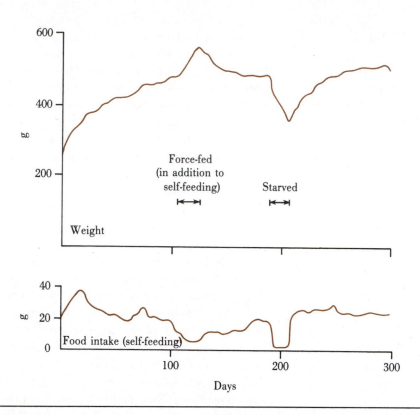

showed that death from LH lesions was not the rule; indeed recovery of feeding typically occurred provided the animal was artificially fed during the immediate postoperative period (Teitelbaum and Epstein 1962). However, feeding and drinking in LH-lesioned rats are in no sense normal, despite considerable recovery. The effects of LH lesions have been shown in a number of mammals in addition to rats. Cats, for example, normally attack and eat mice. But LH-lesioned cats have little interest in mice, and, in fact, when handed a mouse directly, at least one experimental cat went to sleep with the mouse in its jaws (Wolgin et al. 1976).

Despite recovery of the ability to survive on dry food and water, rats with LH lesions control body weight at chronically lower-than-normal levels, apparently indefinitely. Lower-than-normal weight control occurs regardless of whether the diet is highly palatable. It has been suggested that aphagia is a response to chronic lowering in the body weight set point, the target about which weight is somehow controlled.

Powley and Keesey (1970) showed that the duration of aphagia following LH lesions is a function of the rat's relative weight at the time the lesion is made (Figure 17-13). When LH lesions were made in animals that had been eating as much as they wanted, they remained aphagic until they lost enough weight to reach about 80% of normal. In another group that had been preoperatively starved so that their weights were 60% to 65% of normal, lesioned rats resumed eating very soon after surgery. They were, in fact, slightly hyperphagic until they gained weight up to the 80% level. In a third group that was force-fed to make them heavier than normal, the duration of aphagia was prolonged. These animals also resumed eating when the lower-than-normal weight had been attained.

Figure 17-13 Body weights of rats that received small (4 seconds) or larger (7 seconds) LH lesions. Before the lesions the rats were either fed as much as they would eat (ad libitum) or starved to reduce body weight at the time of the lesion before the level at which it would ultimately be regulated. See text for explanation. Modified from Powley, T.L., and R.E. Keesey. 1970. J. Comp. Physiol. Psychol. 70:25-36.

● Control
◉ Moderate lesion
○ Greater lesion
—— Fed ad lib
– – – Initially starved

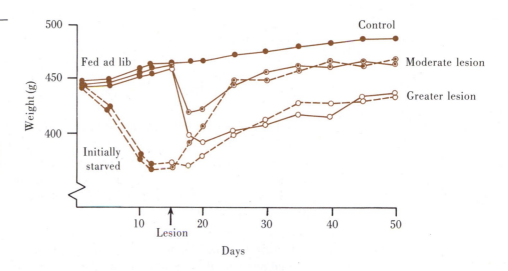

Although there are many persistent (and presumably permanent) deficits in feeding and drinking in animals with LH lesions, they do respond normally to being placed in a cold environment. Like normal rats they increase their food intake under such conditions. They also increase their intake if the caloric density of the diet is reduced and do so in a relatively normal fashion.

The results of VMH and LH lesions led to the dual center model of hypothalamic control of feeding, which envisioned a free-running LH feeding center, the neural output of which translated into the behavioral action of feeding. Most of the time, when the animal was not eating, the activity of the LH was assumed to be restrained by the VMH satiety center. The VMH was thought to monitor one or more correlates of the organism's state of energy balance and release the LH from inhibition when fuel reserves went below a certain level. Note that the model assumes that the VMH influences feeding only by modifying the activity of the LH. Electrophysiological studies revealed that firing rates of VMH and LH units were negatively correlated (as VMH firing increased, LH firing decreased). All of this, plus other lines of evidence, was taken as strong support for the concept of the VMH as a satiety center.

Among other considerations one hypothesis, the *glucostatic hypothesis* proposed by Jean Mayer (1955), postulated that the VMH contained specialized glucoreceptors that monitored the level of glucose and in turn produced satiety. When plasma glucose levels were high (as after feeding), VMH glucoreceptors would be stimulated, leading to satiety. However, the failure of direct infusions of glucose into the VMH to suppress feeding in hungry rats (Epstein 1960) discounted this theory. If there really is a "glucostat," it probably is in the liver.

The interpretation of LH lesioning itself is not as simple and straightforward as was initially assumed. The story is complicated by the fact that both sensory and motor pathways also run through this region (e.g., Marshall et al. 1971, Levitt and Teitelbaum 1975). It is yet another example of the "trained-cricket" problem.

Lesions produced in the conventional manner by passing electric currents through brain tissue destroy not only cell bodies (the hypothetical "centers") in the region around the electrode but also numerous axons that merely are passing through the region en route to some other part of the brain. The notion of two hypothalamic centers controlling feeding has been sternly challenged in recent years by evidence that damaged fibers of passage may be responsible for the classic effects of VMH and LH lesions (e.g., Zeigler and Karten 1974, Grossman 1975).

Gold (1973) argued that VMH lesions confined to the histologically defined boundaries of the VMH nuclei do not produce hyperphagia and obesity. He reported that only lesions which spread laterally from the nucleus cause the classic changes in feeding and body weight. Gold interpreted his findings to mean that the VMH syndrome is not really the result of damage to the VMH but to some other structure. Gold's preference for the other structure was the *ventral noradrenergic bundle* (VNAB), which arises in more posterior areas of the brainstem and projects anteriorly to the hypothalamus (and elsewhere). This pathway is located just lateral to the ventromedial nuclei and without question is damaged by many of the "VMH lesions" that have been described in the literature.

Consistent with Gold's hypothesis is the finding by several different investigators that micro knife cuts anterior, posterior, or lateral to the VMH produce hyperphagia and obesity (Sclafani and Berner 1977). These knife cuts cause minimal damage to cells in the VMH nuclei, but transect fibers in the area. Moreover, destruction of the VNAB near its origin also results in hyperphagia and obesity (Ahlskog and Hoebel 1973).

However, do VMH lesions cause obesity and hyperphagia only because they interrupt the VNAB? The answer turns out to be clearly, "No." There are several lines of evidence. First, the critical experiment compared the effects of VMH, VNAB, and combined VMH-VNAB lesions on food intake and weight gain. If VMH lesions were effective only because they transected the VNAB, then all three lesions should produce hyperphagia and obesity to the same degree. In fact, the effects of VMH and VNAB lesions were shown to be additive, implying two distinct inhibitory mechanisms controlling feeding (Ahlskog et al. 1975).

Other recent evidence implicates yet a third central mechanism in the inhibitory control of feeding. Its exact anatomical locus is not yet clear, but it is an ascending group of neurons that produce serotonin as a neurotransmitter (i.e., a serotonergic pathway). At present the evidence is largely pharmacological, but several workers have reported that depletion or destruction of central serotonergic systems leads to hyperphagia and obesity.

One line of evidence that originally seemed to support the concept of an LH feeding center was the demonstration that electrical stimulation of the lateral hypothalamus could elicit feeding and food-rewarded behaviors. Although some laboratories produced maps that appeared to have identified a center, the exact location of the center varied considerably from one laboratory to another. The reason for this soon became apparent. Electrical stimulation of a particular hypothalamic site could give rise to a variety of behaviors *depending on which goal objects were present in the testing situation* (Valenstein et al. 1970).

Experiments revealed that switching from one electrically elicited behavior to another was not limited to oral activities such as eating and gnawing. Animals could be switched from eating to copulating, behaviors that obviously involve very different motor sequences.

So what is electrical stimulation of the hypothalamus doing if it is not activating discrete centers for this or that bit of behavior? The most reasonable explanation is that the stimulation activates pathways that are essential for all active, motivated behaviors. Given that these pathways are activated, stimuli (goal objects) in the immediate environment direct the appearance of specific behaviors. The broad spectrum of behavioral *deficits* produced by LH lesions (and other treatments) is consistent with the view that the LH area and the pathways that course through it are important to a wide range of active behaviors rather than part of a neuronal circuit that controls only feeding.

But what of the cells in the LH area? Are they completely unimportant to the control of feeding as the preceding discussion seems to indicate? Recent work reveals that these cellular groups may still play an important role. That role may be more directly related to feeding than to other motivated behaviors. Rolls (1978) has identified LH and neighboring cells of monkeys that seem to respond selectively to hunger and food-related stimuli. Whether or not similar cells exist in the rat and other species is not yet known.

The notion of two neural centers is difficult to defend. Although the neurons in the LH and VMH seem to have a role in controlling feeding, the brain mechanisms that control feeding clearly involve much more.

The gravest problem with the satiety center concept of VMH function is that rats with VMH lesions do control their feeding activity with remarkable precision. Given palatable foods, VMH lesioned rats control body weight at a higher-than-normal level, but at a *particular* level. More recent evidence has indicated that rats with VMH lesions can control intake just as precisely as normal rats when feeding intragastrically and intravenously (e.g., Liu and Yin 1974). They can compensate for intragastric loads. Analysis of their meal patterns reveals the usual positive correlation between the size of a given meal and the time to the next meal (Thomas and Mayer 1978). Except for a heightened sensitivity to factors that affect palatability, rats with VMH lesions appear to control intake just as precisely as normal animals. This is hardly what would be expected if the master satiety system had been destroyed.

Recent Interpretations

One contemporary view of the VMH as stated by Powley (1977) emphasizes changes in hypothalamic control of visceral functions. Specifically, Powley suggests that the behavioral changes accompanying VMH lesions arise from an exaggeration of the normal patterns of digestive reflexes triggered by the oropharyngeal sensations associated with food in mouth. Because of a general overresponse of visceral systems to the taste (and other stimuli) associated with food, the VMH animal is in effect locked into a positive feedback loop, and hyperphagia is the result. The vagus nerve, especially its efferent branches, is presumed to be a major pathway in this exaggerated response to food since vagotomy reverses the obesity that develops after VMH lesions (Powley and Opsahl 1974, but see Wampler and Snowdon 1979 for an alternate view).

Another aspect of the problem is that, given minimal access to food, rats with VMH lesions are hyperinsulinemic (Frohman and Bernardis 1968). Insulin drives glucose and amino acids into cells and fosters their conversion to glycogen and fat. Insulin injections in normal animals induce hyperphagia and obesity. Thus another reasonable hypothesis is that the hyperphagia and obesity that accompany VMH lesions are caused by the hyperinsulinemia.

Experiments designed to evaluate this hypothesis indicate that hyperinsulinemia is a factor in VMH obesity, but it is not the whole story. Rats prevented from developing hyperinsulinemia (by destroying the insulin-producing cells in the pancreas and injecting fixed amounts of a long-acting form of insulin) still develop hyperphagia and obesity when given VMH lesions (Vilberg and Beatty 1975). The magnitude of both effects is less in these animals than in animals that do become hyperinsulinemic.

So changes in insulin secretion contribute to the VMH syndrome, but other factors evidently are important as well. Most likely the other responses are other activities of the viscera controlled by the VMH via efferent branches of the vagus nerve. The implication is that stimuli associated with food would trigger exaggerated visceral reflexes including but not limited to hyperinsulinemia and increased gastric acid secretion, which, in turn, would produce an exaggerated absorptive phase of digestion. Too much of the ingested nutrients would be driven into storage, creating a temporary and immediate deficit. Eating would be stimulated, and the animal would be locked into a positive feedback cycle with hyperphagia being the inevitable consequence. The hypothesis can easily explain the excessively long meals taken by the VMH rat. If one assumes that a broad range of visceral reflexes is exaggerated, including reactions to noxious tastes, then it is easy to explain finickiness.

The greatest weakness of the hypothesis is its explanation of the static phase of the VMH syndrome. This is a troublesome problem for many theories because it is hard to explain why the dynamic period of hyperphagia stops and especially why it stops at a precise level of obesity. Powley's explanation is that there are limits to how long the visceral systems can operate in an exaggerated manner; that is, hyperphagia ends because the visceral systems become fatigued. But why does not the fatigue dissipate and the hyperphagia resume? The hypothesis awaits rigorous closer examination. The set point, if it exists, remains elusive.

SUMMARY

In spite of conflicting views and much that remains to be settled, the emerging contemporary view of feeding in mammals, as exemplified by the laboratory rat, places great emphasis on an unfortunately large number of humoral, neural, and metabolic factors that together control feeding. More than likely the system is redundant in the sense that some of the controls could be eliminated without grossly altering the working of the system. This brings one to the point of drawing comparisons with the fly and asking what, in biological terms, is going on with all of this complexity?

There is not a complete picture even for the fly. There still are lingering questions concerning central mechanisms and possible roles of humoral factors. But it is clear that control of feeding in flies is vastly simpler than in mammals. It is simple enough that one can write a short computer program that models the fly system quite nicely. The system appears to be explained largely by peripheral mechanisms.

The fly system also contains redundancy, but it is minimal. As far as is known, satiety is monitored by only two routes. One involves stretch receptors in the region between the crop and midgut that operate through the recurrent nerve. The other, which might be viewed as a backup, apparently uses stretch receptors in the body wall that connect to the ventral nerve cord.

Feeding in flies also is stopped or, more properly, interrupted by temporary adaptation and habituation. But this does not lead to long-term satiation.

The mammalian system involves peripheral mechanisms but, apparently, places far less emphasis on them than does the fly. The nerves that are involved (e.g., vagus in the mammal as opposed to the recurrent in the fly) are unlikely to be homologous (from the same ancestral origin) because of the very distant relationships of the two organisms. Until more is known about the vagus nerve, one might argue that they are hardly even analogous (have the same function).

The mammal appears to rely much more on chemical—metabolic and hormonal—mechanisms in the control of feeding than does the fly. The amount of redundancy in the mammalian system seems to be disconcertingly excessive.

The control of food intake obviously is very important, and at least some backup equipment can be useful or even critical. Given the number of biochemicals involved in food and metabolism, there may be some advantage to fine-tuning the system to different substances, although how an animal achieves a balanced diet could be a whole new problem.

Flies, however, seem to do quite well with much simpler systems than rats, and feeding really does not seem to warrant all that much complexity. The difference between flies and rats may be one more case of neural parsimony in insects (Roeder 1967) (pp. 404 and 405). Perhaps vertebrates have acquired a lot of excess neural and endocrine baggage through the eons and simply have gotten away with it. Perhaps, on the other hand, it really is advantageous. The reader can ponder the question.

Before closing this chapter one must return to its main purpose—to provide an *example* of neural control of functional behavior—and consider the control of *all other behaviors* in an animal's repertoire. Progress has been made in tracing some of the pathways, even of language in humans, through the brain, as seen in Chapter 15. But we still are far from understanding how it is all put together even for single, apparently simple functional behavior (such as starting and stopping feeding), let alone the collection of all behaviors.

As a postscript one might ask a practical question of immense interest to many people: is there any hope for overweight persons, or are they locked into a level of weight and feeding from which they cannot escape? This is a complex question, and many persons should seek the advice of a physician. Studies of other species, however, suggest two practices that may alter the level at which one's body controls its weight: avoid good-tasting items and increase one's level of exercise.

Recommended Additional Reading

Bennet, W., and J. Gurin. 1982. The dieter's dilemma. Basic Books, Inc., Publishers, New York.

Novin, D., W. Wyrwicka, and G.A. Bray, editors. 1976. Hunger: basic mechanisms and clinical implications. Raven Press, New York.

Pfaff, D.W., editor. 1982. The physiological mechanisms of motivation. Springer-Verlag New York, Inc., New York.

Stunkard, A.J., editor. 1980. Obesity. W.B. Saunders Co., Philadelphia.

CHAPTER 18

James W. Grier

Nestling red-shouldered hawk (*Buteo lineatus*).

ONTOGENY OF BEHAVIOR

During growth and development an animal (except protozoans) goes from a single cell through intermediate stages to the point where it is a functional organism consisting of a very large number of differentiated cells. When and how does behavior enter this picture? Even after birth or hatching, animals rarely display all of their behavioral repertoire at the start. Different behaviors become manifest at different times in the individual's life, more or less in orderly sequence. Behaviors shown by young animals may be replaced by or overlap with adult behaviors. Examples are illustrated in Figure 18-1. What determines the changes, and how do they occur? To address these questions one must deal with the intertwined subjects of ontogeny (the developmental history of an individual), learning, and memory.

Unfortunately, these are not easy subjects with which to deal. Nature, in the course of evolutionary branching, has not stuck with one simple plan for development. Different organisms develop in many different ways. They use different schemes of embryological or asexual processes, acquire particular characteristics at different stages and in different sequences, and develop fairly directly into the adult form in some cases and only indirectly—often through quite different larval stages—in other species.

Many persons now view learning as a more diverse phenomenon than was formerly suspected. The internal, molecular workings of learning and memory have proven most intractable and have provided some of the greatest challenges to biology and psychology.

The ability of biological systems to store large amounts of information in the nervous system is awesome. The capacity of animal memory surpasses the wonders of orientation and migration in terms of its incredibility. Consider the role of memory in the daily lives of humans. Regardless of amount of formal education, everyone knows an incredible number of places, things, and skills. Before the advent of writing many persons remembered so much of their genealogy that it would take many hours or days to quote it all. Most persons have had the experience of recognizing faces, voices, or musical phrases, perhaps only from the first few notes, after many years since the last exposure to them.

Are the mechanisms of memory and learning the same in all animals? The way that an ant remembers how to run through a maze may be vastly different from the way a rat does the same task (Schneirla 1959). In either the ant or the rat, the ways that they tackle the spatial problems may be quite different from how they deal with learning and remembering in feeding or social functions.

Because of the complexity of the subjects, the amount of available literature, and the usual limitations of a textbook, the material in this chapter and Chapter 19 represents a limited sampling and overview of the subjects. Ontogeny and a few learning topics will be covered in this chapter. The remainder of the learning and memory material is discussed in Chapter 19.

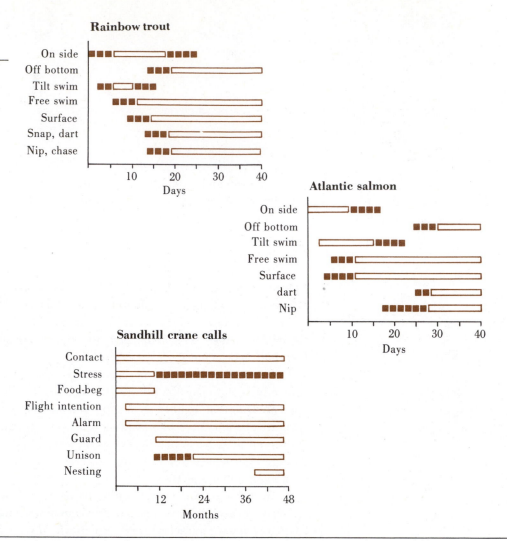

Figure 18-1 Sequential appearance of behaviors during the life of individuals of different species. *Open bars,* Common occurrence in most individuals during the indicated ages; *dark broken bars,* ranges resulting from variability among individuals and preliminary and subsequent traces of the behavior. Modified from Dill, P.A. 1977. Anim. Behav. 25:116-121; Fernald, R.D., and N.R. Hirata. 1979. Z. Tierpsychol. 50:180-187; Candland, D.K. 1971. In Moltz, H. The ontogeny of vertebrate behavior. Academic Press, Inc., New York; Altmann, J., and K. Sudarshan. 1975. Anim. Behav. 23:896-920; Archibald, G.W. 1976. Ph.D. Thesis. Cornell University, Ithaca, N.Y.

EMBRYOLOGICAL AND NEONATAL ROOTS OF BEHAVIOR

The rates of development and stage of development at hatching or birth vary tremendously among animal species. There may be stages variously defined as larval, instar, imago, adult, embryonic, fetal, neonatal, postnatal, immature, juvenile, and subadult. Some young are *altricial* at birth or hatching; that is, they are helpless on their own and require further care and attention before they can leave the site of birth or hatching, such as a nest or den. In other species the young may be *precocial* (Figure 18-2). They can move about, feed, and have at least a limited capacity to deal with the hazards of life on their own. Some precocial young appear to be simply like miniature adults. The terms altricial and precocial only define points on a continuum. Different species may occur almost anywhere between one extreme and the other.

Even the point of birth is somewhat arbitrary; marsupials are "born" essentially as fetuses and must crawl into the mother's pouch to complete their early development. Because of this tremendous diversity, it is difficult to generalize. Behavioral developments that take place before birth or hatching in some species may occur afterward in other species.

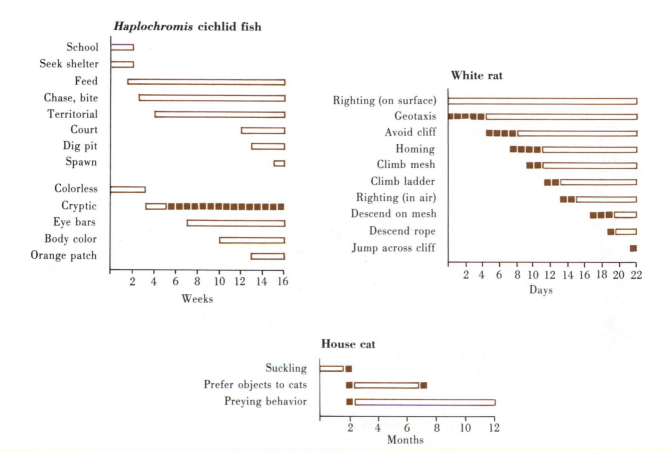

**Figure 18-2 Precocial
young. White-tailed deer
fawn—20 minutes old.**
Photo by Leonard Lee Rue III.

A major difference between vertebrate and invertebrate embryos concerns the contraction of muscles prior to hatching. Embryonic muscles of vertebrates show considerable twitching, jerking, and movement. Among amniotes (reptiles, birds, and mammals), even the amniotic membrane displays contractions and movements. Invertebrate embryos, however, are virtually motionless while packed in the egg and, among those which go through a pupal stage, again in the pupae. Their earliest movements, hence "behaviors," are seen in the larval or instar stages. The earliest beginnings of behavior, thus, usually are within the embryo or soon thereafter in vertebrates and after hatching in invertebrates. Most of the information up until the last few decades came from vertebrates.

Many of the earliest notions about behavioral development, along with development in general, were based on the idea of **preformation.** Under this concept the organism is fully formed and complete even in the sperm or egg, and development amounts primarily to growth. This idea now is of little but historical interest, having been replaced by the basic concept of **epigenesis** (Needham 1959). Epigenesis states that various characteristics, including behavior, are not present initially but form and become visible during development.

Along epigenetic lines, however, there have been several different opinions concerning the early development of behavior. Opposing viewpoints have roughly followed the old nature versus nurture dichotomy. Most of the early workers simply observed embryos. Different persons often focused on different species and then speculated and attempted to generalize to vertebrates or all animals. There rarely were any manipulative or well-designed, critical experiments.

The nature side of the argument, that early behavior is essentially **predetermined,** can be traced to at least 1885 when W. Preyer published a large volume on behavioral embryology. He and later workers observed that early movements begin before the embryo is capable of responding to external stimuli and that there are rhythmic activities that do not respond to outside stimulation. One viewpoint that developed, known as **autogenous motility,** stated that the early motions were intrinsic to the muscle tissue or developing motor neurons. The movements were not believed to be either activated or modified by sensory input or feedback.

A different view, based partly on other species that have earlier development of reflex pathways, became known as the **reflexogenous** concept. Under this viewpoint the neuromuscular pathways still are predetermined, but movements are not shown until the pathways develop; the movements reflect the pathways; and movement is sensitive to external stimulation. Subdivisions of this and some of the other views depended on the orientation of the researcher. Neuroanatomists tended to work conceptually from the inside out, viewing the structure as of primary interest and the behavior as only secondary, whereas behaviorists worked from the outside in, considering the structure as of secondary importance. One of the early neuroanatomists, Coghill (e.g., 1929), proposed that movement begins as a total, undifferentiated pattern, out of which discrete or independent parts develop. Other workers (e.g., Swenson 1929, Windle 1944), still in the reflexogenous camp, suggested that it was the other way around—that parts of the behavior develop first, then add together to form functional patterns.

Evidence that accumulated during the 1930s and 1940s made the autogenous and reflexogenous views difficult to generalize. One way to deal with the diverse information was to propose more abstract generalizations, such as **systemogenous behavior.** This viewpoint (e.g., Carmichael 1963, Anokhin 1964) focused more on the behavioral system and simply stated, in essence, that the movements needed for survival under the particular environmental situation would be developed before birth. Emphasis was not on the particulars of how development occurred but that the outcome could vary from species to species, depending on the environmental factors the species encountered at birth. But the various outcomes for the different species still were predetermined.

Meanwhile, back at the nurture side of the argument, there developed a **probabilistic** viewpoint about behavior in the embryo. The main proponents of this school were Holt (1931) and Kuo (1939, 1967). Building, in part, on early information (during the period from 1910 to the 1930s) about nerve growth, such as dendrites growing toward axons from which they received the most stimulation, the probabilistic view stated that early behavioral development was not predetermined and certain. Rather, it was viewed as being only probable, depending on a host of stimulating factors. These included mechanical agitation such as from the heart beats and amniotic contractions, internal and external sensory input, hormones, and musculoskeletal effects of use in conjunction with previous developmental history. The early behavioral substrate was viewed as a blank slate (the **tabula rasa** associated with Hobbes and Locke, see discussion in Bolles 1979), which became channeled by external and internal events until it developed into an observable outcome. Various outcomes were possible, although some, namely those generally observed, were most probable.

Figure 18-3 Relationships among various schools of thought concerning the ontogeny of behavior. Of the ones shown here, the autogenous school (emphasized) has received the most experimental support.

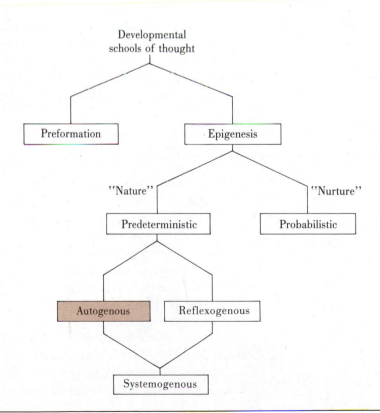

Other major contrasts between this view and that of the predeterminists were whether structure only influenced function (predeterminist) or if effects could go both ways (probabilist) and whether development proceeded in discontinuous jumps (predeterminist) or more smoothly and continuously (probabilist). Gottlieb (1970) provides further details and references. The relationships between these and the other schools of thought are diagramed in Figure 18-3.

Investigation into these issues lessened for a few years, but with improved techniques for working with embryological and neuromuscular tissues, the pace picked up again, and there now is a wealth of recent information. A sampling of these findings follows.

Basic movement patterns in a wide variety of species appear to develop normally, even when they are deprived of the normal sensory input. Examples among arthropods include locust flight (Kutsch 1974) and lobster movement (Davis 1973). Chick embryos with cut sensory nerves displayed fairly normal activity in their leg movements during most of the subsequent embryonic development (Hamburger et al. 1966). These observations support the autogenic viewpoints established earlier on the basis of rhythmic movements and movements that occurred before the establishment of complete nerve-muscle connections.

In a classic demonstration of the presence of developed behavior before any opportunity for sensory feedback or rehearsal, Bentley and Hoy (1970) investigated the species-specific song production of crickets (*Teleogryllus* sp.). The songs are produced by stridulatory movements of the forewings by adult males. Immature crickets go through 9 to 12 instars, with the wings becoming functional only at the final, adult stage. By lesioning an appropriate part of the brain in late (but not yet adult) instars, they discovered that the larval wings, almost too small to be recognized as wings, were set in motion. Electrodes inserted into the wing muscles showed contraction patterns characteristic of those which produced the song in adults (Figure 18-4). The neural program needed to produce the species-specific song was entirely present in the nervous system but normally was inhibited from expression by another part of the nervous system. The inhibiting region apparently was removed by the brain lesion, which permitted the premature release of the behavior.

Tracing movements back to their earliest origins is difficult for three main reasons. First, there may be several movements occurring simultaneously or nearly simultaneously so that it is difficult to know which ones are related and which are only passive. Second, embryos and their muscular movements are so small that the movements may be almost imperceptible and, even if visible, difficult to photograph or otherwise measure. Third, the size, proportions, and postures of the growing embryo change considerably so that it can be difficult to compare measurements from one time to another.

One solution to these problems is to record *electromyograms* (EMGs) from individual muscles. These are similar to recording action potentials from neurons. This permits the contractions to be recorded electronically and directly. One can investigate the developing patterns of contractions and compare them with patterns in subsequent, fully developed behaviors. An operational, functional behavior requires coordination at several levels: intramuscular, intermuscular, intrajoint, interjoint, and interlimb. Accordingly, patterns of contraction can be studied at all these levels.

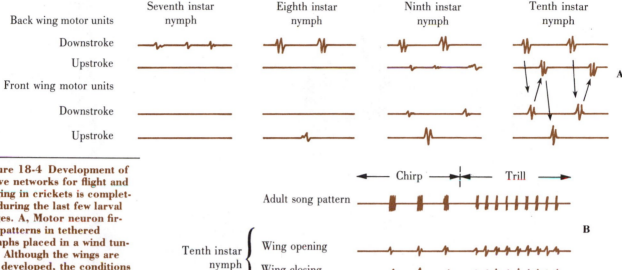

Figure 18-4 Development of nerve networks for flight and singing in crickets is completed during the last few larval stages. A, Motor neuron firing patterns in tethered nymphs placed in a wind tunnel. Although the wings are not developed, the conditions stimulate the motor neurons to fire in typical fashion, as if the wings were functional. At first the pattern is only partial. Then the pattern develops for other muscles. At the final instar, before the adult stage and before the wings emerge fully, the full motor pattern is present. Upstroke and downstroke neurons alternate, and the back wing leads the front wing, as indicated by arrows. B, Muscle impulses during calling song; top trace is adult song pattern; bottom is from a tenth instar nymph with a brain lesion that disinhibits the output patterns. Tenth instar nymphs normally do not attempt to sing. Also see Figures 4-6 and 4-10. Modified from Bentley, D., and R.R. Hoy. 1974. Sci. Am. 231(2):34-44.

A. Bekoff (1976, 1978) studied the development of intrajoint and interjoint coordination in chick embryos. She implanted fine (approximately 50 μm diameter) wire electrodes with long flexible leads into various leg muscles so normal movements were relatively unrestricted. Her recording setup and results are illustrated in Figure 18-5. The results appear to demonstrate that patterns of contraction similar to those shown in coordinated hatching movements are already present by at least the ninth day of development. (Chicks hatch after 21 days of development.) Furthermore, reflex arcs are not yet present and functional. Several things can be concluded: coordination may be present long before it would seem to be "needed"; the coordination appears to develop without the need for sensory feedback; and coordinated movements may be assembled gradually before they become obvious to more superficial observations. There now is much evidence for central pattern generators among a wide variety of vertebrates and invertebrates, as discussed in Chapter 15. Some of these central outputs may be expressed very early and used in some of the first movements of the embryo.

It also would be efficient, as in some invertebrate systems (e.g., Camhi 1977), to use basic patterns of contraction in different behaviors. The basic movements used in hatching may be used again later in walking in slightly different combination. If so, much of the ontogenetic development of behavior might consist of development of coordination patterns at higher levels of organization. Different behaviors, for example, could use different interlimb combinations of the same underlying interjoint patterns. The extent to which different behaviors share pattern generators remains the subject of much ongoing research and debate.

Another approach to analyzing the early development of behavior and locating the cells that are initially responsible for directing the behavior is through a technique called *fate mapping*—following and correlating mutant or abnormal lines of cells from early embryonic stages into the adult. The location of the responsible site is known as the *behavioral focus*. Hotta and Benzer (e.g., 1970,

Figure 18-5 Schematic diagram of apparatus used by Beckoff to investigate muscular output of developing chicken embryos. Recording setup and results: similarity of movement patterns between 9 (A) and 20 (B) days of development. Intensified movements of these types are used during hatching at age 21 days. Based on photos and illustrations in Bekoff, A. 1978. In Burghardt, G.M., and M. Bekoff, editors. The development of behavior. Garland STPM Press, New York.

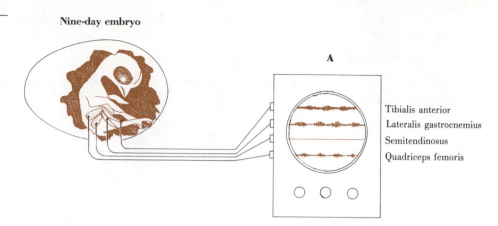

Nine-day embryo

A

Tibialis anterior
Lateralis gastrocnemius
Semitendinosus
Quadriceps femoris

1972) and subsequent workers have constructed fate maps for a number of mutant alleles affecting behavior in *Drosophila melanogaster*. These include abnormal electroretinograms, artificially induced leg movements, flight, and more complex courtship and mating behaviors. In males, for example, courtship orientation and wing vibrating map to locations in the head, whereas copulation movements map to the thorax. More detail is expected when additional mutants and markers become available.

All of these lines of evidence strengthen the inference that many patterns and components of behavior develop early in life and are under very significant central nervous system control. This central control, it seems reasonable to assume, is in turn directed by relatively unmodifiable genetic instruction. Thus there is strong support for autogenetic development *in some behaviors in some organisms*.

MATURATION AND BEHAVIORAL CHANGE

Many of the changes that are observed in an animal's behavior can be traced to maturation of the underlying structures. *Maturation* refers to normal changes during growth and development that cannot be correlated with environmental variability, but they still are *changes* and may give the appearance of learning. (Learning is defined on p. 579.) Subsequent changes, such as refinement and perfection of the behavior, may or may not involve learning.

Practice, as will be seen shortly, is a phenomenon that leads to an improvement in behavior in many cases but not necessarily in all situations. It commonly is thought to involve increasing completion, organization, or reinforcement of neural pathways responsible for the specific movements, perhaps through changes in dendritic synapses. Practice also may produce improvement through the exercise effects of strengthened muscular and skeletal tissue. But practice is not understood very well, and it is difficult to place it in relationship to maturation and learning, just as its role in play (see Chapter 12) has not yet been settled.

Twenty-day embryo: hatching

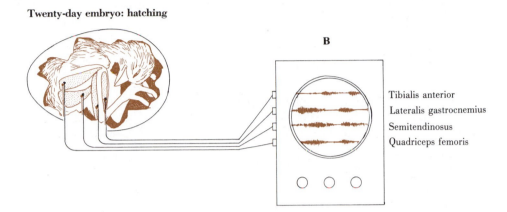

Tibialis anterior
Lateralis gastrocnemius
Semitendinosus
Quadriceps femoris

The effects of hormones on behavior, as described in Chapter 16, can be viewed as maturational changes in many instances. Such would be the case with the neonatal or fetal surges of androgen that masculinize the brain. Other hormonal effects that cause behavioral changes from one growth stage to another include going from one larval stage to another and from larvae to pupae and changes seen in puberty and premolting and molting behaviors. Hormones and, in a sense, maturation also play a large role in starting, developing, and stopping reproductive and migratory behaviors.

Clear examples of maturational changes now are fairly numerous. One involves color preferences in amphibians. Frightened frogs (*Rana* sp.), for example, jump toward water, showing a strong preference for blue and an avoidance of green. Tadpoles of the same genus, however, move toward green. The preferences have been demonstrated behaviorally and correlated with evoked potentials recorded from electrodes placed at different points in the optic pathway, from the retina to the diencephalon (Muntz 1962a,b, 1963a). The change in this response with age (Figure 18-6) corresponds to a change in the sensory cells and visual pigments in the retina of the frog's eye. The first cells to develop absorb green wavelengths of light. Later two additional types of retinal cells develop: one believed to act synergistically with the first set, plus another type, all of which create the new blue preference. It appears that the development of the frog's color preference behavior is strongly influenced by the maturation of the sensory cells. The conclusions are strengthened by similar findings in other amphibians, some of which have similar preferences and sensory cells and some of which have different preferences and correspondingly different visual pigments (Muntz 1963b).

Cuttlefish (*Sepia* sp.), a type of cephalopod mollusc, prey on small shrimp, among other things. Adult cuttlefish will learn from punishment, such as from electric shock, not to attack. But young cuttlefish do not seem to learn to stop attacking. They continue to attack in spite of electric shock or when the shrimp are protected behind glass so that the young cuttlefish fail to catch them (Wells 1958). Young cuttlefish continue to attack until physically exhausted.

But the young individuals do display some changes in their attack behavior. The full behavior has four main components: when first presented with a shrimp, the cuttlefish shows a latency before giving any response; then the eye closer to

Figure 18-6 Changes in the color preference of frogs' (*Rana temporaria*) phototactic responses as the animals matured from tadpoles to frogs. These curves are similar for both behavioral choice and physiological, evoked potential responses; that is, frogs responded by jumping toward different choices in the frequencies as indicated, and electrodes inserted in the visual pathway showed neural firing rates in similar proportions. Modified from Muntz, W.R.A. 1962. J. Neurophysiol. 25:699-720, and Muntz, W.R.A. 1963. J. Exp. Biol. 40:371-379.

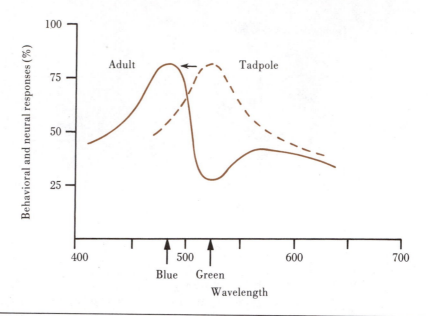

the shrimp turns toward it; next the cuttlefish turns its body and both eyes toward the shrimp; finally it attacks and throws out the tentacles to grab the shrimp. All but the first component are relatively invariable, taking about 10 seconds regardless of age or experience. The first, latency stage, however, shows a marked change with experience. A naive cuttlefish requires about 2 minutes before showing any response to its first shrimp. The second shrimp elicits a response around 40 to 50 seconds. The latency rapidly declines further until, after about five trials, it is 10 seconds or less. Although the improvement in response time might appear superficially to be learning, the change is quite consistent, apparently unmodifiable, and probably amounts to a practice effect or some form of maturation specific to the attack behavior. The change in response time occurs whether the cuttlefish is rewarded or not and whether it is initially well-fed or starved (Wells 1958) (Figure 18-7).

What about the change from young cuttlefish that do not "learn" to stop attacking to adults that do? This also appears to result from maturation. The region of the adult cuttlefish brain known to be involved in learning, the vertical lobe, does not develop until later in the cuttlefish's life.

Another classic example involves chicken behavior. Chickens are ground-dwelling birds that obtain their food by pecking at it. The chicks are not fed by their parents, although hens may peck toward food, indicating its location, or occasionally even hold items in their beaks for the chicks. For the most part, however, chicks must feed themselves from the very start, and they do quite well if artificially raised without an adult.

At first the chicks are not very accurate and miss many of their pecks at food. With time, however, there is a considerable improvement. Cruze (1935) held chicks in the dark to reduce their movement and fed them powdered food by hand for various periods of time. He then tested their accuracy in pecking at grain. After the initial test he allowed them to feed naturally in light and then retested

Figure 18-7 Latency of attack of cuttlefish on shrimp. The latencies decline rapidly with repeated trials, but the pattern of decline does not depend on success of capture, starvation, or age within a range of 1 to 5 days. Modified from Wells, M.J. 1958. Behaviour 13:96.

● Rewarded for attacking

○ Not rewarded for attacking

▲ Starved for 5 days and then rewarded for attacking (1 trial per day)

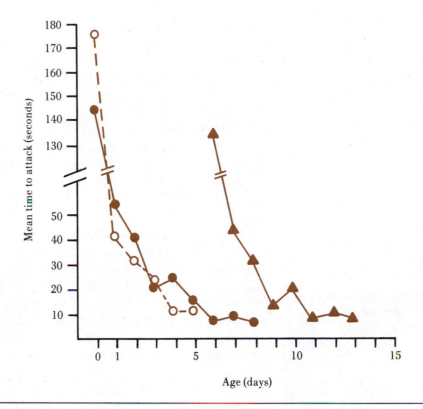

them 12 hours later. Accuracy improved in all cases after practice, suggesting a practice effect. But it also improved among older chicks that had never practiced before (Table 18-1). This suggests an improvement from something maturing, whether within the sensory, neural, or muscular systems. The improvement may have resulted from increased strength and ability of the leg and neck muscles.

Further evidence for the maturation effect plus an indication that reward or sensory feedback was not required was obtained by an ingenious experiment by Hess (1956). He fitted chicks with rubber hoods and prismatic lenses that deflected their vision by 7 degrees. Control chicks wore plastic lenses that did not deflect their vision. Results were recorded in Plasticine with a nail for a target;

Table 18-1 Number of missed food pecks by chicks of different ages before and after a 12-hour practice period*

	Age When Removed from Dark (Hours)				
	24	**48**	**72**	**96**	**120**
First attempt	6.04	4.32	3.00	1.88	1.00
After 12 hours of practice and 12 hours of rest†	—	1.96	1.76	0.76	0.16

Modified from Cruze, W.W. 1935. J. Comp. Psychol. 19:371-409.
*Values represent the mean number of misses out of 25 pecks averaged for 25 chicks.
†Actual age at second test is same as birds in next group at their first test. Birds removed from the dark at age 48 hours, for example, were tested first upon removal from the dark, allowed to eat naturally for 12 hours, held without food for 12 hours, and then retested.

each peck left a dent in the Plasticine. The chicks were first tested and then allowed to feed (while wearing the glasses) for 3 days, then retested. Both groups showed tighter clustering, hence, "accuracy," in their pecks, but those in the treatment group continued to miss the target—they simply missed it in a more consistent manner. The treatment chicks had managed to get some food while on their own, but barely enough, and many were losing weight. They had not learned to compensate for the deflected vision, and the tighter clustering of pecks was interpreted to result from some type of a maturational response. Roughly similar pecking responses and results have been shown in gulls (Hailman 1969).

Another behavior in which the role of maturation has been studied involves wing-flapping "exercises" by nestling birds. The young of most birds will stand and pump or flap their wings before they are able to fly. As early as 1873 Spalding showed that the wing flapping might not be necessary. He put young swallows in cages so small that the young were unable to open their wings. But when released at an age where they should be able to fly, the caged swallows flew as well as other young that had been able to stretch their wings and exercise. Grohmann (1939) obtained similar results by raising an experimental group of pigeons in cardboard tubes that prevented flapping wing movements.

Provine (1979) experimentally investigated wing flapping in chicks of domestic chickens. He amputated the wings of one group and fitted them with limb prostheses made from soda staws. Chicks were dropped a short distance under carefully standardized conditions and photographed stroboscopically (Figure 18-8). The mean number of wing flaps by chicks with wings intact and chicks with the plastic wing prostheses were not only statistically similar (t test), they were virtually identical. Proprioceptive effects from the muscles, located in the breast, and possible effects from neonatal movements before amputation could not be ruled out. But it was clear that postnatal sensory feedback that would be associated normally with an intact wing was not necessary for the flapping patterns.

As a final example of maturational effects on behavior and the differences between species, the development of prey-handling ability in predators will be considered. Mueller (1974) investigated prey recognition and predatory behavior in young American kestrels *(Falco sparverius)*, a type of small falcon. These birds typically prey on mice, among other things, and kill them in the usual falcon fashion by biting the neck soon after capture. Acquisition of components of the behavior might be accomplished by familiarity and learning while under the care of their parents, by simply pursuing any small moving object and being rewarded or reinforced to concentrate on those with which they have success in catching and eating, or by having the patterns of behavior develop by maturation of the nervous system. Young kestrels in the wild seem to feed mostly on insects at first. If given live mice prematurely, they show either fear or no response. This information suggests that the behavior requires maturation.

Mueller hand-reared and hand-fed young kestrels to prevent experience with whole animals or live prey. Then at ages 10 weeks to 1 year the kestrels were tested on four types of items: tissue-paper balls and tissue model mice, stuffed mouse skins, dead mice, and live mice. In each case, except the live mice, which were moving, the objects were presented either stationary or moving. The falcons showed few responses to the paper objects and then only in a playful manner. The

Figure 18-8 Stroboscopic photographs of wing flapping of a 13-day intact chick *(right)* and a wingless chick with plastic prostheses *(left)*. The chicks were dropped gently a short distance to stimulate the wing flapping. This study indicated that sensory feedback from the intact wing was not necessary for the wing-flapping pattern. See text for further explanation. From Provine, R.R. 1979. Behav. Neural Biol. 27:233. Copyright by Academic Press, Inc. Photos courtesy Robert R. Provine.

stuffed mouse likewise received little attention. The dead mice were taken by a third of the kestrels, which attacked and bit them in typical fashion, as if they were live. With live mice, however, eight of nine birds attacked the first time with responses that were "intense, rapid, sustained and well-oriented" with capture and killing being expert and indistinguishable from adult behavior. All birds achieved consistent, expert performance in fewer than six trials. Mueller inferred that experience was unimportant in the recognition, capture, and killing of mice by American kestrels.

Raber (1950), in studying the ontogeny of predatory behavior of owls *(Asio otus)*, obtained similar findings except there appeared to be a **critical period** within which live prey had to be taken. If live prey was not taken during that period, the birds would not attack live prey at a later time if given the opportunity. Thorpe (1948) demonstrated prey-killing behavior in another owl *(Athene noctua)* as a vacuum activity; the owls would display the entire predatory sequence, including killing the imaginary prey, in the absence of any prey stimulus.

The ontogeny of predatory behavior in mammals is less clearly a case of maturation and may involve more learning, depending on the species. Eibl-Eibesfeldt (1956, 1963) and Leyhausen (1956, 1965) have suggested that experience and learning are necessary for orienting bites to the necks of prey by mustelids and cats. Gossow (1970) and Eaton (1970), however, believed instead that more maturation may be involved in at least some of these mammals. Eibl-Eibesfeldt (1951) studied the ontogeny of nut opening in squirrels and showed that learning was not important in that instance. Several workers (e.g., Fox 1969) have noted the importance of movement in eliciting attack by various mammalian predators.

It is clear that the roles of maturation and learning in the development of behavior vary widely. Variation can be seen even among animals in the same class, such as mammals. Furthermore, maturation and learning are not completely separate categories but may be related. Maturation of the most extreme form, with no learning involved, could be viewed as a subtly disguised form of innate behavior (instinct, genetic, nature, etc.). It simply is innate behavior that takes some time to develop. As discussed elsewhere in this book (Chapter 4), most behavior probably lies somewhere between the two extremes.

EFFECTS OF DEPRIVATION AND GENERAL EFFECTS OF EARLY EXPERIENCE ON THE DEVELOPMENT OF BEHAVIOR

Being deprived of various stimuli or opportunities for learning seems to pose few problems in the ontogeny of some behaviors in some species, as indicated in parts of the preceding section. Deprivation may have some effects, however, as in the case of the owls observed by Thorpe (1948). In some instances the effects of deprivation are striking. This section deals with a few examples where deprivation caused significant impact on subsequent behavior. The greatest effects of deprivation have been observed in mammals, from which all of the following examples come. Examples of deprivation effects in birds will be discussed later under the separate section devoted to the development of song in birds.

Kittens have been the subjects of extensive research on the effects of deprivation of vision and visually guided behavior (e.g., Wiesel and Hubel 1963). When deprived of normal visual experiences for the first 2 to 3 months of life, kittens are functionally blind thereafter. In various experiments they have been deprived of sight by being kept in darkness, by wearing diffusing hoods, and by having the eyelids carefully sutured closed, an operation that is easily reversed. In the eye-suturing experiments (e.g., Wiesel and Hubel 1965) either one or both eyes were sutured; sutures were subsequently removed and eyes were opened; or eyes that had been open were subsequently sutured closed. The results were clear in all cases: the visual pathway in the brain deteriorated for all eyes deprived during early life. Effects included reduced size of neurons in the lateral geniculate body and reduction of connections of the visual cortex in the brain on the appropriate side. Kittens with single eyes sutured showed a loss of binocular vision, with the loss occurring on the affected side of the brain. Opening of the eyes after these effects had occurred led to only slight, insignificant recovery. Similar deprivation of vision, for comparable periods of time, in adult cats did not cause similar losses.

Lack of use during the developmental period may affect not only the sensory pathways but the entire sensorimotor response in a general way. This has been studied in several species with a variety of techniques. Perhaps one of the most ingenious and best-known studies involves self-produced responses in kittens in which an active kitten and a passive kitten were exposed to the same visual environment and visual movements (Held and Hein 1963). Through the use of a merry-go-round apparatus (Figure 18-9), one kitten pushed and generated the movements while the other rode in a gondola. Mechanical linkages duplicated all of the active kitten's body movements at the gondola for the passive kitten. The procedure was repeated with 10 pairs of kittens, each pair coming from a different

Figure 18-9 Apparatus for equating motion and consequent visual feedback for actively moving (A) and passively moving (P) kittens. From Held, R., and A. Hein. 1963. J. Comp. Physiol. Psychol. 56:872. Copyright 1963 by the American Psychological Association. Reprinted by permission of the authors.

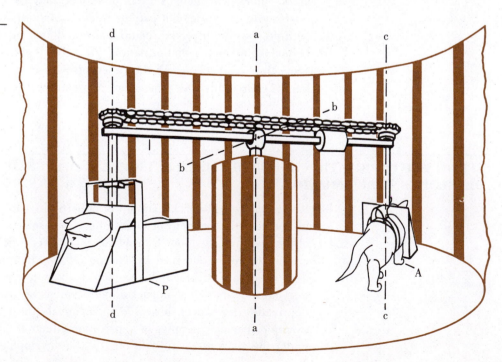

litter. Kittens were placed in the apparatus for 3 hours per day for several weeks. The remainder of the time they were kept in the dark with their mothers and littermates. At the end of the period they were tested in two principal ways: (1) with a visually guided paw placement test in which they were carried toward a horizontal surface (toward which a normal kitten reaches its paws before contact) and (2) a visual cliff test in which kittens were placed on a narrow platform over a piece of glass with a patterned surface immediately under the glass on one side and a similar surface 30 inches below on the other side. At the end of the experiment all of the active kittens responded to the paw-reach and visual cliff tests normally, whereas none of the passive kittens showed the reaching response or discriminated between the shallow and deep sides of the cliff test. Held and Hein concluded that "self-produced movement with its concurrent visual feedback is necessary for the development of visually-guided behavior." They eliminated anatomical or physiological deterioration of the systems by showing normal blinking and pupillary responses in the passive kittens and by obtaining subsequent recovery after the passive kittens were permitted to move normally in a lighted environment.

In experiments involving puppies, Melzack and Scott (1957) showed that environmental deprivation during early life can lead to abnormal behavioral responses associated with pain. They used several litters of Scottish terriers, with each litter divided into one of two groups, a control group where the pups were raised as pets in private homes and the treatment group where pups were raised in diffusely lighted isolation cages, deprived of normal sensory and social experience. This treatment lasted from the time of weaning until about 8 months of age. The dogs were then tested in a variety of ways 3 to 5 weeks after being released from isolation, and some were retested again 2 years later. The normally raised dogs showed normal avoidance responses to pain induced by electric shock, pinpricks, and heat, whereas all of the deprived dogs showed significantly different, nonadaptive responses. The deprived dogs showed local reflex responses to the noxious stimuli, indicating that the senses functioned at least to some extent, but the dogs showed no organized avoidance, attempts to get away, or any signs of emotional distress. The deprived dogs also were abnormal otherwise in their responses to objects in the environment. They would walk repeatedly, for example, into water pipes along the wall of their testing rooms.

Probably the best-known, most widely publicized studies of deprivation involve rhesus monkeys deprived of a normal, live, soft mother and normal peer relationships (Harlow 1959, Chamove et al. 1973). Several aspects of growth, development, and individual and social behavior were investigated experimentally by comparing young monkeys raised under normal (although captive) conditions and those raised in various combinations of soft and wire surrogate mothers, no peers, different categories of peer relationships, and peers but no mothers. The experimental monkeys developed normally from a physical and physiological standpoint but not behaviorally. Young deprived during the first 6 to 12 months of life showed a host of problems and inabilities in emotional and social behaviors. The problems lasted throughout the lives of the affected individuals.

Deprivation is a relative concept. Although some of the cases cited earlier might be viewed as rather severe, deprivation lies at one end of a continuum with environmental enrichment at the other end. The normal situation is somewhere in between. Normal kittens, for example, would be reared with their mother and littermates in surroundings that stay relatively constant day after day. Eventually, however, the kittens are able to move about and explore. They then are exposed to a variety of things by their own movements or by being carried about by their mother. But even with such exploration and movement the kittens are exposed to new things within limits. It would be possible to artificially enrich their experiences by introducing a wider variety of novel objects, sights, and sounds, carrying them in vehicles, and changing the general surroundings. Would such things have the opposite effect of deprivation? And what could be learned about the normal development of behavior?

Environmental Enrichment

The implications of enriched versus deprived rearing are obviously of profound importance and interest in human development, and there is a large body of human-related literature and studies. The findings have been fairly consistent regardless of whether the mammal studied was human, other primate, carnivore, or rodent, however, so discussion here will be confined to a few brief comments from studies of white rats.

Before proceeding and as somewhat of a disclaimer, however, one needs a few notes of caution. The subject is fraught with semantic and methodological differences of approach and opinion. Terms related to deprivation include isolation and restriction. Enrichment, on the other hand, has included environments referred to as free, unrestricted, enriched, or complex. It is rarely clear just what "normal" means, particularly when dealing with the domesticated, laboratory-reared animals commonly used in research. Meyers (1971) attempted to obtain a degree of uniformity in recognizing the polarity by classifying environments simply as restricted or complex. Methodology has varied widely. Various mazes commonly are used to measure behavioral performance. Furthermore, as usual, theories about the underlying mechanisms abound. Thus, although some findings are fairly consistent, closer inspection of the field reveals a morass of unsettled issues. The consistent, if not somewhat superficial, findings will be considered as well as the important anatomical and physiological differences between deprived and enriched forms of rearing.

It appears to be a fairly safe conclusion that an enriched environment between the time of birth and puberty leads to enhanced problem-solving ability in rats (and many other, but not all, species of mammals). On the other hand, a rat's performance can be degraded by a deprived environment. Furthermore, these effects appear to be genuinely related to the performance in question (e.g., ability in a maze) and are not artifacts from differences in sensory ability, emotionality, or simple exploratory behavior. Further discussion and references are provided by Meyers (1971).

The most concrete effects of deprivation versus enrichment can be seen in the nervous tissues of the rats. Most information along these lines stems from the pioneering work of Rosenzweig et al. (e.g., 1972) at the University of California at Berkeley. They reared rats generally in one of three environments: (1) enriched with 10 to 12 animals in a group with various objects, frequent handling, and experience in mazes, (2) social control with 3 animals living together in a standard laboratory cage, and (3) isolated control where animals are housed individually in cages with three solid sides.

Enriched-environment rats showed significant differences in the cortex of the brain compared with the isolated controls. Depending on which neural characteristic was considered, the social-control rats were generally somewhere intermediate. Enriched environments during rearing led to heavier and thicker cortical tissue, larger neurons, increased glial tissue, and increased dendritic branching. Information on acetycholinesterase (AChE) levels implies that there may be increased synaptic activity in the neural tissue of enriched-environment animals compared with the deprived individuals. (AChE is the primary enzyme that deactivates acetylcholine after synaptic transmission.)

One can conclude that the nature of environmental input and opportunities for specific motor responses to that input can importantly affect the early development and organization of the nervous system in at least some species of mammals. As will be seen shortly, much the same can be said also for some birds.

It is thus very clear that early experience in the life of the individual may importantly affect its subsequent behavior. These effects of experience are different from what would be considered as either simple maturation or traditional notions of learning. Learning, as commonly defined, usually refers to more specific changes in behavior. These developmental effects are general.

A closely related area of general, or "psychological" as opposed to specific behavior, involves the ontogeny of emotional behavior (Candland 1971). The conclusions from work on the ontogeny of emotional behavior are quite similar to what was considered earlier: experiences during early life may have significant effects on various emotions later. These effects may be all too familiar in human experience, and they have been amply demonstrated in a variety of animals under objective study. Again, although there may be differences in emotions attributable to other factors such as inheritance, there are clear effects in many cases resulting from early experience. As with the general problem-solving ability, most persons would not associate these effects with what would traditionally be called learning. (Traditional categories of learning are discussed in Chapter 19.)

Development and the Acquisition of Specific Changes in Behavior

Up to this point we have been discussing mostly *general* affects of development in behavior, such as general sensory abilities, emotional and social reactions, and speed of attack in predators. The distinction between *general* and *specific* is not always clear and obviously forms a continuum with much overlap. But some changes in behavior clearly involve quite specific outcomes, such as the type of animal with which an individual will mate or the specific sounds or songs that a bird or other animal makes. The specific aspects often show many similarities with the general aspects discussed earlier and frequently arise during early development of the individual. A few examples will be described next. An important component of the picture, the critical period, first needs to be emphasized.

Critical or Sensitive Periods

Before proceeding, one must stop to recall, from Chapter 16 and the earlier discussions, that certain events in an animal's life often have to occur at a particular time during life for effects in the animal's behavior to be realized. Examples include the sexual organization of male brains in mammals resulting from a surge of testosterone during a brief period around the time of birth, development of visual and sensorimotor pathways in young but not older kittens, and the development of normal emotional, social, and predatory aspects of behavior. The period of time during which the events impose their significant impacts on the behavior of the animal is called the *critical* or *sensitive period* of the animal's life. Bateson (1979) provides a recent discussion of critical periods. The phenomenon of critical period shows up repeatedly also in respect to specific behavioral changes and should be kept clearly in mind in the following sections.

Imprinting

Perhaps the most familiar form of behavioral development in many animals is *imprinting*, the process of a young animal forming an association or identification with another animal, object, or class of items. Imprinting received much attention beginning with Lorenz (1935) and has been the subject of much research and review (e.g., Immelmann 1972, Hess 1973a, Hess and Petrovich 1977). It has been recognized or at least implied for centuries even in nursery rhymes such as "Mary Had a Little Lamb" (also see comments in Chapter 2).

In the best-known examples of imprinting, involving the following of a parental object, or *filial imprinting*, a young animal forms an impression of its parent soon after hatching or birth and follows that object for a period of time thereafter. Under natural circumstances the object is the actual parent, and the following is correct. Under artificial conditions, however, if the true parent is not present, the young animal may imprint on almost any other moving object present nearby, including humans and artificial objects (e.g., Figures 2-4 and 18-10). Two important early findings of imprinting studies were that (1) many species, particularly precocial ones, show a marked critical period during which exposure to the

Figure 18-10 Apparatus used in the study of laboratory imprinting consists primarily of a circular runway around which a decoy duck can be moved. In this drawing, a duckling follows the decoy. The controls of the apparatus are in the foreground. From Hess, E.H. 1959. Science 130:134.

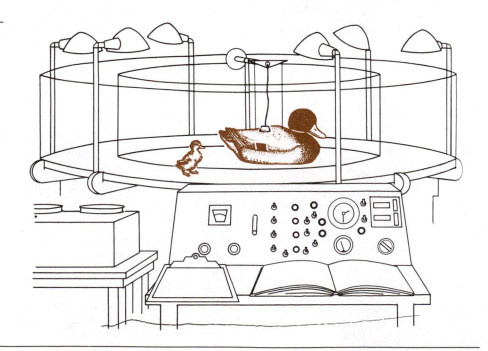

Figure 18-11 Critical age for laboratory imprinting in mallards expressed as the percent of animals making perfect scores. Modified from Hess, E.H. 1973. Imprinting. Van Nostrand Reinhold Co., Inc., New York.

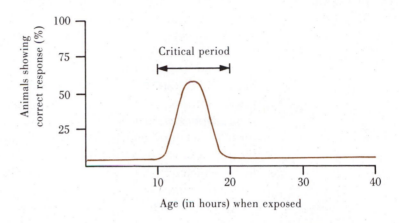

Figure 18-12 Strength of laboratory imprinting as a function of distance traveled by ducklings, with exposure time held constant. Modified from Hess, E.H. 1973. Imprinting. Van Nostrand Reinhold Co., Inc., New York.

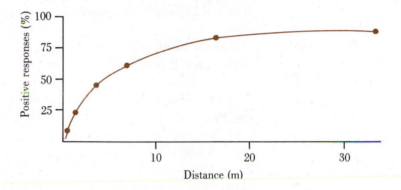

imprinting object must occur (Figure 18-11) and (2) the strength of the imprinting depends on the effort expended by the young animal in following the object (Figure 18-12); that is, the imprinting process is not merely passive; the young animal itself must actively take part.

In addition to filial imprinting, many other categories of imprinting have now been identified. These include imprinting that leads to eventual choice of sexual partner, or *sexual imprinting,* imprinting to types of food that will be eaten, and, it has been proposed, imprinting to specific habitat or nesting substrate that will be occupied later in life. Imprinting to the odors of a home stream is described for salmon in Chapter 7.

As might be expected, there is *much* variability in the roles that imprinting plays in the life of different species. Some, such as whooping cranes (Chapter 21), show filial imprinting but not clear sexual imprinting if later given sexual choices, but they do show a degree of sexual imprinting if the proper species is not available later. Some, such as golden eagles (Durden 1972), show sexual imprinting but will slowly reverse to their own species if given the opportunity. Some show sexual imprinting that is not reversible, and some show no sexual imprinting whatsoever—they appear to choose the correct class of mating partners and the correct sex of the correct species, if such is present, or they simply will not engage in reproductive behavior. Variation in the types and extent of imprinting occurs markedly even among relatively closely related groups such as waterfowl.

Although studied most extensively among birds, imprinting is also well known among other vertebrates, particularly mammals. In the most specific sense, it is responsible for individual recognition, particularly between mothers and off-spring and is based on individual sights, sounds, or odors encountered soon after birth or hatching during the critical period. The mother-offspring bonding is reciprocal in some cases and involves a critical period during which the mother learns which offspring to accept or reject thereafter (also see under parental care in Chapter 10).

An understanding of imprinting has proven quite useful in applied cases of animal behavior (Chapter 21). It forms the basis for "cooperative" (as opposed to forced) artificial insemination of many species of wildlife (e.g., Grier 1973), and it has been used judiciously where sexual imprinting problems are not involved or expected in cross-fostering offspring to other species for reintroduction into the wild.

Persons wanting further information on imprinting should refer to Hess (1973) or Hess and Petrovich (1977).

Development of Bird Vocalizations

The means by which birds acquire the ability to produce specific patterns of sound has attracted much attention and has elucidated the ontogeny of behavior perhaps better than any other single topic. Bird vocalizations are varied and complex, providing extensive detail and means of identification, and the birds themselves are relatively easy to raise and manipulate in captivity. Through the use of magnetic tape recordings and contemporary electronic technology, it is relatively easy to accurately reproduce sounds, create artificial analogs, and experimentally manipulate virtually all aspects of the subject. As a result, there currently exists an immense and continually growing body of literature on the subject. Pioneering work was done by Marler (1952), Konishi (1965), Thorpe (1958), Emlen (1972), and many others. Because of the limitations of space, only a cursory summary of the findings will be presented here. Persons wishing further details should consult any of the many available reviews on the subject (e.g., Marler and Mundinger 1971, plus references in Marler and Peters 1981).

As should now be of no surprise, there is much variability in the vocalizations of different species, of different individuals of a given species, within an individual, and even within how much variation may occur. Some birds have only a few simple vocalizations, and others may have a rich and changing repertoire. In general, the simpler and nonreproductive sounds are referred to as **calls**. The term **song** usually is reserved for the more complex reproductive vocalizations and, in the minds of most ornithologists, just for species in one order, Passeriformes, commonly known as *perching birds* or *songbirds*.

Among other, nonpasserine orders of birds, it appears that the vocalizations, including those used in reproduction, are largely impervious to environmental effects. These birds, when mature and in reproductive condition, generally give species-typical vocalizations regardless of whether raised in isolation, in the presence of other species, or under other types of experimental manipulation such as deafening. Examples of such birds include chickens, doves (Chapter 4), and cranes (Chapter 5).

The Passeriformes, however, present quite a different picture. Some, such as the brown-headed cowbird, have an apparently fixed and species-typical song, similar to the nonpasserine birds. At the other extreme, however, are birds that have a highly variable natural repertoire and, in some cases, can accurately mimic the sounds of other organisms, including humans. Mynah birds and a few of the starlings, for example, can mimic to the extent that a listener can identify the gender and often even the individual that the bird is mimicking.

The group of birds that has been most instructive regarding the development of behavior, however, includes those which are intermediate between the two extremes—birds that modify their songs in the process of development, but only within limits. There are many birds in this group. Various species of sparrows have received the most study. The results of different studies, using different techniques, different species, and by different persons in different locations, have provided different pieces to what seems to be emerging as one puzzle or one

Table 18-2 Generalized results of experiments on song learning among many passerine birds*

Condition and Treatment(s)			Fledgling → Critical Period → Sub-song → Rehearsed Song → Primary Song	Results
Normal wild birds	Male		Normal	Normal song
	Female		Normal	No song
Female hearing own species			Injected with testosterone	Sings like male
Male isolated from own species	Kept in quiet surroundings		← Isolated after critical period	Normal song
			Isolated before critical period → |	Develops own song
	Presented with recordings of song	Normal song from own species	Before → |	Develops own song
			| ← During → |	Normal song
			← After critical period	Develops own song
		Backward song from own species	(During critical period)	Sings backward
		Song from different species	(During critical period)	Develops own song—not normal and not other species
Surgically deafened	Before critical period		Deafen → |	No song
	After critical; before full primary		Deafen	Song deteriorates
	After full primary		Deafen	Normal, lasting song

*See text for explanation.

Figure 18-13 Generalized pattern of events in song development in several species of songbirds. For further explanation and references, see text.

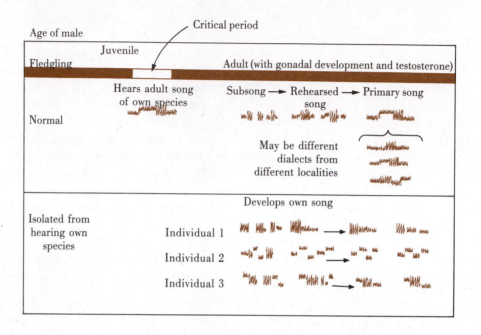

picture. Because of this, I have taken the liberty of synthesizing a general or hypothetical case rather than citing different cases for different parts of the story. Persons wanting specific actual examples from which this was built should refer to the reviews cited earlier. Details vary depending on which species one is considering, but the general pattern appears to fit most cases.

A generalized summary of various research is shown in Table 18-2. From this has been derived the interpretation illustrated in Figure 18-13. The complete, normal sequence in the acquisition and ontogeny of song in this group of birds is as follows. While still juveniles or young adults, the specifics of which vary from species to species, the young male hears the song of its own species from other birds around and close to it. During a critical period some form of memory (sometimes referred to as a *template*) of the sound pattern is established within the bird's nervous system. The bird does not sing or attempt to repeat the sound itself during this time; it simply has captured a memory of the sound. Later, often after a period of several months, when it begins to mature sexually and under the influence of testosterone on the nervous system, the bird begins to vocalize. Initially the song is not complete but consists only of bits and pieces, the **subsong**. These components eventually (over days or weeks) form together and begin to resemble the song that the bird heard as a juvenile. This song is then **rehearsed** and perfected until it becomes the full **primary song**.

Apparently during the rehearsal period the bird is listening to its own song, comparing it with the pattern in its memory, and correcting it until the two match. After this point the song becomes fixed and will not subsequently change. Before becoming fixed, the song pattern is vulnerable. If prevented from hearing their own rehearsals, via deafening, birds slowly lose the correct pattern that was developing, and the song deteriorates. If they never heard a song in the first place, they develop and fix their own concoction. If they hear *another* song, artificial, abnormal, or from other species, whether or not they memorize and subsequently copy it (which or whether both of the processes are involved is not known) depends on *how close the pattern is to the natural song for that species*. In other words, there are limits, or a range, as to the patterns that will work. Generally they have to be from or closely resemble the normal song. There often is much variation in the natural song, however, and local dialects in song patterns may develop. Marler and Tamura (1964), for example, demonstrated dialects in three populations of white-crowned sparrows within a 100-mile distance near San Francisco.

The ranges of variation that different species permit vary considerably. Many species will show some variation but not accept the songs of other species or quite different artificial patterns. Some not only accept but naturally copy the songs of other species. And, at the other extreme, some have a very narrow range and quite stereotyped pattern.

The singing or responding to singing by these birds is clearly under hormonal influence. Normally males sing, and females do not, but the females respond to the male songs in various ways. Females given testosterone, however, will go through the sequence of song development as if they were males and eventually sing. Recent work (e.g., Baker et al. 1981, Searcy and Marler 1981) has shown that *female responsiveness* to male song follows an analogous pattern to that of male singing; that is, the specific songs to which females respond when they become sexually mature depends on the patterns they heard when they were young. Other research, involving males canaries (Nottebohm 1981), has shown that changes in the acquisition of song (including new song in subsequent years—canaries are more flexible than many of their relatives) is accompanied by measurable changes in specific regions of the brain. These changes are believed to involve increases and decreases in numbers of synapses.

There still is more, however, to the development of song in birds. Even in the case of those with the most stereotyped songs, experience and environmental influences may affect the singing. The classic case of stereotyped song is that of the cowbird *(Molothrus ater)*. The species is a brood parasite. Females lay their eggs in the nests of well over 100 other species, which then raise the young. Young cowbirds thus are reared in the presence of other species and do not have early experiences hearing their own species. Thus it has always been assumed that their songs are under strict genetic control. Research with isolated, hand-reared individuals seemed to confirm these beliefs. Isolated males give typical songs, and isolated females respond to them.

But it has been discovered (West et al. 1981) that there nonetheless is variation among cowbird songs and that their potency or effect on other male and female cowbirds varies considerably. The more potent songs stimulate more female responses and also cause a much greater response from other males. Normally the potency varies with the individual male's position in the dominance hierarchy of the cowbird group it joins after leaving its foster parents. Through fighting and other social interactions, the male learns its place in the local social setting and, thereby, is affected in how potent a song to sing; that is, variation in song is influenced by social experiences. In artificial manipulations the most dominant males were moved from their own groups of cowbirds to others. When they gave their most potent songs in the new groups, they were killed by males already established in those groups. Meanwhile the groups from which they came shifted in their absence, and new males became most dominant and began singing the most potent songs.

Birds are not the only organisms in which unexpected cases of learning have been discovered. Environmental influences, critical periods, and related phenomena sometimes show up in species of animals where not anticipated and sometimes do not appear where they might be predicted, at least in those taxonomic groups of animals with the most complex nervous systems. This brings us to a general consideration of learning and the next chapter.

SUMMARY

This is the first of two chapters devoted to the development and change of behavior during the lives of individual animals. Some behaviors occur only early in life, to be replaced subsequently by other behaviors. There is much variability among species in the sequencing of behavior, presence and absence of different behaviors, rates at which they develop, and the influence of environmental variability on behavioral variability. Interpretation and generalization of observations about the development of behavior are complicated by major developmental differences among animals. Vertebrate embryos, for example, generally show movement before hatching, whereas those of invertebrates do not; many invertebrates possess larval stages that are difficult to compare with vertebrate development. Some young hatch or develop into independent ways of living, whereas others may be helpless and require extended parental care.

A major, ongoing controversy in behavioral development has been concerned with intrinsic versus external factors in determining the outcome of development. Several schools of thought and concepts are discussed in the chapter. They include preformation and epigenesis, probabilism and predetermination, and autogenous, reflexogenous, and systemogenous behavior. Much evidence exists among a wide variety of organisms that many behaviors, such as cricket song and bird hatching, can develop without sensory input. The best evidence at present suggests autogenic development, that is, intrinsic within the neural and muscular systems, to be the initial mechanism of behavioral development at least in some behaviors in some species.

Maturation, which involves normal growth and development of underlying anatomical parts and pathways, plays an important role in the initial appearance and changes of many behaviors. Some of the changes that might otherwise be interpreted solely as learning often can be accounted for largely as maturation of underlying systems. Examples include color preferences in amphibians, accuracy of pecking by some birds, flying ability, and attack behavior in cephalopods. Maturation and learning, however, are not mutually exclusive. The effects of environmental variability and experience on development of maturing behaviors vary greatly. Predatory behavior among avian and mammalian predators, for example, appears to depend on the species; in some experience seems important, and in others it is not.

The development of normal behavior has been studied in part through the opposite techniques of environmental deprivation and enrichment. In many cases these appear to impose little effect on the behavior in question. In others, however, particularly in the ontogeny of many mammals, the effects are striking. Deprivation may lead to degeneration or very abnormal outcomes, whereas enrichment can lead to opposite effects. Areas that are affected notably include visual pathways in the nervous system, sensorimotor responses, and a number of emotional and social responses.

Effects of sensory and environmental input, whether normal or resulting from either deprivation or enrichment, frequently occur only during critical periods in the individual's life; that is, particular inputs or deprivations will produce an effect if presented at one time but not at another. Critical periods have been shown most clearly during imprinting phenomena and during the process of song learning among several species of birds. Critical periods also occur, however, in many other, less obvious instances of behavioral development in some species.

As is seen frequently in biology, particularly in the subject of behavior, one of the major conclusions to emerge from the study of behavioral ontogeny is simply that there are many differences among different animals. Plasticity of behavior and the effects of experience and environmental variability differ greatly from species to species, sometimes even among closely related species. Furthermore, what is modifiable in one species may not be in another species, but the latter may have *different* components of its behavior that are modifiable; that is, different species vary in *which* aspects of behavior are plastic. Cowbird song, for example, is not dependent on hearing other cowbirds sing, in contrast to song learning in some other species, but learning and the effects of social experience nonetheless have been found to be very important in how and when cowbirds sing.

Recommended Additional Reading

Aronson, L.R., E. Tobach, D.S. Lehrman, and J.S. Rosenblatt, editors. 1970. Development and evolution of behavior. W.H. Freeman and Co., Publishers, San Francisco.

Bateson, P. 1979. How do sensitive periods arise and what are they for? Anim. Behav. 27:470-486.

Burghardt, G.M., and M. Bekoff, editors. 1978. The development of behavior: comparative and evolutionary aspects. Garland Publishing Co., New York.

Cowan, W.M., editor. 1981. Studies in developmental neurobiology. Oxford University Press, Inc., New York.

Denenberg, V.H. 1963. Early experience and emotional development. Sci. Am. 208:138-146.

Denenberg, V.H. 1972. Readings in the development of behavior. Sinauer Associates, Inc., Sunderland, Mass.

Hailman, J.P. 1969. How an instinct is learned. Sci. Am. 221:98-106.

Harlow, H.F. 1959. Love in infant monkeys. Sci. Am. 200:68-74.

Hinde, R.A. 1970. Animal behaviour, 2nd ed. McGraw-Hill Book Co., New York.

Marler, P., and P. Mundinger. 1971. Vocal learning in birds. In H. Moltz, editor. 1971. The ontogeny of vertebrate behavior. Academic Press, Inc., New York.

Moltz, H., editor. 1971. The ontogeny of vertebrate behavior. Academic Press, Inc., New York.

Nash, J. 1970. Developmental psychology. Prentice-Hall, Inc., Englewood Cliffs, N.J.

Spear, N.E., and B.A. Campbell. 1979. Ontogeny of learning and memory. Halsted Press, New York.

West, M.J., A.P. King, and D.H. Eastzer. 1981. The cowbird: reflections on development from an unlikely source. Am. Sci. 69:56-66.

CHAPTER 19

Mouse in a T maze.

LEARNING AND MEMORY

Chapter 18 ended with discussion of critical periods, imprinting, and the development of bird song. These topics are important to the contemporary view of learning. In fact, the ontogeny of behavior merges inseparably with the larger subject of learning so that this chapter amounts to a continuation of the previous one. In addition to imprinting, of which bird song development in part is an example, there are many other categories of learning that have been identified. This chapter describes briefly these and related phenomena, surveys the presence of learning among taxonomic groups, and then attempts to synthesize the information from both chapters. Memory, a necessary component of *all* learning, is also discussed. The biological mechanism of memory has proven an elusive, fascinating topic in neurobiology.

A universally accepted definition of learning does not exist. For purposes here the following will be used: **learning** *is a specific change or modification of behavior involving the nervous system as a result of experience with an external event or series of events in an individual's life.* By focusing on external events, one can exclude fatigue and maturational and purely developmental processes. By specifying the nervous system, other changes that might affect behavior, such as a broken leg, are excluded. Within the framework of what normally is considered learning, however, there are at least 10 possibly unique types of learning. Whether or not they are really unique or are all merely variations on a single, more general underlying process still is a matter of ongoing debate. At least at the cellular, neuronal level there are many ways for neural processing to be modified.

LEARNING CATEGORIES AND TERMINOLOGY

Learning is a general phenomenon by which internal changes of some kind lead to changes either in the specific nature of motor patterns or in the filtering processes by which the animal responds to particular stimuli. Under various circumstances and in different species the outward expression of learning varies considerably, and this has led to the classification of different categories of learning. Just as there is no generally agreed on definition of learning, there is no one system of classification, and the categories vary widely among different authorities. Thus the classifications that follow are not fixed, and there may be much overlap from one category to another. They simply include types of learning commonly recognized and discussed in the fields of biology and psychology.

Habituation

Habituation involves the gradual fading of an unlearned response to a stimulus that proves to be safe or irrelevant. The initial response usually is one associated with danger, such as fleeing, crouching and becoming immobile, or some form of startle response. The stimulus generally is something new or unfamiliar to the animal. After repeated occurrence without significant meaning to the animal, the stimulus loses its novelty and is ignored. This probably is the most primitive and universal form of learning. It can be seen in virtually all species and is an important means of dealing with an otherwise overwhelming amount of environmental sensory input, most of which is irrelevant to the animal. Habituation is the type of learning involved in the early hawk-goose study of fear responses, as discussed in Chapter 2. In that case birds habituated to the appearance of other birds that were not hawks or to hawk shapes that did not attack.

Sensitization

Sensitization is, in its outward appearance, approximately the opposite of habituation. Internally it is thought to involve different underlying neural processes. With sensitization the animal shows an *increased* response to repeated stimuli. This generally involves highly relevant stimuli, such as encounters with predators, unlike the irrelevant stimuli involved in habituation.

Classical Conditioning

Classical conditioning focuses on changes in the stimuli that elicit behavior and is based on natural or normal stimulus-response systems of species. The basic stimulus and response are referred to as the **unconditioned stimulus (US)**, such as the sight of food, and the **unconditioned response (UR)**, such as salivation. When the US is properly paired with a novel or different stimulus, such as the sound of a bell, the new stimulus, which is referred to as the **conditioned stimulus (CS)**, may elicit the behavior, now referred to as the **conditioned response (CR)**.

The bell-food-salivation example is easily recognized as being associated with the Russian physiologist and psychologist Pavlov. This form of conditioning often is called *Pavlovian conditioning* because of his well-known work on the subject. In the original Russian writing on classical conditioning the terminology should have been translated as **conditional** rather than conditioned. The term *conditional* better reflects the meaning that the conditional stimulus depends on (is conditional on) pairing with the initial (unconditional) stimulus to elicit the response. The first translations, however, used the term *conditioned*, and it has stuck for the most part since, although the term conditional is slowly working its way into the vocabulary.

Although usually studied in the laboratory with artificial CSs such as ringing bells or flashing lights, classical conditioning clearly is a natural phenomenon with real biological value. Under natural conditions animals may learn to associate the presence of food or other items with other normally occurring stimuli. One of the major characteristics of classical conditioning seen in connection with this point is known as **sign tracking** or **autoshaping**; that is, learned behavior will become shaped by or track the associated stimuli or sign of something. Predators, for example, will learn to cue into and follow the auditory, olfactory, or visual signals from potential prey.

The conceptualization of classical conditioning has changed considerably over the years. Initially and for many years the view was that the CS merely substituted for the US. This has been called the *stimulus substitution model* or **S-R (stimulus-response) learning.** The responses (UR and CR), however, are not always identical and may even be opposite with some drug stimuli. Other models such as the *preparatory-response model*, in which the CS is viewed as preparing the animal for the US, and the *compensatory-response model*, in which the CS may lead to a behavioral compensation for the US (particularly when the UR and CR are not the same) also have pitfalls, and a good comprehensive model of classical conditioning does not exist at present. From different outcomes in different situations involving different species about all that can be said is that the conditioning depends on the specific nature of the particular US, CS, and the animal's natural behavioral predispositions. Most of the recent emphasis has been placed on the important pairing between the CS and the US so the phenomenon now is considered as **S-S (stimulus-stimulus) learning.**

In classical conditioning the specific nature of the pairing between CS and US is very important. The strength of the conditioning depends on several factors including (1) the consistency with which the US follows the CS, (2) the latency, interval, or amount of time between the US and the CS, and (3) the particular relevance or strength of the US to the animal at the time it is presented. The strength of the sight of food, for example, is more important to a hungry animal than to one that is satiated.

Several other important terms are used in connection with conditioning. The normal sequence, called **forward conditioning,** is for the CS to be presented before the US. If the reverse occurs (which generally does not lead to any learning), it is called **backward conditioning.** If the CS and US are presented in random sequence with variable intervals so that there is no clear association between the two, the situation is referred to as **pseudoconditioning.** If changes in behavior are observed with backward or pseudoconditioning, they generally are considered as artifacts and not signs of bona fide learning. Thus, by definition, classical conditioning is forward conditioning.

When a clear association between the CS and the US ceases to exist so that the US occurs without being preceded by the CS, the animal's response to the CS may decline and disappear completely. If so, the conditioning is said to have been extinguished, or **extinction** has occurred. This situation, in which the CS no longer occurs with the US, is different from the phenomenon of **forgetting,** which is a decrease in the conditioning resulting from an intervening lapse of *time*. Analogous phenomena, which may or may not share an underlying neural substrate, can be seen in losses of habituation. A loss of habituation as a result of a lapse of time without the stimulus is referred to as **recovery,** which is similar to forgetting. A change of habituation as a result of a stimulus becoming relevant or because of an interaction with a new stimulus is called **disinhibition.**

Operant Conditioning

Operant or instrumental conditioning is a second major form of conditioned learning. In this type of conditioning *responses*, called **operants,** are instrumental in producing certain *consequences* in the environment of the animal. The consequence is called a **reinforcing stimulus, reinforcer,** or **SR.** This type of conditioning is involved in maze learning and what many (e.g., Thorpe 1963) call *trial-and-error learning*.

This type of conditioning has been subject to a vast amount of research by numerous workers using many kinds of methodologies. To connoisseurs of conditioning there are subtle differences between instrumental and operant, and there has arisen an array of confusing terminology related to positive and negative reinforcers and reinforcements, rewards, punishments, escapes, and different types of avoidance. The terms are used in different combinations so that, for example, not all negative reinforcement is "punishment," and there are both negative and positive punishments. The uninitiated student is cautioned to be alert. Persons interested in the details should see a contemporary text on psychology or learning, for example, Domjan and Burkhard (1982) and perhaps consult a local friendly psychology instructor.

Instrumental conditioning differs, at least conceptually, from classical conditioning in that instead of focusing on the connection (contingency or contiguity) between two different stimuli (CS and US), the emphasis is on the contingency between the instrumental response, or behavior of the animal, and the reinforcer, or consequence of that behavior. Instrumental conditioning is at least somewhat similar to classical in that the contingency or pairing of interest can **shape** the animal's behavior. And, as in classical conditioning, there are many important factors such as frequency of pairing and consistency of the association. The instrumental response and the reinforcer can be paired consistently and continuously or only part of the time. If only part of the time, it can be on a **fixed** (e.g., every tenth time) or **variable** schedule. Furthermore, the time interval between behavior and reinforcer may be constant or variable. Interestingly, conditioning under variable schedules is more resistant to extinction than under fixed or continuous and predictable schedules.

From comparative work with different species, it has been discovered that not all are equally capable of learning or becoming conditioned to the same things or at the same rates. Some species seem to learn better under some circumstances, and others are better in other situations. To biologists this fits well with concepts of evolutionary adaptiveness, which varies among species, and the idea of **bio-**

Figure 19-1 Gila monster—a lizard with poisonous skin (and also a poisonous bite). Photo courtesy T. Brakefield.

logical constraints. In psychological parlance these evolved predispositions have gone under names such as **preparedness, relevance,** or **belongingness.**

The adaptiveness of conditioned forms of learning, in animals capable of it, should be obvious. In classical conditioning it would be to an animal's advantage to associate or anticipate noxious, dangerous stimuli that should be avoided. Conversely, positive stimuli may prove advantageous as predictors of useful events or resources. An animal that learns to associate a certain sound, for example, with potential food may become much more efficient and successful in obtaining food. Similarly any animal that can assume some degree of control over its surroundings or at least in how it relates to its surroundings, in a framework of instrumental conditioning, would have a distinct selective advantage over animals that could not.

Taste Aversion

Taste aversion learning appears superficially as another form of classical conditioning involving avoidance. It differs markedly, however, in the opinion of some (not all) learning theorists and perhaps is best viewed as a separate category of learning. With taste aversion the animal associates sickness or severe discomfort in the digestive system with an item that has been ingested. The animal subsequently avoids further ingestion of similar items, even to the point of starvation if that is the only food available. Such learning is highly advantageous because it prevents poisoning. Taste aversion learning appears to be widespread and general at least among vertebrates.

Although widespread, the phenomenon of taste aversion is also quite variable among species and, perhaps, even individuals. Its expression depends in part on the normal food-sensing modalities of the species (e.g., whether by olfaction or vision). It also may come into play at different stages of predation, depending on the species. Some predators, for example, will continue to kill but not eat objects with noxious associations, whereas others will avoid the objects altogether. Different animals also may vary in the strength of the aversion, depending on a host of factors such as species, previous experience with the food (before poisoning), and perhaps other factors such as age, sex, and level of hunger.

Many species of organisms have acquired, through the course of evolution, the ability to produce taste aversion in other animals that feed on them. Well-known examples include monarch butterflies and many other invertebrates plus some amphibians, particularly toads, and the gila monster lizard (Figure 19-1), which have poison glands in their skin. The natural sequence for the acquisition of taste aversion learning is for an animal to ingest an object or substance that produces sickness, usually including nausea and vomiting. Through the coupling of vision or olfaction or perhaps other sensory modalities with the taste of the sickness-producing agent, the animal becomes averted (i.e., learns to avoid future encounters).

The phenomenon can be elicited not only by natural noxious agents but also by a variety of artificial means, including hypodermic injection of sickness-producing chemicals and, apparently, radiation-induced sickness. The organism receiving such treatment becomes averted to food that may have been ingested before the induced sickness. Research on taste aversion in a variety of species has revealed several important differences between this form of learning and other types of avoidance conditioning. Latencies between the stimulus (noxious food) and response (sickness with vomiting, etc.) can be quite long, up to 2 hours or more. The aversion usually is formed with a single encounter, and it is highly resistant to extinction. When taste-averting stimuli and other forms of avoidance-producing stimuli are presented together, the distinction between the two forms of learning becomes particularly noticeable.

There are some similarities between taste aversion learning and other avoidance learning. In both cases the learning is greater with bigger differences between the natures of the CS and US, greater novelty of CS, increased intensity of US, more fear initially of CS, and if repeated pairings occur. Learning is weaker in both cases if the animal has previously been exposed to CS without pairing with US. Finally, there is much variability among species in both cases. The similarities, however, are far outnumbered by the differences (Table 19-1). For further comparisons and references, see Kalat (1977).

Latent or Exploratory Learning

Latent or exploratory learning involves an animal using experience gained at one time in the modification of behavior at a much later time. Information and memories that may not be immediately useful are nonetheless retained and may become useful subsequently. This is particularly important for spatial information. In a laboratory setting, for example, rats that have been allowed to spend time in and explore a maze without food or other reward learn to negotiate the

Table 19-1 Differences between food aversion and other forms of associative learning

Characteristic	Food Aversion Learning	Other Associative Learning
Time and repetition	One-trial learning with CS-US delays up to 24 hours	Requires repetition and delays between CS-US of no more than a few seconds
Relevance	Poisons more readily associated with foods than with lights and sounds	Shock more readily associated with lights and sounds than with tastes
Specificity to training environment	Food aversion generalizes readily to a new environment	Shock avoidance generalizes less readily to a new environment
Maturation of learning abilities	Weanling rats equal to adults	Weanling rats worse than adults
Effect of active versus passive exposure to CS	Less learning if taste is force-fed than if actively ingested	Passive presentation of CS is effective; active participation not considered important
Learning under anesthesia	Yes	No
Synaptic pharmacology	Cholinergic and anticholinergic drugs have little or no effect on learning; aversions may depend on histaminergic synapses	Cholinergic and anticholinergic drugs have large effect; role of histamine unknown

From Kalat, J.W. 1977. In N.W. Milgram, L. Krames, and T.M. Alloway, editors. Food aversion learning. Plenum Press, New York.

maze for reward at a later time much faster than rats that are introduced to the maze for the first time. The advantages of such learning, the prior acquisition of knowledge in a sense, would be quite clear in the wild. It gives an animal more efficient access to resources, an additional advantage of residing in a known home range (see discussion in Chapter 7). This may serve as the basis for much of the exploratory behavior that is observed in animals, including humans and other primates.

Place or Spatial Learning

Place or spatial learning concerns an animal learning its surroundings and familiar place or familiar path (Chapter 7). It involves an ability to become oriented in space and, if necessary, to reorient to new locations. The value of this type of learning is clearly obvious, so much so that it may be taken for granted. It may form one of the most common, if not overlooked, forms of learning and may involve some of the most primitive and ancient parts of nervous systems. Many persons would consider this ability under instrumental, trial-and-error learning, such as in learning to negotiate mazes, but it is not clear whether different forms of learning are involved or not.

Cultural or Observational Learning

Cultural or observational learning has now been documented in many of the higher vertebrates, particularly birds and mammals. In this form of learning one animal learns to do something by watching or otherwise sensing what another animal is doing, including not only overt gross actions but, in many cases, even very subtle mannerisms. Through culture, information can pass very quickly from animal to animal and effectively bypass the long periods required for evolution and inherited acquisition of new behavioral traits. As an analog for the information-carrying agents in inherited traits (genes), Dawkins (1976) proposed the term *meme* for the information elements, at least in human ideas, for cultural transmission. Traits are most likely to be learned culturally from parents or peers, and several social factors may be involved.

Interest in the analogies between cultural and genetic transfer of information and the interaction between the two modes is picking up and may well serve a new direction for much research in the near future. For expanded treatment of the topic, Bonner (1980) gives a readable popular introduction to the subject; Cavalli-Sforza and Feldman (1981) present a more in-depth, technical treatment; Hutchinson (1981) considers the implications for humans.

Culture, at least in rudimentary aspects and as a form of learning, has been found to be quite widespread, at least among mammals and some birds. Examples of cultural transmission include the learning of potato and grain washing by Japanese monkeys, milk-bottle lid removal by European tits, and predator mobbing by passerine birds. In the predator-mobbing example, birds learn to mob objects that they see others mobbing (e.g., Curio et al. 1978a,b) (Figures 19-2 and 19-3). As in other forms of learning, there seems to be a predisposition for mobbing objects of a certain type (such as owls), but via learning and through artificial manipulation, the birds can be taught to mob other things such as nonpredatory birds and even plastic bleach bottles.

Figure 19-2 Experimental setup to investigate acquisition of mobbing by European blackbirds. The presentation box was moved into view by a pulley operated by the experimenter and then rotated to expose the stimuli to the subjects. An owl, which elicits a strong mobbing response, was shown to the mobbing bird ("teacher") while a novel object, either a honeyeater bird, a nonpredatory and unfamiliar species to blackbirds, or a plastic bleach bottle was presented to the observing learner blackbird. Modified from Vieth, W., et al. 1980. Anim. Behav. 28:1217-1229.

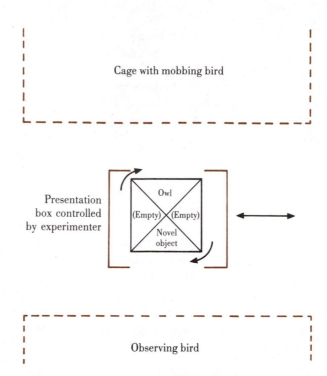

Figure 19-3 Results of European blackbird mobbing studies. Sequence of trials: (1) empty box presented to control for effect of box movement and appearance, (2) presentation of novel object, without teacher seeing the owl, (3) presentation of owl and novel object to teacher and learner, respectively, and simultaneously, and (4) test of learner's response to novel object alone. Mobbing responses were standardized to that initially shown to an empty presentation box. Decreased responses to repeated presentations of empty box or novel subject alone represent habituation. Note that the birds learned to mob even the empty plastic bottle, but the response was not as strong as toward a novel species of harmless bird. Modified from Curio, E., et al. 1978. Z. Tierpsychol. 48:184-202.

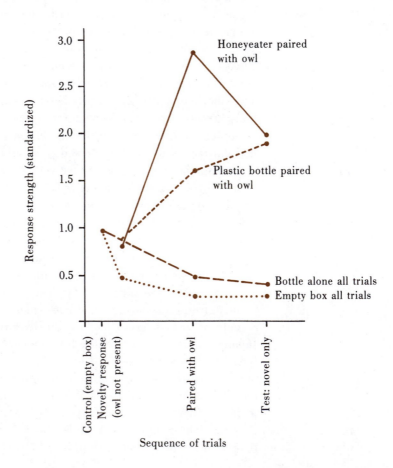

Imprinting

Imprinting is recognized as one of the major forms of learning. Imprinting involves learning the properties of a stimulus object toward which the animal subsequently directs its otherwise normal feeding, social, or other behavior, depending on the species and situation. The learning period generally occurs early in the individual's life, often during only a very narrow critical period. This form of learning is discussed extensively in Chapter 18.

Insight Learning

Insight learning is perhaps the most advanced form of learning, at least from the *Homo*-chauvinistic viewpoint. In this form of learning the individual derives new behavior or solutions to problems by insight or thinking about them. Previous *specific* experience is not involved, although general and closely related experiences may help. Rather, the modified behavior and actions are new to the situation. This probably is the least widespread and least understood form of learning. It may involve processes of mental modeling and simulation; that is, the individual thinks about something and tries it out mentally before putting it into overt form. Scientific hypotheses are formed by insight learning.

Insight learning and mental modeling, however, are by no means confined to scientists or formal education. A familiar but less recognized form of such thinking occurs when one thinks about future personal interactions and relationships with the world and other humans. Examples are "If I bend this lever, I wonder if the machine will work?" and "If I wear such and such clothes to the party next weekend, what will so and so think of me?" Many, if not most, persons are good at rehearsing likely conversations and events in advance. It may be that complex social relationships played a major role in the evolution of insight capabilities.

Mental simulations and hypothesizing form a type of trial-and-error situation that can be repeated, modified, and worked out in the comfort of one's own mind, usually without the risks of the real world. Thus it would seem much more efficient and less hazardous than real trial and error; the advantages seem immense. If the advantages really are great, why are not more animals capable of greater insight learning than they appear to be? If as in computer simulation the ability requires considerable processing and memory capacity, it may be that humans are the only living species that possess sufficient mental ability. We do not, however, have a monopoly on insight learning—just the major share of it. Novel solutions to problems such as reaching places with difficult access can be seen in other primates and other mammals as well as a few birds, including members of the crow family (Corvidae).

Learning-Set Learning

Learning-set learning is a form of learning that may be closely related to insight learning, perhaps differing mostly in degree or just in name. It has been studied fairly extensively in other animals, particularly primates and a variety of other mammals. In learning sets animals generalize from previous learning and solve closely related learning tasks more rapidly. In essence the animals learn to learn or learn how to improve their learning. The learning set involves a set of principles or a **strategy**. For example, if food is found under an object, the animal will continue to search under similar objects, but if not, it switches to a different object. This particular strategy is called a *win-stay, lose-shift strategy*. Many persons consider this type of learning as an extension of instrumental conditioning.

In a related, perhaps converse phenomenon, is a type of learning known as **learned helplessness.** If an animal is faced with an intractable problem for which there is no solution, it may simply give up and, in essence, learn that it cannot learn. As an example with dogs (Seligman 1975), animals were given shock treatment from which some could not escape and some could. Those which could not eventually stopped trying and then would not escape even when provided with the opportunity.

Overlaps and Combinations of Types of Learning

Overlaps and combinations of types of learning occur in many situations, and it is not always clear which category to place particular instances in or whether the categories are really distinct in the first place. Classical and instrumental forms of conditioning share many properties in common, and many persons believe they may just be different aspects of the same process. Other instances of overlap between categories have been implicated earlier.

In perhaps one of the clearest examples involving cultural transmission (from mother to offspring) and instrumental learning (offspring learning to capture prey) with what seems to even involve some elements of insight learning (mother varying her behavior in response to the outcome of the offspring's learning), mother cats aid (teach?) the learning of prey capture by their kittens (Caro 1979, 1980). During the period when kittens are age 4 to about 12 weeks, mother cats go through a sequence where they first bring dead prey to the kittens and eat it in their presence; next they bring dead prey but do not eat it. Then they bring crippled but live prey for the kittens to play with and eat. Finally they bring unharmed prey and release it in front of the kittens. If the prey escapes, the mother recaptures it, brings it back, and releases it again. Then the kittens follow the female into the field and begin capturing their own prey, after which the mother no longer takes part in interactions over prey with the kittens. It should be noted, however, that female cats often engage in similar behavior (as many cat owners will attest!) when no kittens are present, bringing the prey instead to their human associates or as a vacuum activity. Thus it is not certain how much, if any, insight learning is really involved.

PHYLOGENETIC SURVEY OF LEARNING

An attempt to survey the various phyla of animals for the role and extent of learning in different species is difficult. First, there has been a disproportionate amount of study among species. Some, such as the rat and human, have been studied extensively, whereas several other phyla, most orders of animals, and the vast majority of species have gone virtually untouched.

The second major problem is that there are so many differing opinions about what constitutes learning and how to properly measure it. Bitterman, for example, in a scathing attack on studies of invertebrate learning, remarked (1975:144),

"We are seeing . . . a diminishing respect for that tradition among comparative psychologists, who find it fashionable now to denigrate their past (about which they know less and less) and to stand in awe before the loose anthropomorphic and teleological models of the ethologists." Beyond the rhetoric there still are some valid criticisms. The main points that Bitterman stresses in the same article from which the preceding quote was taken are that (1) greater attention to and familiarity with the findings of the intensively studied species, such as the rat, are needed because many of the pitfalls have been recognized and (2) studies must become more efficient and objective to remove observer bias. Lahue and Corning (1975) suggest a third major problem—that even as split as learning has become, we still may not have fine enough resolution on various categories. They believe that when several criteria for a given form of learning are considered, it is found that different groups may be quite different. Using habituation as an example, Thompson and Spencer (1966) have classified 9 (or 10 if the eighth is split further) criteria for identifying the presence of habituation. In their phyletic survey Lahue and Corning show that different criteria are satisfied in different species and that available information also shows marked differences in underlying mechanisms. In view of this divergence in habituation, they remark, "Comparative statements based only upon behavior [are] of little value—it is similar to announcing that two species are the same because both can fly. As Jensen (1967) points out, man could be characterized . . . as a featherless biped that talks and would accordingly be in the same grouping as a plucked trained parrot" (pp. 167-168).

Many persons, however, disagree with this sentiment. There are at least two good reasons for a comparative view: (1) finding similarities in structure and function where they do exist and (2) where underlying structures are not similar, understanding the biological principles that lead to convergence in appearance. With flight, for example, or the external morphology of aquatic species, the structures clearly are not always homologous, but they are analogous, and the biophysical principles of supporting oneself in air or water are of much interest. Most of this comparative survey is synthesized from articles in Corning et al. (1973-75) with expanded treatment where deserving or of interest.

Protozoans

Protozoans have the absolute minimum of equipment with which to learn. Nonetheless, a number of protozoan species have demonstrated a clear ability to habituate that is not simply a matter of temporary fatigue. How such habituation might occur is not yet known. Other findings purporting to show other types of learning, however, remain controversial (Corning and von Burg 1973).

Porifera

Porifera display virtually no behavior and have received little attention from a learning context.

Coelenterates

Coelenterates are simple animals but only deceptively simple from a learning standpoint. The general view of coelenterate behavior since the turn of the century has been that coelenterates, as well as many other lower organisms, behaved in a very straightforward simple reflex manner. Parker (1919), for example, stated that the sea anemone is "a delicately adjusted mechanism whose activities [are] made up of a combination of simple responses to immediate stimulation." As late as 1952 Pantin stated that coelenterates, more than any other animal, brought us close to "a complete analysis of the structural units on which behavior is based." But some, such as Jennings (1905, 1906), knew better and argued to the contrary.

In addition to responses to external stimuli, coelenterates show endogenous, often cyclic, patterns of behavior and much variability from individual to individual and from species to species and group to group. Compounding this is a host of miscellaneous problems. The movements and responses of coelenterates may be very slow—sometimes too slow to be observed by eye—and may tax both the patience of the observer and, seemingly, the capabilities of the animal itself.

Such complications have resulted in many problems and alternate explanations for coelenterate "learning" studies. Early experiments (e.g., Nagel 1894, Parker 1896) suggested that certain anemones would learn to discriminate between filter paper soaked in juice of crab meat and crab meat itself. The anemones would move the pieces of filter paper to the mouth, swallow them, then eject them. Eventually the tentacles would respond more slowly or simply not carry paper to the mouth; then the tentacles would reject the paper altogether. The anemones, however, would continue to accept crab meat. Only the tentacles that were tested would show the effect; other tentacles would still accept filter paper (until they, too, were repeatedly stimulated). The slowing of response by specific tentacles also is shown normally toward pieces of food. Later studies (reviewed in Rushforth 1973) attempting to confirm that the anemones (or their tentacles) would learn to distinguish between flavor-soaked paper and real crab meat, however, produced highly variable and conflicting results. Some workers believed that the findings could be explained by sensory adaptation, habituation, or accumulation of substances on the tentacles or in the surrounding water.

An apparently overlooked aspect of the anemone behavior, incidentally, whether it involves learning or not, is how does a coelenterate recognize an undigestable item and expel it in the first place? With the relative sparseness of the nervous "system," relative lack of specialized neurons, and lack of a well-defined central system, how do anemones discriminate whether something is food or not? Or, if perhaps they monitor the progress of digestion, how do they keep track of the length of time something has been present in the gastrovascular cavity without being either digested or further digested?

Other attempts to show conditioning or more advanced forms of learning have been plagued similarly by complicating factors, conflicting results, and differing interpretations. Perhaps the clearest demonstration of more complex learning in coelenterates involves swimming reactions in a sea anemone of the genus *Stomphia*. But it is not very complex and still not very clear. When exposed to contact or chemicals from particular surface regions of two species of starfish, *Hippasteria spinosa* or *Dermasterias imbicata*, or the nudibranch, the anemone *Stomphia* elongates its body, detaches from the substrate, then "swims" away by whirling or bending its body column (Robson 1961) (Figure 19-4). The response develops over a period of seconds; then, after landing in a new location—if the animal moves at all—the contractions slow down and quit after a few minutes. The

Figure 19-4 Sequence of swimming and resettling by the anemone (Stomphia coccinea) after contact with the starfish Dermasterias. Modified from Rushforth, N.B. 1973. In Corning, W.C., et al., editors. Invertebrate learning, Vol. 1. Plenum Press, New York, and Robson, E.A. 1961. J. Exp. Biol. 38:343-363.

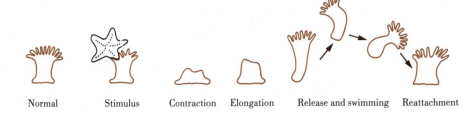

Normal   Stimulus   Contraction   Elongation   Release and swimming   Reattachment

anemone becomes inactive and eventually reattaches to the substrate. The nudibranch is a predator of these anemones, and the reaction seems to be one of escape. Aside from occasionally bumping into the anemones, however, the starfish are not predators of anemones and pose no threat to them; the reaction of the anemone to the starfish is a bit puzzling. Furthermore, the chemicals from the bodies of the nudibranch and starfish are quite different and stimulate the anemones at different regions. Starfish stimulate only the anemone's tentacles, whereas the nudibranch stimulates only the body column of the anemone. Another starfish, *Henricia*, does not stimulate *Stomphia* to swim.

It is not clear why the anemones respond as they do to stimuli from the starfish *Hippasteria* or *Dermasterias*. Nonetheless, this response has been used as a possible conditioning response. Ross (1965) used contact with the starfish *Dermasterias* as the unconditioned stimulus. He tried three conditioned stimuli, that is, stimuli presented prior to the US: (1) a pipe cleaner dipped in food extract (from a clam) and touched to the tentacles, (2) contact with the nonstimulating starfish *Henricia*, and (3) gentle pressure applied to the base of the anemone. There was no evidence of classical conditioning or of any interaction between US and CS for the pipe cleaner and *Henricia*. But pressure to the base led to a reduction in the subsequent swimming response, an effect that Ross termed *conditioned inhibition*. However, pressure at the base may have had some other mechanical or neural effect.

Although studies of more complex learning have not been convincing, there is good evidence (see Rushforth for references and details) for habituation. Animals habituated to mechanical stimulation will still react to other (e.g., photic) stimuli. Habituation to particular stimuli generally last, however, only for a few hours or days at most. As a group coelenterates may represent organisms with the earliest types of neurons (perhaps protoneurons) and nervous systems. Studies of modifiability of these systems still are confusing.

Platyhelminthes

Platyhelminthes, or flatworms, have shown hints of more advanced learning ability. Of the three classes in this phylum, only one, the free-living Turbellaria (which includes the planarians), has received much attention. The other two, Trematoda (flukes) and Cestoda (tapeworms), are internal parasites, difficult to work with, and largely unknown behaviorally.

The results of learning studies with planarians have been variable and conflicting and have engendered much controversy. Many of the problems have stemmed from the use of different species and genera and, even with the same species, quite different techniques by different workers. Many of the early studies and some of the recent ones lacked proper experimental controls. Reviews of learning in planarians are numerous; see, for example, McConnell and Jacobson (1973), and Corning and Kelly (1973) and references contained therein for further details.

Even with the simplest form of learning, habituation, there have been differing results. Several persons have reported evidence for habituation to mechanical or light stimuli (e.g., Walter 1908, Westerman 1963), whereas Bennett and Calvin (1964) were not able to show habituation in their studies. But a number of positive, well-controlled studies clearly show that at least some species under some conditions will habituate. Habituation, particularly to light, has also been used as a first step to experiments with more complex forms of learning.

One of the earliest studies of operant conditioning in planarians was conducted in the Netherlands by van Oye (1920). He showed that planarians could gradually be trained and perform better than untrained controls in tests where they had to travel up the side of their container, across the underside of the water surface, and down a wire to get to food suspended in the middle of the water (Figure 19-5).

The greatest amount of controversy and confusion over planarian learning arose over later experiments with conditioning. At first the arguments were the usual in-house disagreements and bickering that seem to accompany studies in any area of research. The real fireworks started, however, when McConnell et al. (1959) first reported that the specific learning could be passed on to regenerated planarians. If planarians are cut in half (or even smaller pieces), each part regenerates the missing portion. McConnell et al. reported that if trained planarians

Figure 19-5 Conditioning of planarians to reach food. A, Naive animals do not reach food item suspended below the water's surface. B, Animals reach food suspended immediately at the surface by traveling along the undersurface. With gradual lowering of the food the animals learn to travel down the wire suspending the food. C, Animals readily travel to suspended food by using the wire. Modified from Corning, W.C., and S. Kelly. 1973. In Corning, W.C., et al., editors. Invertebrate learning, Vol. 1. Plenum Press, New York.

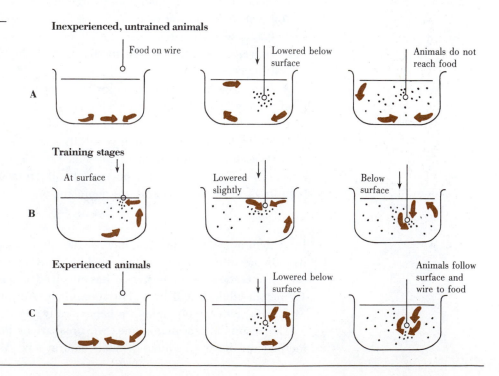

were cut up, the regenerated individuals would learn faster than naive regenerates. This initiated a flurry of worm-running experiments that produced more and more surprising and unbelievable results and a journal, *Worm Runner's Digest*, that were not taken seriously by many biologists.

The most surprising experiments came with subsequent cannibalism studies (for a review, see Fjerdingspad 1971). Trained planarians were chopped up and fed to untrained planarians, and the previously untrained animals were then able to learn much faster. All of this suggested a transfer of learning that had some molecular basis, perhaps even specific memory molecules. This would be most interesting if true, but several persons found it too much to believe, and the research fell under very close scrutiny and criticism. In particular, some of the

Figure 19-6 Double discrimination learning of planarians. The US was shock, and the CSs were either light or vibration. A, Responses of animals given light paired with shock or vibration without shock, followed by extinction (neither paired with shock), then the reverse—vibration paired with shock or light neutral, followed by extinction trials. B, Responses of different subjects given pairings in reverse of those in A; vibration was paired with shock first and then light (with a period for extinction after each). Evidence of conditioned learning is provided by (1) acquisition of response only for appropriate stimulus, (2) more rapid acquisition of response in stage 3 than in stage 1, (3) extinction after shock was stopped, and (4) more rapid extinction in stage 4 than in stage 2. Modified from Block, R.A., and J.V. McConnell. 1967. Nature 215:1465-1466.

○   Light
●   Vibration
〰   Paired with shock
—   No shock given

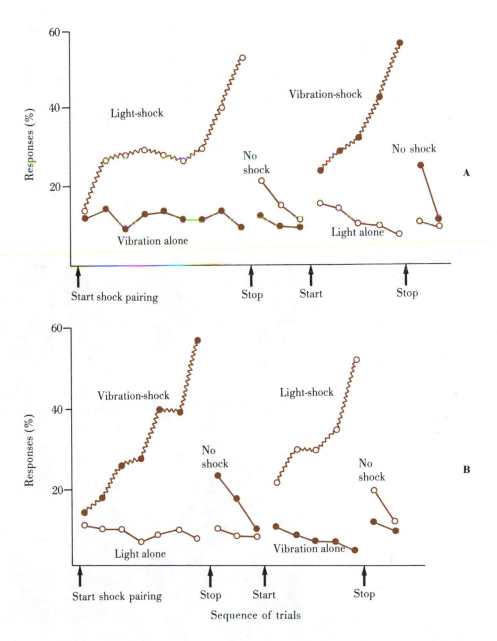

early work had not properly controlled for pseudoconditioning, fatigue of the animals, and other factors.

As a result of the problems, many of the learning experiments were redone more carefully. There were some negative results, but the majority seemed to clearly support an interpretation of bona fide learning. An example of one of the most carefully controlled and clearest studies is shown in Figure 19-6 (Block and McConnell 1967).

Further studies of possible molecular transfer of learning brought even more surprises, seemingly showing very specific learning effects. McConnell and Shelby (1970), for example, trained planarians to go to one arm or another of a T maze; then they chopped them up and fed them to naive animals in four groups. Group I was fed animals trained to go to one arm; group II was fed animals trained to go to the other arm; group III was fed animals from *both* of the trained groups; and group IV was a control group, fed on untrained planarians. Groups I and II, fed on animals trained for one arm or the other, learned to go to the proper arm much faster than the controls, whereas Group III took longer than the controls. They showed more head-waving at the junction of the maze, as if undecided or confused. Other workers, however, were not always (or in some cases ever) able to confirm these results.

The possibility of molecular transfer, focusing primarily on RNA but also with some attention to protein, was further studied such as by blocking RNA synthesis and extending the studies to other animals such as crabs and rats. (Planarians have a simple digestive system that does not destroy macromolecules as in more

Figure 19-7 Conditioned learning in octopus. A, A crab and geometric figure (white square in this case) are presented to an octopus. If it takes the crab to eat but then is shocked by electrodes on the white square (B), it subsequently will not take crabs when presented with the white square (C). If the crab and square are moved toward the octopus, it not only will not attack but retreats (D). Crabs presented alone will be attacked. By using similar and related techniques, octopuses can be trained to take crabs and avoid fish, vice versa, and discriminate among many different objects both by vision and by touch (see Figure 19-8). From Boycott, B.B. 1965. Sci. Am. 212(3):42-50.

A

B

advanced animals. Thus, higher forms have had to rely on injections of material rather than simple, cannibalistic feeding.) The results of these supposed memory transfer experiments still have not been interpreted or understood to the satisfaction of everyone, and the controversy continues today.

Although the issue of memory transfer still is not settled, disagreement over whether or not planarians can learn has largely subsided. Most (not all) persons accept that conditioning exists in at least some species of Turbellaria under some conditions. Planarians served importantly to draw attention to the importance of considering simpler, invertebrate animals in the study of learning. Corning and Kelly (1973) conclude, "The flatworm may not achieve the position in psychology that the fruit fly gained in genetics and the bacterium achieved in molecular biology, but it certainly has stimulated interest in researching animals other than the rat."

Molluscs

Molluscs learn as might be expected from their phylogenetic position. Cephalopods, however, seem to be outstanding exceptions. Octopuses seem more advanced in some respects even than arthropods, and although there are no direct evolutionary relationships, they act much like many vertebrates. These two groups, cephalopods and vertebrates, provide an outstanding example of evolutionary convergence in respect to vision and the outward expression (but not the underlying detailed structure) of learning. The similarities may be related to speed of movement (other molluscs are famous for being slow) and ability (via the

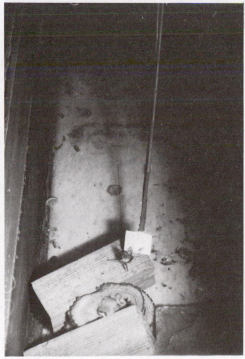

C   D

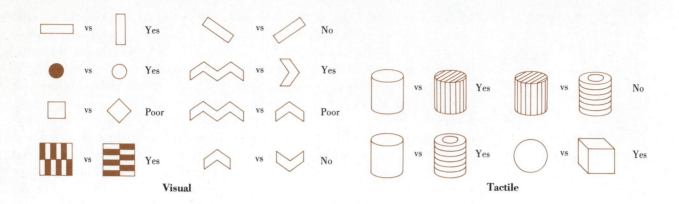

Figure 19-8 Examples of some of the visual and tactile stimuli used to test octopus learning and discrimination abilities. Octopuses could be taught to discriminate between smooth and grooved objects and, with some effort, between round and cube-shaped objects. They would not distinguish grooved objects, however, on the basis of vertical or horizontal grooves. Limits also were found for their ability to distinguish on a visual basis. Octopuses also were tested also on a variety of natural objects, such as smooth and rough-shelled molluscs, which they could easily distinguish. Modified from Sanders, G.D. 1975. In Corning, W.C., et al., editors. Invertebrate learning, Vol. 3. Plenum Press, New York.

advanced eye) to sense events at a distance rather than just at the surface of the body. Cephalopods also have appendages, the tentacles, capable of manipulating things in the environment, like many (not all) vertebrates.

In spite of some severe criticism of the research (Bitterman 1975), higher learning in octopuses, particularly one species *(Octopus vulgaris)*, seems to have been well demonstrated and has been widely accepted. Earlier work involved discrimination learning whereby octopuses, which normally rush out to grab food items, could be negatively conditioned. Octopuses would refuse food that was paired with visual cues followed by shock (Figure 19-7). Subsequent work involved finer details of visual and tactile discrimination (Figure 19-8).

Returning to the other molluscs (reviewed by McConnell and Jacobson 1973 and Willows 1973), there have been few learning studies in other molluscs except the gastropods. There have been many claims of learning in snails, including one rather farfetched one about a pet snail that learned to recognize the voice of its owner (Dall 1881). But close scrutiny of the various evidence reveals convincing support only for habituation and sensitization types of learning. Withdrawal and escape behavior, involving habituation and sensitization, have received much attention in gastropods, particularly *Aplysia*, and much is known even of the specific neural substrate and pathways, as was seen in Chapter 15. There have been numerous reports of higher forms of learning, but all suffer from inadequate methods. Negative conclusions based on the available evidence do not rule out other types of learning; such learning simply has not been demonstrated. Willows (1973) points out that most researchers have been looking in the wrong places by using visual stimuli, whereas gastropods rely primarily on chemical senses and by investigating simple withdrawal or escape patterns rather than more complex homing, reproductive, and feeding behaviors known to exist in gastropods.

Annelids

Annelids (reviewed by Dyal 1973 and McConnell and Jacobson 1973) show evidence of limited, if not primitive, capacity for some higher forms of learning. Aside from cephalopods they show the most learning among the groups thus far surveyed. There are three classes: Polychaeta or marine sedentary and mobile annelids, Oligochaeta including earthworms and freshwater species, and Hirudinea or leeches. As with most previous invertebrate studies, the majority of pub-

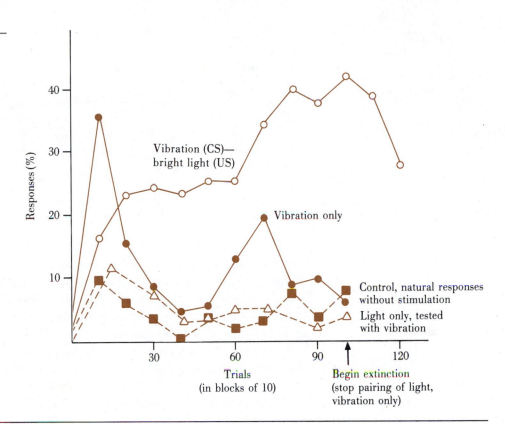

Figure 19-9 Learning versus related phenomena in earthworms (Lumbriculus variegatus). Bright light was used as the US, a mild vibration as the CS. The classical conditioning group received the two stimuli paired in the standard fashion. The sensitization group received vibration only. The pseudoconditioning group received 10 trials of light only, followed by 5 of vibration only, alternating throughout the series of trials. The control group received no bright light. Responses were measured by testing the worms with weak vibrations. Modified from Ratner, S.C., and K.R. Miller. 1959. J. Comp. Physiol. Psychol. 52:102-105.

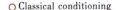

○ Classical conditioning
● Sensitization group
△ Pseudoconditioning group
■ Control group

lished claims for annelid learning involve inadequate methods and lack of control for such things as pseudoconditioning, sensitization, and other problems. In spite of this, there have been some properly designed studies, and habituation has been clearly demonstrated in all three classes. In a classic study (Yerkes 1912) an earthworm learned to turn correctly in a T maze in response to shock for punishment and darkness and moisture for reward. Subsequent studies have been beset with methodological problems, but maze learning has more or less been confirmed. There have been a number of properly controlled experiments that clearly demonstrate classical conditioning in earthworms, using, for example, weak vibrations as the CS and light as the US (Ratner and Miller 1959) (Figure 19-9). A number of differences have been shown between classes, although so few species have been properly studied that it may be premature to generalize very far.

Arthropods

Arthropods comprise most of the species (perhaps 80% to 90%) and possibly most of the individuals of animal life found on Earth. There are so many different kinds that it is convenient to break the phylum into major subphyla rather than just consider classes. There are three subphyla: Trilobita, the individuals of which are extinct and found only in fossil form; Chelicerata, which includes horseshoe crabs, spiders, sea spiders, and ticks, mites, scorpions, and others; and Mandibulata, which includes centipedes, millipedes, crustaceans, insects, and a few other minor groups. Among the two major mandibulate classes, the crustaceans (Crustacea) and insects (Insecta), there are (depending on authority) around 30

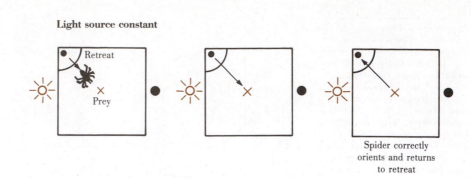

Figure 19-10 Disorientation of spiders (*Agelena* sp.) in response to learned orientation to light. If the direction of a light source is reversed after a spider leaves its retreat and arrives at its food, it travels in the opposite direction in an attempt to return to its retreat. Modified from Bartels, M. 1929. Z. Vergl. Physiol. 10:527-593.

orders of crustaceans and nearly as many orders of insects. One estimate (Elzinga 1978) places the number of species at around 715,000, which means that insects alone comprise 70% to 75% of all known, cataloged species of animals. The entire Chordata phylum, which includes the vertebrates, on the other hand, contains only around 45,000 species.

In the face of such numbers and diversity, it is surprising and discouraging that relatively nothing is known about arthropod learning. Most of the arthropod studies have been confined to spiders, a few familiar crustaceans, and four or five orders of insects, with a disproportionate concentration on hymenoptera (ants, wasps, and bees). Among hymenoptera, the honeybee has been the "white rat" of arthropod studies.

In spite of the paucity of arthropod studies, however, there have been enough that one can at least have some glimpses of learning capability in this phylum. As might be expected, habituation has been clearly demonstrated in both living subphyla, the chelicerates and mandibulates. One of the earliest and more interesting demonstrations of habituation (Peckham and Peckham 1894) involved orb-weaving spiders. If a vibrating tuning fork is held near the web of these spiders, they will drop from the web on a thread of web material. With subsequent presentations of the stimulus, the spiders drop shorter and shorter distances until they cease to drop at all. Habituation of all responses is not universal, however. Palka and Babu (1967), for example, showed that the escape withdrawal response in some scorpions habituates after repeated stimulation, but a striking response does not. Studies of arthropod learning are so few and involve such variable methods that good, quantitative comparisons among species and groups are not yet possible.

One form of learning that has been clearly identified among many, but not all, arthropods that have been investigated is *spatial* or *place* memory. For many species it involves a "home" base of some type—a cavity or den or a resting and hiding place on a web, hive, etc. In addition, in many cases it may involve the position of a food item. Bees are well known for their ability to return to nectar and pollen sources (Chapters 7 and 9). A classic study by Tinbergen (Chapters 2 and 7) involved the ability of a wasp to return to sites of prey burying. Numerous other examples involving insects, such as ants, cockroaches, and others, can be included. Among Crustacea, many lobsters, crayfish, and crabs have been shown able to return to a consistent, familiar hole or hiding place. Several studies with

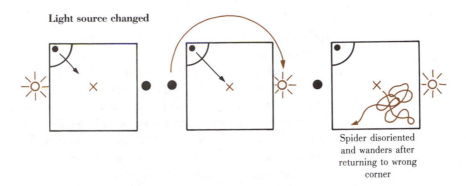

Light source changed

Spider disoriented
and wanders after
returning to wrong
corner

spiders (e.g., Bartels 1929, Peters 1932) showed that spiders learn and remember the locations of prey on their webs; even if it is artificially removed without disturbing the web, the spiders will return to the exact place where the prey was located and search repeatedly for the missing item. Subsequent searches may become less specific, with the spider ranging slowly farther and farther from the site from which the prey had been taken.

The orientation cues that are involved in place learning frequently are tactile or visual. In one genus of spider, *Agelena*, the animals become disoriented if the source of light is changed while they are away from their retreat (e.g., Bartels 1929) (Figure 19-10). Further studies (Gorner 1958) demonstrated much variability in the disorientation (e.g., percentage of spiders becoming disoriented under different circumstances), and, perhaps more importantly, it makes a difference *when* the lights are changed. If the lights are switched before or soon after the spider leaves its retreat, there is little or no disorientation. The disorientation becomes worse and worse, however, the longer the spider has been out when the lights are changed. Disorientation is greatest if the lights are changed after the spider has arrived at its prey item.

Higher levels of conditioning have been demonstrated in spiders. It was shown as early as 1884 (Dahl) that spiders would avoid tackling bees, certain beetles, or otherwise edible insects coated with turpentine. Bays (1962) and Walcott (1969) trained different species of spiders to associate dead flies that had been dipped either in sugar or quinine solutions with sounds of different pitch. After the association had been learned, unflavored glass beads were substituted for the flies, and the spiders either discarded or bit the beads depending on the pitch of the sound that was presented.

Conditioning also has been shown in several other studies of arthropods. The use of conditioning in bees, such as for studies of color discrimination, is classic and well known. The specifics of learning in bees may be less familiar but are quite revealing. As revealed by Frisch (1967), bees learn the color of flowers on which they land only during the 2 seconds before they land. In a different situation involving learning the hive location, honeybees relearn the position of their hive each morning but only with their first trip out. Bees will return to hives that were moved at night before the first trip, as is known to beekeepers around the world. But if the hive is moved after the first trip, even with the bees inside, the bees subsequently will become lost.

In an apparently good example of instrumental conditioning, cockroaches of various species can be trained to leave their preferred dark retreats and stay in the light after receiving electric shocks in the dark (reviewed in Alloway 1973). Razran (1971) discounted the response as being aversive inhibitory conditioning where the normal negative phototaxis response was simply being inhibited by a stronger response to shock. Alloway (1973), on further consideration and detailed observation, however, described the cockroach behavior as much more complex than simple avoidance of shock, and he attributed it to genuine instrumental conditioning. In another interesting twist of learning in insects, grain beetles *(Tenebrio molitor)* have retained learning through metamorphosis (e.g., Alloway 1972); this is intriguing because most of the larval neurons, at least peripheral ones, are replaced during metamorphosis.

Maze learning has been studied extensively in ants by several persons. Studies by Schneirla and associates alone spanned a period of 40 years (e.g., Schneirla 1929, Weiss and Schneirla 1967, with numerous publications in between). The general method has been to use various mazes, runways, and gates through which ants had to travel to get from their nest on one side to food on the other. Visual cues have been controlled by artificial lighting, and chemical cues are manipulated by interchanging or replacing liners in the passageways.

One can barely do justice to this much research in a few short sentences, and there are many considerations and qualifications (such as a trait in ants called *centrifugal swing*) (Schneirla 1929) that cannot be covered here. In brief, however, ants were found to use several cues—visual, chemical, and kinesthetic (movement and sequences of position) in various combinations. But they could get by without some cues, depending on which stage of learning was involved. Chemical cues, for example, and somewhat surprisingly, could be dispensed with as well as visual (as normally happens when ants are underground). Evidence of kinesthetic cues was demonstrated by lengthening segments of well-learned mazes; ants would attempt to turn at the point where previous turns had been located.

Two important generalizations resulted from the ant studies. First, learning progresses through relatively distinct stages. The first one or two stages involve essentially trial and error, with the elimination of entrance into blind passages. The final stage involves integration of the behavior into a more coherent, smooth whole by which the ant efficiently negotiates the passages. Second and perhaps most important for a comparative view of learning, maze learning in the ants is *situation specific*; that is, each part of a maze is learned separately, and new mazes must be learned from start. Even learning a maze on the way to food does not improve learning an identical maze on the return trip. There appears to be little ability to generalize or transfer learning in one situation to a new situation, even when the situations are otherwise identical. Situation specificity in the learning of arthropods has also been shown in several other groups and species; recall, for example, the details of color learning by bees approaching and landing at feeding sites and the possibly similar phenomenon of spiders becoming visually disoriented depending on when the lights are switched.

Another aspect of learning shown by invertebrates is the *compartmentalization* of learning ability. As reviewed by Krasne (1973), for example, crayfish and lobsters are able to avoid adversaries by swimming away by flapping their abdomens. Furthermore, they can sense and avoid obstacles while doing so. However, they seem unable or "unwilling" to learn to use these same patterns of movement to swim to food that they otherwise cannot reach. The characterization of insects as mechanical and robotlike is not completely accurate because they do have capacity for some learning. But relative to the flexibility seen in cephalopods and vertebrates, the robot characterization for insects may not be all that misleading. There appears to be no convincing evidence of learning set, cultural, or insight types of learning in arthropods.

The diversity among organisms and the particular types of learning found in different species of arthropods have raised a very important concept concerning the *biological relevance of particular learning capacity*. This or a similar notion will surface again when the biological constraints of learning among vertebrates are considered. The relevance of different kinds of learning to different natural histories has received increasing emphasis during recent years. Krasne (1973), for example, tabulates differences in morphology, senses, and "ways of living" for 22 groups of crustaceans alone. He says:

> [We] would anticipate that an animal such as the lobster, living in a fixed burrow from which it emerges at night to forage, might well learn something about the topography of its home range and the locus of its burrow. We might therefore anticipate that it would be capable of learning its way through mazes in the laboratory. On the other hand, we would not be surprised at the absence of such an ability in a small planktonic creature such as a copepod, which is forever wafted here and there by water currents.

Echinoderms

Echinoderms have shown little evidence of learning, in spite of studies going back to Jennings (1907). However, they are quite difficult to study. Willows and Corning (1975) summarized the situation: "Definitive conclusions concerning echinoderm learning must be forestalled until better preparations and perhaps more patient investigators appear."

Vertebrates

Vertebrates provide most of the familiar learning studies. Even with vertebrates, however, there has been comparative neglect of the wide diversity of organisms available.

Among vertebrates that have been studied, there are no clear relationships between learning characteristics and phylogenetic position. It appears that all vertebrates, from fish and birds to many species of mammals, are capable of rapid avoidance learning in feeding situations with noxious stimuli (of many sorts). In classical conditioning with light and shock, fish and pigs require about the same, or slightly less, time to learn than rhesus monkeys. Rats surpass monkeys in learning simple mazes, and they are about equal to humans in complex mazes. Fish, chickens, and several mammals have been shown to learn discrimination tasks at roughly the same rate. Rats learn visual discrimination so much faster than chimpanzees that Hebb (1958:454) remarked that a "large brain like a large government may not be able to do simple things in a simple way." In other comparisons there are many differences between species and even between some strains or races of the same species so that in one situation one learns more quickly and in another situation the other learns better. There are several learning tasks that other vertebrates can do better than humans. But humans still are capable of a greater array of complex learning abilities than any other animal.

From a phylogenetic standpoint, incidentally, primates are low on the ladder of mammalian evolution, and mammals are from a lower line of reptiles than are the much more recent (hence, advanced?) lines that led to dinosaurs, birds, and even the modern-day reptiles (see Figure 5-4). Primates may have relatively advanced brains in some respects but not necessarily in all details, and aside from the brain, primates are rather generalized and not very advanced mammals. If it is any consolation, recall that among the behaviorally advanced invertebrates, cephalopods come off the evolutionary tree from a relatively low branch.

If not phylogenetic position, which factors appear to relate to differences in learning among different vertebrates? Warren (1973) lists such things as life history (e.g., predators versus herbivores), sensory dominance (different species are tuned to different sensory modalities—rhesus monkeys to visual cues, which they learn more rapidly than auditory, whereas cats learn better in response to sounds), and "response availability." As an example of the latter, Warren describes differences between two different kinds of fish: in a light-shock avoidance training test, goldfish learn to escape much more rapidly than Siamese fighting fish (Otis and Cerf 1963). Goldfish are bottom-feeding scavengers, and their normal response is to flee in the presence of aversive stimuli. Siamese fighting fish, on the other hand, are predators and are typically aggressive; they normally stay where they are or fight back in response to threats. Thus in the training methods used, goldfish would be expected to flee much more quickly than Siamese fighting fish; the normal responses available in the species' repertoires are quite different.

Warren attempted to compare information on learning sets but concluded that there have been too many confounding and complicating variables (such as species differences in visual cues and in response to visual and nonvisual cues) to "yield uncontaminated measures." As a result (Warren 1973), "One can therefore never safely conclude that any quantitative difference in learning set performance, however large, is a valid indication of a difference in learning capacity, rather than a reflection of species differences in adaptation to the arbitrary demands of the test situation." It is much like the problem of trying to eliminate cultural biases from human I.Q. tests, only more so. Warren concludes (1973:500-501):

The classical approach to the comparative psychology of learning was based on oversimplified and obsolete ideas concerning natural selection and phylogeny. Animal species are subjected to selection for survival and success in reproduction, not for the degree to which they manifest progressively more humanlike capacities for learning and problem solving in the Skinner box or WGTA (Wisconsin General Testing Apparatus).

The basic point of this discussion is that each surviving species has become adapted to survive . . . by whatever means that responded most adequately to selection pressures. Specific learning capacities must have been subjected to the same kind and degree of selection as any other trait of the organism, and therefore, specific learning capacities are no more likely to vary with taxonomic status than any other functional or morphologic trait.

Types of learning that have not been adequately studied and compared include (in addition to the problems with learning sets) cultural and insight learning. Insight, or what amounts in most basic form to problem-solving learning, has received some attention relative to how animals deal with obstacles in the environment. Again, however, differences probably are more closely related to lifestyles and the usual environmental circumstances in which the species is found than in any kind of phylogenetic rank, at least among the vertebrates.

MECHANISMS OF LEARNING AND MEMORY

Learning has been clearly reduced to the cellular, molecular level only for habituation in a type of marine snail, *Aplysia*. The mechanism will be described, after which a closely related topic, the biological mechanism(s) of memory, will be considered.

Habituation and Sensitization in *Aplysia*

The gill withdrawal reflex of *Aplysia* described in Chapter 15 (see Figure 15-26) has been the subject of intense study by E.R. Kandel and his associates (Kandel 1979a,b). These studies have elucidated the mechanisms of basic forms of learn-

Figure 19-11 Habituation in a motor neuron of the marine snail *Aplysia*. Repeated stimulation of a sensory neuron leads to repeated firings of that neuron but a decline in the postsynaptic response of a motor neuron with which it synapses. Long-term (days and weeks) habituation also **can be seen in recordings from the motor neuron.** Modified from Kandel, E.R. 1979. Sci. Am. 241(3):66-76.

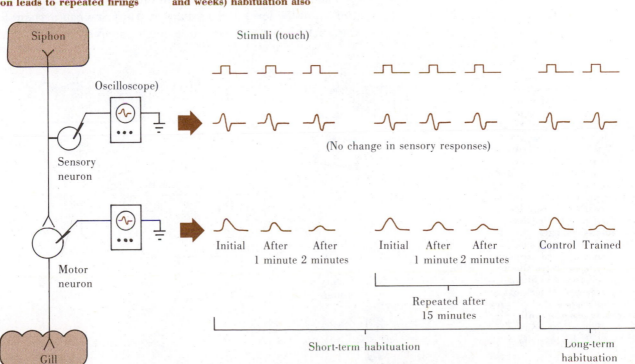

ing in this mollusc and suggest avenues of approach for other organisms. The gill withdrawal reflex shows the learning effects of both habituation and sensitization. Habituation results from repeated simple touching of the siphon; the gill withdrawal reaction decreases and eventually ceases. If a noxious stimulus is applied to the head at the same time the siphon is touched, however, sensitization occurs: the gill withdrawal increases. Habituation effects have been shown to occur over periods ranging from only a few minutes to several days and weeks (Figure 19-11). Sensitization can reverse any previous habituation.

Habituation, at least over the short term, has been shown to involve a reduction of synaptic transmitter release from the sensory neurons. The sensory neurons still fire (Figure 19-11), but the action potential results in less transmitter being released. This is thought to be caused by less calcium inflow during action potentials, with calcium being partly responsible for synaptic vesicles binding to the release sites in the presynaptic neurons. As a result of less transmitter being released into the synapse, there are insufficient EPSPs to fire the motor neurons (e.g., L7), and the response diminishes or habituates.

Sensitization, on the other hand, involves a different mechanism that is mediated by **facilitating interneurons.** These interneurons connect between the sensory neurons of the head and the motor neurons as well as the excitatory interneurons. A simplified diagram is shown in Figure 19-12. The facilitating interneuron makes *presynaptic* connections to the siphon sensory neuron axons. The facilitating interneuron, by release of its transmitter (serotonin), *increases* the amount of transmitter (acetylcholine) that is released in turn by the siphon sensory neuron. This increases the EPSPs in the motor neuron and additional excitatory interneuron. The increased excitation leads to increased muscular contraction and an increase in the gill withdrawal, hence sensitization.

Depending on the nature of the stimuli, one can produce inhibition, disinhibition, and sensitization at the level of the synapses. Thus one can account for changes in behavior, or learning, over time in a single, simple neural pathway

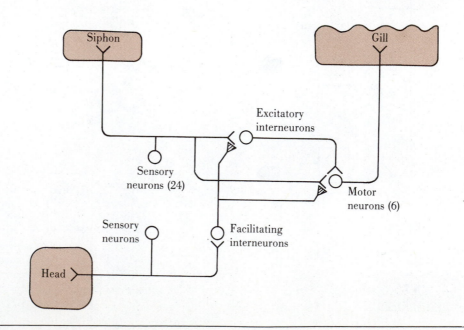

Figure 19-12 Sensitization in the gill-withdrawal reflex of *Aplysia*. **Sensitization is a form of learning and memory in which the response to a stimulus is enhanced. A stimulus to the head activates neurons that excite facilitating interneurons, which end on the synaptic terminals of the sensory neurons. These neurons are plastic, that is, capable of changing the effectiveness of their synapse. The transmitter of the facilitating interneurons, thought to be serotonin, modulates the release of sensory-neuron transmitter to the excitatory interneurons and motor neurons.** Modified from Kandel, E. R. 1979. Sci. Am. 241(3):66-76.

from input to output. Similar modifications of synaptic transmission have been shown in other invertebrates and even in a few isolated vertebrate preparations. The possibility that such mechanisms form the basis for higher orders of learning and memory is promising and intriguing. Kandel and co-workers, plus others in other laboratories, are actively pursuing these leads to higher and more complex parts of the *Aplysia* (and related molluscs) nervous system (e.g., Carew et al. 1983). Much, however, remains to be done. Next the process of memory in general will be considered.

Memory and the Elusive Engram

The ability to learn, regardless of the type of learning involved, depends on the ability of the nervous system to record changes in neural input, that is, *store information*. Any processing of information must follow certain patterns, steps, or sets of instructions, and *that* information has to be stored somewhere. Sagan (1977), based on data from Britten and Davidson (1969), graphed the total amount of information that can theoretically be stored in genes versus nervous systems (Figure 19-13). He calculated that the threshold where brains can store more information than the available amount of DNA is crossed at about the level of the reptiles. If this is the case, more or additional programs can be stored in the brain than in the genes. Furthermore, it seems biologically possible for any given program to be stored *either* place. It is conceivable, given the vagaries of evolution and genetic variability, for storage sites for particular programs to even shift back and forth from one location to the other. It would not be particularly surprising that comparable categories of behavior in closely related organisms might be mostly (in the analysis of variance sense, see Chapter 4) inherited in one group or mostly acquired in another. If a given group could be followed over evolutionarily long periods of time, the behavior conceivably could be largely genetically determined initially, switch to having a greater learned component, and then return to being more genetically fixed.

Figure 19-13 Amount of information (in bits) estimated to be contained in genes versus brains (and extrasomatic sources). Modified from Sagan, C. 1977. The dragons of Eden. Random House, Inc., New York, and derived from Britten, R.J., and E.H. Davidson. 1969. Science 165:349-357.

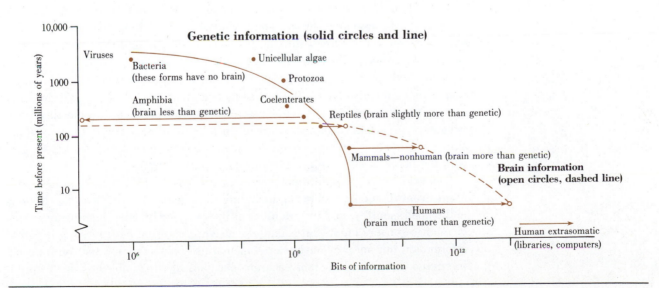

The possibility of the evolution of innate responses from learned responses was proposed in 1896 by J.M. Baldwin and is known as the *Baldwin effect*. More recently the phenomenon has been discussed by G.G. Simpson, who considers that learning may confine individuals to restricted circumstances, in which individuals with genetic predispositions or more innate limitations to stay within those circumstances might be selected over those more likely to stray. Note that this interpretation falls squarely within Darwinian principles of natural selection. It is the evolution of characteristics within populations by shifts in gene frequencies and not a Lamarckian interpretation of inheritance of individually acquired characteristics. Waddington has further elaborated the concept and calls the process *genetic assimilation* by canonization of initial nongenetic traits. Wecker (1964) provides a specific possible example and further discussion involving differences in habitat selection among *Peromyscus* mice.

There is no doubt, however, that much information is not stored in the genes but in the nervous system. Furthermore, the capacity of biological memory is very impressive. Examples of human memory were noted at the start of Chapter 18. Large feats of memory have been shown also for other mammals and some birds. Some food-storing birds, for example, may remember the locations of thousands of stored seeds over periods of several months (Sherry et al. 1981, Shettleworth 1983). How is all of this accomplished? What is the biological mechanism of memory? And is there a specific location in the brain where memory is located? These questions have provided some of biology's greatest continuing challenges. The physical representation that is assumed to exist for each specific memory in the brain is called an **engram,** but we do not know what or where it is (Lashley 1950).

Based on a wealth of relatively consistent information, the basic model for the operation of memory, as a phenomenon if not a process, is as follows. Initially there is a transient sensory trace, which may last less than ½ second or so, followed by **short-term memory (STM)** or *working memory*. Depending on the circumstances and species, the information may reside in the STM for a few minutes to hours and then either disappear or become fixed in **long-term memory (LTM),** or what might be referred to as *library memory*. The process of transferral from STM to LTM is not understood but is thought to involve some kind of *rehearsal* or repetition—perhaps some kind of reverberation whereby the information cycles several times through the same pathways and then becomes **consolidated** into LTM. Recall of the information depends on some type of *retrieval* process, perhaps by a matching process against the stored template of information. Proper recall requires the full sequence of events. When recall does not occur, it may be because of one or more problems: fading of the memory trace (whatever and wherever it is), failure of consolidation, or inability to retrieve the stored information. Improvement of human memory generally attempts to deal with the latter problem (e.g., Ericsson and Chase 1982).

The evidence for two different memories, STM and LTM, is quite widespread. Many events that create amnesia, including electroconvulsive shock, concussions to the head, senility, and interruptions or interference by other significant events or intense sensory input, for example, cause a loss of the *most recently acquired* information but not previously stored memories. This suggests that those events interfered with something that was transient, that is, wiped out the trace before it became stored more permanently. In general, memories appear to become more and more resistant to loss with an increase of time (although recall can also become more difficult with time if memories become misplaced, as it were, among a large number of other memories). The general characteristics of the STM versus LTM phenomenon are found over a wide diversity of organisms, including mammals, fish, and insects, and even in isolated parts of nervous systems such as segments of the spinal cord (e.g., Chamberlain et al. 1963). The commonly accepted model of STM-LTM is not without some problems, however. When tested for different forms of learning occurring simultaneously with stimuli that would reduce STM, it has been found that STM can be interfered with while LTM still takes place (e.g., Maki 1979). Hence, the relationship between STM and LTM may be more correlational than causal.

One of the most intriguing and at the same time frustrating features about memory is that it, or at least many types of it, seems to exist diffusely throughout the nervous system. In attempts to find a location for memory, different parts of the brain have been removed. The extent of loss caused by such removal seems to depend on the amount of neural tissue removed, not its location, as if the memory exists everywhere simultaneously. This phenomenon has been shown in animals as diverse as mammals and cockroaches (Chen et al. 1970). In the case of the cockroaches, for example, intact cockroaches, headless insects, and isolated segments of the insects were classically conditioned to avoid electric shock. Initial learning occurred in all cases but was not very good unless the head and brain were present. Once learning had occurred, however, all segments of the nervous system showed evidence of and retained the learning. These findings support the concepts of both STM and LTM, with STM apparently requiring participation by the brain, and diffuse memory, whereby the learning, once acquired, resides in several parts of the nervous system. Not having a specific focal point or precise location to investigate has made it difficult to study the exact mechanism of memory.

In spite of the difficulties in finding a precise location, much research and thinking have been conducted on just what the mechanism of memory in biological systems might be. The answers still are not in, and there has been much disagreement. But the schools of thought fall into two general categories: molecular and structural mechanisms. The molecular line of thinking received much impetus from the early, controversial planarian studies described earlier in this chapter plus findings that learning seemed to correlate with increased levels of RNA and protein synthesis. According to this view, memory is somehow encoded into "memory molecules," whereby information is stored in a manner perhaps analogous to that which occurs in DNA. Memory recall would then require some sort of mechanism for decoding this information and returning it to a neural basis, which seems to create logistical problems for the model.

The structural line of thinking proposes that there are structural changes in the neural pathways with alterations resulting from use or disuse, which would facilitate future replays over those pathways which become established. Although this view has not been proven and we are far from a complete understanding or being able to demonstrate a specific case, the bulk of recent evidence seems to lean in the direction of structural changes. In particular, it appears with learning and memory that there are significant changes in *dendritic spines* in the synapses

Figure 19-14 Learning and dendritic branching. A, Comparison of interneurons from the optic tectum of the jewel fish *(Hemichromis bimaculatus)* **reared socially in communities versus isolated.** *a,* **Axons. B, Schematic diagram of learning effects on dendritic branching. A** from Coss, R.G., and A. Globus. 1978. Science 200:788. Copyright 1978 by the American Association for the Advancement of Science. Illustration courtesy Richard G. Coss. **B** modified from Eccles, J.C. 1977. The understanding of the brain, 2nd ed. McGraw-Hill Book Co., New York.

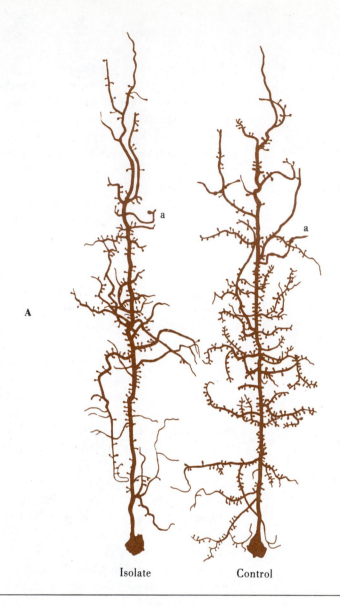

A

Isolate   Control

between different neurons (Figure 19-14). For examples with more information and discussion, see Coss and Globus (1978) and Greenough (1975). Also note that the focus is on synaptic sites, as discussed earlier for *Aplysia*.

The findings of increased RNA and protein synthesis are compatible with the model of structural changes because structures themselves are composed of molecules, and although they would not encode the memory directly, they still would involve protein synthesis.

Individual neurons by themselves, however, would not seem likely candidates for carrying entire memories or the "programs" for behavior. Many neurons must act in concert with each other. Thus it is likely that in the final analysis learning and memory involve similar principles to those found in general neural processing (Chapter 15), such as abstraction. Along these lines, a model for learning and memory in the octopus has been proposed (e.g., Boycott 1965) whereby abstractions lead to "memory cells" in specific regions of the brain and in which new information is compared with previous experience. Then, depending on the out-

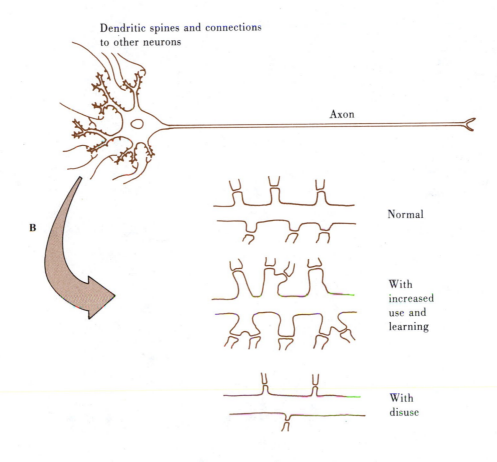

Dendritic spines and connections
to other neurons

Axon

B

Normal

With
increased
use and
learning

With
disuse

come, appropriate motor pathways are stimulated and inappropriate ones are inhibited (Figure 19-15). Similar models are now being pursued in other organisms such as bees (e.g., Menzel and Erber 1978).

Whatever the case, all that can be said for certain is that we remain a very long way from understanding exactly what is occurring in the processes of learning and memory. The topic is fascinating, however, and likely to stimulate intensive research for many years to come.

LEARNING PROCESSES VERSUS LEARNING PHENOMENA: A SYNTHESIS

So what sense can be made of this information distributed so unequally across the spectrum of animal species? First, it is important to review the questions and working assumptions that, implicitly or explicitly, have been deemed important and which have motivated the subject, to the point in some cases of outright

Figure 19-15 Hypothesized memory and learning system in the octopus, based on numerous brain-lesioning experiments and differential interruption of memory. A, Diagram of nervous system, including major parts of the brain. B, Components of the memory and learning system with indications (in parentheses) of parts of the brain where found. According to this scheme, information from the eyes is combined with input from other senses in memory cells. These memories are then compared with future visual input, and, depending on the match, either attack or retreat responses are stimulated while opposite behaviors are simultaneously inhibited (arrows and negative signs in center of illustration). Modified from Young, J.Z. 1965. Proc. R. Soc. Lond. (Biol.) 163:285-320, and Boycott, B.B. 1965. Sci. Am. 212(3):42-50.

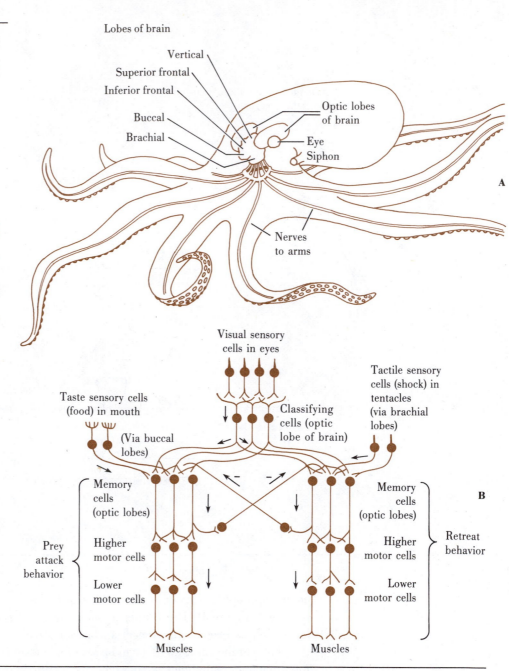

controversy. One general question has concerned whether particular behavior is innate (genetic, instinct, nature) or learned (nurture). This is based on the biologically naive and improper assumption that it must be either-or. All behavior is unavoidably the joint product of both genetic and environmental factors, as discussed in Chapter 4. What really is of interest and the proper way of phrasing the question is "To what extent can the observed *variability* in particular behavior be allotted to genetic versus environmental variability among several different individuals?" or, for a given individual, "To what extent can the behavior of that animal be modified as a result of experience or surrounding, environmental events during its lifetime?" Two separate questions are being dealt with—one concerning variability in populations and the other concerning development in an individual.

Maturation and modifiability of normal developmental processes complicate the picture. Examples of developmental changes include modified tube-webs of spiders restricted initially to abnormal holes and altered courtship patterns of male Salticidae spiders initially forced to perform in abnormal spaces. One might also view the ontogeny of bird song in this light. What is encountered is a biological continuum that merges from what is classically considered as development into what is classically considered as learning. Across species and groups of species there is a continuum in the *extent* to which something such as bird vocalization is modifiable by experience. Nature simply is not obliged to come in neatly divisible packages with unambiguous boundaries as one might like; continuums are a fact of life.

Aside from the nature-nurture problem, the other major set of questions and assumptions about learning has concerned whether the characteristics of learning are universal and whether differences between different animals are only a matter of degree. That is, when modifiability of behavior as a result of experience does occur, is it the same whether it occurs in rats, humans, pigeons, snails, or worms? Can the characteristics of learning that are discovered about rats be applied also to humans (and pigeons, snails, etc.)? At the extreme, if something is learned about the neural details of habituation in *Aplysia*, does that tell anything about habituation in humans or about other forms of learning such as conditioning? There is the constant problem of analogy versus homology. While not rejecting that learning might be universal, many persons are not willing to accept the proposition yet; they instead maintain an agnostic position on the topic.

As has been seen, there are many categories of the outward expression or *phenomena* of learning. Among different groups of animals, learning of a particular type, such as habituation, may have some common attributes but also much variation. Do similar appearances of learning mean that the underlying, neural *processes* are also the same? Or could one simply be seeing convergence, as Bitterman (1975) phrased it, "whether the resemblance between vertebrate and invertebrate learning is any more profound than that between the hand of an ape and the claw of a lobster."

It is at this point that the echinoderms may provide some insight. Look closely again at Figure 5-4, particularly at the main division between the two major trunks of protostomes and deuterostomes. Consider where the major centers of highlights of learning are among the animal groups: cephalopods, arthropods, and vertebrates. What were the common ancestors like, and did these major groups all inherit their capacity for learning from a common ancestor? Or were learning abilities derived independently? These questions cannot be answered for sure. The common ancestors are long gone, and all one has to work with today are a few fossils (which tell nothing about learning) and a few modern-day remnants of ancient groups. We are stuck with inferences at best.

Present-day echinoderms and the lower chordates may be just spectacular degenerates from the glory that was our common past. But they probably are not. The fossil record (for echinoderms) goes back too far, and there are too many different forms to have that many degenerates. Even if they started from one degenerate form, the problems and complexities of a more modern biological world would seem too harsh for the successful radiation of degenerates in a free-living (as opposed to parasitic) state. It seems more reasonable that we are in fact seeing genuine remnants of a time when life was simpler. If so and if the diversity of simple forms with limited learning capacity at the bases of *both* protostome and deuterostome trunks represents the status of learning in the distant past, then it seems reasonably safe to infer that the protostomes and deuterostomes parted company long before advanced forms of learning were acquired. This would mean we are looking at analogies and convergence rather than homologies and common, ancestral relationships.

Some insight can be gathered also from an advanced group, the birds. Perhaps one of the most enlightening conclusions from studies of birds is the *amount of variability* in the extent and manners that song can be modified even within a single order of organisms (Passeriformes). Birds are certainly a diverse lot, but from the perspective of variation in all other forms of animal life, birds constitute an extraordinarily uniform group. It is therefore remarkable that the modifiability of their vocalizations can be so diverse. This implies that learning (development or however one wishes to view bird song) is not necessarily conservative.

When all of this evidence is taken together—similarities in greatly (and anciently) diverged groups, differences in closely related groups, situation-specific learning, compartmentalization, and the growing list of categories of learning—it seems a good *guess* that learning has evolved on numerous occasions and because of different specific advantages. Similarities in *phenomena* probably only represent convergent evolution. There is a high probability that underlying neural processes are different. Knowledge of the mechanisms of habituation in *Aplysia* may or may not tell much about habituation of young birds to the patterns or shapes of other birds flying overhead. This will not be known until we *also* know the detailed mechanism(s) of habituation in birds (and mammals, insects, and others).

It is important to emphasize that a guess is merely a guess, not an answer. In spite of the diversity seen in basic neurons, there are many shared characteristics, including ionic and other biochemical properties such as some transmitters, among virtually all organisms. Many of these common attributes probably resulted from common ancestory. Learning, at least the so-called higher types, results from processes occurring in groups of neurons, hence above the level of the basic neuron. Many of the higher forms of learning appear associated with advanced senses for distant events, perhaps abilities to manipulate objects, and other factors such as predation; many of these aspects of life do not appear to have been present or as prominent among the common ancestors of the protostomes and deuterostomes. Nonetheless, perhaps the roots of higher learning processes were present back then and what is seen today is related. Or perhaps there are constraints on nervous systems so that when modifiability does occur, it occurs only in certain basic ways, and the processes end up being the same (or similar to some extent) whether independently evolved or not. If this is so, then the whole issue becomes merely academic.

All that can be said at the present is that we really do not know what the underlying processes are, and we still are struggling with and sorting out the characteristics of the phenomena. All three of the major behavioral disciplines—

psychology, ethology, and neurophysiology—have contributed significantly to the contemporary understanding of learning. Psychology developed careful designs and controls of learning studies and showed how to measure and differentiate subtle characteristics of learning phenomena. Ethology led to a broad evolutionary perspective, looking at a wide variety of organisms, considering problems of convergence, analogy, and homology, and stressing the roles of natural, ecological, and evolutionary forces in the differences among species. Neurophysiology is providing access to difficult tissues at the miniature level and is opening the door to deal with complexity at or beyond the astronomical order of magnitude.

In the meantime, we continue to deal with learning mostly as *phenomena not processes*. The detailed inner working of the brain as a whole remains a black box. Most progress in understanding the neural substrate of learning amounts to unwrapping only the outermost layers of that black box.

At the level of treating learning as a phenomenon, one of the most satisfying contemporary interpretations of learning was proposed by Mayr (1974). It encompasses all of the variability, similarities, biological constraints, developmental aspects, critical periods, and other matters discussed earlier. It is a simple analogy drawn from understanding of computer programs. Without specifying the underlying mechanisms, Mayr interprets learning as follows:

A genetic program which does not allow appreciable modifications during the process of translation into the phenotype I call a *closed program*—closed because nothing can be inserted in it through experience. Such closed programs are widespread among the so-called lower animals. A genetic program which allows for additional input during the life span of its owner I call an *open program*. Even this improvement in terminological precision does not remove all our difficulties. A particular instinctive behavior act is, of course, never controlled directly by the genotype but rather by a behavior program in the nervous system which resulted from the translation of the original genetic program. It is particularly important to make this distinction for the open program. The new information acquired through experience is inserted into the translated program in the nervous system rather than into the genetic program because, as we know, there is no inheritance of acquired characters.

An open program is by no means a tabula rasa; certain types of information are more easily incorporated than others. . . . Whether natural selection will favor the evolution of an open or a closed program for a given behavior depends on the circumstances. For instance, the shorter the life span of an individual, the smaller the opportunity to learn from experience. . . . The situation is radically different in species with a more or less extended period of parental care. Here, the fixed responses of the newborn can be quite few in number, being limited primarily to adequate responses to the parents.

The longer the period of parental care, the more time will be available for learning, hence, the greater the opportunity to replace the closed genetic program by an open program. The great selective advantage of a capacity for learning is, of course, that it permits storing far more experiences, far more detailed information about the environment, than can be transmitted in the DNA of the fertilized zygote.

Under what circumstances is a closed genetic program favored and under what others an open one? . . . the genetic program for formal signaling must be essentially closed; to state it more generally, selection should favor the evolution of a closed program when there is a reliable relationship between a stimulus and only one correct response. . . . On the other hand, noncommunicative behavior leading to an exploitation of natural resources should be flexible, permitting an opportunistic adjustment to rapid changes in the environment . . . Such flexibility would be impossible if such behavior were too rigidly determined genetically.

The longer the life span of an individual, the greater will be the selective premium on replacing or supplementing closed genetic programs by open ones. . . . The direction of many evolutionary pathways, thus, is clear. It often leads to a gradual opening up of the genetic program, permitting the incorporation of personally acquired information to an even greater extent. . . . There are two prerequisites for this to happen . . . [a] greater storage capacity than is needed for the carefully selected information of a closed genetic program . . . [that is,] a larger central nervous system . . . [and] prolonged parental care. When the young of a species grow up under the guidance of their parents, they have a long period of opportunity to learn from them—to fill their open programs with useful information on enemies, food, shelter, and other important components of their immediate environment.*

All organisms (except the simplest and most sedentary such as the sponges) must have at least *some* openings in their behavioral programs. At the minimum, for example, animals must at least use sensory input to avoid bumping into obstacles. Beyond that, it appears that further opening of behavioral programs to outside input depends on specific requirements and details of the program, specific life-styles, reliability of stimuli or cues, and general attributes such as storage capacity *and* parental care. (Many organisms have parental care, but many do not also have the necessary storage capacity for further outside information.)

Mayr's interpretation of learning may not be perfect in all detail and may need further development for better resolution, such as to accommodate human language. At a simpler level (than human language), the interpretation has not been developed to the level of resolution necessary to handle all the details even of bird communication. Furthermore, the closed-open program model has not been extended yet to such things as insight learning and consciousness. But Mayr's closed-open program model provides a useful, if only heuristic, framework from which to view what is commonly called learning.

SUMMARY

The subject of learning merges with that of ontogeny and the development of behavior; hence, this chapter serves as a continuation of the previous chapter. *Learning* is defined as a specific change or modification of behavior as a result of experience with an external event or series of events in an individual's life. There are, however, many observable forms that behavioral modification can take—each with its own apparent characteristics. Many categories have been named.

*From Mayr, E. 1974. Am. Sci. 62:650-658.

The current list of learning and learninglike phenomena includes habituation, sensitization, classical conditioning, instrumental or operant conditioning, pseudoconditioning, discriminant learning, taste aversion learning, latent or exploratory learning, cultural or observational learning, imprinting, insight learning, and learning-set learning. These various categories are described and compared. Several of these categories overlap, and many can be subdivided into smaller categories.

Just how distinct the various forms of learning are and what their relationships are among different taxonomic groups of animals form the basis of several ongoing controversies. This chapter includes a phylogenetic survey of learning. From such a survey it is clear that learning does not exist on a single, progressive scale simply leading up to a peak in humans. Instead, learning appears as a set of branching and independently evolved phenomena. Some groups with advanced forms of learning, such as cephalopods and primates, actually are relatively low on their respective phylogenetic branches. Instead of depending simply on phylogenetic relationships, learning capabilities appear to depend on life history characteristics, sensory capabilities, particularly ability to sense distant events, and the ability to manipulate objects via extensions of the body.

Learning requires a capability of memory. The characteristics of memory have received much attention and research but remain poorly understood. Several hypotheses have been proposed for the mechanism and location of memory within the nervous system, including both molecular and structural sites. Some progress has been made in understanding simpler forms of learning, such as habituation and memory in invertebrates like the marine snail *Aplysia*. Among vertebrates the most promising basic mechanism for memory appears to be increases, decreases, and changes in particular neural pathways as mediated by dendritic branching and synaptic contacts. There may also be other internal molecular and membrane changes but the engram, or storage site, remains elusive.

In an attempt to synthesize the current information and understanding of learning, different forms of learning can be considered as phenomena (i.e., the observable characteristics) as opposed to processes (i.e., underlying neural mechanisms). Such a comparison has involved two central controversies: the nature-nurture problem and whether learning processes are universal and species differ only in matter of degree or if the underlying processes themselves vary.

The nature-nurture or instinct-learning question is commonly phrased in a mistaken manner and is misleading, as has been stressed elsewhere in this book. All behavior is a joint product of both genetic and environmental factors. One must consider variability of behavior in populations as related to genetic and environmental variability or, alternately, the extent to which specific behavior of individuals can be modified. The question boils down to two separate aspects: variability of behavior in populations and development of behavior in individuals. When phrased in this manner, it becomes clear that there are continuums, such as in song learning among different species of birds, and nature is not neatly divisible into instinct versus learning categories.

Concerning the problem of universality of learning among different species, not enough is known yet about the internal, neural mechanisms. It seems likely, however, from an inspection of the positions of organisms at the base of phylogenetic branches, such as the echinoderms, that many of the advanced forms of learning that appear similar among animals may have been independently derived and simply represent convergent evolution.

Returning to learning as phenomena and ignoring processes for the moment, Mayr has characterized learning as the extent of openness in the translation of genetic information to neural systems in producing information-control systems that are analogous to computer programs. Behavioral programs that do not permit much input are said to be closed, whereas those which permit input are open. Some species have more open points than other species, but even with instances that are most open, the inputs must be at particular times and within particular limits, just as one cannot enter alphabetical information into a computer program requiring numerical data. Closed programs are expected when the relationship between behavior and input is reliable and relatively constant, as in much of communication. Open programs, on the other hand, are advantageous in cases of rapid changes in the environment and such things as the exploration of some resources. Open programs, however, require a larger memory capacity and may need prolonged parental care during the initial stages of learning.

Mayr's conceptualization of open versus closed behavior programs reflects and encompasses the observed characteristics of learning better than the instinct versus learning dichotomy because it better permits a continuum of openness and emphasizes specific times and limits that particular programs open up. However, it remains only a heuristic description of phenomena, not processes. The searches for better understanding of the mechanisms of learning continue.

**Recommended
Additional Reading**

Bitterman, M.E. 1975. The comparative analysis of learning. Science 188:699-709.

Bitterman, M.E., V.M. LoLordo, J.B. Overmier, and M.E. Rashotte. 1979. Animal learning. Plenum Press, New York.

Bonner, J.T. 1980. The evolution of culture in animals. Princeton University Press, Princeton, N.J.

Boycott, B.B. 1965. Learning in the octopus. Sci. Am. 212:42-50.

Corning, W.C., J.A. Dyal, and A.O.D. Willows, editors. 1973, 1975. Invertebrate learning, 3 vols. Plenum Publishing Corp., New York.

Curio, E., U. Ernst, and W. Vieth. 1978. Cultural transmission of enemy recognition: one function of mobbing. Science 202:899-901.

Deutsch, J.A., editor. 1973. The physiological basis of memory. Academic Press, Inc., New York.

Domjan, M., and B. Burkhard. 1982. The principles of learning and behavior. Brooks/Cole Publishing Co., Monterey, Calif.

Drucker-Colin, R.R., and J.L. McGaugh, editors. 1977. Neurobiology of sleep and memory. Academic Press, Inc., New York.

Hess, E.H. 1973. Imprinting. Van Nostrand Reinhold Co., New York.

Hess, E.H., and S.B. Petrovich, editors. 1977. Imprinting. Dowden, Hutchinson, & Ross, Inc., Stroudsburg, Pa.

Hinde, R.A. 1970. Animal behaviour, 2nd ed. McGraw-Hill Book Co., New York.

Mackintosh, N.J. 1974. The psychology of animal learning. Academic Press, Inc., New York.

Mayr, E. 1974. Behavior programs and evolutionary strategies. Am. Sci. 62:650-658.

Medin, D.L., W.A. Roberts, and R.T. Davis. 1976. Processes of animal memory. Lawrence Erlbaum Associates, Inc., Hillsdale, N.J.

Menzel, R., and J. Erber. 1978. Learning and memory in bees. Sci. Am. 239:102-110.

Milgram, N.W., L. Krames, and T.M. Alloway, editors. 1977. Food aversion learning. Plenum Press, New York.

Pribram, K.H., and D.E. Broadbent, editors. 1970. Biology of memory. Academic Press, Inc., New York.

Rosenzweig, M.R., and E.L. Bennett, editors. 1976. Neural mechanisms of learning and memory. MIT Press, Cambridge, Mass.

Rosenzweig, M.R., E.L. Bennett, and M.C. Diamond. 1972. Brain changes in response to experience. Sci. Am. 226:22-29.

Spear, N.E., and R.R. Miller, editors. 1981. Information processing in animals. Lawrence Erlbaum Associates, Inc., Hillsdale, N.J.

Sutherland, N.S., and N.J. Mackintosh. 1971. Mechanisms of animal discrimination learning. Academic Press, Inc., New York.

Ungar, G., editor. 1970. Molecular mechanisms in memory and learning. Plenum Press, New York.

PART FOUR

ADDITIONAL TOPICS

CHAPTER 20

Leonard Lee Rue III

A domestic dog and deer fawn in an amicable relationship.

ABNORMAL BEHAVIOR

Life does not always proceed by the book. Conflicting stimuli or competing interests may arise that cannot always be resolved. Such situations may present the researcher with opportunities to gain insight into the mechanisms of behavior. But conflicts and abnormal situations are not always so beneficial to the animals themselves. Conflict is only one of several problems that may result in what is commonly referred to as *abnormal behavior*.

Abnormal behavior, as recognized by most persons, is best known among human beings or animals living in association with humans. Abnormal behavior is rarely seen under natural conditions in the absence of human influence. This may be partly because it is overlooked, unrecognized, or unreported. However, it is more likely that, as one automobile bumper sticker put it, "Nature takes care of abnormalities." When abnormalities occur, they usually do not survive for long; the abnormal individual becomes someone else's meal very quickly. Or if an unusual characteristic happens to be better than the "normal" in terms of survival and reproduction, the new characteristic may replace the former and become the new normal.

Abnormal behavior is another category that is difficult to define precisely. Based on the discussions and definitions of several persons (Fox 1968), the concept of abnormal behavior possesses the following common denominators:

1. The behavior differs from the usual or normal; that is, it is in some way *uncommon*.
2. It is in some manner *maladaptive*. Other terms that have been used in this context include *inappropriate*, *aberrant*, *undesirable*, and *disorderly*.

Before one can recognize behavior that is different from normal, the normal repertoire must be known. This implies and requires much previous experience with the species. Sackett (1968) takes this point a step further and suggests that abnormal can be determined only in comparison against observations of specific, experimentally comparable behaviors of other individuals. Regardless of how tightly one controls the groups being compared, the circumstances, context, or environment must also be considered. Uncommon refers to the presence of a behavior under a given set of circumstances. Rareness per se is not important. For example, honeybees fanning their comb in the presence of a fire is normal. It is rare, however, because the circumstance (fire near the hive) is rare.

The manner in which a behavior differs from normal may be either by presence or absence of a behavior or by the *degree* or magnitude of a behavior that otherwise would be considered normal. Female bears, pigs, and dogs with their young, for example, normally show protective behavior. Similar behavior in the absence of offspring, however, might be considered abnormal. Or if the guarding is excessive and the female shows aggression toward a distant animal that poses little or no threat, then the behavior could be considered abnormal.

The second test of whether behavior is normal or abnormal is to ask whether or not it is adaptive or functional. It may be, as has been the case with play, that one cannot recognize any function of a behavior; hence, it would be difficult to judge its potential adaptiveness. But in many cases the behavior would seem to be clearly nonadaptive; it may be harmful, represent a wasteful or inefficient use of resources, or otherwise not be conducive to survival or reproduction. Adaptiveness is very difficult to assess.

Consider, for example, a noisy awkward animal in a pack of predators. The animals of this species normally hunt with stealth, cunning, and cooperation among the pack animals. The noisy individual disrupts hunts and is not permitted by the others to join during prehunt activities and subsequent searches and chases. It would seem at first sight that the noisy animal is displaying maladaptive behavior and should be selected against. It may be, however, that the effect of the noisiness and having to stay back at the den proves to be highly adaptive under the circumstances and leads to increases in both individual and inclusive fitness. By staying at the den, the individual may stay in charge of the young. Its particular noisiness warns other animals to stay out of the vicinity. This increases the survival of the offspring, creates a division of labor in the pack, and leads to increases in efficiency and the degree of social behavior.

A judgment of adaptiveness also depends in part on the orientation and perspective of the observer, environmental context, and recent evolutionary history. For example, broody behavior in a wild chicken, whereby the hen stops laying eggs and begins incubation, leads to survival and reproduction. In a chicken at an egg farm, on the other hand, broodiness leads to the stew pot.

The net result of all this is that behavior varies tremendously from species to species, and what is normal for one species or situation may be abnormal for another. Many herbivores normally become quiet and immobile at the appearance of distant danger, whereas many carnivores normally will run at the first detection of a distant threat. These problems leave one without a very precise definition of what is normal and what is abnormal. Be that it may, we will proceed with the intuitive feeling that abnormal behavior depends on how common and adaptive the behavior is under particular circumstances. Given the preceding, several examples of what might be thought of as abnormal behavior among animals both in and out of human-related contexts will be considered next.

EXAMPLES OF ABNORMAL BEHAVIOR IN ANIMALS

Nondomestic Animals under Natural Conditions

Cases of abnormal behavior in wild animals are somewhat arbitrary and, as already mentioned, difficult to observe. The majority probably represent either the fringe or tail end of natural variation or problems resulting from disease or injury that happened to be observed and recorded before the animal was lost from the population. There has been little attempt to synthesize descriptions of abnormal behavior in wild animals. Aside from occasional reviews of specific abnormalities, the information is mostly scattered as anecdotal short notes in the backs of journals. A few examples follow.

Elliot (1977) describes an interesting case of a raven (*Corvus corax*) hanging upside down from branches by its bill and feet. He was unable to find any previous reports of such behavior and could not assign any function or adaptiveness to the behavior. It might be classed as an example of play rather than abnormal behavior. Elliot speculated that it might have been related to normal aerial courtship displays shown by males. After hanging in various manners and repeating the sequences, the bird perched normally and eventually flew away.

Two interesting cases of abnormal nest building by eastern phoebes *(Sayornis phoebe)* were recorded by Weeks (1977). They were observed at different times and locations during a study involving 277 nests, the remainder of which were normal. One nest was formed of fresh moss and mud and was built in a continuous mat along the side of a bridge beam. The nest was 2.2 m (over 6 feet) long and measured 2.5 to 3.75 cm thick for 1 m and 1.25 to 2.5 cm thick for the remainder. A clutch of five eggs was laid at one end of the "nest," with all of the young being hatched and successfully fledged. The second instance involved another larger-than-normal nest that was composed of five separate cups. There was a normal cup in the center with two cups on each side descending in stairstep fashion. The inner two cups contained some lining material, and the outer two were shallow depressions. The normal, center cup was used for a clutch of five eggs.

Weeks commented on the nonadaptive nature of the abnormal nest building, noting the obvious strains placed on the birds' time and energy budgets for the additional building effort. Also the young, believed to be the first clutch at the first nest, were fledged late—about the time when other birds were fledging their second clutches.

Sometimes abnormal mates are chosen. Brown (1977) noted, for example, a male brown-headed cowbird *(Molothrus ater)* that approached and attempted to court two male purple martins *(Progne subis)*. The behavior displayed by the cowbird was typical of the display normally shown by male cowbirds toward female cowbirds.

Animals in Zoos

The problem of wild animals in zoos is classic. It is academic as to whether or not the behavioral problems represent abnormal behavior. It is really the circumstances that are abnormal for the animals, and the responses are so common that the behaviors themselves probably should not be viewed as abnormal. Under natural conditions, however, the behaviors would be quite uncommon and maladaptive.

The information in this subsection has been summarized from a review by Meyer-Holzapfel (1968). The problems result from a number of causes including attempts to escape, fear-causing stimuli, lack of appropriate normal stimuli and boredom, lack of exercise, improper nutrition, inadequate housing and facility needs, and intraspecific social problems caused by artificially forced groupings and inability of the animals to escape from the interactions.

Abnormal Escape or Hiding Reactions

These reactions may go to either of two extremes: (1) animals may become so hyperactive and violent in attempts to escape that they dash against the sides of the facility and injure or kill themselves, or (2) they may enter a stupor and become immobile, often for days or permanently. Much hyperactivity, injury, or death may occur at first capture or during transport. The problems can be enhanced by social factors. A black bear at the bottom of the social ranking at one zoo, for example, was attacked frequently by other bears. As a result it sat motionless in a rock hollow and gnashed its teeth. Many of these problems can be resolved by more humane capture and transport techniques, initial use of tranquilizers, provision of hiding places, and other attention to the animal's artificial and social environment. These problems often depend importantly on the age of an animal when it is captured from the wild. Younger animals tend to adapt to the conditions of captivity better than older ones. With most species it is best not to attempt to place adults in captivity in the first place. In addition to being unfit for captivity, they have proven their ability to survive in the wild.

Refusal to Eat

Many species will not eat after capture or change of housing. Elephant seals, for example, may refuse food for 40 to 100 days, and snakes frequently will not eat for several weeks, months, or, in some instances, even up to 2 years. Refusal to eat often leads to death.

Stereotyped Motor Reactions

The list of stereotyped movements by captive wild animals is long and the causes manifold. There are several common patterns. The most familiar involves pacing. The same path is followed back and forth, and the activity often seems ceaseless or compulsive. Some of the most common paths are straight, figure-eight, circular, and oval.

Pacing usually results from lack of space for normally active species or from a lack of suitable surroundings. Digging animals, such as armadillos, may pace if not provided with suitable substrate to burrow into, and many animals will pace if denied sleeping or hiding places.

A shortened version of pacing is weaving behavior. The animal simply sways back and forth or rocks up and down. It has been noted particularly in horses, bears, elephants, large ungulates, and parrots of various species. In some cases the development of the behavior has been observed. The animal initially paces; after a period of time it simply takes a few steps to the left, then a few to the right, then eventually stands in one position and just swings the head from side to side.

When animals that show stereotyped movements are moved to new and better surroundings, they often will change or cease the patterns. But this is not always the case, particularly if the previous confinement was very cramped or if the animal had been kept that way for long periods.

Exceptional stereotyped behavior was shown by a pair of sloth bears in a case studied by Meyer-Holzapfel (1968). The female displayed an array of complex abnormal behaviors at feeding time. The female subsequently was separated from the male for reasons not associated with the abnormal behavior, and the abnormal behavior ceased within a day. After 6 weeks the female was reunited with the male, and the complex set of abnormal behaviors returned immediately with the first meal. It was discovered that the problem resulted from dominance of the male, which gulped his food and then chased the female from her bowl. By finally realizing this, the zoo personnel were able to eliminate the problem by isolating the female at feeding time and for a period of at least 20 minutes after she had finished.

Displacement and Redirected Behavior

The conflict and stress that often accompany confinement may lead to displacement or redirected behavior. The problems may lead to irrelevant movements otherwise seen in eating, drinking, sleeping, nest building, or grooming, such as body and head scratching in many species of primates (including humans). When unable to reach something they wish to eat, many animals will bite or chew other objects. Some cases of play behavior by zoo animals have been interpreted as displacement behavior.

Self-mutilation

Animals in captivity occasionally will damage themselves. Birds may pluck their own feathers, sometimes to the point where they are virtually naked. Mammals, particularly opossums, carnivores, long-tailed monkeys, and the small South American monkeys, will bite, chew, or scratch themselves, sometimes to the point of death.

Self-mutilation has been interpreted in several ways: (1) pathological maintenance behavior, (2) insufficient activity in otherwise active animals, with subsequent circulation problems in extremities, (3) minor injuries or parasites that initiate licking and scratching, and (4) especially in carnivores and primates, aggression that has been redirected toward themselves.

Aggressiveness

Aggressiveness, whether normal or excessive, is often a problem in zoo animals. It frequently results from overcrowding, hypersexuality, and various other problems that are not well understood. Aggression may be directed against conspecifics, including mates and offspring, or against the human keepers. Brown bear mothers, for example, normally remain with their cubs for over a year under natural conditions. But in captivity they may go into estrus when cubs are only 2 months old or less, and they may then abandon the cubs or maul them.

Aggressiveness toward keepers is a problem and can develop suddenly even in formerly calm animals. Many keepers have been killed as a result. Male animals usually become most aggressive during the rut or courtship period, whereas females may become most dangerous when tending offspring. These should be viewed in many instances as normal and adaptive behavior, except the aggression may be directed toward humans. For reasons not understood, many animals form a strong antagonism toward specific people and not toward others—even when there has been no mistreatment involved.

Serious aggression toward humans often develops, apparently as misdirected intraspecific aggression, by animals that were hand-raised and may be imprinted on humans. This is a common problem with hand-raised birds and ungulates. It is unwise to allow unfamiliar and unprepared persons to enter the confines of such animals kept in captivity; also it is generally a mistake to release such hand-raised animals back to the wild. Perhaps one of the more amusing examples, except to the people concerned, involved a hand-raised great-horned owl *(Bubo virginianus)* that had been released in a rural cemetary and disrupted a subsequent funeral.

Miscellaneous Abnormal Behavior

Meyer-Holzapfel (1968) includes several other categories of abnormal behavior observed in zoos. These include a variety of abnormal sexual behaviors. In one case a male brown hyena frequently copulated with its metal water bowl. A chimpanzee was reported to form a sexual relationship with a cat, and in a similar instance an immature female gorilla showed pseudomale behavior toward a dog. Imprinting by birds, depending on the species, frequently results in sexual behavior being displayed toward other species, including humans.

Various perversions also occur in eating behavior. Common problems involve coprophagy (eating of feces, which, incidentally, is normal for many species such as some lagomorphs and rodents), licking and eating objects and parts of the cage or enclosure, and cannibalism, particularly of young. Parental care is often highly dependent on both the internal hormonal condition of the parent and on specific signals, sign stimuli or releasers, given by the offspring. Either or both of these factors may be disrupted in captivity with the result that the normal parental system breaks down.

Other problems and abnormal behavior include prolonged infantile behavior, especially food begging, and a reversion to juvenile behavior. Reversion to behavior shown normally only by young generally occurs in association with injury, such as broken wings or legs in birds and various injuries in carnivores. It has been suggested that immature behavior shown by invalid elderly persons may share a common basis with such reversions in other animals.

The last category to be mentioned in association with zoo animals is what might best be described as apathy. Social animals kept in isolation or after separation from a mate or partner often become indifferent to their surroundings, have a poor appetite, and show various forms of disturbed, mildly abnormal behavior. This is particularly evident in monogamous birds and great apes (e.g., orangutans and gorillas). In the latter cases even the facial expressions have been reported to be interpretable by humans as expressions of depression and mourning.

Figure 20-1 Unsuitable conditions at a wayside exhibit of animals. Photo by James W. Grier.

Figure 20-2 Example of proper housing facilities at a modern zoo: a lynx exhibit at the Minnesota Zoological Garden. A, Artist's perspective of the naturalistic exhibit. B, Photograph of a lynx in this exhibit. Courtesy Tom Cajacob, Minnesota Zoological Garden.

Many of the problems still remain at older zoo facilities and at smaller, usually private, wayside exhibits of animals (Figure 20-1). Fortunately, however, many of the types of problems discussed in this section are disappearing with enlightened practices of zoo management, construction of new facilities that are specifically tailored to the animals' needs (Figure 20-2), and improved understanding of animal behavior.

Captive Animals Used in Research

Abnormal behavior in captive research animals is important in two very different ways. In the first, such behavior is unwanted, unplanned, and unexpected and may interfere with the main goals and procedures of the research project. At the other extreme, however, abnormal behavior is frequently the actual subject of particular research projects. In such cases the behavior is studied as a model or analog of abnormal behavior in humans or, less frequently, as an attempt to understand the problems of the animals themselves to improve future management of the species.

The problems of accidental abnormal behavior depend to a large extent on whether the animals are nondomestic or domestic. Nondomestic animals in captivity for research can often be considered in much the same position as wild animals brought into captivity for zoos and display. Much of the research conducted on animals is not behavioral, and the researchers may pay little or no attention to behavioral implications of their care and experimental treatments. Depending on management practices and the purposes of the research, the problems, causes, and cures may be very similar to what was discussed in the preceding section.

The most favorable situations for studying wild species under captive conditions involve modern zoos with properly designed facilities and a comprehensive, integrated view of all aspects of the animal's life, health, and welfare. Another equally favorable situation for studying captive wild animals involves institutions specialized for work with one species or taxonomic group. Examples of specialized research institutions include the International Crane Foundation at Baraboo, Wisconsin, Peregrine (falcon) Fund at Cornell University, the Primate Center at the University of Wisconsin at Madison, and several similar facilities directed toward primates, birds of prey, waterfowl, large cats, ungulates, insects, and others.

Accidental, unwanted abnormal behavior also may occur with domestic animals used in research. These problems are similar to those created by abnormal behavior in domestic animals used for other purposes, such as for food production or as pets (discussed later).

Abnormal behavior that is deliberately caused and studied experimentally in research is the subject of a massive body of literature and is an important part of the discipline of psychology. Results of such research may serve as feedback to improve management and care of captive animals being studied for other purposes. But the most frequent interest in such research by far has been to learn what information from other species might possibly explain abnormal behavior in humans.

Perhaps the name best associated with work in the area of experimental neuroses and abnormal behavioral problems in animals is that of the Russian physiologist-phychologist Ivan Pavlov. Pavlov began research early in the twentieth century; he was initially interested in the physiology of digestion. This led him into the subjects of conditioning and learning, for which he is best known and which were discussed earlier in this book. He went on to study abnormal behaviors and was able, for example, to produce nervous breakdowns and "brainwashing" in dogs. These were achieved with such things as very strong stimuli, extending the length of aversive stimuli, frequent presentation of aversive stimuli, switching positive and negative stimuli to cause uncertainty and confusion (some effects referred to as *collision*), and a number of related techniques. This early work has been followed by much research into similar conditioning effects, hormonal effects, neural pathways, corticovisceral or "psychosomatic" interrelationships, and how behavioral problems can be either caused or prevented. The negative and positive aspects of this work, unfortunately or fortunately, respectively, have been applied to humans in some instances (e.g., Sargant 1957).

The effects of early experience and social surroundings, particularly involving parents and peers, have received much research. The work has tended to concentrate, although not exclusively, on primates and canids. The early work of Harlow and colleagues (e.g., Harlow and Zimmerman 1959) has become classic. The implications of this research for normal biological mechanisms are incorporated in earlier portions of this book. An excellent review that focuses on the abnormal behavioral aspects is provided by Sackett (1968), who concluded that behavioral impairments could result from improper or inadequate interactions with either mother or peers. Peer relationships seemed to pose the greatest effects. Deficiencies were mostly emotional, motivational, and social rather than intellectual. Furthermore, there was a more or less critical or vulnerable period consisting of the first 6 months of life. Although the effects can be permanent and extend into adult life, they are not necessarily irreversible. Suomi and Harlow (1972), for example, have treated some of the problems through the use of "therapist" monkeys.

Domestic Livestock

Domestication must be considered as a significant factor in animal behavior, including abnormal behavior. The situation with domestic animals differs significantly from the situations discussed earlier not only by the presence of domestication per se but also in the human element. Human interaction with domestic animals involves many more animals worldwide and a larger diversity of human understanding. Many people understand their animals' needs and behavior very well, but many others do not. In dealing with domestic animals there is much ignorance, faulty tradition, and a tendency to anthropomorphize excessively.

Before proceeding, the meaning of domestication must be clarified, and the general effects of domestication on behavior must be considered. Good discussions are provided by Fox (1968) and Ratner and Boice (1975). The contemporary biological meaning includes an implication of evolutionary and presumably genetic changes resulting from a close relationship with humans. This means that the organisms have been removed, at least partially, from the forces of natural selection and subjected instead to artificial (human-related) selective pressures. These pressures do not have to be deliberate or conscious; that is, the humans do not have to be aware of their effects.

There is a distinction between domestication and taming. *Tameness* refers to the calmness and ease of approach and handling of certain individual animals or strains or species if most of the individuals in the group are similar in this respect. Tameness is also sometimes referred to as *socialization* (to human beings). The opposite of being tame is referred to commonly as being *wild*. Thus wild is used rather loosely in both contexts—nondomestic (referring to recent evolutionary background) and not calm in the presence of humans.

For examples of these points, consider the hunting relationships between humans and birds of prey or between humans and cheetahs. People have been forming an association, as in the sport of falconry, with these animals for thousands of years. But, until just recently, these species have not bred in captivity. Humans simply keep replenishing their needs for individuals from wild populations, and it is unlikely that the association has had much significant impact on the selection and evolution of the wild populations. But individual raptors or cheetahs, in the process of being handled and trained, may become quite tame; that is, they are tame but not domestic.

Among many domestic species, however, individuals that are not handled and raised in close association with humans may become very wild, that is, fearful, excited, and not calm in the presence of humans. Some of the least tame animals are from domestic species that grow up without being properly handled by people. Many individual animals are both domestic and tame.

Domestication also is not necessarily "degenerate," in spite of biologists sometimes referring to white rats and domestic chickens as zoological monstrosities. Domestic animals have been selected under particular, human-associated circumstances, and within this setting they should be viewed as adapted and not as inferior (Boice 1973). Although domestic animals may be very fit, this does not mean that they do not have problems.

Fox (1968) listed some of the common characteristics for which domestic animals, depending on the species, have been selected. Many involve behavior directly or indirectly. They include the following:

1. *Tameness or docility and ease of handling.* Natural tameness may have been a factor of some species entering the domesticated state in the first place. It has been claimed that the difference between the wild and savage Norway rat and the docile domestic black rat resulted from the mutation of a single gene (Keeler and King 1942).

2. *Ability to live under human conditions of housing and nutrition, along with regional and climatic conditions where different groups of people live.* Hill and downs sheep, for example, have different grazing and fleece characteristics, which permit them to live in different climates.

3. *Characteristics of economic importance.* These include high fertility, production of useful items (such as eggs, milk, or wool), and rapid growth and efficient use of food.

4. *Prolongation of infant stages (neoteny) or reduction of some adult characteristics (paedomorphosis).* It has been suggested that this results in animals that are less specialized and more flexible behaviorally, which may facilitate care and even attractiveness to humans. Along with a reduction of secondary sexual characteristics (e.g., reduction in horns, thickness of skin, territorial aggressiveness), it may become easier to crowd the animals, handle them, and process their products. Some breeds of dogs are described as "perpetual puppies." Keeping animals in the immature stages is accomplished sometimes by castration.

5. *Acceleration of evolutionary changes by selective inbreeding and hybridization in addition to rapid artificial selection.* Comparing the pace of evolutionary changes wrought by humans versus those imposed by nature might be debatable, but humans have clearly demonstrated the ability to rapidly select characteristics deemed most desirable.

The effects of domestication on behavior depend on the particular behavior in question. Hence, egg laying may be increased, broodiness decreased, flight distances may be decreased, the threshold of barking responses may be altered, and activity involved in searching for food may be decreased, but other behaviors, such as the movements involved in copulation or maintenance, may be unchanged. Even within a species, such as chickens or dogs, some strains may be selected for fighting ability for sporting purposes, whereas in other strains fighting and aggressiveness are selected against. "Abnormal" behavior may now be considered against this background.

Many of the abnormal behaviors seen in domestic animals, in spite of selection to tolerate captive conditions, nonetheless may be similar to problems of non-domestic animals in captivity. These problems include escape-related behaviors, pacing and weaving, and a number of problems related to overcrowding—including cannibalism of young.

Many of the observed differences between domestic species and their wild counterparts may be environmental rather than genetic. Miller (1977) showed that domesticated mallards reared in an uncrowded environment displayed courtship behavior that looked identical to that of their wild counterparts rather than the "degenerate" displays commonly shown by domestic individuals.

Horses are known for a number of behavioral disorders, often referred to as "vices," including crib biting, where they chew on parts of their stalls, wind sucking, where they pull up on an object such as a board and gulp air, kicking, biting, weaving, and shying. Refusal to eat occurs occasionally in several domestic species. Eating of young and other maternal care problems are seen in some species, particularly swine, cats, and some poultry. Crowding and other forced social situations may lead to serious, often fatal aggressiveness.

One characteristic that is, perhaps, remarkable in some domestic species is the ability to tolerate extremely limited, if not cramped, individual confinement. White rats, hamsters, mice, domestic rabbits, and poultry are routinely subjected to housing where there is barely room to turn around or lie down. Yet many strains have been selected for such a life and apparently do quite well under these conditions with few signs of boredom, pacing, or other behaviors that might be considered abnormal.

Problems of fear and attempted escape can cause bizarre reactions and serious losses under social conditions in domestic species that herd or flock. These include cattle, sheep, turkeys, and some strains of white leghorn chickens. Animals in a group may easily panic and stampede. In chickens, for example, the problem is known as *chicken hysteria syndrome*, as described graphically by Ferguson (1968):

Whenever someone entered the pen a brief period of tense quietness would ensue to be quickly interrupted when a hen would emit a loud squawk and then dash to a new location. This would be followed immediately by blind running and milling around with loud squawking, by the entire flock. Hens that had been concealed within community nests left these to participate in the panic behavior, at the end of which the majority of birds would once again be crowded silently under feeders or in nests and corners [Figure 20-3].

Figure 20-3 Chicken hysteria. Many birds are crowded under the small feeder, and others are in the open nest boxes, some with heads showing above the edge. Most are quite tense, and at a signal from one of the hens they will dash frantically to a new location. From Sanger, V.L., and A.H. Hamdy. 1962. Am. Vet. Med. Assoc. 140:455-459. Courtesy Ohio Agricultural Research and Developmental Center.

The behavior occurs in some flocks and not in others even from the same hatch of chicks.

Many behavioral problems seen in domestic animals apparently are caused by modified environmental cues and related hormonal effects. In addition to effects of conflict and stress on the sympathetic nervous system, these problems may produce several reproductive system anomalies via interactions with epinephrine and adrenocortical steroids from the adrenal gland and ACTH from the pituitary. Males may exhibit a number of reproductive deficiencies such as incomplete copulatory movements, excessive sexual activity, and sexual interest in inappropriate objects. Females are susceptible to a number of problems in estrous and maternal behavior.

Subestrus, also known as *silent heat*, *quiet ovulation*, and *anomalous heat*, occurs when a female in estrus does not act as if she were in estrus. This condition has been seen in mares, ewes, and sows and has been reported in up to 20% to 30% of cattle (references in Fraser 1968). In milder forms there is an array of estrous problems where females display partial or abnormal estrous behavior. At the other extreme, females may show excessive estrus and even nymphomania. This condition, mediated hormonally, is believed to have an underlying genetic basis.

The most common maternal problems are rejection of young and failure in normal caregiving (epimeletic) behavior. These abnormalities include delays in grooming after birth, desertion of young, active and aggressive desertion such as butting or biting, and refusal to allow suckling. Cannibalism occurs occasionally. Other problems include early onset of maternal behavior. Before a female gives birth to her own young, she attempts to adopt others' offspring. These latter problems may result in confusion over young and subsequent rejection.

Domestic animals are also subject to a wide variety of abnormal behaviors resulting from undernutrition and malnourishment. These may lead, in mild cases, to such things as hyperactivity and increased searching behavior or increased fighting and aggressiveness under social conditions. In more severe cases there may be developmental problems and actual interference with neural function. A good introductory review to nutritional problems is provided by Worden (1968).

Domestic Pets

Pets are exposed to the problems mentioned in all of the preceding sections, plus more! We now are dealing with a much larger absolute number of animal-human relationships, and the proportion of persons who really understand animal behavior is much lower. Ignorance, misunderstandings, anthropomorphism, and "old wives' tales" are rampant when it comes to pets. Deliberate mistreatment of pets is relatively uncommon, but the inadvertent problems are quite numerous. People living in close association with social and highly domesticated species, such as the dog, may create subtle, psychological problems in their pets just as they cause similar problems among themselves.

Pets may suffer developmental and social problems such as from inconsistency in training and treatment, particularly when given directions by different people, isolation and lack of social contact and attention, many effects of improper early handling and experience, insufficient stimulation, excessive stimulation and handling, and improper housing and feeding (often of a human nature rather than what the animals need).

All of this results in a vast array of problems including sudden changes in behavior of the pet, reproductive problems of many kinds, including false pregnancies, abnormal care-seeking (et-epimeletic behavior), autism and stereotypy (excessive fear, shyness, and withdrawal), depression, hysteria, epilepsy, climactic fits, vices, many neuroses and psychopathic behaviors, plus psychosomatic conditions. The latter include anorexia nervosa (failure to eat from a nervous condition), gastrointestinal ulcer, heart problems, vomiting, improper defecation or urination, various muscular spasms, other ingestional and digestional problems, respiratory problems, and even stroke and sudden death (reviewed in Schmidt 1968).

Many of the problems depend on the species and may involve causes that are not readily apparent to the owner. Cats, for example, may suddenly stop using their litter box with the introduction of new cats or humans into the house or neighborhood. Male cats may begin spraying urine in the house even if neutered.

A particularly striking case of abnormal behavior in a pet brought in for veterinary treatment was uncovered by Schmidt (1968); it was discovered that the dog had shown sexual interest in its female owner and other women. After being frustrated, however, "The animal became enraged . . . The owner showed us deep wounds on her arms. The dog had mad, dementia-like seizures, in which it tore up the curtains, knocked over furniture, and tore apart armchairs." The dog showed no interest in females of its own species and, in fact, attacked them.

CAUSES OF ABNORMAL BEHAVIOR: A REVIEW

The causes of abnormal behavior are many and diverse. To review the problems discussed in the preceding sections, the causes of behavioral disorders may be categorized as follows:

1. *Extremes in natural behavior at the limits of natural variation.* This is not really a "cause" but a recognition of some behavior that occurs within the limits of normal variability.
2. *Malfunctions.* These include congenital and genetic problems, disease, parasites, and injury, all of which may damage neural, endocrine, and muscular or morphological characteristics and functioning. Genetic problems may result from mutations, from chromosomal aberrations, or, more commonly, from intense selective breeding or inbreeding.
3. *Early experience and developmental and learning problems.* Animals are particularly sensitive at narrow critical periods in some species to exposure to wrong objects (imprinting problems), trauma, lack of peer or parental relationships, and inadequate exposure to novel situations.
4. *Conflict, uncertainty, and "stress."* These problems, operating primarily through the sympathetic nervous system and the hypothalamic-adrenocortical interactions, may cause "nervous breakdown," decreased immune responses, and a variety of neuroses, psychoses, and psychosomatic effects.
5. *Abnormal stimuli or environmental surroundings.* These may lead to "supernormal stimuli," as in birds incubating the wrong eggs or feeding something that is not their offspring. Abnormal environments may lead to excessive stimulation and conflict-stress or may lack necessary stimuli.

Most, although certainly not all, of the abnormal behaviors involve human interactions in one way or another. Many abnormalities undoubtedly occur in nature, but the individuals probably do not survive to pass them on to offspring. If they do continue to exist, they may serve as the foundation for further evolution, and when they are observed by humans, we include them in the normal repertoire of the species. Selection imposed by domestication may reduce the impact of the associations with humans, but at the same time, domestication may expose animals, especially pets, to new human-related problems.

PREVENTING, CORRECTING, AND CURING ABNORMAL BEHAVIOR

What can be done to prevent, correct, and cure behavioral problems in animals? The first, most obvious, and most general statement that can be made is to get to know the normal behavior of the species in question. Increased familiarity with the animals not only allows one to recognize the abnormal in the first place, but it also generally provides one with more insight and understanding of the species' surroundings and needs. Beyond the general recommendation to spend time with the animals, steps to prevent, correct, or cure abnormal behavior depend on the particular species and problems with which one is working. This leads into the subject of applied behavior and the next chapter.

SUMMARY

Abnormal behavior is difficult to define precisely but generally refers to behavior that is either uncommon or maladaptive or both. Such behavior is observed most commonly in human-associated situations, perhaps because abnormal animals do not survive long in nature or because what is seen in nature simply is defined as normal. There is much variability in behavior among different species, and what is abnormal for one may be normal for another.

What abnormal behavior is seen in the wild can be accounted for usually as extremes in the distribution of natural variability or the results of disease, injury, or congenital problems.

In zoos and research facilities abnormal behavior is seen in attempts to escape or hide, refusal to eat, stereotyped motor patterns often resulting from insufficient space or environmental requirements, abnormal conditions of development and social surroundings, displacement and redirected behavior, and a number of problems that may lead to self-mutilation.

Domestic livestock and pets commonly have been selected, deliberately or otherwise, through many generations to live under the usual human-associated conditions. There is a distinction between domestication and tameness, or ease of handling and response to confinement. Some domestic animals may not be tame and vice versa. Domestic animals occasionally show abnormal behavior related to some of the same causes as in zoos and research but, in addition, suffer a host of other problems usually associated with ignorance or misunderstanding of animal behavior.

The best solutions for preventing abnormal behavior in human surroundings are to increase the understanding of the normal biology and behavior of the particular species and to better provide for those requirements.

Recommended Additional Reading

Fox, M.W. 1968. Abnormal behavior in animals. W.B. Saunders Co., Philadelphia.

Sluckin, W. 1979. Fear in animals and man. Van Nostrand Reinhold Co., New York.

CHAPTER 21

James W. Grier

In the waiting room of a veterinary office.

APPLIED
BEHAVIOR

Humans probably have put an understanding of animal behavior to use throughout much of their own evolutionary history. Aside from needing to avoid the larger, dangerous other predators, humans in hunting, fishing, and trapping have needed to know something about animals in order to ambush, trap, and trick them. As a fur trapper once said, "To trap a mink you have to know where it has been before it gets there."

It is unlikely that man's understanding of animal behavior has always been as conscious or deliberate as it is today, although our view now may be largely a contemporary chauvinism. The principles in the relationship may still be similar; they may have just been refined and become more sophisticated with modern technology for studying, measuring, and analyzing behavior. In a sense, much of the application of animal behavior has been (and continues to be) somewhat symbiotic, and at times it is difficult to determine who is training whom. In a Skinner-type interpretation, humans have learned how to give animals their rewards so that we can get ours. Or as the cartoon of two rats in a Skinner box goes, "Boy, do I have this guy trained; every time I hit this bar, he gives me a food pellet." Furthermore, at least with domestication, both proximate and ultimate factors may be involved. Humans selected for care and breeding those animals which provided the most milk, research data, or other rewards with the fewest bites and scratches.

The purposes of this chapter are to discuss a subject, applied behavior, of interest in its own right, to partially synthesize and catalog various topics within the subject, to provide useful information for persons who may be headed toward careers in zoos, research, and wildlife, and to stimulate wider thinking on alternate solutions to problems involving animal ecology and behavior. With recent ecological and aesthetic problems resulting from the use of various pesticide chemicals, for example, alternate behavioral and nonlethal solutions are much preferred by many persons. Applied behavior is receiving increased interest as a subject in its own right, including at least one professional journal *Applied Animal Ethology*.

A useful knowledge of behavior for any given species, regardless of the use to which it may be put, generally requires much time for a person to acquire. Such knowledge may be obtained by reading, by deliberate observation of the species, or by more subconscious absorption from spending much time around the animals. Because of the many behavioral differences between different species and the amount of time needed to become familiar with them, most persons can acquire a useful, subtle understanding of only a few species or groups of related species during their lifetime. Persons who may understand some animals very well may have almost no understanding of or ability to deal with other species even distantly related. Persons who understand waterfowl, for example, may be almost ignorant of the behavior of birds of prey and vice versa. This problem and the movement of personnel from working with one species to another, often for administrative purposes only, have frequently weakened the quality of zoos, wildlife agencies, and other programs. In all cases there is one cardinal dictum pertaining to the successful application of animal behavior: *become as familiar as possible with all aspects of the animal.*

Modern applications of understanding of animal behavior fall into several categories. The remainder of this chapter systematically considers these categories and also demonstrates some of human ingenuity in dealing with other species and their behavior. The major categories of applied behavior are (1) management of animals in captivity, (2) management of animals in the wild, (3) improvement of research on animals, and (4) animal training.

MANAGEMENT OF ANIMALS IN CAPTIVITY

General Considerations

Regardless of the purpose of captivity or species involved, there are several general concerns that must be addressed. Following is a discussion of these concerns, and together they may be viewed as something of a checklist.

Legal Ramifications and Requirements
There now are many legal requirements for the capture or other means of obtaining particular species, their transport, possession, and care. There may be permits or licenses and various forms of inspection that are needed. Minimum standards, such as in cage size and construction, frequently are involved. In addition, there may be local or regional zoning laws that prohibit certain species from being held in certain neighborhoods or locations. All of these legal and regulatory matters must be addressed before considering any of the following.

Housing Facilities
Except for a few species of domestic animals (some people object even there), very few animals can or should be merely placed in a simple cage, room, or form of confinement. Not only size but shape, construction materials, location, and associated furniture and hiding places, if necessary, must be considered. Bigger is not always better, nor must the surroundings always be necessarily complex. House cats, for example, for reasons not completely understood, seem quite content to live in cages if provided with cardboard boxes to sit in, an appropriate, spacious view before them, a few toys with which to play, and, depending on the degree to which the individual has been socialized previously, someone or something to give them occasional attention.

Specialized Handling and Moving Techniques and Equipment
Not all animals can be easily handled or picked up. Many are dangerous, too active or excitable, or simply too large and heavy.

Nutrition and Feeding
A few species have generalized diets and can be fed broad carnivorous or herbivorous diets. But many have very specialized types of food or places, times, or ways of eating that require special attention. Hognose snakes, in spite of being

otherwise interesting, fairly tame, and seemingly suitable for captivity, require a specialized diet of toads or, for most individuals, a long and difficult period of training to shift them to another form of food.

Behavioral Welfare ("Psychological Concerns")

Almost every species has behavioral requirements that need to be met. Often what is a concern for one species is irrelevant to another. Although there are disagreements and differences in philosophy on this point, there are probably few species that need "freedom" per se, that is, cannot tolerate captivity at all, even if all other needs are given attention. Freedom is mostly an abstract and relative human notion that probably is not meaningful to other animals because they are more tuned to proximate concerns such as food, activity, and reproduction. But there clearly are many psychological needs that must be met for most animals before they appear well adjusted and normal in captivity. This is perhaps one of the biggest challenges to keeping animals in captivity and the point over which most humans are likely to disagree, even to the point of going to court or the legislature.

Many species are social, for example, and require either others of their own species or, if raised in certain ways, human surrogates. This is one of the most common problems with house dogs that are owned by people who are rarely at home or who relegate the dog to solitary confinement in a kennel. It can lead to the behavioral syndrome known as *kennelitis* (Fox 1968). At the same time there are a suitable number and a particular complement of individual animals that may be placed together. Not all combinations work. In other examples cats need objects on which to sharpen their claws; armadillos and badgers need a substrate in which they can dig; and most rodents need things on which to chew.

Prevention and Correction of Disease and Injury

There are general precautions and concerns related to animal health that pertain to almost all species. In addition, most species have their own peculiarities that must be known if they are to be kept successfully. There may be, for example, certain diseases or parasites to which they are particularly susceptible. Or there may be particular environmental conditions that predispose them to problems.

Special Problems

Many species have unique needs or weaknesses of which anyone working with them should be aware. Extended periods of inactivity on certain substrates, for example, may lead to special foot or leg problems. Social incompatibilities may be expected under certain restricted circumstances, such as at particular seasons.

References and Specialists

To follow the cardinal dictum of getting to know the animal, there are two important steps: (1) to observe the animal itself and (2) to consult knowledgeable sources. The best sources of information are the professional, specialized literature and specialists in the field. Although some insight and hints can sometimes be obtained from amateurs or from the racks at pet shops, the most useful publications are to be found on the library shelves of large institutions, including major or specialized universities, modern zoos and museums, and a few large city libraries. One of the best sources is the *International Zoo Year Book*, which began publication in 1960 and annually devotes a major section to the special captive management requirements of a particular group of animals, such as reptiles or ungulates. Examples of other good general, introductory references are Crandall (1964) and the Universities Federation for Animal Welfare (UFAW) handbook (1972). Other excellent handbooks are published, for example, by the Chemical Rubber Company in the United States. Young (1973) and Fowler (1978) are excellent sources for information and techniques on capture, restraint, and handling techniques for a large variety of species.

In addition to the published sources, there usually are interested specialists who work with particular species or groups of animals and who may be contacted by letter or telephone. A person interested in starting to work with a species or group of animals for the first time should spend a period of apprenticeship, usually at least 3 months, working with an experienced person.

Some of the more common specialty areas, aside from domestic livestock, common pets, and the more general zoo animals, include bees, tropical fish, marine fish, marine invertebrates, pigeons, gallinaceous game birds, waterfowl, furbearers, ungulates, canids, felids, primates, and birds of prey.

Zoos

The history of the use of animals in menageries and zoos is long and fascinating. Zuckerman (1980) provides a review of the general history and case studies of 24 modern zoos and specialist collections. The term *menagerie* is derived from the French *menager*, which refers to the general management and care of a household. In the earliest use of menagerie, it meant the enclosure attached to a home where the family livestock was kept. Louis XIV, in the seventeenth century, added wild animals to the menagerie attached to his palace at Versailles and established a new custom. The word *zoo* dates from 1847, when it was derived as an abbreviation for the Zoological Gardens of London.

Although our terminology, if it can be called that, is relatively recent, the practice of keeping wild animals in captivity goes back to before recorded history. By the time of ancient Egypt certain wild animals were considered sacred; specimens were kept in temples and parks; and animals were mummified and kept in special mausoleums. The Egyptians and later the Romans held animals at first in high esteem and used them in triumphal processions. With this came the training of camels, elephants, and even lions and such species as the oryx to pull chariots. Collections were set up in Mesopotamia, India, and China. As early as 1100 BC there was a collection in China known as the Garden of Intelligence. The taming and training of animals were developed to a high degree, to the point of teaching baboons to play musical instruments and pick different characters of the alphabet on cue. More and more animals were collected and paraded before the public.

The Romans, however, then proceeded over a few hundred years to transform their interests in wild animals into, as Zuckerman describes it, "a depraved and brutal cult." Professional schools were established for training gladiators, and the public slaughter of wild animals and men became national sport. For centuries people crowded into the stadiums, amphitheaters, and coliseums to watch the butchery—often during shows that went on for months on end. Trajan is recorded as holding games for 4 months, during which 10,000 men and 11,000 wild animals were killed. Nero held one event with a company of horsemen against 400 bears and 300 lions. In some cases the coliseums were flooded, and gladiators in boats fought hippopotamuses, seals, and crocodiles.

All of this obviously required the keeping of large numbers of animals in captivity. Augustus, during a 15-year period, kept 3500 wild beasts. From 79 to 81 AD Titus was said to keep and "use up" 5000 animals.

A temporary stop was put to all of this in 325 by Constantine, the first Roman emperor to become a Christian. But the gladitorial and animal combats were resumed in the sixth century by Justinian, an otherwise scholarly and serious Roman emperor, after he took a mistress, then wife, named Theodora. She was the daughter of a bear keeper in Constantinople and, according to the reports, was rather wild herself. When gladitorial and animal fighting resumed as a sport, the only major difference from the earlier periods was that it became illegal for priests and bishops to watch. These animal contests were held at least into the twelfth century.

Relics of this violent past in today's culture can still be seen in bullfights, cockfights, dogfighting, circus lion-tamers and ringmasters with their whips, and the general machismo of men who can wrestle or otherwise associate with ferocious beasts.

The next phase in the evolution of zoos, although they were not called that then, began more or less when organized gladiatorial games came to an end. It became fashionable for kings and monarchs to have, give, and be accompanied by collections of wild animals. The tradition in the royal collection in England can be traced back to at least 1235 AD. This collection remained until the Zoological Gardens were opened in 1828.

Elsewhere there were royal zoos throughout Europe, including in France, Prague, and Vienna, and there are records of somewhat similar zoos even in Mexico before the arrival of Columbus. Vestiges of the use of animals in performances continued and grew in popularity with the growth of circuses, many of which traveled.

Next came the holding of animals for science and education. Earlier cases of the use of collections of captive wild animals for educational purposes included the Chinese Garden of Intelligence mentioned earlier and the zoo of Ptolemy II in Alexandria around 300 BC. But the use of animals for other than recreation or pomp and power for the owner did not really begin until the Versailles menagerie in 1665. One of the early uses of collections of wild exotic animals was for obtaining fresh specimens (in the absence of refrigeration and preservatives) for anatomical studies, although such studies did not really become important from a zoo standpoint (most of the anatomical specimens were obtained from local, wild species) until about the mid-eighteenth century. Zoos then achieved a new prominence for use in studies of taxonomy, anatomy, and physiology. The Zoological Society of London was granted a royal charter in 1829 as a scientific society, with its associated zoo—this zoo to be scientific from the start—for "the advancement of Zoology and Animal Physiology and the introduction of new and curious subjects of the Animal Kingdom." Such uses of zoos then spread around the world.

Following World War I came a period of decay for zoos. Zoological and biological research for the most part were being undertaken in new settings. The sub-

Figure 21-1 Modern zoological facilities designed for multiple educational, scientific, and recreational purposes: a beaver exhibit at the Minnesota Zoological Garden. A, Beaver cutting aspen trees (which are placed in the exhibit daily by keepers). B, A beaver working on a dam with flowing water in the exhibit. C, Above and below water viewing of the beaver pond. D, Class studying the exhibit. In addition to these viewing facilities, the exhibit includes closed-circuit television from inside the beaver den for public and scientific viewing and has included blinds for extended scientific observation of the animals. The exhibit also is inhabited by other normal members of the pond community including fish, turtles, and waterfowl. Photos courtesy Tom Cajacob, Minnesota Zoological Garden.

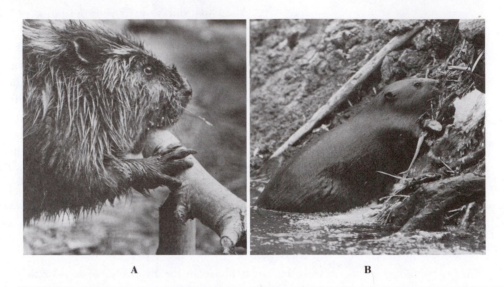

A   B

jects of previous research at zoos, such as comparative anatomy and taxonomy, were considered to be less glamorous, and there was a constant clash between the use of animals for public display and for research. There also were serious economic pressures, and social attention was directed to many other concerns. Zoos, for the most part, slid from what limelight they had and suffered from lack of attention and funding.

After World War II, however, the plight of zoos began to reverse, and the latest phase, one of commerce and conservation, began. Public interest started to increase for new zoos and displays. This increased demand and pressure for collecting opened a new world of animal commerce, and the harvest began to stress the populations of some species. The rarer the animal, the more it was desired. This was accompanied by many other environmental problems, including severe loss of habitat for some species. At the same time there was a rapidly growing public awareness and shift in attitudes in favor of animal welfare, both in the wild and in captivity. Combined with new knowledge about the animals, particularly in the field of animal behavior, and the building (or rebuilding, as in Europe after the destruction by bombings) of new facilities, zoos stepped into a largely new position in two important respects: (1) for the conservation of dwindling populations of wild species—in some instances as the only place until they could be reintroduced into the wild—and (2) for the study of behavior.

Behavior became important both for the design of facilities and management and as a subject for further scientific study. In many cases modern zoos provide the best circumstances for studying animal behavior.

The state of the art in philosophy, planning and design, layout, and operation

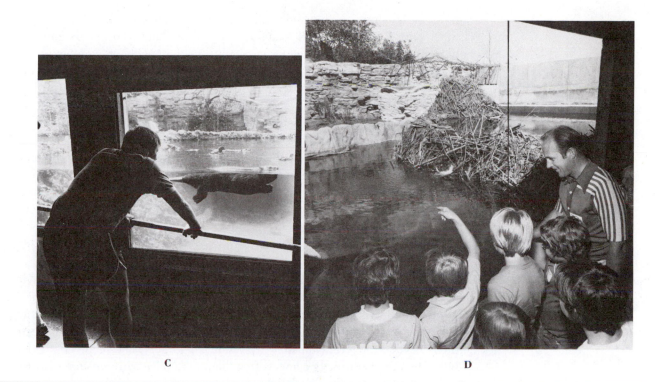

C   D

of zoos today has been elevated to a high level of sophistication (Figure 21-1). Note that the diverse functions of public display and recreation plus education and scientific study, including the opportunity for much behavioral research, have been brought back together in modern zoos and are smoothly integrated at a level not formerly known.

Livestock and Research Animals

The management of livestock and research animals in captivity differs from that of wild animals in two important respects. First, the animal species are generally domestic and hence better adapted for the captive conditions. They have fewer specialized needs and are easier to work with in the first place. Second, there are many more people involved with these activities and these species, making it much easier to learn about the special requirements and particular behaviors of the animals.

For livestock animals the number of qualified people and good literature is vast, although there still is comparatively little deliberate attention paid to the behavior of the animals. Many universities have entire departments and curricula devoted to animal science, that is, livestock production, handling, and management. Several professional journals are available in these fields, and most libraries have entire sections of books devoted to horses, cattle, and poultry, for example. In many rural agricultural communities there are clubs, such as 4-H, where one can learn about the proper care and management of these animals.

Work with livestock often is a family tradition, and in some communities there are probably more genuine experts than nonexperienced people in working with horses, cattle, sheep, swine, and poultry. There are many people who have spent a major part of their lives, for example, with horses and who understand their behavior and all of their subtle movements just as thoroughly or even better than some specialists working with an obscure, rare primate deep in a tropical rain forest. For persons wishing to delve into the behavior and management of livestock animals, Hafez (1975) is an excellent starting point.

With the common animals used in zoological and psychological research, such as mice, rats, hamsters, rabbits, cats, dogs, and pigeons, there are far fewer people working with the animals than in agriculture and livestock. But the numbers are still very high, and one can find good, qualified persons and much literature at most universities and various other research institutions, including health and hospital laboratories. Excellent books on the subject include the UFAW handbook cited earlier.

Pets

What can be done to improve the application of knowledge about animal behavior to pets? The overall picture involves large masses of the general public with many long-term and well-established misunderstandings about animals and their behavior. Far too many veterinarians have a poor understanding of domestic animals beyond the level of anatomy, physiology, surgery, and medications (Fox 1968). Thus the most likely contacts that the public have with the community of animal professionals often do not help. Unfortunately veterinary curricula are already so tightly crammed with requirements that it is unlikely that many will squeeze in courses on animal behavior in the near future. But that would be one good route because veterinarians have the best and most numerous contacts with the general public on issues related to animal behavior.

The need for veterinarians to have formal training in behavior, along with education in anatomy, physiology, and surgery, is perhaps summarized best by Taylor (1962): "If the veterinarian is to understand the whole mechanism of domestic animals (and that is one of the things expected of him) he cannot afford to leave out the functions of the cerebrum."

In spite of occasional problems and misunderstandings, the present level of pet care in the Western world generally is quite good. In fact, some people have remarked that the average pet probably receives better care and nutrition than the average person on this planet.

The association between pets and humans may be very important for both participants. Some people need every friend they can get, and the adage that "man's best friend is a dog" may not be far from the truth. Lorenz (1952) provides a delightful essay on the behavioral relationship between man and dog. Levinson (1968, also see Holden 1981) discusses the relationship further and indicates that pets may be very significant to the mental hygiene of most persons but particularly for children, the elderly, and persons who are isolated or suffer emotional problems. Petting animals may substitute for social grooming, which has been lost in much of contemporary human culture.

As has been echoed as a common theme throughout this book, the more general biology one understands, the better one can understand behavior and manage one's pets. It is useful to know not only what should be done for the proper management of pets, but also—to avoid the mistakes—what should *not* be done and which things create problems in the behavior of the species. Particularly important, as discussed in Chapter 20, are the considerations of social environment, early experience, and lack of conflict and stress on a pet. If one is fully aware of these potential problems, one may be able to properly care for the pets or, perhaps, decide that having such a pet is not in the best interests of the animal in the first place.

Veterinary Management of Animals

The importance of veterinarians to the care and management of captive animals has been already stressed in earlier sections. Veterinarians and their assistants probably work more on a daily basis with animals in captivity than any other group of people. Applied animal behavior is their means of livelihood.

The arguments for increased understanding of animal behavior have been given by Taylor (1962) and Fox (1967). For small animals pet-practice veterinarians should know the best methods for raising and training dogs and cats. This includes an awareness of critical periods, weaning, housebreaking, and various training that may be associated with different breeds.

Veterinarians should be able to recognize signs of abnormal behavior and disturbances and then be able to recommend the appropriate treatment, whether via drugs, deconditioning, or retraining. Whereas many pets, often valuable animals, are simply disposed of because of behavioral problems, it may be possible to save these pets through proper treatment. This is much like animals with broken legs that used to be simply killed. Behavior not only is important for signs of behavioral problems per se, such as from improper rearing, training, or social interactions, but the symptoms of many diseases, parasites, and injuries are also behavioral. Because many of the problems arise from the behavior of the pet *owner*, it is as important for the veterinarian to understand human psychology as it is to understand animal behavior.

By understanding the finer subtleties of animal behavior, veterinarians can also establish better conditions within their clinics and in the manners in which they handle and care for animals. It may be possible, for example, to avert some of the problems associated with the separation of pets and their owners during treatment and boarding. Trauma, fear, pain, and suffering of the animal for the veterinary situation may be reduced. Knowledge of clinical psychology can suggest changes, often simple, that greatly aid and speed recovery from surgery or medical treatment.

Finally, a knowledge of animal behavior may greatly improve advice on initially preventing problems. Some pet owners may have excessively aggressive and dangerous animals that are a threat not only to themselves but to others. Such animals often are disposed of, only to be replaced by another that becomes equally a problem. The veterinarian who recognizes such cases can recommend better rearing and training techniques or, perhaps, a different breed that is more docile and less susceptible to developing the problem.

In large animal practice much of the same can be said as just discussed; plus the understanding of behavior may facilitate capture, handling, and transport of the animals. Many large mammals, for example, often refuse to cooperate because of problems with strange odors, of which the handler (being human and relying more on vision) is not aware. Cattle that otherwise refuse to enter a loading chute or truck may readily move in if some of their familiar bedding, another animal with which they are familiar, or another familiar-smelling object is placed ahead of it.

Many of the problems in large animal practice, particularly involving reproduction (e.g., artificial insemination), arise from improper surroundings, including the social situation, or previous trauma. New stock often must be introduced to a herd or flock for replacement or for genetic considerations. But disruption of the established social hierarchy can create serious, even fatal, effects. The problem is very likely with the introduction of older males. New males usually are introduced best as youngsters or where no other males exist—the others perhaps having been removed. The elimination or reduction of these problems may require an understanding of behavior and the ability to analyze the particular situation.

MANAGEMENT OF NONDOMESTIC ANIMALS

Managing Pests and Nuisance Animals

The common human approach to problems caused by "pest" or "nuisance" animals, usually animals that eat or damage things we want for ourselves, is to simply try to destroy or eliminate them. If possible the animals usually are killed. Reducing or eliminating a population by killing or harvesting works in some cases but not with many pest species. Killing of animals or trapping and transporting them away from the area may be effective with the larger birds and mammals, but it is notoriously ineffective for smaller mammals and many species of insects. Trying to eliminate some species is about as futile as trying to dip all the water out of the ocean with a bucket; everytime you remove a bucketful, more water moves in to take its place.

An old, institutionalized folly along these lines, now largely abandoned, was the concept of bounties, whereby persons were paid to kill the animals. Payment was usually made on presentation of evidence of the kill, such as for the legs, ears, skin, or whole carcass. Several studies (reviewed in Palm et al. 1970), however, showed bounty systems to be *ineffective*. In Michigan, for example, after many years of bounty and the expenditure of several hundred thousand dollars, as many or more wolves, coyotes, and foxes were being brought in as ever. The system is often *fraudulent*. Animals may be imported for collection of bounty. Coyote pelts, for example, were bought in the Dakotas for $3 each and "sold" for bounty in Michigan for $35 each; one such pair of traffickers submitted 248 fraudulent pelts in a 9-month period. It is *wasteful*. Many more animals may be killed than if other approaches were used; furthermore, some techniques, such as trapping, often lead to the deaths of nontarget species. It is *costly*. Payments for bounties have run into the millions of dollars per year and in many cases have greatly exceeded the cost of losses if there had been no attempts to control in the first place.

In addition to not being effective in many cases, lethal techniques of control create other problems. Many persons object to the methods as being cruel and inhumane. Different people may disagree on the value of a species; some persons regard it as desirable, whereas others consider it highly undesirable. Furthermore, there often are negative side effects, particularly with chemical pesticides that may kill or affect other species, including in some cases humans.

Thus the notion that humans can *control* certain species, in the sense of eliminating them, usually is a myth. Instead, one may have to view it as making the best of the situation and search for the most acceptable compromise. This is best referred to as *pest management*. It involves using many different approaches to the problem and finding the best combination of different techniques at different times and places to achieve the smallest population of the problem species or otherwise reducing its impact on our interests. A knowledge of the behavior of the species generally is extremely useful in this endeavor.

Examples of the application of knowledge about animal behavior to pest management are becoming more numerous every day. A classic case involved the reduction of screwworm populations in the southwestern portions of the United States. The screwworm is the larva of a fly *(Cochliomyia hominivorax)*. Adults lay their eggs on cattle, and the maggots burrow beneath the skin, causing a reduction in health and productivity and damaging the skin for later use as leather. Behavioral research, however, showed that the females of the species mate only once. So large numbers of males were raised in the laboratory, sterilized with radiation, then spread through the affected regions by being dropped in cardboard boxes from airplanes. These and other sterile male techniques, including the search for genetic strains that lead to subsequent sterility, have now been explored for a number of pest species.

Another general technique in insect pest management involves the use of pheromones to attract the species. The attractants are used either to bring the pests in where they can be killed or to simply permit the population to be sampled and assessed for density and estimates of economic thresholds, the point at which benefit of control exceeds the cost.

Deer frequently become a nuisance around highways, where the glare of oncoming lights disorients them and they dash into the road, causing collisions. In some areas small mirrors have been placed on posts along the road to deflect the light of approaching vehicles while they are still at a distance, hence, alerting the deer well in advance.

Coyotes have posed an ongoing problem among livestock producers. And in spite of many and varied efforts to control the species, populations of coyotes appear not only to be holding their own but are actually increasing. Present populations of coyotes appear to be well adapted to living with humans. Attempts have been made to find solutions to the coyote-livestock conflict that do not involve poisons or other lethal methods. One approach that has been suggested but not yet accepted by all involved parties uses taste aversion learning (Figure 21-2). If coyotes can be made sick in association with the normal tastes or other sensory cues of the livestock, such as sheep or chickens, then the coyotes might not bother the livestock, much in the same fashion that birds leave monarch butterflies and other animals leave toads alone. The first chemical proposed for use in such applications was lithium chloride. This substance, however, may form aversions to salt taste rather than the taste of the prey; there have been misunderstandings over dosages and experimental protocols and results; and many persons remain unconvinced of its potential. For further discussion see Gustavson et al. (1979) and Burns (1983).

Figure 21-2 The use of taste aversion in attempts to control coyote predation. The coyote in this picture, which was photographed during one of the original experiments, is eating a lamb it has just killed. Subsequently it was fed lamb meat containing lithium chloride, which made it sick. The coyote then refused to kill sheep in future tests. Photo courtesy Carl R. Gustavson.

Other proposals, almost all behavioral, for alleviating the coyote-livestock conflict involve such techniques as using Komondor guard dogs (Linhart et al. 1979) or toxic collars on the sheep (Savarie and Sterner 1979). Perhaps one of the most interesting proposals involves the addition of llamas to flocks of sheep. The llamas are exceptionally curious and, upon the approach of a coyote, go rushing over to investigate. This frightens the coyote, which then goes off to find its meal elsewhere.

Other examples of behavioral management of pest and nuisance species are numerous. It can only be reiterated that each situation usually involves unique behaviors and solutions. This calls for both understanding of the behaviors involved and a good imagination. Again, the more familiar one is with the behavior of the subjects, the more likely will there be success in dealing with the problems. Information on behavior must be combined with knowledge of the overall life history and population circumstances. A behavioral solution that works where animals live densely packed together, such as vampire bats (Linhart 1975), may be totally ineffective among animals in a sparsely distributed species.

Game, Nongame, and Endangered Species Management

At the same time that humans might wish to reduce the populations of some species, there are others which are all too easily reduced and eliminated or which, for some other reason, we wish to increase in the wild. Many of the best solutions involve applications of knowledge about the behavior of the species.

Game and Nongame Management

The field of wildlife management, in the modern sense of scientific study and deliberate attempts to manage on a large scale, is slightly less than 100 years old. Much of the impetus for the field was a response to the excesses of overhunting and overtrapping during the nineteenth century. These problems led to the extinction of several species, including the passenger pigeon, and near extinction of several others, such as the American bison, beaver, and pronghorn antelope.

Although some efforts were made by the ancients and various laws and regulations have been enacted for some hundreds of years, the real establishment and development of wildlife management began during the twentieth century. Aldo Leopold generally is credited as the father of the subject, beginning with his now classic text, *Game Management* (1933). He also started an academic department at the University of Wisconsin.

Since that beginning, wildlife management has been approached through two main channels: ecology and, to a lesser and more recent extent, wildlife behavior. Most of the early attention was directed at defining which species were most suitable for sport killing (the so-called game species) and finding techniques to maintain a sustained yield. Because of the difficulties of working with such species on a large scale, the numerous environmental and biological factors involved in the populations, and the difficulty of accurately and precisely measuring populations in the first place, it rarely has been clear what real effects the management efforts have had on the species. But many of them were logical, and they seem to have worked. Some species, such as white-tailed deer, may be more numerous now than at any previous time, although some of the population growth may have been merely coincidental with effects to manage the species. Regardless of its effects, wildlife management has become a large discipline: it has generated a large body of literature pertaining to wild animals, and there are now agencies at most levels of government.

Changes in the field of wildlife management during the most recent decades, in addition to an increasing awareness of the importance of animal behavior, include increased public awareness and interest toward other nongame species of animals, thus shifting the center of gravity in the subject slightly away from hunting, fishing, and trapping. Another change is an increased realization within the field that much, if not most, of the effort in wildlife management must be directed toward people management rather than just managing the animals themselves. That is, most of the problems one encounters with wildlife populations result from human activities, and one must treat the basic, human causes and not just the symptoms in the wildlife.

Thus the biological side of the problem has slipped slightly into the background as more and more effort is directed to the human-related aspects. But within the biological context, the behavior of the animals is seen as becoming more and more important in dealing with the species. There currently is no single reference devoted solely to the principles of applied behavior in wildlife management, but there is a growing body of behavioral information that could be applied to many species, and a general synthesis may be forthcoming.

A good example involves the Uganda kob (Buechner 1974) (Figure 21-3). The management is based on several years of intensive study of a large number of animals, many of which were individually marked. The techniques allowed a careful analysis of both the structured population dynamics and the animals' behavior. The hunting harvest is based on an understanding of the social behavior.

The managed population is in the Toro Game Reserve in the Semliki Plains of western Uganda. It consists of about 18,000 kob antelope, which are socially organized into 13 subunits of 1000 to 1500 animals each. The gene flow between subunits is believed low; whatever exists is probably being carried by wandering

Figure 21-3 Male Uganda kob *(Adenota kob)* **on its territory in a courtship arena (lek) with several displaying males.** Photo by Leonard Lee Rue III.

young males. Within each subunit there are several categories of social groupings. The orientation of the subunit is around more or less historical territorial breeding grounds.

The territorial grounds, or leks, consist of a cluster of 30 to 40 small male territories. Total size of the lek is only around 200 m in diameter. The lek is where the breeding action takes place with probably fewer than 50 males breeding most of the 500 to 600 adult females. The system works well with nearly 100% of the females being bred.

Spread between the leks are larger, spaced-out single-male territories. These single territories are considered important for maintaining the spacing and social stability of the herd. Although the males holding the single territories are able to successfully copulate only rarely, these single territories are preferred by females for their daily activities. There they are less harassed by young males, and they may be somewhat protected from predation. The bulk of the males are to be found in an all-male, bachelor herd. Turnover on the lek is rapid, and the active males that hold lek territories lose them and move to the bachelor herd. There they recuperate before returning once again to the lek. In addition to the social, age, and sex structure of the population, the species' needs for food and protection from natural predators have been considered in the management of hunting in this system. Indiscriminate shooting of the allowable number of animals was not permitted. Both the males holding lek territories and the integrity of the lek itself were considered of prime importance and, therefore, not to be disturbed. The lek males were considered important for the genetic composition of the population.

The males on single territories were likewise considered important for spacing as "integral components of the structural fabric" of the kob's system. Females, because nearly all were being bred and thus none could really be considered as surplus, were regarded as very important for natural recruitment in the population. Unlike cases of overpopulation with white-tailed deer or elk, it did not seem likely that the shooting of females could stimulate higher rates of reproduction.

This left the all-male bachelor herd as the place for hunting. Although the bachelor herd contained some of the lek males that were recuperating and younger males that could be future breeders, the exploitation of the bachelor herd would still provide the most random losses from the gene pool and the least impact on the genetics of the population. Hence, hunting was permitted only for subadult males in the bachelor herds, with numbers to be spread over all the subunits.

The herd was monitored from 1963 to 1969, during which time the cropping consisted largely of the subadult males taken out of the bachelor herds, a small number of trophy adults taken out of the bachelor herds, and a small percent of females. In addition to the legal take, there was a fair amount of poaching, much of which was fawns that could be approached within spearing distance, and there were natural losses to predators, disease, and injury. As well as could be determined, the population remained stationary during the 6-year period. The regulated, managed exploitation by humans did not appear to impose any serious adverse effects.

Endangered Species Management

Endangered species management, a relatively recent development within the wildlife profession, has tended to give more consideration to behavioral aspects of the problems and their solutions than in most other wildlife management. Endangered species are those believed to be in danger of becoming extinct within the near future, although it is very much of an economic-political problem as well, and we frequently do not know the real size or vulnerability of populations. Working with endangered species often involves much effort, cost, and logistical problems; the work generally is a very labor-intensive field of endeavor for the number of animals involved. This type of work with birds has been dubbed *clinical ornithology*.

A conference on the management of endangered birds was held at Madison, Wisconsin, in 1977 to bring workers in the field together (Temple 1977). Subsequently there have been a number of meetings to discuss management of endangered birds and other groups. Two examples using birds have been chosen to illustrate applied behavior in endangered species management.

The first involves problems between two species. Puerto Rican parrots *(Amazona vittata)* have become endangered, apparently as a result of increased nest predation and competition. Intensive study by Snyder and Taapken (1977) showed that the most serious competitor was another hole-nesting bird, the pearly-eyed thrasher *(Margarops fuscatus)*. These thrashers, which have shown rapidly increased populations during the past several decades, enter the parrots' nests and destroy the eggs and small chicks. Parrot nests were guarded directly by observers who frightened or shot the thrashers, but this required much manpower; some thrashers entered unnoticed; and there was an endless supply of thrashers—26 being shot at one site. Another technique involved removing the parrot eggs for artificial incubation and replacing them with plaster eggs—which the thrashers proceeded to fill with dents.

The solution came, however, after 3 years of experiments. It was found that the parrots and thrashers have slightly different preferences for nesting cavities. The thrashers prefer shallower holes, around 90 cm deep, whereas the parrots will accept deeper holes. The solution further used the thrashers' own social behavior. Preferred thrasher nest boxes were erected near parrot nest sites. This provided a suitable location for a pair of thrashers, which then did not bother the parrots. The intraspecific territorial behavior of the thrashers kept other thrashers out of the vicinity and away from the parrots.

In the next example knowledge of the behavior of cranes has been used in the management and conservation of whooping cranes (Drewien and Bizeau 1977). Until recently the only remnant population of this species nested in Wood Buffalo National Park, Northwest Territories, Canada, and migrated to wintering grounds at the Aransas National Wildlife Refuge in Texas. During the 1970s an imaginative and ambitious program was initiated to introduce the species to new nesting and wintering areas so all of the species' "eggs would not be in one basket," so to speak.

It was known that cranes normally lay two eggs but only raise one chick. Because of this one egg could be taken from wild clutches with little impact on the

Figure 21-4 Whooping crane with sandhill cranes wintering at Los Lunas, New Mexico. The whooping crane is about 1¾ years old, is a male, and was raised by sandhill cranes. Photo courtesy Rod Drewien.

natural reproduction. Eggs were thus collected initially to establish a captive population at Patuxent Wildlife Research Center in Maryland. It also was known that cranes normally migrate from nesting areas as family groups, and apparently the migratory routes and wintering areas are learned by chicks migrating with their parents.

It was hypothesized that, although differences in feeding and other behaviors existed, sandhill cranes might be used as foster parents for whooping cranes. After much consideration and preliminary planning, a population of greater sandhill cranes at Grays Lake National Wildlife Refuge in Idaho was selected for the reintroduction experiments. Eggs were brought in from the Wood Buffalo wild population along with additional eggs from Patuxent, from production by the captive population, and exchanged for the sandhill eggs in pairs known to be good incubators and parents.

Some losses and problems were experienced, but overall the results were very successful. Foster sandhill parents accepted and hatched the whooping crane eggs and then reared the young (Figure 21-4). The young were captured and individually marked for further study and tracking. The sandhills accepted them back after handling. The whooping cranes followed their foster parents to the sandhill cranes' wintering grounds in New Mexico, subsequently left the sandhill families, and now appear to be joining up with other whooping cranes fostered by other sandhill cranes. The birds' behavior has been followed closely to ensure that imprinting problems would not lead to whooping cranes attempting to mate with sandhills when they matured. It looks as though the program is well on its way toward establishing a new population of wild whooping cranes. The cooperation of wildlife agencies and the public in the region has been excellent; there has been an active education and information program; and, as of last report, no whooping cranes are known to have been accidentally or deliberately shot by anyone.

APPLIED ANIMAL BEHAVIOR IN RESEARCH AND PHOTOGRAPHY

Many people receive full or partial employment by *studying* or *photographing* animals, rather than just for displaying them in zoos or for other performance purposes, producing and selling them, and managing their populations. They simply study or photograph them. These activities often are conducted in association with teaching, writing, or other professions. A few well-known names in one or both of these fields include, for example, Konrad Lorenz, Niko Tinbergen, George Schaller, Jacques Cousteau, Jane van Lawick-Goodall, Diane Fossey, Frederick Kent Truslow, Alan and Joan Root, and several whose photographs appear in this book. The Roots photographed such films as *The African Elephant*, *Year of the Wildebeest*, and *Mysterious Castles of Clay* (about termites). Today there are hundreds, if not thousands, of other less widely known professional animal researchers and photographers.

In virtually all cases an intimate understanding of the behavior of their subjects is essential to success. Examples of techniques in the study and photography of behavior are given in Chapter 3.

TRAINING ANIMALS

The training of animals is the last major category of applied behavior. It probably is the first category that would come to the minds of the general public when applied behavior is mentioned. The use of trained animals is familiar in circus and similar performances, as actors in semidocumentary or entertaining motion pictures and television, in work animals, and even in the home.

In more utilitarian and academic uses, trained animals have figured importantly in the field of psychology, where thousands of rats and individuals of other species are trained for study every day. Much knowledge about bird orientation and navigation has been obtained from trained pigeons. Training has been useful in experiments designed to infer sensory and other capabilities of animals.

In military usage there are many instances of trained animals, other than serving simply as beasts of burden or for riding. The use of carrier pigeons for messages is a classic illustration. Recent uses of trained animals for the military includes porpoises used to retrieve torpedoes and bombs from the ocean floor.

In livestock and zoo applications, animals can be trained to drink or eat from certain implements, follow a certain daily routine, and come or do other activities on command. There are probably few people who work with animals who do not use training to some extent.

Training is based on a deliberate or unconscious understanding of the various principles of learning, as discussed in Chapter 19. Because learning was treated in detail there, training techniques will not be discussed further here except to stress the essentials. The basic techniques of training involve repeated rewards (or occasionally punishment) for the desired behavior. The closer in time the reward follows the behavior and the more often it is repeated, the more fixed the training becomes. When training animals, one must consider the natural learning tendencies and abilities of the species. As has been discussed earlier, most primates, for example, cannot be trained to new toilet habits without extensive effort. Many breeds of domestic animals have been selected for enhanced trainability for certain characteristics, such as dogs that retrieve objects, herd sheep, or lead blind people.

SUMMARY

Applied behavior, as a separate field of endeavor, has only recently been recognized as a topic in its own right, although the basic principles may go back for thousands of years in human understanding. Understanding of animal behavior can be applied in numerous areas including zoo practice, livestock and research animal management, veterinary practice, care and enjoyment of pets, research and photography of wild animals, casual enjoyment of wild animals, pest and nuisance control, management of game, nongame, and endangered species, and various training of animals. Examples are provided in these areas.

In addition to the specific considerations for different species and different topics, there are a number of general considerations, particularly concerning captive animals. Such considerations include legal aspects, disturbance, housing facilities, specialized handling techniques and equipment, nutrition, safety, and techniques to increase knowledge about the species. The central dictum of applied behavior, as in most areas of behavior, is that one must understand the species. Because of differences among species, knowledge about the behavior of particular species cannot always be transferred to other species. Most persons in a lifetime can become familiar with only a few species at most.

Recommended Additional Reading

Crandall, L.S. 1964. Management of wild animals in captivity. University of Chicago Press, Chicago.

Fowler, M.E. 1978. Restraint and handling of wild and domestic animals. Iowa State University Press, Ames, Iowa.

Fox, M.W., editor. 1968. Abnormal behavior in animals, W.B. Saunders Co., Philadelphia.

Geist, V., and F. Walther, editors. 1974. Behaviour of ungulates and its relation to management, vols. 1 and 2. International Union for Conservation of Nature (IUCN), Morges, Switzerland.

Hafez, E.S.E., editor. 1975. Behavior of domestic animals. The Williams & Wilkins Co., Baltimore.

Palm, C.E., et al. 1970. Vertebrate pests: problems and control. National Academy of Sciences, Washington, D.C.

Temple, S.A., editor. 1977. Endangered birds, management techniques for preserving threatened species. University of Wisconsin Press, Madison, Wis.

Universities Federation for Animal Welfare. 1972. The UFAW handbook on the care and management of laboratory animals, Churchill Livingstone, Edinburgh.

Young, E., editor. 1973. The capture and care of wild animals. South Africa Nature Foundation/World Wildlife Fund, Ralph Curtis Books, Hollywood, Fla.

Zuckerman, L., editor. 1980. Great zoos of the world. Westview Press, Inc., Boulder, Colo.

APPENDIX A

SUGGESTIONS FOR OBSERVING LIVE ANIMALS

Animals of other species surround humans almost everywhere, even in large cities, and nearly everyone sees them whether intentionally looking or not. Meaningful behavioral observation of animals may not occur, however, without experience or suggestions on how to start. This appendix provides suggestions for watching animals to help engage the observer in seeing things that otherwise might be overlooked. The various projects suggested in this appendix are quite simple and should be viewed only as starters or as bases for course projects during an introductory quarter or semester. Before live animals are used in behavior projects, experiments, and other laboratory or field experiences, however, a number of preliminary concerns need to be addressed.

CAUTION BEFORE OBSERVING ANIMALS

Legal, Ethical, Safety, and Humane Considerations

This currently is a complex world of ethics, rules, regulations, and laws pertaining to live animals, both wild and captive. Furthermore, many animals can be outright dangerous, physically or ecologically, and pose safety problems to persons involved in the project, to innocent bystanders in the vicinity, or to the local environment (such as when exotic species escape). For the welfare and protection of all concerned—instructors, students, property owners, neighbors, and the animals themselves—both instructors and students must be alert to and familiar with the possible legal, ethical, and safety ramifications of *whatever* is being done with the animals, and instructors should carefully supervise projects and clearly identify who is responsible in case of errors, problems, accidents, or violations. All persons working on independent exercises outside formal organized laboratory work or fieldwork are responsible for conducting everything properly, including having permits or licenses when necessary and obtaining permission to enter and work on private property. If in doubt, seek advice. Respect the rights of landowners and also be considerate of the animals themselves. For further information along these lines, see the Animal Behavior Society Animal Care Committee (1981) guidelines.

Choice of Species: the August Krogh Principle

When choosing the species with which to work and circumstances of the exercise, it is important to recognize differences between species and situations. Some species and situations are easier to work with or more appropriate for a given problem (the so-called *August Krogh principle*) (Krebs 1975). Given the limited time during a course, however, some persons may not discover the problems until it is too late to change. Thus one should choose reliable species and circumstances to the best possible extent and avoid unusual, exotic, and unknown situations.

Time and Scheduling Problems

A combination of normal procrastination and lack of appreciation of the time involved in working with live animals, even on simple exercises, can create a real time crunch and unfinished work at the end of a course. Exercises should be organized and structured throughout the course; if not, one should be aware of the likely time problem at the onset and perhaps set intermediate deadlines.

Use of Literature

Depending on the nature of the project, one may need to consult various references. Particularly if done in the context of a course assignment, this should involve original journals, abstracts, and perhaps specialized books. An acceptable literature review should include more than books of "pet shop" quality and more than one can get simply from the card catalog of a library.

SUGGESTED EXERCISES FOR OBSERVING BEHAVIOR

The observation of animals, given the preceding precautions and logistical constraints (limitations of time, travel, equipment, and money), is limited only by one's imagination and the availability of the animals. The following suggestions represent only a few possibilities. A number of other field and laboratory workbooks (e.g., Price and Stokes 1975, *Biological Sciences Curriculum Study* 1975) provide numerous additional exercises. The published, professional literature also provides a wealth of things that can be done; one can repeat or modify original studies that have been reported in behavioral and related journals, such as those on ornithology, mammalogy, ichthyology, and entomology. Most journals report studies on at least some easily accessible organisms, and some studies use a minimum of equipment. Repeated studies will provide insight into the research process, will focus attention on the subject of interest, and may even provide some unexpected surprises (such as different results from those reported). Above all, one's own imagination should not be overlooked. With a little creativity, one can conduct all sorts of interesting and meaningful behavioral observations on live animals. Finally, one should not feel that he or she has to make work out of all of this; it is possible and recommended that one enjoy watching animals. Perhaps one wishes to watch animals only for pure enjoyment and with little or no intent of doing a specific project or obtaining "meaningful" outcomes. Speaking as one who has spent a considerable portion of my life just watching animals, in addition to doing specific projects at other times, I highly recommend watching animals for no other lofty purpose than simply watching animals. For suggestions on specific projects aside from those given earlier (i.e., repeating published studies reported in the journals, using exercises published elsewhere, or devising one's own), the following are offered.

Four-hour Quantification of Behavior

This project is the most straightforward and least likely to encounter unexpected problems. One can use any species that may be reasonably active (i.e., shows a diversity of behavior) and observe it during an active period (not during periods when it is likely to be sleeping, resting, or inactive during the entire period). Almost all animals will be inactive during some of the observation time, and one cannot always predict when they will be inactive, but one should try to choose a 4-hour block of time when the animal is likely to be most active. The animal should be acting "on its own," that is, not in response to the observer or another human being, although it may be responding to other animals of its own or other species. The animal should be in an environment that is not too simple and restricted; that is, it should be able to show a reasonable diversity of behavior. Mice that are alone in simple cages, for example, cannot do very much. Mice that are housed in more complex surroundings with other mice, exercise wheels, and a variety of structures on which to move, on the other hand, have more opportunity for expressing behavior. Single cattle, horses, or other livestock in a small, confined area also cannot be expected to do very much, whereas those in fields with others of their species may show more activities and movements. Also situations should be avoided in which the animal is likely to be doing the same thing the entire time, such as sleeping, trying to escape from confinement, or, as with many species of fish confined to an aquarium, simply moving back and forth.

Animals that might make good subjects include mice and other pets (as long as they can be watched without influencing their behavior), wild or pet birds, many domestic livestock species (a flock of bantam chickens or pigeons is particularly good for choosing one animal and following it), and some fish or other aquarium species such as crabs or crayfish. Many arthropods (insects, crustaceans, spiders, and others, including some that might be readily available as pests) also make good subjects. Fruit flies are excellent subjects if one can get close enough or have some kind of magnifying ability to see them; one or a few pairs can be set up in a small container with plants, food, etc., and the one to be followed can be identified by using a mutant, for example, one with white eyes.

Once the animals, time, and place have been chosen, spend at least three 4-hour blocks of time with it. The first period of 4 hours or longer is simply to familiarize oneself with the animal's behavior and to devise a scheme for watching and recording the behavior, perhaps by speaking into a tape recorder or devising a system of shorthand notation. It is necessary to recognize and name different movements and behavior—something that may be easy with some behaviors and very difficult with others.

One must be conscious of and alert to *all* movements the animal makes and not just those which are familiar. Watch for subtle movements and activities, such as scratching a part of the body, different postures, etc., that might otherwise be considered insignificant. If the animal makes sounds, a system will be needed for describing the different sounds and calls. If preparing a report, describe the initial observations, problems experienced, and how it was decided to identify and record the behaviors, including a key to symbols and each behavior pattern.

Next (perhaps the next day at the same time period) watch the animal and record all behavior for one continuous 4-hour period. The observations should be recorded on a form or sheet that has been devised to maintain a running tally of everything the animal does and when. This can be done with a sheet prepared beforehand that has each minute on a separate line, or a log of everything that happens may be kept; that is, every time the animal changes behavior, write down the time and behavior. Some form of shorthand will almost be necessary because most animals can do a lot of things faster than one can write them down in longhand!

After the continuous 4-hour observation is completed, conduct another session, preferably the next day during the same 4-hour period, in which the behavior is only randomly sampled. (One may, of course, watch the animal between sampling periods, but only the behavior for the predetermined sampling segments of time is *recorded*.) The way to do this is to divide the 4 hours into equal chunks of time and then randomly choose (with a random numbers table, by drawing slips out of a hat, flipping coins, or any other suitable randomizing procedure) a number of segments amounting to a total observation time of 1 hour (one fourth of the total time available). For example, one could choose 60 1-minute segments, 12 5-minute segments, or some other choice (each segment should be no longer than 5 minutes). Or one can spot sample by choosing a large number of instants (at least 360) or, for example, 10-second segments, during which one does not follow the behavior for a period of time but simply records what the animal is doing at the instant of sampling. The particular times of sampling, regardless of length of sampling segments, are to be randomly chosen *before* the observation begins.

After both the continuous and sampling 4-hour observations are finished, quantitatively compare the results of the two ways of observing; that is, from the data, figure out how much actual time the animal spent doing different things and what the proportion or percent of time spent in each activity was. If one has had statistics and wants to make comparisons statistically, this is to be encouraged. The behaviors are to be quantitated by measuring the amount of time or number of instances each behavior is shown, but the results of the two methods (continuous and sampling) may be compared by subjectively describing how one thinks they compare. Did both methods give similar percentages, or were they quite different? Use graphs and tables as appropriate.

Species Ethogram

As described in Chapter 3, determine as complete an ethogram for a species as possible, based on one's observations plus those from the literature. This does not involve sampling or quantitative measurements necessarily but, rather, simply compiling a qualitative list of the various acts that the species engages in.

Determination of Home Range or Territory

This project has the potential of being quite interesting, but it also may end up being more time consuming, difficult, and vulnerable to unexpected problems and not getting data. It may, however, offer a little more adventure. Persons wanting to do this project must know something about home range and will have to do some reading on the subject in general before getting started (see discussion and references in Chapter 7).

Subjects can be wild animals, domestic animals, or pets—as long as they are reasonably unconfined (or confined in an area larger than their natural home range), and they must be unaffected by observation of them, so they are choosing where they go on their own. Individuals may be followed by tracking, marking, or observing at a distance. It may be possible to do this project with animals in a large aquarium or terrarium if the animals do not occupy the whole area. Sufficient time and number of observation periods must be allowed to gain a reasonable picture of the space (or volume, if in water) that the animal is occupying. This will preferably involve sampling the animal's paths or whereabouts over more than a week. One should plan to spend at least 12 hours (total time—not necessarily all at once) observing or tracking the animal. This does not count time for capturing, marking, or otherwise finding the animal or setting up aquaria initially.

Conduct Markov Chain or Other Multiple-behavior Analysis

See the discussion and references in Chapter 3 (pp. 80 and 81) for a guide to performing multiple-behavior analysis.

Plot Numbers of Behaviors Against Observation Time

See the discussion and figure on pp. 69 and 70.

Prepare a List of Species Communication Behaviors

This is similar to the species ethogram but is confined to communication behavior. Compile a list and description of all communication signals that the animals use with others of their own species. Use both personal observations and literature sources. Consider that not all communication signals are visual or auditory.

Modification of Behavior

This project requires some knowledge and understanding of learning before it can be done (see Chapters 2, 18, and 19). In addition, one may need to read other sources, such as an introductory psychology or learning text. Familiar categories of learning include classical and operant conditioning, but there are also many others, including imprinting (if one can hatch some chicks or get other suitable young animals) and taste aversions.

To document the acquisition of modified behavior, one needs data before, during, and after (or at the end). For before and after this can include photographs, motion pictures, videotape, or simple written descriptions. For documenting the acquisition, one needs to keep careful records and present the data in tables, graphs of learning curves, or whatever is most appropriate in the particular situation.

This should be more than simply teaching a dog to do cute tricks. Such tricks are completely legitimate instances of behavioral modification and may indeed by used. But carefully document the trial-by-trial acquisition of the modified behavior. Include a carefully recorded description of the progress (including problems and lack of progress) that was involved.

Behavior Experiment

If one chooses this route for a project, inferential statistics and at least a minimum experimental design will be needed. Statistical tests would involve, for example, t test, ANOVA, or an appropriate nonparametric test (Conover 1980). One also should have some reasonable hypothesis about behavior—either a new one or one someone else has proposed. If one has not had a course in statistics and knows little or nothing about at least simple experimental design and statistics, guidance and instructions will be required. Otherwise one of the other projects might be preferable.

APPENDIX B

GENE FREQUENCIES IN POPULATIONS

The gene pool of a species contains a large number of genes for a great many traits. At any one locus on a chromosome there may be two or more, sometimes many, different genes or alleles in a population. There may be a dominant and a recessive allele symbolized by A and a, respectively. Any individual will have two: A and A, A and a, or a and a. But among different individuals there may also be, in addition to A's and a's, A_1's, A_2's, etc. The **gene frequency** is the frequency, expressed as a proportion, of any particular allele in the entire population.

Consider a simple population of 32 individuals, of which 8 are AA, 16 are Aa, and 8 are aa. There is a total of 64 copies of alleles at that locus in the population: 8 A + 8 A + 16 A + 16 a + 8 a + 8 a or 32 A's and 32 a's. The frequency for A is thus 32/64, 50%, or, as a proportion of the total, 0.5. The frequency of a is the same.

There are many possible ways in which the 32 individual animals might mate and reproduce. For purposes of discussion we will say that they choose mates in proportion to their availability so that 2 AA's mate with each other, 4 choose Aa mates, and 2 choose aa, etc. (Figure B-1). (Figure B-2 or a copy of it may be used as a worksheet.) Of the resulting pairs there are 1 AA × AA, 4 AA × Aa, 2 AA × aa, 1 aa × aa, 4 aa × Aa, and 4 Aa × Aa. If each pair produces 12 offspring, with the possible combinations of alleles in expected proportions, for example, each AA × Aa mating produces 6 AA young and 6 Aa young, then there will be 192 young: 48 AA, 96 Aa, and 48 aa. There will be 384 copies of the alleles: 48 A + 48 A + 96 A, etc. The frequency of A will be 192/384, or 0.5— the *same* as in the parent population, although the population sizes are much different.

The preceding example is simply an illustration of a familiar mathematical equation known as the **Hardy-Weinberg law** or **equilibrium.** The basic idea was known earlier but not published until 1908 when it was set forth independently by Hardy, a British mathematician, and Weinberg, a German physician. The *proportion* of one allele, for example, A, *at that locus in the total population* can be referred to as p, and the proportion of the other, a, is q. The ratios of A : a in the sperms and eggs will be the same as in the population, and they will combine binomially in the next generation into genotypes in the frequency of $p^2 + 2pq + q^2$. *The proportions will remain the same through subsequent generations* regardless of population size *if not subjected to chance fluctuations or external influences.*

Binomial simply means "two names," as applied here to the two alleles. Sometimes there may be more than two possible alleles available for a locus (although any one chromosome would only have one allele and an individual could only carry two at a time). If so or if one considers more than one locus (pending complications from linkage and crossover, etc.), similar multinomial formulas

Figure B-1 Accounting of gene frequencies at one locus for two alleles, A and a, in a single case involving no chance fluctuation and no natural selection, as described in the text. Circled numbers represent 32 individual animals, which pair as indicated and produce 12 offspring per pair. The genotypes of the offspring are shown in the right side of the listing. Totals for each genotype are shown at the bottom and, below them, calculation of p, the proportion of the total represented by allele A. The proportion of a, symbolized by q, could be calculated similarly or simply by subtracting p from 1.00.

♂		♀		Pairs		Number of offspring* AA	Aa	aa
① AA——AA ⑰				① AA × AA ⑰		12		
② AA AA ⑱				② AA × Aa ㉑		6	6	
③ AA AA ⑲				③ AA × Aa ㉕		6	6	
④ AA AA ⑳				④ AA × aa ㉙			12	
⑤ Aa Aa ㉑				⑤ Aa × AA ⑱		6	6	
⑥ Aa Aa ㉒				⑥ Aa × Aa ㉒		3	6	3
⑦ Aa Aa ㉓				⑦ Aa × Aa ㉖		3	6	3
⑧ Aa Aa ㉔				⑧ Aa × aa ㉚			6	6
⑨ Aa Aa ㉕				⑨ Aa × AA ⑲		6	6	
⑩ Aa Aa ㉖				⑩ Aa × Aa ㉓		3	6	3
⑪ Aa Aa ㉗				⑪ Aa × Aa ㉗		3	6	3
⑫ Aa Aa ㉘				⑫ Aa × aa ㉛			6	6
⑬ aa aa ㉙				⑬ aa × AA ⑳			12	
⑭ aa aa ㉚				⑭ aa × Aa ㉔			6	6
⑮ aa aa ㉛				⑮ aa × Aa ㉘			6	6
⑯ aa——aa ㉜				⑯ aa × aa ㉜				12
						48	96	48

*For example, Aa × Aa.

	A ♀ a	
A	AA 3	Aa 3
a	Aa 3	aa 3

♂   Σ = 12

A = 48 + 48 + 96 = 192
a = 96 + 48 + 48 = 192
                           384

p = 192/384 = 0.5

apply. The mathematics get messier, but the result is the same: the gene frequencies remain the same over successive generations. And, in general, the frequency of any particular homozygote is the square of the gene frequency. The frequency of a heterozygote is twice the product of the frequencies of the two genes contributing to it.

The Hardy-Weinberg law, as stated earlier, has two if's tied to it: "if not subjected to chance fluctuations" and "if not subjected to external influences." The problem of *chance*, or stochastic fluctuations, will be considered first. If the simple illustration earlier were a real situation, the numbers would not likely turn out as evenly as they did because of the small population, just as one is unlikely to always get 6 heads and 6 tails if a coin is tossed 12 times. Mating may not occur in exact proportion to availability. The proportions of possible combinations may not be exactly equal, and all pairs are not likely to each have exactly 12 offspring. Perhaps there were, by chance, 140 young: 46 AA, 60 Aa, and 34 aa. The frequency of A would be $(46 + 46 + 60)/280 = 0.54$. The differences that result in the gene frequencies of small populations simply as a result of chance (i.e., stochastic properties) are referred to as *genetic drift*.

In a large population, however, and most populations are larger than most persons realize, genetic drift becomes much less significant. The random fluctuations of frequencies in one direction tend to be cancelled by random fluctuations

Figure B-2 Blank worksheet in the format of Figure B-1 for exercise purposes. Differential survival and reproduction, as in natural or artificial selection, may be imposed on the parents and offspring of the previous simple example to explore the effects on p and q, the gene frequencies.

	♂	♀		Pairs		Number of offspring*		
						AA	Aa	aa
①	AA	AA ⑰	①	AA × AA ⑰				
②	AA	AA ⑱	②	AA × Aa ㉑				
③	AA	AA ⑲	③	AA × Aa ㉕				
④	AA	AA ⑳	④	AA × aa ㉙				
⑤	Aa	Aa ㉑	⑤	Aa × AA ⑱				
⑥	Aa	Aa ㉒	⑥	Aa × Aa ㉒				
⑦	Aa	Aa ㉓	⑦	Aa × Aa ㉖				
⑧	Aa	Aa ㉔	⑧	Aa × aa ㉚				
⑨	Aa	Aa ㉕	⑨	Aa × AA ⑲				
⑩	Aa	Aa ㉖	⑩	Aa × Aa ㉓				
⑪	Aa	Aa ㉗	⑪	Aa × Aa ㉗				
⑫	Aa	Aa ㉘	⑫	Aa × aa ㉛				
⑬	aa	aa ㉙	⑬	aa × AA ⑳				
⑭	aa	aa ㉚	⑭	aa × Aa ㉔				
⑮	aa	aa ㉛	⑮	aa × Aa ㉘				
⑯	aa	aa ㉜	⑯	aa × aa ㉜		—	—	—

*For example, Aa × Aa.

	A ♀ a	
A	AA	Aa
a	Aa	aa

♂   Σ =

A = + + =
a = + + = ___

p =   /   =

in the opposite direction, and the true proportions remain closer to their expected values, if random mating occurs. Thus the stochastic considerations may be more easily ignored, and the Hardy-Weinberg equation fits quite adequately.

The assumption of random mating is not always met. In human populations, for example, tall people tend to mate more often than by chance with tall people, short people with short people, etc. Nonrandom mating is said to be *assortative*. But assortative mating is believed to be relatively insignificant in nature, and random mating is usually a legitimate assumption. If nonrandom, assortative mating were more prominent, it would increase the likelihood of such things as sympatric speciation. But there is little evidence that this occurs. Persons wishing to pursue the matter further should refer to other texts, such as Mayr (1963). For all practical purposes it is safe to assume random mating and that gene frequencies should remain more or less at some equilibrium. New genes that enter a large population as a result of mutations or chromosomal aberrations can be expected to remain rare, except as described next.

The other "if," aside from chance fluctuations, that may upset the Hardy-Weinberg equilibrium involves *external influences*. These are discussed in Chapter 5. Figure B-2 may be used for imposing different levels of selection, from differential survival or reproduction, and for exercise in manipulating and understanding the numbers.

APPENDIX C

NERVOUS SYSTEM RESPONSE TIMES

The times of events in animal nervous systems are usually measured in milliseconds (one thousandth of a second). Most persons, in the course of their daily experiences, however, are not familiar with such small units of time. Few people would readily relate to the difference between 45 and 450 msec, for example, although they differ by a factor of 10. By measuring a simple response for oneself, however, one can begin to comprehend other cases more easily. The response involves seeing an event and then making a simple movement with the hand. The amount of time required for the event to be sensed by the eyes, neural impulses to be sent to the brain and processed, instructions to be sent down the spinal cord and out the arm to the muscles of the hand, and the muscles to contract can be measured easily in a variety of ways.

The easiest way to measure the response time is with a digital timer accurate to one hundreth of a second. Such timers may be found as routine equipment in many research or classroom laboratories, incorporated in other measurement equipment, and in some electronic watches, pocket calculators, and sporting event timers. The response is simply timed by having one person hit the switch or button that starts the timer; then a second person turns it off as quickly as possible. The second person's response time is displayed on the readout. The time can be converted easily to milliseconds.

Another similar method uses a paper, strip-chart recorder with an event marker for an electrode stimulator or other event item. The chart can be run at high speed with one person triggering or starting the event marker and a second person either tripping the marker a second time or stopping it in the case of a continuous marker. The response time can be measured from the recorded distances on the chart in routine time-measurement fashion for paper recorders. Similarly, one could use an oscilloscope. The gist of the method involves measuring the time it takes for one person to respond to the sight of another person's action.

A method that requires a minimum of equipment, only a "poor man's oscilloscope," uses the physics of falling objects. It is based on the old tavern bet that one cannot catch a dollar bill dropped between one's fingers before it has fallen out of reach of the closing fingers. The time required to stop a falling object once one sees it start can be measured using any stiff piece of paper. A standard computer card works fine. The dollar bill trick works (i.e., it cannot be caught by most persons) only if the bill is held so that the middle of it is between the open fingers. If it is held so the fingers are closer to the bottom edge, there is more of the bill to fall through the fingers; it takes longer; and most persons can catch it. For this exercise the paper card should be held so the open fingers are at the bottom edge, as shown in Figure C-1.

The basic technique is for one person to hold the card at the top, with the open fingers of the subject at the bottom. As soon as the subject sees the other person let go of the card, the subject stops it by closing his or her fingers. The card is marked off in centimeters or directly in times from the bottom. The distance can be converted to time from the following equations and Table 1.

$$\text{Distance} = (\tfrac{1}{2})\,(\text{acceleration due to gravity})\,(\text{time}^2)$$

Or

$$\text{Time} = \sqrt{\frac{(2)\,(\text{distance [in centimeters]})\,(\text{seconds}^2)}{980.6\ \text{cm}}}$$

The card can be dropped several times to determine the minimum distance (time) at which the subject can detect the fall and stop it. Regardless of which of these methods is used, the average minimum time required for a simple finger-closing response to a visual stimulus for an average young adult to middle-aged person after a few practice trials is usually around 130 msec.

A visual stimulus involves a relatively uncomplicated pathway for neural impulses: retina to brain to spinal cord to finger muscle. After determining the simple visual response times, one can try an auditory stimulus. The subject can

Figure C-1 Response time for catching a falling object. The marked computer card is dropped by one person and caught by another. At the start (A), the bottom edge of the card is held level with the top edge of the subject's thumb and forefinger. Then the response time can be read at the top edge of the closed fingers (B). The response time photographed here is between 135 and 143 (approximately 141) msec. Photos by James W. Grier.

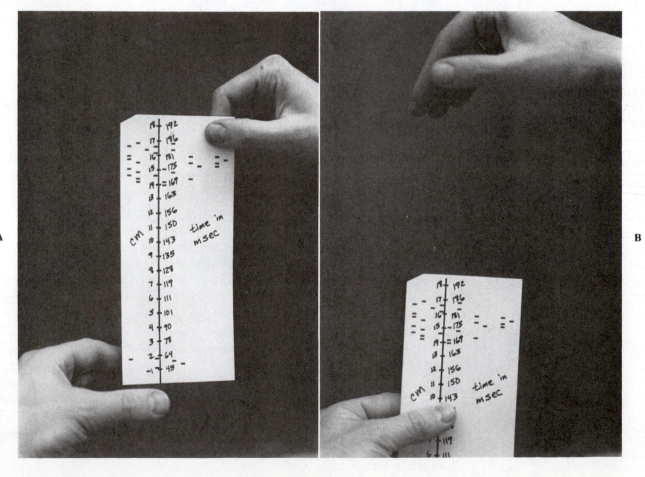

A   B

close his eyes, and the person dropping the card can give a vocal signal at precisely the moment the card is let loose.

Now to complicate matters and involve more neural processing and pathways, the person can add a signal, such as a raised finger by the card at the moment it is released, to indicate whether the card is to be stopped or if it should be allowed to fall. That is, one signal can be used to indicate it should be stopped and another to indicate it should not be stopped. Vocally one can use different signals, such as "yes" or "no." How fast can the subject respond and not make mistakes? The response time, including time to evaluate the instructions, should take much longer. In fact, it may be much more difficult to catch the card at all. With some imagination, this simple exercise can be varied in other ways to explore nervous system response times.

Table 1 Conversion of distance to time for falling objects

Distance (Centimeters)	Time (Milliseconds)	Distance (Centimeters)	Time (Milliseconds)	Distance (Centimeters)	Time (Milliseconds)
1	45	11	150	21	207
2	64	12	156	22	212
3	78	13	163	23	217
4	90	14	169	24	221
5	101	15	175	25	226
6	111	16	181	26	230
7	119	17	186	27	235
8	128	18	192	28	239
9	135	19	197	29	243
10	143	20	202	30	247

LITERATURE CITED

Abeelen, J.H.F. van. 1979. Genetic analysis of locomotor activity in immature mice from two inbred strains. Behav. Biol. 27:214-217.

Able, K.P. 1974. Environmental influences on the orientation of free-flying nocturnal bird migrants. Anim. Behav. 22:224-238.

Adams, J.L. 1974. Conceptual blockbusting. W.H. Freeman & Co., Publishers, San Francisco.

Ahlskog, J.E., and B.G. Hoebel. 1973. Overeating and obesity from damage to a noradrenergic system in the brain. Science 182:166-169.

Ahlskog, J.E., P.K. Randall, and B.G. Hoebel. 1975. Hypothalamic hyperphagia: dissociation from hyperphagia following destruction of noradrenergic neurons. Science 190:399-401.

Aideley, D.J. 1978. The physiology of excitable cells, 2nd ed. Cambridge University Press, New York.

Alcock, J. 1973. Cues used in searching for food by red-winged blackbirds (Agelaius phoeniceus). Behaviour 46:174-188.

Alcock, J. 1975. Animal behavior: an evolutionary approach. Sinauer Associates, Inc., Sunderland, Mass.

Aldis, O. 1975. Play fighting. Academic Press, Inc., New York.

Alexander, R.D. 1974. The evolution of social behavior. Ann. Rev. Ecol. Syst. 5:325-383.

Alexander, R.D., and D.W. Tinkle. 1981. Natural selection and social behavior: recent research and new theory. Chiron Press, New York.

Allee, W.C. 1943. Where angels fear to tread: a contribution from general sociology to human ethics. Science 97:517-525.

Allen, E., et al. 1975. Letter to the editor. NY Rev. Books 22:43-44.

Allen, E., et al. 1976. Sociobiology: another biological determinism. BioScience 26:182-186.

Allison, T., and H. Van Twyver. 1970. The evolution of sleep. Nat. Hist. 79:56-65.

Alloway, T.M. 1972. Retention of learning through metamorphosis in the grain beetle, Tenebrio molitor. Am. Zool. 12:471-477.

Alloway, T.M. 1973. Learning in insects except Apoidea. In W.C. Corning, J.A. Dyal, and A.O.D. Willows, editors. Invertebrate learning. Vol. 2. Plenum Press, New York.

Altmann, J. 1974. Observational study of behavior: sampling methods. Behaviour 49:227-267.

Altmann, J., and K. Sudarshan. 1975. Postnatal development of locomotion in the laboratory rat. Anim. Behav. 23:896-920.

Altmann, S.A., and J. Altmann. 1977. On the analysis of rates of behaviour. Anim. Behav. 25:364-372.

Alvarez, F., F. Braza, and A. Norzagaray. 1976. The use of the rump patch in the fallow deer (D. dama). Behaviour 56:298-308.

Anand, B.K., and J.R. Brobeck. 1951. Hypothalamic control of food intake in rats and cats. Yale J. Biol. Med. 24:123-140.

Anderson, D.J. 1982. The home range: a new nonparametric estimation technique. Ecology 63(1):103-112.

Andersson, M., and J. Krebs. 1978. On the evolution of hoarding behaviour. Anim. Behav. 26(3):707-711.

Andrew, R.J. 1974. Arousal and the causation of behavior. Behaviour 51:135-165.

Animal Behavior Society (ABS) Animal Care Committee. 1981. Guidelines for the use of animals in research. Anim. Behav. 29:1-2.

Anokhin, P.K. 1964. Systemogenesis as a general regulator of brain development. In W.A. Himwich and H.E. Himwich, editors. The developing brain. Elsevier North-Holland, Inc., New York.

Antin, J., J. Gibbs, J. Holt, R.C. Young, and G.P. Smith. 1975. Cholecystokinin elicits the complete behavioral sequence of satiety in rats. J. Comp. Physiol. Psychol. 89:784-790.

Archibald, G.W. 1976a. The unison call of cranes as a useful taxonomic tool. Ph.D. Thesis. Cornell University, Ithaca, N.Y.

Archibald, G.W. 1976b. Crane taxonomy as revealed by the unison call. Proc. Int. Crane Workshop 1:225-251.

Aronson, L.R., E. Toback, D.S. Lehrman, and J.S. Rosenblatt, editors. 1970. Development and evolution of behavior. W.H. Freeman & Co., Publishers, San Francisco.

Aschoff, J. 1979. Circadian rhythms: influences of internal and external factors on the period measured in constant conditions. Z. Tierpsychol. 49:225-249.

Baerends, G.P. 1976. The functional organization of behaviour. Anim. Behav. 24:726-738.

Baerends, G.P., R. Brouwer, and H.T. Waterbolk. 1955. Ethological studies on Lebistes reticulatus (Peters). I. An analysis of the male courtship pattern. Behaviour 8:249-334.

Baker, M.C., K.J. Spitler-Nabors, and D.C. Bradley. 1981. Early experience determines song dialect responsiveness of female sparrows. Science 214:819-821.

Baker, R.R. 1978. The evolutionary ecology of animal migration. Holmes & Meier Publishers, Inc., New York.

Baker, R.R. 1980a. The mystery of migration. The Viking Press, New York.

Baker, R.R. 1980b. Goal orientation by blindfolded humans after long-distance displacement: possible involvement of a magnetic sense. Science 210:555-557.

Balda, R.P. 1980. Recovery of cached seeds by a captive Nucifraga caryocatactes. Z. Tierpsychol. 52:331-346.

Barash, D. 1976. Male response to apparent female adultery in the mountain bluebird (Sialia currucoides): an evolutionary interpretation. Am. Nat. 110:1097-1101.

Barash, D.P. 1977a. Sociobiology and behavior. Elsevier Science Publishing Co., Inc., New York.

Barash, D.P. 1977b. Sociobiology of rape in mallards *(Anas platyrhynchus):* responses of the mated male. Science 197:788-789.

Barash, D.P. 1978. Rape among mallards: reply to technical comments. Science 210:282.

Barber, V.C. 1968. The structure of mollusc statocysts, with particular reference to cephalopods. Symp. Zool. Soc. Lond. 23:37-62.

Barchas, J.D., H. Akil, G.R. Elliott, R.B. Holman, and S.J. Watson. 1978. Behavioral neurochemistry: neuroregulators and behavioral states. Science 200:964-973.

Barker, J.L., and T.G. Smith, Jr., editors. 1980. The role of peptides in neuronal function. Marcel Dekker, Inc., New York.

Barlow, G.W. 1968. Ethological units of behavior. In D. Ingle, editor. The central nervous system and fish behavior. University of Chicago Press, Chicago.

Barlow, G.W. 1974. Hexagonal territories. Anim. Behav. 22:876-878.

Barlow, G.W. 1977. Modal action patterns. In T.A. Sebeok, editor. How animals communicate. Indiana University Press, Bloomington.

Barlow, G.W., and J. Silverberg, editors. 1980. Sociobiology: beyond nature/nurture? AAAS Symposium 35. Westview Press, Inc., Boulder, Colo.

Baron, R.A., D. Byrne, and B. Kantowitz. 1978. Psychology: understanding behavior. W.B. Saunders Co., Philadelphia.

Bartels, M. 1929. Sinnesphysiologische und psychologische Untersuchungen an der trichterspinne *Agelena labyrinthica* (Cl.). Z. Vergl. Physiol. 10:527-593.

Bateson, P. 1979. How do sensitive periods arise and what are they for? Anim. Behav. 27:470-486.

Bays, S.M. 1962. Training possibilities of *Araneus diadematus.* Experientia 18:423.

Beach, F.A. 1950. The snark was a boojum. Am. Psychol. 5:115-124.

Beach, F.A. 1955. The descent of instinct. Psychol. Rev. 62:401-410.

Beer, C.G. 1975. Was professor Lehrman an ethologist? Anim. Behav. 23:957-964.

Beitinger, T.L., J.J. Magnuson, W.H. Neill, and W.R. Shaffer. 1975. Behavioural thermoregulation and activity patterns in the green sunfish, *Lepomis cyanellus.* Anim. Behav. 23:222-229.

Bekoff, A. 1976. Ontogeny of leg motor output in the chick embryo: a neural analysis. Brain Res. 106:271-291.

Bekoff, A. 1978. A neuroethological approach to the study of the ontogeny of coordinated behavior. In G.M. Burghardt and M. Bekoff, editors. The development of behavior: comparative and evolutionary aspects. Garland Publishing, Inc., New York.

Bekoff, M. 1976. Animal play: problems and perspectives. In P.P.G. Bateson and P.H. Klopfer, editors. Perspectives in ethology. Vol. 2. Plenum Press, New York.

Bellrose, F.C. 1958. Celestial orientation in wild mallards. Bird Banding 29:75-90.

Belzer, W.R. 1970. The control of protein ingestion in the black blowfly, *Phormia regina* (Meigen). Ph.D. Thesis. University of Pennsylvania, Philadelphia.

Benderly, B.L. 1980. The great ape debate. Science 80 1(5):61-65.

Bennet, W., and J. Gurin. 1982. The dieter's dilemma. Basic Books, Inc., Publishers, New York.

Bennett, E., and M. Calvin. 1964. Failure to train planarians reliably. Neurosci. Res. Program Bull. 2:3-24.

Bentley, D.R. 1971. Genetic control of an insect network. Science 174:1139-1141.

Bentley, D.R., and R.R. Hoy. 1970. Development of motor patterns in crickets. Science 170:1409-1411.

Bentley, D.R., and R.R. Hoy. 1972. Genetic control of the neuronal network generating cricket *(Teleogryllus gryllus)* song patterns. Anim. Behav. 20:478-492.

Bentley, D.R., and R.R. Hoy. 1974. The neurobiology of cricket song. Sci. Am. 231(2):34-44.

Benzer, S. 1973. Genetic dissection of behavior. Sci. Am. 229(6):24-37.

Berndt, R., and H. Sternberg. 1969. Alters und geschlechtsunterschiede in der dispersion des trauerschnappers *(Ficedula hypoleuca).* J. Ornith. 110:22-26.

Beyer, C. 1979. Endocrine control of sexual behavior. Raven Press, New York.

Biological Sciences Curriculum Study (BSCS). 1976. Investigating behavior. W.B. Saunders Co., Philadelphia.

Bitterman, M.E. 1975. The comparative analysis of learning. Science 188:699-709.

Bitterman, M.E. 1979. Historical introduction. In M.E. Bitterman et al., editors. Animal learning, survey and analysis. Plenum Press and NATO, New York.

Bitterman, M.E., V.M. LoLordo, J.B. Overmier, and M.E. Rashotte. 1979. Animal learning. Plenum Press, New York.

Black, C.H., and R.H. Wiley. 1977. Spatial variation in behavior in relation to territoriality in dwarf cichlids *Apistogramma ramirezi.* Z. Tierpsychol. 45:288-297.

Bligh, J., and K.G. Johnson. 1973. Glossary of terms for thermal physiology. J. Appl. Physiol. 35:941-961.

Block, G.D., and S.F. Wallace. 1982. Localization of a circadian pacemaker in the eye of a mollusc, *Bulla.* Science 217:155-157.

Block, R.A., and J.V. McConnell. 1967. Classically conditioned discrimination in the planarian, *Dugesia dorotocephala*. Nature 215:1465-1466.

Bloom, F.E., editor. 1980. Peptides: integrators of cell and tissue function. Raven Press, New York.

Bloom, F.E. 1981. Neuropeptides. Sci. Am. 245:148-168.

Blough, D.S. 1977. Visual search in the pigeon: hunt and peck method. Science 196:1013-1014.

Blum, M.S., and N.A. Blum, editors. 1979. Sexual selection and reproductive competition in insects. Academic Press, Inc., New York.

Bodian, D. 1972. Neuron junctions: a revolutionary decade. Anat. Rec. 174:73-82.

Boice, R. 1973. Domestication. Psychol. Bull. 80:215-230.

Bolles, R.C. 1975. Theory of motivation. Harper & Row, Publishers, Inc., New York.

Bolles, R.C. 1979. Learning theory. Holt, Rinehart & Winston, New York.

Bolles, R.C. 1980. Some functionalistic thoughts about regulation. In F.M. Toates and T.R. Halliday, editors. Analysis of motivational processes. Academic Press, Inc., New York.

Bond, R.R. 1957. Ecological distribution of breeding birds in the upland forests of southern Wisconsin. Ecol. Monogr. 27:351-384.

Bonner, J.T. 1958. The relation of spore formation to recombination. Am. Nat. 92:193-200.

Bonner, J.T. 1980. The evolution of culture in animals. Princeton University Press, Princeton, N.J.

Boorman, S.A., and P.R. Levitt. 1980. The genetics of altruism. Academic Press, Inc., New York.

Boring, E.G. 1957. A history of experimental psychology, 2nd ed. Appleton-Century-Crofts, New York.

Boudreau, J.C., and C. Tsuchitani. 1973. Sensory neurophysiology. Van Nostrand Reinhold Co., New York.

Boycott, B.B. 1965. Learning in the octopus. Sci. Am. 212(3):42-50.

Brady, J. 1979. Biological clocks. University Park Press, Baltimore.

Breed, M.D., C.M. Hinkle, and W.J. Bell. 1975. Agonistic behavior in the German cockroach, *Blattella germanica*. Z. Tierpsychol. 39:24-32.

Brenowitz, W.A. 1981. "Territorial song" as a flocking signal in red-winged blackbirds. Anim. Behav. 29(2):641-642.

Brines, M.L., and J.L. Gould. 1979. Bees have rules. Science 206:571-573.

Britten, R.J., and E.H. Davidson. 1969. Gene regulation for higher cells: a theory. Science 165:349-357.

Brockmann, H.J., A. Grafen, and R. Dawkins. 1979. Evolutionarily stable nesting strategy in a digger wasp. J. Theor. Biol. 77:473-496.

Brockmann, H.J., and J.P. Hailman. 1976. Fish cleaning symbiosis: notes on juvenile angelfishes (Pomacanthus, Chaetodontidae) and comparisons with other species. Z. Tierpsychol. 42:129-138.

Brodie, E.D., Jr. 1978. Biting and vocalization as antipredator mechanisms in terrestrial salamanders. Copeia 1978:127-129.

Brower, L.P., J.V.Z. Brower, and E.P. Cranston. 1965. Courtship behavior of the queen butterfly, *Danaus gilippus*. Zoologica 50:1-39.

Brown, C.R. 1977. Brown-headed cowbird courting a purple martin. Auk 94:395.

Brown, J.L. 1963. Ecogeographic variation and introgression in an avian visual signal: the crest of the Stellar's jay, *Cyanocitta stelleri*. Evolution 17:23-39.

Brown, J.L. 1964. The integration of agonistic behavior in the Steller's jay *Cyanocitta stelleri* (Gmelin). Univ. Calif. Publ. Zool. 60:223-328.

Brown, J.L. 1975. The evolution of behavior. W.W. Norton & Co., Inc., New York.

Brown, K.T. 1974. Physiology of the retina. In Mountcastle, V.B., editor. Medical physiology, 13th ed. The C.V. Mosby Co., St. Louis.

Bruning, E. 1973. The physiological clock, 3rd ed. Springer-Verlag New York, Inc., New York.

Buckle, G.R., and L. Greenberg. 1981. Nestmate recognition in sweat bees (*Lasioglossum zephyrum*): Does an individual recognize its own odour or only of its nestmates? Anim. Behav. 29:302-809.

Buechner, H.K. 1974. Implications of social behavior in the management of Uganda kob. In V. Geist and F. Walther, editors. The behaviour of ungulates and its relation to management. International Union for Conservation of Nature and Natural Resources, Morges, Switzerland.

Bullock, T.H., R. Orkand, and A. Grinnel. 1977. Introduction to nervous systems. W.H. Freeman & Co., Publishers, San Francisco.

Burghardt, G.M. 1970a. Intraspecific geographical variation in chemical food preferences of newborn garter snakes (*Thamnophis sirtalis*). Behaviour 36:246-257.

Burghardt, G.M. 1970b. Defining "communication." In J.W. Johnston, Jr., D.G. Moulton, and A. Turk, editors. Communication by chemical signals. Appleton-Century-Crofts, New York.

Burghardt, G.M. 1977. Ontogeny of communication. In T.A. Sebeok, editor. How animals communicate. Indiana University Press, Bloomington.

Burghardt, G.M., and M. Bekoff, editors. 1978. The development of behavior: comparative and evolutionary aspects. Garland Publishing Co., New York.

Burns, R.J. 1983. Coyote predation aversion with lithium chloride: management implications and comments. Wildlife Soc. Bull. 11:128-133.

Burtt, E.H., Jr. 1977. Some factors in the timing of parent-chick recognition in swallows. Anim. Behav. 25:231-239.

Butler, S.R., and E.A. Rowe. 1976. A data acquisition and retrieval system for studies of animal social behaviour. Behaviour 57:281-287.

Byers, J.A. 1977. Terrain preferences in the play behavior of Siberian ibex kids (*Capra ibex sibirica*). Z. Tierpsychol. 45:199-209.

Cade, T.J., and G.L. Maclean. 1967. Transport of water by adult sandgrouse to their young. Condor 69:323-343.

Cade, W. 1975. Acoustically orienting parasitoids: fly phonotaxis to cricket song. Science 190:1312-1313.

Caldwell, P.C., and R.D. Keynes. 1957. The utilization of phosphate bond energy for sodium extrusion from giant axons. J. Physiol. 137:12P-13P.

Caldwell, R.L. 1979. Cavity occupation and defensive behaviour in the stomatopod *Gonodactylus festai*: evidence for chemically mediated individual recognition. Anim. Behav. 27:194-201.

Camhi, J.M. 1977. Behavioral switching in cockroaches: transformations of tactile reflexes during righting behavior. J. Comp. Physiol. Psychol. 113:283-301.

Campbell, B., editor. 1972. Sexual selection and the descent of man. Aldine Publishing Co., Hawthorne, N.Y.

Candland, D.K. 1971. The ontogeny of emotional behavior. In H. Moltz, editor. The ontogeny of vertebrate behavior. Academic Press, Inc., New York.

Candland, D.K. 1979. A profound Quaker meeting. Contemp. Psychol. 24:965-967.

Capen, D.E., editor. 1981. The use of multivariate statistics in studies of wildlife habitat. U.S. Department of Agriculture Forest Service, General Technical Report RM-87, Fort Collins, Colo.

Caplan, A.L., editor. 1978. The sociobiology debate. Harper & Row, Publishers, Inc., New York.

Carew, T.J., R.D. Hawkins, and E.R. Kandel. 1983. Differential classical conditioning of a defensive withdrawal reflex in *Aplysia californica*. Science 219:397-400.

Carlson, A.D., and J. Copeland. 1978. Behavioral plasticity in the flash communication systems of fireflies. Am. Sci. 66:340-346.

Carlson, N.R. 1980. Physiology of behavior, 2nd ed. Allyn & Bacon, Inc., Boston.

Carmichael, L. 1963. The onset and early development of behavior. In L. Carmichael, editor. Manual of child psychology. John Wiley & Sons, Inc., New York.

Caro, T.M. 1979. Relations between kitten behaviour and adult predation. Z. Tierpsychol. 51:158-168.

Caro, T.M. 1980. Predatory behaviour in domestic cat mothers. Behaviour 74:128-148.

Caro, T.M. 1981. Predatory behaviour and social play in kittens. Behaviour 76:1-24.

Carr, A. 1965. The navigation of the green turtle. Sci. Am. 212(5):79-86.

Carthy, J.D. 1956. Animal navigation. Charles Scribner's Sons, New York.

Carthy, J.D. 1968. The pectines of scorpions. Symp. Zool. Soc. Lond. 23:251-261.

Carthy, J.D., and G.E. Newell, editors. 1968. Invertebrate receptors. Symp. Zool. Soc. Lond. No. 23.

Caryl, P.G. 1979. Communication by agonistic displays: what can games theory contribute to ethology? Behaviour 68:136-169.

Cassidy, J. 1979. Half a century on the concepts of innateness and instinct: survey, synthesis, and philosophical implications. Z. Tierpsychol. 50:364-386.

Catchpole, C.K. 1980. Sexual selection and the evolution of complex songs among European warblers of the genus *Acrocephalus*. Behaviour 74:149-166.

Cavalli-Sforza, L.L., and M.W. Feldman. 1981. A quantitative approach. Princeton University Press, Princeton, N.J.

Chamberlain, T.J., P. Halick, and R.W. Gerard. 1963. Fixation of experience in the rat spinal cord. J. Neurophysiol. 26:662-673.

Chamove, A.S., L.A. Rosenblum, and H.F. Harlow. 1973. Monkeys (*Macaca mulatta*) raised only with peers: a pilot study. Anim. Behav. 21: 316-325.

Charnov, E.L. 1976. Optimal foraging, the marginal value theorem. Theor. Popul. Biol. 9:129-136.

Chen, W.Y., L.C. Aranda, and J.V. Luco. 1970. Learning and long- and short-term memory in cockroaches. Anim. Behav. 18:725-732.

Chepko, B.D. 1971. A preliminary study of the effects of play deprivation on young goats. Z. Tierpsychol. 28:517-576.

Childs, G., R. Maxson, R.H. Cohn, and L. Kedes. 1981. Orphons: dispersed genetic elements derived from tandem repetitive genes of eucaryotes. Cell 23:651-663.

Clark, G.A., Jr. 1975. Additional records of passerine terrestrial gaits. Wilson Bull. 87:384-389.

Clayton, D. 1976. The effects of pre-test conditions on social facilitation of drinking in ducks. Anim. Behav. 24:125-134.

Cloarec, A. 1980. Post-moult behaviour in the water-stick insect *Ranatra linearis*. Behaviour 73:304-324.

Clutton-Brock, T.H., and P.H. Harvey, editors. 1978. Readings in sociobiology. W.H. Freeman & Co., Publishers, San Francisco.

Cochran, W.W., G.G. Montgomery, and R.R. Graber. 1967. Migratory flights of *Hylocichla* thrushes in spring: a radiotelemetry study. Living Bird 6:213-225.

Cody, M.L. 1968. On the methods of resource division in grassland bird communities. Am. Nat. 102:107-147.

Coghill, G.E. 1929. Anatomy and the problem of behavior. Cambridge University Press, Cambridge, England.

Cohem, D.B. 1979. Sleep and dreaming: origins, nature, and function. Pergamon Press, Inc., New York.

Cold Spring Harbor Symposia on Quantitative Biology. 1952. The neuron. Biological Laboratory, Long Island, N.Y.

Cole, L.C. 1954. The population consequences of life history phenomena. Q. Rev. Biol. 29: 103-137.

Cole, L.C. 1957. Biological clock in the unicorn. Science 125:874-876.

Colgan, P.W., W.A. Nowell, and N.W. Stokes. 1981. Spatial aspects of nest defence by pumpkinseed sunfish (*Lepomis gibbosus*): stimulus features and an application of catastrophe theory. Anim. Behav. 29:433-442.

Collias, N.E. 1960. An ecological and functional classification of animal sounds. In W.E. Lanyon and W.N. Tavolga, editors. Animal sounds and communication. Am. Inst. Biol. Sci. 7:368-391.

Collias, N.E., and E.C. Collias, editors. 1976. External construction by animals. Benchmark papers in animal behavior. Vol. 4. Dowden, Hutchinson, and Ross, Inc., Stroudsburg, Pa.

Collins, B.J., and J.D. Davis. 1978. Long term inhibition of intake by mannitol. Physiol. Behav. 21:957-965.

Conner, R.N., and C.S. Adkisson. 1977. Principal component analysis of woodpecker nesting habitat. Wilson Bull. 89:122-129.

Conner, W.E., and W.M. Masters. 1978. Infrared video viewing. Science 199:1004.

Conover, W.J. 1980. Practical nonparametric statistics, 2nd ed. John Wiley & Sons, Inc., New York.

Cook, A. 1977. Mucus trail following by the slug *Limax grossui* Lupu. Anim. Behav. 25:774-781.

Cook, A. 1979. Homing by the slug *Limax pseudoflavus*. Anim. Behav. 27:545-552.

Corben, C.J., G.J. Ingram, and M.J. Tyler. 1974. Electrophysiological correlates of meaning. Science 186:944-947.

Corning, W.C., J.A. Dyal, and A.O.D. Willows, editors. 1973-1975. Invertebrate learning. Vols. 1 to 3. Plenum Press, New York.

Corning, W.C., and S. Kelly. 1973. Platyhelminthes: the turbellarians. In W.C. Corning, J.A. Dyal, and A.O.D. Willows, editors. Invertebrate learning. Vol. 3. Plenum Press, New York.

Corning W.C., and R. Von Burg. 1973. Protozoa. In W.C. Corning, J.A. Dyal, and A.O.D. Willows, editors. Invertebrate learning. Vol. 3. Plenum Press, New York.

Cornsweet, T.N. 1962. The staircase method in psychophysics. Am. J. Psychol. 75:485-491.

Corrent, G., D.J. McAdoo, and A. Eskin. 1978. Serotonin shifts the phase of the circadian rhythm from the *Aplysia* eye. Science 202:977-979.

Coss, R.G., and A. Globus. 1978. Spine stems on tectal interneurons in jewel fish are shortened by social stimulation. Science 200:787-790.

Count, E.W. 1958. The biological basis of human sociality. Am. Anthropol. 60:1049-1085.

Cousteau, J. 1973. Attack and defense. World Publishing, New York.

Cowan, W.M., editor. 1981. Studies in developmental neurobiology. Oxford University Press, Inc., New York.

Crandall, L.S. 1964. Management of wild animals in captivity. University of Chicago Press, Chicago.

Crews, D. 1975. Psychobiology of reptilian reproduction. Science 189:1059-1065.

Crews, D. 1979. The hormonal control of behavior in a lizard. Sci. Am. 241:180-187.

Crick, F.H.C. 1979. Thinking about the brain. Sci. Am. 241(3):219-232.

Crow, J.F. 1976. Genetics notes, 7th ed. Burgess Publishing Co., Minneapolis.

Cruze, W.W. 1935. Maturation and learning in chicks. J. Comp. Psychol. 19:371-409.

Curio, E. 1976. The ethology of predation. Springer-Verlag New York, Inc., New York.

Curio, E., U. Ernst, and W. Vieth. 1978a. Cultural transmission of enemy recognition: one function of mobbing. Science 202:899-901.

Curio, E., U. Ernst, and W. Vieth. 1978b. The adaptive significance of avian mobbing. II. Cultural transmission of enemy recognition in blackbirds: effectiveness and some constraints. Z. Tierpsychol. 48:184-202.

Czeisler, C.A., M.C. Moore-Ede, and R.M. Coleman. 1982. Rotating shift work schedules that disrupt sleep are improved by applying circadian principles. Science 217:460-462.

Dahl, F. 1884. Das Gehor- und Geruchsorgan der Spinnen. Arch. Mikroskop. Anat. 24:1-10.

Dall, W.H. 1881. Intelligence in a snail. Am. Nat. 15:976-978.

Daly, M., and M. Wilson. 1978. Sex, evolution, and behavior. Duxbury Press, North Scituate, Mass.

Daniken, E. von. 1977. Chariots of the gods? G.P. Putnam's Sons, New York.

Darling, F.F. 1938. Bird flocks and the breeding cycle. Cambridge University Press, Cambridge, England.

Darwin, C. 1859. The origin of species. John Murray Publishers, London.

Darwin, C. 1871. The descent of man, and selection in relation to sex. D. Appleton & Co., New York.

Darwin, C. 1873. Expression of the emotions in man and animals. D. Appleton & Co., New York.

Davidson, D.W. 1978. Experimental tests of the optimal diet in two social insects. Behav. Ecol. Sociobiol. 4:35-41.

Davidson, J.M., and R.J. Davidson, editors. 1980. The psychobiology of consciousness. Plenum Press, New York.

Davies, N.B., and T.R. Halliday. 1979. Competitive mate searching in male common toads, *Bufo bufo*. Anim. Behav. 27:1253-1267.

Davis, W.H., and H.B. Hitchcock. 1965. Biology and migration of the bat *Myotis lucifugus* in New England. J. Mammal. 46:296-313.

Davis, W.J. 1973. Development of locomotor patterns in the absence of peripheral sense organs and muscles. Proc. Natl. Acad. Sci. USA 70:954-958.

Dawkins, M. 1971. Perceptual changes in chicks: another look at the "search image" concept. Anim. Behav. 19:566-574.

Dawkins, M. 1977. Do hens suffer in battery cages? Environmental preferences and welfare. Anim. Behav. 25:1034-1046.

Dawkins, R. 1976. The selfish gene. Oxford University Press, Inc., New York.

Dawkins, R. 1978. Replicator selection and the extended phenotype. Z. Tierpsychol. 47:61-76.

Dawkins, R. 1979. Twelve misunderstandings of kin selection. Z. Tierpsychol. 51:184-200.

Dawkins, R. 1980. Good strategy or evolutionarily stable strategy? In G.W. Barlow and J. Silverberg, editors. Sociobiology: beyond nature/nurture? Westview Press, Inc., Boulder, Colo.

Dawkins, R., and H.J. Brockmann. 1980. Do digger wasps commit the Concorde fallacy? Anim. Behav. 28:892-896.

Dawkins, R., and M. Dawkins. 1976. Hierarchical organization and postural facilitation: rules for grooming in flies. Anim. Behav. 24:739-755.

Dawkins, R., and J.R. Krebs. 1978. Animal signals: information or manipulation? In J.R. Krebs and N.B. Davies, editors. Behavioural ecology, an evolutionary approach. Blackwell Scientific Publications, Ltd., London.

De Robertis, E. 1960. Some observations on the ultrastructure and morphogenesis of photoreceptors. J. Gen. Physiol. 43(Suppl.):1-13.

de Vos, G.J. 1979. Adaptations of arena behavior in black grouse *(Tetrao tetrix)* and other grouse species (Tetraoninae). Behaviour 68:277-314.

de Vries, T. 1973. The Galapagos hawk. Ph.D. Thesis. University of Amsterdam.

Deese, J. 1978. Thought into speech. Am. Sci. 66:314-321.

Deguchi, T. 1981. Rhodopsin-like photosensitivity of isolated chicken pineal gland. Nature 209:706-707.

Dement, W., and N. Kleitman. 1957. Cyclic variations of EEG during sleep and their relations to eye movements, body motility, and dreaming. Electroencephalogr. Clin. Neurophysiol. 9:673.

Demong, N.J., and S.T. Emlen. 1978. Radar tracking of experimentally released migrant birds. Bird Banding 49:342-359.

Denenberg, V.H. 1963. Early experience and emotional development. Sci. Am. 208:138-146.

Denenberg, V.H. 1972a. Biobehavioral bases of development. In V.H. Denenberg, editor. The development of behavior. Sinauer Associates, Inc., Sunderland, Mass.

Denenberg, V.H. 1972b. Readings in the development of behavior. Sinauer Associates, Inc., Sunderland, Mass.

Denny, M. 1980. Locomotion: the cost of gastropod crawling. Science 208:1288-1290.

Dethier, V.G. 1971. A surfeit of stimuli: a paucity of receptors. Am. Sci. 59:706-715.

Dethier, V.G. 1976. The hungry fly. Harvard University Press, Cambridge, Mass.

Deutsch, J.A., editor. 1973. The physiological basis of memory. Academic Press, Inc., New York.

Deutsch, J.A., W.G. Young, and T.J. Kalogeris. 1978. The stomach signals satiety. Science 201:165-167.

Dewsbury, D.A. 1978. What is (was?) the "fixed action pattern"? Anim. Behav. 26:310-311.

Diamond, S., editor. 1974. The roots of psychology. Basic Books, Inc., Publishers, New York.

Dilger, W.C. 1962. The behavior of lovebirds. Sci. Am. 206(1):88-98.

Dill, P.A. 1977. Development of behaviour in alevins of Atlantic salmon, Salmo salar, and rainbow trout, S. gairdneri. Anim. Behav. 25:116-121.

Dingle, H.A. 1969. Statistical and information analysis of aggressive communication in the mantis shrimp Gonodactylus bredini Manning. Anim. Behav. 17:561-575.

Dissanayake, E. 1974. A hypothesis of the evolution of art from play. Leonardo 7:211-217.

Dobzhansky, T. 1951. Genetics and the origin of species. Columbia University Press, New York.

Dodson, P. 1974. Dinosaurs as dinosaurs. Evolution 28:494-497.

Domjan, M., and B. Burkhard. 1982. The principles of learning and behavior. Brooks/Cole Publishing Co., Monterey, Calif.

Drewien, R.C., and E.G. Bizeau. 1977. Cross-fostering whooping cranes to sandhill crane foster parents. In S.A. Temple, editor. Endangered birds. University of Wisconsin Press, Madison.

Drickamer, L.C. 1972. Experience and selection behavior in the food habits of Peromyscus: use of olfaction. Behaviour 41:269-287.

Drucker-Colin, R.R., and J.L. McGaugh, editors. 1977. Neurobiology of sleep and memory. Academic Press, Inc., New York.

Dunbar, M.J. 1960. The evolution of stability in marine environments: natural selection at the level of the ecosystem. Am. Nat. 94:129-136.

Dunbar, R.I.M. 1976. Some aspects of research design and their implications in the observational study of behaviour. Behaviour 58:78-98.

Dunstone, N., and R.J. O'Connor. 1979. Optimal foraging in an amphibious mammal. Anim. Behav. 27:1182-1201.

Durden, K. 1972. Gifts of an eagle. Simon & Schuster, New York.

Dwyer, T.J. 1975. Time budget of breeding gadwalls. Wilson Bull. 87:335-343.

Dyal, J.A. 1973. Behavior modification in annelids. In W.C. Corning, J.A. Dyal, and A.O.D. Willows, editors. Invertebrate learning. Vol. 1. Plenum Press, New York.

Eaton, R.L. 1970. The predatory sequence, with emphasis on killing behavior and its ontogeny, in the cheetah (Acinonyx jubatus Schreber). Z. Tierpsychol. 27:492-504.

Eaton, R.L. 1974. The cheetah: the biology, ecology, and behavior of an endangered species. Van Nostrand Reinhold Co., New York.

Eccles, J.C. 1970. Facing reality: philosophical adventures by a brain scientist. Springer Publishing Co., Inc., New York.

Eccles, J.C. 1977. The understanding of the brain, 2nd ed. McGraw-Hill, Inc., New York.

Eckert, R. 1963. Electrical interaction of paired ganglion cells in the leech. J. Gen. Physiol. 46:575-587.

Ehrman, L., and P.A. Parsons. 1976. The genetics of behavior. Sinauer Associates, Inc., Sunderland, Mass.

Eibl-Eibesfeldt, I. 1951. Beobachtungen zur Fortpflanzungsbiologie und Jungendentwicklung des Eichhornchens. Z. Tierpsychol. 8:370-400.

Eibl-Eibesfeldt, I. 1956. Angebornes und Erworbenes in der technik des Beutetotens (Versuche am Iltis, Putorius putorius L.). Z. Saugtierk 21:135-137.

Eibl-Eibesfeldt, I. 1958. Versuche uber den Nestbau erfahrungsloser Ratten. (Wiss. Film B757.) Inst. wiss. Film, Gottingen.

Eibl-Eibesfeldt, I. 1963. Angebornes und Erworbenes im Verhalten einiger Sauger. Z. Tierpsychol. 20:705-754.

Eibl-Eibesfeldt, I. 1972. Ethology: the biology of behavior, 2nd ed. Holt, Rinehart & Winston, New York. (Translated by E. Klinghammer.)

Eisenberg, J.F. 1967. A comparative study in rodent ethology with emphasis on the evolution of social behavior. Proc. US Natl. Mus. 122:1-51.

Eisenberg, J.F., and P. Leyhausen. 1972. The phylogenesis of predatory behavior in mammals. Z. Tierpsychol. 30:59-93.

Eisner, T., and D.J. Aneshansley. 1982. Spray aiming in bombardier beetles: jet deflection by the coanda effect. Science 215:83-85.

Eisner, T., K. Hicks, M. Eisner, and D.S. Robson. 1978. "Wolf-in-sheep's clothing" strategy of a predaceous insect larva. Science 199:790-794.

Elliot, R.D. 1977. Hanging behavior in common ravens. Auk 94:777-778.

Ellis, D.H. 1979. Development of behavior in the golden eagle. Wildlife Monogr. 70:1-94.

Ellis, M.J. 1973. Why people play. Prentice-Hall, Inc., Englewood Cliffs, N.J.

Elzinga, R.J. 1978. Fundamentals of entomology. Prentice-Hall, Inc., Englewood Cliffs, N.J.

Emlen, J.M. 1966. The role of time and energy in food preference. Am. Nat. 100:611-617.

Emlen, J.M., and M.G. Emlen. 1975. Optimal choice in diet: test of a hypothesis. Am. Nat. 109:427-435.

Emlen, S.T. 1972. An experimental analysis of the parameter of bird song eliciting species recognition. Behaviour 41:130-171.

Emlen, S.T. 1975. The stellar-orientation system of a migratory bird. Sci. Am. 233(2):102-111.

Emlen, S.T. 1980. Ecological determination and sociobiology. In G.W. Barlow and J. Silverberg, editors. Sociobiology: beyond nature/nurture? Westview Press, Inc., Boulder, Colo.

Emlen, S.T., and J.M. Emlen. 1966. A technique for recording migratory orientation of captive birds. Auk 83:361-367.

Emlen, S.T., and L.W. Oring. 1977. Ecology, sexual selection, and the evolution of mating systems. Science 197:215-223.

Emmel, T.C. 1976. Population biology. Harper & Row, Publishers, Inc., New York.

Enright, J.T. 1980. The timing of sleep and wakefulness. Springer-Verlag New York, Inc., New York.

Eoff, M. 1977. Artificial selection in *Drosophila simulans* males for increased and decreased sexual isolation from *D. melanogaster* females. Am. Nat. 3:259-277.

Epstein, A.N. 1960. Reciprocal changes in feeding behavior produced by intrahypothalamic chemical injections. Am. J. Physiol. 199:969-974.

Epstein, R., R.P. Lanza, and B.F. Skinner. 1980. Symbolic communication between two pigeons. Science 207:543-545.

Epstein, R., R.P. Lanza, and B.F. Skinner. 1981. "Self-awareness" in the pigeon. Science 212:695-696.

Ericsson, K.A., and W.G. Chase. 1982. Exceptional memory. Am. Sci. 70:607-617.

Esch, H. 1967. The evolution of bee language. Sci. Am. 216(4):96-104.

Eskin, A. 1971. Some properties of the system controlling the circadian activity rhythm of sparrows. In M. Menaker, editor. Biochronometry. National Academy of Sciences, Washington, D.C.

Evans, D.L. 1980. Multivariate analysis of weather and fall migration of saw-whet owls at Duluth, Minnesota, M.S. Thesis. North Dakota State University, Fargo.

Evans, R.L. 1976. The making of psychology. Alfred A. Knopf, Inc., New York.

Ewert, J.P. 1980. Neuroethology, an introduction to the neurophysiological fundamentals of behavior. Springer-Verlag New York, Inc., New York.

Ewing, A.W., and V. Evans. 1973. Studies on the behaviour of cyprinodent fish. I. The agonistic and sexual behaviour of *Aphyosemion biviltatum* (Lönnberg 1895). Behaviour 46:264-278.

Fagen, R.M. 1976. Exercise, play, and physical training in animals. In P.P.G. Bateson and P.H. Klopfer, editors. Perspectives in ethology. Vol. 2. Plenum Press, New York.

Fagen, R.M. 1981. Animal play behavior. Oxford University Press, Inc., New York.

Fagen, R.M., and R.N. Goldman. 1977. Behavioural catalogue analysis methods. Anim. Behav. 25:261-274.

Falconer, D.S. 1960. Introduction to quantitative genetics. Oliver & Boyd, Edinburgh.

Farish, D.J. 1972. The evolutionary implications of qualitative variation in the grooming behaviour of the hymenoptera (Insecta). Anim. Behav. 20:662-676.

Farkas, S.R., and H.H. Shorey. 1976. Anemotaxis and odour trail following by the terrestrial snail *Helix aspersa*. Anim. Behav. 24:686-689.

Felsenstein, J. 1980. A view of population genetics. Science 208:1253.

Fentress, J.C., editor. 1976. Simpler networks and behavior. Sinaurer Associates, Inc., Sunderland, Mass.

Ferguson, N.B.L., and R.E. Keesey. 1975. Effect of a quinine-adulterated diet upon body weight maintenance in male rats with ventromedial hypothalamic lesions. J. Comp. Physiol. Psychol. 89:478-488.

Ferguson, W. 1968. Abnormal behavior in domestic birds. In M.W. Fox, editor. Abnormal behavior in animals. W.B. Saunders Co., Philadelphia.

Fernald, R.D., and P. Heinecke. 1974. A computer compatible multi-purpose event recorder. Behaviour 48:268-275.

Fernald, R.D., and N.R. Hirata. 1979. The ontogeny of social behavior and body coloration in the African cichlid fish *Haplochrommis burtoni*. Z. Tierpsychol. 50:180-187.

Ferron, J. 1976. Comfort behavior of the red squirrel (*Tamiasciurus hudsonicus*). Z. Tierpsychol. 42:66-85.

Ferron, J. 1981. Comparative ontogeny of behaviour in four species of squirrels (Sciuridae). Z. Tierpsychol. 55:193-216.

Ficken, R.W., M.S. Ficken, and J.P. Hailman. 1978. Differential aggression in genetically different morphs of the white-throated sparrow (*Zonotrichia albicollis*). Z. Tierpsychol. 46:43-57.

Fischer, G.L. 1975. The behaviour of chickens. In E.S.E. Hafez, editor. The behaviour of domestic animals. The Williams & Wilkins Co., Baltimore.

Fischer, R.B. 1981. God did it, but how? Cal Media, La Mirada, Calif.

Fisher, R.A. 1930. The genetical theory of natural selection. Clarendon Press, Oxford.

Fisher, R.A. 1958. The genetical theory of natural selection, 2nd ed. Dover Publications, Inc., New York.

Fisler, G.F. 1977. Interspecific hierarchy at an artificial food source. Anim. Behav. 25:240-244.

Fitzpatrick, J.W. 1981. Search strategies of tyrant flycatchers. Anim. Behav. 29:810-821.

Fjerdingstad, E., editor. 1971. Chemical transfer of learned information. North-Holland Publishing Co., Amsterdam.

Flugel, J.C. 1933. A hundred years of psychology. G. Duckworth & Co., Ltd., London.

Ford, E.B. 1975. Ecological geneticcs, 4th ed. Chapman & Hall, Ltd., London.

Forster, L. 1982. Vision and prey-catching strategies in jumping spiders. Am. Sci. 70:165-175.

Fouts, R.S. 1972. The use of guidance in teaching sign language to a chimpanzee. J. Comp. Physiol. Psychol. 80:515-522.

Fouts, R.S., and R.L. Mellgren. 1976. Language, signs, and cognition in the chimpanzee. Sign Lang. Studies 13:319-346.

Fowler, M.E. 1978. Restraint and handling of wild and domestic animals. Iowa State University Press, Ames.

Fox, L.R., and P.A. Morrow. 1981. Specialization: species property or local phenomenon? Science 211:887-893.

Fox, M.W. 1967. The place and future of animal behavior studies in veterinary medicine. J. Am. Vet. Med. Assoc. 151:609-615.

Fox, M.W., editor. 1968. Abnormal behavior in animals. W.B. Saunders Co., Philadelphia.

Fox, M.W. 1969. Ontogeny of prey-killing behavior in the Canidae. Behaviour 35:259-272.

Fraenkel, G.S., and D.L. Gunn. 1961. The orientation of animals. Dover Publications, Inc., New York.

Frankel, R.B., R.P. Blakemore, and R.S. Wolfe. 1979. Magnetite in freshwater magnetotactic bacteria. Science 203:1355-1356.

Fraser, A.F. 1968. Behavior disorders in domestic animals. In M.W. Fox, editor. Abnormal behavior in animals. W.B. Saunders Co., Philadelphia.

Fraser, D. 1979. Aquatic feeding by a woodchuck. Canad. Field-Naturalist 93:309-310.

Freemon, F.R. 1972. Sleep research. Charles C Thomas, Publisher, Springfield, Ill.

Fretheim, T.E. 1969. Creation, fall, and flood. Augsburg Publishing House, Minneapolis.

Fretwell, S.D., and H.L. Lucas, Jr. 1970. On territorial behavior and other factors influencing habitat distribution in birds. I. Theoretical development. Acta Biotheor. 19:16-36.

Fricke, H.W. 1979. Mating system, resource and defence and sex change in the anemonefish *Amphiprion akallopisos*. Z. Tierpsychol. 50:313-326.

Frisch, K. von. 1914. Demonstration von Versucher zum Nachweis des Farbensinnes bei angeblich total farbenblinder Tieren. Verhandl. d. Deutsch. Zool. Ges. in Freiburg, Berlin.

Frisch, K. von. 1962. Dialects in the language of the bees. Sci. Am. 207(2):78-87.

Frisch, K. von. 1967. The dance language and orientation of bees. Belknap/Harvard University Press, Cambridge, Mass.

Frisch, K. von. 1971. Bees, their vision, chemical senses, and language. Cornell University Press, Ithaca, N.Y.

Frisch, K. von. 1974. Animal architecture. Harcourt Brace Jovanovich, Inc., New York.

Frohman, L.A., and L.L. Bernardis. 1968. Growth hormone and insulin levels in weanling rats with ventromedial hypothalamic lesions. Endocrinology 82:1125-1132.

Gaither, N.S., and B.E. Stein. 1979. Reptiles and mammals use similar sensory organizations in the midbrain. Science 205:595-597.

Galler, S.R., K. Schmidt-Koenig, G.J. Jacobs, and R.E. Belleville, editors. 1972. Animal orientation and navigation. Science and Technology Information Office, NASA Spec. Publ. 262.

Gallup, G.G., Jr. 1970. Chimpanzees: self-recognition. Science 167:86-87.

Gamow, G. 1952. The creation of the universe. The Viking Press, New York.

Gamow, R.I., and J.F. Harris. 1973. The infrared receptors of snakes. Sci. Am. 228(5):94-100.

Garcia, J., W.G. Hankins, and K.W. Rusiniak. 1974. Behavioral regulation of the milieu interne in man and rat. Science 185:824-831.

Gardner, B.T., and R.A. Gardner. 1969. Teaching sign language to a chimpanzee. Science 165:664-672.

Gardner, B.T, and R.A. Gardner. 1978. Comparative psychology and language acquisition. In K. Salzinger and F. Denmark, editors. Psychology: the state of the art. Ann. NY Acad. Sci. 309:37-76.

Gauthreaux, S.A., Jr., editors. 1980. Animal migration, orientation, and navigation. Academic Press, Inc., New York.

Geist, V. 1972. An ecological and behavioural explanation of mammalian characteristics and their implication to therapsid evolution. Z. Säugetierkunde 37:1-15.

Geist, V., and F. Walther, editors. 1974. Behaviour of ungulates and its relation to management. Vols. 1 and 2. International Union for Conservation of Nature and Natural Resources, Morges, Switzerland.

Gentry, R.L. 1973. Thermoregulatory behavior of eared seals. Behaviour 46:73-93.

Gersuni, G.V. 1971. Sensory processes at the neuronal and behavioral levels. Academic Press, Inc., New York.

Getty, T. 1981a. Analysis of central-place space-use patterns: the elastic disc revisited. Ecology 62(4):907-914.

Getty, T. 1981b. Territorial behavior of eastern chipmunks (*Tamias striatus*): encounter avoidance and spatial timesharing. Ecology 62(4):915-921.

Getty, T. 1981c. Structure and dynamics of chipmunk home range. J. Mammal. 62(4):726-737.

Gibb, J.A. 1957. Food requirements and other observations on captive tits. Bird Study 4:207-215.

Gibbs, J., D.J. Fauser, E.A. Rowe, B.J. Rolls, E.T. Rolls, and S. Maddison. 1978. Bombesin suppresses food intake in the rat. Soc. Neurosci. Abstr. 4(529):174.

Gilbert, L.E. 1982. The coevolution of a butterfly and a vine. Sci. Am. 247(2):110-121.

Gilbert, W. 1978. Why genes in pieces? Nature 271:501.

Gold, R.M. 1973. Hypothalamic obesity: the myth of the ventromedial nucleus. Science 82:488-490.

Goldstein, M.C. 1971. Stratification, polyandry, and family structure in central Tibet. Southwest. J. Anthropol. 27:64-74.

Gorner, P. 1958. Die optische und kinasthetische Orientierung der trichterspinne *Agelena labyrinthica* (Cl.). Z. Vergl. Physiol. 41:111-153.

Gossow, H. 1970. Vergleichende Verhaltensstudien an Marderartigen I. Uber LautauBerungen und zum Beuteverhalten. Z. Tierpsychol. 27:405-480.

Gottfried, B.M., and E.C. Franks. 1975. Habitat use and flock activity of dark-eyed juncos in winter. Wilson Bull. 87(3):374-383.

Gottlieb, G. 1970. Conceptions of prenatal behavior. In L.R. Aronson, E. Tobach, D.S. Lehrman, and J.S. Rosenblatt, editors. Development and evolution of behavior. W.H. Freeman & Co., Publishers, San Francisco.

Gottlieb, G. 1971. Development of species identification in birds. University of Chicago Press, Chicago.

Gould, J.L. 1974. Genetics and molecular ethology. Z. Tierpsychol. 36:267-292.

Gould, J.L. 1975. Honey bee recruitment: the dance-language controversy. Science 189:685-693.

Gould, J.L. 1980. Sun compensation by bees. Science 207:545-547.

Gould, J.L., and K.P. Able. 1981. Human homing: an elusive phenomenon. Science 212:1061-1063.

Gould, J.L., J.L. Kirschvink, and K.S. Deffeyes. 1978. Bees have magnetic remanence. Science 201:1026-1028.

Gould, S.J. 1976. Biological potential vs. biological determinism. Nat. Hist. 85(5):12-22.

Gould, S.J. 1980a. Is a new and general theory of evolution emerging? Paleobiology 6(1):119-130.

Gould, S.J. 1980b. Sociobiology and the theory of natural selection. In G.W. Barlow and J. Silverberg, editors. Sociobiology: beyond nature/nurture? Westview Press, Inc., Boulder, Colo.

Gould, S.J., and E.S. Vrba. 1982. Exaptation—a missing term in the science of form. Paleobiology 8(1):4-15.

Gray, J.A.B. 1959. Initiation of impulses at receptors. In J. Field, editor. Handbook of physiology. Section I, Neurophysiology. Vol. I. Waverly Press, Baltimore.

Green, G.W. 1964. The control of spontaneous locomotor activity in *Phormia regina* Meigen. I. Locomotor activity patterns in intact flies. J. Insect Physiol. 10:711-726.

Greenberg, N. 1976. Thermoregulatory aspects of behavior in the blue spiny lizard *Sceloporus cyanogenys* (Sauria, Iguanidae). Behaviour 59:1-21.

Greene, H.W., and G.M. Burghardt. 1978. Behavior and phylogeny: constriction in ancient and modern snakes. Science 200:74-77.

Greene, H.W., and R.W. McDiarmid. 1981. Coral snake mimicry: does it occur? Science 21:1207-1212.

Greenough, W.T. 1975. Experiential modification of the developing brain. Am. Sci. 63:37-46.

Greenwald, O.E. 1978. Kinematics and time relations of prey capture by gopher snakes. Copeia 1978(2):263-268.

Grier, J.W. 1968. Pre-attack behavior of the red-tailed hawk. M.S. Thesis. University of Wisconsin, Madison.

Grier, J.W. 1971. Pre-attack posture of the red-tailed hawk. Wilson Bull. 83:115-123.

Grier, J.W. 1973. Techniques and results of artifical insemination with eagles. Raptor Res. 7:1-12.

Grier, J.W. 1975. Avian spread-winged sunbathing in thermoregulation and drying. Ph.D. Thesis. Cornell University, Ithaca, N.Y.

Griffin, D.R. 1955. Bird navigation. In A. Wolfson, editor. Recent studies in avian biology. University of Illinois Press, Urbana.

Griffin, D.R. 1958. Listening in the dark. Yale University Press, New Haven, Conn.

Griffin, D.R. 1959. Echoes of bats and men. Doubleday & Co., Inc., New York.

Griffin, D.R. 1970. Migrations and homing of bats. In W.A. Winsatt, editor. Biology of bats. Academic Press, Inc., New York.

Griffin, D.R. 1976a, 1981. The question of animal awareness. Rockefeller University Press, New York.

Griffin, D.R. 1976b. A possible window on the minds of animals. Am. Sci. 64:530-535.

Griffin, D.R. 1977. Anthropomorphism. BioScience 27:445-446.

Griffin, D.R., J. Friend, and F. Webster. 1965. Target discrimination by the echolocation of bats. J. Exper. Zool. 158:155-168.

Griffin, D.R., and R.J. Hock. 1949. Airplane observations of homing birds. Ecology 30:176-198.

Grohmann, J. 1939. Modifikation oder Funktionsreifung? Ein Beitrag zur Karung der wechselseitigen Beziehungen zwischen Instinkthandlung und Erfahrung. Z. Tierpsychol. 2:132-144.

Groos, K. 1898. The play of animals. D. Appleton & Co., New York. (Translated by E.L. Baldwin.)

Grossman, S.P. 1975. Role of the hypothalamus in the regulation of food and water intake. Psychol. Rev. 82:200:224.

Grubb, T.C., Jr. 1974. Olfactory navigation to the nesting burrow in Leach's petrel (*Oceanodroma leucorrhoa*). Anim. Behav. 22:192-202.

Grubb, T.C., Jr. 1977. Why ospreys hover. Wilson Bull. 89:149-150.

Gurin, J. 1980. Chemical feelings. Science 80 1(1):28-33.

Gustavson, C.R., L.R. Brett, J. Garcia, and D.J. Kelly. 1979. A working model and experimental solutions to the control to predatory behavior. In R. Markowitz and V. Stevens, editors. Studies of captive wild animals. Nelson-Hall Publishers, Chicago.

Guthrie, E.R., and G.P. Horton. 1946. Cats in a puzzle box. Rinehart, New York.

Hadidian, J. 1980. Yawning in an old world monkey, *Macaca nigra* (primates: Cercopithecidae). Behaviour 75:133-147.

Hafez, E.S.E., editor. 1975. Behavior of domestic animals. The Williams & Wilkins Co., Baltimore.

Hailman, J.P. 1967. The ontogeny of an instinct:the pecking response in chicks of the laughing gull (*Larus atricilla* L.) and related species. Behaviour Supplement 15.

Hailman, J.P. 1969. How an instinct is learned. Sci. Am. 221(6):98-106.

Hailman, J.P. 1977. Optical signals. Indiana University Press, Bloomington.

Hailman, J.P. 1978a. The question of animal awareness: evolutionary continuity of mental experience. Auk 95:614-615.

Hailman, J.P. 1978b. Rape among mallards: technical comments. Science 201:280-281.

Hailman, J.P. 1978c. The behavior of communicating (review). Auk 95:771-774.

Hailman, J.P., and J.J.I. Dzelzkalns. 1974. Mallard tail-wagging: punctuation for animal communication? Am. Nat. 108:236-238.

Hainsworth, F.R., and L.L. Wolf. 1979. Feeding: an ecological approach. Adv. Study Behav. 9:53-96.

Halberg, F. 1973. Laboratory techniques and rhythmometry. In J.N. Mills, editor. Biological aspects of circadian rhythms. Plenum Press, New York.

Haley, T.J., and R.S. Snider, editors. 1964. Responses of the nervous system to ionizing radiation. Second International Symposium of University of California at Los Angeles. Little, Brown, & Co., Boston.

Halliday, T.R. 1975. An observational and experimental study of sexual behaviour in the smooth newt, *Triturus vulgaris* (Amphibia:Salamandridae). Anim. Behav. 23:291-322.

Halliday, T.R. 1976. The libidinous newt: an analysis of variations in the sexual behaviour of the male smooth newt, *Triturus vulgaris*. Anim. Behav. 24:398-414.

Hamburger, V., E. Wenger, and R. Oppenheim. 1966. Motility in the chick embryo in the absence of sensory input. J. Exp. Zool. 162:133-160.

Hamerstrom, F. 1957. The influence of a hawk's appetite on mobbing. Condor 59:192-194.

Hamilton, W.D. 1963. The evolution of altruistic behavior. Am. Nat. 97:354-356.

Hamilton, W.D. 1964. The genetical theory of social behavior. J. Theor. Biol. 7:1-32.

Hamilton, W.D. 1971. Geometry for the selfish herd. J. Theor. Biol. 31:295-311.

Hamilton, W.D. 1972. Altruism and related phenomena, mainly in social insects. Ann. Rev. Ecol. Syst. 3:193-232.

Hamilton, W.J. III. 1973. Life's color code. McGraw-Hill Book Co., New York.

Hansen, E.W. 1966. The development of maternal and infant behavior in the rhesus monkey. Behaviour 27:107-149.

Harcourt, A.H. 1978. Activity periods and patterns of social interaction: a neglected problem. Behaviour 66:121-135.

Harden Jones, F.R. 1968. Fish migration. Edward Arnold (Publishers), Ltd., London.

Hardy, J.W. 1976. Comparative breeding behavior and ecology of the bushy-crested and Nelson san blas jays. Wilson Bull. 88:96-120.

Harlow, H.F. 1959. Love in infant monkeys. Sci. Am. 200:68-74.

Harlow, H.F., and R.R. Zimmerman. 1959. Affectional responses in the infant monkey. Science 130:421-432.

Hasler, A.D., A.T. Scholz, and R.M. Horrall. 1978. Olfactory imprinting and homing in salmon. Am. Sci. 66:347-355.

Hausfater, G. 1976. Predatory behavior of yellow baboons. Behaviour 56:44-68.

Hazen, R.M. 1978. Curve-fitting. Science 202:823.

Hazlett, B.A., editor. 1977. Quantitative methods in the study of animal behavior. Academic Press, Inc., New York.

Heath, J.E. 1965. Temperature regulation and diurnal activity in horned lizards. U. Calif. Publ. Zool. 64:97-136.

Hebb, D.O. 1958. Textbook of psychology. W.B. Saunders Co., Philadelphia.

Hecht, M.K., and W.C. Steere, editors. 1970. Essays in evolution and genetics. Appleton-Century-Crofts, New York.

Heinrich, B. 1975. Thermoregulation in bumblebees. II. Energetics of warm-up and free flight. J. Comp. Physiol. Psychol. 96:155-166.

Heinrich, B. 1979. Bumblebee economics. Harvard Unviersity Press, Cambridge, Mass.

Heinrich, B., and G.A. Bartholomew. 1979. The ecology of the African dung beetle. Sci. Am. 241(5):146-156.

Held, R., and A. Hein. 1963. Movement produced stimulation in the development of visually guided behavior. J. Comp. Physiol. Psychol. 56:872-876.

Hendrichs, H. 1975. Changes in a population of dikdik *Madoqua* (*Rhynchotragus*) *kirki* (Gunther 1880). Z. Tierpsychol. 38:55-69.

Herter, K. 1962. Der Temperatursinn der Tiere. Ziensen Verlag, Wittenberg, Germany.

Hess, E.H. 1956. Space perception in the chick. Sci. Am. 195:71-80.

Hess, E.H. 1962. Ethology: an approach toward the complete analysis of behavior. In New directions in psychology. Holt, Rinehart & Winston, New York.

Hess, E.H. 1973a. Imprinting. Van Nostrand Reinhold Co., New York.

Hess, E.H. 1973b. Comparative sensory processes. In D.A. Dewsbury and D.A. Rethlingshafer, editors. Comparative psychology. McGraw-Hill Book Co., New York.

Hess, E.H., and S.B. Petrovich, editors. 1977. Imprinting. Dowden, Hutchinson, & Ross, Inc., Stroudsburg, Pa.

Hetherington, A.W., and S.W. Ranson. 1940. Hypothalamic lesions and adiposity in the rat. Anat. Rec. 78:149.

Hetherington, A.W., and S.W. Ranson. 1942. Effect of early hypophysectomy on hypothalamic obesity. Endocrinology 31:30-34.

Heymer, A. 1977. Ethologisches Wörterbuch/Ethological Dictionary/Vocabulaire Ethologique. Paul Parey, Berlin.

Hickman, C.P., Jr., L.S. Roberts, and F.M. Hickman. 1984. Integrated principles of zoology, 7th ed. The C.V. Mosby Co., St. Louis.

Hinde, R.A. 1956. Ethological models and the concept of drive. Br. J. Phil. Sci. 6:321.

Hinde, R.A. 1970. Animal Behaviour, 2nd ed. McGraw-Hill, Inc., New York.

Hinde, R.A. 1981. Animal signals: ethological and games-theory approaches are not incompatible. Anim. Behav. 29:535-542.

Hobson, J.A., and R.W. McCarley. 1977. The brain as a dream state generator: an activation-synthesis hypothesis of the dream process. Am. J. Psychol. 134(12):1335-1348.

Hobson, J.A., T. Spagna, and R. Malenka. 1978. Ethology of sleep studied with time-lapse photography: postural immobility and sleep-cycle phase in humans. Science 201:1251-1253.

Hockett, C.P. 1948. Biophysics, linguistics, and the unity of science. Am. Sci. 36:558-572.

Hodges, C.M. 1981. Optimal foraging in bumblebees: hunting by expectation. Anim. Behav. 29:1166-1171.

Hodos, W., and C.B.G. Campbell. 1969. Scala naturae: why there is no theory in comparative psychology. Psychol. Rev. 76:337-350.

Hoebel, B.G., and P. Teitelbaum. 1966. Weight regulation in normal and hypothalamic hyperphagic rats. J. Comp. Physiol. Psychol. 61:189-193.

Holden, C. 1979. Paul MacLean and the triune brain. Science. 204:1066-1068.

Holden, C. 1981. Human-animal relationship under scrutiny. Science 214:418-420.

Hölldobler, B. 1967. Zur Physiologie der Gast-Wirt-Beziehungen (Myrmecophilie) bei Ameisen: I, das Gastverhältnis der *Atemeles* - und *Lomechusa*-Larven (Col. Staphlinidae) zu *Formica* (Hym. Formicidae). Z. Vergl. Physiol. 56:1-21.

Hölldobler, B. 1969. Orientierungsmechanismen des Ameisengastes *Atemeles* (Coleoptera, Staphlinidae) bei der Wirtssuche. Verh. Deutsch. Zool. Gesell. 33:580-585.

Hölldobler, B. 1970. Zur Physiologic der Gast-Wirt-Beziehungen (Myrmecophilie) bei Ameisen:II, das Gastverhaltnis des imaginalen *Atemeles pubicollis* Bris. (Col. Staphylinida) zu *Myrmica* und *Formica* (Hym. Formicidae). Z. Vergl. Physiol. 66:215-250.

Hölldobler, B. 1971a. Recruitment behavior in *Camponotus socius* (Hym. Formicidae). Z. Vergl. Physiol. 75:123-142.

Hölldobler, B. 1971b. Communication between ants and their guests. Sci. Am. 224(3):85-93.

Hölldobler, B. 1976. Tournaments and slavery in a desert ant. Science 192:912-914.

Holling, C.S. 1959. The components of predation as revealed by a study of small-mammal predation of the European pine sawfly. Can. Entomol. 91:293-320.

Holling, C.S. 1966. The functional response of invertebrate predators to prey density. Mem. Entomol. Soc. Can. 48:1-86.

Holst, E. von, and U. von Saint Paul. 1960, 1963. On the functional organization of drives. Anim. Behav. 11:1-20. (Translated from Naturwiss. 18:409-422.)

Holt, E.B. 1931. Animal drive and the learning process. Holt, New York.

Hopkins, C.D. 1974. Electric communication in the reproductive behavior of *Sternopygus macrurus* (Gymnotoidei). Z. Tierpsychol. 35:518-535.

Hopson, J. 1980. Growl, bark, whine, and hiss. Science 80 1:81-85.

Horn, H.S. 1968. The adaptive significance of colonial nesting in the Brewer's blackbird (*Euphagus cyanocephalus*). Ecology 49:682-694.

Horridge, G.A. 1965. Intracellular action potentials associated with the beating of the cilia in ctenophore comb plate cells. Nature 205:602.

Hotta, Y., and S. Benzer. 1970. Genetic dissection of the *Drosophila* nervous system by means of mosaics. Proc. Natl. Acad. Sci. USA 67:1156-1163.

Hotta, Y., and S. Benzer. 1972. Mapping of behavior in *Drosophila* mosaics. Nature 240:527-535.

Hubel, D.H. 1963. The visual cortex of the brain. Sci. Am. 209:54-62.

Hubel, D.H. 1979. The brain. Sci. Am. 241(3):45-53.

Hubel, D.H., and T.N. Wiesel. 1979. Brain mechanisms of vision. Sci. Am. 241(3):150-162.

Huber, P. 1810. Recherches sur les Moeurs des Fourmis Indigenes. J.J. Paschoud, Paris.

Huey, R.B., and E.R. Pianka. 1977. Natural selection for juvenile lizards mimicking noxious beetles. Science 195:201-203.

Hughes, G.M. 1957. The coordination of insect movements. II. The effect of limb amputation and the cutting of commissures in the cockroach (*Blatta orientalis*). J. Exp. Biol. 34:306-333.

Hull, D.L. 1980. Sociobiology: another new synthesis. In G.W. Barlow and J. Silverberg, editors. Sociobiology: beyond nature/nurture? Westview Press, Inc., Boulder, Colo.

Hunsaker, D. 1962. Ethological isolating mechanisms in the *Sceloporus torquatus* group of lizards. Evolution 16:62-74.

Hunt, J.H., editor. 1980. Selected readings in sociobiology. McGraw-Hill Book Co., New York.

Hunter, M.L., Jr. 1980. Microhabitat selection for singing and other behaviour in great tits, *Parus major*: some visual and acoustical considerations. Anim. Behav. 28:468-475.

Hutchinson, G.E. 1981. Random adaptation and imitation in human evolution. Am. Sci. 69:161-165.

Hutt, C. 1966. Exploration and play in children. In P.A. Jewell and C. Loizos, editors. Play, exploration, and territory in mammals. Academic Press, Inc., London.

Huxley, J.S. 1914. The courtship habits of the great crested grebe (*Podiceps cristatus*), with an addition to the theory of sexual selection. Proc. Zool. Soc. Lond. 2:491-562.

Huxley, J.S. 1938. The present standing of the theory of sexual selection. In G.R. de Beer, editor. Evolution. Clarendon Press, Oxford.

Immelmann, K. 1972. Sexual and other long-term aspects of imprinting in birds and other species. Adv. Study Behav. 4:147-174.

Iversen, L.L. 1979. The chemistry of the brain. Sci. Am. 241(3):134-149.

Jacobs, C.H., N.E. Collias, and J.T. Fujimoto. 1978. Nest colour as a factor in nest selection by female village weaverbirds. Anim. Behav. 26:463-469.

Jacobs, J. 1981. How heritable is innate behaviour? Z. Tierpsychol. 55:1-18.

Jaeger, R.G., R.G. Joseph, and D.E. Barnard. 1981. Foraging tactics of a terrestrial salamander: sustained yield in territories. Anim. Behav. 29:1100-1105.

James, F.C. 1971. Ordinations of habitat relationships among breeding birds. Wilson Bull. 83:215-236.

James, W. 1890. Principles of psychology. Macmillan, Inc., New York.

Jennings, H.S. 1905. Modifiability in behavior. I. Behavior of sea anemones. J. Exp. Zool. 2:447-473.

Jennings, H.S. 1906. Behavior of the lower organisms. Columbia University Press, New York.

Jennings, H.S. 1907. Behaviour of starfish *Asterias forreri*. U. Calif. Publ. Zool. 4:53-185.

Jennings, T., and S.M. Evans. 1980. Influence of position in the flock and flock size on vigilance in the starling, *Sturnus vulgaris*. Anim. Behav. 28:634-635.

Jensen, D.D. 1967. Polythetic operationism and the phylogeny of learning. In W.C. Corning and S.C. Ratner, editors. Chemistry of learning. Plenum Press, New York.

Jewell, P.A. 1966. The concept of home range in mammals. Symp. Zool. Soc. Lond. 18:85-109.

Jewell, P.A., and C. Loizos. editors. 1966. Play, exploration and territory in mammals. Symp. Zool. Soc. Lond. 18:1-280.

Johnson, D.H. 1980. The comparison of usage and availability measurements for evaluating resource preference. Ecology 61:65-71.

Johnson, D.H. 1981. The use and misuse of statistics in wildlife habitat studies. In D.E. Capen, editor. The use of multivariate statistics in studies of wildlife habitat. U.S. Department of Agriculture Forest General Technical Report RM-87, Fort Collins, Colo.

Johnston, R.E., and T. Schmidt. 1979. Responses of hamsters to scent marks of different ages. Behav. Neural Biol. 26:64-75.

Johnson, V.R., Jr. 1977. Individual recognition in the banded shrimp *Stenopus hispidus* (Olivier). Anim. Behav. 25:418-428.

Jukes, T.H. 1980. Silent nucleotide substitutions and the molecular evolutionary clock. Science 210:973-978.

Kacelnik, A. 1979. The foraging efficiency of great tits (*Parus major* L.) in relation to light intensity. Anim. Behav. 27:237-241.

Kalat, J.W. 1977. Biological significance of food aversion learning. In N.W. Milgram, L. Krames, and T.M. Alloway, editors. Food aversion learning. Plenum Press, New York.

Kamil, A.C., and T.D. Sargent, editors. 1981. Foraging behavior: ecological, ethological, and psychological appraoches. Garland Publishing, Inc., New York.

Kandel, E.R. 1976. Cellular basis of behavior. W.H. Freeman & Company, Publishers, San Francisco.

Kandel, E.R. 1979a. Behavioral biology of *Aplysia*. W.H. Freeman & Co., Publishers, San Francisco.

Kandel, E.R. 1979b. Small systems of neurons. Sci. Am. 241(3):66-76.

Kandel, E.R., and J.H. Schwartz. 1981. Principles of neural science. Elsevier North-Holland, Inc., New York.

Karplus, I. 1979. The tactile communication between *Cryptocentrus steinitzi* (Pisces, Gobiidae) and *Alpheus purpurilenticularis* (Crustacea, Alpheidae). Z. Tierpsychol. 49:173-196.

Karplus, I., M. Tsurnamal, R. Szlep, and D. Algom. 1979. Film analysis of the tactile communication between *Cryptocentrus steinitzi* (Pisces, Gobiidae) and *Alpheus purpurilenticularis* (Crustacea, Alpheidae). Z. Tierpsychol. 49:337-351.

Keeler, C.E., and H.O. King. 1942. Multiple effects of coat color genes in the Norway rat, with special reference to temperament and domestication. J. Comp. Physiol. Psychol. 34:241-250.

Keeton, W.T. 1972. Biological science. W.W. Norton & Co., Inc., New York.

Keeton, W.T. 1974a. The mystery of pigeon homing. Sci. Am. 231(6):96-107.

Keeton, W.T. 1974b. The orientational and navigational basis of homing in birds. Adv. Study Behav. 5:47-132.

Kelley, D. 1980. Auditory and vocal nuclei in the frog brain concentrate sex hormones. Science 207:553-555.

Kennedy, G.C. 1950. The hypothalamic control of food intake in rats. Proc. R. Soc. Lond. 137:535-548.

Kennedy, R.J. 1969. Sunbathing behaviour in birds. Br. Birds 62:249-258.

Kessel, B. 1953. Distribution and migration of the European starling in North America. Condor 55:49-67.

Kessel, E.L. 1955. Mating activities of balloon flies. Syst. Zool. 4:97-104.

Kevan, P.G. 1975. Sun-tracking solar furnaces in high arctic flowers: significance for pollination and insects. Science 189:723-726.

Kevan, P.G. 1976. Sir Thomas More on imprinting: observations from the sixteenth century. Anim. Behav. 24:16-17.

Kiester, E., Jr. 1980. Images of the night. Science 80 1:36-43.

Kiley, M. 1972. The vocalizations of ungulates, their causation and function. Z. Tierpsychol. 31:171-222.

Kiley-Worthington, M. 1976. The tail movements of ungulates, canids, and felids with particular reference to their causation and function as displays. Behaviour 56:69-115.

Kimura, M. 1979. The neutral theory of molecular evolution. Sci. Am. 241(5):98-126.

King, J.A., D. Maas, and R.G. Weisman. 1964. Geographic variation in nest size among species of Peromyscus. Evolution 18:230-234.

Kinnaman, A.J. Mental life of two Macacus rhesus monkeys in captivity. Am. J. Psychol. 13:173-218.

Klopfer, P.H. 1963. Behavioral aspects of habitat selection: the role of early experience. Wilson Bull. 75(1):15-22.

Klopfer, P.H. 1965. Behavioral aspects of habitat selection: a preliminary report on stereotypy in foliage preferences of birds. Wilson Bull. 77(4):376-381.

Klopfer, P.H. 1969. Habitats and territories. Basic Books, Inc., Publishers, New York.

Klopfer, P.H. 1974. An introduction to animal behavior: ethology's first century, revised ed. Prentice-Hall, Inc., Englewood Cliffs, N.J.

Klopfer, P.H., and J.P. Hailman. 1965. Habitat selection in birds. Adv. Study Behav. 1:279-303.

Knudsen, E.I., and M. Konishi. 1978. A neural map of auditory space in the owl. Science 200:795-797.

Koford, C.B. 1957. The vicuna and the puna. Ecol. Monogr. 27:153-219.

Konig, J.F.R., and R.A. Klippel. 1963. The rat brain. A stereotaxic atlas of the forebrain and lower parts of the brain stem. The Williams & Wilkins Co., Baltimore.

Konishi, M. 1965. The role of auditory feedback in the control of vocalization in the white-crowned sparrow. Z. Tierpsychol. 22:770-783.

Konishi, M., and E.I. Knudsen. 1979. The oilbird: hearing and echolocation. Science 204:425-427.

Kovach, J.K. 1973. Soviet ethology: the first all-union conference on ecological and evolutionary aspects of animal behavior. Behaviour 44:14-211.

Kovach, J.K. 1980. Mendelian units of inheritance control color preferences in quail chicks (Coturnix coturnix japonica). Science 207:549-551.

Kovach, J.K., and G.C. Wilson. 1981. Behaviour and pleiotropy: generalization of gene effects in the colour preferences of Japanese quail chicks (C. coturnix japonica). Anim. Behav. 29:746-759.

Kraly, F.S., W.S. Carty, S. Resnick, and G.P. Smith. 1978. Effect of cholecystokinin on meal size and intermeal interval in the sham-feeding rat. J. Comp. Physiol. Psychol. 92:697-707.

Kramer, B. 1978. Spontaneous discharge rhythms and social signalling in the weakly electric fish Pollimyrus isidori (Cuvier et Valenciennes) (Mormyridae, Teleostei). Behav. Ecol. Sociobiol. 4:61-74.

Kramer, D.L., and W. Nowell. 1980. Central place foraging in the eastern chipmunk, Tamias striatus. Anim. Behav. 28:772-778.

Kramer, G. 1953. Die Sonnenorientiering der Vogel. Verh. Deut. Zool. Ges. Freiburg 1952:72-84.

Kramer, G. 1957. Experiments in bird orientation and their interpretation. Ibis 99:196-227.

Krasne, F.B. 1973. Learning in Crustacea. In W.C. Corning, J.A. Dayal, and A.O.D. Willows, editors. Invertebrate learning. Vol. 2. Plenum Press, New York.

Krebs, C.J. 1978. Ecology. Harper & Row, Publishers, Inc., New York.

Krebs, H.A. 1975. The August Krogh principle: "For many problems there is an animal on which it can be most conveniently studied." J. Exp. Zool. 194:221-226.

Krebs, J.R. 1973. Behavioral aspects of predation. In P.P.G. Bateson and P.H. Klopfer, editors. Perspectives in ethology. Vol. 1. Plenum Press, New York.

Krebs, J.R. 1978. Optimal foraging: decision rules for predators. In J.R. Krebs and N.B. Davies, editors. Behavioural ecology, an evolutionary approach. Blackwell Scientific Publications, Ltd., London.

Krebs, J.R., and N.B. Davies, editors. 1978. Behavioural ecology, an evolutionary approach. Blackwell Scientific Publications, Ltd., London.

Krebs, J.R., J.T. Erichsen, M.L. Webber, and E.L. Charnov. 1977. Optimal prey selection in the great tit (Parus major). Anim. Behav. 25:30-38.

Krebs, J.R., J.C. Ryan, and E.L. Charnov. 1974. Hunting by expectation or optimal foraging? A study of patch use by chickadees. Anim. Behav. 22:953-964.

Kreithen, M.L., and W.T. Keeton. 1974. Detection of polarized light by the homing pigeon, Columbia livia. J. Comp. Physiol. Psychol. 89:83-92.

Kruuk, H. 1972. The spotted hyena. University of Chicago Press, Chicago.

Kummer, H. 1971. Primate societies. Aldine, Atherton, Chicago.

Kung, C., S.Y. Chang, Y. Satow, J. VanHouten, and H. Hansma. 1975. Genetic dissection of behavior in Paramecium. Science 188:898-904.

Kuo, Z.Y. 1939. Studies in the physiology of the embryonic nervous system. II. Experimental evidence on the controversy over the reflex theory in development. J. Comp. Neurol. 70:437-459.

Kuo, Z.Y. 1967. The dynamics of behavior development. Random House, Inc., New York.

Kushlan, J.A. 1978. Nonrigorous foraging by robbing egrets. Ecology 59:649-653.

Kuterbach, D.A., and B. Walcott. 1982. Iron-containing cells in the honey bee (Apis mellifera). Science 218:695-697.

Kutsch, W. 1974. The influence of the wing sense organs on the flight motor pattern in maturing adult locusts. J. Comp. Physiol. Psychol. 88:413-424.

Lack, D. 1933. Habitat selection in birds with special reference to the effects of afforestation on the Breckland avifauna. J. Anim. Ecol. 2:239-262.

Lack, D. 1943. The life of the robin. H.F. & G. Witherby Co., London.

Lack, D. 1966. Population studies of birds. Oxford University Press, Oxford.

Lack, D. 1968. Ecological adaptations for breeding in birds. Methuen, Ltd., London.

Lade, B.I., and W.H. Thorpe. 1964. Dove songs as innately coded patterns of specific behaviour. Nature 202:366-368.

Lahue, R., and W.C. Corning. 1975. Synthesis: a comparative look at vertebrates. In W.C. Corning, J.A. Dyal, and A.O.D. Willows, editors. Invertebrate learning. Vol. 3. Plenum Press, New York.

Lall, A.B., H.H. Seliger, W.H. Biggley, and J.E. Lloyd. 1980. Ecology of colors of firefly bioluminescence. Science 210:560-562.

Lashley, K. 1950. In search of the engram. Soc. Exp. Biol. Symp. 4:454-482.

Laven, H. 1940. Beiträge zur Biologie des Sandregenpfeifers (Charadrius hiaticula L.). J. Ornithol. 88:183-288.

Lawick-Goodall, J. van. 1968. The behaviour of free-living chimpanzees in the Gombe Stream Reserve. Anim. Behav. Monogr. 1:165-311.

Lehner, P.N. 1979. Handbook of ethological methods. Garland Publishing, Inc., New York.

Lehrman, D.S. 1953. A critique of Konrad Lorenz's theory of instinctive behavior. Q. Rev. Biol. 28:337-363.

Lehrman, D.S. 1964. The reproductive behavior of the ring dove. Sci. Am. 211:48-54.

LeMagnen, J. 1971. Advances in studies on the physiological control and regulation of food intake. In E. Stellar and J.M. Sprague, editors. Progress in physiological psychology. Vol. 4. Academic Press, Inc., New York.

Leopold, A. 1933. Game management. Charles Scribner's Sons, New York.

Lerwill, C.J., and P. Makings. 1971. The agonistic behavior of the golden hamster Mesocricetus auratus (Waterhouse). Anim. Behav. 19:714-721.

Leshner, A.I. 1978. An introduction to behavioural endocrinology. Oxford University Press, Inc., New York.

Levine, J.D., and R.J. Wyman. 1973. Neurophysiology of flight in a mutant Drosophila. Proc. Natl. Acad. Sci. 70:1050-1054.

Levine, L. 1958. Studies on sexual selection in mice. Am. Nat. 92:21-26.

Levinson, B.M. 1968. Interpersonal relationships between pet and human being. In M.W. Fox, editor. Abnormal behavior in animals. W.B. Saunders Co., Philadelphia.

Levitt, D.R., and P. Teitelbaum. 1975. Somnolence, akinesia, and sensory activation of motivated behavior in the lateral hypothalamus syndrome. Proc. Natl. Acad. Sci. 72:2819-2823.

Lewin, R. 1981a. Lamarck will not lie down. Science 213:316-321.

Lewin, R. 1981b. Seeds of change in embryonic development. Science 214:42-44.

Lewontin, R.C. 1965. Selection in and of populations. In J.A. Moore, editor. Ideas in modern biology. Natural History Press, New York.

Leyhausen, P. 1956. Verhaltensstudien an Katzen. Z. Tierpsychol. Beiheft 2.

Leyhausen, P. 1965. Uber die Funktion der relativen Stimmungshierarchie. Dargestellt am Beispiel der phylogenetischen und ontogenetischen Entwicklung des Beutefangs von Raubtieren. Z. Tierpsychol. 22:412-494.

Ligon, J.D., and D.J. Martin. 1974. Pinon seed assessment by the pinon jay, Gymnorhinus cyanocephalus. Anim. Behav. 22:421-429.

Lindauer, M. 1952. Ein Beitrag zur Frage der arbeitsteilung in Bienenstaat. Z. Vergl. Physiol. 34:299-345.

Lindauer, M. 1971a. Communication among social bees. Harvard University Press, Cambridge, Mass.

Lindauer, M. 1971b. The functional significance of the honeybee waggle dance. Am. Nat. 105:89-96.

Linhart, S.B. 1975. The biology and control of vampire bats. In G.M. Baer, editor. Natural history of rabies. Vol. 2. Academic Press, Inc., New York.

Linhart, S.B., R.T. Sterner, T.C. Carrigan, and D.R. Henne. 1979. Komondor guard dogs reduce sheep losses to coyotes: a preliminary evaluation. J. Range Manage. 32(3):238-241.

Lissman, H.W. 1958. On the function and evolution of electric organs in fish. J. Exp. Biol. 35:156-191.

Lissman, H.W. 1963. Electric location by fishes. Sci. Am. 208(3):50-59.

Liu, C.M., and T.A. Yin. 1974. Caloric compensation to gastric loads in rats with hypothalamic hyperphagia. Physiol. Behav. 13:231-238.

Lloyd, J.E. 1966. Studies on the flash communication system in Photinus fireflies. Misc. Publ. Museum Zool. U. Mich. 130:1-95.

Lloyd, J.E. 1975. Aggressive mimicry in Photuris fireflies: signal repertoires by femmes fatales. Science 187:452-453.

Lloyd, J.E. 1980. Male Photuris fireflies mimic sexual signals of their females' prey. Science 210:669-671.

Lloyd, J.E. 1981. Mimicry in the sexual signals of fireflies. Sci. Am. 245(1):139-145.

Lockard, R.B. 1971. Reflections on the fall of comparative psychology: is there a lesson for us all? Am. Psychol. 26:168-179.

Lockley, R.M. 1967. Animal navigation. Hart Publishing Co., Inc., New York.

Loewenstein, W.R. 1971. Mechano-electric transduction in the Pacinian corpuscle. Initiation of sensory impulses in mechanorecepters. In W.R. Loewenstein, editor. Handbook of sensory physiology. Vol. 1. Springer-Verlag New York, Inc., New York.

Loizos, C. 1966. Play in mammals. In P.A. Jewell and C. Loizos, editors. Play, exploration, and territory in mammals. Academic Press, Inc., London.

Lorenz, K.Z. 1932. A consideration of methods of identification of species-specific instinctive behaviour patterns in birds. J. Ornithol. 80:50-98. (Translated by R. Martin in Lorenz, K.Z. 1970. Studies in animal and human behavior. Harvard University Press, Cambridge, Mass.)

Lorenz, K.Z. 1935. Der Kumpan in der Umwelt des Vogels. J. Ornithol. 83:137-213, 289-413.

Lorenz, K.Z. 1941. Vergleichende Bewegungsstudien an Anatinen. Suppl. J. Ornith. 89:194-294.

Lorenz, K.Z. 1952. King Solomon's ring. Thomas Y. Crowell Co., New York.

Lorenz, K.Z. 1958. The evolution of behavior. Sci. Am. 199(6):67-78.

Lorenz, K.Z. 1970. Studies in animal and human behaviour. Vol. I. Harvard University Press, Cambridge, Mass. (Translation of earlier papers.)

Lorenz, K.Z. 1981. The Foundations of Ethology. Springer-Verlag New York, Inc., New York.

Lorenz, K.Z., and N. Tinbergen. 1938. Taxis und Instinkthandlung in der Eirollbewegung der Graugans I. Z. Tierpsychol. 2:1-29.

Lott, D.F. 1981. Sexual behavior and intersexual strategies in American bison. Z. Tierpsychol. 56:97-114.

Louw, G. 1979. Biological "strategies." Science 203:955.

Low, W.A., R.L. Tweedie, C.B.H. Edwards, R.M. Hodder, K.W.J. Malafant, and R.B. Cunningham. 1981a. The influence of environment on daily maintenance behaviour of free-ranging shorthorn cows in central Australia. I. General introduction and descriptive analysis of day-long activities. Appl. Anim. Ethol. 7:11-26.

Low, W.A., R.L. Tweedie, C.B.H. Edwards, R.M. Hodder, K.W.J. Malafant, and R.B. Cunningham. 1981b. The influence of environment on daily maintenance behaviour of free-ranging shorthorn cows in central Australia. II. Multivariate analysis of duration and incidence of activities. Appl. Anim. Ethol. 7:27-38.

Ludlow, A.R. 1976. The behaviour of a model animal. Behaviour 58:131-172.

Lustick, S., B. Battersby, and M. Kelty. 1978. Behavioral thermoregulation: orientation toward the sun in herring gulls. Science 200:81-83.

Luttenberger, V.F. 1975. Zum problem des gahnens bei reptilien. Z. Tierpsychol 37:113-137.

MacArthur, R.H. 1958. Population ecology of some warblers of northeastern coniferous forests. Ecology 39:599-619.

MacArthur, R.H., and E.R. Pianka. 1966. On optimal use of a patchy environment. Am. Nat. 100:603-609.

Mackintosh, N.J. 1974. The psychology of animal learning. Academic Press, Inc., New York.

MacLean, P. 1981. Training the brain. Interview on options in education series. National Public Radio Education Services, Washington. (Cassette tape available from N.P.R.)

MacNichol, E.F., Jr. 1964. Three-pigment color vision. Sci. Am. 211(6):48-56.

Maier, N.R.F., and T.C. Schneirla. 1935. Principles of animal psychology. McGraw-Hill Book Co., New York.

Maki, W.S. 1979. Discrimination learning without short-term memory: dissociation of memory processes in pigeons. Science 204: 83-85.

Manning, A. 1967. An introduction to animal behavior. Addison-Wesley Publishing Co., Inc., Reading, Mass.

Marler, P. 1952. Variations in the song of the chaffinch, *Fringilla coelebs*. Ibis 94:458-472.

Marler, P. 1957. Specific distinctiveness in the communication signals of birds. Behaviour 11:13-39.

Marler, P. 1959. Developments in the study of animal communication. In P.R. Bell, editor. Darwin's biological work: some aspects reconsidered. John Wiley & Sons, Inc., New York.

Marler, P., and W.J. Hamilton III. 1966. Mechanisms of animal behavior. John Wiley & Sons, Inc., New York.

Marler, P., and P. Mundinger. 1971. Vocal learning in birds. In H. Holtz, editor. The ontogeny of vertebrate behavior. Academic Press, Inc., New York.

Marler, P., and S. Peters. 1981. Sparrows learn adult song and more from memory. Science 213:780-782.

Marler, P., and M. Tamura. 1964. Culturally transmitted patterns of vocal behavior in sparrows. Science 146:1486.

Marler, P., and J.G. Vandenbergh, editors. 1979. Social behavior and communication. Handbook of behavioral neurobiology. Vol. 3. Plenum Press, New York.

Marshall, J.F., B.H. Turner, and P. Teitelbaum. 1971. Sensory neglect produced by lateral hypothalamic damage. Science 174:523-525.

Marx, J.L. 1980. Ape-language controversy flares up. Science 207:1330-1333.

Marx, J.L. 1981. Genes that control development. Science 213:1485-1488.

Mason, P.R. 1975. Chemo-klino-kinesis in planarian food location. Anim. Behav. 23:460-469.

Mason, W.A. 1976. Windows on other minds. Science 194:930-931.

Mast, S.O. 1911. Light and the behavior of organisms. John Wiley & Sons, Inc., New York.

Matsubara, J.A. 1981. Neural correlates of a nonjammable electrolocation system. Science 211:722-725.

Matthews, G.V.T. 1951a. The sensory basis of bird navigation. J. Inst. Navigation 4:260-275.

Matthews, G.V.T. 1951b. The experimental investigation of navigation in homing pigeons. J. Exp. Biol. 28:508-536.

Matthews, G.V.T. 1955. Bird navigation. Cambridge University Press, Cambridge, England.

Matthews, R.W., and J.R. Matthews. 1978. Insect behavior. John Wiley & Sons, Inc., New York.

Maturana, H.R., J.Y. Lettvin, W.S. McCulloch, and W.H. Pitts. 1960. Anatomy and physiology of vision in the frog *(Rana pipiens)*. J. Gen. Physiol. 43:129-175.

Mayer, J. 1955. Regulation of energy intake and body weight. The glucostatic theory and the lipostatic hypothesis. Ann. Acad. Sci. 63:15-43.

Maynard Smith, J. 1964. Group selection and kin selection. Nature 201:1145-1147.

Maynard Smith, J. 1968. Evolution in sexual and asexual populations. Am. Nat. 102:469-473.

Maynard Smith, J. 1971. What use is sex? J. Theor. Biol. 30:319-335.

Maynard Smith, J. 1972. Game theory and the evolution of fighting. In J. Maynard Smith, editor. On evolution. Edinburgh University Press, Edinburgh.

Maynard Smith, J. 1976. Evolution and the theory of games. Am. Sci. 64:41-45.

Mayr, E. 1963. Animal species and evolution. Belknap/Harvard University Press, Cambridge, Mass.

Mayr, E. 1974. Behavior programs and evolutionary strategies. Am. Sci. 62:650-658.

Mayr, E., and W.B. Provine, editors. 1980. The evolutionary synthesis. Harvard University Press, Cambridge, Mass.

McCarley, R.W., and J.A. Hobson. 1977. The neurobiological origins of psychoanalytic dream theory. Am. J. Psychiatry 134(11):1211-1221.

McConnell, J.V., and A.L. Jacobson. 1973. Learning in invertebrates. In D.A. Dewsbury and D.A. Rethlingshafer, editors. Comparative psychology. McGraw-Hill Book Co., New York.

McConnell, J.V., A.L. Jacobson, and D.P. Kimble. 1959. The effects of regeneration upon retention of a conditioned response in the planarian. J. Comp. Physiol. Psychol. 52:1-5.

McConnell, J.V., and J. Shelby. 1970. Memory transfer in invertebrates. In G. Ungar, editor. Molecular mechanisms in memory and learning. Plenum Press, New York.

McCosker, J.E. 1977. Fright posture of the Plesiopid fish *Calloplesiops altivelis:* an example of batesian mimicry. Science 197:400-401.

McEwen, B.S. 1976. Interactions between hormones and nerve tissue. Sci. Am. 235(1):48-58.

McFarland, D., and R. Sibly. 1972. "Unitary drives" revisited. Anim. Behav. 20:548-563.

McKinney, F. 1975. The behaviour of ducks. In E.S.E. Hafez, editor. The behaviour of domestic animals, 3rd ed. The Williams & Wilkins Co., Baltimore.

Mech, L.D. 1966. The wolves of Isle Royale. Fauna of the National Parks of the U.S.—Fauna Series 7. U.S. Government Printing Office, Washington, D.C.

Mech, L.D. 1970. The wolf: the ecology and behavior of an endangered species. Natural History Press, New York.

Mech, L.D. 1977. Wolf-pack buffer zones as prey reservoirs. Science 198:320-321.

Meddis, R. 1975. On the function of sleep. Anim. Behav. 23:676-691.

Medin, D.L., W.A. Roberts, and R.T. Davis. 1976. Processes of animal memory. Lawrence Erlbaum Associates, Inc., Hillsdale, N.J.

Meisel, R.L., and I.L. Ward. 1981. Fetal female rats are masculinized by male littermates located caudally in the uterus. Science 213:239-242.

Melzack, R., and T.H. Scott. 1957. The effects of early experience on the response to pain. J. Comp. Physiol. Psychol. 50:155-161.

Menaken, M. 1971. Biochronometry. National Academy of Sciences. Washington, D.C.

Menzel, R., and J. Erber. 1978. Learning and memory in bees. Sci. Am. 239(1):102-108.

Mertens, R. 1956. Das Problem der Mimikry bei Korallenschlangen. Zool. Jahrb. Abt. Syst. Oekol. Geogr. Tiere. 84:541-576.

Metzgar, L.H. 1967. An experimental comparison of screech owl predation on resident and transient white-footed mice (*Peromyscus leucopus*). J. Mammal. 48:387-391.

Meyer-Holzapfel, M. 1968. Abnormal behavior in zoo animals. In M.W. Fox, editor. Abnormal behavior in animals. W.B. Saunders Co., Philadelphia.

Meyerriecks, A.J. 1972. Man and birds: evolution and behavior. The Bobbs-Merrill Co., Inc., Indianapolis.

Meyers, B. 1971. Early experience and problem-solving behavior. In H. Moltz, editor. The ontogeny of vertebrate behavior. Academic Press, Inc., New York.

Michelsen, A. 1979. Insect ears as mechanical systems. Am. Sci. 67:696-706.

Milgram, N.W., L. Krames, and T.M. Alloway, editors. 1977. Food aversion learning. Plenum Press, New York.

Miller, D.B. 1977. Social display of mallard ducks (*Anas platyrhynchos):* effects of domestication. J. Comp. Physiol. Psychol. 91:221-232.

Miller, N.E., and M.L. Kessen. 1952. Reward efforts of food via stomach fistula compared with those of food via mouth. J. Comp. Physiol. Psychol. 45:555-564.

Mills, J.N. 1973. Biological aspects of circadian rhythms. Plenum Press, New York.

Milne, L., and M. Milne. 1964. The senses of animals and men. Atheneum Publishers, New York.

Mock, D.W. 1977. Maintenance behavior and communication in the brown pelican (Review of Schreiber 1977). Wilson Bull. 89:639-641.

Mock, D.W. 1980. Behavioral mechanisms in ecology. Am. Sci. 70:325.

Moericke, V. 1950. Ueber das Farbensehen der Pfirsichblattlaus, *Myxodes persical* Sulz. Z. Tierpsychol. 7:265-274.

Moermond, T.C. 1979. The influence of habitat structure on anolis foraging behavior. Behaviour 70:147-167.

Mohl, B. 1968. Auditory sensitivity of the common seal in air and water. J. Auditory Res. 8:27-38.

Molnar, R.E. 1977. Analogies in the evolution of combat and display structures in ornithopods and ungulates. Evol. Theory 3:165-190.

Moltz, H. 1965. Contemporary instinct theory and the fixed action pattern. Psychol. Rev. 72:27-47.

Moltz, H., editor. 1971. The ontogeny of vertebrate behavior. Academic Press, Inc., New York.

Mook, D.C., and E.M. Blas. 1968. Quinine-aversion thresholds and finickiness in hyperphagic rats. J. Comp. Physiol. Psychol. 65:202-207.

Moore, B.R., and S. Stuttard. 1979. Dr. Guthrie and *Felis domesticus* or: tripping over the cat. Science 205:1031-1033.

Moore, R.G. 1978. Seasonal and daily activity patterns and thermoregulation in the southwestern speckled rattlesnake (*Crotalus mitchelli pyrrhus*) and the Colorado desert sidewinder (*Crotaluscerastes laterorepens*). Copeia 3:439-442.

Moore, R.Y. 1978. Central neural control of circadian rhythms. Front. Neuroendocrinol. 5:185-206.

Moore-Ede, M.C., F.M. Sulzman, and C.A. Fuller. 1982. The clocks that time us. Harvard University Press, Cambridge, Mass.

Morgan, C.L. 1894. Introduction to comparative psychology. Scribner, New York.

Morrell, G.M., and J.R. Turner. 1970. Experiments on mimicry. I. The response of wild birds to artificial prey. Behaviour 36:116-130.

Morris, D. 1957. "Typical intensity" and its relation to the problem of ritualization. Behaviour 11:1-12.

Morris, D., editor. 1970. Patterns of reproductive behavior. McGraw-Hill Book Co., New York.

Morrison, G.R. 1974. Alterations in palatability of nutrients for the rat as a result of prior tasting. J. Comp. Physiol. Psychol. 86:56-61.

Morton, E.S. 1977. On the occurrence and significance of motivational-structural rules in some bird and mammal sounds. Am. Nat. 111:855-869.

Mountcastle, V.B., editor. 1980. Medical physiology, 14th ed. Vol. one. The C.V. Mosby Co., St. Louis.

Moynihan, M.H. 1962. The organization and probable evolution of some mixed species flocks of neotropical birds. Smithsonian Misc. Collections 143(7):1-140.

Moynihan, M.H. 1970. Control, suppression, decay, disappearance and replacement of displays. J. Theor. Biol. 29:85-112.

Moynihan, M.H., and A.F. Rodaniche. 1977. Communication, crypsis, and mimicry among cephalopods. In T.A. Sebeok, editor. How animals communicate. Indiana University Press, Bloomington.

Mrosovosky, N., and T. Powley. 1977. Set points for body weight and fat. Behav. Biol. 20:205-233.

Mueller, H.C. 1974. The development of prey recognition and predatory behavior in the American kestrel Falco sparverius. Behaviour 49:313-324.

Mueller, H.C. 1977. Prey selection in the American kestrel: experiments with two species of prey. Am. Nat. 3:25-29.

Mueller, H.C., and D.D. Berger. 1973. The daily rhythm of hawk migration at Cedar Grove, Wisconsin. Auk 90(3):591-596.

Mueller, H.C., and P.G. Parker. 1980. Naive ducklings show different cardiac response to hawk than to goose models. Behaviour 74:1-2.

Muller, H.J. 1932. Some genetic aspects of sex. Am. Nat. 66:118-138.

Müller-Schwarze, D. 1968. Play deprivation in deer. Behaviour 31:144-162.

Müller-Schwarze, D. 1971. Pheromones in black-tailed deer (Odocoileus hemionus columbianus). Anim. Behav. 19:141-152.

Müller-Schwarze, D., and R.M. Silverstein, editors. 1980. Chemical signals, vertebrates and aquatic invertebrates. Plenum Press, New York.

Muntz, W.R.A. 1962a. Microelectrode recordings from the diencephalon of the frog (Rana pipiens), and a blue-sensitive system. J. Neurophysiol. 25:699-711.

Muntz, W.R.A. 1962b. Effectiveness of different colors of light in releasing the positive phototactic behavior of frogs, and a possible function of the retinal projection to the diencephalon. J. Neurophysiol. 25:712-720.

Muntz, W.R.A. 1963a. The development of phototaxis in the frog (Rana temporaria). J. Exp. Biol. 40:371-379.

Muntz, W.R.A. 1963b. Phototaxis and green rods in urodeles. Nature 199:620.

Mykytowycz, R., and E.R. Hesterman. 1975. An experimental study of aggression in captive European rabbits, Oryctolagus cuniculus (L.). Behaviour 52:104-123.

Myrberg, A.A., Jr. 1972. Social dominance and territoriality in the bicolor damselfish, Eupomacentrus partitus (Poey) (Pisces: Pomacentridae). Behaviour 41:207-231.

Mysterud, I., and H. Dunker. 1979. Mammal ear mimicry: a hypothesis on the behavioural function of owl "horns." Anim. Behav. 27:315-317.

Nagel, W.A. 1894. Experimentelle sinnesphysiologische Untersuchungen an Coelenteraten. Arch. Ges. Physiol. 57:493-552.

Nash, J. 1970. Developmental psychology. Prentice-Hall, Inc., Englewood Cliffs, N.J.

Nauta, W.J.H., and M. Feirtag. 1979. The organization of the brain. Sci. Am. 241(3):88-111.

Nebes, D., and R.W. Sperry. 1971. Hemispheric deconnection syndrome with cerebral birth injury in the dominant arm area. Neuropsychologia 9:247-259.

Needham, J. 1959. A history of embryology. Abelard-Schuman, New York.

Neisser, U. 1966. Cognitive psychology. Appleton-Century-Crofts, New York.

Nelson, M.E., and L.D. Mech. 1981. Deer social organization and wolf predation in northeastern Minnesota. Wildlife Monogr. 77:1-53.

Nicolai, V.J. 1976. Evolutive neuerungen in der Balz von haustaubenrassen als ergebnis menschlicher zuchtwahl. Z. Tierpsychol. 40:225-243.

Noble, G.K. 1936. Courtship and sexual selection of the flicker (Colaptes auratus luteus). Auk 53:269-282.

Norton-Griffiths, M. 1969. The organization, control, and development of parental feeding in the oystercatcher (Haematopus ostralegus). Behaviour 34:55-114.

Nottebohm, F. 1981. A brain for all seasons: cyclical anatomical changes in song control nuclei of the canary brain. Science 214:1368-1370.

Novin, D., W. Wyrwicka, and G.A. Bray, editors. 1976. Hunger: basic mechanisms and clinical implications. Raven Press, New York.

Numbers, R.L. 1982. Creationism in 20th-century America. Science 218:538-544.

O'Brien, D.F. 1982. The chemistry of vision. Science 218:961-965.

Ohmart, R.D., and R.C. Lasiewski. 1971. Roadrunners: energy conservation by hypothermia and absorption of sunlight. Science 172:67-69.

Olds, J., and P. Milner. 1954. Positive reinforcement produced by electrical stimulation of septal area and other regions of the rat brain. J. Comp. Physiol. Psychol. 47:419-427.

Oppenheim, R.W. 1972. Prehatching and hatching behaviour in birds: a comparative study of altricial and precocial species. Anim. Behav. 20:644-655.

Orians, G.H., and N.E. Pearson. 1979. On the theory of central place foraging. In D.J. Horn, G.R. Stairs, and R.D. Mitchell, editors. Analysis of ecological systems. Ohio State University Press, Columbus.

Ostrom, J.H. 1979. Bird flight: how did it begin? Am. Sci. 67:46-56.

Ostrom, J.H. 1980. The evidence for endothermy in dinosaurs. In D.K. Thomas and E.C. Olson, editors. A cold look at the warm-blooded dinosaurs. Westview Press, Inc., Boulder, Colo.

Otis, L.S., and J.A. Cerf. 1963. Conditioned avoidance learning in two fish species. Psychol. Rep. 12:679-682.

Otte, D. 1972. Simple versus elaborate behavior in grasshoppers: an analysis of communication in the genus *Syrbula*. Behaviour 42:291-322.

Oye, P. van. 1920. Over het geheugen by fr flatwormen en andere biologische waarnemingen bji deze dieren. Natuurwet. Tydschr. 2:1.

Page, T.L. 1982. Transplantation of the cockroach circadian pacemaker. Science 216:73-75.

Palka, J., and K.S. Babu. 1967. Toward the physiological analysis of defensive responses of scorpions. Z. Vergl. Physiol. 55:286-298.

Palm, C.E., et al. 1970. Vertebrate pests: problems and control. National Academy of Sciences, Washington, D.C.

Panaman, R. 1981. Behaviour and ecology of free-ranging female farm cats (*Felis catus* L.). Z. Tierpsychol. 56:59-73.

Pantin, C.F.A. 1952. The elementary nervous system. Proc. R. Soc. Lond. 140:147-168.

Parker, G.A. 1978. Searching for mates. In J.R. Krebs and N.B. Davies, editors. Behavioural ecology. Blackwell Scientific Publications, Ltd., London.

Parker, G.H. 1896. The reactions of *Metridium* to food and other substances. Bull. Mus. Harvard 29:107-119.

Parker, G.H. 1919. The elementary nervous system. J.B. Lippincott Co., Philadelphia.

Partridge, B.L. 1982. The structure and function of fish schools. Sci. Am. 246(6):114-123.

Partridge, L. 1974. Habitat selection in titmice. Nature 247:573-574.

Partridge, L. 1976a. Individual differences in feeding efficiencies and feeding preferences of captive great tits. Anim. Behav. 24:230-240.

Partridge, L. 1976b. Field and laboratory observations on the foraging and feeding techniques of blue tits (*Parus caeruleus*) and coal tits (*P. ater*) in relation to their habitats. Anim. Behav. 24:534-544.

Patterson, F. 1978. The gestures of a gorilla: sign language acquisition in another pongid species. Brain Lang. 5:72-97.

Pavlov, I.P. 1906, 1927. Conditioned reflex. Oxford University Press, Oxford. (Translated by G.V. Aurep.)

Pearl, R. 1903. The movements and reactions of freshwater planaria. Q. J. Microsc. Sci. 46:509-714.

Pearson, O.P. 1966. The prey of carnivores during one cycle of mouse abundance. J. Anim. Ecol. 35:217-233.

Peck, J.W. 1978. Rats defend different body weights depending on palatability and accessibility of their food. J. Comp. Physiol. Psychol. 92:555-570.

Peckham, G.W., and E.G. Peckham. 1894. The sense of sight in spiders with some observations on the color sense. Trans. Wisc. Acad. Sci. 10:231-261.

Pelkwijk, J., J. Ter, and N. Tinbergen. 1937. Eine reizbiologische Analyse einiger Verhaltensweisen von *Gasterosteus aculeatus* L. Z. Tierpsychol. 1:193-204.

Pengelley, E.T. 1974. Circannual clocks. Academic Press, Inc., New York.

Pengelley, E.T., and S.J. Asmundson. 1971. Annual biological clocks. Sci. Am. 224(4):72-79.

Pennycuick, C.J., and J.A. Rudnai. 1970. A method of identifying individual lions *Panthera leo* with an analysis of the reliability of identification. J. Zool. Lond. 160:497-508.

Pepperberg, I.M. 1981. Functional vocalizations by an African grey parrot. Z. Tierpsychol. 55:139-160.

Perdeck, A.C. 1958. Two types of orientation in migrating starlings, *Sturnus vulgaris* L., and chaffinches, *Fringilla coelebs* L., as revealed by displacement experiments. Ardea 46:1-37.

Peters, H. 1932. Experimente uber die Orientierung der Kreuzspinne *Epeira diademata* Cl. im Netz. Zool. Jahrb. 51:239-288.

Peters, M. 1971. Sensory mechanisms of homing in salmonids: a comment. Behaviour 39:18-19.

Peters, R.P., and L.D. Mech. 1975. Scent marking in wolves. Am. Sci. 63:628-637.

Peterson, S.R., and R.S. Ellarson. 1977. Food habits of oldsquaws wintering on Lake Michigan. Wilson Bull. 89(1):81-91.

Pettingill, O.S., Jr. 1970. Ornithology in laboratory and field. Burgess Publishing Co., Minneapolis, Minn.

Pfaff, D.W., editor. 1982. The physiological mechanisms of motivation. Springer-Verlag New York, Inc., New York.

Phelps, M.E., D.E. Kuhl, and J.C. Mazziotta. 1981. Metabolic mapping of the brain's response to visual stimulation: studies in humans. Science 211:1445-1448.

Pianka, E.R. 1970. On r- and K-selection. Am. Nat. 104:592-597.

Picton, T.W., S.A. Hillyard, H.I. Krausz, and R. Galambos. 1973. Human auditory evoked potentials. I. Evaluation of components. Electroenceph. Clin. Neurophysiol. 36:179-190.

Pierotti, R. 1982. Habitat selection and its effect on reproductive output in the herring gull in Newfoundland. Ecology 63(3):854-868.

Pietrewicz, A.T., and A.C. Kamil. 1979. Search image formation in the blue jay (*Cyanocitta cristata*). Science 204:1332-1333.

Pietsch, T.W., and D.B. Grobecker. 1978. The compleat angler: aggressive mimicry in an antennariid anglerfish. Science 201:369-370.

Pittendrigh, C.S. 1958. Perspectives in the study of biological clocks. In A.A. Buzzati-Traverso, editor. Perspectives in marine biology. Scripps Institution of Oceanography, La Jolla, Calif.

Polyak, S. 1941. The retina. University of Chicago Press, Chicago.

Polyak, S. 1957. In H. Kluver, editor. The vertebrate visual system. University of Chicago Press, Chicago.

Poole, T.B. 1966. Aggressive play in polecats. In P.A. Jewell and C. Loizos, editors. Play, exploration, and territory in mammals. Academic Press, Inc., London.

Popper, A.N., and S. Combs. 1980. Auditory mechanisms in teleost fishes. Am. Sci. 68:429-440.

Porter, J.P. 1904. A preliminary study of the psychology of the English sparrow. Am. J. Psychol. 15:313-346.

Potts, G.W. 1973. The ethology of *Labroides dimidiatus* (Cuv. and Val) (Labridae, Pisces) on aldabra. Anim. Behav. 21:250-291.

Powell, R.A. 1978. A comparison of fisher and weasel hunting behavior. Carnivore 1:28-34.

Powley, T.L. 1977. The ventromedial hypothalamic syndrome, satiety, and a cephalic hypothesis. Psychol. Rev. 84:89-126.

Powley, T.L., and R.E. Keesey. 1970. Relationship of body weight to the lateral hypothalamic syndrome. J. Comp. Physiol. Psychol. 70:25-36.

Powley, T.L., and C.A. Opsahl. 1974. Ventromedial hypothalamic obesity by subdiagraphramatic vagotomy. Am. J. Physiol. 226:25-33.

Premack, D. 1972. Teaching language to the ape. Sci. Am. 227:92-99.

Premack, D., and G. Woodruff. 1978. Chimpanzee problem-solving: a test for comprehension. Science 202:532-535.

Preston, J.L. 1978. Communication systems and social interactions in a goby-shrimp symbiosis. Anim. Behav. 26:791-802.

Preyer, W. 1885. Specielle physiologie des embryo. Grieben, Leipzig.

Pribram, K.H., and D.E. Broadbent, editors. 1970. Biology of memory. Academic Press, Inc., New York.

Price, E.O., and A.W. Stokes. 1975. Animal behavior in laboratory and field. W.H. Freeman & Co., Publishers, San Francisco.

Pringle, J.W.S. 1957. Insect flight. Cambridge University Press, Cambridge, England.

Provine, R.R. 1979. "Wing-flapping" develops in wingless chicks. Behav. Neural Biol. 27:233-237.

Pycraft, W.P. 1912. The infancy of animals. The Hutchinson Publishing Group, Ltd., London.

Pyke, G.H. 1981a. Why hummingbirds hover and honeyeaters perch. Anim. Behav. 29:861-867.

Pyke, G.H. 1981b. Honeyeaters foraging: a test of optimal foraging theory. Anim. Behav. 29:878-888.

Pyke, G.H. 1981c. Optimal foraging in hummingbirds: rule of movement between inflorescences. Anim. Behav. 29:889-896.

Pyke, G.H., H.R. Pulliam, and E.I. Charnov. 1977. Optimal foraging: a selective review of theory and tests. Q. Rev. Biol. 52(2):137-154.

Raber, H. 1950. Das Verhalten gefangenen Waldohreulen (Asio otus otus) and Waldkauze (Strix aluco aluco) zur Beute. Behaviour 2:1-95.

Randall, J.A. 1978. Behavioral mechanisms of habitat segregation between sympatric species of Microtus: habitat preference and interspecific dominance. Behav. Ecol. Sociobiol. 3:187-202.

Randall, J.A. 1981. Comparison of sandbathing and grooming in two species of kangaroo rat. Anim. Behav. 29:1213-1219.

Rasa, O.A.E. 1971. The causal factors and function of "yawning" in Microspathodon chrysurus (Pisces, Pomacentridae). Behaviour 39:39-57.

Rasmussen, T., and W. Penfield. 1947. Further studies of sensory and motor cerebral cortex of man. Fed. Proc. 6:452.

Ratner, S.C., and R. Boice. 1975. Effects of domestication on behaviour. In E.S.E. Hafez, editor. The behavior of domestic animals, 3rd ed. The Williams & Wilkins Co., Baltimore.

Ratner, S.C., and K.R. Miller. 1959. Classical conditioning in earthworms, Lumbricus terrestris. J. Comp. Physiol. Psychol. 52:102-105.

Ray, T.S., and C.C. Andrews. 1980. Antbutterflies: butterflies that follow army ants to feed on antbird droppings. Science 210:1147-1148.

Razran, G.A. 1971. Mind in evolution: an east-west synthesis. Houghton Mifflin Co., Boston.

Regnier, F.E, and E.O. Wilson. 1971. Chemical communication and "propaganda" in slave-maker ants. Science 172:267-269.

Renner, M. 1960. The contribution of the honey bee to the study of time-sense and astronomical orientation. Cold Spring Harbor Symp. Quant. Biol. 25:361-367.

Rhijn, J.G. van. 1977. The patterning of preening and other comfort behaviour in a herring gull. Behaviour 63:71-109.

Richards, O.W. 1927. The specific characters of the British bumblebees (Hymenoptera). Trans. Ent. Soc. Lond. 75:233-268.

Richardson, W.J. 1978. Timing and amount of bird migration in relation to weather: a review. Oikos 30:224-272.

Robinson, D.N. 1981. The psychobiology of consciousness. Am. Sci. 69:463-464.

Robson, E.A. 1961. Some observations on the swimming behaviour of the anemone Stomphia coccinea. J. Exp. Biol. 38:343-363.

Rodgers, J.A., Jr. 1978. Breeding behavior of the Louisiana heron. Wilson Bull. 90:45-59.

Roeder, K.D. 1948. Organization of the ascending giant fiber system in the cockroach (Periplaneta americana L.) J. Exper. Zool. 108:243-262.

Roeder, K.D. 1962. The behaviour of free flying moths in the presence of artificial ultrasonic pulses. Anim. Behav. 10:300-304.

Roeder, K.D. 1963. Echoes of ultrasonic pulses from flying moths. Biol. Bull. 124:200-210.

Roeder, K.D. 1967. Nerve cells and insect behavior. Harvard University Press, Cambridge, Mass.

Roeder, K.D., and A.E. Treat. 1961. The detection and evasion of bats by moths. Am. Sci. 49:135-148.

Rolls, E.T. 1978. Neurophysiology of feeding. Trends Neurosci. 1:1-3.

Romanes, G.J. 1884. Mental evolution in animals. Keegan, Paul, Trench & Co., London.

Romanes, G.J. 1889. Mental evolution in man. D. Appleton & Co., New York.

Rosenzweig, M.R., and E.L. Bennett, editors. 1976. Neural mechanisms of learning and memory. MIT Press, Cambridge, Mass.

Rosenzweig, M.R., E.L. Bennett, and M.C. Diamond. 1972. Brain changes in response to experience. Sci. Am. 226(2):22-29.

Ross, D.M. 1965. The behavior of sessile coelenterates in relation to some conditioning experiments. Anim. Behav. Suppl. 1:43-55.

Ross, W.N., B.M. Salzberg, L.B. Cohen, and H.V. Davila. 1974. A large change in dye absorption during the action potential. Biophys. J. 14:983-986.

Rossler, V.E. 1978. Ubertragung von verhaltensweisen durch transplantation von anlagen neuroanatomischer strukturen bei amphibienlarven. Z. Tierpsychol. 46:1-13.

Rothenbuhler, N. 1964. Behavior genetics of nest cleaning in honey bees. IV. Responses of F_1 and backcross generations to disease-killed brood. Am. Zool. 4:111-123.

Roubik, D.W. 1978a. Competitive interactions between neotropical pollinators and Africanized honey bees. Science 201:1030-1032.

Roubik, D.W. 1978b. Curve-fitting (response to R.M. Hazen). Science 202:823.

Rovner, J.S., G.A. Higashi, and R.F. Foelix. 1973. Maternal behavior in wolf spiders: the role of abdominal hairs. Science 182:1153-1155.

Rowell, T.E., R.A. Hinde, and Y. Spencer-Booth. 1964. "Aunt"-infant interaction in captive rhesus monkeys. Anim. Behav. 12:219-226.

Rumbaugh, D.M., editor. 1977. Language learning by a chimpanzee: the LANA Project. Academic Press, Inc., New York.

Rusak, B., and G. Groos. 1982. Suprachiasmatic stimulation phase shifts rodent circadian rhythms. Science 215:1407-1409.

Ruse, M. 1979. Sociobiology: sense or nonsense. D. Reidel Publishing Co., Boston.

Rushforth, N.B. 1973. Behavioral modifications in coelenterates. In W.C. Corning, J.A. Dyal, and A.O.D. Willows, editors. Invertebrate learning. Vol. 1. Plenum Press, New York.

Sackett, G.P. 1968. Abnormal behavior in laboratory reared rhesus monkeys. In M.W. Fox, editor. Abnormal behavior in animals. W.B. Saunders Co., Philadelphia.

Sade, D.S. 1965. Some aspects of parent-offspring and sibling relations in a group of rhesus monkeys, with a discussion of grooming. Am. J. Physical Anthropol. 23:1-17.

Sagan, C. 1977. The dragons of Eden. Random House, Inc., New York.

Sagan, C. 1979. Broca's brain. Random House, Inc., New York.

Sale, P.F. 1971. Apparent effect of prior experience on a habitat preference exhibited by the reef fish, Dascyllus aruanus (Pisces: Pomacentridae). Anim. Behav. 19:251-256.

Sales, G., and D. Pye. 1974. Ultrasonic communication by animals. Chapman & Hall, Ltd., London.

Sanders, G.D. 1975. The cephalopods. In W.C. Corning, J.A. Dyal, and A.O.D. Willows, editors. Invertebrate learning, Vol. 3. Plenum Press, New York.

Sanger, V.L., and A.H. Hamdy. A strange fright-flight behavior pattern (hysteria) in hens. Am Vet. Med. Assoc. 140:455-459.

Sargant, W.W. 1957. Battle for the mind, a physiology of conversion and brainwashing. Doubleday & Co., Inc., New York.

Sauer, E.G.F. 1954. Die Entwicklung der Lautau Berungen vom Ei ab schalldicht gehaltener Dorngrasmucken (Sylvia c. communis Latham) im Vergleich mit spater isolierten und mit wildlebenden Artgenossen. Z. Tierpsychol. 11:10-93.

Saunders, D.S. 1977. An introduction to biological rhythms. John Wiley & Sons, Inc., New York.

Savage-Rumbaugh, E.S., D.M. Rumbaugh, and S. Boysen. 1980. Do apes use language? Am. Sci. 68:49-61.

Savarie, P.J., and R.T. Sterner. 1979. Evaluation of toxic collars for selective control of coyotes that attack sheep. J. Wildlife Manage. 43(3):1979.

Schaller, G.B. 1967. The deer and the tiger. University of Chicago Press, Chicago.

Schaller, G.B. 1972. The Serengeti lion. University of Chicago Press, Chicago.

Schally, A.V., T.W. Redding, H.W. Lucien, and J. Meyer. 1967. Enterogasterone inhibits eating by fasted mice. Science 157:210-211.

Scheller, R.H., J.F. Jackson, L.B. McAllister, J.H. Schwartz, E.R. Kandel, and R. Axel. 1982. A family of genes that codes for ELH, a neuropeptide eliciting a stereotyped pattern of behavior in Aplysia. Cell 28:709-719.

Schenkel, R. 1966. Play, exploration and territoriality in the wild lion. In P.A. Jewell and C. Loizos, editors. Play, exploration, and territory in mammals. Academic Press, Inc., London.

Scherzinger, W. 1970. Zum Aktionssystem des Sperlingskauzes (Glaucidium passerinum L.) Zoologica 118:1-120.

Schiller, C.H. 1957. Instinctive behavior. International Universities Press, Inc., New York.

Schleidt, W.M. 1961. Reaktionen von truthuhnern auf fliegende Raubvogel und Versuche zur Analyse ihrer AAM's. Z. Tierpsychol. 18:534-560.

Schleidt, W.M. 1974. How "fixed" is the fixed action pattern? Z. Tierpsychol. 36:184-211.

Schleidt, W.M., M. Schleidt, and M. Magg. 1960. Storung der Mutter-Kind-Beiziehung bei Truthuhern durch Gehorverlust. Behaviour 16:254-260.

Schmidt, J.P. 1968. Psychosomatics in veterinary medicine. In M.W. Fox, editor. Abnormal behavior in animals. W.B. Saunders Co., Philadelphia.

Schmidt-Koenig, K. 1975. Migration and homing in animals. Springer-Verlag New York, Inc., New York.

Schmidt-Koenig, K., and W. Keeton. 1978. Animal migration, navigation, and homing. Springer-Verlag New York, Inc., New York.

Schmidt-Koenig, K., and C. Walcott. 1978. Tracks of pigeons homing with frosted lenses. Anim. Behav. 26:480-486.

Schneider, D. 1974. The sex-attractant receptor of moths. Sci. Am. 231(1):28-35.

Schneirla, T.C. 1929. Learning and orientation in ants. Comp. Psychol. Monogr. 6(4):1-143.

Schneirla, T.C. 1959. An evolutionary and developmental theory of biphasic processes underlying approach and withdrawal. Neb. Symp. Motivation 7:1-42.

Schnell, G.D. 1970a. A phenetic study of the suborder Lari (Aves). I. Methods and results of principal components analyses. Syst. Zool. 19:35-57.

Schnell, G.D. 1970b. A phenetic study of the suborder Lari (Aves). II. Phenograms, discussion, and conclusions. Syst. Zool. 19:264-302.

Schoener, T.W. 1971. Theory of feeding strategies. Ann. Rev. Ecol. Syst. 2:369-404.

Schoener, T.W. 1981. An empirically based estimate of home range. Theor. Popul. Biol. 20:281-325.

Schoener, T.W., and A. Schoener. 1982. Intraspecific variation in home-range size in some anolis lizards. Ecology 63(3):809-823.

Schreiber, R.W. 1977. Maintenance behavior and communication in the brown pelican. Ornithol. Monogr. 22:1-78.

Schüz, E. 1971. Grundriss der Vogelzugskunde. Verlag, Paul Parey, Berlin.

Schwartzbaum, J.S., and H.P. Ward. 1958. An osmotic factor in the regulation of food intake in the rat. J. Comp. Physiol. Psychol. 51:555-560.

Scientific American. 1979. The brain (special issue). Sci. Am. 241(3). (September issue.)

Sclafani, A., and C.N. Berner. 1977. Hyperphagia and obesity produced by parasagittal and coronal hypothalamic knife cuts. J. Comp. Physiol. Psychol. 91:1000-1018.

Sclafani, A., and D. Springer. 1976. Dietary obesity: similarities to hypothalamic and human obesity syndromes. Physiol. Behav. 17:461-471.

Searcy, W.A., and P. Marler. 1981. A test for responsiveness to song structure and programming in female sparrows. Science 213:926-928.

Searcy, W.A., and K. Yasukawa. 1983. Sexual selection and red-winged blackbirds. Am. Sci. 71:166-174.

Sebeok, T.A., editor. 1977. How animals communicate. Indiana University Press, Bloomington.

Seely, M.K., and W.J. Hamilton III. 1976. Fog catchment sand trenches constructed by tenebrionid beetles, Lepidochora, from the Namib Desert. Science 193:484-486.

Selander, R.K. 1972. Sexual selection and dimorphism in birds. In B. Campbell, editor. Sexual selection and the descent of man. Aldine Publishing Co., Chicago.

Seligman, M.E.P. 1975. Helplessness: on depression, development, and death. W.H. Freeman & Co., Publishers, San Francisco.

Shapiro, C.M., R. Bortz, and D. Mitchell. 1981. Slow-wave sleep: a recovery period after exercise. Science 214:1253-1254.

Sherry, D.F. 1981. Parental care and the development of thermoregulation in red junglefowl. Behaviour 76:250-279.

Sherry, D.F., J.R. Krebs, and R.J. Cowie. 1981. Memory for the location of stored food in marsh tits. Anim. Behav. 29:1260-1266.

Shettleworth, S.J. 1983. Memory in food-hoarding birds. Sci. Am. 248(3):102-110.

Sibatani, A. 1980. The Japanese brain. Science 80 1(8):22-27.

Sidman, R.L., S.H. Appel, and J.L. Fuller. 1965. Neurological mutants of the mouse. Science 150:513-516.

Sidtis, J.J., B.T. Volpe, J.D. Holtzman, D.H. Wilson, and M.S. Gazzaniga. 1981. Cognitive interaction after staged callosal section: evidence for transfer of semantic activation. Science 212:344-346.

Sih, A. 1980. Optimal behavior: can foragers balance two conflicting demands? Science 210:1041-1043.

Silberglied, R.E. 1979. Communication in the ultraviolet. Ann. Rev. Ecol. Syst. 10:373-398.

Silverberg, J. 1980. Sociobiology, the new synthesis? An anthropologist's perspective. In G.W. Barlow and J. Silverberg, editors. Sociobiology: beyond nature/nurture? Westview Press, Inc., Boulder, Colo.

Simmons, J.A. 1979. Perception of echo phase information in bat sonar. Science 204:1336-1338.

Simmons, J.A., M.B. Fenton, and M.J. O'Farrell. 1979. Echolocation and pursuit of prey by bats. Science 203:14-21.

Simon, D., and R.H. Barth. 1977. Sexual behavior in the cockroach genera Periplaneta and Blatta. I. Descriptive aspects. Z. Tierpsychol. 44:80-107.

Simpson, G.G. 1970. Uniformitarianism: an inquiry into principle, theory, and method in geohistory and biohistory. In M.K. Hecht and W.C. Steere, editors. Essays in evolution and genetics in honor of Theodosius Dobzhansky. Appleton-Century-Crofts, New York.

Simpson, M.J.A., and A.E. Simpson. 1977. One-zero and scan methods for sampling behaviour. Anim. Behav. 25:726-731.

Sims, E.A.H. 1976. Experimental obesity, dietary-induced thermogenesis, and their clinical implications. Clin. Endocrinol. Metab. 5:377-395.

Skinner, B.F. 1938. The behavior of organisms. Appleton-Century-Crofts, New York.

Skinner, B.F. 1974. About behaviorism. Random House, Inc., New York.

Skinner, B.F. 1976. Particulars of my life. Alfred A. Knopf, Inc., New York.

Skinner, B.F. 1979. The shaping of a behaviorist: part two of an autobiography. Alfred A. Knopf, Inc., New York.

Skinner, B.F. 1981. Selection by consequences. Science 212:501-504.

Sluckin, W. 1979. Fear in animals and man. Van Nostrand Reinhold Co., New York.

Small, W.S. 1901. Experimental study of the mental processes of the rat. Am. J. Psychol. 12:206-239.

Smartt, R.A. 1978. A comparison of ecological and morphological overlap in a Peromyscus community. Ecology 59(2):216-220.

Smith, E.O., editor. 1977. Social play in primates. Academic Press, Inc., New York.

Smith, J.N.M., and R. Dawkins. 1971. The hunting behaviour of individual great tits in relation to spatial variations in their food density. Anim. Behav. 19:695-706.

Smith, N.G. 1968. The advantages of being parasitized. Nature 219:690-694.

Smith, R.L. 1979. Repeated copulation and sperm precedence: paternity assurance for a male brooding water bug. Science 205:1029-1031.

Smith, W.J. 1977. The behavior of communicating. Harvard University Press, Cambridge, Mass.

Sneath, P.H.A., and R.R. Sokal. 1973. Numerical taxonomy. W.H. Freeman & Co., Publishers, San Francisco.

Snodgrass, R.E. 1935. Principles of insect morphology. McGraw-Hill Book Co., New York.

Snowdon, C.T. 1969. Motivation, regulation, and the control of meal parameters with oral and intragastric feeding. J. Comp. Physiol. Psychol. 69:91-100.

Snowdon, C.T. 1975. Production of satiety with small intraduodenal infusions in the rat. J. Comp. Physiol. Psychol. 88:231-238.

Snyder, N.F.R., and J.D. Taapken. 1977. Puerto Rican parrots and nest predation by pearly-eyed thrashers. In S.A. Temple, editor. Endangered birds. University of Wisconsin Press, Madison.

Snyder, S.H. 1980. Brain peptides as neurotransmitters. Science 209:976-983.

Spear, N.E., and B.A. Campbell. 1979. Ontogeny of learning and memory. Halsted Press, New York.

Spear, N.E., and R.R. Miller, editors. 1981. Information processing in animals. Lawrence Erlbaum Associates, Inc., Hillsdale, N.J.

Spencer, H. 1855. Principles of psychology. D. Appleton & Co., New York.

Spencer, H. 1892. The principles of ethics. Vol. I. D. Appleton & Company, New York.

Spencer, H. 1896. Principles of psychology, 2nd ed. D. Appleton & Co., New York.

Sperry, R.W. 1970. Perception in the absence of the neocortical commissures. In Perception and its disorders. Res. Publ. Assoc. Res. Nervous and Mental Disease. Vol. 48.

Sperry, R.W. 1974. Lateral specialization in the surgically separated hemispheres. In F.O. Schmitt and F.G. Worden, editors. The neurosciences: third study program. MIT Press, Cambridge, Mass.

Sperry, R.W. 1982. Some effects of disconnecting the cerebral hemispheres. Science 217:1223-1226.

Spieth, H.T. 1974. Courtship behavior in Drosophila. Ann. Rev. Ent. 19:385-405.

Sprague, R.H., and J.J. Anisko. 1973. Elimination patterns in the laboratory beagle. Behaviour 47:257-267.

Srb, A.M., R.D. Owen, and R.S. Edgar. 1965. General genetics. W.H. Freeman & Co., Publishers, San Francisco.

Stanley, S.M. 1974. Relative growth of the titanothere horn: a new approach to an old problem. Evolution 28:447-457.

Stebbins, G.L., and F.J. Ayala. 1981. Is a new evolutionary synthesis necessary? Science 213:967-971.

Stefansson, V. 1944. The friendly arctic. Macmillan, Inc., New York.

Stellar, E. 1954. The physiology of motivation. Psychol. Rev. 61:5-22.

Stephenson, G.R., and T.W. Roberts. 1977. The SSR system 7: a general encoding system with computerized transcription. Behav. Res. Meth. Instrument. 9:434-441.

Stetson, M.H., and M. Watson-Whitmyre. 1976. Nucleus suprachiasmaticus: the biological clock in the hamster? Science 191:197-199.

Stevens, C.F. 1966. Neurophysiology: a primer. John Wiley & Sons, Inc., New York.

Stoddart, D.M. 1976. Mammalian odours and pheromones. Camelot Press, Ltd., London.

Storm, R.M. 1967. Animal orientation and navigation. Oregon State University Press, Corvallis.

Street, P. 1976. Animal migration and navigation. Charles Scribner's Sons, New York.

Stunkard, A.J., editor. 1980. Obesity. W.B. Saunders Co., Philadelphia.

Sulak, K.J. 1975. Cleaning behaviour in the centrarchid fishes, Lepomis macrochirus and Micropterus salmoides. Anim. Behav. 23:331-334.

Suomalainen, E. 1969. Evolution in parthenogenetic Curculionidae. Evol. Biol. 3:261-296.

Suomi, S.J., and H.F. Harlow. 1972. Social rehabilitation of isolate-reared monkeys. Dev. Psychol. 6:487-496.

Sutherland, N.S., and N.J. Mackintosh. 1971. Mechanisms of animal discrimination learning. Academic Press, Inc., New York.

Swenson, E.A. 1929. The active simple movements of the albino rat fetus: the order of their appearance, their qualities, and their significance. Anat. Rec. 42:40.

Takahashi, J.S., and M. Zatz. 1982. Regulation of circadian rhythmicity. Science 217:1104-1111.

Tavolga, W.N., editor. 1976. Sound reception in fishes. Benchmark papers in animal behavior. Vol. 7. Dowden, Hutchinson, & Ross, Inc., Stroudsburg, Pa.

Tavolga, W.N., and J. Wodinsky. 1963. Auditory capacities in fishes: pure tone thresholds in nine species of marine teleosts. Bull. Am. Mus. Nat. Hist. 126:177, 179-239. (Reprinted in Tavolga 1976.)

Taylor, E.L. 1962. The place of animal behavior studies in veterinary science. Vet. Rec. 74:521-524.

Tegner, M.J., and P.K. Dayton. 1976. Sea urchin recruitment patterns and implications of commercial fishing. Science 196:324-326.

Teitelbaum, P., and A.N. Epstein. 1962. The lateral hypothalamic syndrome: recovery of feeding and drinking after lateral hypothalamic lesions. Psychol. Rev. 69:74-90.

Temple, S.A., editor. 1977. Endangered birds: management techniques for preserving threatened species. University of Wisconsin Press, Madison.

Terrace, H.S., L.A. Petitto, R.J. Sanders, and T.G. Bever. 1979. Can an ape create a sentence? Science 206(4421):891-902.

Tets, G.F. Van. 1965. A comparative study of some social communication patterns in the Pelecaniformes. Ornith. Monogr. 2:1-88.

Thiessen, D.D., K. Owen, and M. Whitsett. 1970. Chromosome mapping of behavioral activities. In G. Lindzey and D.D. Thiessen, editors. Contributions to behavior genetic analysis: the mouse as a prototype. Appleton-Century-Crofts, New York.

Thomas, D.W., and J. Mayer. 1978. Meal size as a determinant of food intake in normal and hypothalamic obese rats. Physiol. Behav. 21:113-117.

Thomas, R.D.K., and E.C. Olson. 1980. A cold look at the warm-blooded dinosaurs. American Association for the Advancement of Science (AAAS) Selected Symposium 28. Westview Press, Inc., Boulder, Colo.

Thompson, C.R., and R.M. Church. 1980. An explanation of the language of a chimpanzee. Science 208:313-314.

Thompson, R.F., and W.A. Spencer. 1966. Habituation: a model phenomenon for the study of neuronal substrates of behavior. Psychol. Rev. 73:16-43.

Thorndike, E.L. 1911. Animal intelligence: experimental studies. Macmillan Publishing Co., Inc., New York.

Thorpe, W.H. 1948. The modern concept of instinctive behaviour. Bull. Anim. Behav. 7:2-12.

Thorpe, W.H. 1951. The definition of terms used in animal behaviour studies. Bull. Anim. Behav. 9:34-40.

Thorpe, W.H. 1958. The learning of song patterns by birds, with special reference to the song of the chaffinch, Fringilla coelebs. Ibis 100:535-570.

Thorpe, W.H. 1963. Learning and instinct in animals, 2nd ed. Methuen, London.

Thorpe, W.H. 1979. The origins and rise of ethology. Heinemann Educational Books, Ltd., London.

Tiger, L., and R. Fox. 1971. The human biogram. In L. Tiger and R. Fox, editors. The imperial animal. Holt, Rinehart & Winston, New York.

Tinbergen, L. 1960. The natural control of insects in pinewoods. I. Factors influencing the intensity of predation by song birds. Arch. Neerl. Zool. 13:265-343.

Tinbergen, N. 1942. An objectivistic study of the innate behavior of animals. Biblioth. Biotheor. 1:39-98.

Tinbergen, N. 1948. Social releasers and the experimental method required for their study. Wilson Bull. 60:6-51.

Tinbergen, N. 1951. The study of instinct. Oxford University Press, Inc., New York.

Tinbergen, N. 1952. The curious behavior of the stickleback. Sci. Am. 187(6):22-26.

Tinbergen, N. 1953. The herring gull's world. William Collins Sons & Co., Ltd., London.

Tinbergen, N. 1959. Comparative studies of the behavior of gulls (Laridae): a progress report. Behaviour 15:1-70.

Tinbergen, N. 1960. The evolution of behavior in gulls. Sci. Am. 203(6):118-130.

Tinbergen, N. 1965. Animal behavior. Time-Life Books, Inc., New York.

Toates, F.M., and T.R. Halliday. 1980. Analysis of motivational processes. Academic Press, Inc., New York.

Tobias, J.M. 1952. Some optically detectable consequences of activity in nerve. Cold Spring Harbor Symp. Quant. Biol. 17:15-25.

Tolman, E.C. 1932. Purposive behavior in animals and men. Century Co., New York.

Topoff, H., and J. Mirenda. 1980. Army ants do not eat and run: influence of food supply on emigration behaviour in Neiv amyrmex nigrescens. Anim. Behav. 28:1040-1045.

Trivers, R.L. 1971. The evolution of reciprocal altruism. Rev. Biol. 46:35-51.

Trivers, R.L. 1972. Parental investment and sexual selection. In B. Campbell, editor. Sexual selection and the descent of man. Aldine Publishing Co., Hawthorne, N.Y.

Trivers, R.L. 1974. Parent-offspring conflict. Am. Zool. 14:249-264.

Trumler, E. 1959. Das "Rossigkeitsgesicht" und ähnliches Ausdrucksverhalten bei Einhufern. Z. Tierpsychol. 16:478-488.

Tucker, D.W. 1959. A new solution to the Atlantic eel problem. Nature 183:495-501.

Turner, E.R.A. 1961. Survival values of different methods of camouflage as shown in a model population. Proc. Zool. Soc. Lond. 136:273-284.

Tuttle, M.D., and M.J. Ryan. 1981. Bat predation and the evolution of frog vocalizations in the neotropics. Science 214:677-678.

Tyler, M.J., and D.B. Carter. 1981. Oral birth of the young of the gastric brooding frog Rheobatrachus silus. Anim. Behav. 29:280-282.

Tyler, S. 1979. Time-sampling: a matter of convention. Anim. Behav. 27:801-810.

Uexküll, J. von. 1909. Umwelt und Innerwelt der tiere. Springer-Verlag, Berlin.

Uexküll, J. von. 1934. A stroll through the worlds of animals and men. In C.H. Schiller, editor. 1957. Instinctive behavior. International Universities Press, Inc., New York.

Underwood, H. 1977. Circadian organization in lizards: the role of the pineal organ. Science 195:587-589.

Ungar, G., editor. 1970. Molecular mechanisms in memory and learning. Plenum Press, New York.

Universities Federation for Animal Welfare. 1972. The U.F.A.W. handbook on the care and management of laboratory animals. Churchill Livingstone, London.

Urquhart, F.A. 1976. Found at last: the monarch's winter home. Natl. Geographic 150:160-173.

Urquhart, F.A., and N.R. Urquhart. 1979. Breeding areas and overnight roosting locations in the northern range of the monarch butterfly (Danaus plexippus plexippus) with a summary of associated migratory routes. Can. Field-Naturalist 93:41-47.

Valenstein, E.S., V.C. Cox, and J.W. Kakolewski. 1970. Reexamination of the role of the hypothalamus in motivation. Psychol. Rev. 77:16-31.

Van Der Kloot, W.G. 1956. Brains and cocoons. Sci. Am. 194(4):131-140.

van Tienhoven, A. 1968. Reproductive physiology of vertebrates. W.B. Saunders Co., Philadelphia.

Verplanck, W.S. 1957. A glossary of some terms used in the objective science of behavior. Psychol. Rev. 64(Suppl.):1-42.

Vieth, W., E. Curio, and U. Ernst. 1980. The adaptive significance of avian mobbing. III. Cultural transmission of enemy recognition in blackbirds: cross-species tutoring and properties of learning. Anim. Behav. 28:1217-1229.

Vilberg, T.R., and W.W. Beatty. 1975. Behavioral changes following VMH lesions in rats with controlled insulin levels. Pharmacol. Biochem. Behav. 3:377-384.

Vincent, L.E., and M. Bekoff. 1978. Quantitative analyses of the ontogeny of predatory behaviour in coyotes, Canis latrans. Anim. Behav. 26:225-231.

Visser, M. 1981. Prediction of switching and counterswitching based on optimal foraging. Z. Tierpsychol. 55:129-138.

Wade, M.J. 1980. Wright's view of evolution. Science 207:173-174.

Wade, N. 1976. Sociobiology: troubled birth for new discipline. Science 191:1151-1155.

Wade, N. 1980. Does man alone have language? Apes reply in riddles, and a horse says neigh. Science 208:1349-1351.

Walcott, C. 1969. A spider's vibration receptor: its anatomy and physiology. Am. Zool. 99:133-144.

Walcott, C., J.L. Gould, and J.L. Kirschvink. 1979. Pigeons have magnets. Science 205:1027-1028.

Walcott, C., and R.P. Green. 1974. Orientation of homing pigeons altered by a change in the direction of an applied magnetic field. Science 184:180-182.

Walkinshaw, L.H. 1973. Cranes of the world. Winchester Press, Inc., Tulsa, Okla.

Wallace, R.L. 1980. Ecology of sessile rotifers. Hydrobiologia 73:181-193.

Walter, H.E. 1908. The reactions of planaria to light. J. Exp. Zool. 5:35-163.

Wampler, R.S., and C.T. Snowdon. 1979. Development of VMH obesity in vagotomized rats. Physiol. Behav. 22:85-93.

Ward, P. 1965. Feeding ecology of the black-faced Dioch *(Quelea quelea)* in Nigeria. Ibis 107:173-214.

Warren, J.M. 1973. Learning in vertebrates. In D.A. Dewsbury and D.A. Rethlingshafer, editors. Comparative psychology. McGraw-Hill Book Co., New York.

Watkins, L.R., and D.J. Mayer. 1982. Organization of endogenous opiate and nonopiate pain control systems. Science 216:1185-1192.

Watson, J. 1965. The molecular biology of the gene. W.A. Benjamin, Inc., New York.

Watson, J.B. 1913. Psychology as the behaviorist views it. Psychol. Rev. 20:158-177.

Watson, J.B. 1930. Behaviorism. W.W. Norton & Co., Inc., New York.

Watson, R.I. 1971. The great psychologists, 3rd ed. J.B. Lippincott Co., Philadelphia.

Wecker, S.C. 1963. The role of early experience in habitat selection by the prairie deermouse, *Peromyscus maniculatus bairdi.* Ecol. Monogr. 33:307-325.

Wecker, S.C. 1964. Habitat selection. Sci. Am. 211(4):109-116.

Weeks, H.P., Jr. 1977. Abnormal nest building in the eastern phoebe. Auk 94:367-369.

Wehner, R. 1976. Polarized-light navigation by insects. Sci. Am. 235(1):106-114.

Weiss, B.A., and T.C. Schneirla. 1967. Intersituational transfer in the ant *Formica schaufussi* as tested in a two-phase single choice point maze. Behaviour 28:269-279.

Wells, M.J. 1958. Factors affecting reactions to *Mysis* by newly hatched *Sepia.* Behaviour 13:96-111.

Wells, M.J., and S.K.L. Buckley. 1972. Snails and trails. Anim. Behav. 20:345-355.

Wenner, A.M. 1964. Sound communication in honeybees. Sci. Am. 210(4):116-124.

Wenner, A.M. 1971. The bee language controversy. Educational Programs Improvement Corp., Boulder, Colo.

Werner, E.E., and D.J. Hall. 1976. Niche shifts in sunfishes: experimental evidence and significance. Science 191:404-406.

West, M.J., A.P. King, and D.H. Eastzer. 1981. The cowbird: reflections on development from an unlikely source. Am. Sci. 69:56-66.

Westerman, R.A. 1963. A study of the habituation of responses to light in the planarian *Dugesia dorotocephala.* Worm Runner's Digest 5:6-11.

White, D.H., and D. James. 1978. Differential use of fresh water environments by wintering waterfowl of coastal Texas. Wilson Bull. 90(1):99-111.

White, M.J.D. 1970. Heterozygosity and genetic polymorphism in parthenogenetic animals. In M.K. Hecht and W.C. Steere, editors. Essays in evolution and genetics. Appleton-Century-Crofts, New York.

Whittaker, R.H., and P.P. Feeney. 1971. Allelochemics: chemical interactions between species. Science 171:757-770.

Wickler, W. 1968. Mimicry in plants and animals. McGraw-Hill Book Co., New York. (Translated by R.D. Martin.)

Wicksten, M.K. 1980. Decorator crabs. Sci. Am. 242(2):146-154.

Wiens, J.A. 1966. On group selection and Wynne-Edwards' hypothesis. Am. Sci. 54:273-287.

Wiens, J.A., S.G. Martin, W.R. Holthaus, and F.A. Iwen. 1970. Metronome timing in behavioral ecology studies. Ecology 51:350-352.

Wiesel, T.N., and D.H. Hubel. 1963. Effects of visual deprivation on morphology of cells in the cat's lateral geniculate body. J. Neurophysiol. 26:978-993.

Wiesel, T.N., and D.H. Hubel. 1965. Comparison of the effects of unilateral and bilateral eye closure on cortical unit responses in kittens. J. Neurophysiol. 28:1029-1040.

Wilcox, R.S. 1979. Sex discrimination in *Gerris remigis:* role of a surface wave signal. Science 206:1325-1327.

Wilczek, M. 1967. The distribution and neuroanatomy of the labellar sense organs of the blowfly *Phormia regina* Meigen. J. Morphol. 122:175-201.

Williams, G.C. 1966. Adaptation and natural selection. Princeton University Press, Princeton, N.J.

Williams, G.C. 1975. Sex and evolution. Princeton University Press, Princeton, N.J.

Willows, A.O.D. 1973. Learning in gastropod mollusks. In W.C. Corning, J.A. Dyal, and A.O.D. Willows, editors. Invertebrate learning. Vol. 2. Plenum Press, New York.

Willows, A.O.D., and W.C. Corning. 1975. The echinoderms. In W.C. Corning, J.A. Dyal, and A.O.D. Willows, editors. Invertebrate learning. Vol. 3. Plenum Press, New York.

Wilson, E.O. 1965. Chemical communication in the social insects. Science 149:1064-1071.

Wilson, E.O. 1971. The insect societies. Belknap/Harvard University Press, Cambridge, Mass.

Wilson, E.O. 1972. Animal communication. Sci. Am. 227(3):52-60.

Wilson, E.O. 1975a. *Leptothorax duloticus* and the beginnings of slavery in ants. Evolution 29:108-119.

Wilson, E.O. 1975b. Sociobiology, the new synthesis. Belknap/Harvard University Press, Cambridge, Mass.

Wilson, E.O. 1976. Academic vigilantism and the political significance of sociobiology. BioScience 26:183-190.

Wilson, E.O. 1980. A consideration of the genetic foundation of human social behavior. In G.W. Barlow and J. Silverberg, editors. Sociobiology: beyond nature/nurture? Westview Press, Inc., Boulder, Colo.

Wilson, E.O., and W.H. Bossert. 1963. Chemical communication among animals. Recent Prog. Horm. Res. 19:673-716.

Wilson, E.O., T. Eisner, W.R. Briggs, R.E. Dickerson, R.L. Metzenberg, R.D. O'Brien, M. Susman, and W.E. Boggs. 1978. Life on earth. Sinauer Associates, Inc., Sunderland, Mass.

Wilson, S. 1973. The development of social behavior in the vole *(Microtus agrestis).* Zool. J. Linn. Soc. 52:45-62.

Wiltschko, W. 1972. The influence of magnetic total intensity and inclination on directions preferred by migrating European robins *(Erithacus rubecula).* In S.R. Galler, et al., editors. Animal orientation and navigation. Science and Technical Information Office, NASA Spec. Publ., Washington, D.C.

Wimsatt, W.A. 1970, 1977. Biology of bats. Vols. I, II, and III. Academic Press, Inc., New York.

Windle, W.F. 1944. Genesis of somatic motor function in mammalian embryos: a synthesizing article. Physiol. Zool. 17:247-260.

Winkler, H. 1973. Nahrungserwerb und Konkurrenz des Blutspechts, *Picoides (Dendrocopos) syriacus.* Oecologia 12:193-208.

Wittenberger, J.F. 1981. Animal social behavior. Duxbury Press, Boston.

Wolf, L.L. 1975. "Prostitution" behavior in a tropical hummingbird. Condor 77:140-144.

Wolgin, D.L., J. Cytrawa, and P. Teitelbaum. 1976. The role of activation in the regulation of food intake. In D. Novin, W. Wyrwicka, and G. Bray, editors. Hunger: basic mechanisms and clinical implications. Raven Press, New York.

Wood, D.S. 1979. Phenetic relationships within the family Gruidae. Wilson Bull. 91:384-399.

Woodcock, A., and M. Davis. 1978. Catastrophe theory. Avon Books, New York.

Worden, A.N. 1968. Nutritional factors and abnormal behavior. In M.W. Fox, editor. Abnormal behavior in animals. W.b. Saunders Co., Philadelphia.

Wright, S. 1955. Classification of the factors of evolution. Cold Spring Harbor Symp. Quant. Biol. 20:16-24.

Wright, S. 1968-1978. Evolution and the genetics of populations. Vols. 1 to 4. University of Chicago Press, Chicago.

Wynne-Edwards, V.C. 1962. Animal dispersion in relation to social behavior. Hafner, New York.

Wynne-Edwards, V.C. 1965. Self-regulating systems in populations of animals. Science 147:1543-1548.

Wynne-Edwards, V.C. 1977. Intrinsic population control: an introduction. Inst. Biol. Symp. on Population Control by Social Behavior, London.

Yalden, D.W., and P.A. Morris. 1975. The lives of bats. Quadrangle/New York Times Book Co., New York.

Yerkes, R.M. 1912. The intelligence of earthworms. J. Anim. Behav. 2:332-352.

Young, E., editor. 1973. The capture and care of wild animals. South Africa Nature Foundation/World Wildlife Fund. Ralph Curtis Books, Hollywood, Fla.

Young, J.Z. 1965. The organization of a memory system. Proc. R. Soc. Lond. (Biol.) 163:285-320.

Zach, R. 1978. Selection and dropping of whelks by northwestern crows. Behaviour 67:134-148.

Zach, R. 1979. Shell dropping: decision-making and optimal foraging in northwestern crows. Behaviour 68:106-117.

Zack, S. 1978. Head grooming behavior in the praying mantis. Anim. Behav. 26:1107-1119.

Zahavi, A. 1979. Why shouting? Am. Nat. 113(1):155-156.

Zeigler, H.P. 1976. Feeding behavior of the pigeon. Adv. Study Behav. 7:285-389.

Zeigler, H.P., and H.S. Karten. 1974. Central trigeminal structures and the lateral hypothalamic syndrome. Science 186:636-637.

Zimen, E. 1976. On the regulation of pack size in wolves. Z. Tierpsychol. 40:300-341.

Zimmerman, N.H., and M. Menaker. 1979. The pineal gland: a pacemaker within the circadian system of the house sparrow. Proc. Natl. Acad. Sci. USA 76:999-1003.

Zoeger, J., J.R. Dunn, and M. Fuller. 1981. Magnetic material in the head of the common Pacific dolphin. Science 213:892-894.

Zuckerman, L., editor. 1980. Great zoos of the world. Westview Press, Inc., Boulder, Colo.

Zwislocki, J.J. 1981. Sound analysis in the ear: a history of discoveries. Am. Sci. 69:184-191.

INDEX

a, 663

A, 663

Abedus herberti, male, egg brooding by, 331

Abeelen, J.H.F. van, 101, 670

Aberrant behavior, 621; *see also* Abnormal behavior

Ablation, stimulation, and lesioning in study of neurons, 417-418

Able, K.P., 219, 220, 670, 677

Abnormal behavior, 621-635
  accidental, 628
  in captive animals used in research, 628-629
  causes of, review of, 634
  definition of, 621
  in domestic livestock, 629-633
  in domestic pets, 633
  examples of, 622-633
  miscellaneous, 626
  in nondomestic animals under natural conditions, 622-623
  preventing, correcting, and curing, 634
  in zoo animals, 623-627

Abstraction, 484
  and pattern recognition, 484-491

Acceleration of behavior, developmental, 142

Accidental abnormal behavior, 628

Accidental species, 345

Accidents during play, 391

Accipiter albicilla, hunting success of, 374

Accipiter hawks
  acceptance of new prey by, 369
  activity and inactivity in, 150
  flying patterns of, 162-163
  migration activity of, 152

Accipiter nisus, 374

Accommodation, 448

Acetylcholine, 427, 518
  and sensitization, 604

Acetylcholinesterase levels and enriched environment, 569

Action patterns, 68
  fixed, 36
    defining and secondary attributes of, 41
    and sign stimulus, relationship of innate releasing mechanisms to, 40

Action patterns—cont'd
  modal, 36

Action potential, 421, 424
  compound, 417, 428
  ionic flow during, 424
  and muscle contraction, time sequence of, 437
  repeated, 425

Action-specific energy, 39

Activation-synthesis theory of dreams, 157-158

Active sleep, 155

Activity
  animal, changing patterns of circadian rhythms in, 153
  awake, of human, in cave, 153
  bird, automated recording of, 78-79
  displacement, 42
  and food deprivation in blowflies, relation of, 532-533
  and inactivity, 148-162
  migration, time of, in birds of prey, differences in, 152
  neural, recording, to study senses, 445

Activity patterns in rattlesnakes, 149, 161, 162

Acts, behavioral, 68
  sequences of, 80, 81

Acuity, visual, 460-461

Acupuncture and opiate system, 523

Ad libitum observation, 74

Adams, J.L., 393, 670

Adaptation(s), 104
  of animal, total, 105-106
  and blowfly's feeding behavior, 536
  sensory, by moth ear, 429
  sexual reproduction as means to facilitate, 305

Adaptive behavioral responses, 21

Adaptive radiation, 140

Adaptive structures, 104

Addiction, 523

Adenosine, methylxanthines and, 523-524

Adenota kob, 650
  management of, 649-651

Adkisson, C.S., 83, 673

Adoption, 338-339

Adrenal cortex, hormones of, and feeding behavior, 543

Adrenal medulla, hormones of, and feeding behavior, 543

Adrenocorticotrophic hormone, 518

Advance warning system of prey, 376

Aepyceros malampus, aggression in, 251-252

Afferent neurons, 415

African Elephant, 653

African fruit-eating bat, communal display in, 328

African hunting dogs, social behavior of, 226, 227

African lions, vibrissae patterns in, as identification, 63

African termite
  magnetic senses of, 463
  temperature regulation by, 169

African wild dogs
  prehunting behavior of, 281
  switch in prey in, 366

Afterimage, negative, demonstration of, 446

Agapornis fischeri, carrying behavior in, 98

Agapornis roseicollis, carrying behavior in, 98

Agapornis sp., carrying behavior in, 98

Agelena sp., learned orientation to light in, 598-599

Aggregation
  passive, coincidental, 226
  simple, 342, 343, 344

Aggression
  in domestic animals, 631
  male, and fighting, 319, 320, 323
  sublethal, in agonistic behavior, 237

"Aggressive mimics," 371

"Aggressive play," 387-388

Aggressive threats, 281

Aggressiveness
  in captive animals, 625
  in male cats, 99
    following testosterone treatment, 79
  as single-gene effect, 95

Agonistic behavior, 237, 238

Ahlskog, J.E., 548, 670

Aideley, D.J., 525, 670

Aiello, A., 457

Air movements, orientation based on, 200

Akil, H., 671

Alarm calls, 280-281

Alarm patterns of cephalopods, 287

Albinism in mice and behavior, 95

Alcmaeon, 29

Alcock, J., 41, 180, 670

Alcor, locating, as test of vision, 460

Aldis, O., 395, 670
Alexander, F., 394
Alexander, R.D., 336, 395, 670
Algom, D., 680
Allee, W.C., 253, 670
Alleles, 93; *see also* Genes
    frequency of, 662-665; *see also* Gene
        frequency
Allen, E., 253, 670
Allison, T., 158, 670
Allomaternal care, 337
Allomones, 266
Alloparental care, 337
Allopaternal care, 337
All-or-none principle, 438
Alloway, T.M., 584, 600, 617, 670, 680,
    684
Alpha-fetoprotein and development of "fe-
    male" brain, 521
Alpheus purpurilenticularis and *Cryptocen-
    trus steinitzi*, mutualism be-
    tween, 347
Alpheus rapax and goby, interindividual
    interactions between, 348
Alternate interpretations of data, 82
Altmann, J., 28, 55, 74, 75, 76, 85, 554,
    670
Altmann, S.A., 28, 76, 670
Altricial animals, 322, 323
Altricial young, 555
Altruistic behavior, 117-118, 253-258
    as social behavior, 229
Alvarez, F., 276, 670
Amacrine cell, 412
    of retina, 486
Amazona vittata, management of, 651-
    652
Ambushing, stalking and, by predators,
    371
American bison, 648
American bison bull, *Flehmen* in, 311
American coot, habitat selection by, 198
American foulbrood, 94
American kestrel
    hunting success of, 374
    prey-handling ability of, maturational
        changes in, 564-565
American Sign Language of the deaf,
    teaching of, to apes, 295, 298
American toad
    defense maneuvers of, 382
    dwarf, fertilization of eggs by, 310,
        311
American wigeon, habitat selection by,
    198
Amino acid structure of proteins, 87
Amniotes, embryonic, behavior of, 556
Amphibia; *see* Amphibians
Amphibians
    auditory communication in, 275

Amphibians—cont'd
    care of eggs of, by male, 333
    cerebellums of, 501
    color preference of, maturational
        changes in, 561
    color vision in, 458
    feeding behavior in, genetic control of,
        102
    hearing of, 453
    migration of, 209
    sleep characteristics of, 158
Amphioxus, nervous system of, 411
Amphipods, oldsquaw ducks eating, 177
Amphiprion, sex reversal in, 315
Amplifiers, 65
Amplitude of sound waves, 8-9
Anadromous fish, 206
Analgesia, natural, 522, 523
Analog devices to record data, 77
Analog modulation of signals, 270
Analogous structures, 110
Analogy versus homology in learning,
    611
Analysis
    complex, and computers, 83-84
    multivariate
        computers and, 83-84
        of woodpecker nesting habits, 83
    of possible communication behavior,
        sociometric, 80, 81
Anand, B.K., 545, 670
Anas platyrhynchos, 146
Anatinae, taxonomic relationships of, as
    elucidated by behavioral char-
    acteristics, 130
Anatomical and behavioral events, hor-
    monally dependent, annual se-
    quence of, 516
Anatomists, 48
Anatomy and physiology, medical, and
    understanding of behavior, 30
Andean condors, reproductive cycles of,
    311
Anderson, D.J., 199, 670
Andersson, M., 181, 670
Andrea Doria effect, 200
Andrew, R.J., 39, 670
Andrews, C.C., 346, 687
Androgen
    effect of, on behavior, 561
    uptake of, in brain of frog, 517
Anemia, sickle cell, 113
Anemonefish, sex reversal in, 315
Anemones
    hermit crabs and, commensalism be-
        tween, 346
    learning of, 590
    swimming and resettling response of,
        590-591
Anemotaxis, 200

Angiotensin, 518
Anglerfish, hunting methods of, 370,
    371
Anguilla, migration by, 207-209
Anhinga, taxonomic relationship of, 130
Animal(s)
    activity of, changing patterns of circadi-
        an rhythm in, 153
    altricial, 322, 323
    awareness of, 126-127
    behavior of; *see* Behavior
    behavioral repertoire of, 69-70
    in captivity
        management of, 638-646
        used in research, abnormal behavior
            in, 628-629
    cold-blooded, 168
    construction by, 182-183
    distraction of, by observers, 60-61
    domestic, characteristics of, 630
    hand-raised, 625
    head ornaments of, role of, 110
    horned, play in, 396
    locating and identifying, 62-65
    markings of
        artificial, 63, 64
        natural, 62, 63
    nondomestic
        abnormal behavior in, under natural
            conditions, 622-623
        management of, 646-653
    nuisance, managing, 646-648
    perceptual world of, 21
    precocial, 322-323
    research, management of, 643-644
    semelparous, 107
    social world of, roles in, 230
    suggestions for observing, 657-661
    tame, 629
    total adaptation of, 105-106
    training, 654
    veterinary management of, 645-646
    warm-blooded, 167
    wild, 629
    in zoos, abnormal behavior in, 623-
        627
Animal Behavior Society Animal Care
    Committee, 62, 657, 670
Animal Intelligence: Experimental Studies,
    45
Animal kingdom, phylogenetic relation-
    ship of major groups in, 116
Animal prey, 358
Animal sampling, focal, 75
Anion cell of blowfly, 534
Anisko, J.J., 165, 690
Anisogamy, 318
Annelids
    learning in, 596-597
    light sensitivity of, 455

Annelids—cont'd
nervous system of, 405, 410
Anokhin, P.K., 557, 670
Anolis carolinensis, behavior of, hormones and, 513
Anolis lizards
behavior of, hormones and, 512, 513
foraging behavior in, 179
Anomalous heat, 632
Anorexia nervosa in pets, 633
ANOVA, 661
Ant lion, hunting methods of, 371
Antbutterflies, 345-346
Anteater, spiny; *see* Echidna
Antelope squirrels, encounters of, at feeder, 344
Antelopes
lek behavior in, 328
pronghorn, 648
defecation by, 165
Antennae for hearing, 453
"Antennagrams," 268
Antennarium, hunting methods of, 370, 371
Anthropoides paradisea, 131
Anthropoides virgo, 131
Anthropomorphism, 31, 125-126
Antin, J., 543, 670
Anting, 167
Antipredation cause of grouping, 232-234
Antipredator behavior, 357-383
Antipredator defense tactics, 375-382
Antithesis, 279-280
communication by, 276-280
Antlers, moose, 320
Ants
adoption among, 338
and aphids, commensalism between, 346-347
army
breeding times of, 311, 314
commensals of, 345, 346
combination messages of, 277
desert
polarized light detection in, 460
use of "sun compass" by, 204
domestication among, 352-353
group efficiency in, 236, 237
maze learning in, 600
mound-building, parasitism by, 349
odor trails for, 204
optimal foraging by, 179
parasitism among, 349
pheromone-producing glands in, 273
shelters of, 181
signal to initiate transport by, 281
social behavior of, 230
"thief," 349
weaver, 183

Ants—cont'd
weaver—cont'd
living bridge of, 236, 237
Anurans, orienting behavior of, 209
Aotus trivirgatus, number of displays in, 286
Apathy in zoo animals, 626
Apes
black, yawning in, 166
language in, controversy over, 294-300
symbolic communication in, 296, 297
Aphagia, 545, 546
Aphasia, 497
Broca's, 494-495
Wernicke's, 495
Aphids
and ants, commensalism between, 346-347
peach, color vision in, study of, 446
and stingless bees, commensalism between, 346
wooly, and predaceous lacewing larvae, 371, 372-373
Apidae, 290
Apinae, 290
Apis florea, dance of, 290; *see also* Bees
Apis mellifera, advanced communication in, 288-293; *see also* Honeybees
Apis mellifera linguistica, dance of, 289-290
Aplysia sp.
cardiovascular output of, control of, 502
circadian rhythm in, 155
egg-laying behavior of, genetic control of, 102
gill withdrawal reflex in, neural control of, 503
habituation and sensitization in, 603-605
learning in, 596
and memory, 603-609
memory and elusive engram in, 605-609
neural circuits in, 502, 503
study of nervous system of, 49
Aposematism, 377
Appel, S.H., 689
Appetitive component of behavior, 33
Apple snail as prey of Everglades kite, 366
Applied animal behavior in research and photography, 653
Applied Animal Ethology, 637
Applied behavior, 636-655
Aquatic plankton, "migration" of, 202
Arachnids

Arachnids—cont'd
care of eggs of, by male, 333
migration by, 202
Aranda, L.C., 673
Archibald, G.W., 132, 133, 134, 138, 139, 140, 554, 670
Archipallium, 501
Archistriatum, 501
Arctic insects, temperature regulation by, 170
Arctic terns, migration of, 201, 210
Arctiid moth, 20
Arctiidae, 20
Arcuate fasciculus, 495
Area, familiar, 199
bird migration and, 212, 213, 214
home range and, 199-200
Area-concentrated search in predators, 369
Arena behavior, 231, 326-328
Aristotle, 29
Armadillos
pacing by, 624
psychological needs of, 639
sleep time for, 159
Armor, body, as defense against predator, 376
Army ants
breeding times of, 311, 314
commensals of, 345, 346
optimal foraging by, 179
Armyworms, 5
Aronson, L.R., 143, 577, 670
Arousal/stimulation theory of play, 395
Artemia metanauplii as prey of herring larva, 375
Artemia nauplii as prey of herring larva, 375
Arthropods
as commensals, 345
communication in, 283-284
compound eyes of, 456
hearing in, 454
heat and infrared sensing by, 461
hunting methods of, 371
learning in, 597-601
marine, auditory communication in, 275
nervous system of, 410
study of, 49
polarized light detection of, 460
Artifacts, observational, 56-57
"Artificial bat," 15
Artificial stimulation of brain and cricket calls, 102
Aschoff, J., 152, 153, 670
Asexual reproduction, 92
Asiatic land crab, eyes of, 456
Asio flammeus, 378
Asio otus, 378

Asio otus—cont'd
  predatory behavior of, maturational changes in, 565
Asmundson, S.J., 154, 686
Aspartate, 518
Assimilation, genetic, 606
Association neurons, 415
Associative learning, food aversion and other, differences between, 584
Assortative mating, 665
Atemeles pubicollis, use of ant signals by, 354
Athene noctua, predatory behavior of, maturational changes in, 565
Atlantic salmon, sequential appearance of behavior in, 554
Atlases, stereotaxic, of brain, 417, 419
Attack, response to, by coordinated group, 234
Attack behavior of cuttlefish, maturational changes in, 561-562, 563
Attack perimeter, 240
Attendant species, 345
Attention, "shifts of," 368
Attitudes and experiments, 55
Auditory communication, 274-275
Auditory ossicles, 452
Auditory pathway of human brain, 491
August Krogh principle, 657
Augustus, 641
Aurep, G.V., 686
Austerman, M., 390
Australian brush turkey, temperature sense of, 447
Australian honeyeaters, foraging methods of, 176
Australopithecus africanus, 254
Australopithecus habilis, 254
Autism in pets, 633
Autogamy, 100
Autogenous motility, 556
Autoinhibitors, 266
Automated recording
  of bird activity, 78-79
  of pigeon pecking responses, 77
Automatic systems to record data, 77
Autoshaping, 580
Autotoxins and wastes, 266
Autotrophs, 104
Average variation, 99
Aversion
  food, and other associative learning, differences between, 584
  taste, 583-584
    forms of learning using, 48
Aversive conditioning of predator, 369-370
Avoidance learning
  taste aversion as, 583, 584

Avoidance learning—cont'd
  by vertebrates, 601
Awake activity of human in cave, 153
Awareness
  of animals, 126-127
  definition of, 126
Axel, R., 688
Axo-axonic synapses, 415
Axon, 413
  structure of, 414
  terminals of, 414
Axo-somatic synapses, 415
Ayala, F.J., 125, 690

B

Baboons
  capture of vervet monkeys by, 391
  hamadryas, aggression in, 251
  leopard attacking, 342
  olive, 56
    mating of, 310, 311
    as prey of chimpanzee, 374
Babu, K.S., 598, 685
Bacillus larvae, 94
Backswimmer insects, optimal behavior in, 180
Backward conditioning, 581
Bacteria, magnetic senses of, 463
Badgers
  psychological needs of, 639
  social behavior of, 226
Badis badis, number of displays in, 286
Baer, G.M., 682
Baerends, G.P., 28, 74, 492, 670
Bahama duck, 130
Bain, A., 28, 30
Baker, M.C., 575, 670
Baker, R.R., 202, 220, 223, 670
Balanus nubilis, muscle fibers of, 438
Bald eagles
  feeding nestling, 59
  migration of, 201
  social behavior of, 226
Balda, R.P., 181, 670
Baldwin, J.M., 606
Baldwin effect, 606
Balearica pavonina, 131
Balearica pavonina ceciliae, 131
Balearica pavonina pavonina, 131
Balearica regulorum, 131
Balearica regulorum gibbercepst, 131
Balearica regulorum regulorum, 131
Balearicinae, 131
Balloon used to study bird migration, 66-67
Banding to identify birds, 65, 211
Barash, D., 256, 328, 671
Barbary dove, cooing song pattern of, 97
Barber, V.C., 470, 671
Barchas, J.D., 517, 518, 671

Bark, 285
Barker, J.L., 525, 671
Barlow, G.W., 28, 36, 41, 239, 253, 256, 259, 671, 674, 675, 677, 680, 689
Barn owl, brain of, map of, 483
Barnacle, muscle fibers of, 438
Barnard, D.E., 680
Baron, R.A., 23, 51, 671
Bartels, M., 598, 599, 671
Barth, R.H., 308, 689
Bartholomew, G.A., 176, 679
Basic behavior of animals
  definition of, 147
  examples of, 147
  maintenance behavior and, 146-184
Basket cells, 414, 500
Bass, largemouth, hunting success of, 374
Bat(s)
  African fruit-eating, communal displays in, 328
  "artificial," 15
  calls of, 10-12
    to detect insects, 14
    identification of, by moth, 431
    response of Arctiid moth to, 20
    simulation of, 15
  chirping, 10
  echo accentuation by, 12
  echoes of flying moth and, 14-15
  echolocation and hunting behavior in, 7-15
  echolocation calls of, types of, 11
  energy-emitted senses of, 466
  familiar areas of, 200
  flight paths of, 17
  FM, 10
    little brown bats as, 13
  frog-eating, 355
  fruit, echolocation by, 8
  hearing of, 448, 453
    suppression and amplification in, 12
  horseshoe, echolocation in, 8
  larynx of, 6
  little brown; see Little brown bat
  and moths, 4-20
  moth's response to call of, 17-19
  play in, 390
  roosting, mixed-species groupings in, 344
  rousette, echolocation by, 8
  sensory orientation by, 200
  sleep time for, 158, 159, 162
  sounds of, 7
  temperature regulation by, 168
  use of Doppler effect by, 11
  use of naturally occurring sounds by, 200

Bat(s)—cont'd
    vision of, 7, 8
Bat detectors, 13
Batesian mimicry, 377, 379, 380
Bateson, P., 570, 577, 671, 676, 681
Bathing, 166-167
Battersby, B., 683
Bay-breasted warbler, feeding habitat of,
    193
Bays, S.M., 599, 671
Beach, F.A., 28, 41, 47, 51, 671
Bear
    black
        abnormal escape reactions in, 624
        hunting success of, 374
        play in cubs of, 391
    play in, 396
    polar, migration by, 220
    as predators, 360
    sloth, abnormal behavior in, 624-625
    weaving behavior in, 624
Beatty, W.W., 550, 691
Beaver, 648
    cutting aspen trees, 642
    mountain, sleep time for, 159
    shelters of, 181, 182
Bedbugs, location of prey by, 449
Beer, C.G., 28, 32, 51, 671
Bees
    altruistic behavior of, 118
    color vision in, 458
    dances of, 286, 288-293
    familiar paths for, 200
    inactivity in, 148
    incitement to hunt in, 281
    learning in, 598, 599
    magnetic senses of, 463
    nest cleaning by, 94-95
    polarized light detection in, 460
    "queen substance" in, role of, 271
    shelters for, 181, 182
    social behavior of, 230
    stingless
        and aphids, commensalism between,
            346
        communication in, 290-291
        mutualism-parasitism involving, 351
    temperature regulation by, 168, 169
    temperature sense of, 447
    use of sun compass by, 205, 206
Beetles
    Batesian mimicry by, 379, 380, 381
    bombardier, defense by, 377
    as commensals, 345
    dung, foraging behavior in, 176-177
    grain, learning in, 600
    head ornament of, 110
    Hercules, head ornamentation of, 320
    leaf-rolling, 183
    lucanid, head structures of, 320

Beetles—cont'd
    temperature regulation by, 169
    tenebrionid, trapping of moisture by,
        172
    use of ant signals by, 354
    whirligig, energy-emitted senses of,
        466-467
Begging responses in newly hatched her-
        ring gulls, 38
Behavior(s), 68
    abnormal, 621-635; see also Abnormal
        behavior
    academic interest in, division of, 27
    adaptiveness of, 622
    agonistic, 237, 238
    albinism in mice and, 95
    "altruistic," 117-118, 253-258
        as social behavior, 229
    analysis of, categories of, 80, 81
    antipredator, 357-383
    appetitive component of, 33
    applied, 636-655
        in research and photography, 653
    arena, 231, 326-328
    attack, of cuttlefish, maturational
        changes in, 561-562, 563
    basic and maintenance, 146-184
    biological structures and, 21
    and biology, 3-23
    branches of study of, development of,
        31-33
    caching, 181
    cannibalism-avoiding, in spiders, 308-
        309
    carrying, in lovebirds, 98
    categories of, 142
    causes of, introduction to, 1-143
    change in
        hormones and, 561
        maturation and, 560-565
        and nervous system disorders, 48-
            49
        specific, development and, 569-576
    chemicals involved in regulation of, list-
        ing of, 518
    communication, sociometric analysis of
        possible, 80, 81
    components of, 68-69
    compulsive, 523
    conflict, 41-42
        in evolution of behavior, 142
    consistent, 102
    constancy of, 102
    consummatory component of, 33
    convergence, 140
    convergent, 110
    courtship, of empid fly, 141
    Darwin on, 30-31
    definition of, 3, 53

Behavior(s)—cont'd
    describing, recording, and cataloging,
        68-77
    development of, effects of deprivation
        and general effects of early
        experience on, 565-576
    developmental acceleration of, 142
    displacement and redirected, in captive
        animals, 625
    eating, abnormal, 626
    ecological and evolutionary forces af-
        fecting, 4
    and ecology, 103-107
    effect on
        of captivity, 61
        of domestication, 631
        of predation, 360
    egg-laying, in Aplysia, genetic control
        of, 102
    elimination, 165-166
    embryological and neonatal roots of,
        554-560
    emotional, ontogeny of, 569
    environment and, 90-92
    environmental factors affecting, 128
    epimeletic, 632
    et-epimeletic, in pets, 633
    ethological and ecological aspects of,
        144-399
    evolution of, 128-143
        comparative method to study, 128-
            129
        consequences of genetic and ecologi-
            cal variability in, 108-143
    evolutionary characteristics of, 140-
        143
    evolutionary foundation of understand-
        ing of, 30-31
    "exapted," 142
    extremes of, as abnormal behavior,
        634
    familiarity with species when studying,
        55-56
    feeding, 70
        in amphibians, genetic control of,
            102
        in blowfly, 532-533
        of rat, 540-541
    focal sampling of, 75
    foraging; see Foraging
    frequency of, alterations in, as result of
        signal, 263
    genetic control of, additional evidence
        for, 100-102
    and genetics, 92
    genetics and ecology of, 86-107
    group selection and, 117-118
    grouping of, 70
    group-selected, 118
    guard, 234

Behavior(s)—cont'd
hearing and, 4-20
history of study of, early, 29-33
homologous
   courtship displays in ducks as, 129
   head scratching as, 129
and hormones
   interdependence of, 507
   neurochemicals and, 506-525
hunting, and echolocation in bats, 7-15
incubation
   of herring gull, interruptions in, 492
   hormones and, 511
insect, and hormones, 509
interanimal and intraanimal, and chemicals, relationships of, 508
internal control of, 400-617
internal control mechanisms of, 3
   Darwin on, 30-31
internal integration of, 472-505
interspecific, 70, 340-383
lek, 231, 326-328
maintenance, 70
malfunction of, causing abnormal behavior, 634
mechanical view of, 126
medical anatomy and physiology and, 30
modification of, 661
nature-nurture problem regarding, 610-611
navigation, 70
of noctuid moth, hearing and, 15-20
number of, plotting, 661
observation of
   and measurement of, 52-85
   suggested exercises for, 658-661
   and time spent observing, relationship of, 69
as observed output, 144-399
ontogeny of, 552-577
   schools of thought concerning, 557
optimal, 180
orientation, 70
outward, Watson and, 46
pecking
   of birds, maturational changes in, 563-564
   of chicken, maturational changes in, 562-563
physiological and psychological aspects of, 400-617
play, examples of, 386-390
preadapted, 142
predator, 357-383
predator-mobbing, of birds, 585, 586
predatory
   evolution of, 359-361

Behavior(s)—cont'd
predatory—cont'd
   maturational changes in, 564-565
prehunting, ritualized, 281
prey, 360
proximate factors affecting, 20-21, 128
psychological, 48
purposeful, 125-126
quantification of, 4-hour, 658-660
quantitative genetics, and heritability, 96-100
radio telemetry to study, 63, 64, 65
recording of, 65-66
   methods for, 76-77
redirected, 41-42
reductionist view of, 68
regression in, by zoo animals, 626
and reproduction, 303-339
reproductive
   definition of, 303
   sequences of, 306-310
rhesus monkey, interobserver reliability in measurement of, 68
role of future events in, 125-127
sampling, 73-76
   random, 75-76
   statistical, 74-76
and scientific methods, 54-55
scientific view of, beginning of, 30
selfish genes and, 119-121
sequential appearance of, 554-555
sexual, abnormal, in zoo animals, 626
shaping of, 582
single genes and, 94-95
social; see Social behavior
sound and, 4-20
species communication, listing, 661
species-specific, 48
study of, 3
   history of, 53
   methods of, suggestions for learning, 84
survival and reproduction and, 127-128
systemogenous, of embryo, 557
territorial, 238, 240
   as epideictic phenomenon, 118
and ultimate factors, 21-22, 128
uniformitarianism and, 110
units of, 68-71
vacuum, 38
   of predators, 364
variations of, among different individuals, 80, 81
vertebrate, and hormones, 509-516
voluntary control of, 481
weaning, 337
weaving, in captive animals, 624

Behavior of Organisms, 46-47
Behavior experiment, 661
Behavior patterns, 68
   adaptive, 104
   evasive, of moths toward bat calls, 17-19
   innate or instinctive, 3
   "learned," 4
Behavioral acts, 68
   methods of classifying, 69
   methods of describing, 69
   sequences of, 80, 81
Behavioral and anatomical events, hormonally dependent, annual sequence of, 516
Behavioral biology
   descent of, abstraction of, 28
   relationships among other disciplines and, 26
Behavioral characteristics, taxonomic relationships of Anatinae as elucidated by, 130
Behavioral data, interpreting and presenting, 77-84
Behavioral development in embryo, 556-558
Behavioral ecology, relationship of, to other disciplines, 26
Behavioral focus, 559
Behavioral mutations, 100-101
Behavioral repertoire
   of animal, 69-70
   of mallard duck, portion of, 70-71
   of species, 56
Behavioral responses
   adaptive, 21
   to proximate conditions, 21
Behavioral welfare of captive animals, 639
Behaviorism, 46
Behaviorism, 46-48
Beitinger, T.L., 170, 671
Bekoff, A., 391, 394, 395, 399, 559, 560, 671
Bekoff, M., 395, 560, 577, 671, 672, 691
Bell, P.R., 281, 683
Bell, W.J., 672
Belleville, R.E., 223, 677
Bell-food-salivation conditioning, 580
Bellrose, F.C., 214, 671
Belongingness and learning, 583
Belzer, W.R., 538, 671
Benderly, B.L., 295, 297, 299, 671
Bennet, W., 551, 671
Bennett, E., 592, 671
Bennett, E.L., 617, 687
Bentley, D.R., 97, 102, 558, 559, 671
Benzer, S., 100, 101, 559, 671, 680
Berger, D.D., 150, 685

Bernardis, L.L., 549, 677
Berndt, R., 210, 671
Berner, C.N., 548, 689
Beta-endorphin, 518
Bever, T.G., 690
Beyer, C., 525, 671
Bicolor damselfish, dominance hierarchy
    in, 242
Biggley, W.H., 682
Bighorn sheep
    male fighting in, 319
    migration by, 220
    parental teaching by, 332
Binaural hearing, 448
Binocular vision, 448
Binoculars, 65
Binomial, 663-664
Bioelectric events, recording of, to study
    neurons, 416-417
Biological constraints, 582-583
Biological determinism, 257
Biological relevance of learning capacity,
    601
Biological Sciences Curriculum Study,
    658, 671
Biological structures and behavior, 21
Biological view of evolution, 122-123
Biologists, neurophysiological, 49
Biology
    and behavior, 3-23
    behavioral
        descent of, abstraction of, 28
        relationship among other disciplines
            and, 26
    cellular relationship of, to other disci-
        plines, 26
    population, relationship of, to other dis-
        ciplines, 26
Bipolar cells of retina, 486
Bipolar neurons, 412, 414
Bird(s)
    abnormal sexual behavior by, 626
    activity of, automated recording of, 78-
        79
    advanced forms of communication in,
        293-294
    alarm calls of, similarity of, 281
    alloparental helping in, 338
    banding of, 65
    body temperatures of, 168-169
    bower, 142
    breeding times of, 311, 314
    calls of, 293, 572
    care of eggs of, by males, 333
    cerebellum of, 501
    circadian rhythms in, 155
    color vision in, 458
    commensals of, 345
    communal displays in, 328
    cooperative foraging by, 234

Bird(s)—cont'd
    displays in, numbers of, 286
    drinking by, 171-172
    dust bathing by, 167
    ears of, 453
    elimination behavior of, 165-166
    embryonic behavior of, 556
    escape responses of, to flying birds of
        prey, 39
    food robbing by, 348-349
    habitat preference of, 196
    habitat selection of, 193
    hand-raised, 625
    head scratching by, 128, 129
    hunting methods of, 371
    insectivorous
        switch in prey of, 366
        vacuum behavior in, 38
    insight into learning from, 612
    insight learning in, 587
    Kaspar Hauser, 195, 196
    learning in, to increase food supplies,
        179
    locomotion of, 162-163
    magnetic senses of, 462-463
    and mammals, reproductive behavior
        of, differences in, 329-330
    mating patterns in
        versus parental investment, 329
        and sexual dimorphism, 317
    memory of, 606
    migration of, 210-219
        study of, 66-67
        weather variables affecting, 82-83
    mimicking by, 573
    mixed-species grouping in, 344
    navigation by, genetics and, 212
    nervous system of, 409
    nest of, 183
    nestling, wing-flapping "exercises" of,
        maturational changes in, 564
    observation of, 659
    odor following in, 191
    passerine, see Passeriformes
    pecking responses of, maturational
        changes in, 563-564
    perching, songs of, development of,
        572-576
    predator-mobbing behavior of, 585,
        586
    preening by, 167
    as prey of raptors, 374
    of prey
        activity and inactivity in, 150
        difference in migration activity times
            of, 152
        flying, escape responses of birds to,
            39
    reciprocal feeding in, 275
    recognition by, 229

Bird(s)—cont'd
    reproductive sequences of, hormones
        of, 510-511
    role of sensory channels in, 279
    self-mutilation by, 625
    shelters of, 181, 182, 183
    sleep characteristics of, 158
    songs of, 293, 572
        dialects in, 575
        functions of, 293-294
        of male, female responsiveness and,
            575
        primary, 574
        rehearsal of, 574
    spread-winged postures of, 170-171
    storage of prey by, 363
    stretching by, 166
    temporal pattern recognition of, 448
    trained, 654
    use of sounds for orientation of, 200
    variability in, 612
    vocalizations of, 275
        development of, 572-576
        meanings of, 271
    widow, young, markings of, 351
Bird leks, 326-327
Bird and mammal calls, convergence in,
    284, 285
Bishop, J., 387, 390, 397
Bison, American, 648
    Flehmen shown by bull of, 311
    migration by, 220
Bitterman, M.E., 28, 29, 44, 51, 588,
    589, 596, 611, 617, 671
Bivalves, nervous system of, 405
Bizeau, E.G., 652, 675
Black, C.H., 239, 671
Black apes, yawning in, 166
Black bear
    abnormal escape reaction in, 624
    hunting success of, 374
    play in cubs of, 391
Black geese, 130
Black rat, 630
Black widow spiders, cannibalism of,
    306
Black-and-white warbler, niche-gestalt
    for, 197
Blackbirds
    alarm call of, 281
    European, mobbing behavior by, 586
    red-winged
        sexual dimorphism in, 317
        social behavior of, 226, 227
        vocalization of, 271
Blackburnian warbler, feeding habitat of,
    193
Black-footed ferrets, 357
Black-headed duck, nest parasitism by,
    350

Black-necked crane, 131, 140
Black-tailed deer
    pheromone-producing glands of, 272
    play in, 391, 394
Bladder, swim, and sound, 452, 453
Blakemore, R.P., 676
Blarina, responses of, to increases in prey, 368
Blas, E.M., 545, 684
Bligh, J., 167, 671
Blind spot, 486
Blinds to study animals, 58-60
Block, G.D., 154, 671
Block, R.A., 593, 594, 672
Blocking of sense organs or neural connections to study senses, 445
Blood sugar levels and feeding in fly, 532, 533
Bloom, F.E., 525, 672
Blough, D.S., 77, 672
Blowfly
    control of feeding in, 532-540
        flowchart of, 539
    digestive system of, 536-537
    labellum of, sensory receptors on, 534
    parabiotic, 538
Blue geese, migration of, 210
Blue titmouse, alarm call of, 281
Blue tits
    foraging behavior in, 179
    habitat preference of, 196
Blue whale, motor neurons of, 413
Blue-winged teal, habitat selection by, 198
Blum, M.S., 339, 672
Blum, N.A., 339, 672
Boas
    care of eggs of, by male, 333
    heat and infrared sensing by, 461
    incubation of eggs by, 333
Bobcat with rabbit, 359
Bodian, D., 414, 672
Body(ies)
    cell, of neuron, 413
    cleaning of, by commensals, 346
    markings on, false, on butterflies, 379
    Nissl, 414
    orientation of, 468
Body lice, localization of prey by, 449
Body posture, changes in, in drinking ducks and hawks, 75
Body shape displays and coloration in octopus, 287
Body weight
    control of, hypothetical feeding system for, 530
    set-point mechanism for, 529
Boggs, W.E., 107, 143, 692
Boice, R., 629, 630, 672, 687

Boidae, heat and infrared sensing by, 461
Bolles, R.C., 28, 527, 530, 531, 557, 672
Bombardier beetle
    Batesian mimicry by, 380
    defense by, 377
Bombesin, 518
    and satiety, 543
Bombinae, 290
Bombus fervidus, foraging paths of, 175
Bombykol, 465-466
Bombyx mori, chemical senses of, 465-466
Bonasa umbellus, 260
Bond, R.R., 197, 672
Bonding, mother-offspring, 572
Bonner, J.T., 585, 617, 672
Bony fishes, gill movements of, 165
Boorman, S.A., 247, 259, 672
Boring, E.G., 51, 672
Bortz, R., 689
Bossert, W.H., 274, 692
Botfly, infestation of nests by, 351
Bottle-nosed dolphin, sleep time for, 158, 159
Boudreau, J.C., 471, 672
Boundary marking, territorial, pheromones and, 272
Bounties, 646
Boutons, 414
Bowerbirds, 142
    communal displays in, 328
Boxes
    "problem," Thorndike's, 44, 45
    Skinner, 46, 47
Boycott, B.B., 594, 608, 610, 617, 672
Boysen, S., 688
Bradley, D.C., 670
Brady, J., 185, 672
Bradykinin, 518
Brain
    artificial stimulation of, and cricket calls, 102
    of barn owl, map of portion of, 483
    changes in, from enriched environment, 569
    complexity of, 473-475
    deuterostome, 404
    hierarchies of centers in, 493
    human; *see* Humans, brain of
    of iguana, map of portion of, 482
    information storage in, 605
    insect, 410
    invertebrate, pathways through, 502
    language pathways in, 493-496
    lateralization of, 497-500
    left side of, functions of, 498, 499
    of mammals, 475, 501
        sexual differentiation of, 519-521

Brain—cont'd
    mapping, 481-484
    motor centers in, 491-493
    pathway of information from, to motor neuron, 480
    pathways through, 472-505
    pleasure centers in, 521
    protostome, 403
    of rat
        placement of electrodes in, 418
        sections involved in feeding behavior in, 544, 545-549
        stereotaxic atlas of, 419
    reptilian, 501
    right side of, functions of, 498
    stereotaxic atlases of, 417, 419
    vertebrate, 411
    visual pathways from retina to, 487, 488
Brain volume of hominid species, increase in, 254
"Brainwashing," 628
Branchial columella, 453
Bray, G.A., 551, 685, 693
Braza, F., 670
Breathing, 165
Breed, M.D., 238, 672
Breeding
    seasonal patterns of, 310-311, 314
    selective, and mutations, 101
Brenowitz, W.A., 271, 672
Brett, L.R., 678
Briggs, W.R., 107, 143, 692
Brightening detectors, 488
Brimes, M.L., 286, 672
Britten, R.J., 605, 672
Brittle star, nervous system of, 408
Broadbent, D.E., 617, 687
Brobeck, J.R., 545, 670
Broca, P., 28, 49, 494, 497
Broca's aphasia, 494-495
Broca's area, 494, 495
Brockmann, H.J., 181, 248, 346, 672, 674
Brodie, E.D., Jr., 376, 672
Brolga crane, 131, 135, 139, 140
Brooding
    egg, by water bug, 331
    gastric, by frog, 331
    of young, parental, and thermoregulatory ability, 332
Brouwer, R., 670
Brower, J.V.Z., 672
Brower, L.P., 308, 672
Brown, C.R., 623, 672
Brown, J.L., 140, 226, 259, 264, 271, 280, 301, 672
Brown, K.T., 469, 672
Brown bat; *see* Bat(s); Little brown bat

Brown hyena, abnormal sexual behavior in, 626
Brown pelicans
   communal hunting by, 372
   signals of, 286
Brown-headed cowbird
   abnormal mate selection by, 623
   song of, 573
Bruce effect, 271
Brugger, C., 188
Brugger, K., 188
Bruning, E., 185, 672
Brush turkey
   Australian, temperature sense of, 447
   egg laying by, 333
Bry, E., 226, 317, 326
Bryozoans, social behavior in, 230
Bubo bubo, 378
Bubo virginianus, 625
Buckle, G.R., 228, 672
Buckley, S.K.L., 202, 692
Buechner, H.K., 649, 672
Bufo bufo, male fighting in, 323
Bufo sp., as prey of hognose snake, 366
Bufotenin, 518
Bug, water, male, egg brooding by, 331
Bugeranus carunculatus, 131
Bullock, T.H., 27, 28, 29, 51, 414, 441, 490, 505, 672
Bumblebees, 290
   body temperatures of, 168, 169
   foraging paths of, 175
   optimal foraging by, 179
Buntings
   indigo
      activity of, recording of, 77, 78
      study of communication in, 268
      use of stars by, for navigation, 214
   reed, alarm call of, 281
Burghardt, G.M., 140, 261, 267, 365, 375, 560, 577, 671, 672, 678
Burkhard, B., 582, 614, 675
Burns, R.J., 647, 672
Burrowing, shelters made by, 182
Burrowing movements, 163
Burtt, E.H., Jr., 338, 672
Bushy-crested jay, territory of, 239
Busycon carica, hunting success of, 374
Buteo galapagoensis, predation by, 363
Buteo lineatus, 552
Buteo sp.
   activity and inactivity in, 150
   migration activity of, 152
Butler, S.R., 77, 672
Butterflies
   false body markings on, 379
   Heleconius, larvae of, 265
   mating of, 310
   messages sent by, 262
   monarch; *see* Monarch butterflies

Butterflies—cont'd
   plant "messages" and, coevolution of, 265
   queen, courtship sequence of, 308
   role of sensory channels in, 279
   swallowtail, aggression in, 251
Buzzati-Traverso, A.A., 686
Byers, J.A., 391, 672
Byrne, D., 51, 671

C
2-^{14}C deoxy-D-glucose, 484
Cabbage looper, 5
Caching of excess food, 343
Caching behavior, 181
Cacique, mutualism-parasitism involving, 351
Caddis fly, larvae of, shelter for, 183
Cade, T.J., 171, 172, 672
Cade, W., 355, 672
Caffeine, 523-524
Cajacob, T., 627, 642
Cajal, 28
Calcium ions in neurons, 420
Caldwell, P.C., 427, 672
Caldwell, R.L., 228, 229, 672
Calloplesiops altivelis, protective mimicry by, 379
Calls
   alarm, 280-281
   of bats, 10-12
      to detect insects, 14
      identification of, by moth, 431
      moth's response to, 17-19
      response of Arctiid moth to, 20
      simulation of, 15
      types of, 11
   of birds, 293, 572
      and mammals, convergence in, 284, 285
   of cranes
      guard, 132
         sonograms of, 133
         in unison call, 136
      unison, 132, 134, 135
         characteristics of, 136-137
         shared characteristics of, 138, 139
   cricket, and artificial stimulation of brain, 102
   distress, 280-281
   of hunting little brown bat contrasted with human speech, 13
   sandhill crane, sequential appearance of, 554
Calvin, M., 592, 671
Camhi, J.M., 559, 673
Camouflage as defense against predators, 377-378
cAMP, 517

Campbell, B., 339, 673, 688
Campbell, B.A., 577, 690
Campbell, C.B.G., 47, 679
Canada geese during migration, 186
Canadian sandhill crane, 131
Canals, semicircular, 450-451
Canaries, song of, changes in acquisition of, 575
Candland, D.K., 261, 554, 569, 673
Canids
   communal hunting by, 372
   play in, 391
   as predators, 360
   role of sensory channels in, 279
Canis latrans, play in, 391
Cannibalism, 626
   with planarians, 593, 594
   and reproduction, 306, 308
Cannibalism-avoiding behavior in spiders, 308-309
Cannon, W., 542
Canon, Morgan's, 44
Cantoring, 163
Canvasback duck, habitat selection by, 198
Capen, D.E., 197, 673, 680
Caplan, A.L., 253, 259, 673
Capra hircus, play in, 394
Capra ibex, injuries to, during play, 391
Capreolus capreolus, 375
Captive animals used in research, abnormal behavior in, 628-629
Captivity
   abnormal reaction during, 623-626
   animals in, management of, 638-646
   effect of, on animal behavior, 61
   unsuitable conditions of, 626
Capture of animals, legal ramifications and requirements for, 638
Capture and hunting methods of predators, 371-374
Carbidae, 377
Cardiovascular output in *Aplysia*, control of, 502
Care
   alloparental, 337
   parental, 330-337
      extended, 332
      factors conducive to development of, 333-334
      and sexual selection, 335
   of young, reproduction and, cooperation in, 337-339
Carew, T.J., 605, 673
Caribou
   defensive maneuvers of, 380
   and foxes, play between, 388-389
   migration by, 220
Carlson, A.D., 356, 673
Carlson, N.R., 541, 673

Carmichael, L., 557, 673
Carnivores
  alloparental helping in, 338
  brains of, 475
  color vision in, 458
  matriarchal group of, 330
  self-mutilation by, 625
Carnivorous, 357; see also Predator
Caro, T.M., 392, 588, 673
Carolina duck, 130
Carp, hearing in, 453
Carpenter bee, shelter of, 182
Carr, A., 209, 673
Carrigan, T.C., 682
Carrying behavior in lovebirds, 98
Carter, D.B., 331, 691
Carthy, J.D., 190, 223, 445, 471, 673
Cartilaginous fishes, respiratory movements of, 165
Carty, W.S., 681
Caryl, P.G., 267, 673
Casmerodius albus, nonoptimal foraging in, 180
Cassidy, J., 41, 673
Caste, 228
Catadromous fish, 207
Cataloging, describing, and recording behavior, 68-77
Catastrophe theory, 240
"Catch" muscles of molluscs, 438
Catchpole, C.K., 318, 673
Catfish, electric, 375
Catharus ustulatus, transmitter on, 64
Catocala sp., 5
Cats
  brain of, 411
    electrode study of, 481
  captivity and behavior of, 61
  color vision in, 458
  defense threat posture of, 281
  domestic
    abnormal behavior in, 631, 633
    free-ranging, time budgets for, 151
  dominance in, 241
  ear of, 452
  elimination behavior of, 165
  fighting by, and testosterone, 54, 55
  "greeting" reaction in, 56, 57
  hearing of, 448
  house
    housing facilities for, 638
    male, aggressiveness in, following testosterone treatment, 79
    sequential appearance of behavior in, 555
  hunger threshold of, 362
  hunting methods of, 371
  and ionizing radiation, 463
  lateral hypothalamic lesions in, feeding behavior and, 546

Cats—cont'd
  male, aggressiveness in, 79
  nervous system of, 408
  play in, 396
  play intent movements in, 278
  as predators, 360
  predatory behavior of, learning and, 565
  psychological needs of, 639
  selective breeding of, 101
  sleep time for, 158, 159, 160
  teaching of predatory behavior to kittens by, 588
  visual pathway from retina to brain in, 487
Cats in a Puzzle Box, 57
Cattle, free-ranging, in Australia
  grazing patterns of, 150
  shorthorn, time budgets of, 148
Causation, immediate, classifying behavior by, 69
Cavalli-Sforza, L.L., 585, 673
Cell(s)
  amacrine, 412
  anion, of blowfly, 534
  basket, 414, 500
  "daughter," 92
  ganglion, of frog retina, output of, 487
  glial, 428
  light-sensitive, and vision, 455
  neurosecretory, 507
  Purkinje, 412, 414, 500
  pyramidal, 412, 414
  of retina, 485, 486
  sense, visual, structure of, 469
  visual, operation of, 486
Cell body of neuron, 413
Cellular biology, relationship of, to other disciplines, 26
Cellular membrane, 420
Centers, 416
Central control hypothesis of muscle contraction, 438
Central nervous system, 404
  of invertebrates, identified neurons in, 404
  pain pathway in, in vertebrates, hypothesized, 522
Central nervous system mechanisms and control of feeding in rat, 543-549
Central stimulus filtering, 443
Centrifugal swing in ants, 600
Centrocercus urophasianus, 302
Centrosomes, 413
Cephalochordates, nervous system of, 411-412
Cephalopods
  advanced forms of communication in, 287-288

Cephalopods—cont'd
  color vision in, 458
  communication in, 283-284
  learning in, 595-596
  movements of, 163
  nervous system of, 405
Cerci, cockroach, neural research on, 432-435
Cercopithecus aethiops and play, 391
Cercopithecus campbelli, play in, 386
Cerebellar cortex, neuronal types in, 412
Cerebellum, role of, 500-501
Cerebral cortex, human, map of, 482
Cerebrum, right versus left sides of, 497-500
Cerf, J.A., 686
Cestoda, 591
Cetaceans
  brains of, 474
  migration by, 220
Chaffinch, alarm call of, 281
Chains
  Markov, 80, 81
  of synapses, 415
Chamberlain, T.J., 607, 673
Chambers, pressure, for hearing, 453
Chameleons, true, hunting methods of, 371
Chamove, A.S., 567, 673
Chance
  and experimentation, 55
  in gene frequencies, 664
Chance encounters, 342, 343, 344
Chang, S.Y., 681
Changula hyemalis, food choices of, 177
Chappell, M.A., 189
Characteristics, new, origins of, 114-117
Charged particles, 419
Charnov, E.L., 174, 175, 185, 673, 681, 687
Chase, W.G., 606, 676
Chase and pursuit by predators, 371
Cheetah, hunting success of, 374, 375
Chelicerata, 597
  learning in, 598-601
Chelonia mydas, homing ability of, 209-210
Chemical communication, 271-274
Chemical cues, identification of, by combinations of sensory input, 464
Chemical defenses against predators, 376-377
Chemical interactions between organisms, classification of, 266
Chemical senses, 463-466
Chemical synapses, 424
Chemicals
  exogenous mood-altering, pleasure and pain and, 521-524
  interspecific, 266

Chemicals—cont'd
  and intraanimal and interanimal components of behavior, relationships of, 508
  intraspecific, 266
  involved in regulation of animal behavior, listing of, 518
  transmitter, 507, 508
Chemoreceptors and feeding behavior of fly, 534, 535
Chemosensory hairs of fly, 526, 535
Chen, W.Y., 673
Chepko, B.D., 394, 673
Chestnut-breasted teal, 130
Chiasognathus granti, head structures of, 320
Chick, pecking response of, maturational changes in, 563-564
Chick embryos, development of, 558
Chicken
  brain of, motor centers in, 491-493
  distress calls of, 281
  domestic, courtship sequence of, 309
  observation of, 659
  pecking behavior of, maturational changes in, 562-563
  prairie, leks in, 326-327
  wing flapping in, 564
Chicken hysteria syndrome, 631-632
Childbirth and opiate system, 522
Childs, G., 90, 673
Chilee pintail, 130
Chilee teal, 130
Chilee wigeon, 130
Chimpanzee
  abnormal sexual behavior by, 626
  brain of, 411
  hearing of, 453
  hunting success of, 374
  language in, controversy over, 294-300
  play-face in, 398
  "self-awareness" of, 127
  sleep time for, 159
  study of, 60
  tools used by, 371
Chinchilla, sleep time for, 159
"Chip" calls of alarm, 281
Chipmunks, foraging methods of, 176
Chipping sparrows
  habitat preference of, 196
  hand raising of, 195
Chiroptera, 5
  play in, 390
Chirp, 285
Chirping bats, 10
Chloride ions in neurons, 420
Cholecystokinin, 518
  and satiety, 543
Chondrichthyes

Chondrichthyes—cont'd
  gill respiratory movements of, 165
  use of electrical shock by, 375
Chordates, lower, nervous system of, 411-412
Chromatographs, 66
Chromosomes
  in fruit flies, 95
  physical maps of, 96
  statistical maps of, 95-96
Chrysopa slossonae, 372
Chthonolasius, parasitism by, 349
Church, R.M., 297, 690
"Cicada" principle, 234
Cicadas
  communal displays in, 327, 328
  role of sensory channels in, 279
Cichlid fish
  dwarf, territory of, 239
  Haplochromis, sequential appearance of behavior of, 555
  reproductive relationships in, 330
  visual signals in, 284
Cilia, 470
Circadian pacemakers, 154-155
Circadian rhythms, 151
  of animal activity, changing patterns in, 153
Circannual rhythms, 151
  of ground squirrels, 154
Cirphis unipuncta, 5
Cistrons, 89
  as regulator genes, 90
Cladocera, nervous system of, 406
Clamp, voltage, 418
Clams, nervous system of, 405, 406
Clark, G.A., Jr., 163, 673
Classic Darwinian view of evolution, 111
Classical conditioning, 45, 580-581
Classification, historical, of behavior, 69
Clawed frogs, androgen uptake in brain of, 517
Clayton, D., 74, 673
Cleaner fish, 346
  sex reversal in, 315
Cleaner shrimp cleaning zebra moray, 346
Cleaners
  facultative, 346
  obligate, 346
Cleaning
  body, by commensals, 346
  nest, by honeybees, 94-95
Cleft, synaptic, 426, 427
Clever Hans effect, 298
"Click mechanism," 438
Cliff swallows, 317
  nest of, 182
Climbing, 163
Clines, 114

Clines—cont'd
  behavior traits of, 140
Clinical ornithology, 651
Cloarec, A., 171, 673
Clock-shifting, 205
Cloning, 92
Closed-circuit television, 65
Clutton-Brock, T.H., 259, 673
CNS; see Central nervous system
Coal tits
  foraging behavior in, 179
  habitat preference of, 196
Cocaine, 524
Cochlea, 451
Cochliomyia hominivorax, 647
Cochran, W.W., 210, 673
Cockroaches
  agonistic behavior in, 238
  cerci of, neural research on, 432-435
  circadian pacemakers of, 154, 155
  courtship sequence of, 309
  instrumental conditioning in, 600
  memory in, 607
  startle response of, 432-435
  ventral nerve cord of, 434
Cocoon-spinning patterns, 182
  hormonal control and, 509
Cody, M.L., 197, 673
Coefficient of relationship, 243-244
Coelenterates
  hearing in, 454
  learning in, 589-591
  light sensitivity of, 455
  movement by, 163
  nervous system of, 405
  poisoning by, 375
  simple reflexes in, 475
Coghill, G.E., 556, 673
Cognition, 45
Cognitive map, 45
Cognitive theories, 45
Cohen, D.B., 673
Cohen, L.B., 687
Cohn, R.H., 673
Coincidental passive aggregations, 226
Cold Spring Harbor Symposia on Quantitative Biology, 673
Cold-blooded animals, 168
Cole, L.C., 152, 174, 673
Coleman, R.M., 674
Colgan, P.W., 240, 673
Collared dove, cooing song pattern of, 97
Collateral pathways and crossing of sensory information to other motor neurons, 479
Collembolans as commensals, 345
Collias, E.C., 185, 673
Collias, N.E., 185, 284, 673, 680
Collins, B.J., 542, 673
Collision, 629

Colonial insects, social behavior of, 230-231

Colonial lower vertebrates, social behavior of, 230

Color
  discrimination of, 458
  spectrum of, 455

Color preferences
  in amphibians, maturational changes in, 561
  artificial genetic selection of, in coturnix quail, 101

Color vision, 447, 455-458

Coloration and body shape displays in octopus, 287

Columbiformes, drinking behavior of, 171

Columella
  branchial, 453
  tympanic, 453

Combasson, young, markings of, 351

Combat, cranial, and displays in ungulates, evolution of, 323

Combs, S., 454, 686

Command neurons, 491

Commensalism, 342, 344-346

Commensals, trophic, 345-346

Commissural neurons, 416

Common crane, 131, 134, 139, 140

Common pintail, 130

Common teal, 130

Common toads, male fighting in, 323

Communal displays, 326-328

Communal hunting by predators, 372

Communication
  advanced forms of, 287-294
  advantages and functions of, 280-282
  as advertising, 267
  in arthropods, 283-284
  auditory, 274-275
  basic characteristics of, 262
  in cephalopods, 283-284
  chemical, 271-274
  common patterns in, 285
  conditional properties of, 267
  contexts of, observed, 280
  convergence in, 284
  definition of, 261-268
  electrical, 276
  evolution of, 280-294
  informational aspects of, 266
  of invertebrates, 283-284
  manipulative aspects of, 267
  mathematical formulations for, 263
  methods of studying, 268-269
  mimicking of, by another species, 294
  modes and channels of, comparison of, 271-276
  modes and mechanisms of, 269-280

Communication—cont'd
  modes and mechanisms of—cont'd
    and ecological considerations in, 260-300
  modulated, 269
  by multiple channels and antithesis, 276-280
  in nervous system, 413
  operational definition of, 263
  phylogenetic comparisons of, 282-286
  reciprocal, as social behavior, 228
  role of different sensory channels in, 279
  signals as, 261-262
  sound for, 274-275
  symbolic, 296, 297
  tactile, 275-276
  in vertebrates, 283-285
  visual, 276, 277
  in water striders, study of, 67

Communication behavior, sociometric analysis of possible, 80, 81

Communication lines of host, use of, by parasites, 353-357

Communication repertoire, total, 279

Communication signals, mantis shrimps and, 264

Communication system of fireflies, 355-356

Comparative methods, 54
  and historical inference in biology, 109-110
  to study evolution of behavior, 128-129

Comparative psychology, 31, 43-48, 50
  development of discipline of, 32
  ethology, neurobiology and, 22-51

Compartmentalization of learning by invertebrates, 600-601

Compass(es)
  navigational, used by birds, 214
  sun, 205, 206, 214

Compensatory-response model for learning, 581

Competition
  over food, 343
  and habitat, 193
  over nesting places, 343

Competitive ability of group, increased, 236

Competitive exclusion, 193

Complex analyses and computers, 83-84

Compound action potential, 417, 428

Compound eyes, 456
  of insect, vision through, 458

Compulsive behavior, 523

Computer-compatible techniques for recording data, 77

Computers, complex analyses and, 83-84

Concentration gradients, ionic, 427

"Concorde fallacy," 181

Conditional ESS, 250

Conditional properties of communication, 267

Conditional stimulus, 580

Conditioned inhibition in anemone, 591

Conditioned learning in octopus, 594-595

Conditioned response, 580

Conditioned stimulus, 45, 580

Conditioning
  aversive, of predator, 369-370
  backward, 581
  classical, 45, 580-581
  forward, 581
  instrumental, 581-583
  operant, 47, 581-583
    in planarians, 592
  Pavlovian, 580
  of planarians, 592

Condors, Andean, reproductive cycles of, 311

Conductance, ionic, in neuron, changes in, 427

Conduction, saltatory, 426

Condylodera tricondyloides, 380

Cone
  photosensitive part of, 469
  and vision, 459

Cone snails, poisoning by, 375

Conflict
  and abnormal behavior, 621
  causing abnormal behavior, 634
  parent-offspring, 335-337
  weaning, 337

Conflict behavior, 41-42
  in evolution of behavior, 142

Confusion of predator by group, 233

Connecting neurons, 415

Conner, R.N., 83, 673

Conner, W.E., 673

Conover, W.J., 661, 673

Consciousness
  animal, 127
  and language, 299, 499

Consequence
  naming behavior by, 69
  "selection by," 128

Consistent behavior, 102

Consolidation and memory, 606

Constancy of behavior, 102

Constant frequency bat calls, 11

Constantine, 641

Constitutive enzymes, 90

Constraints, biological, 582-583

Constriction patterns of snakes, 376

Constructive interference with sound, 10

Consummatory component of behavior, 33

Contact perimeter of territory, 240

Conte, M., 387

Contexts of communication, observed, 280

Continuous signals, 271

Contraction, muscular, 436
  action potential and, time sequence of, 437
  before hatching of embryo, 556
  hypothesis of, 438

Contralateral neurons, 416

Control
  of behavior
    internal, 400-617
    voluntary, 481
  of feeding
    in insect, 532-540
    in mammal, 540-550
  weight, lipostatic theory of, 529

Convergence
  behavior, 140
  in bird and mammal calls, 284, 285
  in communication, 284
  in nervous system, 477

Convergent behavior, 110

Convergent evolution, 110

Cooing song patterns of doves, inheritance of, 97

Cook, A., 673

Cooperation in reproduction and care of young, 337-339

Cooperative defense against predators, 232-233

Cooperative foraging of group, 234

Coordinated group, response to attack by, 234

Coordination, development of, 559

Coordination displays, 267

Coot, American, habitat selection by, 198

Copeland, J., 356, 673

Coprophagy, 626

Copulation, 306, 310-311
  forced, 328

Corben, C.J., 127, 674

Corepressors, complexes of, 91

Corn earworm, 5

Corn snake killing mouse, 376

Corning, W.C., 589, 591, 592, 595, 596, 617, 670, 674, 675, 680, 681, 682, 688, 692

Cornsweet, T.N., 446, 674

Corpora allata, 508, 509

Corpora cardiaca, 508, 509

Corpus callosum, 497

Corpuscle, pacinian, 469

Correlations and the "trained cricket," 81-83

Corrent, G., 155, 674

Cortex
  adrenal, hormones of, and feeding behavior, 543
  cerebellar, neuronal types in, 412
  cerebral, human, map of, 482
  of telencephalon, 501
  visual, 488, 489
    visual processing in, 490

Corti, organ of, hair cells of, 452

Corticosteroids, 518

Corticosterone and feeding behavior, 543

Corvidae
  insight learning in, 587
  social behavior of, 231

Corvus corax, abnormal behavior in, 623

Coss, R.G., 608, 674

Cottontail, encounters of, at feeder, 344

Coturnix coturnix japonica, color preference in, 101

Coturnix quail chicks, artificial selection in, 101

Coughing, 165

Count, E.W., 674

Courtship
  darkening of male guppy during, 74
  and fighting in three-spined sticklebacks, 37
  sequences of, 310-311

Courtship behavior of empid fly, 141

Courtship display
  of European great-crested grebe, 282
  in Sceloporus lizard, 270
  of surface-feeding ducks, 129

Courtship pattern of spiders, altered, and developmental change, 611

Courtship sequences, 306-307, 309-310
  complexity and variability in, 312-313
  of various species, flowcharts of, 308-309

Cousteau, J., 383, 653, 674

Cow, sleep time for, 159

Cowan, W.M., 577, 674

Cowbird
  brown-headed, 573
    abnormal mate selection by, 623
  giant, parasitism-mutualism in, 351
  nest parasitism by, 350
  protection against botflies by, 351
  songs of, 575-576

Cowie, R.J., 689

Cox, V.C., 691

Coyote-canids, New England, play in, 391

Coyote-livestock conflict, 647, 648

Coyotes
  breeding with red wolves by, 117
  play in, 395
  taste aversion learning by, 647, 648

Crabs

Crabs—cont'd
  decorator, camouflage by, 378
  hermit, and anemones, commensalism between, 346
  horseshoe, egg-laying and fertilization in, 310
  land, Asiatic, eyes of, 456
  learning in, 598
  observation of, 659

Craig, W., 28, 33

Crandall, L.S., 640, 655, 674

Cranes
  classification of, 131
  "dance" of, 132, 134
    characters of, 135, 136-137
  duetting of, 134, 275
  guard calls of, 132
    sonograms of, 133
    in unison call, 136
  inferred family tree of, 139
  parental teaching by, 332
  sandhill
    calls of, sequential appearance of, 554
    as foster parents of whooping crane, 653
  species of, 132
    distribution of, 133
  unison calls of, 132, 134, 135
    characteristics of, 136-137
    shared, 138, 139
  vocalizations of, evolution of, 130-140
  whooping
    imprinting in, 571
    management of, 652-653

Cranial combat and displays in ungulates, evolution of, 323

Cranston, E.P., 672

Crawling, slug, cost of, 163-164

Crayfish
  claw opener muscle of, 437-438
  compartmentalization of learning in, 600-601
  eye of, 456
  learning in, 598
  nervous system of, 406
  observation of, 659
  shelter of, 182
  study of nervous system of, 49
  tail flexion in, neural circuit controlling, 502

Creation, special, 123-124

Creationism, 123-124

Creativity and play, 393

Crest, raising of, in Steller's jay, 271

Crews, D., 512, 513, 517, 525, 674

Crick, F.H.C., 499, 674

Cricket, 380
  calls of, and artificial stimulation of brain, 102

Cricket—cont'd
   ears of, 452
   nerve networks for flight in, development of, 559
   song of
      larval development and, 558, 559
      patterns of, inheritance of, 97
   tachinid fly attacking, 355
   temporal pattern recognition of, 488
   "trained," correlations and, 81-83
Critical periods, 44, 569-570
Crocodilians
   care of eggs of, by male, 333
   incubation of eggs by, 333
   as predators, 360
Crocuta crocuta
   gorging by, 362
   play in, 395
"Crop milk" of doves, 511
Crossarchus obscurus, 389
Crotalidae, heat and infrared sensing by, 461
Crotalus cerastes, activity pattern in, 161, 162
Crotalus mitchelli, activity pattern in, 161, 162
Crotophaginae, social behavior of, 231
Crow, J.F., 107, 143, 674
Crowned cranes, 131
Crown-of-thorns starfish, defense from predators by, 376
Crows
   group defense by, 233
   insight learning in, 587
   learning in, to increase food supplies, 179
   northwestern, whelk dropping by, 177, 178, 179
   social behavior of, 231
Crustaceans (Crustacea), 597
   care of eggs of, by male, 333
   gastric mill of, neural circuit controlling, 502
   hearing in, 454
   learning in, 598-601
   "migration" of, 202
   observation of, 659
Cruze, W.W., 562, 563, 674
Cryptocentrus steinitzi and *Alpheus purpurilenticularis*, mutualism between, 347
Cuban sandhill crane, 131
Cuckoldry, 328
Cuckoos
   European
      mimicry of, 350
      nest parasitism by, 350
   mimicking behavior of, 294
   social behavior of, 231
Cultural learning, 585

Culturally stable strategy, 251
Cunningham, R.B., 683
Curio, E., 358, 360, 362, 363, 369, 372, 373, 374, 375, 376, 383, 585, 586, 617, 674, 691
Current, 420
Cuttlefish
   attack behavior of, maturational changes in, 561-562, 563
   hunting success of, 374
Cutworms, 5
Cuvier, G., 7, 23
Cycle
   estrous, normal, and behavior of female white rat, 515
   menstrual, of human female, 514
   reproductive, of ringed plover, 515
   of sound, 9
Cyclic AMP, 517
Cyclic AMP complex and transcription, 91
Cytoplasm, 420
Cytrawa, J., 693
Czeisler, C.A., 152, 674

D

Dahl, F., 599, 674
Dale, H.H., 28
Dall, W.H., 596, 674
Dall's porpoise, sleep time for, 158, 159
Daly, M., 248, 305, 315, 329, 339, 674
Damselfish, dominance hierarchy in, 242
Dance(s)
   of bees, 286, 288-293
   of cranes, 132, 134
      characters of, 135, 136-137
   "penguin," of European great-crested grebe, 282
   round, of honeybee, 288
   sickle, of honeybee, 290
   waggle, of honeybee, 288-289
Daniken, E. von, 124, 674
Daphnia sp. as prey of three-spined sticklebacks, 371
Darkening of male guppy during courtship, 74
Darling, F.F., 312, 674
Darwin, C., 24, 28, 30, 31, 50, 93, 109, 111, 117, 118, 128, 174, 252, 253, 279, 319, 353, 674
   on behavior, 30-31
Darwinian view of evolution, 111
Darwin's finches, adaptive radiation in, 140
Data
   alternate interpretations of, 82
   behavioral, interpreting and presenting, 77-84
   correlations of, 81-83
   recording of, 76-77

Data forms to record behavior, 76
"Daughter" cells, 92
Davidson, D.W., 179, 674
Davidson, E.H., 605, 672
Davidson, J.M., 127, 674
Davidson, R.J., 127, 674
Davies, N.B., 28, 176, 185, 259, 323, 674, 681, 686
Davila, H.V., 687
Davis, J.D., 542, 673
Davis, M., 240, 693
Davis, R.T., 617, 684
Davis, W.H., 221, 674
Davis, W.J., 558, 674
Dawkins, M., 62, 167, 674
Dawkins, R., 28, 119, 120, 121, 143, 167, 179, 180, 181, 246, 248, 250, 251, 259, 267, 301, 585, 672, 674, 689
Dayal, J.A., 681
Dayton, P.K., 331, 690
De Humani Corporis Fabrica, 30
de Robertis, E., 469, 674
de Vos, G.J., 325, 674
de Vries, T., 363, 674
Death, 107
deBeer, G.R., 680
Decisions, transmissions of, by neurons, 422-427
Decorator crabs, camouflage by, 378
Deer
   black-tailed
      pheromone-producing glands in, 272
      play in, 391, 394
   head ornament of, 110
   as nuisances, management of, 647
   pheromones to mark territories of, 272
   as prey of puma, 374
   red, harems of, 322
   roe, 375
   tail-flagging in, 276
   use of neutral zones by, 382
   white-tailed, 649
      fighting by, 59
      young of, 555
Deer mouse
   number of displays in, 286
   responses of, to increase in prey, 368
Deese, J., 299, 499, 674
Defecation, 165-166
Defense
   as cause of grouping, 232-234
   offensive, of prey, 376-377
   against predator, cooperative, 232-233
Defense ring, musk-oxen, 233
Defense tactics, antipredator, 375-382
Defensive threats, 281
Deffeyes, K.S., 677
Defining attributes of fixed action patterns, 41

Deguchi, T., 155, 674
Deme, 106
    differences in gene frequency in, 114
    of rabbits, distinct, 106
Dement, W., 157, 674
Demoiselle crane, 131, 139
Demong, N.J., 64, 66, 67, 211, 674
Dendrites, 413, 414
    spines of, 414
        and memory, 607-608
Dendritic potentials, 422
Dendro-dendritic synapses, 415
Dendrograms, phylogenetic, 138
Dendrolasius, parasitism by, 349
Denenberg, V.H., 48, 577, 675
Denmark, F., 677
Denny, M., 163, 164, 675
Deoxyribonucleic acid, 88, 89, 107
    copying of, 92
    eukaryotic, 90
    "junk-," "gibberish-," or "nonsense-,"
        90
    prokaryotic, 90
    recombinant, 89-90
    ribonucleic acid, protein and, 87-90
    roles of, 92
    transcription of, environmental influ-
        ences in, 91
Dependents and predation, 363-364
Depolarization, 420, 421
    spontaneous, 421, 425
Depolarized neuron, 420, 421
Depressants and wastes, 266
Deprivation
    and blowfly's feeding behavior, 536
    effects of, on development of behavior,
        565-576
    versus enrichment, neural differences
        in, 568-569
    environmental, effects of, 566-567
    food, and activity in blowflies, relation
        of, 532-533
Dermasterias imbicata, contact of anemone
    with, 590, 591
Descent of Man, 30
Describing, recording, and cataloging be-
    havior, 68-77
Description
    of behavior, physical, 69
    pure, 54
Description statistics, 78
Desert ants
    polarized light detection in, 460
    use of "sun compass" by, 204
Desert rattlesnakes, activity patterns in,
    161, 162
Desert species, reproductive cycles of,
    311, 314
Destructive interference with sound, 10

Detectors, brightening and dimming, 488
Determinism, biological, 257
Dethier, V.G., 465, 532, 533, 535, 537,
    538, 675
Deuterostome pattern of nervous system,
    404
Deuterostomes, communication in, 283
Deutsch, J.A., 542, 617, 675
Development
    and acquisition of specific changes in
        behavior, 569-576
    of behavior, effects of deprivation and
        general effects of early experi-
        ence on, 565-576
    of bird vocalizations, 572-576
    of embryo
        behavioral, 556-558
        and evolution, 125
    problems in, causing abnormal behav-
        ior, 634
Developmental acceleration of behavior,
    142
Developmental processes, modifiability of,
    611
Dewsbury, D.A., 41, 471, 675, 679, 683,
    691
Dialects, local, in bird songs, 293-294
Diamond, M.C., 617, 687
Diamond, S., 51, 675
Diapause in insects, hormonal control of,
    509
Dickerson, R.E., 107, 143, 692
Diets, specialized, and parental care,
    334
Difference, potential, 419-420
Differential survival and reproduction,
    111
Diffusion of pheromones, 272, 275
Digestive system of blowfly, 536-537
Digger wasps
    ESS in, 248
    optimality and, 181
    use of landmarks by, 203
Digging movement, 163
Digital devices to record data, 77
Digital modulation of signals, 270
Digital signals, 269
Dijkgraaf, S., 8
Dikdiks, ears and horns of, individual
    markings in, 62
Dilger, W.C., 98, 675
Dill, P.A., 554, 675
Dimethoxyphenylethylamine, 518
Dimethyltryptamine, 518
Dimming detectors, 488
Dimorphism
    and polygyny, 255
    sexual, and mating patterns in birds,
        317
Dinosaurs

Dinosaurs—cont'd
    head ornament of, 110
    nervous system of, 409
    phylogenetic grouping of, 117
    as predators, 360
Direction, discrimination of, 448
Directional information from signals, 270
Disadaptation in blowfly, 536
Discrete signals, 269-270
Discrimination
    color, 458
    learning of, by vertebrates, 601-602
    of mimics by predators, 380, 381
    by nervous system, 413
    by neurons, 422
    spatial or temporal pattern of, 448
Disease
    American foulbrood, 94
    in captive animals, prevention and cor-
        rection of, 639
    Gilles de la Tourette's, 495
Disinhibition, 581
Disorderly behavior, 621; *see also* Abnor-
    mal behavior
Displacement activity, 42
Displacement behavior in captive animals,
    625
Displays, 261
    body shape, and coloration in octopus,
        287
    communal, 326-328
    coordination, 267
    courtship
        of European great-crested grebe,
            282
        in *Sceloporus* lizard, 270
    cranial combat and, in ungulates, evo-
        lution of, 323
    evolutionary origins of, 282
    numbers of, in various species, 286
    persuasion, 267
    protean, 376
    speculum-flashing, in mallard, 148
    transfer of, in evolution of behavior,
        142
    visual, 276
    warning, 377
Dissanayake, E., 393, 675
Distance, discrimination of, 448
Distraction of animals by observer, 60-
    61
Distress calls, 280-281
"Diurnal migration" of aquatic plankton,
    202
Diurnal patterns of predation, 363
Divergence in nervous system, 477
Diversive exploration, 393
Dizziness, 468
DNA; *see* Deoxyribonucleic acid
Dobzhansky, T., 143, 675

Documentation and historical inference in biology, 109-110

Dodson, P., 117, 675

Dog pups, play in, 387

Dogs
abnormal behavior in, 633
antithesis in, 279
"brainwashing" in, 628
environmental deprivation in, results of, 567
hunting, African, social behavior of, 226, 227
learned helplessness in, 588
play intent movements in, 278
play-bow of, 397
as predators, 360
rolling in strong-smelling substances by, 167
salivation of, Pavlov's study using, 45
selective breeding of, 101
social needs of, 639
"thrill-seeking" play in, 393
trained, 654
urination by, 165
vacuum behavior in, 38
wild, African
prehunting behavior of, 281
switch in prey in, 366

Dolphins
bottle-nosed, sleep time for, 158, 159
convergent evolution and, 110
magnetic senses of, 463
mixed-species groupings in, 344
social behavior of, 230, 231
use of sonar by, 466

Domestic animals, characteristics of, 630

Domestic livestock, abnormal behavior in, 629-633

Domestic pets, abnormal behavior in, 633

Domestication, 629-633
ant, 352-353
definition of, 629
effect of, on behavior, 631

Dominance hierarchies, 240, 241, 242

Dominance-based monogamy, 316

Dominant genes, 93

Dominis, J., 342

Domjan, M., 582, 617, 675

Dopamine, 518

Doppler effect, 10
use of, by bats, 11

Doppler shifts, 11

Double discrimination learning of planarians, 593

Douc langur, play-face of, 397

Doves
cooing song patterns of, inheritance of, 97
drinking behavior of, 171

Doves—cont'd
maternal role of, 329
ring, reproductive sequence in, 511, 512

Downy woodpecker, nesting habits of, 83

Dragonflies, temperature regulation by, 169

Dreaming, 157-158

Dreaming sleep, 155

Drewien, R.C., 652, 675

Drickamer, L.C., 179, 675

Drift, genetic, 664

Drifting and habitat, 192

Drinking, 171-172

Drive, 527
problems with concept of, 39-40

Drosophila melanogaster
behavioral mutations among, 100-101
fate maps for, 560
laying an egg, 86

Drosophila sp.
chromosomes in, 95
evolution of, 304
selective breeding of, 101

Drucker-Colin, R.R., 617, 675

Drug addiction, 523

Drying behavior, 167

Dual center model of feeding in mammals, 544-545, 547

Ducklings
imprinting in, 570, 571
innate responses of, to hawk silhouette, 41

Ducks
black-headed, nest parasitism by, 350
drinking by, body position changes in, 75
gadwall; see Gadwall duck
goldeneye, 56
habitat selection by, 198
mallard
behavioral repertoire of, portion of, 70-71
speculum-flashing display in, 148
oldsquaw, food choices of, 177
surface-feeding, courtship displays in, 129
taxonomic relationships of, as elucidated by behavioral characteristics, 130

Duets, 275

Dufour's gland, 273

Dulosis, 352-353

"Dump nests," 351

Dunbar, M.J., 118, 675

Dunbar, R.I.M., 76, 675

Dung beetle, foraging behavior in, 176-177

Dung flies, ESS in, 248

Dunker, H., 378, 379, 685

Dunn, J.R., 693

Dunstone, N., 180, 675

Duodenal hormones and satiety, 543

Duplicative theory of vision, 459

Durden, K., 251, 571, 675

Dust bathing, 167

Dwarf American toad, fertilization of eggs by, 310, 311

Dwarf cichlids, territory of, 239

Dwarf honeybee, dance of, 290

Dwyer, T.J., 149, 675

Dyal, J.A., 596, 617, 670, 674, 675, 682, 688, 692

Dzelzkalns, J.J.I., 147, 276, 678

E

Eagles
bald
feeding nestling by, 59
migration of, 201
social behavior of, 226
geese defending against, 233
golden
aggression in, 251
imprinting in, 571
inactivity in, 150
martial, clipping feathers of, to mark, 64
neophobia in, 369
play in, 385
as predators, 360
time without eating for, 362
tree blind to photograph, 60

Ear(s), 451-453
of Arctiid moth, 20
external, of little brown bat, 12
of dikdiks, individual markings in, 62
inner, 453
moth, neural research on, 428-432
of noctuid moth, 5, 16, 17
structures of, in mammals, 452

Eardrum, 452, 453

Earthworms
breeding times of, 311, 314
food choice of, 177
hearing in, 454
learning in, 597
nervous system of, 407

East crowned cranes, 131, 139

Eastern phoebes, abnormal nest building by, 623

Eastern sarus crane, 131, 134

Eastzer, D.H., 577, 692

Eat, refusal to, after capture, 624

Eating; see also Feeding
function of, 172-173
and group behavior, 232

Eating behavior, abnormal, 626

Eating disorders
in domestic animals, 631

Eating disorders—cont'd
in pets, 633
Eaton, R.L., 383, 565, 675
Eccles, J.C., 28, 441, 480, 499, 500, 505, 608, 675
Ecdysis, 171
Echidna
care of eggs by, by male, 333
sleep time for, 159
Echinoderms
insight into learning from, 611-612
learning in, 601
nervous system of, 406, 411
Echoes
accentuation of, by bats, 12
of CF bat calls, 11
characteristics of, 11
of flying moths, bats and, 14-15
of FM bat calls, 10-11
of moths with wings in different positions, appearance and, 16
suppression of, by humans, 12
Echolocation
calls of bats and, types of, 11
and hunting behavior in bats, 7-15
use of, 466-467
by bats, 10-12
Eckert, R., 414, 675
Ecological aspects of animal behavior, 144-399
Ecological and genetic variability, consequences of, in evolution of behavior, 108-143
Ecology
and behavior, 103-107
of behavior
genetics and, 86-107
relationship of, to other disciplines, 26
Ectoparasites, 348
Ectothermy, 168, 169, 170
Edgar, R.S., 107, 690
Edwards, C.B.H., 683
Eels
electric, 375
generation of electricity by, 462
glass, 208
migration by, 207-209
moray, and octopus, 383
yellow, 208
Effect
Andrea Doria, 200
Baldwin, 606
Bruce, 271
Clever Hans, 298
"fountain," 234
Fraser Darling, 312
law of, Thorndike's, 45
placebo, 523
Efferent neurons, 415

Efficiency
energetic, of groups, 237
feeding, with group, 234-235
of movement through fluid by group, 237
Egg brooding by male water bug, 331
Egg dumping, 351
Egg retrieval by greylag goose, 34, 35
Egg-laying behavior in Aplysia, genetic control of, 102
Eggs
care of, by male, 333
of noctuid moth, 5
Egrets, great, nonoptimal foraging in, 180
Ehrman, L., 99, 100, 107, 675
Eibl-Eibesfeldt, I., 28, 38, 41, 565, 675
Eisenberg, J.F., 72, 359, 383, 675
Eisner, M., 675
Eisner, T., 107, 143, 372, 457, 675, 692
Electric catfish, 375
Electric eels, 375
Electric fish
electrical sensing of, 462
energy-emitted senses of, 466
Electric rays, 375
Electric senses, 462
Electrical communication, 276
Electrical field surrounding fish, 462
Electrical shock to kill prey, 375
Electrodes
positioning of, in brain, 417
to study neurons, 416-417
Electromagnetic spectrum, 455
Electromyograms to study embryonic behavior, 558-559, 560-561
Electron microscope for neurophysiological study, 416
Electronic counters to record data, 77
Electrons, 419
Electronic potential, 422
Elephant seals, refusal of, to eat after capture, 624
Elephants
alloparental helping in, 338
matriarchal groups of, 330
reproductive cycles of, 311
sleep time for, 158, 159, 162
social behavior of, 230, 231
visual communication in, 276
weaving behavior in, 624
Elimination, postures for, 165
Elimination behavior, 165-166
Elk as prey of puma, 374
Ellarson, R.S., 177, 686
Elliott, G.R., 671
Elliott, R.D., 622, 623, 675
Ellis, D.H., 68, 675
Ellis, M.J., 395, 675

Elvers, 208
Elzinga, R.J., 598, 675
Embryo
behavioral development of, 556-558
contraction of muscles by, 556
Embryological roots of behavior, 554-560
Embryonic development and evolution, 125
EMG; see Electromyograms
Emitted light, use of, by fish, 467
Emitted-energy senses, 466-467
Emlen, J.M., 28, 77, 78, 174, 180, 185, 324, 675, 676
Emlen, S.T., 28, 64, 66, 67, 77, 78, 180, 211, 212, 226, 268, 572, 674, 675, 676
Emmel, T.C., 334, 676
Emotional behavior, ontogeny of, 569
Empid fly, courtship behavior of, 141
Encounters, chance, 342, 343, 344
Endangered species, management of, 651-653
Endocrine glands, 507
Endocrine system
information processing by, 3-4
and movement, 33
and protein synthesis, 92, 96
Endoparasites, 348
Endoplasmic reticulum, 413
Endorphins, 521
in fish, 450
Endothermy, 168
End-plates, motor, 436, 437
Energetic efficiency of groups, 237
Energy
action-specific, 39
need for, 104
shortages of, 104-105
Engram, 606
Enkephalins, 518, 521
Enrichment
versus deprivation, neural differences in, 568-569
environmental, 568-569
Enright, J.T., 185, 676
Enterogastrone and satiety, 543
Entrained rhythms, 150
Environment
causing abnormal behavior, 634
and behavior, 90-92
genes and, 90-91
natural, and behavior, 103
and parental care, 333-334
role of, in gene frequencies, 113
spatial differences in, 103
Environmental considerations in reproductive behavior, 310-312
Environmental deprivation, effects of, 566-567
Environmental enrichment, 568-569

Environmental factors affecting behavior, 128

Environmental influences in transcription of DNA, 91

Environmental interaction, 90-91

Environmental variability, 103

Environmental variance, 99

Environmental variation, 103
  at global level, 104
  at local level, 105

Enzymes, constitutive, 90

Eoff, M., 101, 676

Ephemeroptera, reproduction and, 107

Epideictic phenomena, 118

Epidermis, shedding of, 171

Epigamic selection, 319

Epigenesis, 556

Epilepsy, 494
  cutting corpus callosum of person with, 497
  in pets, 633

Epimeletic behavior, 632

Epinephrine, 507, 518
  and feeding behavior, 543
  molecular structure of, 518

Epstein, A.N., 546, 547, 675, 690

Epstein, R., 127, 296, 297, 299, 676

Equal sex ratios, 248

Equilibrium
  Hardy-Weinberg, 663, 664, 665
  in neurons, 420
  punctuated, 125

Erber, J., 609, 617, 684

Eremias lugubris, 379
  protective mimicry by, 379

Erichsen, J.T., 681

Ericsson, K.A., 606, 676

Erithacus rubecula, navigation by, and magnetic field, 215

Ernst, U., 617, 674, 691

Error, types I and II, 55

Escape reactions
  abnormal, 623-624
  of turkey chick to hawk silhouette, 41

Escape responses of birds to flying birds of prey, 39

Esch, H., 288, 289, 291, 292, 676

Eskin, A., 153, 674, 676

ESS; see Evolutionarily stable strategy

Estradiol
  and development of "male" brain, 520-521
  and feeding behavior, 543
  tritium-labeled, to locate receptor area, 517

β-Estradiol, molecular structure of, 519

Estrogens, 518
  and incubation behavior in doves, 511

Estrous cycle, normal, and behavior of female white rat, 515

Estrus, 514, 515

Et-epimeletic behavior in pets, 633

Ethical considerations in observation of animals, 657

Ethogram, 56, 69-70
  division of, 70
  rodent, 72-73
  species, determination of, 660

Ethological aspects of animal behavior, 144-399

Ethology, 31, 33-43, 50
  comparative psychology, neurobiology and, 22-51
  development of discipline of, 32
  development of term, 31-32
  early concepts and terms in, 33-39
  and learning, 613
  relationships of, to other disciplines, 26

Ethons, 68

Euglena, light sensitivity of, 455

Eukaryotic DNA, 90

Eukaryotic organisms
  introns in, 88, 89
  transcription in, 89-90

Euphasiopteryx ochracea attacking cricket, 355

European blackbirds, mobbing behavior by, 586

European common crane, 131

European cuckoos
  egg mimicry of, 350
  nest parasitism by, 350

European great-crested grebe, courtship display of, 82

European hare, 375

European polecats, aggressive play in, 387-388

European robins
  magnetic field effects on, 215
  navigation by, 215
  territorial threat responses of, 37

European starling, competition over nesting places by, 343

European storks, migration of, 212

European teal, courtship displays in, 129

European tits, cultural learning of, 585

European warblers
  songs of, and female-choice sexual selection, 318
  use of stars by, for navigation, 214

Evans, D., 219, 676

Evans, R.L., 51, 676

Evans, S.M., 232, 680

Evans, V., 242, 676

Evasive behavior patterns of moths toward bat calls, 17-19, 20

Everglades kite, prey of, 366

Evoked potentials, 417

Evoked potentials—cont'd
  use of, to study senses, 445

Evolution
  of behavior, 128-143
    comparative method to study, 128-129
    consequences of genetic and ecological variability in, 108-143
  biological view of, 122-123
  classic Darwinian view of, 111
  of communication, 280-294
  convergent, 110
  of crane vocalizations, 130-140
  embryonic development and, 125
  historical inferences about, 109
  of mating patterns, 324-325
    proposed model of, 324
  methods of studying, 109-111
  versus other considerations, 122-125
  parallel, 140
  of predatory behavior, 359-361
  process of, 111-122
  "random walk" theory of, 124
  recent considerations concerning, 124-125
  role of genes in, 124-125
  of social behavior, 118
    kin selection and, 243-247
  theory of, development of, 30

Evolutionarily stable strategy, 121, 181, 247-252
  play as, 398

Evolutionary characteristics of behavior, 140-143

Evolutionary foundation of understanding of behavior, 30-31

Evolutionary origins
  of displays, 282
  of new species, 115

Evolutionary theory, 123; see also Evolution

Ewert, J.P., 28, 428, 441, 505, 676

Ewing, A.W., 242, 676

Exaggeration in evaluation of behavior, 142

"Exapted" behavior, 142

Excitatory postsynaptic potentials, 421, 422

Excitatory receptive field, 486

Exclusion, competitive, 193

Exercise
  law of, Thorndike's, 45
  and sleep, 160

Exercise theory of play, 396

Exhaustion of prey by predator, 371

"Existence, struggle for," 111

Exocrine glands, 507

Exogenous mood-altering chemicals, pleasure and pain and, 521-524

Exons, 90

Exoskeleton, shedding of, 171
Expansion, "flash," 234
Experience, early
    causing abnormal behavior, 34
    general effects of, on development of
        behavior, 565-576
Experiment(s)
    attitudes and, 55
    basis of, 54
    behavior, 661
    split-brain, 494, 497-500
    variation and chance in, 55
Experimental manipulation and analysis of
        neurons, 418
Experimental method, 54
    significance of results of, 55
Exploration, 162, 393
Exploratory learning, 584-585
Expression of the Emotions in Man and
        Animals, 30
Extension-type movements, 163
External influences and Hardy-Weinberg
        law, 665
Extinction, 581
Extracellular fluid, 420
Extraterrestrial origin of species, 124
Eyes
    compound, 456
        vision through, 458
    examples of, 456
    of frog, 472
    rapid movement of, 155
    of vertebrates, reflective surface of,
        459

F

F, 100
Facial recognition areas of human brain,
        494
Facilitating interneurons, sensitization in
        Aplysia and, 604
Facilitation, social, and predation, 369
Facultative cleaners, 346
Facultative parasitism, evolution of, 351
Fagen, R.M., 53, 69, 385, 386, 387, 389,
        390, 391, 392, 393, 394, 396,
        397, 398, 399, 676
"Fairness" in play, 396-397
Falco columbarius, hunting success of,
        374
Falco peregrinus, hunting success of,
        374
Falco sp., migration activity of, 152
Falco sparverius
    diets of, 334
    prey-handling ability of, maturational
        changes in, 564-565
Falconer, D.S., 143, 676
Falconiformes, predation by, 359
Falcons as predators, 360

"Fallacy, Concorde," 181
Falling objects, catching, response time
        for, 667-669
False body markings on butterflies, 379
False head as defense against predators,
        379
Familiar area, 199
    bird migration and, 212, 213, 214
    home range and, 199-200
Familiar path, 199
Familiarity with species, 74
    when studying animal behavior, 55-56
Farish, D.J., 166, 676
Farkas, S.R., 676
Fasciculus, arcuate, 495
Fast sleep, 155
Fate mapping to study embryonic develop-
        ment, 559-560
Fatigue, olfactory, 447
Fauser, D.J., 677
Fechner, G.T., 28
Fecundity, high or low, 334
Feeding; see also Eating
    and blood sugar levels in fly, 532,
        533
    in blowfly, cessation of, 535-540
    of captive animals, 638-639
    control of
        in insect, 532-540
        in mammal, 540-550
    and group behavior, 232
    initiation of, in blowfly, 533-535
    in insects, hormonal control and, 509
    internal regulation of, 527-531
    in mammals, dual center model of, 544-
        545, 547
    neural mechanisms and, 526-551
    optimal foraging and, 172-181
    peripheral controls for, in rat, 541-
        543
    problems associated with, 173
    reciprocal, as tactile communication,
        275
    of young, 329
Feeding behaviors, 70
    in amphibians, changes in, genetic con-
        trol of, 102
    in blowfly, 532-533
    of rat, 540-541
Feeding efficiency with group, 234-235
"Feeding" hormones, 510
Feeding system for control of body weight,
        hypothetical, 530
Feeney, P.P., 265, 266, 692
Feirtag, M., 473, 685
Feldman, M.W., 585, 673
Felidae, elimination behavior of, 165
Felsenstein, J., 124, 676
Female-choice sexual selection, 318, 319-
        320

Female-defense monogamy, 316
Females
    care of offspring by, 329
    fighting between males over, 319, 320,
        323
    responsiveness of, to male song, 575
Femme fatales, 356-357
Fenton, M.B., 23, 689
Fentress, J.C., 28, 441, 505, 676
Ferguson, N.B.L., 544, 676
Ferguson, W., 677
Fernald, R.D., 77, 554, 676
Ferrets
    aggressive play in, 387-388
    black-footed, 357
Ferron, J., 676
Fertilization, 92
    despite cannibalism, 306, 308
Fibers, muscle, types of, 438
Fibrillar structures of neurons, 414
Ficken, M.S., 676
Ficken, R.W., 95, 676
Field, J., 677
Field
    electrical, surrounding fish, 462
    magnetic, detection of, 447
    receptive, excitatory, 486
Fighting
    in cats, and testosterone, 54, 55
    courtship and, in three-spined stickle-
        backs, 37
    male aggression and, 319, 320, 323
    between males for females, 319, 320,
        323
    by white-tailed deer, 59
Fighting fish, Siamese, learning of, 602
Filial imprinting, 570-571
Filtering, stimulus, 443
Finches
    Darwin's, adaptive radiation in, 140
    Viduine, song mimicry by, 351
    woodpecker, tools used by, 371
Fireflies
    communal displays in, 327
    communication system of, 355-356
    female, eating male firefly of another
        species, 356
    flash patterns of, 269
    luminescence in, 467
    luminescent signals of, 356
    qualitative differences in light of, 270
    role of sensory channels in, 279
    temporal pattern recognition of, 448
Firings, 424
    rate of
        in neuron, 421, 425
        resting, 425
        spontaneous, 425
Fischer, G.L., 308, 676
Fischer, R.B., 124, 676

Fischer's lovebird, carrying behavior in, 98

Fish
  anadromous, 206
  auditory communication in, 275
  avoidance learning of, 602
  body orientation in, 468
  bony, gill movements of, 165
  care of eggs by, by male, 333
  cartilaginous, respiratory movements of, 165
  catadromous, 207
  cichlid; see Cichlid fish
  cleaner, 346
    sex reversal in, 315
  color vision in, 458, 459
  detection of sound vibration by, 452-453
  displays in, numbers of, 286
  electric
    electrical sensing of, 462
    energy-emitted senses of, 466
  electrical field surrounding, 462
  flashlight, use of emitted light by, 467
  food robbing by, 349
  gorging by, 362
  hearing in, 454
  heat detection by, 462
  insectivorous, switch in prey of, 366
  learning in, 601
  memory in, 607
  migration of, 206-209
  mixed-species groupings in, 344
  nervous system of, 409
  observation of, 659
  odor following in, 191
  pain in, 450
  pleisiopid, protective mimicry by, 379
  porcupine, defense from predators by, 376
  as predators, 360
  as prey, 374
  reef, habitat selection in, 196
  role of sensory channels in, 279
  schools of, 237
  scratching by, 167
  sleep characteristics of, 158
  teleost, intensity threshold of, determination of, 446
  temperature regulation by, 170
  temperature sense of, 447
  Tilapia, territory of, 239
  use of sonar by, 466
  yawning in, 166
"Fish ladders," 206
Fisher, R.A., 248, 303, 676
Fisher and porcupine, 383
Fisler, G.F., 344, 676
Fitness
  gene, 119-120

Fitness—cont'd
  genetic, 243
  inclusive, 243, 245, 246
  individual, 243
"Fittest, survival of," 111
Fitzpatrick, J.W., 179, 676
Fixed action pattern, 36
  defining and secondary attributes of, 41
  and sign stimulus, relationship of innate releasing mechanisms to, 40
Fixed reinforcement schedule, 582
Fjerdingstad, E., 593, 676
Flank rubbing by cats, 56, 57
"Flash expansion," 234
Flash patterns of fireflies, 355-356
Flashlight fish, use of emitted light by, 467
Flatworms
  learning in, 591-595
  movement by, 163
  nervous system of, 405
Fleas as parasites, 348
Fleeing from predator, 375
Flehmen, 310
  by American bison bull, 311
Flicker woodpecker, nesting habits of, 83
Flickers, common, mustache of, 268-269
Flight, 163
Flight mechanism, 438
Flight muscles, 438-439
Flight path
  interception of, by predator, 371
  of moths and bats, 17
Flipper motions, 163
Flocks, foraging, mixed, 345
Florida sandhill crane, 131
Flowers, appearance of, under different light, 457
Flugel, J.C., 29, 51, 676
Fluid, efficiency of movement through, by group, 237
Flukes, 591
  nervous system of, 407
Flushing, groping and, as hunting method, 371
Fly(flies)
  Batesian mimicry by, 379, 381
  caddis, larvae of, shelter for, 183
  as commensals, 345
  dung, ESS in, 248
  empid, courtship behavior of, 141
  fruit; see Fruit flies
  grooming in, 166
  house-; see Houseflies
  jumping spider with, 358
  and mantids, interactions of, 435
  Mertensian mimicry by, 377
  preening by, 167

Fly(flies)—cont'd
  as prey of Mantidae, 375
  tachinid, attacking cricket, 355
  temperature regulation by, 169
Flycatchers
  foraging behavior in, 179
  pied, migration of, 210
Flying, cost of, 163
Flying birds of prey, escape responses of birds to, 39
Flying patterns, 162-163
FM bats, 10
  little brown bats as, 13
Focal animal sampling, 75
Foelix, R.F., 688
Follicle-stimulating hormone, 510, 518
Follicle-stimulating hormone–releasing factor, 518
Food
  clashes over, 342, 343
  exaggerated responses to; see Hyperphagia
  excess, caching of, 343
  intake of, factors affecting, 529-531
  need for, 104
  selection of, 177-179
  stealing of, 348-349
Food aversion and other associative learning, differences between, 584
Food deprivation and activity in blowflies, relation of, 532-533
Food type, optimal, 177-179
Foraging
  cooperative, of group, 234
  improved, with mixed-species groups, 345
  methods for, 175-176
    costs of, 175, 176
  optimal
    description and examples of, 174-181
    feeding and, 172-181
    group nesting and, 235
    among patches, 174, 175-176
    role of learning in, 179
    and search image, 179-180
Foraging behavior
  alteration of, 179
  of Syrian woodpecker, 364
Foraging patterns, 175
Forced copulation, 328
Ford, E.B., 121, 676
Forgetting, 581
Form vision, 460-461
Formica exsecta, parasitism by, 349
Formica fusca, 349
Formica sp., parasitism by Atemeles beetle of, 354
Formica subintegra, raiding by, 353
Forster, L., 358, 676

Forster's tern, hunting success of, 374
Forward conditioning, 581
Fossey, D., 653
Foulbrood, American, 94
"Fountain effect," 234
Fouts, R.S., 295, 298, 676
Fowler, M.E., 640, 655, 676
Fox, L.R., 179, 676
Fox, M.W., 68, 565, 621, 629, 630, 635, 639, 644, 645, 655, 676, 682, 684, 687, 688, 692
Fox, R., 253, 691
Fox
    and caribou, play between, 388-389
    as predator, 360
    red, hunting success of, 375
Fraenkel, G.S., 190, 191, 676
Frankel, R.B., 463, 676
Franks, E.C., 195, 677
Fraser, A.F., 632, 676
Fraser, D., 179, 677
Fraser Darling effect, 312
Freemon, F.R., 159, 185, 677
Free-running period, 152
Fregata, 130
Frequency
    of behavior, alterations in, as result of signal, 263
    gene; see Gene frequency
    of sound, 9
        relationship between sound intensity and, 10
    spike, 425
Frequency distribution of heritability values, 103
Frequency modulated bat calls, 10-11
Fretheim, T.E., 123, 124, 677
Fretwell, S.D., 193, 677
Freud, S., 28, 44
Fricke, H.W., 315, 677
Friction movements, 163
Friend, J., 678
Frilled lizards, defensive threat postures of, 281
Frisch, K. von, 28, 35, 36, 109, 185, 204, 205, 288-293, 301, 458, 599, 677
Frog-eating bats, 355
Frogs
    brain of, 411
    clawed, androgen uptake in brain of, 517
    color preferences of, maturational changes in, 561, 562
    eye of, 472
    gliding behavior of, 140
    gorging by, 362
    nervous system of, 409
    oral birth from gastric brooding by, 331

Frogs—cont'd
    retina of, ganglion cells of, output of, 487
    role of sensory channels in, 279
    visual pathway from retina to brain in, 487
    visual processing by, 486-487
Frohman, L.A., 549, 677
Fruit bat, echolocation by, 8
Fruit flies
    behavioral mutations among, 100-101
    chromosomes in, 95
    communal displays in, 327
    genetic control of behavior in, 101
    observation of, 659
    vibration-sensitive antennae of, 453
FSH; see Follicle-stimulating hormone
Fujimoto, J.T., 680
Fuller, C.A., 185, 684
Fuller, J.L., 689
Fuller, M., 693
Fulvous duck, habitat selection by, 198
Function, 21-22
    classifying and naming behavior by, 69
Functional categories of movement, 162
Functional responses in predators, 368
Fur seal, harems of, 322
Future events, role of, in behavior, 125-127

G

Gadwall duck, 130
    courtship displays in, 129
    habitat selection by, 198
    time budgets for, during breeding season, 149
Gage, P., 48, 49
Gait patterns, 163
Gaither, N.S., 482, 677
Galambos, R., 686
Galapagos hawk, predation by, 363
Galler, S.R., 223, 677, 692
Galliformes, "dump nests" of, 351
Gallup, G.G., Jr., 127, 677
Gallus gallus, parental brooding of, 332
Galvani, L., 30
Gambel's quail, encounters of, at feeder, 344
Game management, 648-651
Game Management, 649
Gametes, 92
Game-theory view of communication, 267
Gamma-aminobutyric acid, 518
Gamma-hydroxybutyrate, 518
Gamow, G., 123, 677
Gamow, R.I., 461, 462, 677
Ganglia, 403, 404, 416
    of frog, output of, 487

Ganglia—cont'd
    of retina, 486
Gannets, use of landmarks by, 212, 213
Gap junctions, 424
Garcia, J., 28, 47, 677, 678
Garden of Intelligence, 640, 641
Garden peas, Mendel's study of, 93-94
Gardner, A., 294, 295
Gardner, B., 294, 295, 677
Gardner, R.A., 677
Garganey duck, 130
Garter snakes
    feeding behavior of, 140
    prey preferences of, 365
Gastric brooding, oral birth from, by frog, 331
Gastropods
    learning in, 596
    movement by, 163
    nervous system of, 405
Gauthreaux, S.A., Jr., 223, 677
Gazelle, Thomson's, as prey
    of cheetah, 374, 375
    of lions, 373
    of wild dogs, 375
Gazzaniga, M.S., 689
Geese
    black, 130
    blue, migration of, 210
    brain of, 411
    defense mechanism of, against eagle, 380
    formation flying by, 237
    greylag
        egg retrieval by, 34, 35
        postures of, and communication, 277
    group defense by, 233, 380
    monogamy of, 329
    true, 130
Geist, V., 27, 323, 655, 672, 677
Gene(s), 93
    dominant, 93
    and environment, 90-91
    fitness of, 119-120
    frequencies of; see Gene frequency
    information storage in, 605
    interactions among, 95
    pleiotropic effects of, 95
    random dispersal of, in population and gene frequency, 114
    recessive, 93
        maladaptive, 112
    regulator, 90
    role of, in evolution, 124-125
    "selfish," 117-122
        and behavior, 119-121
        competing, 121
    single
        and bees, 94

Gene(s)—cont'd
  single—cont'd
    and behavior, 94-95
    and peas, 93
    survival of, 119
Gene frequency, 111
  changes in, gradual accumulation of,
        and origin of new species, 114,
        115
  in populations, 662-665
  reproduction and, 113
  role of environment in, 113
  shifts in, 111-114
  survival and, 112-113
Gene pool, changes in, 111-122
Generalization of mimics by predators,
        380, 381
Generalization gradients, "sharpened
        peaks" of, 368
Generation(s)
  overlap of, and social behavior, 229
  spontaneous, 123
Generator potentials, 422
Genetic assimilation, 606
Genetic control of behavior, additional
        evidence for, 100-102
Genetic drift, 111, 664
  and neutralism, 124
Genetic and ecological variability, conse-
        quences of, in evolution of
        behavior, 108-143
Genetic expression, environmental influ-
        ences in, 91
Genetic fitness, 243
Genetic mosaics of fruit flies, 101
Genetic variation, 92-93
Genetics
  and behavior, 92
  and ecology of behavior, 86-107
  Mendelian, 93-94
  overview of, 87-94
  quantitative, heritability, and behavior,
        96-100
  and variability, 102-103
Genotype
  fitness of, 243
  of population, 113-114
  variability among, 100
Genotypic variance, 99
  zero, 100
Gens (gentes), 350
Gentry, R.L., 170, 677
Geographical isolation and origin of new
        species, 114-117
Geotaxis, 191
Gerard, J.H., 310, 472
Gerard, R.W., 673
Gerris remigis, study of communication in,
        67

Gersuni, G.V., 471, 677
Getty, T., 677
Giant cowbird, parasitism-mutualism in,
        351
Giant squid as predator, 361
Gibb, J.A., 196, 677
Gibbs, J., 543, 670, 677
Gila monster, 582
  noxious taste of, 583
Gilbert, L.E., 265, 677
Gilbert, W., 90, 677
Gill movements, 165
Gill withdrawal reflex in Aplysia, 603,
        604
  neural control of, 503
Gilles de la Tourette's disease, 495
Giraffes
  Masai, male fighting in, 319
  sleep time for, 159
Gland
  Dufour's, 273
  endocrine, 507
  exocrine, 507
  in insects, major, 508, 509
  pheromone-producing, 272, 273
  pineal, circadian pacemaker in, 155
  pituitary, hormones of, and feeding
        behavior, 543
Glass eels, 208
Glial cells, 428
Gliding, 163
Gliding movements, 163
Globus, A., 608, 674
Glomeruli of synapses, 415
Glucagon and feeding behavior, 543
Glucocorticoids and feeding behavior,
        543
Glucostatic hypothesis of feeding behav-
        ior, 547
Glutamate, 518
Glycine, 518
Goats, Toggenberg, play in, 394
Goby
  and Alpheus rapax, interindividual in-
        teractions between, 348
  neon, cleaning green moray, 346
  and shrimp, mutualism between, 347
Gold, R.M., 548, 677
Golden eagles
  aggression in, 251
  imprinting in, 571
Golden hamsters, agonistic behavior in,
        238
Golden plovers, migration of, 210
Goldeneye ducks, 56
Goldfish
  avoidance learning by, 602
  hearing in, 453
  three-point cone sensitivity in, 459
Goldman, R.N., 69, 676

Goldstein, M.C., 677
Golgi, C., 28
Golgi bodies, 413
Golgi's stain, 49
Gonadal hormones and feeding behavior,
        543
Gonadotrophic hormones, 510
Gonodactylus festaei, individual recogni-
        tion by, 229
Gorging, 362
Gorilla
  abnormal sexual behavior by, 626
  language in, controversy over, 294-
        300
  lowland, play in, 390
Gorner, P., 599, 677
Gossow, H., 565, 677
Gottfried, B.M., 195, 677
Gottleib, G., 301, 588, 677
Gould, J.L., 67, 100, 216, 220, 286, 292,
        293, 463, 672, 677, 691
Gould, S.J., 125, 142, 255, 256, 257,
        677
Graber, R.R., 673
Graded potentials, 422
Gradualism, 125
Grafen, A., 248, 672
Grain beetles, learning in, 600
Grasshopper, 380
  brain of, 410
  courtship in, 312
Gray, J.A.B., 469, 679
Gray matter, 477
Gray seal, sleeping, 156
Graziadei, P.P.C., 456
Grazing patterns of free-ranging cattle in
        Australia, 150
Great egrets, nonoptimal foraging in, 180
Great titmouse, alarm call of, 281
Great tits
  microhabitat selection in, 195, 196
  optimal behavior in, 180
  optimal foraging of, learning and, 179
Great-crested grebe, European, courtship
        display of, 282
Greater sandhill crane, 131
Great-horned owl, hand-raised, 625
Grebe, great-crested, European, courtship
        display of, 282
Green, G.W., 532, 678
Green, R.P., 216, 691
Green moray, neon goby cleaning, 346
Green turtle, homing ability of, 209-210
Greenberg, L., 228, 672
Greenberg, N., 678
Greene, H.W., 375, 377, 678
Greenough, W.T., 608, 678
Greenwald, O.E., 375, 678
Green-winged teal, habitat selection of,
        198

"Greeting" reaction in cats, 56, 57
Greylag goose
  egg retrieval by, 34, 35
  postures of, and communication, 277
Grier, J.W., 16, 59, 61, 150, 320, 364,
      385, 418, 436, 453, 461, 572,
      626, 636, 668, 678
Griffin, D.R., 7, 14, 23, 126, 127, 140,
      200, 213, 215, 221, 466, 678
Grinnel, A., 51, 441, 505, 672
Grizzly bears as predators, 360
Grobecker, D.B., 370, 686
Grohmann, J., 564, 678
Grooming, 166-167
  social, 227, 228
    function of, 282
  as tactile communication, 275
Groos, G., 155, 688
Groos, K., 395, 678
Groping and flushing as hunting method,
    371
Grosbeaks, competition for food by, 343
Grossman, S.P., 548, 678
Ground squirrels, circannual rhythms of,
    154
Group
  confusion of predator by, 233
  cooperative defense against predator by,
    232-233
  cooperative foraging by, 234
  coordinated response to attack by, 234
  as defense against predators, 380
  detection of predators by, 232
  division of labor in, 236-237
  efficiency of movement of, through flu-
    id, 237
  facilitation of reproduction in, 235-
    236
  increased competitive ability of, 236
  interspecific, 344
  matriarchal or matrifocal, 324
    among mammals, 329-330
  mixed, 344
  population stability of, 236
  protection of young by, 233
  social, number of animals in, 226-227
  specialization in, 236-237
  survival in, 236
Group nesting and optimal foraging, 235
Group selection
  and behavior, 117-118
  and evaluation of social behavior, 247
Group selection theory of Wynne-
    Edwards, 120
Grouping
  causes of, 231-237
    hypotheses on, 232-237
  mixed-species, 344-345
  phenomenon of, 224-258
Group-selected behaviors, 118

Grouse
  leks in, 326-327, 328
  mating patterns of, 325
Growl, 285
Growth hormone and feeding behavior,
    543
Grubb, T.C., Jr., 176, 678
Gruidae, 130, 131
Grus americana, 131
Grus antigone, 131
Grus antigone antigone, 131
Grus antigone sharpii, 131
Grus canadensis, 131
Grus canadensis canadensis, 131
Grus canadensis nesiotes, 131
Grus canadensis pratensis, 131
Grus canadensis pulla, 131
Grus canadensis rowani, 131
Grus canadensis tabida, 131
Grus grus, 131
Grus grus grus, 131
Grus grus lilfordi, 131
Grus japonensis, 108, 131
Grus leucogeranus, 131
Grus monacha, 131
Grus nigricollis, 131
Grus rubicunda, 131
Grus vipio, 131, 134
Gryllacris sp., 380
Gryllus integer, tachinid fly attacking,
    355
Guard behavior, 234
Guard calls of crane, 132
  sonograms of, 133
  in unison call, 136
Guardian-censorship theory of dreams,
    157
Guenon, Lowe's, play in, 386, 389
Guinea pig, sleep time for, 159
Gulls
  antithesis in, 279
  herring
    grooming in, 167
    habitat selection and, 193
    interruptions in incubation behavior
      of, 492
    newly hatched, begging responses in,
      38
    temperature regulation by, 170
  optimal foraging of, learning and, 179
  reciprocal feeding in, 275
Gunn, D.L., 190, 191, 676
Guppy, male, darkening of, during court-
    ship, 74
Gurin, J., 518, 551, 671, 678
Gustavson, C.R., 647, 648, 678
Guthrie, E.R., 57, 678
Gymnorhinus cyanocephalus, food choice
    of, 177
Gymnothorax meleagris, 379

Gymnotiformes, electrical field surround-
    ing, 462
Gynandromorphs of fruit flies, 101
Gypsy moth, pheromones of, 274
Gyrinus sp., energy-emitted senses of,
    466-467

H
h^2, 99
Habit, "inhibition by," 43
Habitat, requirements of, for species,
    194
Habitat preference of Microtus mice, 192
Habitat selection, 192
  homing, migration and, 186-222
  shifts in preference in, 197
  simple forms of, 190-191
Habitat use, 192
  multivariate analysis for, 197, 198
  and selection, 191-200
Habituation, 580
  in annelids, 597
  in Aplysia sp., 603-605
  in arthropods, 598
  and blowfly's feeding behavior, 536
  in coelenterates, 591
  in molluscs, 596
  in planarians, 592
  in protozoans, 589
Hadidian, J., 166, 678
Hafez, E.S.E., 70, 148, 308, 644, 655,
    676, 678, 684
Hailman, J.P., 28, 41, 102, 127, 147,
    164, 276, 280, 328, 564, 577
    672, 676, 678, 681
Hainsworth, F.R., 678
Hairs, chemosensory, of fly, 526, 535
Hairy woodpecker, nesting habits of, 83
Halberg, F., 153, 678
Haley, T.J., 463, 678
Haliaetus albicilla, 374
Halicitine bees, shelter of, 182
Halick, P., 673
Hall, D.J., 193, 692
Halliday, T.R., 306, 308, 323, 530, 672,
    674, 678, 691
Halperin, S.D., 310
Hamadryas baboons, aggression in, 251
Hamberger, V., 558, 678
Hamdy, A.H., 632, 688
Hamerstrom, F., 364, 678
Hamilton, 127
Hamilton, W.D., 28, 118, 225, 233, 267,
    353, 678
Hamilton, W.J., III, 172, 301, 683, 689
Hampton Court maze, 45
Hamsters
  golden, agonistic behavior in, 238
  pheromones of, 274
  sleep time for, 159

Handling
and killing tactics of predators, 374-375
specialized, for captive animals, 638
Hand-raised animals, 625
Hankins, W.G., 677
Hansen, E.W., 68, 69, 678
Hansma, H., 681
Haplochromis cichlid fish, sequential appearance of behavior of, 555
Haploid number, 92
Haploploidy, 244, 246, 247
Harcourt, A.H., 152, 679
Harden Jones, F.R., 209, 223, 679
Hardy, 663
Hardy, J.W., 239, 679
Hardy-Weinberg law, 111, 663, 664, 665
Hare, European, 375
Harems, 322
Harlow, H.F., 567, 577, 629, 673, 679, 690
Harris, J.F., 461, 462, 677
Hartridge, H., 7
Harvest mouse, shelter built by, 183
Harvester ants, optimal foraging by, 179
Harvey, P.H., 259, 673
Hasler, A.D., 207, 679
Hausfater, G., 391, 679
Hawkins, R.D., 673
Hawkmoths, hearing in, 453
Hawks
  Accipiter; see Accipiter hawks
  activity and inactivity in, 150
  drinking, body posture changes in, 75
  escape reaction of turkey chick to silhouette of, 41
  flying patterns of, 162-163
  Galapagos, predation by, 363
  innate responses of ducklings to silhouette of, 41
  as predators, 360
  readiness to hunt in, 364
  sharp-shinned, with starling, 358
Hazen, R.M., 82, 83, 679
Hazlett, B.A., 55, 85, 679
Head, false, as defense against predators, 379
Head ornaments of animals
  in males, 319, 320
  role of, 110
Head rubbing by cats, 56, 57
Head scratching invertebrates, 128, 129
Head structures of males, similarities in, 321
Head-bobbing patterns in courtship displays in *Sciloporus* lizards, 270
Headshaking in European great-crested grebe courtship display, 282
Hearing, 447

Hearing—cont'd
of amphibians, 453
of bat, suppression and amplification in, 12
and behavior, 4-20
binaural, 448
in invertebrates, 454
moth, 428-432
of noctuid moth and behavior, 15-20
sense of, 447
sensory lateralization in, 500
in vertebrates, 451-452
Hearing area in brain of owl, 483
Hearing organs of insects, 453, 454
"Heat," 514
anomalous, 632
and infrared detection, 461-462
silent, 632
Heath, J.E., 170, 679
Hebb, D.O., 3, 28, 678
Hecht, M.K., 143, 679, 689, 691
Hedgehogs
defense from predators by, 376
sleep time for, 159
Hein, A., 566, 567, 679
Heinecke, P., 77, 676
Heinrich, B., 175, 176, 179, 185, 679
Heinroth, O., 28, 33
Held, R., 566, 567, 679
Heliconius butterfly larvae, 265
Heliosciurus gamblanus, 389
Heliothermy, 168
Heliothis armigera, 5
Helmholtz, H., 30
Helpers, alloparental, 337-338
Helplessness, learned, 588
Hendrichs, H., 62, 679
Henne, D.R., 682
Henricia and anemone, 591
Herbivory as predation, 358
Hercules beetle, head ornamentation of, 320
Heredity, physical versus statistical (inferential) bases of, 95-102
Heritability, 99
degree of, 103
quantitative genetics, and behavior, 96-100
Heritability values, frequency distribution of, 103
Hermaphroditism, 314-315
Hermit crabs and anemones, commensalism between, 346
Herons
Louisiana, courtship in, 312, 313
polarization of light by, 460
Herring gulls
grooming in, 167
habitat selection and, 193

Herring gulls—cont'd
interruptions in incubation behavior of, 492
newly hatched, begging response in, 38
temperature regulation by, 170
Herring larva, hunting success of, 375
Herter, K., 447, 679
Hess, E.H., 28, 41, 447, 449, 453, 454, 460, 471, 563, 570, 571, 617, 679
Hesterman, E.R., 319, 685
Heteronetta atricapilla, nest parasitism by, 350
Heteropolar neurons, 412
Heterosis, 117
Heterothermy, 168
Heterotrophs, 104
Heterotrophy as predation, 358
Heterozygosity, 93
Heterozygote superiority, example of, 113
Hetherington, A.W., 545, 679
Heymer, A., 41, 679
Hibernacula of bats, 5
little brown, migration to, 221
Hiccuping, 165
Hickman, C.P., Jr., 35, 36, 107, 143, 452, 482, 679
Hickman, F.M., 35, 36, 107, 143, 452, 482, 679
Hicks, K., 675
Hiding
to avoid detection, by experimental animals, 57-60
as means of protection, 380
Hiding reactions, abnormal, 623-624
Hierarchical organization of instinct as envisioned by Tinbergen, 42-43
Hierarchies, dominance, 240, 241, 242
Hierodula crassa, hunting success of, 375
Higashi, G.A., 688
High fecundity, 334
High-speed photography, 65
Hilara sartor, 141
Hillyard, S.A., 686
Himwich, H.E., 670
Himwich, W.A. 670
Hinde, R.A., 28, 39, 40, 41, 51, 69, 267, 301, 577, 617, 679, 688
Hinkle, C.M., 672
Hippasteria spinosa, contact of anemone with, 590, 591
Hirata, N.R., 554, 676
Hirudinea, learning in, 596
Histamine, 518
Histologie du Système Nerveux de l'Homme et des Vertébrés, 49

Historical classification of behavior, 69
Historical inquiry, 54
History, 22-51
    of study of behavior, early beginnings
        of, 29-33
Hitchcock, H.B., 221, 674
Hoar, W.S., 35, 36
Hoarding, 181
Hobbes, T., 557
Hobson, J.A., 157, 158, 679, 684
Hock, R.J., 213, 678
Hockett, C.F., 225
Hockett, C.P., 679
Hodder, R.M., 683
Hodges, C.M., 175, 679
Hodgkin, A.L., 28, 49, 424
Hodos, W., 47, 679
Hoebel, B.G., 545, 546, 548, 670, 679
Hognose snakes
    diet for, 638-639
    prey of, 366
Holden, C., 404, 644, 679
Hölldobler, B., 28, 236, 277, 352, 354,
        383
Holling, C.S., 362, 363, 368, 375, 679
Holman, R.B., 671
Holst, E. von, 28, 29, 491, 679
Holt, E.B., 557, 679
Holt, G.G., 404
Holt, J., 670
Holthaus, W.R., 692
Holtz, H., 577, 683
Holtzman, J.D., 689
Home range, 192, 199
    determination of, 660-661
    and familiar area, 199-200
Home territory, 192
Homeotherms, 168
    theoretical relationship between body
        and ambient temperatures in,
        167
Homeothermy, 168
Homing
    habitat selection, migration and, 186-
        222
    migration, proximate factors in, 201-
        202
Homing pigeons, 210
    effects of electromagnetic coils on,
        216
    effects of magnetic bars on, 216
    navigation by, 216-217
Hominid species, brain volume of, in-
        crease in, 254
Homo erectus, 254
Homo sapiens, 254
Homologous behavior, 129
Homologous structures, 110
Homology versus analogy in learning,
        611

Homozygosity, 93, 100
Honeybees
    advanced forms of communication in,
        288-293
    displays in, numbers of, 286
    dwarf, dance of, 290
    homing ability of, 204-205
    Italian, dance of, 289-290
    language of, 286
    nest cleaning by, 94-95
    pheromone-producing glands in, 273
    temperature of hive of, 447
    temperature regulation by, 169
    thermal conservation of, 237
    use of landmarks by, 204, 205
    visual acuity of, 460
Honeybee language controversy, 291-293
Honeydew of aphids, ants feeding on, 346-
        347
Honeyeaters, Australian, foraging meth-
        ods of, 176
Honeyguide, nest parasitism by, 350
Hooded crane, 131, 134, 139, 140
Hooded warbler, niche-gestalt for, 197
Hopkins, C.D., 276, 679
Hopson, J., 285, 680
Horall, R.M., 678
Horizontal cells of retina, 486
Hormonally dependent behavioral and an-
        atomical events, annual se-
        quence of, 516
Hormones, 508
    actions of, pathways and mechanisms
        of, 517-521
    adrenocorticotrophic, 518
    and behavior, interdependence of, 507
    and behavioral change, 561
    duodenal, and satiety, 543
    and environment, 92
    and feeding behavior, 543
    follicle-stimulating, 510, 518
    gonadotrophic, 510
    growth, and feeding behavior, 543
    and incubation behavior in birds, 511
    insect behavior and, 509
    luteinizing, 510, 518
    neurochemical, and behavior, 506-525
    peptide-type, 517
        action of, 517, 519
        molecular structure of, 518
    proteinaceous, 517
        action of, 517, 519
        molecular structure of, 518
    reproductive, of human female, 512-
        514
    somatotrophic, and feeding behavior,
        543
    steroid, 517
        action of, 517, 519
        molecular structure of, 519

Hormones—cont'd
    target neurons of, 517
    thyroid, 518
    affecting transcription, 90, 91
    trophic ("feeding"), 510
    vertebrate behavior and, 509-516
Horn, H.S., 235, 680
Hornbills and guenons, play in, 389
Horned animals, play in, 396
Horned lizard, temperature regulation by,
        170
Horns of dikdiks, individual markings in,
        62
Hornworm, tobacco, larva of, chemical
        senses of, 464
Horrall, R.M., 679
Horridge, G.A., 470, 680
Horse-and-Rider, locating, as test of
        vision, 460
Horses
    behavioral disorders in, 631
    color vision in, 458
    sleep time for, 159
    substrus in, 632
    weaving behavior in, 624
Horseshoe bat, echolocation in, 8
Horseshoe crab, egg-laying and fertiliza-
        tion in, 310
Horton, G.P., 57, 678
Host, communication lines of, use of, by
        parasites, 353-357
Hotta, Y., 101, 559, 680
House cats
    male, aggressiveness in, following tes-
        tosterone treatment, 79
    sequential appearance of behavior in,
        555
House sparrow
    circadian pacemakers in, 155
    competition for food by, 343
    locomotor activity of, and circadian
        rhythms, 153
Houseflies
    breeding times of, 311, 314
    head of, 442
Housing facilities for captive animals,
        638
Hoy, R.R., 102, 558, 559, 671
Hubel, D.H., 28, 49, 51, 473, 489, 490,
        566, 680, 692
Huber, P., 353, 680
Huey, R.B., 379, 680
Hughes, G.M., 147, 680
Hull, D.L., 28, 225, 253, 680
Humane considerations in observation of
        animals, 657
Humans
    aggression toward, by zoo animals,
        625
    body orientation in, 468

Humans—cont'd
body temperature of, 168
brain of
abstraction in, 488-489
auditory pathway of, 491
language and facial recognition areas
of, 494
model of, 473, 474
number of neurons in, 473
pathways of language information to,
496
positron-emission map of, 484
sexual differentiation of, 519-521
in cave, awake activity of, 153
cerebral cortex of, map of, 482
chemical senses of, 463-464
color vision of, 447, 455
distress call of, 281
diversive exploration by, 393
echo suppression by, 12
female
menstrual cycle of, 514
reproductive hormones of, 512-514
hearing of, 447
heat detection by, 462
homing instinct of, 220-221
influence of, and abnormal behavior,
621
injuries to, during play, 391
insight learning in, 587
language of, 282-283
and magnetism, 462-463
nervous system of, 48, 49, 408
neural pathway of, 475-502
play in, 385, 391
costs of, 392
elements of surprise in, 392
play intent movements of, 278
play-wrestling in, 390
recognition by, 229
reproductive cycles of, 311
role of sensory channels in, 279
sensory endings and nerve pathways of,
449
sensory orientation by, 200
sleep in, 155-157
need for, 159-160
sleep time for, 159
social behavior of, 230, 231
sound and pitch discrimination of, 448
vocalization of
contrasted to calls of little brown bat,
13
sonogram of, 14
Hummingbirds
foraging methods of, 176, 180
purple-throated carib, "prostitution" in,
328
temperature regulation by, 168, 169
Hunger

Hunger—cont'd
and predation, 361-363
theories of, 542
"Hunger pangs," 542
Hunger thresholds of predators, 362
Hunsaker, D., 270, 680
Hunt, J.H., 259, 680
Hunt, readiness to, 364-365
Hunter, M.L., Jr., 195, 680
Hunting
communal, by predators, 372
profitability of, 369
ritualized behavior before, 281
Hunting behavior and echolocation in
bats, 7-15
Hunting and capture methods of predators,
371-374
Hunting dogs, African, social behavior of,
226, 227
Hunting success of various predators, 374,
375
Hutchinson, G.E., 585, 680
Hutt, C., 391, 393, 399, 680
Huxley, A.F., 424
Huxley, J.S., 28, 49, 282, 319, 680
Hybrid vigor, 117
Hybrids
Agapornis sp., carrying behavior in,
98
of species, genetics of, 117
sterile or defective, 117
Hydra, nervous system of, 407
Hydrophones, 65
Hydrotaxis, 191
5-Hydroxydimethyltryptamine, 518
5-Hydroxytryptamine, 518
Hyena
brown, abnormal sexual behavior in,
626
food robbing by, 348
play in, 395
spotted
gorging by, 362
hunting success of, 374
switch in prey of, 366
time without eating for, 362
Hymenocera picta, 340
Hymenopteran insects (Hymenoptera)
grooming behavior of, 166
haplodiploidy in, 246, 247
kin selection in, 244
reproductive behavior of, 330
shelters of, 181, 182
social behavior of, 230-231
temperature regulation by, 169
Hyperinsulinemia and hyperphagia, 549-
550
Hyperparasites, 349
Hyperphagia, 545-549
and hyperinsulinemia, 549-550

Hyperphagia—cont'd
and hypothalamus, 545
Hyperpolarization, 420, 421
Hyperpolarized neurons, 420, 421
Hypothalamus
and central mechanism for control of
feeding in rat, 543-549
lesions of, and feeding behavior, 545,
546
role of, in feeding behavior, 549
ventromedial, lesions of, recent interpretations of, 549
Hypothesis, 54
glucostatic, of feeding behavior, 547
grouping, 232-237
learning-to-mother, 338
map-compass, for bird navigation, 218-
219
null, 54
rejection of, 55
sun-arc, for pigeon homing, 217-219
Hypotheticodeductive method, 54
Hysteria in pets, 633

I

Ibex, injuries to, during play, 391
Identification, individual characteristics of
animals as means of, 62, 63
Identified neurons in central nervous sys-
tem of invertebrates, 404
Identifying and locating individual ani-
mals, 62-65
Iguana
brain of, 411
optic tectum of, map of, 482
Iguana iguana, optic tectum of, map of,
482
Image, search
and optimal foraging, 179-180
of predators, 367-370
Immelmann, K., 570, 680
Immobility hypothesis of sleep, 159
Impala, aggression in, 251-252
Imprinting, 570-572, 587
early reference to, 29
Impulses, neural, 424
conflicting, 480
Inactivity and activity, 148-162
Inappropriate behavior, 621; see also
Abnormal behavior
Inbreeding, 100
Inbreeding coefficient, 100
"Inchworms," movement by, 163
Inclusive fitness, 243, 245, 246
of gene, 119
Incubation behavior
in birds, hormones and, 511
of herring gull, interruptions in, 492
Independence, law of, 93-94
Indian sarus crane, 131

Indicator indicator, nest parasitism by, 350

Indigo buntings
activity of, recording, 77, 78
study of communication in, 268
use of stars by, for navigation, 214

Individual(s)
defining, 246
variability among, study of, 98

Individual fitness, 243
of gene, 119

Individual recognition, 228-229
bird songs as means of, 294

Inertial senses, 450-451

Inferences in science, 54

Inferential versus physical bases of heredity, 95-102

Inferential statistics, 78

Information
pathway of, from brain to motor neuron, 480
storage of, and learning, 605

Informational aspects of communication, 266

Infrared and heat detection, 461-462

Infrared light, 455, 457, 458

Ingenuity in study of behavior, 65-67

Ingle, D., 671

Ingram, G.J., 674

Inheritance
of cooing song patterns of doves, 97
of cricket song patterns, 97

Inherited variation, 111

Inhibition, 493
conditioned, in anemone, 591
"by habit," 43
in neural integration, 480

Inhibitory postsynaptic potentials, 421, 422

"Injective knowledge," 31

Injury in captive animals, prevention and correction of, 639

Innate behavior patterns, 3

Innate releasing mechanism, 39
application of, to levels of instinct, 43
relationship of, to sign stimulus and fixed action pattern, 40

Innate responses, 41

Innateness, degree of, 103

Inner ear, 453

Inner telencephalon, 501

Innervation
polyneuronal, of muscle, 438
specificity of, 439

Inquilinism, 352-353

Inquiry, historical, 54

Insectivorous birds, vacuum behavior in, 38

Insects (Insecta), 597
alloparental helping in, 338

Insects (Insecta)—cont'd
aquatic, use of water surface waves for sensing by, 466-467
arctic, temperature regulation by, 170
auditory communication in, 275
backswimmer, optimal behavior in, 180
behavior of, and hormones, 509
brain of, 410
calls of bats to detect, 14
care of eggs by male, 333
colonial, social behavior of, 230-231
color vision in, 458
communal displays in, 327-328
compound eye of, vision through, 458
control of feeding in, 532-540
diet specialization in, 179
excesses of, and hyperphagia, 549-550
family characteristics of, 330
flight muscles of, contraction of, 475
glands in, major, 508, 509
group defense by, 233
hearing organs of, 453, 454
hymenopteran; *see* Hymenopteran insects (Hymenoptera)
initiation of physical transport in, 275
learning in, 598-601
memory in, 607
migration by, 203-206
neural-endocrine integration in, 508
observation of, 659
odor following in, 191
odor trails for, 203, 204
offensive defenses of, 376
optimal foraging by, 179
pheromone-producing glands in, 273
pheromones in, 509
polarized light detection in, 460
predatory, 361
reciprocal feeding in, 275
rocking in, 291
shedding of epidermis by, 171
sleep characteristics of, 158
social; *see* Social insects
study of nervous system of, 49
temperature regulation by, 169
temporal pattern recognition of, 448
temporary parasitism among, 349-350
visual cells of, 469
walking by, neural circuit controlling, 502

Insight learning, 587

Instar stage of development, 558, 559

Instinct, 41
application of innate releasing mechanism to levels of, 43
hierarchical organization of, as envisioned by Tinbergen, 42-43
James on, 43-44

Instinctive behavior patterns, 3

Instrumental conditioning, 581-583

Instrumental learning, 47

Instruments, stereotaxic, 417, 418

Insulin and feeding behavior, 543

Integration, neural, inhibition in, 480

Integrative neurophysiology, relationship of, to other disciplines, 26

Intensity of sound, 9
relationship between pitch and, 10

Intensity differences in signals, 270

Intention movements, 164
list of, 164
locomotion and, 162-164
in evolution of behavior, 142

Interactions
chemicals, between organisms, classification of, 266
environmental, 90-91
among genes, 95
interspecific, 263
classification of, 343
social, involving resources, outcomes of, 243

Interference
observational, 56-57
reducing, techniques for, 57-60
with sound, 10

Interfering, observing without, 56-62

Internal control of behavior, 400-617

Internal integration of behavior, 472-505

Internal mechanisms of behavior, Darwin on, 30-31

Internal regulation of feeding, 527-531

Internal rhythms, 150

International Crane Foundation, 628

International Zoo Year Book, 640

Interneurons, 415-416, 477
facilitating, and sensitization in *Aplysia*, 604

Internuncial neurons, 415

Interobserver reliability in measurement of rhesus monkey behavior, 68

Interpretation of data, alternate, 82

Interpreting and presenting behavioral data, 77-84

Intersexual selection, 319

Interspecific behavior, 70, 340-383

Interspecific chemicals, 266

Interspecific groups, 344

Interspecific interactions, 263
classification of, 343

Intracellular fluid, 420

Intrasexual selection, 319
sexual, 320

Intraspecific chemicals, 266

Intraspecific social behaviors, 70

Introduction to Comparative Psychology, 44

Introgression, 117

Introns, 90
  in eukaryotes, 88, 89
Introspectionism, 44
Invertebrates
  brains of, 474
    pathways through, 502
  communication of, 283-284
  compartmentalization of learning in, 600-601
  embryonic stage of, 556
  hearing in, 454
  learning studies using, 47
  light sensitivity of, 455
  magnetic senses of, 462
  movement by, 163
  nervous system of
    basic, 406
    divergence in, 404-405
  pain in, 450
  reproductive behavior of, 330
  respiratory movements of, 165
  social, recognition by, 228
  study of nervous system of, 49
Ionic concentration gradients, 427
Ionic conductance in neuron, changes in, 427
Ionic flow during action potential, 424
Ionic pump, 420
Ions, 419
  movement of, 420
Ipsilateral neurons, 416
Isogamy, 316
Isolation
  geographical, and origin of new species, 114-117
  species, and reproduction, 306
Isopolar neurons, 412
Isoptera, social behavior of, 230
Italian honeybee, dance of, 289-290
Iteroparity, 316
Iversen, L.L., 525, 680
Iwen, F.A., 692

J

Jacanas, polyandry in, 324
Jackson, J.F., 688
Jacobs, C.H., 198, 680
Jacobs, G.J., 223, 677
Jacobs, J., 103, 680
Jacobson, A.L., 592, 596, 684
Jaeger, R.G., 180, 680
Jaguar, sleep time for, 159
James, D., 198, 692
James, F.C., 197, 680
James, W., 28, 36, 680
  on instincts, 43-44
Japanese Brain: Brain Function and East-West Culture, 499
Japanese crane, 139

Japanese monkeys, cultural learning of, 585
Japanese pattern of sensory lateralization, 500
Jays
  bushy-crested, territory of, 239
  displays of, 140
  pinon, food choice of, 177
  scrub, encounters of, at feeder, 344
  Steller's
    crest raising in, 271
    inferred signals in, 264
Jellyfish
  nervous system of, 407
  "true," 230
Jennings, H.S., 25, 590, 680
Jennings, T., 232, 680
Jensen, D.D., 589, 680
Jet-propulsion movements, 163
Jewell, P.A., 388, 399, 680, 683, 686, 688
Jobaley, M.E., 449
Jogging and opiate system, 523
Johnson, D.H., 197, 680
Johnson, K.G., 167, 671
Johnson, R.E., 274, 680
Johnson, U.R., Jr., 229, 680
Johnston, J.W., Jr., 672
Joseph, R.G., 680
Jukes, T.H., 124, 680
Jumping, 163
Jumping spider with fly, 358
Junco hyemalis
  habitat selection by, 195
  optimal behavior in, 180
Juncos
  habitat selection by, 195
  optimal behavior in, 180
Junctions
  gap, 424
  reciprocal, 415
Junglefowl, red, parental brooding of, 332
Jurine, 7, 23
Justinian, 641

K

k, 243
K, 333-334
Kacelnik, A., 180, 680
Kairomones, 266
Kakolewski, J.W., 691
Kalat, J.W., 584, 680
Kalogeris, T.J., 675
"Kamikaze copulation," 357
Kamil, A.C., 174, 180, 185, 680, 686
Kandel, E.R., 3, 28, 33, 51, 441, 502, 503, 505, 525, 603, 604, 605, 673, 680, 688
Kangaroo rats, grooming in, 167

Kantowitz, B., 51, 671
Karplus, I., 347, 680
Karten, H.S., 548, 693
Kaspar Hauser buds, 195, 196
Katydids, temperature regulation by, 169
Kedes, L., 673
Keeler, C.E., 630, 681
Keenleyside, M.H.A., 284
Keesey, R.E., 544, 546, 547, 676, 687
Keeton, W., 107, 143, 212, 216, 217, 218, 223, 460, 681, 688
Kelley, D., 517, 681
Kelly, D.J., 678
Kelly, S., 592, 595, 674
Kelty, M., 683
Kemp, L., 252
Kennedy, G.C., 528, 681
Kennedy, R.J., 170, 681
Kennelitis, 639
Kenyan crowned cranes, 131
Kerr, W.E., 288
Kerstitch, A., 467
Kessel, B., 212, 681
Kessel, E.L., 141, 681
Kessen, M.L., 542, 684
Kestrel
  American
    hunting success of, 374
    prey-handling ability of, maturational changes in, 564-565
  diets of, 334
Kevan, P.G., 29, 170, 681
Keynes, R.D., 427, 672
Kheper aegyptiorum, foraging behavior in, 176-177
Kiester, E., Jr., 157, 158, 681
Kiley, M., 284, 681
Kiley-Worthington, M., 276, 681
Killer whales
  communal hunting by, 372
  as predators, 361
Killfish, normal versus submissive postures in, 242
Killing and handling tactics of predators, 374-375
Kimble, D.P., 684
Kimura, M., 124, 681
Kin selection, 243, 244
  and evolution of social behavior, 243-247
Kinesis, 190
  in animal orientation, 190
King, A.P., 577, 692
King, H.O., 630, 681
King, J.A., 140, 681
Kinnaman, A.J., 45, 681
Kirschvink, J.L., 677, 691
Kite, Everglades, prey of, 366
Kittens

Kittens—cont'd
  deprivation of vision in, effects of, 566
  self-produced movement of, need for, 566-567
  "thrill-seeking" play in, 393
Kleitman, N., 157, 674
Kleptoparasitism, 348-349
Klinokinesis, 190
Klinotaxis, 190, 191
Klippel, R.A., 544, 681
Kloot, van der, 182, 691
Klopfer, P.H., 36, 51, 195, 196, 223, 671, 676, 681
Kluver, H., 488, 686
Knee-jerk reflex, 476
Knowledge, "injective," 31
Knudsen, E.I., 467, 483, 681
Kob, Uganda, 650
  lek behavior in, 328
  management of, 649-651
Koford, C.B., 386, 681
Komondor guard dogs to frighten coyotes, 648
Konig, J.F.R., 544, 681
Konishi, M., 28, 467, 483, 572, 681
Kopfchengeben, 57
Kovach, J.K., 27, 101, 681
Kraly, F.S., 543, 681
Kramer, B., 276, 681
Kramer, D.L., 176, 681
Kramer, G., 218, 681
Krames, L., 584, 617, 680, 684
Krasne, F.B., 600, 601, 681
Krausz, H.I., 686
Krebs, C.J., 28, 107, 681
Krebs, J., 181, 670
Krebs, J.R., 174, 175, 176, 177, 179, 180, 181, 185, 194, 259, 267, 301, 368, 674, 681, 686, 689
Kreithen, M.L., 460, 681
Krogh, A., 657
Kruuk, H., 362, 366, 374, 395, 681
Kuhl, D.E., 686
Kummer, H., 251, 681
Kung, C., 100, 681
Kuo, Z.Y., 557, 682
Kushlan, J.A., 180, 682
Kuterbach, D.A., 682
Kutsch, W., 558, 682

L

Labellum of fly, sensory receptors on, 534
Labor, division of, in groups, 236-237
Laboratory rats, learning studies using, 47
Labroides, cleaning by, 346
Lacewing larvae, wooly aphids and, 371, 372-373

Lack, D., 118, 193, 350, 682
Lade, B.I., 97, 682
Lady beetle, Batesian mimicry by, 381
Lagena, 451
Lahue, R., 589, 682
Lall, A.B., 270, 682
Lamarck, J.B., de, 124
Lamarckianism, 124
Lampyridae, communication system of, 355-356
Lana, 295, 297
Land crab, Asiatic, eyes of, 456
Landmarks, 199
  use of
    by digger wasps, 203
    by honeybees, 204, 205
Language; see also Communication
  consciousness and, 299, 499
  controversy over, 294-300
  honeybee, 286
    controversy over, 291-293
  human, 282-283
Language areas of human brain, 494
Language information, pathways of, to human brain, 496
Language pathways in brain, 493-496
Langur, douc, play-face of, 397
Lanyon, W.E., 672
Lanza, R.P., 296, 676
Largemouth bass, hunting success of, 374
Larval stage of development, behaviors in, 556
Larynx, 6
Lashley, K., 28, 46, 606, 682
Lasiewski, R.C., 170, 685
Lasius, 349
Latent learning, 584-585
  cognition and, 45
Lateral hypothalamus
  lesions of, and feeding behavior, 546
  in mammals, role of, 544, 545-549
  role of, in feeding behavior, 549
Lateralization
  of brain, 497-500
  in hearing, sensory, 500
Laven, H., 515, 682
Law
  of Effect, Thorndike's, 45
  of Exercise, Thorndike's, 45
  Hardy-Weinberg, 111, 663, 664, 665
  of independence, 93-94
  of segregation, 93
  of "transitoriness," 44
Lawick, H. van, 398
Lawick-Goodall, J. van, 60, 338, 375, 653, 682
Le Magnen, J., 540, 682
Leaf-rolling beetle, 183
"Learned" behavior patterns, 4

Learned helplessness, 588
Learned responses from sensory stimuli, study of, 445-446
Learning
  in annelids, 596-597
  in arthropods, 597-601
  associative, food aversion and other, differences between, 584
  and avian migration, 212
  avoidance
    taste aversion as, 583, 584
    by vertebrates, 601
  categories of, 579-588
  in coelenterates, 589-591
  combination of types of, 588
  conditioned, in octopus, 594-595
  cultural or observational, 585
  definition of, 579
  double-discrimination, of planarians, 593
  in echinoderms, 601
  higher forms of, 612
  insight, 587
  instrumental, 47
  latent or exploratory, 584-585
    cognition and, 45
  learning-set, 587-588
  maze, in ants, 600
  and memory, 578-617
    mechanisms of, 603-609
  in molluscs, 595-596
  overlaps in, 588
  phylogenetic survey of, 588-603
  place or spatial, 585
  in Platyhelminthes, 591-595
  in porifera, 589
  problems in, causing abnormal behavior, 634
  in protozoans, 589
  and reinforcements, 47
  role of, in optimal foraging, 179
  role of parent in, 332
  Schneirla's view of, 50
  "to see," 368
  similarity of, between species, 45
  stimulus-response (S-R), 581
  stimulus-stimulus (S-S), 581
  taste aversion forms of, 48
  terminology for, 579-588
  transfer of, in planarians, 593-594
  trial-and-error, 582
  universality of, 611
  in vertebrates, 601-603
Learning capacity, biological relevance of, 601
Learning curves for rat, sparrow, and monkey, 45
Learning phenomena versus learning processes, 609-614

Learning studies using laboratory rats, numbers of, 47
Learning-set learning, 587-588
Learning-to-mother hypothesis, 338
Leech, nervous system of, 407
Leen, N., 46
Left versus right sides of cerebrum, 497-500
Legal considerations in observation of animals, 657
Legal ramifications and requirements for capture of animals, 638
Lehner, P.N., 55, 63, 68, 81, 85, 199, 682
Lehrman, D.S., 28, 33, 39, 143, 511, 525, 577, 670, 677, 682
Leis dunlopi, Batesian mimicry by, 381
Lek behavior, 231, 326-328
Leks, 326
   bird, 326-327
   for Uganda kob, 650
Lemur, sleep time for, 159
Leopard attacking baboon, 342
Leopoa ocellata, temperature sense of, 447
Leopold, A., 649, 682
Lepidoptera, 5
Leptocephali, 207
Lepus europaeus, 375
Lerwill, C.J., 238, 682
Leshner, A.I., 525, 682
Lesioning, ablation, and stimulation in study of neurons, 417-418
Lesions
   of lateral hypothalamus, food intake and, 546
   of ventromedial hypothalamus
      food intake and, 545, 546
      recent interpretations of, 549
Lesser sandhill crane, 131
Lesser scaup, habitat selection by, 198
Leszczynski, Z., 346, 456
Lettvin, J.Y., 683
Levine, J.D., 101, 682
Levine, L., 95, 682
Levinson, B.M., 644, 682
Levitt, D.R., 547, 682
Levitt, P.R., 247, 259, 672
Lewin, R., 124, 125, 682
Lewis, E., 125
Lewontin, R.C., 118, 682
Lexigrams and language in apes, 295, 297
Leyhausen, P., 359, 363, 383, 565, 675, 682
Library memory, 606
Lice, body, localization of prey by, 449
Licking, abnormal, 626
Light

Light—cont'd
   appearance of flowers under different types of, 457
   emitted, use of, by fish, 467
   infrared, 455, 457, 458
   polarized, detection of, 460
   properties of, 454
   red, 455, 457, 458
   rhythms involving control of, 155
   sensitivity to, 455
   speed of, 454
   ultraviolet, 458
   and vision, 454, 455
Light microscopes for neurophysiological study, 416
Light-sensitive cells and vision, 455
Ligon, J.D., 177, 682
Lilford common crane, 131
Limbic system, function of, 501
Limpets, return to familiar area by, 202
Lindauer, M., 56, 204, 288, 290, 301, 682
Lindzey, G., 690
Linear dominance relationships, 240, 241, 242
Linhart, S.B., 648, 682
Lion cubs, play in, 387
Lions
   communal hunting by, 372, 373
   food intake by, 530
   gorging by, 362
   play in, 387, 388, 391, 395
   play intent movements of, 278
   play-bow of, 397
   Serengeti, 56
   vibrissae patterns in, as identification, 63
Lipostat, 529
Lipostatic theory of weight control, 529
Lissman, H.W., 682
Lithium chloride for taste aversion studies, 647
Little brown bat, 4, 5; see also Bat(s)
   echolocation in, 8
   external ear of, 12
   hunting, calls of, contrasted with human speech, 13
   navigation by, 221
   and noctuid moths, 13-20
Liu, C.M., 549, 682
Live animals, suggestions for observing, 657-661
Livestock
   domestic, abnormal behavior in, 629-633
   management of, 643-644
   observation of, 659
Living places, finding, 186-222
Lizards
   Anolis

Lizards—cont'd
   Anolis—cont'd
      behavior of, hormones and, 512, 513
      foraging behavior in, 179
      frilled, defensive threat postures of, 281
      gila monster, 582
         noxious taste of, 583
      gliding behavior of, 140
      horned, temperature regulation by, 170
      protective mimicry by, 379
      Sceloporus, head-bobbing pattern in courtship displays of, 270
      temporal pattern recognition of, 448
Llamas to frighten coyotes, 648
Lloyd, J.E., 269, 356, 357, 682
Lobsters
   compartmentalization of learning in, 600-601
   development of movement pattern in, 558
   learning in, 598
Local dialects in bird songs, 293-294
Localization of source, ability for, 448-449
Locating and identifying individual animals, 62-65
Lockard, R.B., 32, 682
Locke, J., 557
Lockley, R.M., 223, 682
Locomotion
   basic types of, 163
   costs of, 163-164
   intention movements for, 162-164
      in evolution of behavior, 142
   patterns of, 162-163
   spinal cord and, 481
Locusts
   drifting by, 192
   flight of, development of movement pattern in, 558
   "migrations" of, 203
   nervous system of, 406
Loeb, J., 28
Loewenstein, W.R., 469, 682
Loewi, O., 28
Logbooks, 76
Loizos, C., 388, 391, 399, 680, 683, 686, 688
LoLordo, V.M., 617, 671
Long-distance migration, origins of, 201
Long-tailed monkeys, self-mutilation by, 625
Long-term memory, 606-607
Long-term predator-prey relationships, 383
Loons, eggs of, 333

Lordosis reflex, 514
Lorenz, K.Z., 28, 33, 34, 35, 36, 38, 39, 40, 41, 42, 51, 109, 127, 128, 129, 130, 138, 165, 288, 471, 486, 570, 644, 653, 683
    psycho-hydraulic model of motivation of, 40
Lott, D.F., 311, 683
Loudness of sound waves, 8-9
Louis XIV, 640
Louisiana herons, courtship in, 312, 313
Louse, kinesis in, 190
Louw, G., 247, 683
Lovebirds, carrying behavior in, 98
Low, W.A., 148, 150, 683
Low fecundity, 334
Lowe's guenon, play in, 386, 389
Lowland gorillas, play in, 390
LSD, 523
Lucanid beetles, head structures of, 320
Lucas, H.L., Jr., 193, 677
Lucien, H.W., 688
Luco, J.V., 673
Ludlow, A.R., 54, 683
Luminescent signals of fireflies, 356
Lustick, S., 170, 683
Luteinizing hormone, 510, 518
Luteinizing hormone–releasing factor, 518
    molecular structure of, 519
Luttenberger, V.F., 166, 683
Lycosid spider, parental care by, 331
Lynx, naturalistic exhibit of, 627
Lynx lynx, 378
Lysosomes, 413

M

Macaca nemestrina, circadian activity of, 153
Macaca nigra, yawning in, 166
Macaque, pig-tailed, circadian activity of, 153
MacArthur, R.H., 174, 175, 185, 193, 195, 683
Mackintosh, N.J., 617, 683, 690
Maclean, G.L., 171, 172, 672
MacLean, P., 404, 473, 501, 683
MacNichol, E.F., Jr., 459, 683
Macromeris violacea, 381
Macrotermes, temperature regulation by, 169
Magg, M., 688
Maggot, taxis in, 191
Magnetic fields, detection of, 447
Magnetic senses, 462-463
Magnetism, use of, by birds, 215-216
Magnuson, J.J., 671
Maier, N.R.F., 32, 683
Maier, R., 310, 322
Maintenance behaviors, 70, 164-172

Maintenance behaviors—cont'd
    and basic behaviors of animals, 146-184
Maki, W.S., 607, 683
Makings, P., 238, 682
Maladaptive behavior, 621; see also Abnormal behavior
Maladaptive recessive genes, 112
Malafant, K.W.J., 683
Malenka, R., 679
Males
    aggression and fighting in, 319, 320, 323
    care of offspring by, 329
    fighting between, for females, 319, 320, 323
    head structures of, similarities in, 321
    responsibility of, for care of eggs, 333
Mallard, 130
    behavioral repertoire of, portion of, 70-71
    courtship displays in, 129
    imprinting in, critical period for, 571
    navigation by, 214
    speculum-flashing display in, 148
    use of stars by, for navigation, 214
Malpighi, 416
Mammal and bird calls, convergence in, 284, 285
"Mammalian brain," 501
Mammals
    auditory communication of, 275
    and birds, reproductive behavior of, differences in, 329-330
    brains of, 474-475
        sexual differentiation of, 519-521
    burrow of, 182
    care of developing eggs by males of, 333
    circadian rhythms in, 155
    color vision in, 458
    commensals of, 345
    control of feeding in, 540-550
    displays in, numbers of, 286
    dust bathing by, 167
    ears of, 453
        structures of, 452
    embryonic behavior of, 556
    feeding in, dual center model of, 544-545, 547
    hunting methods of, 371
    insight learning in, 587
    learning studies using, 47
    learning-set learning in, 587
    mating patterns versus parental investment in, 329
    memory in, 607
    migration of, 220-221
    nonhuman, social behavior of, 231
    odor following in, 191

Mammals—cont'd
    ovulation patterns in, 512
    play-face of, 397, 398
    self-mutilation by, 625
    sleep in, 155-157
    sleep characteristics of, 158
    sleep time for, 159
    storage of prey by, 363
    stretching by, 166
    temperature regulation by, 168
    temporal pattern recognition of, 448
    tree-climbing, cerebellum of, 501
    visual cortex of, visual processing in, 490
Manakins, communal displays in, 328
Manchurian crane, 131, 140
Mandarin duck, 130
Mandibulata, 597
    learning in, 598-601
Manduca sexta, larva of, chemical senses of, 464
Manipulative aspects of communication, 267
Manning, A., 28, 51, 683
Mantidae, hunting success of, 375
Mantid (mantis)
    central nervous system controlling behavior in, 481
    and flies, interaction of, 435
    grooming in, 167
    hunger threshold of, 362
    hunting methods of, 371
    preying, cannibalism in, 306, 308
    strikes of, neural circuit controlling, 502
Mantis shrimps
    communication signals and, 264
    individual recognition by, 229
Manual recording of data, 76-77
Map, cognitive, 45
Map-compass hypothesis for bird navigation, 218-219
Margarops fuscatus, 651
Marine arthropods, auditory communication in, 275
Marine snail, study of nervous system of, 49
Markers
    on animals, 63
    pulse, 477
    train, 477
Markings
    animal
        artificial, 63, 64
        natural, 62, 63
    body, false, on butterflies, 379
Markov chains, 80, 81
    analysis of, 661
Markowitz, R., 678

Marler, P., 28, 132, 266, 281, 301, 572, 575, 577, 683, 689
Marmota monax, learning in, to increase food supplies, 179
Marshall, J.F., 574, 683
Martes martes, 378
Martial eagle, clipping feathers of, to mark, 64
Martin, D.J., 177, 682
Martin, R., 682
Martin, R.D., 691
Martin, S.G., 692
Marx, J.L., 127, 297, 683
Masai giraffes, male fighting in, 319
Mason, P.R., 190, 683
Mason, W.A., 127, 683
Mass, D., 681
Mast, S.O., 683
Masters, W.M., 673
Materials
    need for, 104
    shortage of, 104-105
Maternal care problems in domestic animals, 631, 632
Mates, abnormal, 623
Mating
    assortative, 665
    nonrandom, 665
Mating patterns
    evolution of, 324-325
    versus parental investment in birds and mammals, 329
    in reproduction, 313-328
    and sexual dimorphism in birds, 317
    similarities in, 325
Matriarchal groups, 324
    among mammals, 329-330
Matrices, sociometric, 81
Matrifocal groups, 324
Matsubara, J.A., 462, 683
Matthew, R.W., 410, 682
Matthews, G.V.T., 217, 223, 683
Matthews, J.R., 410, 682
Maturana, H.R., 487, 683
Maturation, 560
    and behavioral change, 560-565
Maxson, R., 673
Mayer, D.J., 523, 525, 692
Mayer, J., 549, 690
Mayflies
    communal displays in, 327
    feeding and, 173
Maynard Smith, J., 28, 118, 248, 251, 683, 684
Mayr, E., 121, 125, 128, 143, 304, 613, 614, 616, 617, 665, 684
Maze, Hampton Court, 45
Maze learning in ants, 600
Mazziotta, J.C., 686
McAdoo, D.J., 674

McAllister, L.B., 688
McAvoy, T., 34
McCarley, R.W., 157, 158, 679, 684
McConnell, J.V., 592, 593, 594, 596, 672, 684
McCosker, J.E., 379, 684
McCulloch, W.S., 683
McDiarmid, R.W., 377, 678
McEwen, B.S., 28, 515, 517, 521, 525, 684
McFarland, D., 28, 39, 684
McGaugh, J.L., 28, 617, 675
McKinney, F., 70, 148, 684
Measurement of behavior, observation and, 52-85
Mech, L.D., 27, 273, 362, 374, 382, 383, 389, 684, 685, 686
Mechanical behavior of animals, 126
Mechanical counters to record data, 77
Mechanoreception, 449-450
Meddis, R., 158, 159, 160, 161, 185, 684
Medical anatomy and physiology and understanding of behavior, 30
Medin, D.L., 617, 684
Medulla, adrenal, hormones of, and feeding behavior, 543
Megachiroptera, echolocation in, 8
Megapodes
    egg laying by, 333
    heat and infrared sensing by, 461
Megapodiidae, temperature sense of, 447
Meiosis, 92
    segregation during, complications of, 95
Meisel, R.L., 520, 684
Melatonin, 518
Meliphagidae, foraging methods of, 176
Meliponinae, 290
Mellgren, R.L., 295, 676
Melzack, R., 567, 684
Membrane
    cellular, 420
    of neuron, 420
    tympanic, 452
Meme, 585
Memory
    in *Aplysia* sp., 605-609
    biological mechanism of, 606
    and learning, 578-617
        mechanisms of, 603-609
    library, 606
    location of, 607
    long-term, 606-607
    loss of, 606-607
    short-term, 606-607
    spatial, 199
    working, 606
"Memory molecules," 607
Menagerie, 640

Menaken, M., 185, 684
Menaker, M., 153, 155, 676, 693
Mendel, G., 54, 93, 94, 95
Mendelian genetics, 93-94
Menstrual cycle of human female, 514
Mental Evolution in Animals, 31
Mental Evolution in Man, 31
Mental modeling, 587
Menzel, R., 609, 617, 684
Mercator projection of moth hearing, 430, 431
Mertens, R., 377, 684
Mertensian mimicry, 377
Mescaline, 523
Message, 262
    molecular, 271
    from plants, 263, 265
    sent by butterflies, 262
Messenger ribonucleic acid, 88, 89, 90
    interaction of, with environment, 90-92
    manufacture of, 91
Metacommunication, 278
5-Methoxydimethyltryptamine, 518
5-Methoxytryptamine, 518
Methylxanthines, 523-524
Metzenberg, R.L., 107, 143, 692
Metzgar, L.H., 199, 238, 684
Meyer, J., 547, 683, 688
Meyer-Holzapfel, M., 623, 624, 626, 684
Meyerriecks, A.J., 350, 684
Meyers, B., 568, 684
Mice; *see* Mouse (mice)
Michelsen, A., 453, 684
Microchiroptera, echolocation in, 8
Microelectrodes to study neurons, 416-417
Microhabitat selection in great tits, 195, 196
Microphones, 65
Microscopes for neurophysiological study, 416
Microtubules, 414
Microtus agrestis, play-smells of, 398
Microtus longicaudus, habitat preference of, 192
Microtus montanus, habitat preference of, 192
Micturition, 165-166
Midas sp., Batesian mimicry by, 381
Migration
    of amphibians, 209
    of aquatic plankton, 202
    of arachnids, 202
    of birds, 210-219
        study of, 66-67
        weather variables affecting, 82-83
    definition of, 201-202
    of fish, 206-209

Migration—cont'd
  homing, habitat selection and, 186-222
  and homing, proximate factors in, 201-202
  of insects, 203-206
  long-distance, origins of, 201
  of mammals, 220-221
  of molluscs and crustaceans, 202
  of monarch butterflies, 187-190
  orientation, navigation and, phylogenetic survey of, 202-221
  of reptiles, 209-210
  ultimate factors in, 200-201
Migration activity time of birds of prey, differences in, 152
"Migratory restlessness," 211
Migratory songbirds, north temperate, annual sequence of hormonally dependent events in, 516
Milgram, N.W., 584, 617, 680, 684
Milk, "crop," of doves, 511
Mill, J.S., 28, 30
Miller, D.B., 631, 684
Miller, K.R., 597, 687
Miller, N.E., 542, 684
Miller, R.R., 617, 690
Mills, J.N., 153, 185, 678, 684
Milne, L., 449, 471, 684
Milne, M., 449, 471, 684
Milner, P., 521, 685
Mimicry
  by avian parasites, 350-351
  Batesian, 377, 379-380
  Mertensian, 377
  Müllerian, 379-380
  protective, 377, 379
Mimics
  "aggressive," 371
  of communications, 294
  generalization and discrimination of, by predators, 380, 381
Mink
  with mouse, 359
  optimal behavior in, 180
Minnows, hearing in, 452-453
Mirenda, J., 179, 691
Mississippi sandhill crane, 131
Mistle thrush, alarm call of, 281
Mitchell, D., 689
Mites as commensals, 345
Mitochondria, 413
Mitosis, 92
Mixed ESS, 250
Mixed foraging flocks, 345
Mixed groups, 344
Mixed-species groupings, 344-345
Mizar, localizing, as test of vision, 460
Mobbing behavior of birds, learning of, 585, 586

Mock, D.W., 286
Modal action pattern, 36
Modeling, 54
Modification of behavior, 661
Modulated communication, 269
Moericke, V., 446, 684
Moermond, T.C., 179, 684
Mohl, B., 453, 684
Mole, sleep time for, 159
Mole rat, sleep time for, 159
Molecular mechanisms of memory, 607
Molecular messages, 271
Molluscs
  brain of, 475
  "catch" muscles of, 438
  circadian pacemakers of, 154
  escape from predators by, 375
  hearing in, 454
  learning in, 595-596
  "migration" of, 202
  nervous system of, 405
  odor following in, 191
  offensive defense of, 376
  sleep characteristics of, 158
  study of nervous system of, 49
Molnar, R.E., 110, 323, 684
Molothrus ater
  abnormal mate selection by, 623
  songs of, development of, 575-576
Molting, 171
  in insects, hormonal control of, 509
Moltz, H., 41, 554, 577, 673, 684
Monarch butterflies
  breeding ranges of, 188
  defense mechanisms of, 376, 377
  homing ability of, 203
  messages sent by, 262
  migration of, 187-190, 201
  noxious taste of, 583
  overwintering roosting of, 189
  tagging of, 64
  warning displays of, 377
  wintering sites of, 188, 189
Mongoose and guenons, play in, 389
Monkeys
  elimination behavior of, 166
  Japanese, cultural learning of, 585
  learning in, 601
  learning curve for, 45
  nervous system of, 408
  rhesus
    behavior of, interobserver reliability in measurement of, 68
    deprivation studies involving, 567
    and ionizing radiation, 463
    number of displays in, 286
    play in, 387
  self-mutilation by, 625
  "therapist," 629
  vervet, 391

Monogamy, 315, 316
  convergence of, 328-329
  and differences in reproductive behavior, 329
  dominance-based, 316
  female-defense, 316
  perennial, 316
  seasonal, 316
  serial, 316, 317
  territorial, 316
Monotremes, sleep characteristics of, 158
Montgomery, G.G., 673
Mood-altering chemicals, exogenous, pleasure and pain and, 521-524
Mook, D.C., 545, 684
Moon, monthly movement of, 150
Moore, B.R., 56, 57, 684
Moore, J.A., 682
Moore, R.G., 149, 155, 161, 684
Moore, R.Y., 684
Moore-Ede, M.C., 155, 185, 674, 684
Moose
  antlers of, 320
  as prey of wolf, 374
Moray
  green, neon goby cleaning, 346
  and octopus, 383
  zebra, cleaner shrimp cleaning, 346
More, T., 29
Morgan, C.L., 28, 44, 45, 684
Morgan's canon, 44
Mormyridae, electrical communication in, 276
Morphines, naturally occurring, 521
Morrell, G.M., 380, 381, 684
Morris, D., 339, 685
Morris, P.A., 6, 23, 693
Morrison, G.R., 367, 685
Morrow, P.A., 179, 676
Morton, E.S., 285, 299, 685
Morus, 130
Morus bassanus, use of familiar area by, 213
Mosaics, genetic, of fruit flies, 101
Mosquitoes
  localization of prey by, 449
  role of sensory channels in, 279
  temperature sense of, 447
Mother-infant relationship and abnormal behavior, 629
Mother-offspring bonding, 572
Moths
  Arctiid, 20
  and bats, 4-20
  ears of, neural research on, 428-432
  echo and appearance of, with wings in different positions, 16

Moths—cont'd
  evasive behavior patterns of, toward bat calls, 17-19
  flight paths of, 17
  flying, echoes of, and bats, 14-15
  gypsy, pheromones of, 274
  identification of bat calls by, 431
  night, infrared receptors in, 461
  noctuid; see Noctuid moths
  response of, to call of bat, 17-19
  rocking in, 291
  role of sensory channels in, 279
  silkworm, chemical senses of, 465-466
  temperature regulation by, 169
  tobacco hornworm, identified neurons in supraesophageal ganglion of, 404
Motility, autogenous, 556
Motivation, 527
  psycho-hydraulic models of, Lorenz's, 40
Motivational variation of bird songs, 294
Motor centers in brain, 491-493
Motor end-plates, 436, 437
Motor neurons, 415
  collateral pathways and crossing of sensory information to, 479
  pathway of information from brain to, 480
  vertebrate, 413
Motor reactions, stereotyped, in captive animals, 624-625
Motor region of human cerebral cortex, 482
Mottled duck, habitat selection by, 198
Moulton, D.G., 672
Mound-building ant, parasitism by, 349
Mountain beaver, sleep time for, 159
Mountcastle, V.B., 441, 449, 469, 472, 685
Mouse (mice)
  albinism in, and behavior, 95
  Bruce effect in, 271
  corn snake killing, 376
  deer, number of displays in, 286
  familiar area of, 199
  harvest, shelter built by, 183
  hearing of, 453
  and ionizing radiation, 463
  laboratory, behavioral mutations among, 101
  mink with, 359
  nervous system of, 408
  nesting behavior of, 140
  observation of, 659
  Peromyscus; see Peromyscus mice
  role of sensory channels in, 279
  selective breeding of, 101
  single-gene effects in, 95

Mouse (mice)—cont'd
  sleep time for, 158, 159, 162
  in T maze, 578
Mouthpart palps for hearing, 453
Movements
  air, orientation based on, 200
  anatomy of animal and, 3
  efficiency of, through fluid, by group, 237
  endocrine system and, 3
  eye, rapid, 155
  functional categories of, 162
  intention, 164
    locomotion and, 162-164
      in evolution of behavior, 142
  mechanisms aiding, 3
  nervous system and, 3
  protective, in evolution of behavior, 142
  respiratory, 165
  wing, rapid, muscles and, 438-439
Moynihan, M.H., 286, 287, 288, 299, 345, 685
mRNA; see Messenger ribonucleic acid
Mrosovosky, N., 529, 685
Mud building, shelter made by, 182
Mud wasp, shelter of, 182
Mueller, H.C., 41, 150, 334, 370, 564, 565, 685
Mule as hybrid, 117
Muller, H.J., 304
Müllerian mimicry, 379-380
Mullerian theory, 304
Müller-Schwarze, D., 272, 301, 394, 685
Multiple-behavior analysis, 661
  categories of, 80, 81
Multiple-key paper chart recorders for data, 77
Multipolar neurons, 412, 414
Multivariate analyses
  for animal habitat use, 197, 198
  computers and, 83-84
  of woodpecker nesting habits, 83
Multivariate statistics, 81, 83-84
Mundinger, P., 572, 577, 583
Muntz, W.R.A., 561, 562, 685
Musca domestica
  head of, 442
  proboscis of, 526
Muscle stretch receptors, direct synapses by, 476
Muscles, 436-439
  "catch," of molluscs, 438
  contraction of, 436
    action potential and, time sequence of, 437
    before hatching of embryo, 556
    hypotheses of, 438
  fibers of, types of, 438

Muscles—cont'd
  flight, 438-439
  nervous systems, neurons and, 402-441
  twitch, 438
Muscovy duck, 130
Muscular output, 436-439
Musk-oxen
  defense ring of, 233
  group defense by, 380
Mustela vison, optimal behavior in, 180
Mustelids (Mustelidae)
  neophobia in, 369
  predatory behavior of, 359
  learning and, 565
Mutations, 92-93
  behavioral, 100-101
  selective breeding and, 101
Mutualism, 342, 346-348
Myelin, 425
Myelination, 426
Mykytowycz, R., 319, 685
Mynah birds, mimicking by, 573
Myotis lucifugus, 4
  echolocation in, 8
Myrberg, A.A., Jr., 242, 685
Myrmecocystus mimicus, true slaves in, 352
Myrmica sp., parasitism by Atemeles beetle of, 354
Myrtle warbler, feeding habitat of, 193
Mysterious Castles of Clay, 653
Mysterud, I., 378, 379, 685

N

N, 92
2N, 92
Nagel, W.A., 590, 685
Naloxone, 522-523
Namaqua sandgrouse, soaking of water by, 171-172
Nash, J., 577, 685
Natural analgesia, 522, 523
Natural environment and behavior, 103
Natural responses to experimental stimuli to study senses, 446
Natural selection, 111
  level of operation of, 117-122
Nature versus nurture dichotomy, 556
Nauta, W.J.H., 473, 685
Nautilus sp., communication in, 288
Navigation, 202
  by birds, genetics and, 212
  and orientation
    complex systems of, 200-221
    difference between, 202
    and migration, phylogenetic survey of, 202-221
  star-compass, by birds, 214-215
Navigation behaviors, 70

Navigational compasses used by birds, 214
Nebes, D., 497, 685
Necklace dove, cooing song pattern of, 97
Needham, J., 556, 685
Negative afterimage, demonstration of, 446
Negatively charged particles, 419
Neill, W.H., 671
Neisser, U., 367, 685
Nelson, M.E., 382, 685
Nematode, nervous system of, 407
Neocortex, 501
Neon goby cleaning green moray, 346
Neonatal roots of behavior, 554-560
Neopallium, 501
Neophobia
  in predators, 369
  in rats, 540
Neostriatum, 501
Neoteny, 630
Nero, 641
Nerve impulses, 424
Nerve net, 405
Nerve-muscle synapses, pattern-sensitive, 437-438
Nerves, 416
Nervous systems, 403-412
  of annelids, 405, 410
  of arthropods, 410
  central; see Central nervous system
  of coelenterates, 405
  conceptual division of, 403
  deuterostome pattern of, 404
  disorders of, and behavior changes, 48-49
  of echinoderms, 411
  evolutionary divergence in, 404
  of flatworms, 405
  human, 48, 49
  information processing by, 3-4
  of lower chordates, 411-412
  of molluscs, 405
  and movement, 3
  neurons, muscles and, 402-441
  nonneuronal components of, 428
  peripheral, 404
  of Platyhelminthes, 405
  and protein synthesis, 92, 96
  protostome pattern of, 403
  of protozoans, 405
  response times of, 666-669
  of sponges, 405
  of vertebrates, 412
Nest parasitism, 350
Nesting, group, and optimal foraging, 235
Nesting habits of woodpeckers, multivariate analysis of, 83

Nesting places, competition over, 343
Nests, 182, 183
  abnormal building of, 623
  cleaning of, by honeybees, 94-95
  "dump," 351
  of synapses, 415
Neural activity, recording of, to study senses, 445
Neural control of gill withdrawal reflex in Aplysia, 503
Neural differences from enrichment versus deprivation, 568-569
Neural filtering process, 480
Neural impulses, conflicting, 480
Neural integration, 475
  inhibition in, 480
Neural mechanisms and feeding, 526-551
Neural parsimony, 405, 437
Neural pathway
  simplest, 475
  vertebrate, 475-502
Neural research
  on cockroach cerci, 432-435
  on moth ears, 428-432
Neural response, specificity and selectivity of, 476-477
Neural-endocrine integration
  in insects, 508
  in vertebrates, 509
Neurite, 414
Neurobiology, 31, 48-49, 50, 51
  development of discipline of, 32
  ethology, comparative psychology and, 22-51
Neurochemicals, hormones, and behavior, 506-525
Neuroendocrine system, 403, 507; see also Nervous systems
Neuroethology, 428
Neurofilaments, 414
Neuroglia, 428
Neuronal types in cerebellar cortex, 412
Neurons, 412-428
  afferent or sensory, 415
  bipolar, 412, 414
  classification of, 414
  command, 491
  commissural, 416
  contralateral, 416
  discrimination by, 422
  efferent or motor, 415
  functions of, 413
  heteropolar, 412
  in human brain, number of, 473
  identified, in central nervous system of invertebrates, 404
  internuncial, connecting, or association, 415
  ipsilateral, 416

Neurons—cont'd
  isopolar, 412
  maintenance of, 427
  membrane of, 420
  and memory, 608-609
  motor
    collateral pathways and crossing of sensory information to, 479
    pathway of information from brain to, 480
    vertebrate, 413
  multipolar, 412, 414
  naming of, 415-416
  nervous systems, muscles and, 402-441
  operations of, 418-428
  polarized, 420
  projection, 416
  "pulse-marker," of moth, 432
  recharging of, 427
  recycling of, 427
  "repeater," of moth, 431
  of retina, 485, 486
  sensory, 412
  structures of, 413-416
  study of, techniques for, 416-418
  target, of hormones, 517
  "train-marker," of moth, 432
  transmission of decisions by, 422-427
  unipolar, 414
  variety of, 412
Neurophysiological biologists, 49
Neurophysiological methods, 416-418
Neurophysiology
  integrative, relationship of, to other disciplines, 26
  and learning, 613
Neurosecretory cells, 507
Neuroses in pets, 633
Neurotensin, 518
Neutral theory of evolution, 124
Neutral zone, use of, as defense, 382
Neutralism, 124
Newell, G.E., 471, 673
Newts
  role of sensory channels in, 279
  smooth, courtship sequence in, 307, 308
Nexuses, 424
Niche-gestalts for warblers, 197
Nicolai, V.J., 101, 685
Nigerian crowned cranes, 131
Night moths, infrared receptors in, 461
Night vision, 458-459
Night-viewing devices, 65
Nim, 294, 295, 296, 298
Nissl bodies, 414
Noble, G.K., 268, 685
Noctuid moth, 5
  abstraction by, 484

Noctuid moth—cont'd
  ear of, on thorax, 16, 17
  evasive behavior patterns of, toward bat calls, 20
  eye of, 456
  forms of, 5
  hearing in, 447
    and behavior, 15-20
    little brown bats and, 13-20
Noctuidae, 4
Nodes of Ranvier, 426
No-function hypothesis of sleep, 158-159
Nondomestic animals
  abnormal behavior in, under natural conditions, 622-623
  management of, 646-653
Nongame management, 648-651
Nonhuman mammals, social behavior of, 231
Nonlinear dominance hierarchy, 241
Nonmammalian vertebrates, learning studies using, 47
Nonneuronal components of nervous system, 428
Nonrandom mating, 665
Non-REM sleep, 155
Noradrenergic bundle, ventral, and feeding disorders, 548
Norepinephrine, 427, 518
  molecular structure of, 518
North temperate migratory songbirds, annual sequence of hormonally dependent events in, 516
Northern shoveler, habitat selection by, 198
Northwestern crows, whelk dropping by, 177, 178, 179
Norton-Griffiths, M., 363-364, 685
Norway rat, 630
  learning studies using, 47
Norzagaray, A., 670
Notonecta hoffmanni, optimal behavior in, 180
Notopteris macdonaldi, echolocation by, 8
Nottebohm, F., 28, 575, 685
Novin, D., 551, 685, 693
Nowell, W., 176, 673, 681
NREM sleep, 155
Nuclear species, 345
Nucleoli, 413
Nucleus(i), 416
  suprachiasmatic, circadian pacemaker in, 154
Nudibranch, contact of anemone with, 591
Nuisance animals, managing, 646-648
Null hypothesis, 54
  rejection of, 55
Number, haploid, 92

Numbers, R.L., 123, 685
Numerical responses to increases in prey animals, 368
Nurture versus nature dichotomy, 556
Nutrient detectors in stomach, 542
Nutrition for captive animals, 638-639
Nyctalus sp., larynx of, 6

O

Obesity, 545-549
  and hypothalamus, 545
  in rats, 528
Obesity syndromes in rat, 540
Obligate cleaners, 346
O'Brien, D.F., 459, 470, 685
O'Brien, R.D., 107, 143, 692
Observation
  ad libitum, 74
  of animals
    caution before, 657-658
    choice of species for, 657
    legal, ethical, safety, and humane considerations in, 657
    suggestions for, 657-661
    time and scheduling problems in, 658
    use of literature in, 658
  of behavior
    suggested exercises for, 658-661
    and time spent observing, relationship of, 69
  and measurement of behavior, 52-85
Observational artifacts, 56-67
Observational interference, 56-57
  reducing, techniques for, 57-60
Observational learning, 585
Observed contexts of animal communication, 280
Observing without interfering, 56-62
Ocelli of bees and communication, 292
O'Connor, R.J., 180, 675
Octopamine, 518
Octopus
  care of eggs of, by males, 333
  coloration and body shape displays in, 287
  conditioned learning in, 594-595
  defense mechanisms of, 380
  hunting methods of, 371
  memory and learning system in, 610
  and moray eel, 383
  nervous system of, 405
  as predator, 361
  reproduction and, 107
Octopus vulgaris, learning in, 596
Odor following and home finding, 191
Odor trails for insects, 203, 204
Odors
  homing of fish to, 207
  orientation based on, 200

Odors—cont'd
  and taste transduction, 465
Oecophylla longinodis, 389
O'Farrell, M.J., 23, 689
Off sensors, 488
Offensive defenses of prey, 376-377
Offspring-parent conflicts, 335-337
Ohmart, R.D., 170, 685
Oilbird, echolocation used by, 466, 467
Okapi, sleep time for, 159
Olds, J., 521, 685
Oldsquaw ducks, food choices of, 177
Olfaction, 463
Olfactory fatigue, 447
Oligochaeta, learning in, 596
Olive baboons, 56
  mating of, 310, 311
  as prey of chimpanzee, 374
Olson, E.C., 117, 168, 685, 690
Ommatidia, 458, 460
On detectors, 488
Ontogeny, 96
  of behavior, 552-577
    schools of thought concerning, 557
  of emotional behavior, 569
Operant conditioning, 47, 581-583
  in planarians, 592
Operants, 47, 581
Operator, 89
Operon
  making mRNA along, 91
  repressed, 91
Opiates, naturally occurring, 521
Opossums
  self-mutilation by, 625
  sleep time for, 159
Oppenheim, R., 171, 678, 685
Opsahl, C.A., 549, 687
Optic tectum of iguana, map of, 482
Optimal behavior, 180
Optimal food type, 177-179
Optimal foraging
  description and examples of, 174-181
  feeding and, 172-181
  group nesting and, 235
  among patches, 174, 175-176
  role of learning in, 179
  and search image, 179-180
Optimal patches, choice of, 175-176
Optimal searching, 175
Optimality theory, 174, 247
  and food choice, 177
  problems with, 180-181
Oral birth from gastric brooding by frog, 331
Orb weavers, male, cannibalism-avoiding behavior of, 308
Orb-weaving spiders, habituation and, 598
Oregonia gracilis, camouflage by, 378

Organ(s)
of Corti, hair cells of, 452
hearing, of insects, 453, 454
sense, blocking, interruption or removal of, to study senses, 445
Organisms
chemical interactions between, classification of, 266
eukaryotic
introns in, 88, 89
transcription in, 89-90
prokaryotic, transcription in, 89
Orians, G.H., 176, 685
Orientation
body, 468
definition of, 202
navigation and
complex systems of, 200-221
difference between, 202
and migration, phylogenetic survey of, 202-221
Orientation behaviors, 70
Origin
classifying behavior by, 69
of new species and characteristics, 114-117
of species, extraterrestrial, 124
Origin of Species by Means of Natural Selection, 30, 111, 353
Oring, L.W., 14, 324, 676
Orkand, R., 51, 441, 505, 672
Ornaments, head, of animals
in males, 319, 320
role of, 110
Ornithology, clinical, 651
Oropendolas, mutualism-parasitism involving, 351
"Orphons," 90
Orthokinesis, 190
Orthotaxis, 190
Ortstreue, 210
Oscilloscope to measure response time, 667
Ospreys
foraging methods of, 175-176
hunting success of, 374
Ossicles, weberian or auditory, 452
Ostariophysi, hearing in, 452-453
Osteichthyes
electricity-producing, 375
gill movements of, 165
stretching by, 166
Ostrom, J.H., 141, 167, 685
Otis, L.S., 686
Otte, D., 312, 686
Output, muscular, 436-439
Outward behavior, Watson and, 46
Ovaries of anole lizard, development of, hormones and, 512, 513
Oven bird, nest of, 182

Overeating and weight gain, 529
Overmier, J.B., 617, 671
Ovulation
patterns of, in mammals, 512
quiet, 632
Ovum, 318-319
Owen, K., 690
Owen, R.D., 107, 690
Owl
barn, brain of, map of, 483
defensive threat postures of, 281
ears of, 453
facial mimicry of, 378, 379
great-horned, hand-raised, 625
hearing of, 448
mobbing responses to, 585, 586
neophobia in, 369
pigmy, predation by, differences in, 367
as predators, 360
predatory behavior of, maturational changes in, 565
saw-whet, migration of, 219
screech, postures of, 61
short-eared, predation by, 363
Oxytocin, 510
molecular structure of, 518
Oye, P. van, 592, 686
Oyster catchers
parental feeding in, 364
supernormal stimulus for, 39
Oysters, nervous system of, 405

P

p, 663
p, 55
Pacemaker rhythms, free-running, *Zeitgebers* and, 155
Pacemakers, circadian, 154-155
Pacing by captive animals, 624
Pacinian corpuscle, 469
Paddling movements, 163
Paedomorphosis, 630
Page, T.L., 154, 155, 686
Pain
natural suppressant of, 522-523
pleasure, exogenous mood-altering chemicals and, 521-524
Pain pathway of vertebrates, hypothesized, 522
Pain sensations, 449-450
Palatability
and blowfly's feeding behavior, 536
and food intake, 530-531
Paleopallium, 501
Paleostriatum, 501
Palka, J., 598, 686
Pallium of telencephalon, 501
Palm, C.E., 646, 655, 686
Palolo worms, mating time of, 310, 314

Palps, mouthpart, for hearing, 453
Panaman, R., 151, 686
Pancreatic hormones and feeding behavior, 543
Pandion haliaetus, foraging methods of, 175-176
Panthera leo; see Lions
Pantin, C.F.A., 590, 686
Papilio zelicaon, aggression in, 251
Papio anubis as prey of chimpanzee, 374
Papio cynocephalus, capture of vervet monkeys by, 391
Papio hamadryas, aggression in, 251
Parabiosis, 348
Parabiotic blowflies, 538
Paradoxical sleep, 155
Parallax, 448
Parallel evolution, 140
Parallel paths in nervous system, 477
Paramecium sp.
behavioral mutations among, 100
as prey of *Woodruffia metabolica*, 374
Parasites
trophic, 348
use of hosts' communication line by, 353-357
Parasitism, 342, 348-357
facultative, evolution of, 351
nest, 350
social, temporary, 349-350
Parasitoid, 342
Parental brooding of young and thermoregulatory ability, 332
Parental care, 330-337
extended, 332
factors conducive to development of, 333-334
and sexual selection, 335
Parental investment versus mating patterns in birds and mammals, 329
Parent-offspring conflicts, 335-337
Parker, G.A., 248, 686
Parker, G.H., 590, 686
Parker, P.G., 41, 685
Parrots
Puerto Rican, management of, 651-652
weaving behavior in, 624
Parsimony, neural, 405
Parsons, P.A., 99, 100, 107, 675
Parthe, L., 219
Parthenogenesis, 306
Particles, charged, 419
Particulate factors in Mendelian genetics, 93
Partridge, B.L., 224, 686
Partridge, L., 179, 196, 686
Parus ater
foraging behavior in, 179

Parus ater—cont'd
habitat preference of, 196
Parus caerulus
foraging behavior in, 179
habitat preference of, 196
Parus major
microhabitat selection in, 195, 196
optimal foraging in, 179
Passenger pigeon, extinction of, 648
Passeriformes
cultural learning in, 585
songs of, development of, 572-576
variability in, 612
Passerine birds; see Passeriformes
Passiflora vines, 265
Passive aggregation, coincidental, 226
Patches
optimal, choice of, 175-176
optimal foraging among, 174, 175-176
Path
familiar, 199
parallel, in nervous system, 477
Pathway
auditory, of human brain, 491
collateral, and crossing of sensory information and other motor neurons, 479
of hormone action, mechanisms, and, 517
of information from brain to motor neuron, 480
through invertebrate brains, 502
language, in brain, 493-496
of language information to human brain, 496
neural
simplest, 475
vertebrate, 475-502
pain, of vertebrates, hypothesized, 522
reflex, inhibition of portion of, 478-479
synaptic, principles of, 477
visual, from retina to brain, 487, 488
Patterns
action, 68
activity, in rattlesnakes, 149, 161, 162
alarm, of cephalopods, 287
behavior, 68
adaptive, 104
constriction, of snakes, 376
courtship, of spiders, altered, and developmental change, 611
feeding
in blowfly, 532-533
of rat, 540-541
flash, of fireflies, 355-356
flying, 162-163
gate, 163

Patterns—cont'd
grazing, in free-ranging cattle in Australia, 150
mating; see Mating patterns
ovulation, in mammals, 512
recognition of, and abstraction, 484-491
of signals, 270
of smells, 465
song, inheritance of, 97
spatial or temporal, discrimination of, 448
Pattern-sensitive nerve-muscle synapses, 437-438
Patterson, F., 295, 686
Paul, U. von, 491, 679
Paussidae, 377
Pavlov, I.P., 28, 45, 580, 628, 686
Pavlovian conditioning, 580
Peach aphids, color vision in, study of, 446
Peach-faced lovebirds, carrying behavior in, 98
Pearl, R., 190, 686
Pearly-eyed thrasher competing with parrots, 651-652
Pearson, N.E., 176, 685
Pearson, O.P., 371, 686
Peas, garden, Mendel's study of, 93-94
Peck, J.W., 528, 686
Peckham, E.G., 598, 686
Peckham, G.W., 598, 686
Pecking behavior of chicken, maturational changes in, 562-563
Pecking responses
of birds, maturational changes in, 563-564
pigeon, automated recording of, 77
Pectines of scorpions, study of, 445
Pediocetes phasianellus, leks in, 326-327
Peer relationships and abnormal behavior, 629
Pelecaniformes, taxonomic relationships of, as revealed by comparative behavioral analysis, 130
Pelecanus, 130
Pelicans
brown
communal hunting by, 372
signals of, 286
taxonomic relationships of, 130
Pelkwijk, J., 686
Penfield, W., 482
Pengelley, E.T., 154, 185, 686
"Penguin dance" of European great-crested grebe, 282
Penguins, vocal recognition by, 496
Pennycuick, C.J., 686
Pepperberg, I.M., 297, 686
Peptide-type hormones, 517

Peptide-type hormones—cont'd
action of, 517, 519
molecular structure of, 518
Perception, selective, 367
Perceptual world of animal, 21
Perch, predation by, 359
Perching birds, songs of, development of, 572-576
Perdeck, A.C., 212, 213, 686
Peregrine Fund, 628
Perennial monogamy, 316
Perimeter
attack, 240
contact, of territory, 240
Periods, critical or sensitive, 44, 569-570
Peripheral control hypotheses of muscle contraction, 438
Peripheral feeding controls in rat, 541-543
Peripheral nervous system, 404
Peripheral stimulus filtering, 443
Periplaneta, courtship sequence of, 309
Peristalsis-like motions, 163
Peromyscus leucopus, foraging behavior in, 179
Peromyscus maniculatus, foraging behavior in, 179
Peromyscus maniculatus bairdi, habitat preference of, 195-196
Peromyscus mice
foraging behavior in, 179
habitat preference of, 195-196
number of displays in, 286
responses of, to increase in prey, 368
Persuasion displays, 267
Pests, managing, 646-648
Peters, H., 599, 686
Peters, M., 207, 686
Peters, R.P., 273, 686
Peters, S., 572, 683
Peterson, S.R., 177, 686
Petitto, L.A., 690
Petrovich, S.B., 570, 572, 617, 679
Pets
care of, improvements in, 644-645
domestic, abnormal behavior in, 633
observation of, 659
significance of, 644
Pettingill, O.S., Jr., 58, 686
Pfaff, D.W., 551, 686
Pfungst, O., 298
Phaëthon, 130
Phalacrocorax, 130
Phalanger, sleep time for, 159
Pheasants, "dump nests" of, 351
Phelps, M.E., 484, 686
Phenotype, 93
of population, changes in, 113-114
Phenotypic variance, 99

Phenylethanolamine, 518
Phenylethylamine, 518
Pheromones, 266, 271, 507, 508
    diffusion of, 272, 275
    and environment, 92
    fading of signal of, 272, 274
    glands producing, 272, 273
    in pest management, 647
    and territorial boundary marking, 272
Pherosophus agnathus, Batesian mimicry
        by, 380
Philanthus trangulum, use of landmarks
        by, 203
Philornis sp., infestation of nests by,
        351
Phoebes, Eastern, abnormal nest building
        by, 623
Phormia regina, feeding behavior of, 532-
        540
Photinus macdermotti, 356
Photinus sp., flash patterns of, 269
Photinus tanytoxus, 356
Photoblepharon palpebratus, use of emit-
        ted light by, 467
Photography
    applied animal research in, 653
    high-speed, 65
    time-lapse, 65
Phototaxis, 191, 600
Photuris congener, 356
Photuris sp., females of, mimicry by, 356,
        357
Photuris versicolor, female, eating male
        firefly of another species, 356
Phrynosoma sp., temperature regulation
        by, 170
Phylogenetic comparisons
    of communication, 282-286
    of reproductive behavior, 328-330
Phylogenetic dendrograms, 138
Phylogenetic relationships of major groups
        in animal kingdom, 116
Phylogenetic survey of learning, 588-603
Phylogenetic "trees," 138
    of cranes, 139, 140
Physical description of behavior, 69
Physical maps of chromosomes, 96
Physical versus statistical bases of heredi-
        ty, 95-102
Physical transport and tactile communica-
        tion, 275
Physiological aspects of dreaming, 157-
        158
Physiological and psychological aspects of
        behaviors, 400-617
Physiological psychology
    development of, 32
    relationship of, to other disciplines,
        26
Physiologists, 48

Physiology and anatomy, medical, and
        understanding of behavior, 30
Pianka, E.R., 174, 175, 185, 193, 195,
        333, 379, 680, 683, 686
Piciformes, flying patterns of, 162
Picton, T.W., 491, 686
Pied flycatcher, migration of, 210
Pierce, G.W., 8
Pierotti, R., 193, 686
Pietrewicz, A.T., 180, 686
Pietsch, T.W., 370, 686
Pigeons
    drinking behavior of, 171
    homing; see Homing pigeons
    and ionizing radiation, 463
    magnetic senses of, 463
    maternal role of, 329
    navigation by, 214, 215
    observation of, 659
    passenger, extinction of, 648
    pecking responses of, automated re-
        cording of, 77
    and polarized light, 460
    selective breeding of, 101
    symbolic communication between, 296,
        297
    wing flapping of, 564
Pigmy owls, predation by, differences in,
        367
Pigs
    color vision in, 458
    learning in, 601
    sleep time for, 159
Pig-tailed macaque, circadian activity of,
        153
Pileated woodpecker, nesting habits of,
        83
Pilot whale, sleep time for, 159
Pineal gland, circadian pacemaker in,
        155
Pinnipeds, migration by, 220
Pinon jays, food choices of, 177
Pintail duck, 130
    habitat selection by, 198
Pinus edulis, 177
Pipefish, role reversal in, 324
Pit vipers, heat and infrared sensing by,
        461
Pitch of sound, 9
    relationship between sound intensity
        and, 10
Pits
    to catch prey, 371
    snake, 461, 462
Pittendrigh, C.S., 127, 686
Pitts, W.H., 683
Pituitary gland, hormones of, and feeding
        behavior, 543
Place learning, 585
Place memory in arthropods, 598-599

Placebo effect, 523
Planarians
    kinesis in, 190
    learning in, 591-595
    nervous system of, 407
Plankton
    aquatic, "migration" of, 202
    as prey of herring larva, 375
Plants, "messages" of, 263, 265
Plasmodium falciparum, 113
Platyhelminthes
    learning in, 591-595
    nervous system of, 405
Platypus, care of eggs of, by male, 333
Play, 384-399
    accidents during, 391
    "aggressive," 387-388
    cost of, 391-392
    descriptions of, 386-393
    and exploration, 393
    "fairness" in, 396-397
    general attributes of, 390-393
    invitation to, 278
    movements of, 391
    pleasure from, 392
    reasons for, 394-396
    in social context, 396-399
    solo, 386
    theories of, 394-396
    "thrills" or surprise from, 392-393
    vulnerability to predation during, 391-
        392
Play behavior, examples of, 386-390
Play signals, 393, 397-398
Play-bow, 397
Play-face, mammalian, 397, 398
Play-smells, 397-398
Play-wrestling, 390
Pleasure
    pain, exogenous mood-altering chemi-
        cals and, 521-524
    from play, 392
Pleasure centers in brain, 521
Pleasure theory of play, 394
Pleiotropic effects of genes, 95
Pleisiopid fish, protective mimicry by,
        379
Plesiobiotic relationship, 344
Plethodon cinereus, foraging by, 180
Plovers
    golden, migration of, 210
    ringed, reproductive cycle of, 515
Poikilotherms, 168
    theoretical relationship between body
        and ambient temperatures in,
        167
Poison
    to kill prey, 375
    as offensive defense, 376
Polar bears

Polar bears—cont'd
  migration by 220
  as predators, 360
Polarization, 420
Polarized light detection, 460
Polarized neuron, 420
Polecats
  European, aggressive play in, 387-388
  play in, 391, 395
Polyak, S., 485, 488, 686
Polyandry, 315, 317
  evolution of, 324-325
Polychaeta, learning in, 596
Polyclad, nervous system of, 407
Polygamy, 315
  serial, 316, 317
  simultaneous, 316
Polygynandry, 315-316, 317
Polygyny, 315, 317
  communal displays as form of, 326-
    327
  evolution of, 324
  and sexual dimorphism, 255
Polymorphism, 351
Polyneuronal innervation of muscle, 438
Polynomial, quadratic, in interpretation of
    data, 82
Polypeptide, vasoactive intestinal, 518
Polyploidy and origin of new species,
    117
Poole, T.B., 387, 388, 391, 395, 686
Popper, A.N., 454, 686
Population, 106
  genotypes of, 113-114
  phenotype of, changes in, 113-114
  random dispersal of genes in, and gene
    frequency, 114
  size of, 106-107
Population biology, relationship of, to oth-
    er disciplines, 26
Population stability of groups, 236
Porcupine fish, defense from predators by,
    376
Porcupines
  defense from predators by, 376
  and fisher, 383
Porifera, learning in, 589
Porpoises
  alloparental helping in, 338
  Dall's, sleep time for, 158, 159
  hearing of, 453
  individual differences in appearance of,
    63
  trained, 654
Porter, J.P., 45, 686
Portuguese man-of-war, social behavior
    in, 230
Positively charged particles, 419
Positron-emission map of human brain,
    484

Postsynaptic potentials, 422
Postural reflexes, 476
Posture
  body, changes in, in drinking ducks and
    hawks, 75
  indicating social status, 278
Potassium ions in neurons, 420
Potential
  action; see Action potential
  electrotonic, 422
  evoked, 417
    use of, to study senses, 445
  generator, 422
  graded (dendritic), 422
  postsynaptic, 421, 422
  resting, 420, 421
  zero, 419
Potential difference, 419-420
Potts, G.W., 346, 687
Poultry, abnormal behavior in, 631
Powell, R.A., 359, 687
Powley, T., 529, 546, 547, 549, 685,
    687
Practice, 560
Practice theory of play, 395-396
Prairie chicken, leks in, 326-327
Preadapted behavior, 142
Precocial animals, 322-323
Precocial young, 555, 556
Predation
  common denominators of, 358
  definition of, 358
  and dependents, 363-364
  effect of, on animal behavior, 360
  and habitat, 192-193
  hunger and, 361-363
  patterns in, 363
  proximate factors of, 361-365
  reduced, and mixed-species grouping,
    345
  social facilitation and, 369
  symbioses and, 340-383
  training bias in, 369
  vulnerability to, during play, 391-392
Predation responses to increase in prey
    animals, 368
Predators
  area-concentrated search in, 369
  aversive conditioning of, 369-370
  commensals of, 345
  competing species of, 363
  confusion of, by group, 233
  cooperative defense against, 232-233
  defense against, as cause of grouping,
    232-234
  fleeing from, 375
  functional responses in, 368
  generalization and discrimination of
    mimics by, 380, 381

Predators—cont'd
  handling and killing tactics of, 374-
    375
  humans killing, 362
  hunger thresholds of, 362
  hunting and capture methods of, 371-
    374
  hunting success of, 374, 375
  neophobia in, 369
  parental care of young by, 334
  pressure by, and parental care, 334-
    335
  prey handling of, maturational changes
    in, 564-565
  protection from, evolution of, 341
  search image of, 367-370
  starvation of, 362
  storage of prey by, 363
Predator behavior, 357-383
Predator-mobbing behavior of birds, 585,
    586
Predator-prey relationships, long-term,
    383
Predatory act, components of, 363
Predatory behavior, evolution of, 359-
    361
Predictions, testing of, 54
Preen-behind-wing display in mallard,
    148
Preening, 167
Preformation, 556
Pregnancy and opiate system, 522
Prehunting behavior, ritualized, 281
Premack, D., 127, 295, 687
Preparatory-response model for learning,
    581
Preparedness to learn, 583
Pressure
  predator, and parental care, 334, 335
  selective, 21
Pressure chambers for hearing, 453
Preston, J.L., 348, 687
Prey
  animal, 358
  bad-tasting, 369-370
  behavior of, 360
  birds of; see Birds of prey
  clashes over, 342, 343
  excess, storage of, 363
  exhaustion of, by predator, 371
  finding, recognizing, and selecting,
    problems of, 365-371
  handling of, by predators, maturational
    changes in, 564-565
  increases in, predation responses to,
    368
  preference in, of garter snakes, 365
  selection of, 366, 370-371
  switching of, 366-367
  tools used to get, 371

Preyer, W., 556, 687
Preying mantis, cannibalism in, 306, 308
Pribam, K.H., 617, 687
Price, E.O., 658, 687
Priesner, E., 466
Primary song of bird, 574
Primate Center of University of Wisconsin at Madison, 628
Primates
  alloparental helping in, 338
  brains of, 475
    electrode study of, 481
  color vision in, 458
  insight learning in, 587
  learning in, 602
  learning-set learning in, 587
  matriarchal groups in, 330
  mixed-species groupings in, 344
  play in, 390
  reproductive cycles of, 311
  sleep characteristics of, 158
  social behavior of, 230, 231
  social grooming in, 228
  visual pathway from retina to brain in, 488
Primers, messages as, 271
Principle
  all-or-none, 438
  August Krogh, 657
  "cicada," 234
  "you-first," 233
Principles of Animal Psychology, 32
Principles of Psychology, 30, 43
Pringle, J.W.S., 438, 687
"Private" events, internal, mental process-es as, 46
Probabilistic view of embryonic behavior, 557
Probability, 55
"Problem boxes," Thorndike's, 44, 45
Problem-solving ability and enriched envi-ronment, 568
Prociphilus tesselatus, 372
Profitability of hunting, 369
Progesterone
  and incubation behavior in doves, 511
  molecular structure of, 519
Progne subis, 623
Projection neurons, 416
Prokaryotic DNA, 90
Prokaryotic organisms, transcription in, 89
Prolactin, 510
  in doves, 511
Promiscuity, 315, 316, 324
Promoter, 89
Pronghorn antelope, 648
  defecation by, 165
Propaganda substances, 353
Propagation, 424-425
Proprioceptive senses, 449

Prosopagnosia, 495-496
Prosoplecta semperi, 381
Prostaglandins, 518
Prostitution, 328
Protean displays, 376
Protection of young by group, 233
Protective mimicry, 377, 379
Protective movements in evolution of behavior, 142
Proteinaceous hormones, 517
Proteins
  amino acid structure of, 87
  functions of, 87
  instructions for, 88-89
  RNA and DNA, 87-90
  synthesis of
    nervous and endocrine systems and, 92, 96
    steps in, 88-89
Protogamous hermaphroditism, 314-315
Protostome pattern of nervous system, 403
Protostomes, communication in, 283-284
Protozoans
  learning in, 589
  movement by, 163
  nervous system of, 405
  odor following in, 191
  as predators, 361
  role of sensory channels in, 279
Provine, R.R., 564, 687
Provine, W.B., 128, 684
Proximate factors
  affecting behavior, 20-21, 128
  ultimate and, relationship between, 22
Pseudoconditioning, 581
Psilocybin, 523
Psycho-hydraulic models of motivation, Lorenz's, 40
Psychological behaviors, 48
Psychological concerns for captive ani-mals, 639
Psychological and physiological aspects of behavior, 400-617
Psychology
  comparative, 31, 43-48, 50
    development of discipline of, 32
    and ethology and neurobiology, 22-51
  and learning, 613
  physiological
    development of, 32
    relationship of, to other disciplines, 26
Pterocles namagua, soaking up water by, 171-172
Ptolemy II, 641
"Public" events, outward behavior as, 46
Puerto Rican parrots, management of, 651-652
Pulliam, H.R., 185, 687
Pulse markers, 477

"Pulse-marker" neurons of moth, 432
Puma, hunting success of, 374
Pump
  ionic, 420
  sodium-potassium, 420, 421
Pumping movements, 163
Punctualism, 125
Punctuated equilibria, 125
Punishment, 582
  and learning, 47
Pupa of noctuid moth, 5
Pupal stage of development, 556
Pure ESS, 250
Purine, 518
Purkinje cell, 412, 414, 500
Purple martins, 623
Purple-throated carib hummingbird, "prostitution" in, 328
Purpose, 22
Purposefulness, apparent, 127
Pursuit, chase and, by predators, 371
Putorious putorius, aggressive play in, 387-388
Pycraft, W.P., 394, 687
Pye, D., 301, 688
Pygathrix nemaeus, play-face of, 397
Pyke, G.H., 174, 176, 180, 185, 687
Pyramidal cells, 412, 414
Pythons, heat and infrared sensing by, 461

Q
q, 663
Q/K ratio, 272, 274
"Quacking" of developing queen bees, 291
Quadratic polynomial in interpretation of data, 82
Quail, Gambel's, encounters of, at feeder, 344
Quail chicks, coturnix, artificial selection in, 101
Quantification of behavior, 4-hour, 658-660
Quantitative differences in signals, 270
Quantitative genetics, heritability, and behavior, 96-100
Queen bee, replacement of, 291
Queen butterfly, courtship sequences of, 308
Queen substance, 291
  in bees, 271
Quiet ovulation, 632
Quiet sleep, 155

R
r, 243, 333-334
Rabbits
  bobcat with, 359
  demes of, distinct, 106

Rabbits—cont'd
    distress calls of, 281
    male fighting in, 319
    pheromones to mark territories of, 272
    sleep time for, 159
    visual system of, 488
Raber, H., 565, 687
Raccoon, hunting methods of, 371
Radiation
    adaptive, 140
    of species, 115
Radio telemetry to study animal behavior,
        63, 64, 65
Radiowave sensing, 463
Rainbow trout, sequential appearance of
        behavior in, 554
Ramapithecus punjabicus, 254
Ramon y Cajal, 49
Rana catesbeiana, eye of, 472
Rana sp., color preference of, maturation-
        al changes in, 561
Rana temporaria, color preference of, mat-
        urational changes in, 562
Randall, J.A., 167, 192, 193, 687
Randall, P.K., 670
Random dispersal of genes in population,
        gene frequency and, 114
Random sampling of behavior, 75-76
Random variation, 103
"Random walk" theory of evolution, 124
Range, home, 192, 199
    and familiar area, 199-200
Rangifer tarandus, play in, 388-389
Ranson, S.W., 545, 679
Ranvier, nodes of, 426
Rape, 328
Rapid eye movement, 155
Rapid wing movements, muscles and,
        438-439
Raptorial, 357; *see also* Predator
Raptors
    feeding of dependents by, 364
    hunting success of, 374
    as predators, 360
    readiness to hunt in, 364
Rasa, O.A.E., 166, 687
Rashotte, M.E., 617, 671
Rasmussen, T., 482, 687
Ratio
    Q/K, 272, 274
    sex, equal, 248
Ratner, S.C., 597, 629, 680, 687
Rats
    black, 630
    brain of
        changes in, from enriched environ-
            ment, 569
        placement of electrodes in, 418
        selections of, involved in feeding
            behavior, 544, 545-549
        sexual differentiation of, 519-521

Rats—cont'd
    brain of—cont'd
        stereotaxic atlas of, 419
    control of feeding in, 540-550
    in enriched environment, studies of,
        568-569
    and ionizing radiation, 463
    kangaroo, grooming in, 167
    laboratory, drinking behavior of, 367
    larynx of, 6
    learning in, 601
    learning curve for, 45
    mole, sleep time for, 159
    neophobia in, 540
    Norway, 630
        learning studies using, 47
    obesity in, 528
    play in, 391
    in Skinner box, 46
    sleep time for, 159
    trained, 654
    vacuum behavior in, 38
    weight control by, 527-528
    white
        female, normal estrous cycle and
            behavior of, 515
        sequential appearance of behavior in,
            555
    wood, encounters of, at feeder, 344
Rattlesnakes, 357
    activity patterns in, 149
    desert, activity patterns in, 161, 162
    infrared detecting pits in, 461
    speckled, activity pattern in, 161, 162
Rattus norvegicus, control of feeding in,
        540-550
Rattus sp., larynx of, 6
Raven
    abnormal behavior in, 622-623
    and wolves, play between, 389
Ray, T.S., 346, 687
Rays
    electric, 375
        generation of electricity by, 462
    torpedo, 375
Razran, G.A., 600, 687
Reactions
    escape or hiding
        abnormal, 623-624
        of turkey chick to hawk silhouette,
            41
    "greeting," in cats, 56, 57
    motor, stereotyped, in captive animals,
        624-625
Recall and memory, 606
Receiver, 262
Receptive field, excitatory, 486
Receptors
    muscle stretch, direct synapse by,
        476

Receptors—cont'd
    pain, 449-450
    sensory, on labellum of blowfly, 534
    stretch, and satiety, 542
Recessive genes, 93
    maladaptive, 112
Recharging of neurons, 427
Reciprocal communication as social be-
        havior, 228
Reciprocal feeding as tactile communica-
        tion, 275
Reciprocal junction, 415
Reciprocity selection and evolution of
        social behavior, 247
Recognition
    bird songs as means of, 294
    individual, 228-229
    pattern, and abstraction, 484-491
    among vertebrates, 229
Recombinant DNA techniques, 89-90
Recorder, strip-chart, to measure response
        time, 667
Recording
    of animal behavior, 65-66
    of bioelectric events to study neurons,
        416-417
    of bird activity, automated, 78-79
    of data, 76-77
    describing, and cataloging behavior,
        68-77
    of pigeon pecking responses, automat-
        ed, 77
Recovery, 581
Recuperation hypothesis of sleep, 159
Recycling by neurons, 427
Red deer, harems of, 322
Red fox, hunting success of, 375
Red junglefowl, parental brooding of,
        332
Red light, 455, 457, 458
Red squirrel, 375
Red wolves breeding with coyotes, 117
Red-backed salamanders, foraging by,
        180
Red-billed duck, 130
Redding, T.W., 688
Redhead duck, habitat selection by, 198
Red-headed woodpecker, nesting habits
        of, 83
Redirected behavior, 41-42
    in captive animals, 625
Redstart warbler, *niche-gestalt* for, 197
Red-winged blackbirds
    sexual dimorphism in, 317
    social behavior of, 226, 227
    vocalization of, 271
Reed bunting, alarm call of, 281
Reef fish, habitat selection in, 196
References on captive animals, 639-640
Reflective surface in eyes of vertebrates,
        459

Reflex(es), 475-481
    gill withdrawal, in *Aplysia*, 603, 604
        neural control of, 503
    knee-jerk, 476
    lordosis, 514
    postural, 476
    simple, 475-476
    stretch, two-neuron, 476
Reflex pathways, inhibiting of portion of,
        478-479
Reflexogenous view of embryonic behav-
        ior, 556
Refractory period, 421, 424
Refusal to eat after capture, 624
Regnier, F.E., 353, 687
Regression in behavior by zoo animals,
        626
Regulation
    of animal behavior, chemicals involved
        in, listing of, 518
    internal, of feeding, 527-531
Regulator genes, 90
Rehearsal and memory, 606
Reinforcement
    and learning, 47
    schedules of, 582
Reinforcement theories, 45
Reinforcer, 581, 582
Reinforcing stimulus, 581
Reiter, J., 337
Relationship
    coefficient of, 243-244
    mother-infant, and abnormal behavior,
        629
    peer, and abnormal behavior, 629
    plesiobiotic, 344
    predator-prey, long-term, 383
Releasers, 38
    messages as, 271
    supernormal, 38, 39
Releasing factors, 510
Releasing mechanism, innate, 39
    application of, to levels of instinct, 43
    relationship of, to sign stimulus and
        fixed action pattern, 40
Relevance
    biological, of learning capacity, 601
    and learning, 583
Reliability, interobserver, in measurement
        of rhesus monkey behavior, 68
REM sleep, 155, 157
    functions of, 158
Remoras, commensalism in, 345
Renner, M., 206, 687
Repeated action potentials, 425
"Repeater" neurons of moth, 431
Repetition and memory, 606
Replicator selection, 120
Repressed operon, 91
Reproduction, 92-93
    asexual, 92

Reproduction—cont'd
    behavior and, 303-339
    biological considerations in, 306-310
    and cannibalism, 306, 308
    and care of young, cooperation in, 337-
        339
    differential investment in, by sexes,
        316-321
    disorders of, in pets, 633
    environmental considerations and tim-
        ing of, 310-312
    facilitation of, by grouping, 235-236
    "favorable" times for, 311-312
    and gene frequency, 113
    mating patterns in, 313-328
    readiness for, 309
    regulation of, to conserve resources,
        118
    seasonal patterns of, 310-311, 314
    sexual; *see* Sexual reproduction
    spatial distribution of resources and,
        322-323
    and survival
        and behavior, 127-128
        differential, 111
Reproduction signals, 281
Reproductive behaviors
    definition of, 303
    nonmodal forms of, 328
    phylogenetic comparisons of, 328-330
    sequences of, 306-310
Reproductive cycle of ringed plover, 515
Reproductive hormones of human female,
        512-514
Reptiles
    auditory communication in, 275
    diurnal, color vision in, 458
    drinking behavior by, 171
    ears of, 453
    embryonic behavior of, 556
    migration by, 209-210
    predation by, 359
    role of sensory channels in, 279
    sleep characteristics of, 158
    temperature regulation by, 170
Reptilian brain, 501
Research
    applied animal behavior in, 653
    captive animals used in, abnormal
        behavior in, 628-629
    neural
        on cockroach cerci, 432-435
        on moth ears, 428-432
Research animals, management of, 643-
        644
Resnick, S., 681
Resolution, visual, 460
Resources
    competition for, 104-105
    social interactions involving, outcomes
        of, 243

Resources—cont'd
    spatial distribution of, and reproductive
        behavior, 322-323
Respiratory movements, 165
Response(s)
    begging, in newly hatched herring gulls,
        38
    behavioral; *see* Behavioral responses
    conditioned, 580
    escape, of birds to flying birds of prey,
        39
    functional, in predators, 368
    innate, 41
    learned, from sensory stimuli, study of,
        445-446
    natural, to experimental stimuli to study
        senses, 446
    neural, selectivity and specificity of,
        476-477
    pigeon pecking, automated recording
        of, 77
    predation, to increases in prey animals,
        368
    startle, of cockroaches, 432-435
    stimulus and
        in classical conditioning, 45
        in operant conditioning, 47
    threat, territorial, in European robins,
        37
    unconditioned, 580
Response times, nervous system, 666-
        669
Responsiveness, sexual, of female, 514-
        516
Resting potential, 420, 421
Resting rate of firing, 425
Results, significant, using experimental
        method, 55
Rethlingshafer, D.A., 471, 679, 683,
        691
Reticular activating system, function of,
        501
Retina, 460-461, 485-486
    cells of, 485, 486
    frog, ganglion cells of, output of, 487
    neurons of, 485, 486
    processing of visual information by,
        485
    visual pathways from, to brain, 487,
        488
Retrieval and memory, 606
Rewards, 582
    and learning, 47
Rheobatrachus silus, oral birth from gastric
        brooding by, 331
Rhesus monkey
    behavior of, interobserver reliability in
        measurement of, 68
    deprivation studies involving, 567
    and ionizing radiation, 463

Rhesus monkey—cont'd
　number of displays in, 286
　play in, 387
Rhijn, J.G. van, 167, 687
Rhinolophus ferrumequinum, echolocation
　in, 1
Rhombencephalic sleep, 155
Rhythms, 150-155
　circadian, 151
　　of animal activity, changing patterns
　　　in, 153
　circannual, 151
　　of ground squirrels, 154
　entrained, 150
　internal, 150
　involving light, control of, 155
　pacemaker, free-running, *Zeitgebers*
　　and, 155
　serotonin and, 155
Ribonucleic acid
　deoxyribonucleic acid, and protein, 87-
　　90
　messenger, 88, 89, 90
　　interaction of, with environment, 90-
　　　92
　　manufacture of, 91
　　transfer, 89
Ribonucleic acid polymerase, 91
Ribonucleic acid virus, 88
Ribosomes, 413
Richards, O.W., 353, 687
Richardson, W.J., 219, 687
Ridgway, S.D., 156
Right versus left sides of cerebrum, 497-
　500
Ring doves, reproductive sequence in,
　511, 512
Ringed plover, reproductive cycle of,
　515
Ringing to identify birds, 211
Ring-necked duck, habitat selection by,
　198
Ritualization, 140-141
　in displays, 282
　sources of, 141-142
RNA; *see* Ribonucleic acid
Roach, 381
Roadrunners, temperature regulation by,
　170
Roberts, L.S., 35, 36, 107, 143, 452,
　482, 679
Roberts, T.W., 77, 690
Roberts, W.A., 617, 684
Robins
　European
　　magnetic field effects on, 215
　　navigation by, 215
　　territorial threat responses in, 37
　　sexual dimorphism in, 317
Robinson, D.N., 127, 687

Robson, D.S., 675
Robson, E.A., 591, 687
Robynson, R., 29
Rocking after landing in birds, 291
Rod
　photosensitive part of, 469
　and vision, 459
Rodaniche, A.F., 287, 288, 685
Rodents
　ethogram of, 72-73
　exploration by, 393
　offensive defenses of, 376
　as prey
　　of American kestrel, 374
　　of red fox, 375
　psychological needs of, 639
　temperature sense of, 447
Rodgers, J.A., Jr., 312, 687
Roe deer, 375
Roeder, K.D., 14, 15, 16, 17, 18, 19, 20,
　　23, 49, 405, 428, 431, 432,
　　433, 434, 435, 441, 551, 687
Roles in social groups, 228-229
Rolling and weaving, shelter made by,
　183
Rolls, B.J., 677
Rolls, E.T., 549, 677, 687
Romanes, G.J., 28, 31, 687
Romano, P.L., 156
Root, A., 653
Root, J., 653
Rosenblatt, J.S., 143, 577, 670, 677
Rosenblum, L.A., 673
Rosenzweig, M.R., 568, 617, 687
Ross, D.M., 591, 687
Ross, W.N., 417, 687
Rossler, V.E., 102, 688
Rothenbuhler, N., 94, 95, 688
Rotifers
　habitat selection by, 194
　nervous system of, 407
Roubik, D.W., 82, 83, 688
Round dance of honeybee, 288
Rousette bat, echolocation by, 8
Rousettus sp., echolocation by, 8
Rovner, J.S., 331, 688
Rowe, E.A., 77, 672, 677
Rowell, T.E., 337, 688
Rowing motions, 163
Rubbing by cats, 56, 57
Ruddy duck, habitat selection by, 198
Ruddy sheld duck, 130
Rudnai, J.A., 686
Ruff, communal displays in, 328
Rumbaugh, D.M., 295, 297, 298, 688
Running, 163
　cost of, 163
Rusak, B., 155, 688
Ruse, M., 253, 255, 259, 688
Rushforth, N.B., 405, 590, 591, 688
Rusiniak, K.W., 677

r-versus-k selection, 333-334
Ryan, J.C., 681
Ryan, M.J., 355, 691

S
Sackett, G.P., 69, 621, 629, 688
Sade, D.S., 338, 688
Safety considerations in observation of ani-
　mals, 657
Sagan, C., 124, 159, 283, 404, 498, 501,
　505, 605, 688
Salamanders
　offensive defenses of, 376
　red-backed, foraging by, 180
Sale, P.F., 196, 688
Sales, G., 301, 688
Salivation, dog, Pavlov's study using, 45
Salmon
　Atlantic, sequential appearances of be-
　　havior of, 554
　imprinting in, 571
　migration by, 206-207
　as prey of black bear, 374
　reproduction and, 107
Saltatory conduction, 426
Salticid spider
　hunger threshold of, 362
　male, altered courtship patterns of,
　　611
Salzberg, B.M., 687
Salzinger, K., 677
Sampling
　animal, focal, 75
　of behavior, 73-76
　　random, 75-76
　　statistical, 74-76
Sanders, G.D., 596, 688
Sanders, R.J., 690
Sandgrouse, Namaqua, soaking up water
　by, 171-172
Sandhill crane, 131, 134, 139
　calls of, sequential appearance of, 554
　as foster parents of whooping cranes,
　　653
Sanger, V.L., 632, 688
Sargant, W.W., 629, 688
Sargent, T.D., 174, 185, 680
Sarus crane, 131, 139, 140
Satiety
　in blowfly, 535-540
　and duodenal hormones, 543
　stretch receptors and, 542
Sauer, E.G.F., 364, 688
Saunders, D.S., 185, 688
Savage-Rumbaugh, E.S., 295, 297, 298,
　688
Savarie, P.J., 648, 688
Saw-whet owls, migration of, 219
Sayornis phoebe, abnormal nest building
　by, 623

Scallops
  escape from predators by, 375
  eyes of, 456
  movement by, 163
Scaphidura oryzivora, parasitism-mutualism in, 351
Scatter of sound, 10, 11
Scaup, lesser, habitat selection by, 198
Scavenging, 358
Sceloporus lizard, head-bobbing pattern in courtship displays of, 270
Schaller, G.B., 27, 56, 278, 360, 362, 366, 372, 373, 374, 375, 383, 395, 653, 688
Schally, A.V., 543, 688
Scheduling problems in observation of animals, 658
Scheller, R.H., 102, 688
Schenkel, R., 388, 391, 688
Scherzinger, W., 367, 688
Schiller, C.H., 21, 179, 443, 688, 691
Schiller, F., 394
Schleidt, M., 688
Schleidt, W.M., 36, 41, 126, 688
Schmidt, J., 208, 209
Schmidt, J.P., 633, 688
Schmidt, T., 274, 680
Schmidt-Koenig, K., 223, 677, 688
Schmitt, F.O., 689
Schneider, D., 465, 466, 688
Schneirla, T.C., 28, 32, 33, 47, 50, 553, 600, 683, 688, 692
Schnell, G.D., 138, 688
Schoener, A., 199, 689
Schoener, T.W., 174, 185, 199, 689
Scholz, A.T., 679
Schreiber, R.W., 286, 689
Schüz, E., 212, 689
Schwartz, J.H., 525, 680, 688
Schwartzbaum, J.S., 542, 689
Science, 54
Scientific methods, behavior and, 54-55
Scientific view of behavior, beginning of, 30
Sclafani, A., 528, 548, 689
Scorpions
  nervous system of, 406
  pectines of, study of, 445
  stinging of, 375
Scott, P., 70
Scott, T.H., 567, 684
Scratching, 167
  head, in vertebrates, 128, 129
Screech owl, postures of, 61
Screwworms, control of numbers of, 647
Scrub jay, encounters of, at feeder, 344
Scyphozoans, 230
Sea urchin, nervous system of, 409
Seahorses, role reversal in, 324
Seals
  fur, harems of, 322

Seals—cont'd
  gray, sleeping, 156
  hearing of, 453
  refusal of, to eat after capture, 624
  temperature regulation by, 170
Search, area-concentrated, in predators, 369
Search image
  optimal foraging and, 179-180
  of predators, 367-370
Searching, optimal, 175
Searcy, W.A., 575, 689
Seasonal monogamy, 316
Seasonal patterns of breeding, 310-311, 314
Seasonal rhythms and predation, 363
Sebeok, T.A., 283, 287, 297, 298, 301, 671, 685, 689
Secondary attributes of fixed action patterns, 41
Seely, M.K., 172, 689
Segregation
  law of, 93
  during meiosis, complications of, 95
Selander, R.K., 689
Selection
  "by consequence," 128
  definition of, 122
  epigamic, 319
  group
    and behavior, 117-118
    and evolution of social behavior, 247
  intersexual, 319
  kin, 243, 244
    and evolution of social behavior, 243-247
  natural, 111
    level of operation of, 117-122
  of prey, 370-371
  reciprocity, and evolution of social behavior, 247
  replicator, 120
  r-versus-k, 333-334
  sexual, 319
    female choice in, 318, 319-320
    intrasexual, 320
    parental care and, 335
Selective breeding and mutations, 101
Selective perception, 367
Selective pressures, 21
Selectivity of neural responses, 476-477
Self-awareness, 127
Self-handicapping during play, 396, 397
Selfish Gene, 119
Selfish gene theory, 338
"Selfish genes," 117-122
  and behavior, 119-121
  competing, 121
Self-mutilation by captive animals, 625
Seliger, H.H., 682

Seligman, M.E.P., 588, 689
Semelparity, 316, 324
Semelparous animals, 107
Semicircular canals, 450-451
Semiochemicals, 266
Sender, 262
Senegal dove, cooing song pattern of, 97
Sensations, pair, 449-450
Sense cells
  of human, 449
  visual, structure of, 469
Sense organs, blocking, interruption or removal of, to study senses, 445
Senses
  brief survey of, 449-466
  chemical, 463-466
  electric, 462
  emitted-energy, 466-467
  inertial, 450-451
  magnetic, 462-463
  methods of investigating, 445-446
  proprioceptive, 449
  sensitivity of, 447
  temperature, 447
  vibrational, 451-454
Sensing
  heat and infrared, 461-462
  x-ray and radio wave, 463
Sensitive periods, 569-570
Sensitivity
  of senses, 447
  visual, three-pigment, 458, 459
Sensitization, 580
  in *Aplysia* sp., 603-605
Sensors, off, 488
Sensory adaptation by moth ear, 429
Sensory attributes, 446-449
Sensory cells of retina, 485, 486
Sensory channels, role of, in total communication repertoire, 279
Sensory information, crossing of, and collateral pathways to other motor neurons, 479
Sensory input, combination of, 468
Sensory lateralization in hearing, 500
Sensory neurons, 412, 415
Sensory receptors on labellum of blowfly, 534
Sensory region of human cerebral cortex, 482
Sensory stimuli, learned responses from, study of, 445-446
Sensory world, different, 442-471
Sepia sp.
  attack behavior of, maturational changes in, 561-562, 563
  nervous system of, 407
Sepioteuthis sepioidea, advanced communication in, 287

Sequences
    of behavioral acts, 80, 81
    courtship, 306-307, 309-310
        complexity and variability in, 312-313
        of various species, flowcharts of, 308-309
Sequential appearance of behaviors, 554-555
Serengeti lions, 56
Serial monogamy, 316, 317
Serial polygamy, 316, 317
Serial synapses, 415
Serotonin, 518
    and feeding disorders, 548
    and rhythms, 155
    and sensitization, 604
Set-point mechanism for body weight, 529
Sex ratios, equal, 248
Sex reversal, natural instances of, 314-315
Sex steroids, 510
Sexual behaviors, abnormal, in zoo animals, 626
Sexual differentiation of brain in mammals, 519-520
Sexual dimorphism
    and mating patterns in birds, 317
    and polygyny, 255
Sexual disorders in domestic animals, 632
Sexual imprinting, 571
Sexual reproduction, 92
    controversy over need for, 303-306
    costs of, 304, 305
    as means of adaptation, 305
Sexual responsiveness of female, 514-516
Sexual selection, 319
    female choice, 318, 319-320
    intrasexual, 320
    parental care and, 335
Shaffer, W.R., 671
Shaping of behavior, 582
Shapiro, C.M., 160, 689
"Shark suckers," commensalism in, 345
Sharks
    brain of, 411
    electrical senses of, 462
    predation by, 359
    as predators, 361
    remoras and, 345
"Sharpened peaks of generalization gradients," 368
"Sharp-set," 364
Sharp-shinned hawk with starling, 358
Sharp-tailed grouse, leks in, 326-327
Shedding of epidermis, 171

Sheep
    bighorn
        male fighting in, 319
        migration by, 220
        parental teaching by, 332
    sleep time for, 159
    subestrus in, 632
Shelby, J., 594, 684
Sheld duck, 130
Shelter seeking and construction, 181-183
Sherrington, C.S., 28
Sherry, D.F., 332, 606, 689
Shettleworth, S.J., 606, 689
Shifting balance theory, Wright's, 120, 121-122
Shifts
    "of attention," 368
    Doppler, 11
Shock, electrical, to kill prey, 375
Shoney, H.H., 676
Short-eared owls, predation by, 363
Shorthorn cattle in Australia, free-ranging, time budgets of, 148
Short-term memory, 606-607
Shouting, 285
Shoveler, Northern, habitat selection by, 198
Shoveler duck, 130
Shrews
    hearing of, 453
    inactivity in, 148
    poisoning by, 375
    as predators, 361
    responses of, to increases in prey, 368
    role of sensory channels in, 279
    sleep time for, 158, 159, 162
    tree, sleep time for, 159
Shrimp
    cleaner, cleaning zebra moray, 346
    and goby, mutualism between, 347
    individual, recognition by, 229
    mantis, communication signals and, 264
    as prey of cuttlefish, 374
Siamese fighting fish, learning of, 602
Sibatani, A., 499, 500, 689
Siberian crane, 131, 140
Sibly, R., 39, 684
Sickle cell anemia, 113
Sickle dance of honeybee, 290
Sidewinder, activity pattern in, 161, 162
Sidman, R.L., 101, 689
Sidtis, J.J., 498, 689
Sign stimuli, 38
    and fixed action pattern, relationship of innate releasing mechanism to, 40
Sign tracking, 580

Signal, 262
    alterations in frequency of behavior as result of, 263
    characteristics of, 269-271
    communication, mantis shrimps and, 264
    as communication, 261-262
    continuous, 271
    digital, 269
    discrete, 269-270
    inferred, in Steller's jays, 264
    luminescent, of fireflies, 356
    modification of, 270-271
    play, 393, 397-398
    reproduction, 281
    visual, of cichlid fishes, 284
Significant results using experimental method, 55
Signing approach to language in apes, 295, 297, 298
Sih, A., 180, 689
Silberglied, R.E., 457, 689
Silent heat, 632
Silk and web building, 182
Silkworm moth, chemical senses of, 465-466
Silverberg, J., 253, 255, 256, 259, 671, 674, 675, 677, 680, 689
Silverfish as commensal, 345
Silverstein, R.M., 301, 685
Simmons, J.A., 7, 11, 12, 23, 689
Simon, D., 308, 689
Simple aggregations, 342, 343, 344
Simple reflex, 475-476
Simple synapses, 415
Simpson, A.E., 76, 689
Simpson, G.G., 109, 255, 606, 689
Simpson, M.J.A., 76, 689
Sims, E.A.H., 529, 689
Simulation, 587
Simultaneous hermaphroditism, 314-315
Simultaneous polygamy, 316
Single gene and behavior, 94-95
Siphonophora, social behavior in, 230
"Site tenacity" of birds, 210
Situation specificity in learning of arthropods, 600
Skates, electric, generation of electricity by, 462
Skepticism, 55
Skinner, B.F., 3, 28, 46, 47, 51, 126, 128, 297, 298, 676, 689
Skinner box, 46, 47
Skunks
    chemical defenses of, 376
    predator protection of, 341
    warning displays of, 377
Slave-making, 352-353
Sleep, 155-162
    active, 155
    characteristics of, 158

Sleep—cont'd
dreaming, 155
and exercise, 160
fast, 155
function of, 158-162
hypotheses about, 158-159
need or compulsion for, 158-162
non-REM, 155
paradoxical, 155
quiet, 155
REM, 155, 157
function of, 158
rhombencephalic, 155
stages of, 155, 157
time for, in mammals, 159
Sleep factor delta, 518
Sleep factor S, 518
Sloth, two-toed, sleep time for, 159
Sloth bears, abnormal behavior in, 624-625
Sluckin, W., 635, 689
Slugs
crawling of, cost of, 163-164
moisture conservation of, 237
nervous system of, 405
Small, W.S., 45, 689
Smartt, R.A., 689
Smells, patterns of, 465
Smith, E.O., 392, 689
Smith, G.P., 670, 681
Smith, J.N.M., 179, 689
Smith, N.G., 351, 689
Smith, R.L., 331, 689
Smith, T.G., Jr., 525, 671
Smith, W.J., 265, 266, 280, 289
Smooth newt, courtship sequence in, 307, 308
Snail
apple, as prey of Everglades kite, 366
circadian rhythm in, 155
cone, poisoning by, 375
learning in, 596
marine, study of nervous system of, 49
nervous system of, 405, 407
return to familiar area by, 202
Snakes
constriction patterns of, 376
corn, killing mouse, 376
ears of, 453
garter
feeding behavior of, 140
prey preferences of, 365
gorging by, 362
hearing in, 454
heat focusing by, 448
hognose, diet of, 366, 638-639
infrared heat sensors in, 461
Mertensian mimicry by, 377
movement by, 163
navigation by, 210
nervous system of, 409

Snakes—cont'd
pits in, 461, 462
poisonous, 375
as predators, 360
refusal of, to eat after capture, 624
Sneath, P.H.A., 138, 689
Sneezing, 165
Snelling, J., 64
Snider, R.S., 463, 678
Snodgrass, R.E., 410, 689
Snowdon, C.T., 542, 549, 689, 691
Snyder, N.F.R., 651, 689
Snyder, S.H., 525, 690
Social behavior; see also Sociobiology
elements of, 226-231
evolution of, 118
group selection and, 247
kin selection and, 243-247
reciprocity selection and, 247
indicators of, 226-229
intraspecific, 70
pinnacles of, 230-231
time or energy spent on, 227-228
Social facilitation and predation, 369
Social grooming, 227, 228
function of, 282
Social groups, number of animals in, 226-227
Social insects
commensals of, 345
family characteristics of, 330
food stealing by, 349
group defense by, 233
parasitism of, 354-355
plesiobiotic relationship of, 344
Social interactions involving resources, outcomes of, 243
Social parasitism, temporary 349-350
Social play, 396-399; see also Play
Social status, postures indicating, 278
Social structure
of groups, 228-229
and spacing, maintenance of, 237-242
Social world, animal who's who in, 230
Sociobiological controversy, 252-258
Sociobiology, 53; see also Social behavior
controversy over, 252-258
definition of, 225
introduction to, 224-258
relationship of, to other disciplines, 26
Sociobiology, The New Synthesis, 225, 252, 253
Sociometric analysis of possible communication behavior, 80, 81
Sociometric matrices, 81
Sodium ions in neurons, 420
Sodium-potassium pump, 420, 421
Sokal, R.R., 138, 689
Solitary species, 227
Solo play, 386
Soma, 413-414

Somato-somatic synapses, 415
Somatostatin, 518
Somatotrophic hormone and feeding behavior, 543
Sonar, use of, by animals, 466-467
Song
bird; see Birds, songs of
of cowbirds, development of, 575-576
of crickets, larval development and, 558, 559
of European warblers and female-choice sexual selection, 318
Song patterns, inheritance of, 97
Songbirds
migratory, north temperate, annual sequence of hormonally dependent events in, 516
songs of, development of, 572-576
Sonograms, 66
of guard calls of cranes, 133
of human speech, 14
Sorex sp., responses of, to increases in prey, 368
Sound(s)
of bats, 7
and behavior, 4-20
for communication, 274-275
cycle of, 9
interference with, 10
orientation based on, 200
properties of, 8-10
scatter of, 10, 11
Sound vibration, detection of, 452-453
Sound waves, 8-9
Source, ability to localize, 448-449
South American monkeys, self-mutilation by, 625
Southern crowned cranes, 131
Spacing, social structure and, maintenance of, 237-242
Spagna, T., 679
Spallanzani, L., 7, 23, 215, 445
Sparrows
chipping, hand raising of, 195
habitat preference of, 196
house
circadian pacemakers in, 155
competition for food by, 343
locomotor activity of, and circadian rhythm, 153
learning curve for, 45
white-crowned, dialects in songs of, 575
white-throated, single-gene effects in, 95
Spatial differences in environment, 103
Spatial distribution of resources and reproductive behavior, 322-323
Spatial learning, 585

Spatial memory, 199
in arthropods, 598-599
Spatial pattern discrimination, 448
Spatial summation, 422, 423
Spear, N.E., 577, 617, 690
Special creation, 123-124
Specialists on captive animals, 640
Specialization in groups, 236-237
Species
accidental, 345
attendant, 345
behavioral repertoire of, 56
choice of, for observation, 657
of cranes, 132
distribution of, 133
endangered, 651
familiarity with, 74
when studying animal behavior, 55-56
hominid, brain volume of, increase in, 254
hybrids of, genetics of, 117
isolation of, and reproduction, 306
new, origin of, 114-117
nuclear, 345
origin of, extraterrestrial, 124
radiation of, 115
solitary, 227
Species communication behaviors, listing, 661
Species ethogram, determination of, 660
"Species memory," 190
Species recognition, bird songs as means of, 293
Species-specific behaviors, 48
Specific exploration, 393
Specificity
of innervation, 439
of neural responses, 476-477
situation, in learning of arthropods, 600
Speckled rattlesnake, activity pattern in, 161, 162
Speculum-flashing display in mallard, 148
Spencer, H., 28, 39, 111, 253, 394, 690
Spencer, W.A., 589, 690
Spencer-Booth, Y., 688
Sperm, changes in, 316, 318
Sperry, R.W., 28, 49, 497, 498, 685, 690
Sphex ichneumoneus, optimality and, 181
Sphingidae, hearing in, 453
Spiders
auditory communication in, 275
black widow, cannibalism of, 306
cannibalism-avoiding behavior in, 308-309
communal hunting by, 372
dendritic, 414

Spiders—cont'd
developmental changes in, 611
familiar paths for, 200
hearing in, 454
homing ability of, 202
jumping, with fly, 358
learning by, 599
lycosid, parental care by, 331
nervous system of, 406
observation of, 659
orb-weaving, habituation and, 598
poisoning by, 375
predation by, 359
as predators, 360
salticid, hunger threshold of, 362
storage of prey by, 363
web of, 182
Spieth, H.T., 101, 690
Spike frequency, 425
Spikes, 424
Spinal cord
cross section of, 478
and locomotion, 481
pathway of information from brain through, to motor neurons, 480
Spines
as defense against predator, 376
dendritic, and memory, 607-608
Spinning of cocoon, 182
Spiny echidna (anteater)
care of eggs of, by male, 333
sleep time for, 159
Spitler-Nabors, K.J., 670
Split-brain experiments, 494, 497-500
Sponges, nervous system of, 405
Spontaneous depolarization, 421, 425
Spontaneous generation, 123
Spontaneous rate of firing, 425
Spontaneous variability, 103
Spot, blind, 486
Spotted hyenas
gorging by, 362
hunting success of, 374
Sprague, J.M., 682
Sprague, R.H., 165, 690
Spread-winged postures of birds, 170-171
Springer, D., 528, 689
Squid
advanced communication by, 287-288
giant, as predator, 361
nervous system of, 405
pain in, 450
study of nervous system of, 49
Squirrels
antelope, encounters of, at feeder, 344
color vision in, 458
gliding behavior of, 140
ground, circannual rhythms of, 154
and guenons, play in, 389
nut opening by, 565

Squirrels—cont'd
red, 375
sleep time for, 159
visual system of, 488
SR, 581
S-R learning, 581
Srb, A.M., 107, 690
S-S learning, 581
Stability, population, of groups, 236
Stack, T., 467
Stain, Golgi's, 49
Staircase method of study, 446
Stalking and ambushing by predators, 371
Stanley, S.M., 321, 690
Stanley crane, 131, 139
Star, brittle, nervous system of, 408
Star-compass navigation by birds, 214-215
Starfish
contact of anemone with, 590, 591
crown-of-thorns, defense from predators by, 376
hunting methods of, 371
nervous system of, 409
Starlings
European, competition over nesting places by, 343
migrations of, 212, 213
mimicking by, 573
navigation by, 214
sharp-shinned hawk with, 358
vigilance time and flock size in, 232
Stars, use of, by birds for navigation, 214
Startle response of cockroaches, 432-435
Static obese plateau, 545
Statistical maps of chromosomes, 95-96
Statistical versus physical bases of heredity, 95-102
Statistical sampling of behavior, 74-76
Statistical tests of behavior, 661
Statistics
descriptive, 78
inferential, 78
multivariate, 81, 83-84
Statocyst, 450
Stealing of food, 348-349
Steatornis caripensis, use of echolocation by, 467
Stebbins, G.L., 125, 690
Steere, W.C., 143, 679, 689, 691
Stefansson, V., 388, 690
Stein, B.E., 482, 677
Stellar, E., 544, 682, 690
Steller's jay
crest raising in, 271
inferred signals in, 264
Stenopus hispidus, individual recognition by, 229
Stephenson, G.R., 77, 690

Stereoscopic vision, 448
Stereotaxic atlases of brain, 417, 419
Stereotaxic instruments, 417, 418
Stereotyped motor reactions in captive animals, 624-625
Stereotypy in pets, 633
Sterile hybrids, 117
Sterile male techniques for insect pest control, 647
Sterna forsteri, hunting success of, 374
Sternberg, H., 210, 671
Sterner, R.T., 648, 682, 688
Sternopygus, electrical field surrounding, 462
Steroid hormones, 517, 519
Steroids, sex, 510
Stetson, M.H., 690
Stevens, C.F., 441, 690
Stevens, V., 678
Stick building, shelter made by, 183
Sticklebacks
    care of eggs of, by male, 333
    hunger threshold of, 362
    three-spined; *see* Three-spined stickleback
Stimulation
    ablation, lesioning and, in study of neurons, 417-418
    of brain, artificial, and cricket calls, 102
Stimulus
    abnormal, causing abnormal behavior, 634
    conditioned, 45, 580
    filtering of, 443
    reinforcing, 581
    and response
        in classical conditioning, 45
        in operant conditioning, 47
    sensory, learned responses from, study of, 445-446
    sign, 38
        and fixed action pattern, relationship of innate releasing mechanism to, 40
    supernormal, 38, 39
    unconditioned, 45, 580
Stimulus substitution model for learning, 581
Stimulus-response learning, 581
Stimulus-stimulus learning, 581
Stinging to kill prey, 375
Stingless bees
    and aphids, commensalism between, 346
    communication in, 290-291
    mutualism-parasitism involving, 351
Stingrays, stinging by, 375
Stochastic fluctuations in gene frequencies, 664
Stoddart, D.M., 301, 690

Stokes, A.W., 658, 687
Stokes, N.W., 673
Stomphia, swimming and resettling response of, 590-591
Stonechat, alarm call of, 281
Stopwatch to record response time, 667
Storing of food, 181
Storks, European, migration of, 212
Storm, R.M., 223, 690
Strategy, 247-248
    culturally stable, 251
    evolutionarily stable, 181, 247-252
    involved in hunting for food, 173
    in learning-set learning, 587
    win-stay, lose-shift, 588
Street, P., 223, 690
Stress causing abnormal behavior, 634
Stretch receptors and satiety, 542
Stretch reflex, two-neuron, 476
Stretching, 166
Striatum of telencephalon, 501
Striders, water, communication in, 67
Strip-chart recorder to measure response time, 667
Strongylocentrotus franciscanus, parental care by, 331
Structural mechanisms of memory, 607-608
Structuralism, 44
Structure, social, and spacing, maintenance of, 237-242
"Struggle for existence," 111
Study of Instinct, 36
Stunkard, A.J., 551, 690
Stuttard, S., 56, 57, 684
Sublethal aggression in agonistic behavior, 237
Submission, advantages of, 241, 242
Subestrus in domestic animals, 632
Subsong of bird, 574
Substance P, 518, 522
    in fish, 450
Sucking breathing, 165
Sudan crowned cranes, 131
Sudarshan, K., 554, 670
Sugar, concentration of, and blowfly's feeding behavior, 536
Sula sp., 130
Sulak, K.J., 346, 690
Sulzman, F.M., 185, 684
Summation, 421
    spatial, 422, 423
    temporal, 422, 423
Sun compass, 205, 206
    use of, by birds, 214
Sun-arc hypothesis for pigeon homing, 217-219
Sunfish, contact perimeter for, changes in, 240
Suomalainen, E., 304, 690

Suomi, S.J., 629, 690
Supernormal releasers, 38, 39
Suprachiasmatic nucleus, circadian pacemaker in, 154
Surplus energy theory of play, 394
Surprise in play, 392-393
Survival
    "of fittest," 111
    and gene frequency, 112-113
    of genes, 119
    and grouping, 236
    and reproduction
        and behavior, 127-128
        differential, 111
Susman, M., 107, 143, 692
Sutherland, N.S., 617, 690
Swainson's thrush, transmitter on, 64
Swallows
    cliff, 317
    wing flapping of, 564
Swallowtail butterflies, aggression in, 251
Swans
    monogamy of, 329
    navigation by, 215
Swenson, E.A., 556, 690
Swim bladder and sound, 452, 453
Swimming, cost of, 163
Swine
    abnormal behavior in, 631
    subestrus in, 632
Swing, centrifugal, in ants, 600
Sylvia communis, feeding of dependents by, 364
Symbiosis, 341
    and predation, 340-383
Symbolic communication, 296, 297
Synapses, 414, 415
    chains of, 415
    chemical, 424
    nerve-muscle, pattern-sensitive, 437-438
    nests or glomeruli of, 415
    simple, 415
Synaptic cleft, 426, 427
Synaptic pathways, principles of, 477
Synaptic vesicles, 415, 426
Syndrome
    chicken hysteria, 631-632
    ventromedial hypothalamus, 548; *see also* Hyperphagia; Ventromedial hypothalamus
Synthesis, protein
    nervous and endocrine systems and, 92, 96
    steps in, 88-89
Syrbula sp., courtship in, 312
Syrian woodpecker, foraging behavior of, 364

Systemogenous behavior of embryo, 557
Szlep, R., 680

T

t test, 661
Taapken, J.D., 651, 689
Tabula rasa, 557
Tachinid fly attacking cricket, 355
Tactile communication, 275-276
Tail, movement of, and communication, 276
Tail wagging in waterfowl, 147
Takahashi, J.S., 155, 690
Tameness, 629, 630
Tamias striatus, foraging methods of, 176
Tamura, M., 575, 682
Tape recording of data, 76
Tapetum lucidum, 459
Tapeworms, 591
Tapir, sleep time for, 159
Tarantulas, male, cannibalism-avoiding behavior of, 308
Target neurons of hormones, 517
Taste, 463
    of food, and rat feeding behavior, 541-542
    and odor transduction, 465
Taste aversion, 583-584
Taste aversion forms of learning, 48
Taurine, 518
Tavolga, W.N., 446, 471, 672, 690
Taxis, 190, 191
Taxonomic relationships
    of Anatinae as elucidated by behavioral characteristics, 130
    of Pelecaniformes as revealed by comparative behavioral analysis, 130
Taylor, E.L., 644, 645, 690
Teaching by parents, 332
Teal
    chestnut-breasted, 130
    chilee, 130
    common, 130
    European, courtship displays in, 129
    habitat selection by, 198
Technology in study of behavior, 65-67
Tectum, optic, of iguana, map of, 482
Tegner, M.J., 331, 690
Teitelbaum, P., 545, 546, 547, 679, 682, 83, 690, 693
Telencephalon, 501
Teleogryllus commodus, song patterns of, 97
Teleogryllus oceanicus, song patterns of, 97
Teleogryllus sp., song of, larval development and, 558, 559
Teleonomy, 127
Teleost fish, intensity thresholds of, determination of, 446

Telescopes, 65
Teleutomyrmex schneideri, inquilinism of, 353
Television, closed-circuit, 65
Temperature, sense for, 447
Temperature conformer, 168
Temperature regulation, 167-171
Template of sound pattern, 574
Temple, S.A., 651, 655, 675, 689, 690
Temporal pattern discrimination, 448
Temporal summation, 422, 423
Temporal variation, 104
Temporary social parasitism, 349-350
Tenebrio molitor, learning in, 600
Tenebrionid beetles, trapping of moisture by, 172
Ter, J., 686
Terminals, axon, 414
Termites
    African, magnetic senses of, 463
    reproductive behavior of, 330
    shelters of, 181
    social behavior of, 230
    temperature regulation by, 169, 447
Terns
    arctic, migration of, 201, 210
    Forster's, hunting success of, 374
Terrace, H.S., 294, 295, 296, 297, 298, 690
Territorial behavior, 238, 240
    as epideictic phenomenon, 118
Territorial boundary marking, pheromones and, 272
Territorial monogamy, 316
Territorial threat responses in European robins, 37
Territory, 192, 238
    determination of, 660-661
    examples of, 239
    home, 192
Test(s)
    statistical, of behavior, 661
    t, 661
Testing of predictions, 54
Testosterone, 518
    aggressiveness in male house cats following treatment with, 79
    and development of "male" brain, 520
    and feeding behavior, 543
    and fighting in cats, 54, 55, 79
    and incubation behavior in doves, 511
    molecular structure of, 519
    tritium-labeled, to locate receptor area, 517
Testudo hermanni, yawning and stretching in, 166
Tetrahydroisoquinolines, 518
Tets, G.F. van, 130, 690
Thamonphis sirtalis, prey preference of, 365

Thecla sp., false body markings of, 319
Theodora, 641
Theophylline, 523-524
Theory(ies)
    catastrophe, 240
    cognitive, 45
    dream, 157
    duplicative, of vision, 459
    of evolution, Darwin-Wallace, 111
    evolutionary, 123; *see also* Evolution
    game, and communication, 267
    group selection, of Wynne-Edwards, 120
    of hunger, 542
    of kin selection, 243, 244
    lipostatic, of weight control, 529
    Mullerian, 304
    optimality, 174, 247
        and food choice, 177
        problems with, 180-181
    of play, 394-396
    reinforcement, 45
    selfish gene, 338
    shifting balance, Wright's, 120, 121-122
"Therapist monkeys," 629
Thermal conservation by groups, 237
Thermoregulation
    in evolution of behavior, 141
    and parental care, 332-333
Thermoregulatory movements, 167-171
"Thief ants," 349
Thiessen, D.D., 95, 690
Thomas, D.K., 685
Thomas, D.W., 549, 690
Thomas, R.D.K., 117, 168, 690
Thompson, C.R., 297, 690
Thompson, R.F., 589, 690
Thomson's gazelle as prey
    of cheetah, 374, 375
    of lions, 373
    of wild dog, 375
Thorax, ear of noctuid moth on, 16, 17
Thorndike, E.L., 28, 45, 46, 690
    law of Effect of, 45
    law of Exercise of, 45
    problem boxes of, 44, 45
Thorpe, W.H., 28, 41, 51, 97, 565, 582, 682, 690, 691
Thrasher, pearly-eyed, competing with parrots, 651-652
Threat responses, territorial, in European robins, 37
Threats, 281
Three-pigment visual sensitivity, 458, 459
Three-spined stickleback
    courtship sequence in, 306-307
        and fighting, 37
    hierarchical organization of instinct in, 42

Three-spined stickleback—cont'd
prey selection by, 371
"Thrills" in play, 389, 392-393
Thrush
mistle, alarm call of, 281
Swainson's, transmitter on, 64
Thyroid hormone, 518
Thyroid-releasing factor, 518
Tick, woods, as parasite, 348
Tides, daily, 150
Tiger, L., 253, 691
Tiger beetles, Batesian mimicry by, 380
Tilapia fish, territory of, 239
Timbal of Arctiid moth, 20
Time spent observing and observation of
behavior, relationship of, 69
Time budgets
for free-ranging female domestic cats,
151
of gadwall ducks during breeding sea-
son, 149
of shorthorn cattle in Australia, 148
Time problems in observation of animals,
658
Time-lapse photography, 65
Timing in reproductive behavior, 310-
312
Timing mechanisms, 154-155
Timms, A.M., 284
Tinamous, polyandry in, 324
Tinbergen, L., 179, 690
Tinbergen, N., 28, 34, 35, 36, 37, 38, 39,
42, 51, 109, 179, 203, 277,
279, 288, 306, 367, 487, 493,
598, 653, 683, 686, 691
Tinkle, D.W., 670
Titanotheres, head structure of, 321
Titmouse, alarm call of, 281
Tits
European, cultural learning of, 585
foraging behavior in, 179
great, optimal behavior in, 180
habitat preference of, 196
microhabitat selection in, 195, 196
optimal foraging of, learning and, 179
Titus, 641
Toad
American
defensive maneuvers of, 382
dwarf, fertilization of eggs by, 310,
311
common, male fighting in, 323
noxious taste of, 583
predator protection of, 341
as prey of hognose snake, 366
Toates, F.M., 530, 672, 691
Tobacco hornworm, larva of, chemical
senses of, 464
Tobacco hornworm moth, identified neurons in supraesophageal ganglion of, 404

Toback, E., 143, 577, 670, 677
Tobias, J.M., 417, 691
Tockus fasciatus, 389
Toggenberg goats, play in, 394
Tolman, E.C., 28, 45, 691
Tools used to get prey, 371
"Tooting" of developing queen bees, 291
Topoff, H., 179, 691
Torpedo rays, 375
Tortoise, yawning and stretching in, 166
Touch, 450
Tracking, sign, 580
Tracts, 416
Tragus of little brown bat, 12
Trails, odor, for insects, 203, 204
Train markers, 477
"Trained cricket," correlations and, 81-
83
Training of animals, 654
Training bias in predation, 369
"Train-marker" neurons of moth, 432
Trajan, 640
Transcription, 88, 89
cyclic AMP complex and, 91
of DNA, environmental influences in,
91
in eukaryotic organisms, 89-90
hormones affecting, 90, 91
in prokaryotic organisms, 89
Transduction, 422
mechanisms of, 469-470
odor and taste, mechanism of, 465
Transfer ribonucleic acid, 89
Transitoriness, "law" of, 44
Translation, 88, 89
Transmission
of decisions by neurons, 422-427
of nerve impulse, speed of, 425
by nervous system, 413
Transmitter, 426
of synaptic vesicles, 415
Transmitter chemicals, 507, 508
Transport
of captive animals, 638
physical, and tactile communication,
275
Treat, A.E., 20, 687
Tree blind, 58, 60
Tree shrew, sleep time for, 159
Trees
phylogenetic, 138
of cranes, 139, 140
use of, in habitat selection, 193
Trematodes, 591
Trial-and-error learning, 582
Trichoplusia ni, 5
Tricondyla gibba, Batesian mimicry by,
380
Trilobita, 597
Tritium-labeled hormones, use of, to iden-
tify target neurons, 517

Triturus vulgaris, courtship sequence in,
307
Trivers, R.L., 28, 118, 334, 335, 336,
691
tRNA; see Transfer ribonucleic acid
Trophic commensals, 345-346
Trophic hormones, 510
Trophic parasites, 348
Trophobiosis, 346
Trotting, 163
Trout, rainbow, sequential appearance of
behavior in, 554
True geese, 130
"True jellyfish," 230
True slaves in ants, 352
Trumler, E., 277, 691
Truslow, F.K., 653
Tryptamine, 518
Tryptolines, 518
"Tseee" call of alarm, 281
Tsuchitani, C., 471, 672
Tsurnamal, M., 680
Tube-webs of spiders as developmental
change, 611
Tucker, D.W., 209, 691
Tunneling movement, 163
Tupaia, sleep time for, 159
Turbellaria, learning in, 591, 595
Turk, A., 672
Turkey
Australian brush, temperature sense of,
447
brush, egg laying by, 333
Turkey chick, escape reactions of, to hawk
silhouettes, 41
Turkey hens, care of chicks by, 125,
126
Turner, B.H., 683
Turner, E.R.A., 192, 691
Turner, J.R., 380, 381, 684
Tursiops truncatus, individual differences
in appearance of, 63
Turtle dove, cooing song pattern of, 97
Turtles
migration by, 209-210
navigation by, 210
Tuttle, M.D., 355, 691
Tweedie, R.L., 683
Twitch muscles, 438
Two-neuron stretch reflex, 476
Two-toed sloth, sleep time for, 159
Tyler, M.J., 331, 674, 691
Tyler, S., 76, 691
Tympanic columella, 453
Tympanic membrane, 452
Tympanuchus cupido, leks in, 326-327
Type I error, 55
Type II error, 55
Tyramine, 518
Tyrannidae, foraging behavior in, 179
Tyto alba, brain of, map of, 483

U

Uexküll, J. von, 21, 28, 179, 199, 367, 443, 691
Uganda kob, 650
  lek behavior in, 328
  management of, 649-651
Ultimate factors
  affecting behavior, 128
  behavior and, 21-22
  proximate and, relationship between, 22
Ultrasonic sounds of bats, 7
Ultraviolet light, 458
Umwelt, 21, 443, 444
Umwelt und Innenwelt der Tiere, 21
Uncertainty causing abnormal behavior, 634
Unconditioned response, 580
Unconditioned stimulus, 45, 580
Underwings, 5
Underwood, H., 155, 691
Undesirable behavior, 621; see also Abnormal behavior
Ungar, G., 617, 684, 691
Ungulates
  cranial combat and displays in, evolution of, 323
  distress call of, 281
  elimination behavior of, 166
  group defense by, 233
  hand-raised, 625
  matriarchal groups in, 330
  mixed-group foraging by, 345
  mixed-species groupings of, 344
  play in, 386
  weaving behavior in, 624
Uniformitarianism, 109
  and animal behavior, 110
Unipolar neurons, 414
Unison call of crane, 132, 134, 135
  characteristics of, 136-137
    shared, 138, 139
Universities Federation for Animal Welfare, 640, 691
Urchin, parental care by, 331
Urination, 165-166
Urochordates, nervous system of, 411
Urohydrosis, 166
Urquhart, F.A., 64, 188, 190, 201, 691
Urquhart, N.R., 188, 190, 691
Ursus americanus, play in, 391; see also Bear
Utopia, 29

V

V; see Volts
V_e, 99
V_g, 99
V_p, 99
V_t, 99
Vacuum behaviors, 38
  of predators, 364
Valenstein, E.S., 548, 691
Van Twyver, H., 158, 670
Vandenbergh, J.G., 301, 683
Vandermolen, I., 226
Van Houten, J., 681
Variability
  of behavioral responses, 21
  in bird development, 612
  environmental, 103
  and genetics, 102-103
  among genotypes, 100
  among individuals, study of, 98
  spontaneous, 103
Variable reinforcement schedule, 582
Variables, weather, affecting bird migration, 82-83
Variance, 98-99
  zero genotypic, 100
Variation, 87
  average, 99
  of behavior among different individuals, 80-81
  environmental, 103
    at global level, 104
    at local level, 105
  and experimentation, 55
  genetic, 92-93
  inherited, 111
  lack of, 102
  random, 103
  temporal, 104
Vasoactive intestinal polypeptide, 518
Vasopressin, molecular structure of, 518
Ventral noradrenergic bundles and feeding disorders, 548
Ventromedial hypothalamus
  lesion of
    food intake and, 545, 546
    recent interpretations of, 549
  in mammals, role of, 544, 545-549
  role of, in feeding behavior, 549
Ventromedial hypothalamus syndrome, 548
Venus mercenaria as prey of Busycon carica, 374
Verplanck, W.S., 41, 691
Vertebrates
  behavior of, and hormones, 509-516
  brains of, 411, 474
  circadian pacemakers of, 154-155
  communication in, 283-285
  development of, and predation, 360
  displays in, numbers of, 286
  embryonic muscles of, movement of, 556
  eyes of, reflective surface of, 459
  familiar paths for, 200
Vertebrates—cont'd
  head scratching in, 128, 129
  hearing in, 451-452
  learning in, 595-596, 601-603
  lower, colonial, social behavior of, 230
  motor neuron of, 413
  movement by, 163
  nervous system of, 412
    basic, 406
  neural pathway of, 475-502
  neural-endocrine integration in, 509
  nonmammalian, learning studies using, 47
  pain pathway of, hypothesized, 522
  recognition among, 229
  respiratory movements of, 165
  retina of, 485-486
  rods and cones in, 459
  stretching by, 166
  terrestrial, simple reflexes in, 476
  visual cells of, 469
Vertigo, 468
Vervet monkeys
  capture of, while playing, 391
  play in, 391
Vesalius, A., 30
Vesicles, synaptic, 415, 426
Veterinarians, studies of animal behavior by, 644
Veterinary management of animals, 645-646
Vibration
  sensing of, 454
  sound, detection of, 452-453
Vibrational senses, 451-454
Vibrissae patterns in lions as identification, 63
Vicugna vicugna, play in, 386
Vicuna, play in, 386
Viduine finches, song mimicry by, 351
Vieth, W., 586, 617, 674, 691
Vigilance time in starlings and flock size, 232
Vigor, hybrid, 117
Vilberg, T.R., 550, 691
Vincent, L.E., 395, 691
Vipers, pit, heat and infrared sensing by, 461
Virus, ribonucleic acid, 88
Vision, 454-461
  of bats, 7, 8
  binocular, 448
  color, 447, 455-458
  deprivation of, results of, 566
  duplicative theory of, 459
  form, 460-461
  night, 458-459
  orientation based on, 200
  stereoscopic, 448
Visser, M., 691

Visual acuity, 460-461
Visual cells, operation of, 486
Visual communication, 276, 277
Visual cortex, 488, 489
   visual processing in, 490
Visual displays, 276
Visual information, processing of, by retina, 485
Visual pathways from retina to brain, 487, 488
Visual processing in mammalian visual cortex, 490
Visual sense cells, structure of, 469
Visual sensitivity, three-pigment, 458, 459
Visual signals
   of cichlid fishes, 284
   as communication, 261
Visual techniques for studying neurons, 416
Viviparity and parental care, 333
VMH syndrome; see Hyperphagia; Ventromedial hypothalamus syndrome
Vocalizations, 274-275
   bird
      development of, 572-576
      meaning of, 271
   crane, evolution of, 130-140
   human
      contrasted to call of hunting little brown bat, 13
      sonograms of, 14
Voles
   play-smells of, 398
   thermal conservation of, 237
Volpe, B.T., 689
Voltage clamp, 418
Volts, 419-420
Voluntary control of behavior, 481
Von Burg, R., 589, 674
Vrba, E.S., 142, 677
Vulpes vulpes, 378, 384
Vultures
   social behavior of, 231
   tools used by, 371
   urohydrosis in, 166

W

Wade, M.J., 121, 122, 691
Wade, N., 295, 298, 691
Waggle dance of honeybee, 288-289
Walcott, B., 682
Walcott, C., 216, 463, 599, 688, 691
Walking, 163
Walkinshaw, L.H., 132, 691
Wallace, A.R., 30, 93, 111, 128
Wallace, R.L., 194, 691
Wallace, S.F., 154, 671
Walter, H.E., 592, 691
Walther, F., 655, 672, 677
Wampler, R.S., 549, 691

Warblers
   European
      songs of, and female-choice sexual selection, 318
      use of stars by, for navigation, 214
   feeding habitats of, 193
   niche-gestalts for, 197
Ward, H.P., 542, 689
Ward, I.L., 520, 684
Ward, P., 234, 692
Warm-blooded animals, 167
Warning displays, 377
Warning systems of prey, 376
Warren, J.M., 602, 692
Washing, 166-167
Washoe, 294, 295
Wasps, 381
   digger
      ESS in, 248
      optimality and, 181
      use of landmarks by, 203
   learning in, 598
   mutualism-parasitism involving, 351
   shelters of, 181, 182
   social behavior of, 230
Water boatman, hearing organs of, 453
Water bug, male, egg brooding by, 331
Water striders
   hearing in, 454
   study of communication in, 67
Waterbolk, H.T., 670
Waterfowl
   "dump nests" of, 351
   tail wagging in, 147
Watkins, L.R., 523, 525, 692
Watson, J., 107, 692
Watson, J.B., 28, 46, 692
Watson, R.I., 51, 692
Watson, S.J., 671
Watson-Whitmyre, M., 690
Wattled crane, 131, 139, 140
Wavelength of sound, 9
Waves, sound, 8-9
Weaning conflict, 337
Weather variables affecting bird migration, 82-83
Weaver ants, 183
   living bridge of, 236, 237
Weaverbirds, nest acceptance of, 198
Weaving and rolling, shelter made by, 183
Weaving behavior in captive animals, 624
Weaving birds, nests of, 183
Web building, 182
Webber, M.L., 681
Weberian ossicles, 452
Webs to catch prey, 371
Webster, F., 678
Wecker, S.C., 195, 606, 692
Weeks, H.P., Jr., 623, 692

Weevils
   eyes of, 456
   reproduction of, 304
Wehner, R., 469, 692
Weight
   body
      control of, hypothetical feeding system for, 530
      set-point mechanism for, 529
   control of, 527-529
      lipostatic theory of, 529
Weight gain and overeating, 529
Weinberg, W., 663
Weisman, R.G., 681
Weiss, B.A., 600, 692
Wells, M.J., 202, 561, 562, 563, 692
Wenger, E., 678
Wenner, A.M., 289, 292, 301, 692
Werner, E.E., 193, 692
Wernicke, K., 28, 49, 495
Wernicke's aphasia, 495
Wernicke's area, 494, 495
West, M.J., 575, 577, 692
West crowned cranes, 131, 139
Westerman, R.A., 592, 692
Western pattern of sensory lateralization, 500
Whale
   blue, motor neurons of, 413
   killer
      communal hunting by, 372
      as predators, 361
   pilot, sleep time for, 159
   vocalizations of, 275
Whelk dropping by northwestern crows, 177, 178, 179
Whine, 285
Whirligig beetles, energy-emitted senses of, 466-467
White, D.H., 198, 692
White, M.J.D., 306, 692
White matter, 477
White rat
   female, normal estrous cycle and behavior of, 515
   sequential appearance of behavior in, 555
White-crowned sparrows, dialects in songs of, 575
White-naped crane, 131, 139
   unison call and dance movements of, 134
White-tailed deer, 649
   fighting by, 59
   young of, 555
White-throated sparrows
   habitat preference of, 196
   single-gene effects in, 95
Whitethroats, feeding of dependents by, 364
Whitman, C., 28, 33

Whitsett, M., 690
Whittaker, R.H., 265, 266, 692
Whooping crane, 131, 139, 140
    imprinting in, 571
    management of, 652-653
Wickler, W., 377, 379, 380, 692
Wicksten, M.K., 378, 692
Widow birds, young, markings of, 351
Wiens, J.A., 76, 118, 692
Wiesel, T.N., 28, 49, 473, 490, 566,
        680, 692
Wigeon, 130
    American, habitat selection by, 198
    Chilee, 130
Wilcox, R.S., 67, 692
Wilczek, M., 534, 692
Wild animal, 629
Wild dogs
    African
        prehunting behavior of, 281
        switch in prey of, 366
    hunting success of, 375
Wildebeest
    hunting of, by lions, 373
    migration by, 220
    as prey of hyenas, 366, 374
Wildlife management, 648-651
Wiley, R.H., 239, 671
Willard, 374
Williams, G.C., 225, 304, 305, 334, 339,
        692
Willows, A.O.D., 596, 617, 670, 674,
        675, 681, 682, 688, 692
Wilson, D.H., 689
Wilson, E.O., 26, 27, 28, 56, 107, 123,
        143, 225, 226, 229, 230, 231,
        245, 252, 253, 254, 255, 257,
        259, 263, 266, 273, 274, 277,
        279, 286, 301, 319, 330, 333,
        336, 338, 341, 345, 349, 350,
        353, 354, 355, 383, 397, 687,
        692
Wilson, G.C., 101, 681
Wilson, M., 248, 305, 315, 329, 339,
        674
Wilson, O., 236
Wilson, S., 692
Wiltschko, W., 215, 692
Wimsatt, W.A., 23, 221, 692
Windle, W.F., 556, 692
Wing movement, rapid, muscles and, 438-
        439
Wing-flapping "exercises" of nestling
        birds, maturational changes in,
        564
Wings of moths in different positions, echo
        and appearance of, 16
Winkler, H., 364, 692
Win-stay, lose-shift strategy, 588
Wish fulfillment—disguise theory of
        dreams, 157

Wittenberger, J.F., 259, 316, 335, 693
Wodinsky, J., 446, 690
Wolf, L.L., 678, 693
Wolf spiders, homing ability of, 202
Wolfe, R.S., 676
Wolfson, A., 213, 678
Wolgin, D.L., 546, 693
Wolves
    communal hunting by, 372
    cooperative foraging by, 234
    hunting success of, 374
    neutral zones between packs of, 382
    pheromones to mark territories of, 272,
        273
    as predators, 360
    and ravens, play between, 389
    red, breeding with coyotes, 117
    social status of, postural indications of,
        278
    time without eating for, 362
Wood, D.S., 132, 140, 693
Wood rat, encounters of, at feeder, 344
Wood ticks as parasites, 348
Woodchucks, learning in, to increase food
        supplies, 179
Woodcock, A., 240, 693
Woodpecker finches, tools used by, 371
Woodpeckers
    distress calls of, 281
    flying patterns of, 162
    nesting habits of, multivariate analyses
        of, 83
    Syrian, foraging behavior of, 364
Woodruff, G., 127, 687
Woodruffia metabolica, hunting success
        of, 374
Wooly aphids and predaceous lacewing
        larvae, 371, 372-373
Worden, A.N., 633, 693
Worden, F.G., 689
Working memory, 606
Worm Runner's Digest, 593
Worms, palolo, mating time of, 310, 314
Wren, alarm call of, 281
Wright, P., 387
Wright, S., 118, 120, 121, 122, 693
Writing, muscles and, 436-437
Writing styles, 497
Wundt, W., 28, 30
Wyman, R.J., 102, 682
Wynne-Edwards, V.C., 118, 120, 122,
        247, 693
Wyricka, W., 551, 685, 693

X

Xenobiosis, 349
Xenopus laevis, androgen uptake in brain
        of, 517
X-ray sensing, 463

Y

Yalden, D.W., 6, 23, 693
Yarak, 364
Yasukawa, K., 689
Yawning, 166
Year of the Wildebeest, 653
Yellow eels, 208
Yellowthroat warbler, niche-gestalt for,
        197
Yerkes, R.M., 597, 693
Yin, T.A., 549, 682
"You-first" principle, 233
Young, E., 640, 655, 693
Young, J.Z., 610, 693
Young, R.C., 670
Young, W.G. 675
Young
    altricial, 555
    care of, reproduction and, cooperation
        in, 337-339
    eating of, by domestic animals, 631
    feeding of, 329
    parental brooding of, and thermoregula-
        tory ability, 332
    precocial, 555, 556
    protection of, by group, 233

Z

Zach, R., 167, 177, 178, 179, 693
Zack, S., 167, 693
Zahavi, A., 285, 693
Zatz, M., 155, 690
Zebra
    combination messages of, 277
    hunting of, by lions, 373
    as prey of hyenas, 366
Zebra moray, cleaner shrimp cleaning,
        346
Zeigler, H.P., 548, 693
Zeitgebers, 150
    blocking of, 150, 152
    and free-running pacemaker rhythms,
        155
Zero genotypic variance, 100
Zero potential, 419
Zimen, E., 278, 693
Zimmerman, N.H., 155, 693
Zimmerman, R.R., 629, 679
Zoeger, J., 463, 693
Zoos
    animals in, abnormal behavior in, 623-
        627
    history of, 640-643
    modern facilities of, 627, 642-643
Zuckerman, L., 640, 655, 693
Zugunruke, 211
Zwislocki, J.J., 452, 693
Zygote, 92